telephony's DICTIONARY

SECOND EDITION

telephony's DICTIONARY
SECOND EDITION

Defining 16,000 telecommunication words and terms

By Graham Langley

telephony®

Publishing Corporation
55 East Jackson Blvd.
Chicago, Illinois 60604 USA

Telephony's Dictionary, Second Edition
Defining 16,000 telecommunication words and terms

First Edition: June, 1982
Second Edition: April, 1986

Copyright © 1986 by Telephony Publishing Corp.

ISBN 0-917845-04-8

PRINTED IN THE UNITED STATES OF AMERICA

To Lee, Julia and Peter

FOREWORD

Much has been and much more will be written about the revolution in telecommunications and the coming of the "Information Age". Today, the major topic in conferences and the technical press is ISDN. A short time ago it was "divestiture". These are but two examples of advances or changes that are filling the literature with new terms, abbreviations, acronyms and even old familiar terms being used in new ways.

Readers, especially those new to telecommunications, need a convenient, easy-to-use dictionary of terms. The First Edition of TELEPHONY'S DICTIONARY successfully fulfilled this need. However, the march of time has dictated that an update is needed and Graham Langley has again provided it in this Second Edition.

There are other dictionaries available, but I am not familiar with another that provides such a broad coverage in a form as convenient for the general reader. The IEEE *Standard Dictionary of Electrical and Electronic Terms* is, of course, a definitive text especially of North American usage. The *International Electrotechnical Vocabulary* (IEV) published by the International Electrotechnical Commission (IEC) is a multi-volume, multi-lingual text. Both of these, however, contain much more than telecommunication terms and because of their size are not as convenient to use.

Users of the International Telecommunications Convention, the International Regulations, CCITT or CCIR Recommendations, ANSI T-1 or other standards documents will still need the specialized glossaries or vocabularies that apply to each. These documents tend to be legalistic and the precise meaning of a term defined for one of them may have a significant effect on its interpretation. A precise interpretation of a term in a standard could affect the design of equipment and its eventual compatibility with other equipment.

There are at least three major dialects of English technical terminology — British, American and CCITT/CCIR. The CCITT/CCIR may be a compromise between the British and American, or it may differ from both. Sometimes it may even be a translation from the French. Whatever the source, it has tended to be different from the other two. Happily, now that the CCITT is working in the forefront of systems development, there is a tendency for both the British and American standards organizations to use the CCITT terms. This can only improve international understanding.

Within each of these dialects there are subdialects. Telephone and telegraph people have traditionally used different terms for the same things, or the same terms to mean different things. Since data people came from the telegraph side, they built on the telegraph vocabulary. Within each of these areas, transmission, switching, and traffic people had their own individualistic meaning for some terms. For example, the term "signal" can mean something different in switching and transmission. With the introduction of integrated switching and transmission for IDNs, and with the further integration of services in the ISDN, a common terminology must be found. CCITT and the various standards organizations are working hard to harmonize the dialects.

Because of his unique background, Graham Langley has been able to bridge these dialects and to identify the various meanings of a term, so that the user knowing the context in which it is used, or the source of the text, can find the correct interpretation.

After graduation from London University with first-class honors in 1940, Graham Langley joined British Army Signals and served in the Middle East, North Africa, Sicily and Italy. After the war he joined the United Kingdom overseas civil service working in Singapore and Malaysia for 18 years. He was the last European Director of Telecommunications for the then Federation of Malaya. He then joined Cable & Wireless where he served in engineering posts before he was appointed head of C & W's telephone-related consultancy services. In this capacity he supervised major projects over the world. In 1980, he retired from Cable & Wireless and joined British Telecommunications Systems, Ltd. (BTS) — a company formed to market System X worldwide. When System X ceased to be a U.K. cooperative undertaking but became instead a Plessey/GEC switching system, BTS ceased its activities and he became an independent consultant.

Mr. Langley is ideally suited to prepare a dictionary of terms used in telecommunications, because of his experience, knowledge, and enthusiasm for the subject. The new edition will be welcomed by all in the industry. Let us hope that Graham Langley will retain his enthusiasm to prepare future revisions to take care of the continuing flood of new terms.

John S. Ryan
Chairman
CCITT Study Group XI
AT&T Bell Laboratories
Holmdel, NJ
February, 1986

PREFACE

I want to thank those who wrote to Telephony Publishing or contacted me personally and suggested new entries or pointed out errors in the First Edition. Fortunately, there were remarkably few errors considering the scale of the production and that this Dictionary was the first in its field. The preparation of the Second Edition has given me the opportunity of correcting all the noticed errors.

In the three years since the First Edition was prepared, our industry seems to have brought into common usage almost as many new buzzwords and acronyms as were produced in the previous three decades. Local Area Networks, for example, did not even earn a mention in Edition One but in a great many cities every senior schoolboy now recognizes a LAN when he sees one.

Technical developments and inventions have been the main causes of this word explosion — not merely telematics and information technology (the coming-together of telecommunications and computers) but also the greatly increased use of optic fibers and of digital techniques both for transmission and for switching, the proliferation of satellite-based services and the attention now being given all over the world to interactive cable systems capable of providing home entertainment and business services as well as ordinary telephones.

The cessation of the monopoly previously enjoyed by British Telecom and BT's transition from Government Dept. (part of the Post Office) to a publicly-owned, limited liability company, operating in competition with Mercury Communications Ltd (now a subsidiary of Cable & Wireless plc) has also produced a crop of new words — or given new and strange meanings to old ones. Who would have thought a few years ago that privatization would mean a transfer to publicly-held, shareholding ownership. But even bigger generators of new English language acronyms have been the organizational and regulatory changes in the U.S. which have followed divestiture, the breaking up of the Bell System and deregulation.

I have, therefore, done my best to include in this Second Edition as many new definitions as I can in the hope that TELEPHONY'S DICTIONARY OF TELECOMMUNICATIONS will not merely continue to be a popular work of reference but will also find its way into the personal book collections of every technician, engineer, businessman and diplomat who needs to know what on earth his colleagues are talking about.

Graham Langley
Penang, Malaysia
February, 1986

CONTENTS

A and B leads Wires taken from the center-taps of coils across each of two pairs making up a four-wire circuit. These are used for dc signaling, particularly E & M signaling.

A and B signaling "Bit stealing" signaling procedure used in 24-channel PCM (T1 transmission).

A battery Battery to provide current for heating filaments of electron tubes/thermionic valves.

A board Manual operating position handling calls to nearby towns from local telephone subscribers.

A-law The method of encoding sampled audio waveforms used in the 2048 kbit/s 30 channel pcm primary system, widely used outside America.

A operator Operator who answers subscribers' lines or trunk signals and completes local calls.

A position Operating position in a manual central office where subscribers' lines and operator-controlled trunks are terminated and calls originated by local subscribers are handled. Sometimes called an A board.

AB pack A U.S. power source including the A battery (for heating filaments) and the B battery (high tension for anode voltage).

ab- Prefix to now obsolete cgs electromagnetic units, e.g. abvolt, abampere.

abandoned calls Call during which the calling party goes back on-hook without waiting for the call to be connected or for advice such as busy tone, or information from an operator that the call cannot be connected.

abbreviated dialing System by which a telephone user may dial a short code which instructs the central office to connect the calling line with the called line. A two- or three-digit code is used instead of the normal seven digits.

abbreviated dialing, pool Abbreviated dialing in which the same short code may be used by several different customers in order to obtain connections to a particular line.

abbreviated dialing, unique Abbreviated dialing in which each customer has control of his own series of short codes by which he can be routed to particular lines.

abbreviated dialing prefix The non-numerical code indicating that the information following is an abbreviated number.

abbreviated number The numerical code sent by a telephone caller using abbreviated dialing which identifies the telephone number of the party to whom he wishes to be connected.

abnormal termination Condition encountered when an error is detected so that a series of actions cannot be completed as expected.

abort In data transmission, a function invoked by a primary or secondary sending station causing the recipient to discard (and ignore) all bit sequences transmitted by the sender since the preceding flag sequence.

abort timer Automatic on-hook device for a data modem if no answer-back is received, or the distant end is busy; normally set for 15-second delay.

absence of ground Condition when a telephone switch bank contact is either open or connected to negative battery.

absent-subscriber service At a subscriber's request, calls made to his number during a specified period can be intercepted at the central office/exchange and transferred to: a) absent-subscriber service operators; or b) an answering machine at the central office/exchange, giving appropriate information to the caller and alternately periods of special information tone; c) or the special information tone.

absolute address An address in machine language that identifies a storage or a device without the use of any intermediate reference.

absolute delay The amount of time a signal is delayed. It may be expressed in time (i.e., milliseconds) or in number of characters, pulse times, word times, major cycles, or minor cycles.

absolute zero The lowest temperature theoretically possible, -273.16°C, at which the energy of random motion of particles at thermal equilibrium is minimum. This is zero Kelvin, the thermodynamic temperature scale.

absorber, digit An item of telephone switching equipment into which specified digits can be dialed without causing any response except to make the equipment ready to act upon the next digit received. In Britain these selectors are called discriminating selectors or discriminating selector repeaters.

absorption Attenuation caused by dissipation of energy, e.g., when radio waves lose power going through foliage, rain or snow; when light pulses traveling along an optic fiber encounter unwanted impurities or when sound waves are reflected by soft, energy-absorbing material.

absorption, auroral The loss of energy in a radio wave passing through an area affected by solar auroral activity.

absorption fading Variation in the absorption of radiowave propagated through the ionosphere due to changes in the densities of ionization.

absorption index 1. The ratio of the electromagnetic radiation absorption constant to the refractive index as given by the relation:

$$K' = K\lambda/4\pi n$$

where K is the coefficient of absorption, λ is the wavelength in vacuum, and n is the index of refraction of the absorptive material. 2. The functional relationship between the sun's angle (at any latitude and local time) and the ionospheric absorption.

absorption loss Loss of energy in one circuit caused by conversion into another form of energy or transfer to another circuit.

absorption losses (Optic fiber) Losses caused by impurities, principally iron, cobalt, nickel or water.

absorption peak Greater than average loss at a particular frequency due to greater absorption at that frequency.

AC Alternating current.

AC spark over AC spark-over voltage of a gas discharge protector is the rms value of the sinusoidal voltage at a frequency of 15Hz to 62Hz, the rise of which causes the protector to spark over when the voltage is slowly increased. It is used to indicate the range of application of a protector in case of direct contact with, or magnetic induction from, ac power lines.

AC/DC ringing Telephone signaling method using AC to operate a ringer and DC to operate a relay which breaks the ringing circuit as soon as the called subscriber answers.

accelerated life test A special form of life test for components with high repetition rate of expected input load. Tests may be completed in a small fraction of the time which the component has been designed to endure in normal service.

accelerating electrode Electrode which causes electrons emitted from an electron gun to accelerate in their journey to the screen of a cathode ray tube.

accelerating voltage A voltage applied to an electrode which accelerates a beam of electrons or other charged particles.

acceleration time The time taken by a mechanical input/output device, e.g. a magnetic tape transport, to reach its operating speed after the instruction to act has been interpreted.

accept In data transmission, the condition assumed by a primary or secondary station upon accepting a correctly received frame for processing.

acceptable quality level Percentage of defects which will be accepted a predetermined percentage of the time by a sampling plan during inspection or test.

acceptable reliability level Permitted percentage of failures allowed per thousand operating hours.

acceptance angle (for a uniformly illuminated optical waveguide) Half the vertex angle of that cone within which the power coupled into a waveguide is equal to a specified fraction of the total power coupled into a waveguide. The sine of the acceptance angle is called the numerical aperture.

acceptance cone (Optic fiber) Cone defined by an angle equal to twice the acceptance angle.

acceptance test Test made to ensure equipment as supplied and installed is fully compliant with the issued specifications.

acceptance trials Trials carried out to determine if the specified performance and characteristics have been met.

accepting station A destination station that accepts a message.

acceptor 1. A series resonant circuit. 2. An impurity element that increases the number of holes in a semiconductor crystal.

access 1. Point at which entry is gained to a circuit or facility. 2. Ability of user to enter a given network.

access, random memory Access time is independent of the address and any desired section of the memory may be accessed.

access, sequential memory Items are read off in sequence, the order of which is determined by key values of the items.

access attempt The process by which one or more users interact with a telecommunication system in order to initiate user information transfer.

access-barred signal A signal sent back indicating that the call will not be completed because of a called or calling customer user facility.

access charge Payments made by a long distance carrier to a U.S. local telephone company for use of local facilities, and by telephone subscribers to obtain access to local networks.

access circuit Relay set or junction which routes a call into a particular item of equipment.

access code The preliminary digits that a user must dial to be connected to a particular outgoing trunk group or line.

access control Action taken to permit or deny use of the components of a communications system.

access cost factor (ACF) Formula proposed to be used in calculating value of non-traffic sensitive (NTS) investment in exchange plant.

access denial Access failure due either to the issuing of a system blocking signal by the telecommunication system, or to exceeding maximum access time and nominal system access time fraction during an access attempt. Syn.: system blocking

access-denial probability Ratio of access attempts which result in access denial to total access attempts.

access-denial time Elapsed time between the start of an access attempt and access failure due to access denial.

access failure Termination of an access attempt in any manner, other than initiation of user information transfer between the intended source and destination.

access level Terminals connected to a central processor are not permitted access to all levels of program and data.

access line A circuit between a subscriber and a switching center. Any line giving access to a larger system or network. Also, the private lines feeding a common control switching arrangement or enhanced private switched communications service switch from a PBX.

access originator The functional entity responsible for initiating a particular access attempt.

access permission   Authority given after access controls have checked that an attempt to access a processor has the correct status with no security objections.

access phase   In an information transfer transaction, the phase during which an individual access attempt is made.

access points   A class of junction points in dedicated outside plant. They are semi-permanent splice points, at the junction of a branch feeder cable and distribution cables; points used to make connections for testing or for use of particular communications circuits.

access prefix to the international network   The origin country's single common access prefix both for the intercontinental and the international network.

access request   A control message issued by an access originator for the purpose of initiating an access attempt.

access time   1. In a telecommunication system, the elapsed time between the start of an access attempt and successful access. 2. In a computer, the time interval between the instant when an instruction control unit initiates a call for data and the instant at which delivery of the data is completed. 3. The time interval between the instant at which data are requested to be stored and the instant at which storage is started.

access to supplementary services   Information used to give instruction to the switching equipment that the associated information relates to a supplementary service.

accessibility   The number of appropriate outlets in a switching network which can be reached from an inlet.

accessibility, constant   The number of outlets which can be reached by a source via a switching network is independent of the state of that switching network and its other sources.

accessibility, controlled   The number of circuits which can be tested for a free or busy condition by a source via a switching network is controlled.

accessibility, equivalent   For a link system, the accessibility of a single stage reference switching network which, with the same number of trunks, handles the same traffic with the same probability of congestion.

accessibility, full   Any free inlet can reach any free outlet of the desired route regardless of the state of the system.

accessibility, limited   Access is given only to a limited number of trunks.

accessibility, variable   The number of circuits which can be reached by a source via a switching network is affected by the state of occupation of the switching network and its other sources.

account code   1. Numerical code for classifying revenue or expenses. 2. A multidigit code which may be optionally associated with the user dialing sequence. It is used as an additional means to control and allocate costs. As an example, an OCC customer could purchase one authorization code. When placing a call each employee would use the authorization code, but identify his department by dialing additional digits which represent an account code. The billing information from the OCC to the customer would contain the data on each call as well as the account code.

accounting rate   The rate per traffic unit agreed upon to establish international telecommunications accounts.

accounting rate quote   The part of the total accounting rate per traffic unit corresponding to the facilities made available in each country; this quota is fixed by agreement.

Accounts, Uniform System of   In the U.S.A., grouping of units of telephone plant, identified by standard numerical codes specified by the Federal Communications Commission.

accumulator   1. A computing device which records the cumulative total of signals input. 2. A storage battery.

acid, battery   Dilute sulfhuric acid used as electrolyte in a lead-acid storage battery/accumulator.

acknowledge character   1. A transmission control character transmitted by a station as an affirmative response to a connecting station. 2. A transmission control character transmitted by a receiver as an affirmative response to a sender. An acknowledge character may also be used as an accuracy control character.

acknowledgement signal unit   A signal unit which carries information about whether the signal units in the block indicated were received correctly.

acoustic coupling   Coupling a data terminal or similar device to a telephone line by means of transducers which utilize sound waves to or from a telephone handset.

acoustic delay line   Circuit element which introduces time delay based on the time of propagation of sound waves. Syn.: sonic delay line

acoustic noise   A disturbance in the audio frequency range.

acoustic shock   Shock produced by hearing a sudden loud sound.

acoustic shock reducer   A limiter which cuts off peaks of noise power thereby eliminating shock. Used on operators' headsets.

acoustic wave   A wave transmitted through solid, liquid or gas as a result of mechanical vibrations of a sound source. The frequency range of acoustic waves is usually taken as the effective range of a normal human ear, from 20 Hz up to about 20 kHz.

acoustics   1. The physics and study of sound. 2. Qualities of a room or enclosure which determine the behavior and effect of sounds.

acousto-optic   The interaction of light with a sound wave in a medium via a change of refractive index.

acquisition, satellite   The procedure by which a ground station searches for a satellite and locks itself into alignment.

3

acquisition time 1. In a communication system, the amount of time required to attain synchronism. 2. In satellite control communications, the time required for locking tracking equipment on a signal from a communications satellite.

actinometry The science of measurement of radiant energy, particularly that of the sun.

active Any device or circuit which introduces gain or uses a source of energy other than that inherent in the signal.

active current The component of current that is in phase with the voltage.

active laser medium The material within a laser, such as crystal, gas, glass, liquid, or semiconductor, that emits coherent radiation (or exhibits gain phenomena) as the result of stimulated electronic or molecular transitions to lower energy states. Syn.: laser medium

active line rotation Scrambling procedure which may be used with television transmission for subscription systems in which programs are only usable by authorized receivers.

active load A load formed by an active device.

active material Any chemical substance which plays an active part in a reaction. Compounds used in the plates of lead-acid storage batteries/accumulators and the thorium compounds coated on hot cathodes in electron tubes to improve electron emission.

active network A network that includes a source of power.

active satellite A satellite carrying a station intended to transmit or retransmit radiocommunication signals.

active signaling link A signaling link which has successfully completed the initial alignment procedures and can carry signaling traffic.

active station Station that is currently eligible to enter or accept messages.

active time Time spent in the information transfer phase within the service time interval of an information transfer transaction. NOTE: A user's active time excludes all time spent in the access phase and disengagement phase, all time spent in the idle state and the exit state, and all time outside the service time interval. Active time is comprised of consecutive and alternating operational service periods and outage periods.

active volt-amps The product of voltage times active or in-phase current, which equals the power in watts.

activity In critical path analysis and in PERT (project evaluation and review techniques), an "activity" is normally a task requiring time to complete and needing input resources before leading on to the next event in the program.

activity factor A decimal fraction less than one which represents the fraction of the busy hour that a single voice channel is likely to be actively in use.

activity ratio For a processor, the ratio of the number of records which have been moved in a data file being updated to the total number of records in the file.

actual address See absolute address.

actual measured loss Actual loss reading when test is made under specified conditions.

actual work time (AWT) Time in seconds for a telephone assistance operator to handle a particular type of call or query.

actuator A device used to bring electronic equipment into operation, to calibrate it, or to cause a switch to operate to activate a circuit.

ADA A high-level computer language devised by the French CII-Honeywell Bull and now widely used by the U.S. Department of Defense. ADA is not an acronym, but the name of the girl friend of the 19th Century theoretical inventor of the computer, Charles Babbage.

adapter A fitting used between two sets of equipment to provide a transition from one to another when they cannot be directly interconnected.

adapter, four-wire extension lines A bridge which enables up to six four-wire circuits to be interconnected, with no impedance mismatches and specified levels. Also called a multi-point four-wire bridge.

adapter, four-wire line A device which enables a telephone to be connected to a four-wire circuit with E&M signaling on center points of the two pairs. The adapter accepts on- and off-hook signals.

adapter, long line A loop extender. A higher voltage than normal is made available so that lines of greater resistance may be utilized thereby enlarging the radius of the area served by the central office/exchange.

adapter, touch-tone A device which converts received multi-frequency tones into ten impulses per second loop disconnect signals.

adapter, videotex A device that enables a TV set or personal computer to act as a teletext or viewdata terminal.

adapter dial Plate used in fixing a telephone dial.

adaptive channel allocation A method of multiplexing where the information capacities of channels are not predetermined but are assigned on demand.

adaptive differential pulse code modulation (ADPCM) Technique which reduces storage requirements for speech signals stored in memories by utilizing 3- or 4-bit speech samples instead of the usual 8 bits per sample.

adaptive predictive coding A narrowband analog-to-digital conversion technique employing a one-level or multi-level sampling system in which the value of the signal at each sample time is adaptively predicted to be a linear function of the past values of the quantized signals.

adaptive pulse code modulation A technique which effectively reduces occupied bandwidth per active speaker by reducing sampling rates during periods of overflow peak traffic. See also VRAM, TASI, CELTIC.

adaptive speed device Device that senses speed and code of incoming data and automatically adjusts to it.

4

adaptive system   A general name for a system which is capable of changing itself to meet new requirements.

added bit   A bit delivered to the intended user in addition to intended user information bits and delivered overhead bits. Syn.: extra bit

added block   Any block or other delimited bit group delivered to the intended destination user in addition to intended user information bits and delivered overhead bits. Syn.: extra block

added block probability   The ratio of added blocks to total blocks received at a specified destination during a measurement period. Syn.: extra block probability

adder   A device whose output represents the sum of its inputs.

adder, full   A device which performs binary addition; it has three inputs (including a "carry" input) and two outputs ("sum" and "carry forward").

adder, half   A device which performs lowest level binary addition; it has two inputs and two outputs ("sum" and "carry forward").

additional period   Unit of time used for charging telephone calls longer than the minimum chargeable period.

additive white Gaussian noise   See white noise.

add-on   A facility which enables a recipient of an incoming call to add a third line to the connection, making it a three-way conference call.

address   1. The destination of a message in a communication system. 2. One or a sequence of data characters designating the terminal equipment which is the origin or destination of data. 3. The location in storage of information in a data processing system, i.e., a character or group of characters that identifies a register, a particular part of storage, or some other data source or destination.

address, call   In North America, the last four digits of the telephone number.

address, international telephone   A code which specifies a unique address for any telephone in the world. It consists of (a) a country or regional identity code of one, two, or three digits, (b) a three digit numbering plan area code, (c) a two or three digit central office code, and (d) a four digit station number.

address, memory   The location in a memory store of a byte or word of information.

address, message   That portion of the message that specifies the destination and the handling.

address, relative   An address defined relative to an absolute address or a point in a program.

address, symbolic   A convenient name for an address as used by a programmer. It must be translated into the absolute address before actual use.

address, telephone   The complete 10-digit number which (in North America) specifies the location of a particular telephone. It consists of a three digit area code, plus a three digit central office code, plus a four digit station number.

address field   In data transmission, the sequence of bits immediately following the opening flag of a frame identifying the secondary station sending, or designated to receive, the frame.

address field extension   In data transmission, an enlargement of the address field to include more addressing information.

address format   The way the instruction's address is arranged.

address incomplete signal   A signal sent in the backward direction indicating that the number of address signals received is not sufficient for setting up the call.

address messages   A message sent in the forward direction containing the signaling information required to route and connect the call. It includes address information, class of service information, and additional information relating to user and network facilities. It also may contain the calling customer identity.

address separator   The character which separates the different addresses in the selection signals.

address signal   A signal containing one element of the part of the selection signals which indicates the destination of an initiated call.

address signal complete   A signal sent in the backward direction indicating that signals required for routing the call to the called party have been received, and that no called party's line condition signals will be sent.

addressee   The intended recipient of a message.

addresser, push button   A device which, upon selecting and pushing one of a group of push buttons, will transmit one of a group of pre-set teletypewriter addresses. Used with teletypewriter exchange service (telex).

adjacent channel   Any of two TV channels are considered adjacent when their video carriers, either off-air or on a cable system, are 6 MHz (8 MHz in Europe) apart. For FM signals on a cable system, two channels are adjacent when their carriers are 400 to 600 kHz apart. In any FDM or TDM system, channels are adjacent if they use adjoining bandwidths or time slots.

adjacent-channel interference   Interference caused by a transmitter or carrier system operating in an adjacent channel.

adjacent signaling points   Two signaling points that are directly interconnected by a signaling link.

Administration Management Domain   The set of message handling system entities managed by a telecommunications administration.

administration-selected service restriction   The possibility to prevent all or certain types of outgoing calls and/or service control operations from a telephone line.

administrative processor   A centralized processor for administrative purposes which serves several switching centers.

administrative program   This is part of an application program for an SPC central office/ exchange, specifically relating to software areas such as (a) "soft data" operations, (b) central office/exchange configuration control and extension, (c) traffic measurement and system monitoring, and (d) central office/exchange maintenance and testing.

admittance A measure of the ease with which alternating current flows in a circuit. It is the reciprocal of impedance, and is expressed in siemens. Admittance is the vector sum of a resistive component called "conductance" and a reactive component called "susceptance."

admittance, surge The reciprocal of surge impedance.

admittance, transfer The reciprocal of transfer impedance.

adsorption The adhesion of a thin layer of one substance to the surface of another, e.g., gas or liquid molecules adhering to the surface of a solid.

Advanced Communication Function (ACF) Family of IBM software products incorporating Systems Network Architecture (SNA) and governing intercommunication between computers.

Advanced Communication System Bell System service which enables voice (56 kbit/s) and data (using X25 protocol) to be carried on the same metallic circuit on a shared basis.

Advanced Information System (AIS) Switched data network offered by AT&T Information Systems.

Advanced Mobile Phone Service (AMPS) Bell's cellular radio system for mobile telephones.

aerial 1. In the air, as aerial cable. 2. A radio antenna.

aerial cable A communications cable installed on or suspended from a pole or other overhead structure.

aging The change in properties of a material with time.

air, off the With emphasis on "off": radio transmitter not switched on. With emphasis on "air": received from a radio transmission.

air, on the With emphasis on "on": radio transmitter switched on.

air core An inductor with no magnetic material in its core.

air gap A small gap in a magnetic circuit, e.g., that between pole piece and armature of an electromagnetic relay, even when the relay is in the operated position.

air portable Denotes material which is suitable for transport by an aircraft loaded internally or externally, with no more than minor dismantling and reassembling within the capabilities of user units.

air sounding Measuring atmospheric phenomena or determining atmospheric conditions usually by means of apparatus carried by balloons or rockets.

AIRPAP (air pressurization analysis program) A Bell System software program used to determine the best way of pressurizing and controlling a given cable network.

alarm Flashing light, ringing bell or other visual or audible signal to draw the staff's attention to a condition requiring attention.

alarm, battery Alarm indicating that there is a fault in a main battery circuit.

alarm, carrier An alarm indicating when the carrier supply of a carrier system is interrupted or drops in level by a specified amount.

alarm, contactor Alarm indicating an abnormal condition in a pressurized cable system.

alarm, emergency An urgent alarm indicating a serious fault which must receive immediate attention.

alarm, equipment An alarm indicating a fault in equipment.

alarm, high voltage An alarm indicating that the dc voltage being fed to the switching equipment (normally 50v) has risen above a preset warning level.

alarm, low-high voltage An alarm indicating that a central office battery voltage is outside its permitted and preset range of values.

alarm, low voltage An alarm indicating that a central office battery voltage has fallen below its specified minimum level.

alarm, major carrier Alarm indicating the failure of more than one primary group of carrier channels (usually 12, 16, 24, or 30 voice channels).

alarm, MDF Alarm lamps distributed along a main distribution frame indicating operation of a heat coil, an indicating fuze, or other protective device in a particular section of the frame.

alarm, minor carrier Alarm indicating the failure of multiplexing equipment which affects only a single group of channels.

alarm, non-emergency Alarm indicating a fault which can be dealt with on a routine basis. With modern SPC equipment, such faults are normally diagnosed by the equipment itself and a teletypewriter record made for the central office staff.

alarm, office Alarm indicating a fault in a central office.

alarm, pilot Alarm indicating the interruption of a pilot tone.

alarm, remote Alarm indicating a fault condition at a distant central office.

alarm, transfer Alarm indicating a non-scheduled transfer of a load from a normal to a standby unit.

alarm, urgent Same as emergency alarm.

alarm call service An operator or automatic device calls a given telephone number, at a time specified in advance by the subscriber, and makes an appropriate announcement.

alarm center A location generally within a technical control facility that receives local and remote alarms.

alarm panel, gas pressure A panel on which cable pressure alarms are mounted. Several types are available, with alarm modules providing audible and visual indication of pressure drop in cables. Some types of such panels work to remote contactors located at many points in the cable distribution network, enabling faults to be located with ease and rapidity.

alarm sender A unit at an unattended office which analyzes locally generated alarms and sends a signal to the parent attended central office, enabling maintenance staff there to classify the fault and arrange for it to be given appropriate attention at a suitable time.

alarm sensors   See variation monitors.

alerting lamp   A lamp on a manual operating suite indicating that a particular circuit is carrying an incoming call requiring the operator's attention.

alerting signal   A ringing signal sent to customers to indicate they should answer their telephones.

ALGOL   (algorithmic language) A programming language which uses algorithms to describe mathematical procedures. Usually written in self-contained blocks which can be assembled in such a way that data storage is shared between many blocks.

algorithm   A prescribed finite set of well-defined rules or processes for the solution of a problem in a finite number of steps; the underlying numerical or computational method behind a piece of code.

align   To adjust timing or tuning or gain of a unit or part thereof to ensure that the unit functions correctly.

aligned bundle   A bundle of optical fibers in which the distribution of relative spatial coordinates of each fiber is the same at the two ends of the bundle, as opposed to the random orientation of fibers in bundles typically employed as optical communication transmission lines. An image formed on the input face of an aligned bundle will be seen on the output face.

aligner (data)   A device used to align the elements of one data structure to particular elements of another structure and, in some cases, also to change between the two structures.

alignment   Checking a circuit to ensure it is operating at specified levels at all points and that distortion and other undesired features are within permitted limits.

alignment chart   A chart having three or more scales across which a straightedge can be placed to provide a graphical solution for a particular problem. Also called a "nomograph" or "nomogram."

alignment error rate monitoring   A procedure by which the error rate of a signaling link is measured during the initial alignment.

alive   Either charged with dc potential different from the earth's or connected to a source of ac or dc power. Acoustically, an alive room is one with reflecting rather than absorbent walls.

alkaline cell   A primary cell using an alkaline electrolyte, often potassium hydroxide. Gives higher current drain than ordinary carbon-zinc dry cells.

all-channel amplifier   See amplifier, all channel CATV.

all-number calling   Calling using telephone numbers consisting of numerals only, e.g., seven numerals rather than two letters plus five numerals.

all-purpose computer   Same as general purpose computer; one not designed specifically for telecommunications on-line use.

all-relay automatic system   A small switching unit which only utilizes electromechanical relays with simple contact assemblies and no rotating or electronic components.

all trunks busy   When all the circuits in a group are occupied.

alley arm   Same as a sidearm.

alligator clip   A test clip having long, narrow jaws.

allocated circuit   A circuit designed and reserved for the use of a particular customer.

allocation, frequency   A band of radio frequencies designated for a particular service. International allocations were made by the World Administrative Radio Conference held in Geneva in 1979. Sub-allocations in the U.S. are made by the Federal Communications Commission. For frequency allocation purposes the world is divided into three regions: Region 1 - Europe, Africa, Arabia and Russia; Region 2 - North and South America and Greenland; Region 3 - Southeast Asia and Australasia.

allocations   The assignments of frequencies (by the FCC in the U.S.) for various communications uses (television, radio, land-mobile, defense, microwave, etc.) to achieve a fair division of the available spectrum and to minimize interference among users.

allotter   A switch which parcels out an incoming load to different operating units.

allotting, PBX   A method by which calls to a single directory number are distributed to different number-groups in the office. This spreads the load and increases traffic handling efficiency.

almost differential quasi-ternary code   A Russian proposal for line signals in digital systems.

alnico   An alloy used for permanent magnets usually made up of nickel 18%, aluminium 10%, cobalt 12%, copper 6%, and iron 54%.

alpeth   A cable sheath technology utilizing a corrugated aluminum tape with polyethelene oversheath.

alpha particle   The nucleus of a Helium atom. It is a stable, positively charged particle containing two protons and two neutrons.

alphabet   1. An ordered set of all the letters used in a language, including letters with diacritical signs where appropriate, but not including punctuation marks. 2. An ordered set of symbols used in a language, e.g., the Morse Code alphabet.

alphabet, phonetic   If there is a possibility of words being misheard over the telephone codes are used for each letter, the codes beginning with the letter in question.

alphabet translation   The process converting the meaning in a particular alphabet to one or more different alphabets in the same or different code.

alphabetic   Referring to data using only the letters of the alphabet.

alphageometric   Videotex technique building up letters and pictures by geometric colored shapes as used in Canadian Telidon.

alphameric display   A rectangle divided into a total of 14 segments which form all capital letters of the alphabet and all numbers from 0 to 9. It is commonly used to display information.

alphamosaic Videotex technique building up letters and pictures by small blocks of color as used in British Prestel and French Teletel.

alphanumeric/alphameric A generic term for alphabetic letters, numerical digits and special characters which are machine processable. Used to designate a characterset which contains letters, figures and punctuation marks.

alphaphotographic Videotex technique producing high quality still pictures on the TV screen or on a portion of the screen, as used by "Picture Prestel".

alternate mark inversion signal A pseudo-ternary signal, conveying binary digits, in which successive "marks" are normally of alternative polarity but equal in amplitude, and in which "space" is of zero amplitude. Sometimes called bipolar signal.

alternate mark inversion violation A "mark" which has the same polarity as the previous "mark" in the transmission of AMI signals. Sometimes called bipolar violation.

alternate route Syn.: alternative route, second choice route

alternate route cancel Network management feature which restricts a given percentage of traffic to direct routes only.

alternate route pattern The way a switch has been programmed to hunt for an available circuit to reach a required destination.

alternate routing Directing telephone long distance traffic to another route when the first choice route is busy. Engineering long-distance circuits to achieve this enables considerable economies to be made because direct, high-usage circuit groups are used extremely efficiently and overflow routes are shared by several different parcels of traffic.

alternating current Continuously variable current, rising to a maximum in one direction, falling to zero, then reversing direction and rising to a maximum in the other direction, then falling to zero and repeating the cycle. Usually follows a sinusoidal growth and decay curve.

alternating discharge current through a gas discharge protector is the rms value of an approximately sinusoidal alternating current flowing through the protector.

alternation One of the halves of a complete ac cycle.

alternative (alternate) route A second, or subsequent choice route between two reference points usually consisting of two or more circuit groups in tandem.

alternative routing (of signaling) The routing of a given signaling traffic flow in case of failures of signaling links or routes normally used.

alternative-routing indicator Information sent indicating that the call has been subjected to alternative routing.

alternator A generator of alternating electric power.

alternator, tone A unit in a central office or exchange which generates the advice tones (busy tone, dial tone, re-order tone, etc.) sent out to customers.

alt-route To employ alternate routing techniques.

alt-route automatic The action of a central office/exchange in sequential testing of trunks over several alternate routes in attempting to complete a call.

ALU Arithmetic and logic unit. The hardware part of a processor where arithmetic and logical operations are performed.

alumina Aluminum oxide, used in solid-state components as a dielectric.

aluminum Lightweight metal sometimes used as conducting wire in telephone cables. Its resistance is higher than that of copper wire of the same diameter.

Alvey UK collaborative research program in advanced information technology. (Named after the British Telecoms chairman of the committee responsible.)

ambient Surrounding; often used as shorthand for surrounding air.

ambient noise level The level of acoustic noise existing in a room or other location, as measured with a sound level meter. It is usually measured in decibels above a reference level of 0.00002 newton per square meter in SI units, or 0.0002 dyne per square centimeter in cgs units. Syn.: room noise level

ambient temperature Temperature of surrounding fluid, air, or gas.

ambiguity error An error in the reading of a number when digital representation is changing. All the digits may not change at exactly the same instant, i.e., a photograph could give an incorrect figure, such as 699 changing to 700 might be photographed as 600.

American Bell Inc. (ABI) Name used temporarily by AT&T unit offering enhanced services for business and residential customers. Now under control of AT&T Information Systems.

American Council for Competitive Telecommunications (ACCT) Washington, D.C.-based group which represents specialized common carriers, and value-added, resale and video carriers.

American Morse code A variant of International Morse code.

American National Standards Institute Organization supported by U.S. industry to establish uniformity of standards.

American Standard Code for Information Interchange (ASCII) A code with seven information signals plus one parity check signal, designed for interworking between computers.

American Wire Gauge Standard American method of classifying wire diameter; the Brown & Sharpe (B&S) gauge.

Ameritech Regional holding company for Midwestern states Bell Operating Companies.

ammeter Instrument which measures and records the amount of current in amperes flowing in a circuit.

ammeter, clamp-on An alternating current ammeter which does not have to be inserted in the circuit in order to measure the flow through the meter. It has a coil which can be wrapped around a conductor carrying the current which is measured then by its inductive effect.

amp Abbreviated form of ampere.

ampacity Used by power engineers for the current carrying capacity of a power cable. This is determined by the maximum continuous-performance temperature of the insulation, by the heat generated in the cable (as a result of conductor and insulation losses), and by the heat-dissipating properties of the cable and its environment.

amperage The strength of an electric current, in amperes.

ampere The ampere is the Base SI unit of electric current. The constant current which, if maintained in two straight parallel conductors of infinite length, of negligible circular cross-section, and placed 1 meter apart in vacuum, would produce between these conductors a force equal to 2×10^{-7} newton per meter length.

ampere-hour When the current is one ampere, the quantity of electricity that flows in one hour.

ampere per meter The SI unit of magnetic field strength. The field strength in the interior of an elongated uniformly wound solenoid that is excited with a linear current density in its winding of one ampere per meter of axial distance.

ampere-turns Product of the number of turns of a coil and the current in amperes flowing through the coil. This gives a measure of the magnetomotive force.

amperes rule The relationship between the direction of an electric current and its associated magnetic field.

amplidyne A rotating machine which in effect acts as a power amplifier. A small change in the excited field produces a rapid change in output voltage. Used in servo systems.

amplification The process which results in an output being an enlarged reproduction of the input. Amplifiers may be designed to give amplification of voltage, current or power, or a combination of these.

amplification, current Increase in current magnitude between one point and another.

amplification factor In a device giving constant current to its load, the ratio of the incremental change in output voltage required to maintain the constancy of the current to a corresponding incremental change in the input voltage.

amplifier A device which receives an input signal in wave form and gives as an output a magnified signal.

amplifier, all-channel CATV A broadband amplifier capable of amplifying without distortion television and broadcast radio signals.

amplifier, audio An amplifier designed to cover the normal audio frequency range.

amplifier, balanced A push-pull amplifier with two identical connected signal branches which operate in phase opposition, with input and output connections each balanced to ground.

amplifier, bridging A high input impedance amplifier which introduces a very small loss to the through circuit. Can be used for monitoring.

amplifier, broadband An amplifier used for a broad band of frequencies.

amplifier, buffer An amplifier stage used to isolate a frequency-sensitive circuit from variations in the load presented by following stages.

amplifier, CATV main line An amplifier in a case suitable for external use, able to amplify a broad band of radio frequencies, usually $50 - 220$ MHz, and powered by low-voltage mains frequency power over the same cable.

amplifier, Class A An amplifier in which output current flows during the whole of the input current cycle.

amplifier, Class AB An amplifier in which the output current flows for more than half but less than the whole of the input cycle.

amplifier, Class B An amplifier in which output current is cut off at zero input signal, i.e., a half-wave rectified output is produced.

amplifier, Class C An amplifier in which output current flows for less than half the input cycle.

amplifier, Class D An amplifier operating with pulse-width modulation.

amplifier, compensated An amplifier with consistent performance over a broad band width because its pass-band ends have been specially compensated to counter and cancel out defects in other circuitry parts.

amplifier, compression An amplifier which accepts as input a signal with wide variations in amplitude and produces an output with much smaller variations.

amplifier, differential An amplifier with two inputs but only one output. If inputs are identical the signal is suppressed and output is nil.

amplifier, direct current An amplifier able to amplify direct voltages.

amplifier, hard-of-hearing A deaf aid. A unit in a telephone handset which allows deaf persons to use a telephone instrument without difficulty.

amplifier, heterodyne CATV An amplifier which works on the heterodyne principle, i.e., converts the incoming TV signal to a lower frequency, amplifies this, and reconverts it to a normal TV frequency.

amplifier, high impedance See bridging amplifier.

amplifier, intermediate frequency The high efficiency fixed-frequency amplifiers in a super-heterodyne receiver. These amplify the received signals after their conversion to IF and before their demodulation to the audio range of frequencies.

amplifier, isolation A buffer amplifier, used to minimize the effects of a following circuit on a preceding circuit.

amplifier, line An amplifier for all signals on a line to compensate for line loss in the preceding section.

amplifier, linear An amplifier in which the instantaneous output signal is a linear function of the corresponding input signal.

amplifier, magnetic An amplifier with a control device dependent on magnetic saturation. A small DC signal on a control circuit triggers a large change in the impedance and hence in the output of the circuit.

amplifier, microphone A pre-amplifier. Amplifies the small output from a microphone to make it sufficient to be used as input signal to a power amplifier or another stage in a modulation circuit.

amplifier, monitoring A high impedance bridging amplifier which has no significant effect on the observed received level but enables service to be observed.

amplifier, operator's headset Small amplifier on the receive side of an operator's headset circuit.

amplifier, parametric A reactance amplifier. An inverting parametric device depending essentially on three frequencies: a harmonic of a pump frequency and two signal frequencies.

amplifier, power Amplifier designed to provide an increase in power, rather than just a voltage or current gain.

amplifier, program Amplifier with high quality components and careful design which produces an output signal with the desired frequency-gain characteristics over a wide audio frequency band, i.e., a high quality music band width, normally at least 10kHz.

amplifier, pulse Amplifier able to amplify received pulses and maintain their shape.

amplifier, push-pull A balanced amplifier. Two similar amplifying units connected in phase opposition in order to cancel out undesired harmonics and minimize distortion.

amplifier, sense Amplifier able to regenerate a weak sensing pulse to give it sufficient power to be recognized and acted upon by the next circuit stage.

amplifier, single-channel CATV Amplifier tuned to a particular television channel and able to pass the TV bandwidth without introducing distortion.

amplifier, tuned radio frequency Amplifier tuned to a particular radio frequency or band so that only selected frequencies are amplified.

amplifier, voltage Amplifier which produces an output with an increased voltage, usually across a high impedance.

amplifier stage An amplifier coupled to other devices in cascade so that the output of one stage provides the input to the next.

amplify To increase the amplitude of a signal.

amplitude The peak value of a varying quantity. The actual amplitude of a quantity at a particular instant often varies sinusoidally.

amplitude, pulse The maximum amplitude of a pulse signal.

amplitude distortion Distortion occurring in an amplifier or other device when the output amplitude is not a linear function of the input amplitude under specified conditions.

amplitude equalizer A corrective network which is designed to modify the amplitude characteristics of a circuit or system over a desired frequency range.

amplitude frequency response See insertion loss vs frequency characteristic.

amplitude hit In a data transmission channel, a momentary disturbance caused by a sudden change in the amplitude of the signal.

amplitude modulation A type of modulation in which the amplitude of the continuous or carrier wave is varied above and below its unmodulated level by an amount proportional to the amplitude and frequency of the information signal.

amplitude modulation equivalent See compatible sideband transmission.

amplitude quantized control A synchronization control system in which the functional relationship between actual phase error and derived error signal includes discontinuities.

amplitude quantized synchronization A synchronization control system in which the functional relationship between actual phase error and derived error signal includes discontinuities. Syn.: amplitude quantized control

amplitude-shift keying Data signals which produce a number of different amplitude levels of a sine-wave carrier.

amplitude-versus-frequency distortion That distortion in a transmission system caused by the nonuniform attenuation or gain of the system with respect to frequency under specified terminal conditions. Syn.: frequency distortion

ampoule, carbon monoxide detector Detection device used by external plant workers to check for dangerous concentrations of carbon monoxide in a manhole or cable jointing chamber.

analog Representation by continuously variable physical quantities.

analog channel A channel on which transmitted data can be between specified limits, i.e., there are no discrete packages or pulses of information.

analog circuit switch Real-time switching of an analog signal.

analog computer A computer in which quantities of variable size are represented by another physical quantity such as voltage or position.

analog control A synchronization control system in which the relationship between the actual phase error between clocks and the error signal device is a continuous function.

analog data A physical quantity that is considered to be continuously variable and whose magnitude is made directly proportional to the data or to a suitable function of the data.

analog decoding A process whereby one of a set of reconstructed analog signal samples is generated from the digital signal representing a sample.

analog/digital converter A device which converts an analog signal into a digital representation.

analog encoding A process in which digital signals are generated, representing the sample taken of an analog signal value at a given instant.

analog exchange   An exchange that switches analog signals.

analog junction   A transmission path designed to pass analog signals.

analog loop-back   Method of testing modem and business machines by disconnecting the telephone line and looping back the transmitted line signal into the local receiver.

analog signal   A nominally continuous electrical signal that varies in amplitude or frequency in response to changes of sound, light, heat, position, or pressure impressed on a transducer.

analog switch   A switching equipment designed, designated or used to connect circuits between users for real-time transmission of analog signals.

analog synchronization   A synchronization control system in which the relationship between the actual phase error between clocks and the error signal device is a continuous function over a given range. Syn.: analog control

analog-to-digital coder   See analog-to-digital encoder.

analog-to-digital converter   A device which converts an analog input signal to a digital output signal carrying equivalent information.

analog-to-digital encoder   A device for encoding analog signal samples. Synonyms: analog-to-digital coder; coder

analog transmission   Transmission of a continuously variable signal as opposed to a discretely variable signal.

analog value   A continuously variable value; with no discrete value steps.

analysis, systems   The technique of determining the most effective way of using modern computers to process large amounts of information very rapidly.

analyzer, circuit   A meter with switchable accessories so that different ranges of current, voltage, and resistance may be measured.

analyzer, line fault   A measuring instrument which uses the radar principle to detect and locate any line fault which causes an impedance irregularity. Pulses are transmitted out and reflected back by the fault and their travel time is analyzed to give the distance to the fault. Also called pulse return pattern or time domain reflectometry.

analyzer, network   Circuit elements made up into adjustable complexes of nodes to simulate a real life network so that real performance may be forecast from measurements taken of the model.

analyzer, spectrum   A calibrated visual display unit (cathode ray tube) combined with a tuneable narrow-band filter assembly which enables individual frequencies present in a spectrum to be separated out for measurement of amplitude or power.

analyzer, traffic   Device which scans groups of trunks and common equipment in a central office/exchange and periodically provides a printout indicating the traffic patterns for each circuit or device.

analyzer, wave   An instrument which studies complex wave signals and measures the frequency, amplitude and phase of its components.

anchor   To fasten securely. The device to which an item is fastened. In pole route construction and antenna mast erection, anchors of various types are buried in the ground and rods or guy lines attached to give stability to the above-ground structure.

anchor, cone   A small truncated cone made of sheet steel used as an underground anchor for poles or masts.

anchor, cross-plate   An anchor made of two rectangular steel plates, fitted to a single anchor rod. The plates are twisted to provide a large effective surface to handle the anchor's strain.

anchor, expanding   A type of anchor with blades which are forced out to cut into virgin soil and so improve stability.

anchor, guy   A buried anchor with an anchor rod extending above ground level for attachment of a guy.

anchor, hammer drive   A soft metal tube which is placed in a drilled hole in a masonry wall to securely hold a nail driven into the tube.

anchor, hollow wall   A metal tube, threaded internally at one end, is placed in a drilled hole. A screw is then inserted and tightened so that the tube deforms until it has expanded in effective diameter on the inside of the wall, providing a secure fitting for the screw.

anchor, log   A buried anchor made of one or two wooden logs and an anchor rod.

anchor, plate   A steel plate, either square or bent into a section of a cylinder, with an anchor rod bolted to its center.

anchor, pole key   Steel plates which may be placed against the buried portion of a pole to provide stability.

anchor, rock   A type of anchor designed to be inserted in a hole drilled into a rock and then expanded by turning to provide a secure grip.

anchor, screw   Type of anchor which may be screwed into the ground like a large corkscrew, with a steel strip helix welded to the guy rod.

anchor, swamp   A large form of screw anchor designed for use in very loose soil conditions.

anchor, wood screw   Small tube made of wood fiber, plastic or soft lead. Placed in a drilled hole in a masonry wall it expands into a tight fit when a wood screw is screwed in, and holds the screw securely.

AND gate   Solid state device which only gives a "1" output if all its inputs are also "1".

anechoic chamber   An acoustically dead room, i.e., one with no echoes or reverberations.

angle, radio take-off   The angle between the horizontal from an antenna and the line to the radio horizon.

angle modulation   The modulation of the phase angle of a carrier as in frequency modulation and phase modulation.

angle of arrival   Vertical angle between the horizontal plane and the path by which a radio signal arrives at a receiving antenna.

angle of departure Vertical angle between the horizontal plane and the path by which a radio signal leaves a transmitting antenna.

angle of deviation In optics, the net angular deflection experienced by a light ray after it passes through a refractive medium.

angle of incidence The acute angle between a ray and the normal to the surface on which the ray is incident.

angle of reflection Angle between a reflected ray and the normal to the surface by which the ray has been reflected.

Angstrom unit Unit for measuring very short wavelengths of light. It is 10^{-10} meter, (0.1 nm).

angular frequency The frequency of a periodic phenomenon expressed in radians per second. This is equal to the frequency in hertz multiplied by 2π.

angular misalignment loss An optical power loss caused by angular deviation from the optimum alignment of source to optical waveguide, waveguide to waveguide, or waveguide to detector.

anhydrous Containing no water.

Anik Canadian domestic satellite system.

anion A negatively charged atom, so one with a surplus of electrons which is attracted to an anode or positive electrode.

anisochronous A signal is anisochronous if the time interval separating any two significant instants is not necessarily related to the time interval separating any other two significant instants, e.g., a start-stop transmission as used by teletypes.

anisochronous data channel A communications channel capable of transmitting data but not timing information. Sometimes called an "asynchronous" data channel.

anisotropic Showing different electrical properties in different planes or along different axes.

anneal To heat a metal to a particular temperature, then let it cool slowly. Usually removes brittleness.

announcement machine A recording machine which gives voice guidance to a telephone user if the call cannot be connected, or the person requires information.

announcement trunk Trunk within a central office/exchange which provides access to an announcement machine.

announcer, automatic time A machine which advises callers of the correct time.

announcer, automatic time-temperature An announcing machine which combines the functions of a time announcer with information about local air temperature.

announcer, automatic weather A machine which advises callers of the latest weather reports for local areas.

announcer, intercept An announcing machine which is connected to vacant blocks of numbers in a central office.

annual charge ratio The ratio of the annual charge of one additional circuit on the alternative route to the annual charge of one additional circuit on the high-usage route. The first of these is calculated by summing the annual charge per circuit of each link comprising the alternative route, plus the annual charge of switching one circuit at each intermediate switching center.

annual transmission variations Changes in transmission due to climate changes. In modern equipment arrangements are usually made for these variations to be offset or compensated by built-in devices.

annular Doughnut or ring shaped.

annulling network Components added to filters to improve characteristics at the two ends of pass bands.

annunciator Original name for the indicator on magneto switchboards which indicated the particular line which was calling the exchange.

anode 1. Positive pole or element. 2. The outermost positive element in a vacuum tube, also called the plate. 3. The positive element of battery or cell. 4. A galvanic anode.

anode, galvanic Metal rods buried near underground telephone cable in areas where electrolytic corrosion is likely for metal cable sheath. These rods are connected electrically to the sheath with the moist soil completing an electrical loop, making the sheath a cathode and preventing unwanted electrolytic corrosion of the sheath.

anode, magnesium A galvanic anode used to provide cathodic protection of the cable sheath.

anodize Formation of a thin film of an oxide on a metallic surface, usually to produce an insulating layer.

anolyte Electrolyte with the distribution of ions appropriate to areas near an anode.

anomalistic period The time elapsing between two successive passages of a satellite through its periastron.

anomalous propagation Abnormal propagation due to discontinuities in the propagation medium and often resulting in the reception of signals well beyond their normal range.

anomaly A deviation beyond normal variations.

answer, night trunk Calls to a particular number may be answered from a different telephone during unattended periods by dialing a special code.

answer, universal night Feature provided by most PABXs. Incoming calls may be answered by any extension when attendant's console is unmanned.

answer signal Signal sent in the backward direction to indicate the called line has answered. Answer signals are usually loops to central offices from called subscribers, and battery reversals between offices.

answer supervision The off-hook indication sent back to the originating end when the called line answers. For calls controlled by manual operating positions, the cord circuit supervisory lamp normally dims on receipt of this signal.

answerback A signal sent by a data receiver to a data transmitter indicating it is ready to receive data, or is acknowledging the receipt of data. The answerback is typically part of "handshaking" between devices. For telex, the answerback is usually a code which readily identifies the called company.

answerback unit Unit which sends an answerback signal.

answering delay Time interval between the setting-up of an end-to-end connection between the calling and called stations, and the detection of an answer signal.

answering jack The jack on a manual operating position, associated with a particular line or trunk, into which an operator inserts a plug to answer an incoming call.

answering service, telephone Service provided when a subscriber is unable to answer his own phone. The line is switched to a central bureau which takes messages.

answering set, telephone Recording device attached to a telephone line. When the user is absent the device answers calls, gives a recorded message, and records a short message from the caller.

answering time of operators At the outgoing exchange, the interval between the end of the transmission of the calling signal and its answer by an operator at the distant exchange. At the incoming exchange, the interval between the appearance of a calling signal on a position and its answer by an operator.

antenna A device for transmitting or receiving electromagnetic radiation at radio frequencies.

antenna, all-channel A domestic TV or FM radio receiving antenna able to receive efficiently all the bands in local use.

antenna, aperiodic An antenna which is not periodic or resonant at particular frequencies and so can be used over a wide band of frequencies.

antenna, artificial A device which behaves, so far as the transmitter is concerned, like a proper antenna, but it does not radiate any power at radio frequencies.

antenna, beam A directional antenna.

antenna, billboard Antenna with a large flat or nearly flat reflector. Antennae used with troposcatter systems are often of this type.

antenna, boresight Small antenna used in setting up and calibrating a large directional antenna, such as a ground station working to a satellite system.

antenna, broadband An antenna which operates with acceptable efficiency over a wide band of frequencies without requiring retuning for each individual frequency.

antenna, Cassegrain A double reflecting antenna, often used for ground stations in satellite systems.

antenna, coaxial A dipole antenna made by folding back on itself a quarter wavelength of the outer conductor of a coaxial line, leaving a quarter wavelength of the inner conductor exposed.

antenna, corner An antenna within the angle formed by two plane reflecting surfaces.

antenna, dielectric lens An arrangement of elements in front of the radiating element in order to concentrate the effective beam.

antenna, dipole A center-fed antenna, one half-wavelength long.

antenna, directional An antenna designed to receive or emit radiation more efficiently in a particular direction.

antenna, discone A biconical antenna, omni-directional, fed at the apexes of the cones.

antenna, double doublet A pair of ordinary doublet antennae, at right angles to each other and crossing at their centers.

antenna, doublet A half wavelength long antenna, center fed.

antenna, dummy An artificial antenna, designed to accept power from the transmitter but not to radiate it.

antenna, ferrite Common receiving antenna, a small coil mounted on a short rod of a ferrite material.

antenna, flat top Antenna in which all the horizontal components are in the same horizontal plane.

antenna, folded dipole Two ordinary half-wave dipoles joined at their outer ends and fed at the center of one of the dipoles.

antenna, horn An antenna made by opening up a waveguide into the shape of a horn.

antenna, horn reflector An antenna in which the feed horn extends into a parabolic reflector, and the power is radiated through a window in the horn. Sometimes used instead of ordinary microwave dishes.

antenna, image The imaginary equivalent of an actual antenna, located as far below the ground plane as the actual antenna is above it.

antenna, isotropic A theoretical antenna in free space which transmits or receives with the same efficiency in all directions.

antenna, lens An antenna with metallic or dielectric elements placed in front of the radiating element in order to produce a required emission pattern.

antenna, log-periodic A broad band directional antenna with an array of dipoles of different lengths, with lengths and spacings between dipoles increasing logarithmically away from the feeder.

antenna, long wire An antenna made up of one or more conductors in a straight line pointing in the required direction with total length of several wavelengths at the operating frequency.

antenna, loop One or more turns of wire in the same or parallel planes.

antenna, Luneburg lens A circular antenna designed so that a feed connected at one point produces a major lobe in the direction diametrically opposite.

antenna, MUSA Multiple unit steerable array antenna. Several antennae, often rhombics, the outputs from which may be combined by adjusting phase relationships so as to steer the major lobe of the whole combination to a required direction.

antenna, non-resonant A non-periodic antenna which may be used for different frequencies.

antenna, omni-directional An antenna whose radiating or receiving properties are the same in all horizontal plane directions.

antenna, periodic A resonant antenna, designed for use at a particular frequency.

antenna, periscope Antenna used at some microwave stations, with dish mounted at ground level pointing to a plane reflector mounted at the top of a tower or mast, reflecting the beam to the required bearing.

antenna, pop-up a mast-mounted antenna, normally stored in a vertical shaft below ground level: when needed it is pushed up to its operating height.

antenna, quarter-wave A dipole antenna, with length of one quarter of a wavelength.

antenna, resonant An antenna operated at one of its frequencies of resonance due to standing waves.

antenna, rhombic Large diamond-shaped antenna, with sides of the diamond several wavelengths long. Fed at one of the corners, with directional efficiency in the direction of the diagonal.

antenna, series fed Vertical antenna, fed at its lower end.

antenna, shunt fed Vertical antenna with grounded base and fed at a given point above ground.

antenna, single channel An antenna, usually a dipole, cut to a precise length to give maximum efficiency at a particular frequency.

antenna, steerable An antenna so constructed that its major lobe may readily be changed in direction.

antenna, top-loaded A vertical antenna capacitively loaded at its upper end, often by simple enlargement or the attachment of a disc or plate.

antenna, turnstile An antenna with one or more tiers of horizontal dipoles, crossed at right angles to each other and with excitation of the dipoles in phase quadrature. Used for omni-directional broadcasting, particularly of television.

antenna, uni-directional Directional antenna with one major lobe.

antenna, wave Horizontal long-wire receiving antenna, several wavelengths long, with directivity in the direction of its length.

antenna, whip Thin semi-flexible metal rod or tube, fed at its base.

antenna, Yagi Directional antenna made up of a series of dipoles cut to different lengths. Directors are placed in front of the active dipole and reflectors behind the active element.

antenna array Several antennae coupled together to give a required degree of directivity.

antenna directivity factor The ratio of the power flux-density in the wanted direction to the average value of power flux density at crests in the antenna directivity pattern in the interference section. This is equivalent to the average improvement in signal-to-interference ratio achieved by using the actual antenna in place of an isotropic radiator in free space.

antenna gain-to-noise temperature For satellite earth terminal receiving systems, a figure of merit which equals G/T, where G is the gain in dB of the earth terminal antenna at the receive frequency, and T is the equivalent noise temperature of the receiving system in kelvins.

antenna height above average terrain An FCC rating for radio transmitting stations based on the average antenna height above the terrain in a circular band from 2 to 10 miles from the antenna.

antenna lobe A three-dimensional section of the radiation pattern of a directional antenna bounded by one or two cones of nulls or regions of diminished intensity.

antenna matching The process of adjusting impedance so that the input impedance of an antenna equals the characteristic impedance of its transmission line.

antenna noise temperature The temperature of a resistor having an available noise power per unit bandwidth equal to that at the antenna output at a specified frequency.

antenna pattern A diagram showing the efficiency of radiation in all directions from the antenna.

antenna preamplifier (CAVT) A small amplifier, usually mast-mounted, for amplifying weak signals to a level sufficient to compensate for down-lead losses and to supply sufficient input to system control devices.

antennas, nested rhombic Two rhombic antennae, one smaller than the other, so that the complete diamond-shaped antenna fits inside the area occupied by the larger unit.

anti-jamming Counter measures taken to reduce the effects of any jamming of a radio communication system.

antimony Metal much used in lead alloys and storage cell plates.

antimony battery Name for storage cell which utilizes lead-antimony alloy plates.

anti-node Point at which a voltage or other variable attains a maximum value in a standing wave pattern.

Antiope French version of teletext. A digital data broadcast service associated with a TV signal providing pages of text on a cyclic repetition basis.

antireflection coating A thin, dielectric or metallic film applied to an optical surface to reduce the reflectance and increase the transmittance.

anti-resonance Parallel resonance.

antisidetone circuit Circuit designed to produce a limited reproduction in one's own earpiece of the output from one's own microphone in a telephone instrument.

antisidetone telephone set Telephone instrument so designed that only a very small part of the transmitted speech passes through the earpiece.

anti-siphoning FCC rules which prevent cable systems from "siphoning off" programming for pay cable channels that otherwise would be seen on conventional broadcast TV. "Anti-siphoning" rules state that only movies no older than three years and sports events not ordinarily seen on television can be cablecast.

anti-stuffing device Item to enable pay phones to be modified to eliminate this difficulty.

antivox Use of a voice actuated circuit to prevent operation of a transmitter when an associated receiver is in use.

aperiodic Not resonant nor periodic.

aperiodic antenna An antenna designed to have an approximately constant input impedance over a wide range of frequencies. Syn.: nonresonant antenna

aperture That portion of a plane surface near a unidirectional antenna, perpendicular to the direction of maximum radiation intensity, through which the major part of the radiation passes. The clear diameter of the parabolic reflector of a microwave antenna.

aperture-to-medium coupling loss The difference between the theoretical gain of a very large antenna (as used in beyond-the-horizon microwave links) and the gain that can be realized in operation. It is related to the ratio of the scatter angle to the antenna beamwidth.

APL (a programming language) A computer programming language developed for mathematical, economics and business problems.

apoastron The point in the orbit of a satellite or planet which is a maximum distance from the center of mass of the primary body.

apogee The point in the orbit of an earth satellite which is at a maximum distance from the center of the Earth; the apogee is the apoastron of an earth satellite.

APP (application date) Abbreviation used in Bell System's Universal System Service Order. The date on which the order's requester provides a firm commitment and sufficient information to proceed with writing the primary order.

apparatus, station The equipment installed in a telephone subscriber's premises. Also called subscriber's apparatus or customer's apparatus.

appearance The point at which a circuit is terminated and access is gained to the circuit, with special reference to terminations on boards attended by operators.

Appleton layers Reflecting layers in the ionosphere, the F1 and F2 layers.

application layer The top layer of the Open Systems Interconnection seven-layer logical structure for data services; the end-user layer.

application package A computer program designed to perform a particular type of work.

application program A computer program which enables the processor (including both the hardware and the operating system software) to be applied to a task.

applique A small addition to a standard circuit to enable it to perform additional functions or provide additional features.

applique circuit Circuit modified to provide additional facilities.

Arabian Satellite Communications Organization Organization controlled by Arab governments: Arabsat.

ARAEN Reference apparatus for the determination of transmission performance ratings.

arc A sustained luminous discharge between electrodes.

arc-back Reverse flow of current during what is normally the non-conducting half-cycle of applied alternating voltage in a rectifying tube.

arc chute Device which ensures rapid extinguishment of an arc between contacts in a circuit-breaker by enclosing the arc between plates.

arc current through a gas-discharge protector The current which flows after spark-over when the gap of the protector is bridged by an arc.

architectural protection Relates to facilities built into a system to ensure security is maintained, e.g., that one program cannot interfere with another or access be given to data which should not be available to the other user. Both hardware and software design are involved in this.

architecture 1. The interaction between hardware and software in a computing system to achieve the most economic, efficient, secure, rapid, or easiest-to-maintain system. 2. The arrangement and design of exchange or computer subsystems, or of elements of the local memory capability available to a processor.

archived file A data file which is not available for immediate on-line access. Magnetic tapes are the most common form of archive at present.

archiving The process of maintaining archived files.

arcing Sustained luminous passage of current between contacts or electrodes.

area, accounting Defined territorial area for which accounting records are kept in separate files.

area, base rate That part of an area served by a central office in which all services are provided without an additional distance-related charge.

area, central office Defined territorial area served from a particular central office.

area, cross-sectional The sum of the cross-sectional areas of all the wires in a particular cable serving one function such as a power cable or a stranded wire. Must be measured at right angles to the length of the cable or conductors.

area, exchange Defined territorial area served by one or more central offices in which a single schedule of charges is applicable.

area, foreign Area other than that in which the calling subscriber's line is located. Outside the U.S., a different country is involved; in the U.S. a different numbering plan area is involved.

area, heavy loading Territory in which the weather history is such that overhead wires may be subjected to very low temperatures, severe winds and heavy icing.

area, home The numbering plan area of the calling subscriber.

area, light loading Territory in which the weather history is such that overhead wires may be subjected to light to moderate conditions, with no icing at all.

area, local telephone service Territorial area in which a local tariff applies.

area, maintenance Territorial area in which one group of maintenance personnel are responsible for repair services.

area, medium loading Territory in which the weather history is such that overhead wires may be subjected to moderate conditions with some icing.

area, multi-office, or multi-exchange An exchange area, typically in a metropolitan area, served by several central offices.

area, numbering plan In the U.S. and Canada, a territorial area in which central office codes are not duplicated.

area, operating Territorial area in which construction, operation and maintenance services are allocated by the telephone company management to one organizational group, often with a measure of financial autonomy.

area, primary service Area in which ground-wave reception of transmitted radio signals is dependable and satisfactory.

area, secondary service Area in which received signals from a broadcast transmitter are not always dependable, and fading and interference can occur.

area, service Territorial area served from a single central office or several central offices all located in one building complex.

area, testing Territorial area assigned to a group which performs all trouble testing.

area code The numbering plan area code in North America. The first digit is never 1 or 0, and the second digit is always a 1 or 0.

Arecibo The site in Puerto Rico of the world's largest parabolic antennae, which is constructed in a huge amphitheater-shaped valley.

Ariane European Space Agency rocket used for launching of geostationary satellites.

Ariel Range of UK research satellites.

arithmetic and logic unit The hardware part of a processor which deals with arithmetical and logical operations.

arithmetic expression A combination of arithmetic delimiters, numerals (decimal, hexa-decimal octal or binary) and identifiers enclosed by parenthesis.

arithmetic unit A unit within a processor which only performs arithmetical operations. Sometimes the ALU is called the arithmetic unit, in which case logical operations also are carried out.

arm A wooden rectangular cross-section attached to a pole which carries insulators supporting telephone wires. Sometimes called crossarms. Different lengths are used, depending on the numbers of circuits to be carried.

arm, cable Steel arm attached to a pole, used to support aerial cables rather than open wires.

arm, cable extension Short steel arm used to support an aerial cable and suspend it farther away from the pole.

arm, cable form A branch bundle of wires leaving a main cable form.

arm, extension A crossarm used to extend an existing arm to carry extra circuits or to give extra height.

arm, guard Crossarm erected on the pole in line with the route to give protection to a messenger cable or to open wires.

armature Movable part of a magnetic circuit. Typically relay armatures, which hinge towards the core when the relay operates and force contacts in the springset to open or close, or the armatures in microphones or earpieces which affect or are affected by sound pressure variations via a diaphragm.

armature winding The winding of an electrical machine, either motor or generator, in which current is induced.

armor Steel wires or tapes wrapped around a telephone cable to provide physical strength or protection.

armor, caged Cable armor in which the armor wires are inside a polyethelene jacket. Used on some submarine cables.

armor wire Zinc-coated mild or high-tensile steel wire of various gauges to protect and provide tensile strength for cables.

ARQ Automatic repeat request. A method of error correction whereby a transmitting terminal is commanded to retransmit any data blocks found by the receiving terminal to contain errors.

array, antenna Several directional antennae so placed and interconnected that directivity may be enhanced.

array, broadside An antenna array whose elements are all in the same plane, producing a major lobe perpendicular to the plane.

array, co-linear An antenna array whose elements are in the same line, either horizontal or vertical.

array, end-fire An antenna array whose elements are in parallel rows, one behind the other, producing a major lobe perpendicular to the plane in which individual elements are placed.

array, linear Antenna array whose elements are arranged end-to-end.

array, stacked An antenna array whose elements are stacked, one above the other.

arrester A protective device which provides a low impedance path for dangerous voltages such as lightning, bypassing the equipment by grounding the circuit.

arrester, gas-filled Protective device fitted in central offices threatened by lightning. Two electrodes are sealed in a small gas-filled tube, the gas ionizes when high voltages are received and provides a rapid low-resistance path to ground, thereby protecting the apparatus.

arrester, gas tube surge A gas-filled protective device.

Arrhenius equation A mathematical formula used in designing accelerated life tests where operating

temperatures are likely to affect component degradation.

articulated Describes some early types of submarine cable repeaters made up of jointed strings of watertight containers, intended to behave mechanically in the same way as the submarine cable itself.

articulation The ability to produce clear and distinct speech.

articulation index To measure the articulation index, the speech spectrum is divided into several unequal bands which contribute equally to intelligibility (in terms of a subjectively-measured articulation score). The intensity of speech varies according to the band so weighting factors are introduced, according to the ratio of the speech energy in the band to the hearing threshold.

articulation reference equivalent (AEN) From the French "affaiblissement equivalent pour la nettete." The AEN of a system is measured by carrying out articulation tests alternately on the system under test and on the standard reference system for the determination of AEN, the SRAEN.

articulation score The AS is the basic measure of the intelligibility of a voice system, in terms of the percentage of words correctly understood over a channel perturbed by interference. AS figures do however vary with bandwidth, the number of syllables in the words, speaker-listener familiarity and signal/noise ratio.

artificial ear A device which has the same acoustic impedance as the average external human ear. The artificial ear comprises an acoustic network and a measurement microphone which permit calibration of earphones used in audiometry.

artificial intelligence (AI) The use of knowledge-based computing systems instead of Von Neumann type information-based systems.

artificial line A complex of resistances, inductances, etc., which provides the same overall parameters as a real transmission line or route, including its characteristic impedance and frequency versus attenuation characteristics.

ascending (descending) node The point at which the orbit of a satellite or planet intersects the reference plane.

ASKY Pronounciation of ASCII, the American Standard Code for Information Interchange.

ASR (automatic send and receive) A teletypewriter terminal with paper tape, magnetic or solid-state storage.

assemble To translate a program expressed in an assembly language and to link subroutines.

assembler A program used to assemble; a program capable of translating assembly code into object code. Also called assembly program.

assembly 1. The production of a complete program in machine language. 2. Putting together equipment in order to carry out a task.

assembly, cable A factory-made cable length, complete with connecters at each end.

assembly language A computer-oriented programming language whose statements may be instructions or declarations. The instructions usually have a one-to-one correspondence with machine instructions. Syn.: computer-dependent language

assigned frequency A radio frequency which has been reserved for use by a particular organization.

assigner The circuit elements or logic units which distribute traffic to different items of common control equipment. Also, the operation staff personnel who allocate work loads, equipment or plant to specific users or requirements.

assigning The allocation duties performed by an assigner.

assignment, frequency The process by which authority is obtained for the use of particular radio frequencies for particular purposes.

assignment date Abbreviation used in Bell System's Universal System Service Order. The date on which information required to design the service is to be available. All required facilities and equipment are to be reserved, assigned and/or ordered.

associated mode The mode where messages for a signaling relation involving two adjacent signaling points are conveyed over a directly interconnected signaling link.

associated signaling A mode of operation of a common channel signaling system in which the signals carried by the system relate to a group of speech circuits which terminate in the same exchanges as the signaling system.

Association of Data Processing Service Organizations (ADAPSO) An association representing all sides of this new industry.

Association of Long Distance Telephone Companies (ALTEL) Trade group representing all the major and many of the minor long distance carriers.

assumed values A range of values, parameters, levels, etc., assumed for a mathematical model, hypothetical circuit, or network, from which analyses, additional estimates, or calculations will be made. The range of values, while not measured, represents the best engineering judgment and is generally derived from values found or measured in real circuits or networks of the same generic type, and includes projected improvements.

astatic Term used in mechanical telegraph days to indicate a relay or other system which had no bias or normal rest position in neutral equilibrium.

asymmetrical modulator See unbalanced modulator.

asynchronous Not synchronous.

asynchronous communication system A data communication system in which extra signal elements are appended to the data for the purpose of synchronizing individual data characters or blocks. Syn.: start-stop system

asynchronous data channel See anisochronous data channel.

asynchronous modem A modem which cannot transmit timing information in addition to data: a modem which does not require synchronization with the associated terminal equipment. Sometimes called an "anisochronous" modem.

asynchronous network See nonsynchronous network.

asynchronous operation 1. A sequence in which operations are executed out of time coincidence with any event. 2. An operation that occurs without a regular or predictable time relationship to a specified event. Syn.: asynchronous working

asynchronous tape Recorded bit lengths and spacings on magnetic tape which are variable because of tape transport differences.

asynchronous terminal A terminal using a start-stop data transmission method.

asynchronous time-division multiplexing An asynchronous transmission mode that makes use of time-division multiplexing.

asynchronous transmission 1. A transmission process wherein between any two significant instants in the same group there is always an integral number of unit intervals. 2. Transmission in which each information character is individually synchronized, usually by the use of start and stop elements. Syn.: start-stop transmission

asynchronous working See asynchronous operation.

ATIC (time assignment with sample interpolation) A TASI system for pulse code transmission systems (from the Italian name).

ATM Reciprocity ability to use cards from one credit card organization or bank to obtain services from the Automatic Teller Machines of another body.

atmosphere The gaseous envelope surrounding the Earth, composed of 78% nitrogen, 21% oxygen, 0.9% argon, plus some carbon dioxide and water vapor. The atmosphere is divided into several layers: troposphere, stratosphere, ionosphere, and exosphere.

atmosphere, explosive Air into which other gases or dust powder concentrations have been leaked, making the area dangerous and capable of explosive ignition.

atmospheric duct A layer in the lower atmosphere, occasionally of great horizontal extent, in which the vertical refractivity gradients are such that radio signals are guided or focused within the duct and tend to follow the curvature of the earth with much less than normal attenuation.

atmospheric noise Radio noise caused by natural atmospheric processes, primarily by lightning.

atmospherics Noise on circuits, either line or radio, resulting from lightning or other natural phenomena.

atomic time scale A time scale based on the periodicities of atomic or molecular phenomena.

attachments General name for all outside plant items used in conjunction with poles and route erection.

attachments, authorized Equipment which, although not provided by the telephone company, has access to telephone company services.

attachments, foreign Any items of equipment fixed to a telephone company's pole line by another authority or organization, or items connected to telephone lines without due authority.

attachments, tree Brackets, etc., used on trees used to help support telephone cables or wires.

attack time In an echo suppressor, the time interval between the instant that the signal level at the receive "in" port exceeds the suppression activate point and the instant when the suppression loss is introduced.

attempt The initiation of a telephone call. The repeat of the action is a "retrial" or a "subsequent attempt."

attendant The operator at a PABX console or operating position.

attendant exclusion PABX feature which bars the attendant from monitoring an established call.

attendant recall PABX feature which brings the attendant back in to a circuit to provide assistance when needed.

attendant's cabinet PABX console or operating position.

attendant's switchboard A manual operating position in an automatic exchange/central office.

attended A central office or radio station which normally has maintenance staff on duty.

attended pay station (APS) See attended public telephone.

attended public telephone A public telephone supervised by an attendant who can collect monies due, obtain long distance numbers, and provide general help.

attenuate To reduce in quantity or power.

attenuation The reduction in power level due to line resistance, leakages, induction, etc., resulting in the received signal being lower in volume than the original transmitted signal. In optical fiber systems there are other causes of attenuation, such as absorption, scattering, and losses into radiation modes.

attenuation, free space The theoretical loss between transmitting and receiving antennae.

attenuation, rain Loss of power due to passage of radio signals through heavily moisture-laden clouds or storms.

attenuation coefficient Measure of the attenuation resulting from propagation of a signal along a line. If the voltage at the sending end of a line λkm in length is V_s, and voltage at the receiving end is V_r, then

$$\alpha\lambda = 20 \log \frac{V_s}{V_r}$$

where α is the attenuation coefficient in db per kilometer.

attenuation constant The real part of the propagation constant.

attenuation equalizer Circuit components which enable the frequency and attenuation characteristics of a route to be adjusted to a common standard.

attenuation limited operation The condition prevailing when the signal amplitude (rather than distortion) limits the communication capacity.

attenuation slope on each side of a radio receiver's passband is the ratio of the difference in the attenuations corresponding to two different frequencies beyond the passband to the difference between these frequencies. It is expressed in dB/Hz or dB/kHz. This property is important where an excessively large attenuation slope may lead to serious distortion of the phase/frequency characteristics in the passband.

attenuator Circuit component which introduces loss into the circuit with no distortion.

attitude-stabilized satellite A satellite with at least one axis maintained in a specified direction.

attraction The attractive force between two unlike magnetic poles (N/S) or electrically charged bodies (+/−).

audibility The quality of being able to be heard. For most humans the frequency range of audibility extends roughly from 15 Hz to 15 kHz

audible Range of sound which can be heard by a normal human ear.

audible ringing tone The information tone sent back to the calling telephone subscriber as an indication that the called line is being rung.

audible signals Information tones sent to calling subscribers to advise on the status of the call attempt.

audio Signals or equipment which use or produce signals in the human audio range of about 15 Hz to 15 kHz.

audio frequency The human ear's frequency band, normally from about 15 Hz to about 15 kHz.

audio frequency protection ratio Agreed minimum value of the audio frequency signal-to-interference ratio considered necessary to achieve a subjectively defined reception quality.

audio frequency (AF) signal-to-interference ratio Ratio (expressed in dB) between the values of the voltage of the wanted signal and the voltage of the interference. This ratio corresponds closely to the difference in volume of sound (expressed in dB) between the wanted program and the interference.

audio response unit Output device which provides a spoken response to digital inquiries from a telephone or other device. The response is usually assembled by a computer from a prerecorded vocabulary of words.

audiogram Plot of hearing loss against frequency, for each ear.

audiometer Instrument used in the measurement of hearing.

auditing 1. Checking compliance with authorized procedures and financial regulations; also checking and confirming the accuracy of recorded bookkeeping and accounting records. 2. Verifying the integrity of a system, and attempting to correct errors when detected.

auger, earth A drill used to bore holes in the ground to erect telephone poles.

aural Pertaining to audibility and the sense of hearing.

aurora (or aurora borealis) Flashes of light and luminous bands seen in the night sky in polar latitudes. They are caused by sun-spot activity and usually disrupt radio communications.

authentication Checks included in computer programs to ensure that only those authorized to extract information or amend instructions in a computer are permitted to take such action.

authenticator Words or numbers which must be inserted in the instructions at an appropriate point before a processor will continue with action.

authorization code (auth/code) A unique multidigit code identifying an authorized subscriber. The authorization code is validated via a translation and the call is allowed only if the authorization code is valid. The authorization code is also generally used to identify the routing to be used to place the call and may also identify call restriction and class of service data.

authorized frequency A frequency that is allocated and assigned by a competent authority to a specific user for a specific purpose.

authorized power Maximum power which may be broadcast by a licensed radio station.

authorized user (of satellite channels) Organization (U.S.) permitted to deal directly with COMSAT with regard to satellite facilities.

autoband Automatic data rate determination. The first character to be transmitted following completion of modem hand-shake is an autoband identity character.

autodialer Device which will dial a subscriber's number when a single key or button is pressed.

AUTODIN A world-wide data communications network of the U.S. Defense Communications System. Acronym for automatic digital network.

AUTODIN operation 1. Mode I. A duplex synchronous operation with automatic error and channel controls allowing independent and simultaneous two-way operation. 2. Mode II. A duplex asynchronous operation, without automatic error and channel controls, allowing simultaneous two-way operation. 3. Mode III. A duplex synchronous operation with automatic error and channel controls utilizing one-way message transmission. The return direction is used exclusively for error control and channel coordination purposes. The Mode III channel is reversible on a message basis. 4. Mode IV. A unidirectional asynchronous operation, send or receive only, without error control and channel coordination. The Mode IV channel is nonreversible. 5. Mode V. A duplex asynchronous operation allowing independent and simultaneous two-way transmission. Control characters are used to acknowledge receipt of messages and perform limited channel coordination.

autofax A public facsimile service, renamed Fonofax, operated by British Telecom.

automanual center A center with switchboards for handling subscribers' traffic to operators, including enquiry and directory enquiry traffic.

automated coin toll service A service which relieves operators of handling routine toll call charges from coin boxes. ACTS computes charges, announces these to the customer, counts coins as they are deposited, and sets up the call.

automatic   Action taken by machines (including computers) on receipt of instructions from humans.

automatic alternative routing   During the setting up of a call, the automatic selection of an alternative routing where the originally selected out-going route is unavailable.

automatic answer   The answering of a telephone call by a tape recorder which can pass on a short message and record an incoming message.

automatic answering   A facility by which the called data terminal equipment automatically responds to the calling signal and the call is established.

automatic booked call   With prior indication from a subscriber, a call may be made automatically from his telephone termination to a particular telephone number at a specific date and time.

automatic call distribution   A system designed to evenly distribute heavy incoming traffic among clerks or attendants.

automatic calling   A facility by which selection signals must be entered continuously at the full character rate. The address characters will be generated in the data terminal equipment.

automatic calling card service   Service enabling long distance calls to be originated and charged to a customer's own account, based on the use of a coded card and special telephone sets.

automatic calling unit   A dialing device supplied by a common carrier that permits a business machine to dial calls automatically over the public switched telephone network.

automatic credit card service   Payment of call charges is made by placing a credit card in a specially adapted telephone.

automatic data processing   An interacting assembly of procedures, processes, methods, personnel, and equipment to perform automatically a series of data processing operations that result in a change in the semantic content of the data.

automatic dialing unit   Device capable of automatically transmitting a predetermined set of dialed digits.

automatic exchange   A telephone system in which communications between subscribers is effected without the aid of an operator by means of switches set in operation by the originating subscriber's equipment.

automatic frequency control   Circuitry which controls the frequency of oscillators in equipment within specified limits.

automatic gain correction   An electronic circuit which compares the level of an incoming signal with a previously defined standard and automatically amplifies or attenuates that signal so it arrives at its destination at the correct level.

automatic identified outward dialing   PBX feature which identifies the extension originating a call and enables charges to be allocated to the station concerned.

automatic line testing (ALT)   Facility provided by switching centers/central offices; lines including subscribers' loops are tested during low traffic periods. ALT often enables faults to be found (and cleared) before notification by the subscriber.

automatic message accounting   A system which automatically records all the data of a dialed long distance call needed for preparation of an itemized bill for subscribers. This system also can provide information for inter-administration accounting.

automatic message switching center   Location at which messages are automatically directed according to routing information within the message.

automatic number identification   Equipment which identifies the telephone number of the line initiating a call in order to send this information to the message accounting apparatus.

automatic outward identified dialing   Ability of a switching system to identify the telephone number of the line originating a call, without operator intervention.

automatic personal call   With prior indication from a subscriber, an operator may be associated with an automatically dialed call at the appropriate stage to determine if the called person is available.

automatic repeat attempt   When difficulty is encountered in setting up a call, arrangements can be provided to make another attempt.

automatic request-repeat   A system of error control for data transmission in which the receiving terminal can detect a transmission error and automatically transmit a request-repeat signal to the transmit terminal. The transmit terminal then retransmits the character, code block, or message until it is either correctly received or the error persists beyond a predetermined number of transmittals. Syn.: decision-feedback system; error-detecting-and-feedback system

automatic re-routing   During the setting up of a call, the automatic selection of alternative routing where a congestion signal is received over the originally selected route from a distant switching center.

automatic route selection   Device (or software) which chooses the lowest cost route for long distance calls from WATS lines, leased lines, lines from specialized non-Bell common carriers (e.g., MCI, GTE Sprint), ordinary DDD lines. Syn.: least cost routing

automatic send/receive   A teletypewriter with keyboard and printer plus papertape reader/transmitter and papertape punch (or their electronic equivalents).

automatic sensitivity control   Equipment which controls and adjusts the sensitivity within specified limits.

automatic sequential connection   A facility provided by a public data service to automatically connect, in a predetermined sequence, the data terminal equipment at various addresses to a single data terminal.

automatic signal trunk   A method of controlling the signaling on circuits terminated on manual operating positions. Also called straightforward trunk or plug-supervision trunk.

automatic slope control Automatic correction of changes in slope by compensating circuitry.

automatic switching equipment Equipment in which switching operations are performed by electrically controlled apparatus without the intervention of operators.

automatic system A system in which the switching operations are performed by electrically controlled devices without the intervention of operators.

automatic telephone payment Payment of bills or fund transfer via an ordinary telephone line, (using keypad, not dial). The user calls the computer as for an ordinary telephone call, then keys in code numbers to identify himself, the transaction required to be made, and the amount. The computer does the rest.

automatic test equipment (ATE) Equipment which tests performance degradation (e.g., of long distance circuits) with minimum reliance on human intervention and is able to isolate sections or components found to be outside permitted tolerances.

automatic toll ticketing A system which automatically produces a punched toll ticket or magnetic tape record for every chargeable long distance call originated by subscribers on a direct distance dialing or subscriber trunk dialing basis. This enables bills to be prepared on an itemized call basis.

automatic transferred charge call The automatic debiting to a called subscriber's account of relevant charges for calls made to his telephone number.

automatic transferred debiting of call charges The automatic debiting to a subscriber's account of charges for calls made from any telephone by persons identified by the use of a secret code.

automatic transmission measuring and signaling testing equipment Equipment able to make transmission measurements and signal functional tests on circuits terminating on long distance switching centers. Used almost exclusively on long-distance international circuits.

automatic volume control A gain control, at radio or audio frequency stages, which enables output to be kept effectively constant despite changes in received signal level.

autoquote (AQ) Service provided for hotels by which billing information for calls orginated from guest rooms is automatically transmitted to a receive-only teletypewriter at the hotel.

AUTOSEVOCOM (automatic secure voice communications) A Defense Communications Agency system utilizing crypto-secure telephone instruments and high quality trunks.

auto-start A standby power system which starts up when the public supply fails.

autotransformer Transformer in which both the primary and secondary currents flow through one common part of the coil.

AUTOVON (automatic voice network) The worldwide voice system of the U.S. Department of Defense. High quality trunks are used and different grades of priority pre-emption are provided. Autovon also carries message-switched digital traffic of the U.S. AUTODIN system. See entry under triple.

auxiliary channel In data transmission, a secondary channel whose direction of transmission is independent of the primary channel and is controlled by an appropriate set of secondary control interchange circuits.

auxiliary operation An off-line operation performed by equipment not under control of the central processing unit.

auxiliary operator services system System enabling operators to provide services such as directory assistance and call intercept.

auxiliary power An alternate source of electric power, serving as back-up for the primary power at the station main bus or prescribed sub-bus.

auxiliary route A traffic route within a traffic routing hierarchy which is not a basic route.

availability 1. A measure of the degree to which a system, subsystem or equipment is operable and not in a stage of congestion or failure at any given point in time. 2. The percentage of total trunks in a group which can be accessed by a particular switch.

availability, full A method of providing switched services in which every input to a switching stage has or can have access to every output from the stage.

availability, satellite The existence of paths at a particular time from one ground station to the satellite and back to another ground station.

available line In facsimile transmission, the portion of the scanning line which can be used specifically for picture signals. Syn.: useful line

avalanche effect The effect obtained when the electric field across a barrier region is strong enough for electrons to collide with valence electrons, thereby releasing more electrons and giving a cumulative multiplication effect in a semi-conductor.

avalanche noise Noise contributed to a signal during photocurrent amplification in an avalanche photodiode.

avalanche photodiode A photodiode designed to take advantage of avalanche multiplication of photocurrent. As the reverse-bias voltage approaches the breakdown voltage, hole-electron pairs created by absorbed photons acquire sufficient energy to create additional hole-electron pairs when they collide with ions; thus a multiplication (signal gain) effect is achieved.

average block length The average value of the total number of bits in blocks transferred across a source-user/telecommunication system functional interface. The average block length is specified by the telecommunication system operator and is used in determining values for the block-oriented performance parameters. In systems where the information transferred across the functional interface is not delimited into blocks, the average block length is defined as numerically equal to the data signaling rate.

average busy season (ABS) The three months, not necessarily consecutive, having the highest average time-consistent busy hour traffic.

average busy season busy hour (ABSBH) A time-consistent hour having the highest average business-day load throughout the busy season.

average call duration Obtained by dividing the total number of minutes of conversation recorded by the recorded number of effective calls.

average holding time The sum of the durations of all call attempts made by users during the mean busy hour, divided by the total number of call attempts. The average length of time for which the equipment is in use for call attempts.

axial lead A connecting lead from a resistor or capacitor which comes out from an end, along the axis.

axial ray A light ray that travels along the optical axis.

ayame Japanese geostationary telecommunications satellite.

azimuth The horizontal angle between the north-south line and a particular direction. Scale is given going eastwards from North.

B board Manual switchboard position dealing with incoming calls from other nearby areas to be connected to local subscribers.

B operator Operator at a manual system operating position who answers calls from a distant A board and connects these directly to subscribers terminated on the B board.

B position Manually operated switchboard position dealing with incoming calls from other toll centers.

babble Crosstalk from several interfering channels.

babbling idiot Peripheral equipment which, due to a fault condition, sends repeated messages as fast as it can, usually operating in a loop. Syn.: Intermittent message overflow

babyphone A service providing for a call to be made to a telephone in the off-hook condition for the purpose of audible supervision at the called subscriber's premises.

back bias A feedback signal, in particular a reverse or negative bias.

back board Board mounted on wall to facilitate the mounting of equipment.

back contact Contact which opens when a relay operates.

back cord On a manual operating position the cord nearer the face of the position.

back emf (also counter emf) Voltage induced in reverse direction when current flows through an inductance.

back-haul Routing of a call which appears to take an illogical path through a network. Also, a link between a central office and a main microwave or multiplex station, or between a ground satellite station or submarine cable terminal and switching center controlling the circuits concerned.

back lobe The lobe directly behind the major lobe in a directivity pattern of a radio antenna.

back porch For color TV, that part of the signal between the color burst and the trailing edge of the corresponding blanking pulse. For monochrome TV, the signal part between trailing edge of the horizontal synch pulse and the trailing edge of the blanking pulse.

back scattering 1. In radio wave propagation, that form of wave scattering in which at least one component of the scattered wave is deflected opposite to the direction of propagation of the incident wave. 2. In optic fiber transmission, the deflection of light from its original direction by scattering which reverses the direction of propagation in the optical waveguide.

back-to-back connection Connection between the output of a receiver and the input of a transmitter. Also used when multiplex systems are directly connected together or when a circuit goes through a central office train of switches in a non-hierarchic fashion.

back-to-back coupling Relationship between the power emitted by a transmitting antenna and the power received by a receiving antenna on the same tower but facing in the opposite directions.

back-up A circuit element or facility used to replace an element which has failed.

backbone The high-density portion of any communications network.

backfill Material used to refill a trench after ducts or cables have been laid therein. Also used as a verb, to refill trench.

background noise The total system noise in the absence of information transmission, independent of the presence or absence of a signal.

background processing Processing, usually batch-type, that is not subject to real-time constraints and can be executed whenever computing facilities are not required by real-time programs or those of higher priority.

backing store A large capacity store with slightly longer access time than the immediate access stores used for most calculations.

backoff Procedure used in a data network after a collision and before a second attempt is made to access the network.

backoff, multiple access satellite Avoidance of intermodulation interference in a satellite by reducing the power of a ground station.

backplane Connector blocks and special wiring on rear of shelf. Printed circuit board modules normally mount in front of the backplane.

backscatter 1. The deflection by reflection or refraction of an electromagnetic wave or signal in such a manner that a component of the wave is deflected opposite to the direction of propagation of the incident wave or signal. 2. The component of an electromagnetic wave or signal that is deflected by reflection or refraction opposite to the direction of propagation of the incident wave or signal. 3. To deflect, by reflection or refraction, an electromagnetic wave or signal in such a manner that a component of the wave or signal is deflected opposite to the direction of propagation of the incident wave or signal. 4. That small fraction of light which is deflected out of its original direction along an optic fiber by scattering and suffers a reversal of direction; i.e., it propagates in the optical waveguide back towards the transmitter.

backspace To move backwards one step at a time. Teletypes and teleprinters are not usually able to backspace; however, the effect is obtained by carriage return followed by spacing signals up to the required point in the text.

backstop Mechanical limit on the movement of a relay armature.

backup storage Storage section to which ordinary users are denied access; used for security copies of specified files.

backward In the direction from the called to the calling station.

backward channel 1. In data transmission, a secondary channel whose direction of transmission is constrained to be opposite to that of the primary channel. 2. That channel of a data circuit which passes data in a direction opposite to that of its associated forward channel. NOTE: The backward channel is usually used for transmission of supervisory, acknowledgement, or error-control signals.

backward indicator bit A bit in a signal unit which is used to request a retransmission when a signal unit is received out of sequence.

backward sequence number A field in a signal unit which contains the forward sequence number of a correctly received signal unit being acknowledged.

backward supervision The use of supervisory sequences from a secondary to a primary station.

baffle A mounting board or screen which separates acoustic power at the back of a loudspeaker cone from the power directly emitted from the front of the loudspeaker. Baffles improve bass response by increasing the length of the path difference between the two sources of sound.

balance To equalize loads, currents, etc., between two circuit elements.

balance, active Test of the accuracy of balance network at a two-wire/four-wire hybrid point.

balance, longitudinal A measure of the degree of symmetry of impedance of a network.

balance, terminal The adjustment of balance networks associated with two-wire toll-connecting trunks

coming in to a switching center to ensure optimum operating conditions are obtained.

balance, through The adjustment of intertoll trunk impedances to ensure optimum matching when circuits are interconnected.

balanced Pertaining to electrical symmetry.

balanced amplifier See push-pull amplifier.

balanced circuit Circuit whose two sides are electrically equal in all transmission respects.

balanced code A PCM code whose digital sum variation is finite.

balanced line A transmission line consisting of two conductors in the presence of ground capable of being operated in such a way that when the voltages of the two conductors at all transverse planes are equal in magnitude and opposite in polarity with respect to ground, the currents in the two conductors are equal in magnitude and opposite in direction. Syn.: balanced signal pair.

balanced signal pair See balanced line.

balanced station The station responsible for performing balanced link-level operation. It generates commands and interprets responses, and interprets received commands and generates responses.

balanced three-wire system Power distribution system using three conductors, one of which is balanced to have a potential midway between the potentials of the other two.

balanced-to-ground Condition when impedance-to-ground on one wire of a two-wire circuit equals the impedance-to-ground on the other wire.

balancing, terminal Adjustment of impedances and balances of toll terminating and toll connecting trunks to obtain acceptable return loss characteristics at a toll switching center.

balancing network 1. A circuit used to simulate the impedance of a uniform two-wire cable or open-wire circuit over a selected range of frequencies. 2. A device used between a balanced device or line and an unbalanced device or line to transform from balanced to unbalanced or from unbalanced to balanced.

balcony Small security platform on a utility pole where staff can work safely.

balcony, distributing frame Elevated platform with a guard rail to facilitate working on the upper part of the distributing frame.

balun Balanced-unbalanced. Device used when interconnecting balanced circuits with unbalanced circuits, such as coaxial feed cables with balanced antennae.

banana jack/plug Single conductor connecting arrangement, with a banana-shaped metal spring contact.

band 1. A range of frequencies between upper and lower limits. 2. A group of tracks on a magnetic drum or on one side of a magnetic disc.

band, amateur radio Any of the bands of frequencies authorized for use by radio amateurs in a particular country.

23

band, broadcast The several bands of frequencies authorized for use for broadcasting for public reception, sound and television.

band, citizen's One of two radio-frequency bands used for low power radio transmissions in the U.S. without a license; either 26.965 — 27.225 or 462.55 — 469.95 megahertz. Citizen's band radio is not permitted in some countries; in others different frequencies are authorized.

band, communication satellite Any of the bands of radio frequencies allocated for use by communication satellites:

Up-links to satellites, MHz
5925 — 6425
12500 — 12750
14000 — 14500
27500 — 31000
Down-links from satellites, MHz
3700 — 4200
10700 — 10950
11200 — 11450
17700 — 21200

band, frequency Frequencies between upper and lower specified limits.

band, guard A radio frequency band left unallocated or unused between two active frequencies to give a margin of security against interference.

band, microwave frequency Any of the commonly used frequency band designations (see table) originally based on World War II radar band designations.

P band	0.225	— 0.390 GHz
L band	0.390	— 1.550 GHz
S band	1.55	— 5.20 GHz
X band	5.2	— 10.9 GHz
K band	10.9	— 36 GHz
Q band	36	— 46 GHz
V band	46	— 56 GHz
W band	56	— 100 GHz

band, radio frequency The North American practice is to designate radio frequency bands by a two- or three-letter abbreviation; the International Telecommunication Union uses a number:

Below 300 Hz	ELF = extremely low frequency	—
300 — 3000 Hz	ILF = infra low frequency	—
3 — 30 kHz	VLF = very low frequency	4
30 — 300 kHz	LF = low frequency	5
300 — 3000 kHz	MF = medium frequency	6
3 — 30 MHz	HF = high frequency	7
30 — 300 MHz	VHF = very high frequency	8
300 — 3000 MHz	UHF = ultra high frequency	9
3 — 30 GHz	SHF = super high frequency	10
30 — 300 GHz	EHF = extremely high frequency	11
300 — 3000 GHz	THF = tremendously high frequency	12

Lower limits are exclusive, upper limits are inclusive in each case.

band, transmission Band of frequencies which is passed by a filter or by transmission equipment.

band elimination filter A filter having a single continuous attenuation band, with neither the upper nor lower cut-off frequencies being zero or infinite.

band gap The difference in energy between the valence and conduction bands of a semiconductor. The band gap of optoelectronic components determines their operating wavelengths.

band number A subdivision of the address label in a common channel signaling system containing the most significant bits. Used for routing a signal message and for identifying the circuit group containing the traffic circuit concerned.

band rejection filter See band elimination filter.

band spreading The widening of tuning scales in heavily used radio bands to facilitate tuning in crowded bands.

bandage, cable splicer's Rubber ribbon, about four inches wide, used to wrap a cable splice to protect it temporarily from moisture.

bandpass filter A filter having a single continuous transmission band with neither the upper nor the lower cut-off frequencies being zero or infinite.

bands, cable splice Oversize hose clamps to be placed circumferentially around large cable splice sleeves to maintain a gas-tight cable sheath when the cable is pressurized with gas.

bandwidth 1. The difference between the limiting frequencies of a continuous frequency band. 2. The range of signal frequencies that can be transmitted by a communications channel with defined maximum loss or distortion. Bandwidth indicates the information-carrying capacity of a channel. For analog transmission this is usually expressed in kHz or MHz; for digital transmission the transmission rate, e.g., bit/s, kbit/s, Mbit/s, is used. For an optic fiber system bandwidth is usually given as its capacity to transmit information in a specific time period for a specific length, e.g., 10 Mbit/sec/Km.

bandwidth, modulation acceptance Of a radio receiver for frequency or phase modulated signals (other than those used for broadcast reception) is twice the frequency deviation of an input signal which, when applied at a level 6 dB higher than the maximum usable sensitivity level will produce a ratio (signal + noise + distortion)/(noise + distortion) equal to that specified for the maximum usable sensitivity level.

bandwidth, necessary For a given class of emission, the width of the frequency band which is just sufficient to ensure the transmission of information at the rate and quality required under specified conditions.

bandwidth, nominal Total bandwidth per channel. Four kHz for normal commercial speech channels of 300 to 3400 Hz; 3 kHz for the narrower channels used on some intercontinental submarine cable systems which use 16 voice channels per 48 kHz group instead of the usual 12 channels per group on FDM multiplexing.

bandwidth, occupied The bandwidth of a modulated signal excluding components weaker than 5% of the carrier, or 5% of the strongest side frequency, if higher than that of the carrier.

bandwidth, x dB A bandwidth such that beyond its lower and upper limits any discrete spectrum component or continuous spectral power density is attenuated by at least x dB relative to a given and predetermined zero dB reference level.

bandwidth expansion ratio The ratio of the necessary bandwidth to baseband bandwidth.

bandwidth limited operation The condition prevailing when the frequency spectrum or bandwidth, rather than the amplitude (or power) of the signal, is the limiting factor in communication capability. The condition is reached when the system distorts the shape of the waveform beyond tolerable limits.

bandwidth ratio In a radio communication system, the ratio, in decibels, equal to ten times the common logarithm of a quantity consisting of the occupied bandwidth of the entire baseband, divided by the assigned channel bandwidth.

bank A group of similar items connected together in a specified manner and used in cooperation. In electromechanical telephony switches, a bank is a group of fixed contacts which a brush or wiper travels over to select and complete a circuit.

bank, line The multipled bank of contacts leading to the line or talking circuit in a Strowger-type exchange.

bank, sleeve The multipled bank of contacts in a Strowger exchange used for testing and metering. This is normally the upper bank.

bank, vertical A vertical bank of contacts on a two-motion switch indicating the level reached by the wiper in stepping up its vertical levels.

bank assembly The multipled assembly of banks mounted on a switch frame; a switch can be plugged in to access the banks.

bank points The total number of contacts which have to be made at each position of a two motion or uniselector switch, such as line contacts, metering contacts and testing contacts.

bar The cgs (centimeter-gram-second) unit of pressure.

bar, expanding Long wooden bar with steel slot which fits over an expansion anchor rod and is turned to open up the anchor to grip the soil.

bar, pinch Steel bar with curved end, used for prying.

bar, tamping Steel bar with square end, for tamping backfill around a pole.

bar access To deny access to a facility.

bar code A code used for identifying an item, consisting of a sequence of thick and thin lines which can be read by an optoelectronic scanning device.

bar code scanner Optical character reader. A special optical pen is moved over the printed bar code, the characters are read and translated into digital signals which are sent into a processor.

bare Conductor wire which is not enameled or enclosed in an insulating sheath.

BARITT diode (barrier injection transit time) A microwave diode which can be used as an oscillator in the gigahertz range.

barkhausen effect Magnetization effects which do not change smoothly but in quantum steps as individual molecular magnets change axes.

barnacle A manually-added "fix" to a hardware unit; not incorporated in a normal manufacturing process.

barretter A metallic resistance wire, usually enclosed in a vacuum tube, with a positive temperature coefficient of resistance. It acts as a voltage regulator when in series with the load.

barrier 1. A partition for the insulation or isolation of electric circuits. 2. A blockage within a telephone cable which stops gas or water flowing past the point. Pressurized cables normally have a barrier on the exchange side of the connection to the pressurization system.

barrier, water A specialized type of cable sheath, often thin aluminum, bonded to a plastic main sheath to make the sheath impervious to water.

barrier code An additional digit inserted to prevent misdialed local calls from being charged at toll rates.

barrier layer In an optical waveguide, a localized minimum in refractive index in the region of the core/cladding interface.

base The region between emitter and collector of a transistor. Also, the radix of a numbering system, such as 10 for decimals, 2 for binary, 3 for ternary.

base address A numeric value used as a reference in the calculation of addresses in the execution of a computer program.

base attribute set A minimum set of attributes whose values unambiguously identify a particular management domain in a message handling system.

base station A land station in a land mobile service (radio system providing telephone service for automobiles, etc.)

baseband In the process of modulation, the frequency band occupied by the aggregate of the transmitted signals when first used to modulate a carrier, and in demodulation, the recovered aggregate of the transmitted signals. In the context of local area networks this refers to those employing digital baseband transmission to provide a single high-speed information channel shared by all network nodes.

baseband adaptive transversal equalizer A method of reducing adverse effects of dispersive multipath fading of microwave radio signals. (Developed by the former Bell System.)

baseband bandwidth The band of frequencies occupied by the information channel(s) which is to be conveyed by a radio transmission system.

baseband signaling Transmission of a signal at its original frequency; a frequency not changed by modulation.

BASIC (Beginners all-purpose symbolic instruction code) A computer language designed to be easy to learn and suitable for use in schools.

basic (error correction) method A non-compelled, positive/negative acknowledgement, retransmission error control system.

basic route A traffic route which is provided as a necessary part of a base traffic routing hierachy.

basic services The telephone services offered to all subscribers in return for a basic rental and/or call fees.

basic status In data transmission, a secondary station's capability to send or receive a frame containing an information field.

bass Low audio frequencies, typically below 300Hz.

batch processing A processing method in which groups of input documents or data are collected together and handled as a single unit. Batch working can be compared with real-time working in which transactions are dealt with as they arise; in batch working there can be delay between occurence of the event and the eventual processing of relevant data.

bath tub Shape of typical graph of component failure rates: high during initial period of operation falls to acceptable low level during whole of normal usage period; then rises again as components become time-expired.

Bat-Tap Proprietary device giving voltage division of a central office main battery, e.g., 24v supply from a 50v battery.

battery A group of several cells connected together to furnish current by conversion of chemical, thermal, solar or nuclear energy into electrical energy. A single cell is itself sometimes called a battery.

battery, A See battery, filament.

battery, alkaline Battery using an alkaline electrolyte such as potassium hydroxide.

battery, B An anode or plate battery.

battery, C A grid battery.

battery, central office A group of storage cells connected in series to provide a 48 volt direct current supply to a central office. A lead-acid storage cell has a nominal voltage of 2 volts, which increases to 2.15 volts when being charged and drops to 1.85 volts when being discharged. The normal central office battery therefore consists of 23 cells, increased to 26 cells when the chargers are inoperative. Although 48/50 volt systems are the most common, some national switching systems operate with different voltages, e.g., West Germany uses 60v power in its exchanges.

battery, common A battery which supplies power to several different loads and circuits.

battery, filament Battery which provides current to heat filaments in electron tubes (thermionic valves). An A battery.

battery, floating A low impedance battery, fully charged, connected across the output of central office rectifiers to smooth the output and serve as standby source of power during peak load periods or faults.

battery, grid Battery supplying voltage for biasing grids of electron tubes.

battery, lead-acid A rechargeable secondary cell used in central offices.

battery, local Dry or wet cells used to provide microphone current in unsophisticated telephone systems.

battery, nuclear A source of direct current derived from the radiations of radioactive material.

battery, PBX Source of dc power for private branch exchange. Usually on the same premises as the PBX, though PBXs are sometimes fed from the central office over special conductors.

battery, plate Battery providing anode or plate voltage for electron tubes. A 'B' battery.

battery, signaling Battery providing power to operate relays or light lamps. Not common as a separate unit.

battery, stationary Battery designed to be used in a fixed location.

battery, storage A central office main battery or other secondary cell installation.

battery, talking A low impedance battery suitable for use as a centralized source of speech currents in a telephone office.

battery, testing High voltage battery available at testboards to operate test equipment.

battery and ground pulsing Method of pulsing over high resistance junction circuits using the batteries of the offices at both ends in series. Not widely used outside North America.

battery eliminator A device consisting of a rectifier plus filter capable of supplying dc power at the required voltage without the use of a battery. To substitute for a battery, e.g., in PBX installations, it must have an output of low impedance and low noise level.

battery-ground (BG) signaling Signaling method where the presence or absence of battery between the signaling circuit and ground indicates an off-hook or on-hook condition.

battery tap A device for obtaining a lower voltage from a central office battery without the use of dropping resistors or counter-emf cells. Consists of a tapped stack of rectifier units. See also Bat-Tap.

baud The reciprocal of the length in seconds of the shortest element of the code. If a teletypewriter is operating at 60 wpm, the length of the shortest element of the 7.42-unit code is 22 milli-seconds. The baud rate is therefore $1/0.022 = 45.45$ bauds.

baud rate The number of bits per second of a data signal (baud rate is same as bit rate if all bits have the same time length).

Baudot code Five-unit code used in teletypewriter signaling. See also code, ITA #2.

bay Row or suite of racks on which transmission or switching equipment is mounted.

bay, equal level patching Transmission equipment test panel on which all circuits are lined up to operate at specified standard levels. Facilitates patching of circuits and equipment.

bayonet base A cylindrical lamp base having two locking pins spaced 180 degrees. Used on automotive lamps, ballast lamps, etc., and in some countries for normal domestic lighting fittings.

bayonet neill concelman (BNC) A type of miniature co-axial plug and socket connector.

bazooka A balun, to join together balanced and unbalanced circuits.

BCH code An error correcting code sometimes used with data transmission equipment.

beach anchor The hold-fast at the shore-line for a submarine cable, usually consisting of an earth-embedded strong cross member to which the armor of the shore-end cable is connected inside the beach pit.

beach pit The excavation made to accommodate the joining of the sea and land portions of a cable system. It is sometimes finished as a concrete vault at the landing point, to accommodate the beach anchor and the beach splice.

beach splice The junction of the land cable and shore-end cable above the water's edge at the cable landing point.

beacon A radio transmitter or the signal emitted by it when the emission is used as a directional guide, such as homing beacons or localizer beacons.

bead 1. Glass, ceramic or plastic insulator through which the inner conductor of a coaxial transmission line passes, and by means of which the inner conductor is supported in a position coaxial with the outer conductor. 2. A small program module. Sometimes beads of program are strung together on "threads," to complete the analogy with necklaces.

beam, radio A major lobe of radio power emitted by a transmitter in a specified direction.

beam antenna A directional radio antenna.

beam diameter In optoelectronics, the diameter of a circle, concentric with a beam, through which passes a specified fraction of the total power in the beam. This term is useful only for beams that are or are assumed to be circular in cross section.

beam divergence angle In optoelectronics, half the vertex angle of that cone that encompasses a circle of diameter equal to the beam diameter at all points in the far field. Beam divergence angle is meaningful only when describing the far field of beams that are or that are assumed to be circular in cross section.

beam maser A type of maser (microwave amplification by stimulated emission of radiation) sometimes used for high stability oscillators.

beam spread In optoelectronics, the angle between two planes within which passes a specified fraction of the total power in an optical beam. Beam spread is generally specified for two orthogonal orientations.

beamsplitter In optoelectronics, a device for dividing an optical beam into two or more separate beams.

beamwidth In optoelectronics, the linear dimension of the region over which the beam irradiance falls within specified limits.

beamwidth, antenna The angle between the half-power points (3 dB points) of the main lobe of the antenna pattern when referenced to the peak power point of the antenna pattern. It is measured in degrees and normally taken of the horizontal radiation pattern.

bearer Any basic communications facility from which subscriber services are derived.

beat To beat or mix one radio or voice-frequency signal against another so as to produce a signal of a different frequency; this procedure is used for tuning, for zero beat.

beat frequency Frequency produced when two different frequencies are mixed or combined in a special non-linear circuit; the beat frequency is the difference between the two applied frequencies.

beat frequency oscillator Oscillator designed to produce a signal which beats with an incoming signal so as to produce an audible tone.

beat in To tune one oscillator against another by a zero beat method. Also, to beat a lead cable jointing sleeve to make it fit closely around a cable sheath.

beat note Audible tone generated by combining two different frequencies in a non-linear circuit or as sound waves in air.

beat reception A method of radio reception in which the received radio frequency signal is combined with a locally generated signal on a frequency similar to the incoming signal to produce beats, usually at audio frequency.

beating The action of one periodic signal on another producing a resultant with different periodicity and amplitude.

beats The resultant produced by beating two signals together.

beep tone An intermittent audio tone (commonly a 1/5 second spurt of 1400 Hz) superimposed on a voice circuit as a warning, e.g., that the conversation is being recorded.

Beers law (optics) Absorption coefficient is proportional to molar density, the proportionality constant is the molar absorption coefficient.

beginning-of-tape (BOT) Marker used on magnetic tape to locate the beginning of the permissible recording area.

Bel The common logarithm of the ratio of two powers. One-tenth of a Bel, the decibel, is the commonly used unit.

bell, loud-ringing A telephone bell equipped with gongs of at least 150 mm usually in a weatherproof case. For use in noisy locations, or where the subscriber is likely to be some distance from the phone.

bell, single stroke A bell which chimes once when operated.

bell, weatherproof A telephone bell having all parts except the two gongs inside a weatherproof case, and usually with oversize 150 mm gongs. For installation out of doors.

Bell Atlantic Bell Regional Holding Company for the mid-Atlantic coast region of the U.S.

Bell Communications Research (BELLCORE) The new name for the central service organization providing centralized technical and management services for the Regional Holding Companies.

bell mouth Name given to a device mounted above a cable tank with an opening in the form of a hyperboloid of revolution, through which cable is led from the tank to the deck of a cable ship.

Bell System The whole pre-1984 AT&T organization including Bell Labs, Long Lines, Western Electric and the 23 Bell operating companies. In 1984 the wholly-owned operating companies were separated from the other parts of AT&T and managed on a regional basis. These new regional operating companies (Bell Operating Companies, BOCs) (or Regional Holding Companies, RHCs) are Nynex (New York and New England), Bell Atlantic, BellSouth, Southwestern Bell, US West, Pacific Telesis and Ameritech. The two main subsidiaries left with AT&T are AT&T Communications and AT&T Technologies (successor to Western Electric Co., the manufacturing arm of the group).

AT&T Communications (AT&T-C or ATTCOM) provides long distance voice, data & video transmission services both within the U.S. and between the U.S. and other countries.

AT&T Information Systems (ATTIS) is the principal marketing unit of AT&T Technologies. It provides information processing and distribution systems for business customers in addition to consumer products and services.

AT&T International provides the company's main overseas presence and markets AT&T products and services outside the U.S.

Other sectors of AT&T include AT&T Network Systems, AT&T Technology Systems and AT&T Bell Laboratories.

AT&T continues to hold a minority position in two of the original Bell Operating Companies (Southern New England Tel and Cincinnati Bell).

bell-tapping The intermittent jingling produced by a telephone ringer when a dial is operated on the same line and precautions have not been taken to eliminate it.

Bellboy A Bell System service which provides a paging signal originating at any dial telephone and carried over a radio channel to a miniature pocket radio receiver carried by a customer. When the customer's Bellboy number is dialed, the pocket receiver buzzes three times at 30-second intervals, indicating that he should call his home office. Similar radio paging systems are in use in many areas, using a variety of names.

belt, body A safety belt.

belt, cable An extruded layer of insulation within a cable.

belt, integral A cable belt of extruded insulant.

belt, safety A strong leather or nylon belt designed to give workers security when working on a telephone pole or radio antenna tower or mast. Usually has clips and hooks to hold tools.

benchmark A specified and standardized task designed so that the performance of different processors can be compared; an evaluation process.

bend, cast iron pipe Pipe, usually three inches diameter with right angle bend and radius of about three feet, to enable cables to be brought up to a building or pole from an underground jointing chamber or duct route.

bend, E plane A waveguide bend with the longitudinal axis in a plane parallel to the plane of the electric field vector.

bend, H plane A waveguide bend with the longitudinal axis in a plane parallel to the plane of the magnetic field vector.

bend loss (optic fibers) Increased attenuation because of the loss of high-order modes by radiation (a) when fiber is curved around a restrictive radius of curvature or (b) at microbends caused by small distortions of the fiber imposed by externally induced perturbations.

bender, cable Tool which enables smooth bends to be made in large diameter cables.

bender, piezoelectric Two small piezoelectric elements bonded to a metal strip, but bonded in opposition so that the application of a signal makes the strip bend.

bender, spring A small tool with steel pins which adjusts relays and keys for tension and pressure.

bending radius The limitation of allowable curvature of a cable.

Beverage antenna A directional wave antenna.

Bhaskara Indian telecommunications satellite used for television.

bias 1. A systematic deviation of a value from a reference value. 2. The amount by which the average of a set of values departs from a reference value. 3. Electrical, mechanical or magnetic force that is applied to a relay, vacuum tube or other device to establish an electrical or mechanical reference level to operate the device. 4. Effect on telegraph signals produced by the electrical characteristics of the terminal equipment.

bias, cathode Electron-tube bias by inserting a voltage-dropping resistor in the cathode circuit.

bias, forward Bias which aids the movement of carriers in a transistor.

bias, grid Electron tube bias achieved by applying a negative voltage to the control grid in order to establish the desired operating point.

bias, internal The mechanical or electric bias internal to a particular machine, e.g., a teletypewriter.

bias, magnetic A steady magnetic field applied to bias an electromechanical relay.

bias, marking Lengthening of mark signals and consequent shortening of space signals in telegraph equipment.

bias, mechanical Bias applied by a spring which has to be overcome by an operating signal.

bias, negative Electron tube grid bias to bring the tube to its required operating point.

bias, positive The lengthening of mark pulses.

bias, reverse Bias which discourages the movement of carriers in a transistor.

bias, self Cathode bias; the bias on an electron tube due to the voltage drop across a resistor in series with the cathode and carrying the cathode current.

bias, spacing Lengthening of spacing signals in telegraph equipment.

bias distortion Distortion affecting a two-condition (or binary) modulation in which all the significant intervals corresponding to one of the two significant conditions have uniformly longer or shorter durations than the corresponding theoretical durations.

bias testing A method of testing of equipment, especially line printers, tele-printers etc. An incoming test signal is modified by changing operating characteristics so that one machine's performance in dealing with imperfect signals may be compared with another's.

biax memory A type of two-hole ferrite memory with the read and write wires running at right angles to each other.

bid A single attempt to obtain the service of a resource.

bidirectional A qualification which implies that the transmission of information occurs in both directions.

bifilar Two wires. Bifilar windings can be used to produce balanced circuits or, when connected together at one end, resistances with very low inductance, or to give maximum coupling between windings.

bifurcated Splitting the end of a spring set into two, giving two contacts in parallel, to reduce noise and improve efficiency.

bifurcation Name given to the two-phase approach to the "detariffing" of customer premises equipment (CPE).

BIGFON German wideband integrated service system using optic fibers in local networks (Breitbandiges Integriertes Glasfaser Fernmelde Orts Netz).

bilateral control A synchronization control system between exchanges A and B is bilateral if the clock at exchange A controls that at exchange B and the clock at exchange B controls that at exchange A. Syn.: bilateral synchronization

billboard antenna A broadside antenna array with flat reflectors.

billibit One (US) billion bits or one kilomegabit (10^9 bits).

binary coded decimal Groups of binary digits representing decimal numbers, with each decimal number allocated four binary digits. This system is widely used in telecommunications computer projects.

binary digit Unit of information in two-level digital notation which may be 0 or 1. A member selected from a binary set.

binary figure One of the two figures (0 or 1) used in the representation of numbers in binary notation.

binary modulation The process of varying a parameter of a carrier as a function of two finite and discrete states.

binary notation In the ordinary decimal system displacing a digit one position to the left means multiplying it by a factor of 10. In the binary notation displacement to the left multiplies by a factor of 2. Only the two digits 0 and 1 are used. The binary number "1101" represents 1 unit, 0 x 2, 1 x 4, 1 x 8, i.e. a decimal equivalent of 13.

binary numeral A numeral in the binary (base 2) numbering system represented by the characters 0; 1 and optionally preceded by B.

binary state Either of the two possible conditions of a device which has only two stable conditions.

binary synchronous communication Data transmission with synchronization controlled by timing signals generated at both ends.

binary to decimal conversion Conversion from base 2 to base 10.

binaural Two eared; stereophonic.

binaural effect Directivity effect; the use of two ears enables the direction of a sound source to be determined.

binder Thread or tape wrapped around groups of conductors in multi-pair cables.

binding post A screw terminal making contact with conductors in wires and cables.

binding wire/tape Soft copper wire wrapped around an insulator on an overhead open-wire route and around the line wire itself, to hold the latter in position.

biochip A device fabricated using organic molecules or genetically engineered proteins, with greater computational ability than a silicon chip.

bipolar Having two poles, electric or magnetic.

bipolar binary (ternary) Binary ("to the base 2") refers to a two state code usually 1 or 0, e.g. pulse or no pulse. If the pulse is polarized, i.e. can be $+ = 1$ or -1, then there are clearly three states ($+ = 1$, 0, -1) so the coding is really ternary ("to the base 3") but for transmission purposes a $+ = 1$, 0, -1 code is called bipolar binary, because the information is coded in binary; successive '1' signals normally alternate between positive and negative polarity. Sometimes called bipolar pulse, or pseudo ternary. See Alternate Mark Inversion.

bipolar integrated circuit A type of monolithic integrated circuit based on the use of bipolar transistors, which utilize both electrons and positrons in their operation.

bipolar transistor A transistor in which both p-type material and n-type material are used.

bipolar transmission Signaling method predominantly used for digital services. The signal carrying binary values alternate between positive and negative polarities.

biquinary notation A system similar to that used in the ancient abacus and also in some computers; each decimal digit is represented by two symbols, first is 0 or 1, the 1 representing decimal 5; the second is 0, 1, 2, 3 or 4. Decimal 64 is therefore 1104. Some forms of biquinary use 7 digits for each decimal figure; 01 for less than 5, 10 for 5 — 9 followed by 00001, 00010, 00100, 01000, 10000 for 0 — 4 and 5 — 9 so that each decimal figure is represented by 7 bits of which two only are always 1s, the others always 0s.

bird Slang term for a telecommunications satellite.

birefringence The double bending of light by crystalline materials (such as silica in optic fibers). Increased birefringence (the difference between least and greatest refractive indices) results in reduced coupling between polarization in different planes in an optic fiber, and ensures that polarization is effectively in one plane only, thereby improving transmission properties.

birefringent medium   A material that exhibits different indices of refraction for orthogonal linear polarizations.

Birmingham wire gauge   Scale used for measuring iron and steel wires.

bisector, angle   Movable frame device used to determine correct angle to set a guy when a telephone pole route turns a corner.

bistable latch   A flip-flop which is able to store a 1 or a 0. An S-R flip-flop (S = set, R = re-set).

bistable multivibrator   A flip-flop; a circuit with two stable conditions able to remain in either of these for an indefinite period. It will change over only on receipt of an appropriate trigger pulse.

bit   Abbreviation for binary digit. A pulse whose presence or absence indicates data.

bit, check   An additional bit included in a digital signal used as an error check; a parity bit.

bit, overhead   A bit in a signal which does not carry normal coded information.

bit, parity   A check bit added to an ordinary digital code signal so as to make the number of code 1 bits either even or odd.

bit, service   An overhead bit other than a check bit with a service function, such as request for repetition.

bit, start   Bit preceding character code in start-stop transmission.

bit, stop   Bit following character code in start-stop transmission.

bit-by-bit asynchronous operation   A mode of operation in which rapid manual, semiautomatic, or automatic shifts in the data modulation rate are accomplished by gating or slewing the clock modulation rate. The equipment may be operated at 50 b/s one moment and at 1200 b/s the next moment.

bit count integrity   See character-count

bit duration   Time that it takes one encoded bit to pass a point on the transmission medium.

bit error   When the value of an encoded bit is changed in transmission and interpreted incorrectly by the receiver.

bit error rate   The number of erroneous bits divided by the total number of bits over a stipulated period of time. Two examples of bit error rate are: a) transmission BER — number of erroneous bits received divided by the total number of bits transmitted; and b) information BER — number of erroneous decoded (corrected) bits divided by the total number of decoded (corrected) bits. The BER is usually expressed as a number and a power of 10, e.g., 2.5 erroneous bits out of 100,000 bits transmitted would be 2.5 in 10^5 or 2.5 x 10^{-5}.

bit frame   Synchronizing pulse sent during every 193rd time slot in 24 channel PCM.

bit inversion   The deliberate or fortuitous changing of the state of a bit to the opposite state.

bit oriented   A communications protocol or transmission procedure in which control information is encoded in fields of one or more bits (compared with character or byte oriented protocols).

bit pairing   The practice of establishing within a code set a number of subsets (of two characters each) that have an identical bit representation except for the state of a specified bit.

bit parallel   A method of sending digital code information in which as many paths are used as there are bits in the word being considered.

bit position   The digit position of a bit in a word or byte.

bit rate   The speed at which signal bits are transmitted, usually expressed in bits per second, kbit/s or Mbit/s.

bit-sequence independence   In a digital path or digital section of a PCM system operating at a specified bit rate, a measure of the capability to permit any sequence of bits at that rate, or the equivalent, to be transmitted.

bit sequential   A type of transmission in which the elements of a signal are successive in time.

bit stepped   Operational control of digital equipment in which a device is stepped one bit at a time at the applicable modulation rate.

bit stream transmission   The transmission of characters at fixed time intervals without stop and start elements; the bits that make up the characters following each other in sequence without interruption.

bit string   A linear sequence of bits.

bit stuffing   A synchronization method used in time division multiplexing to handle received bit streams over which the multiplexer clock has no control. Syn.: positive justification, pulse stuffing, stuffing

bit stuffing rate   See nominal bit stuffing rate.

bit synchronous operation   A mode of operation in which data circuit-terminating equipment, data terminal equipment, and transmitting circuits are all operated in bit synchronism with an accurate clocking system. Clock timing is delivered at twice the modulation rate, and one bit is released during one clock cycle.

biternary transmission   A method of digital transmission in which two binary pulse trains are combined for transmission over a system in which the available bandwidth is only sufficient for transmission of one of the two pulse trains when in binary form. The biternary signal is generated from two synchronous binary signals operating at the same bit rate. The two binary signals are adjusted in time to have a relative time difference of one-half the binary interval and are combined by linear addition to form the biternary signal. A decoding process returns the signals to the binary state.

bits per inch (BPI)   Density of data recorded on magnetic tape.

bits per second   The number of bits passing a point per second. A measure of the speed of transmission of digital information; used to describe the information transfer rate on a circuit.

black box   A circuit element with defined functions but with no details given of its internal design.

black circuit   Security classification for a circuit which may be routed over unprotected paths because all information is either encrypted or unclassified.

black designation   A designation applied to all wirelines, components, equipment and systems which handle only encrypted or unclassified signals, and to areas in which no unencrypted, classified signals occur.

black facsimile transmission   In an amplitude modulation facsimile system, that form of transmission in which the maximum transmitted power corresponds to the maximum density of the subject copy. In a frequency modulation system, that form of transmission in which the lowest transmitted frequency corresponds to the maximum density of the subject copy.

black recording   1. In facsimile systems using amplitude modulation, that form of recording in which the maximum received power corresponds to the maximum density of the record medium. 2. In a facsimile system using frequency modulation, that form of recording in which the lowest received frequency corresponds to the maximum density of the record medium.

black signal   In facsimile, the signal resulting from the scanning of a maximum-density area of the subject copy.

blackbody   A totally absorbing body that reflects no radiation.

blackout   A close down of radiocommunications for security purposes. Also, a fade-out of radiocommunications caused by ionospheric activity.

blade, switch   The flat metal moving part of a knife switch.

Blaise   A computerized information service operated by the British Library.

blank   Period during transmission of information during which no characters are recorded.

blank, apparatus   A dummy unit or shelf on an equipment rack filling a space which will be required for additional functional units as the network or system grows.

blanketing   Swamping an area with powerful radio signals in order to prevent reception of specific signals.

blanking (picture)   The portion of the composite video signal whose instantaneous amplitude makes the vertical and horizontal retrace invisible.

blanking   Cutting off a device for a period of time.

blasting   Overloading an audio frequency device and producing distortion.

bleeder   A very high resistance connected in parallel with smoothing capacitors in a high voltage dc system; if power supply is interrupted the capacitors discharge through the bleeder.

bleeding   Preservative which has oozed out to the surface of a pole.

blind   To make a device nonreceptive to unwanted data.

block   1. A group of bits, or N-ary digits, transmitted as a unit. 2. A string of records, a string of words, or a character string to be treated as an entity. 3. A set of things, such as words, characters or digits, handled as a unit. 4. A collection of contiguous records recorded as a unit. Blocks are separated by interblock gaps and each block may contain one or more records.

block, carbon   A protective device against lightning on a main frame in a central office or at a test or distribution point.

block, connecting   A base with terminals or connecting strips used to interconnect wires, cables or cords.

block, discharge   A protective device through which unwanted voltages discharge to ground.

block, fuse   A unit on which fuses and other protective devices are mounted.

block, ocean   An equalization section of an undersea cable.

block, protector   A rectangular carbon block used on main distribution frames as lightning protection. The block is divided into two halves separated by a thin mica spacer or by a self-restoring insulating surface film. Lightning voltages jump the gap and discharge to ground. Also, sealed glass tube with same external dimensions as carbon block, containing electrodes which operate as more efficient lightning arresters than "traditional" carbon blocks.

block, terminal   An insulating base with binding posts to make connections where sets of terminals are mounted.

block, wooden   A dummy carbon block used on main distribution frame where protection against lightning is not required.

block-acknowledged counter   (Signaling System CCITT No. 6). A cyclic counter provided within the signaling terminal to count the number of blocks acknowledged as received at the distant end.

block check   A system of error control based on the observance of preset rules for the formation of blocks.

block check character   A character added at the end of a message or transmission block to facilitate error detection.

block-completed counter   (Signaling System CCITT No. 6). A cyclic counter provided within the signaling terminal to count the number of completed blocks transmitted.

block correction efficiency factor   In data transmission systems employing error control, a performance evaluation parameter given by the ratio of the number of blocks that have been received without error in a given time period (times 100) to the total number of blocks received in the same time period.

block diagram   A diagram of a system, program, switching process or other complex project in which selected portions are represented by annotated boxes with the necessary interconnecting lines to indicate interfacing or relevance.

block efficiency   The average ratio of user information bits to total bits in successfully transferred blocks.

block-error probability The ratio of incorrect blocks to the total number of successful block deliveries during a measurement period.

block-error rate Number of blocks received with erroneous bits compared with the number of blocks received correctly during a stipulated period of time. See bit error rate: note that if a single bit is incorrectly received in an error-correcting data system, the whole block is automatically retransmitted so the block error rate is a good indication of the through-put of a data system.

block-loss probability The ratio of lost blocks to total number of block transfer attempts counted during a measurement period.

block-misdelivery probability The ratio of misdelivered blocks to total number of block transfer attempts counted during a measurement period.

block rate efficiency The ratio of the product of the block transfer rate and the average block length to the signaling rate.

block separation Information indicating that the next character in a telephone service signal is the first character of a block of supplementary information.

block separator The character indicating that the next character in a telephone service signal is the first of a block of supplementary information.

block sum check Syn.: Longitudinal redundancy check

block tilt (cable TV) A method of setting the output levels of all low-band channels at a given number of dB lower than high-band channels.

block transfer attempt A coordinated sequence undertaken to transfer an individual block from a source user to a destination user.

block transfer failure Failure to deliver a block successfully; an unsuccessful block transfer.

block transfer rate The number of successful user information block transfers during a performance measurement period divided by the duration of the period.

block transfer time The elapsed time between the start of a user information block transfer attempt and successful block transfer.

blocked impedance Input impedance of a transducer when its output is connected to a load of infinite impedance.

blocking 1. The forming of blocks for purposes of transmission, storage, checking or other functions. 2. In reference to call traffic flow, blocking occurs when the immediate establishment of a new call is impossible owing to the inaccessibility of facilities in the system being considered. Syn.: congestion 3. In connection with maintenance operations, the deliberate withdrawal from service of a circuit or other resource.

blocking, Erlang B Traffic administrative assumption that a blocked call is abandoned immediately.

blocking, Erlang C Traffic administrative assumption that a blocked call is willing to wait for an idle trunk.

blocking, external When referring to a switching stage, the condition in which no suitable resource connected to that switching stage is available.

blocking, internal In a conditional selection system, the condition in which a connection cannot be made between a given inlet and any suitable free outlet owing to the impossibility of establishing a path.

blocking, Poisson Traffic administrative assumption that a blocked call is held up to the total intended holding time and that if a trunk becomes idle, the call will seize it and use it for the remaining of the holding time.

blocking capacitor A capacitor included in a circuit to stop the passage of direct current.

blockout Indication that a particular set of data is to be disregarded.

blower Slang for microphone.

blower, manhole An airpump with wide-diameter air hose (typically six inches) feeding fresh air into a below-ground manhole or jointing chamber.

blowout, magnetic Strong magnetic field across the contacts of an electric power circuit-breaker which rapidly extinguishes any arc which builds up as contacts open.

board General term for any manually attended or operating position for test or for traffic handling.

board, inward toll Sections of a toll switchboard where incoming toll traffic is handled.

board, linen test A strip of tough linen with numbered holes where cable conductors are placed after testing.

board, outward toll Sections of a toll switchboard where local toll traffic is handled.

board, patch A jack frame on which circuits are terminated and may be patched one to another.

board, power Control and supervisory equipment for central office main power supplies.

board, printed wiring An insulating board which holds a printed circuit with associated components. Often arranged with terminals along one edge so it can be plugged into a shelf connector.

board, service observation A monitoring board from which a senior operator or supervisor may observe and time the speed and accuracy of the work of operators.

board, tag A terminal strip on which cables have been terminated, usually on a soldered or wire-wrapped tag. Binding posts also are called tags.

board, terminal An insulating base on which terminals for wires or cables have been mounted.

board, test A test desk or test panel.

board, test and control A test board similar to an outward toll board but with all facilities for testing transmission and signaling of toll circuits terminated thereon.

board, toll A switchboard where long distance telephone connections are made, supervised and timed.

board, transposition running Simple frame used when constructing long open-wire routes requiring transposition. Wires run through swivels which can be turned as necessary at transposition points.

body In a message handling system: that part of an interpersonal message giving the information the user wishes to communicate.

body capacitance High frequency capacitive path to ground through a human body. Can be used in intruder alarm systems.

boiling out To immerse a paper-insulated cable core in melted paraffin wax to boil off any moisture that may be in the insulation.

bolometer Detector used for the measurement of radio frequency power or current. The bolometer operates on the balancing of a bridge, one arm of which is a resistor whose resistance varies with temperature.

bolt, double-arming A long headless bolt used for connecting two crossarms.

bolt, thimble-eye Galvanized bolt with thimble-eye instead of normal square head; permits direct attachment of a guy line.

bolt, twin-eye Galvanized bolt with two-grooved thimble-eye for two separate guy lines.

bond An electrical connection using a low resistance path.

bond, temporary cable A two-meter length of insulated wire with clips at both ends used to provide sheath continuity while a lead cable is being spliced.

bonding Connecting together (by copper strip) the sheaths of all metal-sheathed cables in a cable vault under a main distribution frame or in a jointing chamber or manhole.

Boolean algebra A calculus using algebraic notation to express logical relationships. Statements are either true or false, and values may be represented by the two binary states 1 and 0; truth tables have been constructed for the various Boolean operations. Since the various electrical states within a computer also have one of two values, circuits can be designed to simulate Boolean operations. These circuit elements are called logic elements or gates: an "and" gate, for example, corresponds to a Boolean "and" operation.

booster, negative impedance A line-powered transistor network inserted in telephone distribution cable systems in order to reduce audio frequency losses and improve high frequency characteristics of cable pairs normally used only for audio frequency services.

booth, telephone An enclosure, big enough for one adult, about 1m square by 2.5m high, which houses a pay telephone, shuts out noise and protects its user from the weather.

bootstrap 1. A technique or device designed to bring itself into a desired state by means of its own action. 2. An existing version, perhaps a primitive version, of a computer program that is used to establish another version of the program. 3. The technique of loading a program into a computer by giving it a simple instruction to load itself. The implication is that the computer can pull itself up by its own bootstraps, though someone has to write all the programs first.

borer, earth An auger designed for boring holes for telephone poles.

borer, increment Small tool used to bore holes in wooden poles to take samples of the wood at different depths below the surface.

boresight, antenna Small antenna used when setting up and calibrating a ground station dish antenna.

borscht Acronym for features to be provided in subscribers' line circuit in a digital exchange b battery feed o overvoltage protection r ringing signal sending s supervisory c codec h hybrid t test (The 'c' feature is not needed in a space-division exchange.)

boson Atomic particle with an integral spin, e.g., a helium atom.

both-way A qualification applying to traffic which implies that the call set-ups occur in both directions.

both-way traffic Traffic in a circuit where call set-up can occur in either direction.

bottom-up forecast Forecast of demand for telephone service based on field surveys and detailed expectation of future land usage in an area.

boule In the manufacture of optical fibers, a synonym for preform.

bounce, contact Momentary undesired condition in which a relay contact rebounds open after closing.

bound mode In optoelectronics, a propagating mode whose power is predominantly in the core of the waveguide. Syn.: core mode, guided mode

boundary node Protocol support node in IBM's Systems Network Architecture.

bow sheaves Large free-running wheels with grooved or flat circumferences, with one, two, or more wheels side by side, mounted on a thwartship axis at the bow of cable ships, over which cable is led aboard (or overboard) to or from the cable engine and the ship's cable tanks.

bow thruster A powerful motor-driven propeller mounted in a tunnel across the bow of a cable ship below the water line to swing the bow to port or starboard without any headway on the ship; also aids in maneuvering at very low speeds or astern.

bow-tie fiber Optic fiber in which the cladding has introduced artificial stresses which result in increased birefringence and improved transmission.

box, apparatus Box providing protection against dust, mechanical damage, and weather.

box, battery Small box holding batteries which provide primary power in local battery or magneto telephone systems.

box, bell Separate bell mechanism used in conjunction with some types of telephone instruments.

box, black Interface or special equipment; input and output parameters are defined but circuit details are not given.

box, cable Terminal box where cable pairs are terminated.

box, jack Box mounted on main distribution frame or other test frame, into which test cords may be plugged to extend lines or circuits to a test panel for measurement or observation. Sometimes called test-boxes.

box, stuffing Device using a sealing washer around a main cable sheath to provide water tight entrance to manholes or cable vaults.

Box-Jenkins technique A U.S.-designed forecasting technique which uses past data and mathematical models to give predicted future levels. Box-Jenkins methods rely heavily on computer technology, many calculations have to be performed to establish optimum conditions.

boxcar averager A device which selects (or "gates") portions of a received waveform, and averages (or integrates) them, thereby increasing the signal-noise ratio of repetitive signals. Syn.: boxcar integrator, gated integrator

brace A metal strap used to give extra rigidity to crossarms mounted on a pole.

brace, crossarm Metal strap from crossarm diagonally down to pole to hold the arm horizontal.

brace, pole A second pole set with its top bolted to the line pole to give stability where a guy line cannot be used to balance strain. Sometimes called struts.

brace, push A pole brace or strut.

brace, vertical Metal strap joining a crossarm to the one above for stability.

bracket, dead-end Bracket with insulator which bolts to a crossarm and is used for deadending open wire conductors.

bracket, phantom transposition A bracket for transposing a phantom circuit of four wires.

bracket, point transposition Bracket which mounts on crossarm in place of normal spindles and insulators, and permits wires to be interchanged in position in accordance with a transposition scheme.

bracket, span transposition A light bracket which enables a point transposition to be made in mid-span.

bracket, transposition Bracket used in place of normal straight spindle to allow transposition of wires at a point.

braid Fine wires woven into a ribbon or tube to give flexibility.

braid, glass Glass fibers woven with a braid ribbon or tube to provide mechanical or electrical protection.

brake, ladder Pressure-operated brake which prevents a mobile ladder from moving when anyone is on the ladder.

branch 1. A set of computer instructions that is executed between two successive decision instructions. 2. In the execution of a computer program, to select one from a number of alternative sets of instructions. 3. To select a branch, as in 1. 4. A direct path joining two nodes of a network or graph. 5. Loosely, a conditional jump or departure from the implicit or declared order in which instructions are being executed.

branch point Point in a program where a branch takes off. Also see node.

branching element A coupler in an optic fiber system which interconnects a transmitter with more than one receiver. See coupler, star; coupler, tee.

branching networks Electrical networks used for transmission or reception of signals over two or more channels.

branching repeater A repeater with two or more outputs for each input.

brand, pole Type designation burnt into a pole after manufacture.

brassboard Syn.: breadboard

breadboard 1. An assembly of circuits or parts used to prove the feasibility of a device, circuit, system, or principle with little or no regard to the final configuration or packaging of the parts. 2. An experimental or prototype model. Syn.: brassboard

break An interruption in continuity or in a circuit.

break, percent For a loop-disconnect system, the percentage of time the circuit is broken, typically 60% during dial pulsing.

break contacts Contacts which open when a relay or key is operated.

breakdown Failure, in particular failure of insulation.

breaker, circuit A cut-out device which breaks a circuit when preset limits of current are exceeded.

break-in The interruption of a teletypewritten circuit by opening the circuit momentarily, or the changing of direction of an echo-suppressor controlled circuit by an interruption from the receiving end.

break-in, attendant The ability of a PABX operator to break in to an established connection.

break-in hangover time (of an echo suppressor) The time interval between the instant when defined test signals, applied to the send and/or receive ports, are altered in a defined manner to restore suppression and the instant when suppression is restored. Removal of loss in the receive path will occur practically simultaneously.

break-in operate time (of an echo suppressor) The time interval between the instant when defined test signals, applied to the send and/or receive ports, are altered in a defined manner to remove suppression and the instant when suppression is removed. Insertion of loss in the receive path will occur practically simultaneously.

break-out Points in a cable, or channels or groups in a transmission system which are terminated at a point other than the end of cable or transmission system.

breakpoint Point in a program where normal sequence is interrupted by external intervention, e.g., an operator's signal or a debugging monitor routine.

Brewster's angle For light incident on a plane boundary between two regions having different refractive indices, that angle of incidence at which the reflectivity of light having its electric field vector in the plane defined by the direction of propagation and the normal to the surface is zero. A window mounted at Brewster's angle with respect to an incident beam is often used as an output window in lasers.

bridge 1. The connection of one circuit in parallel with another without interrupting the continuity of the first. 2. A Wheatstone bridge or other type of measuring instrument operating on a "balanced arms" principle. 3. A strap which connects two adjacent terminals.

bridge, capacitance An ac bridge instrument used for measuring the capacitance of a circuit or component.

bridge, conference Network enabling three or more circuits to be interconnected without upsetting impedance balancing relationships.

bridge, extension Network used to connect an extension telephone instrument to a line.

bridge, impedance An ac bridge instrument used for the measurement of impedance.

bridge, transmission Circuit element in a telephone central office which feeds power separately to the two parties holding a conversation but permits speech power to travel straight from one party to the other.

bridge, Wheatstone A four-armed bridge used to measure resistance by varying an adjustable arm until the bridge has been balanced.

bridge, Wien An ac bridge used for the measurement of capacitance and inductance.

bridge lifters A device which isolates bridged pairs, often by the use of relays or saturable inductors.

bridge tap An undetermined length of wire attached to both legs of a physical circuit between the normal end points of the circuit; this introduces unwanted impedance unbalances, important when circuit is to be used for fast data.

bridge transformer See hybrid coil.

bridged multiple The usual method of connecting up jack appearances on a multiple-appearance manual operating switchboard, with all jacks in parallel.

bridged tap A cable pair continued beyond the point at which the pair is connected to an instrument.

bridging The shunting or paralleling of one circuit with another.

bridging amplifier or bridger An amplifier which is connected directly into the main trunk of a CATV system but isolated from it. It provides services into the distribution or feeder systems.

bridging connection A parallel connection through which some of the signal energy in a circuit may be withdrawn, usually with imperceptible effect on the normal operation of the circuit.

bridging loss The loss at a given frequency resulting from connecting an impedance across a transmission line.

bridging wiper A wiper in a Strowger or step-by-step type telephone exchange which connects with the next outlet before breaking connection with an earlier outlet.

bridle wire Insulated wires leading from an open-wire pair to a terminal box or to other pole-mounted equipment.

brightness A non-quantitative term; the appearance of more or less light from a source. The human eye is not equally sensitive to all colors so the brightest light is not always the most powerful.

broadband Describes a transmission facility having a bandwidth greater than 20 kHz, i.e., a bandwidth sufficient to carry several voice channels. In the local area network context this describes those employing high bandwidth analog transmission, usually over a coaxial cable and in conjunction with frequency division multiplexing.

broadcast 1. A radio or television transmission intended for reception by the public. 2. Dissemination of information to a number of stations simultaneously. 3. To initiate a broadcast transmission.

broadcast band Radio frequency band used for broadcast services.

broadcast conference A type of multi-party telephone conference call in which one station transmits to all and the others can receive but can only talk back to the originator.

broadcast videotex See teletext.

broadcaster's service area Geographical area encompassed by a station's signal. (See predicted grade A contour and predicted grade B contour).

broadcasting Transmitting electromagnetic signals in a multidirectional pattern over the air.

broadcasting-satellite service A radiocommunication service in which signals transmitted or retransmitted by space stations are intended for direct reception by the general public.

broadcasting-satellite space station A space station in the broadcasting-satellite service on an earth satellite.

broadside array See array, broadside.

bronze Alloy of copper and tin with small percentages of other metals; used where high tensile strength wires are required.

Browne & Sharpe gauge A measurement scale for non-iron wires. Now called American Wire Gauge.

Brownian-motion noise Thermal noise; Brown discovered that molecules are always in motion at normal temperatures.

brush Carbon block, metal braid or metal leaf spring used to make contact with rotating elements such as those in an electric motor.

brush, carding Brush with steel bristles used to clean lead cable sheaths before sleeve joints are soldered.

bubble A technique of storing digital information, the magnetic state of a molecular "bubble," usually in a synthetic garnet crystal, is changed by a "write" head and read out by a "read" head in much the same way as a tape recorder except that no physical movement is involved, only the magnetic state moves round a tiny loop. Bubbles can be moved along at 100,000 bits per second, so fast access is possible to large stores of information and power failures do not mean that all the stored information is lost.

buckarm Crossarm fixed at right angles to other arms on pole to carry side leads or branch routes.

bucket 1. In data: a storage unit. 2. In outside plant: open-topped personnel container on extension arm of pole line truck; workers are lifted in it and work from it.

bucket brigade device An integrated circuit used as an analog delay line. The analogy relates to a row of men passing buckets of water along a line and each bucket (which holds information) takes a finite time to finish the journey.

buckling, plate Deformation of lead plates in a storage battery or secondary cell, usually due to high discharge rates, inadequate rectifier power to deal with daily peak loads, or extreme climatic conditions.

buffer 1. An interface unit designed to link circuits together but not allow excessive or unwanted variations in one to affect the other. 2. A routine or storage used to compensate for a difference in rate of flow of data, or time of occurrence of events, when transferring data from one device to another. 3. An area of storage reserved for temporary use in performing input/output operation. Data may be read from or written into the buffer storage area.

bug Something which causes a computer-controlled program to malfunction. Also, to provide equipment which monitors or eavesdrops.

build state The overall configuration of a system. i.e. A complete statement of every hardware and software part contained in the system together with a complete statement of the way in which they are interconnected.

building industry consulting service Service provided by most U.S. telephone companies to give builders and architects professional advice to help ensure that telephone services be installed rapidly and cheaply in any desired location in a building.

building-out Adding capacitance or resistance at a circuit termination in order to bring the impedance of the circuit to its specified value.

building steel (BS) The structural steel columns and horizontal beam framework supporting, for example, a central office building.

built-in check Provision in a process for automatic verification of accuracy of signals being handled.

built-up circuit A toll circuit made up temporarily by connecting together two or more normal toll circuits.

built-up connection A telephone connection utilizing a temporarily built-up circuit.

bulge Nonlinear relationship between frequency and attenuation in a transmission system.

bulging, earth The anomalous propagation condition where microwave radio signals are bent away from the earth's surface.

bulk billing Preparation of telephone bill with a charge for the total number of call units made; there is no itemized list of calls.

bulk-channel charge-coupled device An array of metal-oxide semiconductor capacitors which acts as a charge-transfer device.

bulk encryption The process whereby two or more channels of a telecommunications system are encrypted by a single crypto-equipment.

bulk metering Registration of total call units for all calls dialed by a subscriber, without provision for itemized charging data.

bunched frame alignment signal A frame alignment signal in which the signal elements occupy consecutive digit time slots.

bunching The grouping together of electrons.

bundled The combining together of several services into a single tariff offer.

burden test Procedure used in the U.S. to determine whether or not prices would go up if new services were offered.

bureau, centralized intercept Office where operators deal with call requests which cannot be satisfied by automatic announcement machines.

bureaufax A public-to-public service where both the acceptance and delivery of facsimile messages is effected from a public office and terminal equipment is provided by the telephone or telegraph company or administration.

burial, direct The installation of land cable or shore-end cable directly into the ground without conduits.

burn-in Operation of a device, sometimes under extreme conditions, to stabilize its characteristics and get rid of early failure components before bringing the device into normal service.

burnisher, contact Extremely fine steel file used to clean contacts on relay spring-sets.

burst 1. Interference to, or interruption of, a digital signal. 2. Transmission of a packet of data. 3. In data communications, a sequence of signals counted as a unit in accordance with some specific criterion or measure.

burst mode 1. A system of obtaining full duplex bidirectional transmission of digital speech signals over a two-wire circuit. The two directions of transmission are separated in time by alternating the two directions. The penalty which has to be paid is the use of higher frequencies: if data enters the system at frequency D a typical burst frequency is 2.25 x D. Burst Mode working is sometimes called Ping Pong Mode, or Time Compression Multiplexing (TCM). Burst Mode's principal rival way of transmitting digital speech signals in both directions on a 2-wire circuit is the hybrid system, this separates the two directions by hybrid balance circuits and the use of echo cancellers to ensure that there is sufficient near-end loss to enable proper detection of the far end signal. Yet another way of tackling this problem is for traditional carrier modulation techniques to be used so that one direction of transmission uses a different frequency band from the other direction of transmission, but suitable filters usually mean that this is an expensive alternative. 2. The transfer of a packet of data from a peripheral unit to a central processing unit, with a start signal at the beginning and an end-of-burst signal at the end. The length of the burst can be varied to suit particular requirements.

burst noise   Short periods of unwanted noise.

bus   1. In transmission, part of a circuit which is used in common from one of several, to one of several circuit functions or modules. Sometimes called a highway, because much traffic flows along it. 2. The name for a main power lead between rectifiers/batteries and telephone switching equipment. A bus can be a 50 mm diameter cable of stranded copper, or a rigid framework of copper or aluminum bus bars up near the ceiling, or just a track on a printed circuit board.

bus driver   Circuit which amplifies a bus data or control signal sufficiently to ensure a valid reception.

bus master   Circuit controlling transactions in a bus structure where access to this is shared.

bus network   A network with linear topology in which all nodes branch from a single shared communication line; a multipoint line.

busbar   A main dc power bus.

busdriver, output   An output signal amplifier.

bust   To terminate a message prematurely, usually because of an error made; to abort.

busy   Condition in which a telephone transmission path is in use.

busy back   A busy tone.

busy count   The number of times all the trunks in a group are busy.

busy hour   (of a group of circuits, a group of switches, or an exchange. etc). 1. The uninterrupted period of 60 minutes for which the average intensity of traffic is at the maximum. 2. The busiest hour of the busiest day of a normal week, excluding holidays, weekends, and special event days.

busy hour, peak   The busy hour each day; it usually is not the same over a number of days. Syn.: bouncing busy hour, post-selected busy hour

busy hour, time consistent   The 60 consecutive minutes commencing at the same time each day for which the average traffic volume of the observed exchange or circuit group is greatest over the days of observation. Syn.: mean busy hour

busy key   Key which, when operated, busies-out a switch in a central office so that work may be done on it without the switch being seized for traffic.

busy lamp   Lamp on a switchboard to indicate that a circuit is busy.

busy-out   To take action which causes the line or equipment to test busy to an incoming call.

busy period   The time interval between the seizure of the last available circuit in a system and the next release and resultant idle state of a circuit in that system.

busy season   The three months with the highest busy-hour traffic.

busy signal   A signal indicating that the called line is busy.

busy switch   See busy key.

busy test   A procedure for determining whether a traffic-carrying device is free and available for use.

busy visual   Small indicator lamps used in manual operating positions to show that the associated toll circuits are busy.

butt wrap   Tape wrapping around the core of conductors in a telephone cable, with no overlapping of layers.

buttinsky   Slang for hand-held one piece telephone used in central offices by technicians who can interrupt any line with it.

button, transmitter   The carbon granule capsule used in some telephone transmitters.

buzz   To check continuity of cable circuits by using a buzzer.

bylink   Special termination; particularly of automatic toll circuits, which gives rapid connection to registers thereby avoiding the use of delay signals or second dial tones.

bypass   Any arrangement of circuits which enables communications to be established directly between two organizations (in particular a telephone subscriber and a specialized common carrier) which would normally be expected to use the switched facilities of the local telephone company.

bypass capacitor   Capacitor which provides an ac path, effectively shunting or by-passing other components.

byte   A group of eight bits makes a byte. Typically a 16-bit "word" is itself divided up into two bytes for handling. A byte is usually the smallest addressable unit of information in a data store or memory.

byte mode   A data transfer method between peripheral and central processor in which the unit is the byte. Compare "burst mode," in which a variable length package of data is sent in one burst.

byte multiplexing   A form of time division multiplexing (TDM) in which the whole of a byte from one subchannel is sent as a unit, interlacing in successive time-slots with complete bytes from other subchannels.

C band   Frequency band used for uplinks to satellites (6GHz) and downlinks from satellites (4GHz).

c lead   The third wire of three (+ — and c) which constitute a trunk between switches of a dial central office. This is the wire which controls the guarding, holding, and releasing of switches. When grounded it usually indicates a busy trunk, when open or ungrounded it usually indicates an idle trunk. In Britain, this third wire is called the 'p' wire (from 'private').

cabinet An access point, protected by a case or box, in the local cable network where a main cable or exchange pair can be connected to a distribution cable pair. Syn.: flexibility point

cabinet, apparatus Special case in which telephone equipment is mounted, particularly PABXs on customer's premises.

cabinet, attendant's Console position of a PABX.

cabinet, test Table top case containing basic test gear. Used in small end offices.

cabinet area Geographic area served with telephones from a particular cabinet or flexibility point.

cable A number of insulated conductors assembled in a compact form and covered by a flexible, waterproof protective sheath.

cable, aerial See aerial cable.

cable, air spaced coaxial Coaxial cable in which the central conductor is kept accurately in place by plastic discs, bamboo type extrusions, beads, or helical windings, with air as the main dielectric.

cable, alpeth Cable with extruded aluminum and polyethylene sheath.

cable, armored Cable in which layers of steel tape or wire covering the cablesheath give mechanical protection; used for direct burial.

cable, bamboo Coaxial cable with pinched plastic tube holding the central conductor.

cable, banded Cables bound together by steel strip tapes.

cable, beaded Coaxial cable in which ceramic or plastic beads position the central conductor.

cable, blockwiring Distribution cable on buildings or poles within a city block.

cable, branch Cable which is spliced to and runs away from a main cable.

cable, building out Cable bridged across a loaded cable to bring circuit characteristics to standard values.

cable, buried Cable placed directly in the ground without using conduits.

cable, coaxial Cable with one or more coaxial pairs (tubes) in one sheath.

cable, color-coded Cable having color-coded insulation on the conductors to aid identification.

cable, combination A specially-made cable with some pairs (normally twin-type) and some quads.

cable, composite A specially-made cable with conductors of different types or gauges in one sheath.

cable, connector Cable in standard lengths and sizes, terminated in the factory on standard connectors.

cable, corrosion proof Metallic-sheathed cable with a special oversheath, often polyethylene, to protect against electrolysis.

cable, deep-sea, submarine Unarmored solid insulant coaxial cable used in major submarine cable sys-

tems. Tensile strength is provided by steel wires at the center of the cable within the inner conductor of the coaxial pair.

cable, distribution Secondary cable in a city's distribution network which has terminals from which leads or drops go to subscribers' premises.

cable, entrance Special cable, used to bring circuits from an open wire line to the central office building and avoid bringing open wire and its hundreds of crossarms right to the building.

cable, evencount Cable with the exact number of pairs specified, that is without extra pairs to make up for possibly faulty ones.

cable, feeder Large cable leaving a central office to serve a specific area. Also called main or primary cable.

cable, figure-8 A self-supporting aerial cable with the messenger wire so bound to the conductor cable that the combined cross section is in the shape of a figure 8.

cable, gopher-protected Armored cable specially protected for use in areas where gophers are active.

cable, grease filled Plastic insulated distribution cable with interstices filled with greasy compound, usually petroleum based, to reduce effects of faults.

cable, house Distribution cable within a building or a complex of buildings.

cable, inside wiring Color coded 24 AWG cable, thermoplastic insulated and jacketed, used primarily as customer premises wiring.

cable, integral messenger See cable, Figure-8.

cable, intermediate Short length of cable inserted in an open wire route, e.g., where the route crosses a high voltage power system.

cable, jute-protected Cable for direct burial wrapped in tarred jute (burlap) for protection.

cable, land The portion of a submarine cable reaching from the water's edge to the terminal station.

cable, lashed An aerial cable attached to its messenger wire by a continuously wrapped steel wire.

cable, lateral A branch distribution cable.

cable, layer type Multi-conductor cable with pairs made up in concentric layers, alternate layers having opposite direction of lay.

cable, lead covered Cable protected by an extruded sheath of lead/antimony or lead. In little use today, being superseded by plastic sheathed cable.

cable, loaded Cable having inductance increased (and attenuation at low audio frequencies reduced substantially) by insertion of lumped loading at specified intervals. Sometimes called Pupinised cable, after the inventor of lumped loading

cable, local Distribution cable used in local networks.

cable, L1, L3, L4 coaxial The Bell System's early coaxial cable systems: L1 (1940) provided up to 600 circuits, with repeaters every 8 miles; L2 (1953) gave 1860 circuits, with repeaters every 4 miles; and L4 (1965) 3600 circuits with repeaters every 2 miles.

cable, lossy Cable designed to have high attenuation per unit length.

cable, multisheath Cables provided with more than one type of overall protective sheath.

cable, optical fiber Cable made up of glass fibers protected by plastic coverings; sometimes metallic wires are included as strength members.

cable, paired Cable whose conductors are made up in pairs twisted together.

cable, paper insulated Cable whose conductors are insulated with dry, spirally wrapped paper tape.

cable, plastic insulated Cables using plastic insulation around their conductors. These cables usually have plastic oversheathing also.

cable, pressurized Cable protected against moisture entry by pumping in dry air or nitrogen under pressure from the central office. Leaks produce changes in pressure followed by increased pumping rate to maintain service. This allows faults to be found rapidly without affecting service.

cable, pre-wiring Loosely cabled groups of pairs of wires without an outside sheath. Available for installation in buildings during construction to make telephone installation simple at a later date.

cable, protective Short lengths of fine gauge cable used when heavy gauge cables are brought in to a central office. Provides a fusible link without having to equip all pairs with protective fuses.

cable, pulp insulated Cables with conductors insulated by extruded paper pulp.

cable, quadded Cable with conductors made up in sets of four wires in square formation; diagonally opposite wires are pairs.

cable, radio frequency Cable with low loss at high frequency; used for antenna feeds.

cable, ribbon Cable in a flat configuration with conductors side by side.

cable, riser Cable running vertically in a building to serve upper floors.

cable, self supporting See cable, figure-8.

cable, service Low pair count cable used to provide service to a small group of single subscribers.

cable, shielded Cable with metal tape shield wrapped around the insulated conductors.

cable, silk and cotton Cable used for terminating on frames; now largely superseded by plastic insulant cables.

cable, solid fill Cable with conductors insulated by what appears to be a solid mass of plastic, which is usually of foam-type construction.

cable, spiral-four Quad type cable once used for tactical carrier routes.

cable, stalpeth Cable with sheath made of corrugated aluminum tape, followed by corrugated steel tape, then a polyethelene oversheath.

cable, strip Ribbon-like cable with conductors side by side; used within computers, for cabling between back-plane units, and for house wiring.

cable stub Short length of cable brought out of a main cable and made up of pairs not needed initially but which can be readily accessed at the stub as required.

cable, subfluvial Underwater cable with spirally-wrapped wire armor to protect from abrasion in shallow water.

cable, submarine Telecommunication transmission cable installed on sea, lake or river beds, consisting of conductors and insulants, usually in coaxial form, with provisions for tensile strength to permit laying in the sea, and exterior protection as required by ambient conditions.

cable, submarine, armored Cable with an outside layer of armor wires or tapes to provide tensile strength during the laying operation and protection while resting on the sea bottom.

cable, submarine, armorless Cable having a stranded steel tensile strength member inside the center conductor, with no outer armor.

cable, submarine, main The major portion of a total cable system represented by the part laid in the deeper reaches of the route; to distinguish it from the cable configurations used to come ashore or those that are laid across land.

cable, submarine, repair Cable of lesser attenuation than the main cable used in deep water repairs to permit lengthening of a repeater section without upsetting the repeater gain/cable loss relationship.

cable, submarine, shore-end Seacable with heavy armor (single or double armor, depending on water depth and other conditions) for mechanical protection in shallow water, and containing shielding to reduce electromagnetic interference.

cable, submarine, Type SA A coaxial deep-sea cable having a 0.460-inch core diameter and 1.12-inch outside diameter.

cable, submarine, Type SB A coaxial deep-sea cable having a 0.620-inch core diameter and 1.25-inch outside diameter.

cable, submarine, Type SD A coaxial deep-sea cable having a one-inch core diameter and a 1.25-inch outside diameter.

cable, submarine, Type SF A coaxial deep-sea cable having a 1.5-inch core diameter and a 1.75-inch outside diameter.

cable, submarine, Type SG A coaxial deep-sea cable having a 1.7-inch core diameter and 2.08-inch outside diameter.

cable, swept coaxial Cable lengths which have been tested by the manufacturer and certified to be within specified limits for attenuation, impedance, etc., at all relevant frequencies.

cable, switchboard Cables used for interconnecting racks of transmission or switching equipment in a central office.

cable, tape armored Cable with steel tape wrapped around the sheath for special protection.

cable, terminating Length of cable with textile or enamel insulation on conductors. Used to connect lead-sheathed, paper-insulated cable to MDF; its use prevented moisture from migrating up the paper insulation. Type of cable now rarely encountered: replaced by plastic insulated cable.

cable, tie Cable between two central offices or a central office and a major PBX.

cable, tip Silk and cotton insulated cable used as terminating cable.

cable, toll Low capacitance cable used for toll circuits.

cable, trunk Loaded cable used for voice frequency trunk circuits.

cable, twin A cable composed of two insulated conductors laid parallel and attached to each other by the insulation or by a common covering.

cable, underground In North America, cable pulled in to ducts which can be replaced without excavation. In Europe, any below ground cable, either direct buried or in a duct.

cable, unit type A type of distribution cable with pairs made up into groups, typically of 50 or 100 pairs, facilitating separating out such groups.

cable, video Cable suitable for transmission of signals, including color television.

cable, water blocked Cable with periodic blocking points (a petroleum jelly and polyethelene compound is commonly used) to stop the passage of moisture along the inside of the cable sheath.

cable, wire-armored See cable, subfluvial.

cable attenuation (submarine cable) The measure of the loss in electrical strength encountered by signals sent through submarine cable, usually expressed as a function of frequency and measured between reactance-free resistors representing the resistive component of the cable impedance at high frequencies.

cable core The part of a submarine cable consisting of the center conductor and the dielectric.

cable drum 1. Motor-driven and power-braked large cylinder with a smooth circumference; used to pick up and pay out cable on a cable ship; achieves friction between the cable and the drum surface by several turns of cable held back or drawn off. 2. A cable reel on which cable is wound for storage.

cable engine The main element of the cable machinery, the powered drum or other device on a cable ship which performs the pick-up and controls the pay-out of cable. Also see linear engine.

cable fault location The process of electrical testing at a terminal station to find the location, and sometimes the cause, of a malfunction.

cable floats Inflatable pillow-shaped plastic bags or empty metal drums which, during the installation of a shore-end cable, enable the cable to be kept afloat so it can be pulled from the ship toward the beach. Floats are connected to the cable at regular intervals; once the cable has reached its correct position, the pennants are cut and the cable sinks to its intended resting place.

cable joint (submarine cable) The connection between two pieces of core in a submarine cable. Joints are made in the factory to make up repeater section lengths, and on board ship when cable sections have to be joined either during installation or repair. Also see cable splice.

cable layer See cable ship.

cable machinery The mechanical devices on board a cable ship used for picking up and paying out cable, repeaters, and equalizers and for repair operations.

cable mile Distance measurement in submarine systems: 1,855.3 meters, or 6,087 feet. Also see mile, nautical.

cable pan A large container for holding up to 20 miles of armorless submarine cable in cable ships. Pans are loaded at the factory and transported to the cable ship for final loading.

cable pilot (submarine cable) The designation for the transmission of a supervisory pilot frequency which is inserted as close to the sea cable termination as practicable and extracted likewise at the opposite terminal. Also see pilot, system.

cable plow 1. A strong steel blade towed by a tractor; used to place cable — and occasionally flexible conduit — directly into the earth, the cable passing through a tunnel in the plowshare. 2. A device which is towed over the ocean bottom, making an opening into which the sea cable is placed and buried.

cable powering A method of supplying power to solid-state CATV equipment by using the coaxial cable to carry both signal and power simultaneously.

cable ship A ship specially equipped for cable laying or cable repair; its basic equipment comprises cable engines, hold-back gear, dynamometer, cable tanks, bow sheaves, and fault-location and test equipment.

cable splice 1. The connection of two pieces of cable by joining them mechanically on a pair-for-pair basis and closing the joint with a weather-tight case or sleeve. 2. In submarine operations, connection of two pieces of cable including the armor wires (as distinct from a cable joint which connects only the core).

cable tank The cylindrically-configured holds of cable ships which can store up to 600 miles of cable; also the cable storage facilities in factories and depots.

cable transition The junction between submarine cable ends of different designs, e.g., from single armor to double armor or from shielded to unshielded cable.

cable transporter A large movable device with a motor-driven V-grooved cable sheave and a jockey wheel to keep the cable in the groove; or a small linear engine; used in submarine cable laying to transfer cable from one tank to another or to haul cable toward and into the ship during the cable loading operation.

cable TV Previously called community antenna television (CATV). A communications system which distributes broadcast programs and original programs and services by means of optic fiber, coaxial cables or other cable facilities. In some cable TV systems there is provision for interactive (both way) working, including telephone service.

cable vault Room below ground level in a central office/telephone exchange or cable terminal/repeater station where cables enter the building, usually in conduits.

cable vault ground bar (CVGB) The copper ground bar located in the cable vault (usually under a central office), for bonding cable sheaths.

cablecasting To originate programming over a cable system. Includes public access programming.

cabling The installation of new distribution cables in an area.

calcium battery A long-life lead-acid cell using lead-calcium alloy plates.

Calculagraph Trademarked name for a special time clock used at manual telephone operating positions to stamp call tickets with time of day and elapsed time of the call.

calculating circuits Any computing circuit which is able to perform a mathematical calculation.

calibrate To check, and reset if necessary, a test instrument against one known to be set correctly.

calibration The process of identifying and measuring errors in instruments and/or procedures.

call The setting up of a connection between two stations.

call, collect A telephone call to be paid for by the called subscriber.

call, credit card A telephone call to be charged to a credit card number.

call, free-code A telephone call to a called number which is connected without charge, e.g., directory assistance.

call, interzone A telephone call from one part of a major metropolitan area to another part of the same area.

call, local A telephone call within the local service area.

call, long distance A telephone call beyond the local service area and any area served by multimetering; a toll call.

call, lost A telephone call which cannot be completed because all relevant equipment was busy.

call, no hunt A type of test call made from test equipment: if the dialed number is busy the equipment does not hunt automatically for another line serving the same subscriber.

call, operator-assisted A telephone call such as credit card or collect requiring the help of an operator; usually involves higher charges than a direct dialed call.

call, person-to-person A toll telephone call made to a named person.

call, revertive A subscriber-dialed call between two subscribers served by party line numbers on the same pair of wires.

call, service code A telephone call to a service facility such as a repair desk.

call, station-to-station A telephone call to a customer's number,

call, toll A chargeable telephone call to an area outside the local service area.

call accepted A call control signal sent by called data terminal equipment to indicate that it accepts the incoming call.

call accepted signal In common channel signaling data transmission, a signal sent in the backward direction indicating that the call can be completed.

call address The last four digits of a telephone number.

call announcing Facility on some PABXs and key phone systems by which internal calls may be connected to a small loudspeaker on the telephone instrument (instead of ringing the called line's bell).

call attempt An effort by a subscriber to achieve a connection to another subscriber.

call attempt, abandoned A call attempt aborted by the calling party.

call attempt, completed A successful call attempt answered by the called station.

call attempt, lost A call attempt that cannot be further advanced towards its destination due to an equipment shortage or failure in the network. Sometimes called blocked or rejected call attempt.

call attempt, successful A call attempt, in which the calling station is either switched through to the exchange line terminating unit of the dialed number, or receives busy tone when the dialed number is busy. Note that a successful call attempt does not necessarily result in a successful call.

call-back In data transmission, a facility which enables the originator of a call attempt to a busy terminal to request the network to establish the call when the busy terminal becomes free. Syn.: automatic recall

call barring The ability to prevent all or certain calls to or from a telephone line.

call barring, subscriber selected The capability of a subscriber to prevent all or certain types of outgoing calls and/or service control operations from his telephone line.

call circuit A talking circuit directly from one operator to another. Also called order wire.

call circuit operation The use of a call circuit or order wire to establish toll calls through manually operated toll switchboards.

call collision Contention that occurs when data terminal equipment and data circuit equipment simultaneously transfer a call request packet and an incoming call packet specifying the same logical channel. The DCE will proceed with the call request and cancel the incoming call.

call congestion The probability that a call attempt encounters congestion.

call control character In data operations, a character of an alphabet, or a part of it, which is used for call control. It may be used in conjunction with defined signal conditions on other interchange circuits.

call control procedure The entire set of interactive signals necessary to establish, maintain and release a data call.

call delay The delay suffered when a call reaches automatic switching equipment which is busy processing another call. In the U.S., this delay is considered to be acceptable if not over 1 ½% of the calls are delayed by three seconds during the busy hour.

call detail recording Feature on some PABXs by which all outgoing calls are logged so that charges may be allocated to the telephone extension or department responsible.

call detail recording and reporting An AT&T Technologies enhancement to its Dimension PABXs which provides a communications cost-accounting system to collect and process station message data.

call directing code A routing and identifying code in a message switching system which ensures that a message is routed to the required terminal station and automatically starts up the printer at the station.

call distribution Queuing device which allocates callers to the first available free extension.

call duration The interval between the instant the call is actually established between the calling and the called stations and the instant when the clearing signal is given.

call-failure signal In data transmission common channel signaling: a signal sent in the backward direction indicating that the call cannot be completed due to time out, fault or some other condition.

call forwarding A service feature available in some switching systems where calls can be rerouted automatically from one line to another or to an attendant.

call forwarding busy line A service feature in which an incoming call is automatically routed to the attendant when both the called and alternate lines are busy.

call forwarding don't answer A service feature in which an incoming call is automatically routed to an attendant after a predetermined number of rings or seconds.

call indicator A lamp or LED display on some incoming manual operating positions to show the operator the number which the calling line dialed. This enables the operator to connect the call to the wanted line without speaking to the calling party.

call intensity The ratio of the number of call attempts during a period to the duration of the period.

call intent The desire to establish a call. It is manifested by one or more successive call attempts.

call not accepted A call control signal sent by the called data terminal equipment to indicate that it does not accept the incoming call.

call packing A systematic method of ordering the allocation of calls to free paths. Most commonly, new connections are routed via the most heavily loaded part of the switching network.

call pickup Service feature on some PBXs which permits a station (telephone extension) to answer incoming calls to another station in the same pickup group.

call progress signal A call control signal transmitted by the circuit-terminating data control equipment to the calling data terminal equipment to inform it about the progression of a call (positive call progress signal) or the reason why the connection could not be established (negative call progress signal).

call restriction Device (or software) which bars outgoing service or prevents individual lines (or extensions on a PABX) from making long-distance calls or reaching a toll operator.

call release time In telephone systems, the time from initiation of a clearing signal by a terminal installation until the free circuit condition appears on originating terminal equipment.

call-second A unit of communication traffic, such that one call-second may be defined as one user making one call of one second duration.

call set-up time 1. The overall length of time required to establish a circuit-switched call between users. 2. For data communications, the overall length of time required to establish a circuit-switched call between data terminals, i.e., the time from the initiation of a call request to the beginning of the call message.

call string All the call attempts related to a single call intent.

call transfer Service whereby a call to a subscriber's number is automatically transferred to one or more predetermined alternative numbers when the called number is busy or doesn't answer.

call waiting Service whereby a subscriber will receive a beep tone when he is on the phone, indicating another caller is trying to reach him.

called line Telephone line to which a call is directed.

called-line free indication Capability in some exchanges of automatically informing a user that a line called earlier and found busy is now free.

called line identification facility A facility provided by the network transmitting data which enables a calling terminal to be notified by the network of the address to which the call has been connected.

called-line-identification-request indicator In common channel signaling for data transmission: information sent in the forward direction indicating whether or not the called line identity should be included in the response message.

called line identification signal A sequence of characters transmitted to the calling data terminal equipment to permit identification of the called line.

called line identity In data transmission common channel signaling: information sent in the backward direction consisting of address signals indicating the complete called line identity.

called number The telephone number to which a call is directed.

called office The central office serving the called number.

called party The person on the receiving end of a telephone call.

called-party camp-on See camp-on.

called place The city or locality where the called party is located.

called station The telephone to which a call is made.

calling, selective The ability of a transmitting station to direct a call to the desired station, and of that station to receive it.

calling, three-way A service feature by which a third party can be brought in to a telephone call converting it to a conference-type call.

calling, touch Multifrequency tone signaling by push buttons.

calling card service Service enabling long-distance calls to be charged to a credit card account.

calling device The dial or keypad which generates the pulses or tones to signal the customer's requirements to the central office.

calling line The telephone line where a call is originated.

calling line identification facility In data transmission, a facility provided by the network which enables a called terminal to be notified by the network of the address from which the call has originated.

calling-line-identification-request indicator In data transmission common channel signaling: information sent in the backward direction indicating whether or not the calling line identity should be sent forward in a calling line identity message.

calling line identification signal In data transmission, a sequence of characters transmitted to the called data terminal equipment to permit identification of the calling line.

calling-line identity In data transmission common channel signaling: information sent in the forward direction consisting of a number of address signals indicating the complete calling line identity.

calling line identity message A message sent in the forward direction containing the identity of the calling line. This message, when applicable, is sent subsequent to an address message which does not contain the identity of the calling line.

calling number The number from which a telephone call originates.

calling number indication A service to identify the calling subscriber's number by means of a visual or verbal indication at the called terminal.

calling party The person who originates a telephone call.

calling party's category indicator Information sent in the forward direction denoting the category of the calling party, which is used, together with other call set-up information, to select the appropriate call treatment.

calling place The city or locality where a call originates.

calling rate The average number of originated calls per line during a given period.

calling rate, subscriber The call intensity of a subscriber line.

calling signal The signal transmitted over a circuit to indicate that a connection is desired.

calling station The telephone where a call originates.

calling subscriber The subscriber who originates a call.

calls barred facility A facility that permits a data terminal to either make outgoing calls or to receive incoming calls, but not both.

camp-on Service feature by which a switch observes that the wanted line is busy, waits until it is free, then automatically and immediately connects the calling line which has been waiting.

cancel character A control character which indicates that an error has been made and the data concerned should be disregarded.

cancel from (Can F) Network management control that diverts a preset percentage of traffic attempts overflowing from selected groups.

cancel to (Can T) Network management control that limits a preset percentage of traffic offered to selected groups.

canceling, loss Negative impedance repeaters used to cancel out losses introduced by circuit elements.

candela The base SI unit of luminous intensity. It is the luminous intensity, in the perpendicular direction, of a surface of 1/600,000 square meter of a black body at the temperature of freezing platinum under a pressure of 101,325 newtons per square meter.

candle, stearin A stearic acid candle used as a flux by lead sheath cable jointers.

cannibalize To keep obsolete equipment working by using components from similar items which have already been taken out of service.

cap, binding post Insulating cover put over binding posts in cable terminals.

cap, cable Closed tube fixed over the end of a cable to prevent moisture from entering during storage or transit.

cap, lamp Small colored lamp cap with brass rim, used for the lamps in lamp strips on a switchboard

capability The access right of a device to access a memory store for the transfer of program/data. It limits the access of various memory blocks to specific devices to minimize the possibility of program/data corruption being input to the system.

capacitance The property of a system of conductors and dielectrics that permits the storage of electrically separated charges when potential differences exist between the conductors.

capacitance, battery The capacitance of a central office battery, which acts as a noise-removing shunt filter to improve speech quality through an office.

capacitance, distributed The electrical capacitance between components and wires (other than that of pure capacitors).

capacitance, office cabling Sum of the capacitances of the cabling in a central office.

capacitive turning screw Adjustable screw projecting into a waveguide to provide a variable reactance, in a tunable filter.

capacitor A device which introduces capacitance into an electric circuit.

capacitor, blocking A capacitor, the function of which is to block the flow of direct current; it allows alternating current to pass.

capacitor, building-out A capacitor, the function of which is to increase an electrical circuit's capacitance to a standard value.

capacitor, bypass A capacitor required to provide a low-impedance path around resitors or similar circuit elements for high frequency alternating currents.

capacitor, ceramic A capacitor which uses a ceramic dielectric.

capacitor, coupling A capacitor, the purpose of which is to couple two circuits or two amplifier stages.

capacitor, mylar A fixed capacitor made up of two strips of metal foil separated by a very thin sheet of mylar plastic and then rolled into a cylinder.

capacitor, padder A trimmer capacitor in the oscillator circuit which permits calibration of a superheterodyne radio receiver's frequency range.

capacitor, paper A fixed capacitor made the same as a mylar capacitor but using waxed paper instead of mylar.

capacitor, trimmer A small adjustable capacitor placed in parallel with a larger one and used to tune a circuit precisely.

capacitor, variable A capacitor used to tune radio frequency circuits. Its capacitance can be varied by adjusting the separation between a pair of plates, or by varying the depth or insertion of interleaved plates.

capacity, ampere-hour A measurement of a storage battery's ability to furnish a steady current for a specified period. Battery capacity is usually specified on an 8-hour rate or a 10-hour rate, e.g., a fully charged 1000 AH Battery can be expected to supply a steady 100 Amps for 10 hours.

capacity, current-carrying A measure of the maximum current which can be carried continuously without damage to components or devices in a circuit.

capstan The shaft on which a reel of magnetic tape is mounted.

Captain Character and Pattern Telephone Access Information Network System. Japanese version of videotex, an interactive data retrieval service operated through the public telephone network and giving page displays on a TV receiver.

capture effect An effect associated with the reception of frequency modulated signals in which, if two signals are received on the same frequency, only the stronger of the two will appear in the output. The complete suppression of the weaker carrier occurs at the receiver limiter, where it is treated as noise and rejected.

car, cable A small, wooden seat, from which line technicians work on aerial cables. It is suspended by chains from a pair of wheels which roll along the top of the cable strand.

carbons Small carbon blocks separated by air gaps which are used as lightning arrestors on main distribution frames.

carbon transmitter A telephone microphone using carbon granules and a diaphragm, which varies the pressure on these granules and hence their net resistance.

card, circuit layout record Card maintained by the telephone company or administration listing the facilities and equipment used to provide each special circuit transitting the facility concerned.

card, data A card with punched holes, readable as letters or figures by a computer.

card, dial number Cards inserted in the center of the telephone dial, giving the telephone number.

card, printed circuit Piece of rigid material on which electronic components are mounted and connected by printed circuitry. Also called printed circuit board or printed wiring board.

card, repair Record card maintained by telephone company or administration giving fault history and other relevant information about routing and equipment on a subscriber's line. Now frequently stored in a computer rather than in a physical card index.

card, trouble history See card, repair.

card dialer An attachment to a telephone instrument which dials the required numbers automatically when a selected card is inserted.

card feed Mechanism which places data cards one at a time into the correct position for reading or punching.

card field Group of consecutive columns in a data card which contain a single unit of information.

card punch A machine which punches holes in a data card, either on receipt of instructions from a processor or as instructed by an operator.

card reader Machine which reads punched cards and is able to present the information to another stage of computation, e.g., bill preparation.

card verifying Process using two key punch operators, using the same source information, to verify correctness of the final punched card.

carriage A cable system's procedure of carrying the signals of television stations on its various channels. FCC rules determine which signals cable systems must or may carry.

carrier 1. A wave having at least one characteristic that may be varied from a known reference value by modulation. 2. In a semiconductor, the mobile conduction electron or hole.

carrier, commercial telecommunications An organization whose function is to provide telecommunications service.

carrier, common A commercial telecommunications carrier.

carrier, equivalent four-wire Carrier system designed for use on a two-wire line, using different frequencies for the two directions of transmission.

carrier, exchange Carrier system rarely encountered outside the U.S., producing 24 voice channels in the band 21 — 403kHz, on a frequency division multiplexing basis.

carrier, full In radio transmission, carrier power emitted at a level of 6 dB or less below the peak envelope power.

carrier, handline A light steel spring attached to a belt used by line technicians working on pole routes; it will safely carry the light weight of a handline but releases if additional tension is placed upon it.

carrier, majority In a semiconductor, the type of carrier which predominates, either electrons (in n-type material) or holes (in p-type material).

carrier, minority In a semiconductor, the type of carrier which constitutes less than 50% of the total.

carrier, record The term used to distinguish the organizations dealing only with the transmission of the printed or written word from those engaged in the telephone business.

carrier, reduced In radio transmission, carrier power emitted at a level of 6 to 32 dB (preferably, 16 to 26 dB) below the peak envelope power.

carrier, short-haul A carrier system designed for use over distances of 10 — 200 miles.

carrier, subscriber's Carrier system designed to provide subscribers' loops to a central office on a pair-saving basis. Systems commonly provide one carrier-derived circuit, but systems for 8, 24 or more circuits are also readily available.

carrier, suppressed In radio transmission, carrier power restricted to a level more than 32 dB (preferably, 40 dB or more) below the peak envelope power.

carrier, voice frequency telegraph A wire carrier system which divides a 300-3300 hertz voice channel into 24 frequency-shift telegraph channels spaced at 120 hertz.

carrier dropout A short duration carrier signal loss.

carrier frequency 1. The frequency of a carrier wave. 2. A frequency capable of being modulated or impressed with a second (information carrying) signal. In frequency modulation, the carrier frequency also is referred to as the center frequency.

carrier leak The balance may not be perfect in balanced modulators, as used in suppressed-carrier transmission systems. The remaining small amount of carrier power in the modulated signal is called carrier leak.

carrier level The power of a carrier signal at a particular point in a system, expressed in decibels in relation to some reference level.

carrier line A transmission line used to carry a carrier circuit.

carrier noise level The noise level resulting from undesired variations of a carrier in the absence of any intended modulation. Syn.: residual modulation

carrier-operated device, anti-noise (CODAN) Device which mutes a radio receiver until the carrier signal expected is received.

carrier power of a radio transmitter The average power supplied to the antenna transmission line by a radio transmitter during one radio-frequency cycle under conditions of no modulation; for each class of emission the condition of no modulation should be specified.

carrier sense multiple access/CD (CD = collision detection) Method of controlling access to a shared transmission path, particularly in local area networks. A generally similar procedure is proposed to be used in ISDNs. Terminals listen to a signaling bus and wait for it to be free before they signal. In the event of simultaneous seizure a garbled echo will cause the terminal to stop transmission. Also CSMA/CR (= contention resolution) and CSMA/CA (= collision avoidance).

carrier shift 1. A method of keying a radio carrier for transmitting binary data or teletypewriter signals which consists of shifting the carrier frequency in one direction for a marking signal and in the opposite direction for a spacing signal. 2. A condition resulting from imperfect modulation whereby the positive and negative excursions of the envelope pattern are unequal, thus effecting a change in the power associated with the carrier. There can be positive or negative carrier shift.

carrier signaling Voice circuits on carrier sometimes use a telegraph carrier on one of the voice channels to carry all the associated supervision and signaling circuits. Others may use a signaling frequency within the voice band (in-band signaling), and others a frequency in the guard band between the voice channels (out-of-band signaling). Common-channel signaling (with complete separation of voice from signaling) is becoming standard practice in digital networks.

carrier system A system whereby many channels of electrical intelligence can be carried over a single transmission channel accomplished by modulating the intelligence onto a higher frequency carrier wave, then recovering it at the receiving end through a reverse process of demodulation.

carrier system, four-wire System using two separate wire pairs which transmit in opposite directions.

carrier system (frequency division multiplex) Means of transmitting more than one voice channel over a wideband path by modulating each channel on a different carrier frequency. Standard twelve-channel group occupies the frequency band from 60 to 108 kHz.

carrier system, K type Early twelve-channel FDM carrier system originated by the Bell System. Uses the band 12-60 kHz.

carrier system, L type A wideband system providing 155 twelve-channel groups (1860 channels) using radio bearers or high frequency cables.

carrier system, N2 type Carrier system providing a single twelve-channel group, using pairs in toll cables, one for each direction of transmission.

carrier system, N3 type Carrier system providing two twelve-channel groups, using pairs in toll cables, one for each direction of transmission.

carrier system, O An early carrier system for open wire working in rural areas, providing 16 voice channels.

carrier system, T1 The first of the current generation of time division carrier systems. Uses 24-channel PCM and 1.544 Mbit/s digital transmission.

carrier system, T2 Provides 96 voice channels (the equivalent of four T1 systems) using a 6.3 Mbit/s PCM system.

carrier system, T4 Higher level PCM carrier system providing 4,032 voice channels (168 T1 systems) using 281 Mbit/s PCM and needing a wideband bearer, usually a coaxial cable.

carrier system (time division multiplex) Means of transmitting more than one voice channel over a wideband path by sampling the speech, coding the sample levels (PCM: pulse code modulation) and using digital techniques. There are two basic systems: the 24-channel system as used in America and Japan and the 30-channel system as used in Europe.

carrier system, two-wire System using only a single pair of wires. Different frequency allocations permit transmissions in both directions.

carrier telegraphy The transmission of telegraph signals by means of discrete frequencies allowing the utilization of a voice grade channel for many simultaneous telegraph transmissions.

carrier telephony Form of telephony using carrier transmission, the modulating wave being a voice frequency signal.

carrier-to-noise ratio The ratio, in decibels, of the value of the carrier to that of the noise in the receiver IF bandwidth before any non-linear process such as amplitude limiting and detection.

carrier-to-receiver noise density In satellite communications, the ratio, expressed in dB, of the received carrier power (C) to the received noise power density (kT), where k is Boltzmann's constant and T is the receiver system noise temperature in kelvin. The ratio also is referred to as C/kT.

carrier transmission Form of electrical transmission in which the transmitted electrical wave is a complex wave resulting from the modulation of one or more single-frequency waves by one or more modulating waves.

CARS (community antenna relay service) The 12.75 — 12.95 GHz microwave frequency band which the FCC has assigned to the CATV industry for use in transporting television signals.

cart, cable splicer's A jointer's trailer; small cart carrying a complete set of splicer's tools and materials.

Carterfone A trade-marked acoustically-coupled device intended to couple a two-way radio circuit with local or long distance telephone facilities. The "Carterfone decision," in which the U.S. Federal Communications Commission permitted interconnection of the device opened up the telephone terminal equipment market by permitting subscribers to buy or lease such equipment from firms other than telephone companies.

cartridge (cable TV) Container for recorded programming designed to be shown on a television receiver. The cartridge contains a reel of motion picture film, videotape or electronically embossed vinyl tape, blank or recorded, and uses an external take-up reel.

cascade connection A tandem arrangement of two or more similar component devices with the output of one connected to the input of the next.

case, coil A protective case for loading coils.

case, splice A split case designed to fit around a spliced cable joint providing mechanical protection.

case, strand-mounted A light loading coil case designed to be used mid-span, supported by the aerial cable's messenger strand.

Cassegrain antenna A high-gain parabolic antenna in which the main feed is not at the focus of the dish but near its surface, pointing towards a small subreflector located near the focus.

cassette A self-contained package of reel-to-reel blank or recorded film, videotape or electronically embossed vinyl tape which is continuous and self-rewinding.

catena Series of items in a chained list in a computer store.

catenate To arrange a series of items in a catena.

Caterpillar gear Cable-ship machinery equipped with endless belts like Caterpillar tractor treads which work together to grip submarine cable being paid out or hauled in.

cathode 1. Electron tube: the electrode through which a primary stream of electrons enters the interelectrode space. 2. Electrolytic: the electrode where positive ions are discharged or negative ions are formed or where other reducing reactions occur.

cathode, cold Cathode which functions without heat.

cathode, heated Electron tube cathode with heating circuit to produce electron emission.

cathode, indirectly heated Electron tube cathode with a separate heating circuit to produce electron emission.

cathode follower Electron tube circuit with high input and low output impedances; output load is connected in the cathode circuit of the tube and input applied between control grid and the end of the cathode load.

cathodic protection Protection given to metal cable sheaths against electrolytic corrosion. The outer sheath is connected to a dc supply so that any currents that flow between cable sheath and ground are in the direction which ensures that the atoms of the cable sheath are not encouraged to migrate to the soil, so holes are not produced by local electrical corrosion of the sheath.

cation Positively charged ion or radical that migrates towards the cathode under the influence of a potential gradient.

CATLAS An American Telephone and Telegraph software routine: Centralized Automatic Trouble Locating and Analysis System: used as maintenance tool for locating and diagnosing faults in Bell System electronic central offices/exchanges.

CATT Controlled avalanche transit-time triode: a microwave device which uses IMPATT diode technology.

catwhisker A small, sharp-pointed wire used to make contact with a sensitive point on the surface of a semiconductor. Early domestic radio receivers used crystal detectors with catwhiskers.

cavity coupling Method of extracting or introducing microwave energy from or to a resonant cavity; a common method is slot coupling between the resonator and a waveguide.

cavity resonator A dimensional enclosure or box in which oscillating electromagnetic energy is stored and whose resonant frequency is determined by physical dimensions.

CB Citizens Band; radio frequencies available for public use without formal licensing after technical examination as is needed before amateur transmission is authorized on other frequencies. The 27 MHz band is the most commonly used for CB in most countries.

CCIR From the French for International Radio Consultative Committee. The CCIR is one of the four permanent organs of the International Telecommunication Union. The CCIR does for radio services what the CCITT does for telephony and telegraphy. (See CCITT.)

CCITT From the French for International Telegraph and Telephone Consultative Committee (Committe Consultatif International Telegraphe et Telephone) The CCITT is one of the four permanent organs of the International Telecommunication Union (the ITU). The CCITT deals with technical problems relating to telephone and telegraph services. All member countries of the ITU and certain private operating companies (e.g. AT&T, C&W) can participate in the work. A Plenary Assembly is held every few years, usually at three- or four-year intervals. The Assembly draws up a list of technical questions, the study of which would lead to improvements in international telephony and telegraphy. These questions are entrusted to a number of Study Groups, composed of experts from different countries. The Study Groups draw up Recommendations which are submitted to the next Plenary Assembly. If the Assembly adopts the Recommendations they are published and have an important influence on telecommunications engineers, scientists, designers and manufacturers of equipment, and on operating administrations throughout the world.

CCITT no. 4 signaling Two-tone voice frequency signaling system used on international telephone circuits in Europe.

CCITT no. 5 signaling Inter-register signaling system using pulsed voice frequencies, widely used on intercontinental and international telephone circuits.

CCITT no. 6 signaling Common channel signaling system designed for the control of telephone switches, operating in an analog environment in which signaling information may be sent at 2.4 kbit/s. Error rates which can result from the use of modems have necessitated fixed length messages in SS No. 6; these are 28 bits long, of which eight are allocated to error detection.

CCITT no. 7 signaling Common channel signaling system designed for the control of voice and non-voice services and for use in a digital environment, where signaling information may be sent at 64 kbit/s. Lower error rates than are possible using modems in an analog environment result in a word length in international use of up to 69 octets (552 bits) of which only two are used for error detection. For use within a national system the word length can be up to 279 octets, still using only two octets for error detection.

CCITT R1 signaling Inter-register signaling system used in national systems and internationally on a regional basis especially in North, Central and South America.

CCITT R2 signaling Inter-register signaling system used in many national systems and internationally on a regional basis in Europe, Asia and Africa.

CCS Hundred Call Seconds - this is the US unit of telephone traffic. Europe and the rest of the world uses the Erlang, one circuit continuously occupied for one hour, so there are 36 CCS to 1 Erlang.

CDLRD Abbreviation used in Bell System Universal System Service Orders: on an Other Common Carrier or OCC order, the date by which the confirming design layout report (CDLR) should be received by the responsible system unit.

CEE (From the French) International Commission on Rules for the Approval of Electrical Equipment: basic objectives include development of worldwide specifications for electrical products to safeguard public against electrical accidents and to achieve uniformity of regulations in member countries. The U.S. is not a member country of the CEE.

Ceefax Digital data "Teletext" broadcasting service associated with TV signals transmitted by the British Broadcasting Corp.

CEIRD Confirming engineering information report date. Abbreviation used in Bell System Universal System Service Orders: the date on which all local design groups are to receive a confirming report from the design control group.

celestial guidance Nothing to do with religion: this is the guidance of a rocket or missile by an in-board radio navigation device which observes the apparent position of stars or other stellar bodies.

cell An elementary unit — of data storage, of power supply, or of equipment.

cell, alkaline Primary cell using a potassium hydroxide electrolyte and giving 1.5 volts potential difference per cell.

cell, counter EMF A nickel and sodium hydroxide electrolytic cell used in series opposition with the central office battery to provide the correct net voltage for the equipment. Each cell introduces a drop of 1.5 volts.

cell, dry A small portable battery. The cell operates on Leclanche principles, with a central (positive) carbon electrode and a metal outer case for the negative electrode.

cell, dry counter EMF Stacks of iron/selenium plates are used to give voltage drops in opposition to a main exchange battery.

cell, fuel An electrochemical cell which produces electrical energy from the chemical energy of a fuel and an oxidant.

cell, lead-acid storage An individual cell used in central office batteries. The plates are made of lead, lead/antimony or lead/calcium and the electrolyte is dilute sulfuric acid. The charge/discharge action is reversible; cells may be charged and discharged many times or kept "floating" in a fully charged condition. Nominal voltage is 2 volts per cell, rising to 2.15 if fully charged and falling to 1.85 volts when discharged.

cell, lead-antimony A lead-acid storage cell with plates made of a lead-antimony alloy.

cell, lead-calcium A lead-acid storage cell with lead-calcium alloy used for plates. These cells have a longer life than antimony cells if they are floated, they should not be used on a charge/discharge cycle basis.

cell, Leclanche One of the original non-reversible primary cells, with carbon positive, zinc negative, an ammonium chloride (sal ammoniac) electrolyte, plus manganese dioxide to act as a depolarizer, in close contact with the carbon rod. Gives about 1.5 volts output; it is the predecessor of the ordinary dry cell.

cell, magnesium Primary cell with a magnesium alloy negative electrode.

cell, memory Basic storage unit in a semiconductor or magnetic data store.

cell, mercury Type of dry cell, often used in watches, with a zinc anode, a mercuric oxide cathode, and a potassium hydroxide electrolyte.

cell, nickel-cadmium Storage cell with nickel-steel tubes as plates, the negatives are filled with sponge cadmium and metallic iron, and the positives are filled with nickel oxide. The electrolyte is potassium hydroxide; each cell gives about 1.25 volts.

cell, nickel-iron Storage cell with nickel-steel tubes as plates. The negatives are filled with ferrous oxide and iron, and the positives are filled with nickel oxide. The electrolyte is potassium hydroxide; each cell gives about 1.25 volts.

cell, photoconductive Semiconductor device with resistance that is affected by the amount of light falling on it.

cell, photovoltaic Semiconductor device which generates a voltage when light falls upon it. Used as solar cells, e.g., to power spacecraft.

cell, primary A nonreversible, nonrechargeable cell which converts chemical energy into electrical energy.

cell, secondary A rechargeable cell; a storage battery.

cell, silver-cadmium A long-life sealed storage cell, used on some spacecraft.

cell, silver oxide Type of dry cell often used in hearing aids and cameras, with a zinc anode, a silver oxide cathode, and a sodium or potassium hydroxide electrolyte.

cell, solar Silicon diode used as a photovoltaic cell, e.g., in spacecraft.

cell, storage A rechargeable, secondary cell used to power most central offices. Semiconductor cells used for data storage are sometimes called storage cells.

cellular geographic area The area a cellular radio system operator is authorized to cover and into which another co-channel system may not encroach.

cellular mobile radio telecommunications system/service (CMRS) Automobile radio system using low-power transmitters providing coverage in a limited area but linked (usually by cable) with other similar transmitters operating at different frequencies. The whole system enables radio frequencies to be reused economically, and enables good telephone service to be provided in moving automobiles by switching control from one transmitter cell to another as the mobile moves across an area. In the U.S. the FCC has ruled that each market may be served by two cellular franchises: 'wireline' (usually the local telephone company) and "non-wireline" (controlled by non-telco organizations).

cellulose acetate A tough thermoplastic used to impregnate cotton insulation on conductors to improve handling ability.

Centel Corporation A major U.S. independent telephone company group.

center, automatic intercept Facility that automatically processes intercepted calls and connects to a vacant number announcement or to an announcement machine which gives the correct number, or transfers the call to an intercept operator for handling.

center, automatic teletypewriter switching A switching center which routes Telex teletypewriter calls to their destination.

center, load A power distribution control center in a telecommunications building. Large telecom buildings may have several such centers.

center, nonautomatic relay A message switching center which relays messages toward their address destinations by manual operating procedures.

center, switching The complete set of switching equipment at a location, including related power supplies, transmission equipment, automatic switching equipment, manual switchboard equipment, maintenance and test equipment.

center, tape relay A message switching center where incoming paper tapes are detached from the receiving teletypewriter and taken to an outgoing machine for transmission.

center, telephone switching A telephone company building complex containing one or more central offices.

center, test Location where fault reports are received, lines are tested and fault repair technicians' work is controlled.

center conductor (submarine cable) The inner conductor of the coaxial structure of seacables. They either can be solid-copper wire for armored cables; solid copper wire plus surrounding copper tapes (now obsolete); or a copper tube over a stranded steel member in armorless cable.

center-feed tape   Paper tape with feed holes aligned precisely with the centers of the character holes.

center frequency   1. In frequency modulation, the resting frequency or initial frequency of the carrier before modulation. 2. In facsimile, the frequency midway between picture black and picture white frequencies.

center sampling   A method of sampling a digital data stream at the center of each signal element.

center tap   A physical connection at the midpoint, electrically, of a transformer winding, coil or similar device.

centers, toll switching   The hierarchy of centers which make up the control switching points of the North American long distance dialing network. They are (in order of importance): Class 1 - Regional Centers; Class 2 - Sectional Centers; and Class 3 - Primary Centers. They receive calls from, and switch calls to, the lower class offices in the hierarchy. These are: Class 4C - Toll Center; Class 4P - Toll Point; and Class 5 - End Office.

centigrade   Temperature scale (degrees C) with water freezing at 0°C and boiling at 100°C. Also called the Celsius scale. To convert to degrees Fahrenheit multiply by 9/5 and add 32. To convert to Kelvin add 273 (approximately).

central control   Control equipment (which may be replicated) that is common to an entire exchange.

central office   In telephone operations, the facility housing the switching system and related equipment that provides telephone service for customers in the immediate geographical area.

central office, unigauge   A central office equipped with range extenders to serve customers further away from the central office than was previously possible with the original technology.

central office ground (COG)   The assumed arbitrary point of zero potential in a telephone office.

central office equipment   Apparatus, including switching and signaling equipment, installed in a central office.

central processing unit (CPU)   That part of a computer which contains the logic, computation and control circuits. It controls the interpretation and execution of instructions and, sometimes, contains memory. In telephone switching systems, memory usually is elsewhere. Until the late '70s in computer-controlled telephone exchanges, it was usual to have two CPUs running the exchange — with one working and the other on hot standby, or both working on a load-sharing basis. With the wider use of microprocessors and distributed control, CPUs now tend to be used for administrative tasks and the more complex diagnostic tasks only, typically those requiring access to large data stores. Sometimes CPU's are used on a cluster or multiprocessor basis. There could be a dozen such units in an exchange. One advantage claimed for this approach is that the

traffic handling capacity of the exchange can be increased in stages as the network grows.

centralized automatic message accounting   System which produces itemized billing details for long distance calls dialed by customers; details are recorded at a central facility serving a number of exchanges. In exchanges not equipped for automatic number identification (ANI), calls are routed to a CAMA operator who obtains the calling number and keys it into the computer for billing. Also, the type of trunk used for forwarding CAMA calls.

centralized automatic reporting on trunks (CAROT)   A system for automatically reporting faults identified on toll trunk circuits.

centralized control   A control system in which a significant part of the equipment is in a location remote from the exchange; this system may serve a number of exchanges in an area. Syn.: area control

centralized operation   Operation of a communication network in which transmission may occur between the control station and tributary stations, but not between tributary stations.

CENTREX   Service providing a business telephone customer with direct inward dialing to its phone extensions and direct outward dialing from them.

CENTREX, C.O.   Where the service-providing switch is in the central office.

CENTREX, C.U.   Where the switch is on the customer's premises.

certificate of compliance   The approval of the FCC that must be obtained before a cable system can carry television broadcast signals.

chad   The small circle of paper removed when a hole is punched in a tape or card.

chad   tape   Punched   tape   used   in telegraphy/teletypewriter operation.

chadless tape   Tape in which the chad holes are not punched out completely. This tape is sometimes given an overprinting of the characters so that messages may be read by operators without having to decode the five or seven unit code characters.

chain circuit   An information circuit with several contacts in series so that the circuit is complete only when all the contacts in the chain have been operated. Used for lampsignals, such as those indicating all circuits to a particular destination are busy.

chain printer   A type of line printer in which the characters are carried on a belt or chain which moves across the paper.

chained list   Set of data in which each entry ends by giving the address of the next item in the list. Entries do not need to be entered consecutively or on to consecutive addresses.

chaining search   A search in which each item contains the means for locating the next item to be considered in the search.

chair, lineman's safety   Chair designed to be suspended midspan to permit work on aerial equipment.

chair, operator's   Special chair designed for use by telephone operators, particularly those working in older installations. Modern operating positions are not greatly different from office desks, so high swivel chairs are seldom needed.

challenge Words to be used by operators when they enter a circuit, e.g., "What number are you calling, please?"

chamber, splicing In North America, the area below main distributing frame where external cables can be spliced onto terminating or tip cables. In Europe, a splicing or jointing chamber is usually a manhole in a duct network.

changeback The procedure of transferring signaling traffic from one or more alternative signaling links to a signaling link which has become available.

changeback code A field included in the signaling network management messages used in the changeback procedure; it is used to discriminate messages relating to different changeback procedures performed at the same time towards the same signaling link.

changed number signal A signal sent in the backward direction indicating that the call cannot be completed because the called customer's number has been changed recently.

changeover The procedure of transferring signaling traffic from one signaling link to one or more different signaling links, when the link in use fails or is required to be cleared of traffic.

channel 1. The smallest subdivision of a circuit which provides a single type of communication service, e.g., a voice channel, a telegraph channel, a data channel. Note however that a channel is a unidirectional path, so a telephone circuit includes both directions of transmission and all the associated relay sets etc whereas "channel" is the one-direction transmission path only. Several channels may share a common carrier, as in frequency division and time division systems; in these cases, each channel is allocated a particular frequency band or a particular time slot which is reserved for it. 2. A path along which signals can be sent, e.g., data channel, output channel. 3. The portion of a storage medium that is accessible to a given reading or writing station, e.g., track, band. 4. In information theory, that part of a communications system that connects the message source with the message sink.

channel, adjacent The bands of frequencies on either side of the channel concerned.

channel, AM broadcast Any of the radio frequency bands used for amplitude-modulated broadcasting.

channel, crossbar A path switched through a crossbar central office.

channel, dropped A channel in a multichannel system which is dropped off before reaching the terminal where the main body of groups and channels are terminated.

channel, FM broadcast Any of the radio frequency bands used for frequency-modulated broadcasting.

channel, outboard Channel in an independent sideband system which is furthest from the carrier frequency.

channel, pilot A single frequency signal used to control transmission levels.

channel, public music A CATV service in which a music channel is provided without an extra fee.

channel, radio A band of radio frequencies wide enough to permit radio communication to be established.

channel, SCA Subsidiary carrier authorized. A channel carrying program material for a CATV or FM broadcast system.

channel, service An order wire channel used for telemetry or conversation between engineering control centers.

channel, subscription music A continuous music system carried by a CATV system; available only to customers who pay extra for it.

channel, telegraph A channel capable of carrying telegraph frequency signals.

channel, telephone A channel capable of carrying voice frequency signals.

channel, television Numbered frequency bands used for the broadcasting of television signals. Low numbered channels utilize VHF frequencies between 54 and 88 MHz; higher numbers utilize higher UHF frequencies in the 200 MHz and 500 to 900 MHz ranges.

channel, voice A channel capable of carrying voice frequency signals.

channel-associated signaling A signaling method in which the signals necessary for the traffic carried by a single channel are transmitted in the channel itself or in a signaling channel permanently associated with it.

channel bank A part of a carrier-multiplex terminal that performs the first step of modulation. It multiplexes a group of channels into a higher frequency band and, conversely, demultiplexes the higher frequency band into individual channels.

channel capacity A measure of the maximum possible information rate through a channel, subject to specified constraints.

channel capacity (cable TV) The maximum number of 6MHz (8MHz in Europe) channels which can be simultaneously carried on a CATV system.

channel gate A device for connecting a channel to a highway, or a highway to a channel, at specified times.

channel group Twelve voice channels, frequency-division multiplexed into a single frequency band, 48 kHz wide.

channel modulation First stage of modulation in a frequency division multiplex carrier system; the translation of a voiceband to its appropriate frequency band in the basic twelve-channel group

channel noise level 1. The ratio of the channel noise at any point in a transmission system to some arbitrary amount of circuit noise chosen as a reference. This ratio is usually expressed in decibels above reference noise, abbreviated dBrn, signifying the reading of a circuit noise meter, or in adjusted decibels, abbreviated dBa, signifying circuit noise meter reading adjusted to represent an interfering effect under specified conditions. 2. The noise power density spectrum in the frequency range of interest. 3. The average noise power in the frequency range of interest. 4. The indication on a specified instrument. The characteristics of the instrument are determined by the type of noise to be measured and the application of the results thereof.

channel packing A technique for maximizing the utilization of voice frequency channels used for data transmission by multiplexing a number of lower speed data signals into a single higher speed data stream for transmission on a single voice frequency channel.

channel reliability The percent of time a channel is available for use in a specific direction during a specified period.

channel switching The switching of single channels for the exclusive use of the connection for the duration of a call.

channel time slot A time slot starting at a particular phase in a frame and allocated to a channel for transmitting a character signal, in-slot signaling or other information. Where appropriate a description may be added, for example, "telephone channel time slot."

channelization 1. The allocation of traffic circuits to channels and the forming of these channels into groups and higher order multiplexing. 2. The method of using a single wideband facility to transmit many relatively narrow bandwidth channels by subdividing the wideband channel.

character A member of the character set which is used for the organization, control or representation of data. A single specific symbol, number or letter used to designate the dialable signal caused by a command.

character, magnetic Characters printed with magnetic ink which are machine-recognizable.

character check A method of error detection utilizing the preset rules for the formulation of characters.

character-count and bit-count integrity The preservation of the precise number of characters or bits that are originated in a message, in the case of message communication, or per unit time, in the case of a user-to-user connection.

character interval The total number of unit intervals, including synchronizing, information, error checking, and control bits, required to transmit any given character in any given communications system. Extra signals not associated with individual characters are not included.

character oriented Protocol or procedure that carries control information encoded in fields of one or more bytes.

character printer Output device, like a teleprinter, which prints one character at a time.

character reader A device which can read typescript and produce a machine-readable output without needing an operator to interface the change. Also called an optical reader.

character recognition The process of a machine's recognizing typescript. The two main methods used for machine recognition of characters are: optical character recognition (OCR) and magnetic ink character recognition (MICR). OCR is cheap to produce (ordinary typewriters can be used) but prone to error, especially if documents are soiled or creased. MICR needs special print fonts but is more trustworthy. MICR characters are printed using an ink impregnated with magnetizable particles. The documents go through a magnetizing stage before they are presented to an automatic reader. There are no international standards for print font for use with either procedures, as yet.

character-serial Method of transmission in which characters are sent one at a time, sequentially.

character set 1. A finite set of different characters which are agreed upon and considered to be complete for some purpose, e.g., each of the character sets in ISO Recommendation R646 "six- and seven-bit Coded Character Sets for Information Processing Interchange". 2. An ordered set of unique representations called characters, e.g., the 26 letters of the English alphabet.

character signal A set of digital signal elements representing a character, or in PCM, representing the quantized value of a sample. In PCM, the term "PCM word" may be used in this same sense.

character stepped Pertaining to operational control of start-stop teletypewriter equipment in which a device is stepped one character at a time; the step interval is equal to or greater than the character interval at the applicable modulation rate.

characteristic Properties of a circuit or component.

characteristic, impedance A graph representing a circuit's impedance plotted against frequency.

characteristic distortion Distortion caused by transients which, as a result of modulation, are present in the transmission channel. Its effects are not consistent; its influence upon a given transition is to some degree dependent upon the remnants of transients affecting previous signal elements.

characteristic frequency A frequency which can be easily identified and measured in a given emission.

characteristic impedance The impedance that an infinitely long transmission line would have at its input terminals. A line will appear to be infinitely long if terminated in its characteristic impedance.

characters per inch (CPI) Measurement of density of data recorded on magnetic tape.

charge 1. To replenish or replace the electrical charge in a secondary cell or storage battery. 2. To store electrical energy in a capacitor.

charge, constant voltage Method of charging a secondary cell or storage battery during which the terminal voltage is kept at a constant value.

charge, electric Electric energy stored in a material.

charge, equalizing A special extra charge given to a storage battery to bring all the cells to the same specific gravity reading.

charge, initial period The minimum charge on a manually-controlled toll call; permits conversation for an initial period (usually 3 minutes).

charge, overtime The charge made for telephone calls exceeding the initial period, and for calls made on a WATS tariff when the time limit paid for has been exceeded.

charge, space The effect on the cathode in an electron tube of electrons emitted by a cathode in excess of those which immediately move towards the anode or plate.

charge, trickle A continuous charge of a storage battery using a small current.

charge computer Computer which is fed with rate information about a call and which provides the charge to be made for the call.

charge coupled device (CCD) A device which allows charges to be transferred from one point to another in a controlled manner. Television cameras often use an array of CCDs as their light-sensitive target area.

charge key supervision For manually controlled long distance calls, a method by which an operator signals back to the calling subscriber's central office that the called subscriber has answered, so charging should be commenced.

charge transfer device Semiconductor device in which discrete packets of electric charge are transferred from one location to another.

chargeable duration In international telephone services: 1. The time interval on which the charge for a call is based. 2. In manual or semiautomatic service, the chargeable duration is equal to the duration of the call reduced, if necessary, to make allowance for any interruptions or other difficulties which might have occurred during the call. 3. The duration of a call for which the charge is paid by the calling subscriber (or the called subscriber in the case of a collect call). 4. In the case of manual or semiautomatic operation: (a) either to a 3-minute charge, if the chargeable duration of the conversation is less than 3 minutes; or (b) to the whole number of minutes if the chargeable duration is greater than 3 minutes. Syn: charged duration

charger A rectifier or battery charger, ac input, dc output.

charger, constant voltage Battery charger which maintains constant output voltage, i.e., it starts with a large charging current (if the battery is discharged), then reduces current output as the battery becomes charged, and then floats the battery.

charger-eliminator A battery eliminator; a rectifier which provides low noise, low impedance dc output which can be used to power a PBX without the necessity of a storage battery.

chart, earth's radius Profile chart for radio transmission purposes with base line conforming precisely to the earth's surface; the chart shows the profile for radio waves when there is no refraction of the waves by the atmosphere.

chart, 4/3 earth radius Charts used by radio system planners, with base lines curved to represent a radi-

us 4/3 times greater than the actual earth radius. Straight lines may be drawn on the charts to check clearance for microwave paths.

chart, profile Chart showing clearance over terrain along a proposed microwave path, using a 4/3 earth radius chart.

chassis Metal box or frame where components are mounted.

chatter The multiple rebound of relay contacts or armatures.

Cheapernet A low-cost Local Area Network (LAN) design using CSMA/CD protocol: has many features in common with Ethernet.

check, parity Method of checking that a transmitted digital signal has been correctly received: an extra bit is added to the coded information bits to make the total number of 1s odd (for an odd parity scheme). If the received total is incorrect an automatic request is made for retransmission.

check bit See bit, check.

check character A character used for checking purposes, but often redundant.

check digit An extra digit added to a code number by an internal calculation at the stage when codes are first put into the processor.

check indicator An alarm signal to show the operator an error made has been spotted as the result of a check digit or a parity check operation.

check list A concise list of activities used as an aide-memoire to ensure that all are accomplished.

check loop A device which is attached to interconnect the go and return paths of a circuit at the incoming end of a circuit to permit the outgoing end to make a continuity check on a loop basis.

check out To test a circuit or component.

checking multiple A multiple of subscriber lines provided at a toll operating position so an operator can test the calling line number to see if the number given by the caller is correct.

checking operator service Service which is fully automatic but has provision for operators to be brought in when assistance is needed or checks need to be made.

checking service Service involving an operator's calling in on a customer-dialed call to obtain the calling number and key it into the charging equipment.

checksum The sum of a group of data items associated with the group for checking purposes.

chemical vapor deposition technique A method of fabricating optical waveguide preforms by causing vapors to react and form a deposit that may be in the form of glass oxides.

CHILL A high level programming language for programming SPC telephone exchanges, developed by the CCITT.

chimney, manhole The passage between a manhole cover at road surface level and the roof of the manhole. This permits road level changes to be made without having to rebuild manholes.

chip, semiconductor A small rectangular slice of material, usually silicon and 4 mm^2 or less on which a complete semiconductor device has been built. It can be either a simple single function like a transistor used as an amplifier, or a complex integrated circuit replacing thousands of discrete components. Chips are often called integrated circuits.

chirping A rapid change (as opposed to long term drift) in the emission wavelength of an optical source. Chirping is most often observed in pulsed operation of a source.

choke A coil with high impedance at high frequencies but low dc resistance.

choke, audio frequency A coil with high impedance over the audio frequency band but low dc resistance.

choke, smoothing Coil used as part of a low-pass noise filter through which a dc power supply is fed to equipment.

choke, swinging A coil with a saturated core used in a dc power supply filter to provide voltage regulation.

cholesteric A type of liquid crystal with normally nematic material but molecules arranged in a helical formation. Used widely in liquid crystal display units and watches.

chopper A device for interrupting a light beam or a current in order to produce a pulsating signal, sometimes used for timing pulses.

chopper amplifier If noise on the circuit makes a chopper signal difficult to detect, it can be amplified by a tuned amplifier to ensure recognition and action.

chromatic dispersion Combination (in a fiber optic system) of material dispersion and waveguide dispersion.

chrominance A comparison of a particular color and another color with equal luminance, expressed colorimetrically.

chrominance signal In color TV, the signal which gives the color information. The signal is obtained by combining, in a color coding circuit, specified fractions of the separate video signals into sum and difference signals, which are used to modulate a chrominance subcarrier.

chronopher A device which provides time-of-day and date electrically.

chronotron Electronic device which measures the time interval between events, such as the time between the transmission of a pulse and the arrival of its echo back from a fault.

chute, coin The portion of a coin telephone which accepts coins and conducts them to the coin relay hopper.

chute, stern A structure or device located at the stern of a cable layer. Provides a guide for the passage of cable and submerged devices being overboarded, and has a radius of curvature consistent with the allowable bending radius of the cable.

cipher The pseudo-random signal resulting from the combining of an intelligence signal with a key signal

ciphony equipment Encryption equipment for telephone calls.

circuit 1. A pair of complementary channels, which provide bidirectional communication, with associated equipment terminating in two exchanges. 2. A network of circuit elements (resistances, reactances and semiconductors) which perform specific functions. 3. A schematic diagram of a circuit.

circuit, allocated A circuit set aside for one user's specific use.

circuit, approved A circuit which has been approved to carry classified information without encryption.

circuit, carrier A transmission path used in a carrier system.

circuit, closed An audio or video program circuit not used for broadcast.

circuit, common-battery A circuit which obtains its power from a centralized battery facility which serves several circuits.

circuit, common user A communications circuit used on an equal basis by several users.

circuit, cord The circuit which interconnects switchboard cords to provide talking battery, ringing power and supervision.

circuit, data The electrical means for two-way transmission of binary signals.

circuit, duplex A circuit which permits simultaneous independent communications in both directions.

circuit, electric A path of conductors and/or circuit elements which permits an electric current to flow.

circuit, engineered A standby circuit designed to meet a specific customer requirement.

circuit, four-wire A circuit which uses electrically separate paths for the two directions of transmission.

circuit, free service A telephone message circuit giving toll service with no direct toll charge.

circuit, ground return A single-wire circuit using the earth as its return path. At one time used for dc telegraph circuits. Too noisy for reliable telephone use.

circuit, grounded A circuit which is connected to ground or ground potential at one or more points

circuit, grouping A tie circuit between operating positions which permits one operator to handle traffic offered to several positions during light traffic periods.

circuit, hold and trace Facility on manually operated toll boards permitting operators to apply a special hold and trace tone to faulty toll circuits.

circuit, longitudinal Circuit with one or more wires in parallel and with its return through the earth or through another group of conductors.

circuit, magnetic A closed loop for magnetic flux.

circuit, message Circuit used for toll calls.

circuit, metallic Circuit routed on physical wires, not via fibers or carrier channels or a radio system.

circuit, multiple Circuit with several similar elements in parallel.

circuit, on-call Circuit which is activated only when asked for by the user.

circuit, open-wire Circuit routed on non-insulated wires suspended from insulators on pole-mounted crossarms.

circuit, oscillatory Circuit for electrical oscillations.

circuit, parallel Components or circuits connected to the same input and output terminals so that one path is in parallel with the other.

circuit, phantom Circuit derived from center-taps on two side circuits, so that two pairs of wires can provide three speech paths.

circuit, plugging-up The circuit involved in the use of a plugging-up cord.

circuit, polar A circuit used in the technique of polar telegraphy.

circuit, private line Circuit provided for private use of one user and with no access to the public switched telephone network.

circuit, resonant Circuit with inductance and capacitance which is resonant at a particular frequency.

circuit, ring-down Circuit on which signaling is by a ringing current. If the circuit is short and metallic the ringing current will be dc or at 17 — 20 Hz, if the circuit is long or utilizes carrier facilities a voice-frequency signal is used, typically 1000/20 or 2600 Hz.

circuit, series Components connected so that the same current flows through all in sequence.

circuit, short-period A circuit used only for a specified period daily. Also see on-call circuit.

circuit, sidetone A circuit element in common-battery telephone sets. The talker hears room noise and his own voice reproduced by his own receiver.

circuit, straightforward Circuit on which signaling is in one direction only.

circuit, telecommunication A complete circuit with a specified bandwidth to enable instruments at each end to communicate one with the other.

circuit, telegraph Circuit suitable for transmission of binary signals such as teleprinter signals, usually obtained by subdividing a voice circuit into 24 or more telegraph circuits.

circuit, thin film A circuit whose elements are films a few molecules thick, formed on an insulating substrate.

circuit, three-phase An electric power circuit with three alternating voltages differing in phase by one-third of a cycle (120 degrees).

circuit, toll Circuit between two toll offices.

circuit, transfer Circuit linking two different networks carrying traffic between them.

circuit, tuned An electrical circuit whose components are adjusted or selected to provide desired facilities for signals of a specified frequency.

circuit, two-wire A circuit in which information signals in both directions are carried by the same two-wire path.

circuit, vibrating An audio frequency current, sent through a telegraph relay, insufficient to cause operation but strong enough to make the relay tend to vibrate. A weak incoming telegraph signal, assisted by this vibration, results in positive action by the relay.

circuit breaker See breaker, circuit.

circuit board Plastic base on which components (including integrated circuits) are mounted. Syn.: printed circuit board (PCB)

circuit concentration bay A type of main distribution frame (in a telephone central office) using plug-in jumper cords for cross-connections.

circuit control station That point within the general maintenance organization that fulfills the control responsibilities for leased and special circuits assigned to it.

circuit element A single active or passive functional item in an electronic circuit such as one diode, one transistor, one resistor, etc.

circuit grade The information-carrying capability of a circuit in terms of speed or signal type.

circuit group A group of circuits between two exchanges which are traffic-engineered as a unit.

circuit group, equivalent random A number of theoretical circuits used in conjunction with an equivalent random traffic intensity to permit traffic theories that do not explicitly recognize peakedness to be used in peakedness engineering.

circuit group, first choice In a switching system, the trunk group to which a traffic item is initially offered.

circuit noise level At any point in a transmission system, the ratio of the circuit noise at that point to some arbitrary amount of circuit noise (usually expressed in dBm) chosen as a reference. The ratio is usually expressed in decibels above reference noise, abbreviated dBrn, signifying the reading of a circuit noise meter, or in dBrn adjusted, abbreviated dBa, signifying circuit noise meter reading adjusted to represent an interfering effect under specified conditions.

circuit-released-acknowledgement signal In data common channel signaling: a signal sent in both directions in response to a circuit-released signal indicating that an interexchange data circuit has been released.

circuit-released signal In data common channel signaling: a signal sent in both directions indicating that the interexchange data circuit has been released.

circuit reliability The percentage of time a circuit was available to the user during a specified period of scheduled availability.

$$CR = 100(1 - TO/TS) = 100 \, TA/TS$$

where **TO** is the circuit total outage time, **TS** is the circuit total scheduled time, and **TA** is the circuit total available time. Syn.: time availability

circuit (specific function) Part of an installation forming (or able to form part of) an electric circuit traversed by a current having a definite function, specified in each case.

circuit subgroup A number of circuits with similar characteristics which provide service, protection, equipment limitation or maintenance. The subgroup is not engineered as a unit, but as part of a circuit group.

circuit switched data transmission service A service requiring the establishment of a circuit switched data connection before data can be transferred between data terminals.

circuit switched digital capability (CSDC) The ability to establish an end-to-end switched circuit which can be used for high-speed data transmission.

circuit switched network A switched network which provides a communications channel for exclusive use by connected parties until the connection is released.

circuit switching The switching of circuits for the exclusive use of the connection for the duration of a call.

circuit switching center A communications-electronics complex of circuits, equipment, and supporting facilities used for establishing a connection between two compatible subscribers.

circuit switching unit The equipment used for directly connecting two compatible data terminals for end-to-end data exchanges. Also used to connect a data terminal to a store-and-forward switch.

circuit test access point In telephony, four-wire test-access points located so as much as possible of the international circuit is included between corresponding pairs of these access points at the two centers concerned.

circuit usage for a group of international telephone circuits The percentage ratio between the sum of the holding times during a specified period equal to at least 60 consecutive minutes and the total length of that specified period. In the case of a group of circuits, the circuit usage corresponds to the average traffic density per circuit during the specified period, which unless otherwise indicated, is based on the busy hour.

circular buffer A form of queue in which items of data are placed in successive locations in a memory store, and are later called forward from these locations in the same sequence.

circular hunting A form of station or free-line hunting in which switching equipment hunts over all stations in a directory number hunt group regardless of the starting point.

circular mil The measurement unit of the cross-sectional area of circular conductors. The area of a circle whose diameter is one mil, or 0.001 inch.

circular orbit The path of a satellite in which the distance between the centers of mass of the satellite and the primary body is constant.

circularly-polarized Characteristic of a radio or light wave or other electromagnetic wave whose plane of polarization rotates as the wave propagates in a forward direction.

circulator A passive junction of three or more ports in which the ports can be listed in such an order that when power is fed into any port it is transferred to the next port on the list, the first port being counted as following the last in order.

circulator, radio frequency A coupler which enables more than one transmitter or receiver, at microwave frequency, to use the same antenna.

CISPR International Special Committee on Radio Interference. This body has studied methods of measuring interference and has produced authoritative publications on the subject.

City-Call Inter-city (trunk call) service established in the U.S. by an ITT subsidiary, U.S. Transmission Systems Inc. (USTS).

CIVISION 1. The cryptography of television signals. 2. Enciphered television signals.

cladding 1. Welding together of one metal over another metal, particularly with wires. 2. In an optical waveguide, a homogeneous dielectric fused to and concentrically surrounding the core. Cladding has a lower refractive index than the core, so light is internally reflected along the core.

cladding diameter Diameter of the smallest circle that encloses the surface of the cladding around the core of an optic fiber.

cladding mode In optoelectronics, a propagating mode whose power is predominantly in the cladding.

cladding mode stripper In optoelectronics, a device that employs a material having an index equal to or greater than that of the waveguide cladding which, when applied to the waveguide, provides escape for cladding modes.

clamp Device used to couple parts together mechanically.

clamp, armor Device which grips the armor of a protected cable so it may be secured to a fitting in a manhole or jointing chamber.

clamp, beam A clamp designed for easy fixing to a steel beam. Used to carry insulator spindles.

clamp, bonding ribbon A metal strap used to fasten a bonding ribbon inside a manhole.

clamp, cable Metal clamp used to secure cables in position on a flat surface.

clamp, cable lashing Metal clamp used to secure lashing wire or messenger wire (as used with aerial cables).

clamp, cable strain relief Anti-migration clamp securing aerial cable to messenger wire.

clamp, corner suspension Special clamp for use when messenger wires form sharp corners but the aerial cable takes the bend more smoothly.

clamp, crossover Clamp used to secure together two messenger wires for aerial cables.

clamp, drop wire Clamp used to secure drop wire to a pole or to a building.

clamp, drop wire mid-span Clamp with a supporting hook, used to support a drop wire which leaves an aerial cable route mid-span.

clamp, grade Clamps used near poles and mid-span on grades and slopes to prevent aerial cables from slipping in relation to their messenger wires.

clamp, ground   Clamp to which a ground wire lead is attached and which in turn is attached to a grounded item, such as a metal water pipe.

clamp, ground rod   Clamp used to attach a ground wire lead to a ground rod.

clamp, guy   Galvanized steel clamp used for gripping guy wires. Consists of three bolts holding two shaped clamp units together.

clamp, lashing wire   Clamp used for securing a lashing wire or messenger wire at the end of a span.

clamp, one-hole, cable   A cable clamp which wraps around a cable and is secured by a single fastening bolt.

clamp, pressure valve   A pressure testing valve and gasket used to take gas pressure readings before closing a cable joint.

clamp, span   Clamp designed for securing together messenger strands, mid-span, together with a hook to permit drop wires to leave the cable at the same point.

clamp, strand   A straightforward two-piece, three-bolt clamp used for securing two messenger strands together.

clamp, strand crossover   A special two-piece, two-bolt clamp used for securing together two messenger strands supporting aerial cables which cross each other's paths at right angles.

clamp, strand ground   Clamp used to fix a cable bonding ribbon to an aerial cable messenger strand.

clamp, suspension   Clamp used for securing and supporting a messenger strand or suspension wire at a pole.

clamper   Device which reinserts low frequency signals not faithfully transmitted in a broadband system.

clamping   Holding a pulsed signal in step with a reference waveform.

clamps, splicer's platform   Heavy clamps which permit a splicer's platform to be suspended from cable messenger wires. Rarely encountered outside North America.

clasp, cable   Form of cleating used for surface mounting of small cables; a soft metal strip is screwed to the surface and its ends wrapped around the cable to be secured.

class of office   A rank assigned to switching offices in the telephone network, assigned according to its switching functions, its relationship to other switching offices and transmission requirements.

class of pole   Classification system for strength and top diameter of telephone poles.

class of service   Subgrouping of telephone customers' lines to distinguish different classes from others; for example, those to be given highest priority and others, between business and residential customers, between government lines and others, and between those permitted to make unrestricted international dialed calls and others.

class of service tones   Special tones sent automatically to manual operating positions to indicate the class of service appropriate to the calling subscriber's line.

Class 1 office   A regional toll center; the highest level toll office in the North American switching hierarchy. There are ten such offices in the U.S. and two in Canada; together they form the basic upper level network for long distance traffic in North America. The network, including lower level toll offices, is programmed to handle calls in a systematic and economical way, with alternate routes provided when the first choice is busy. Long distance calls entering the network will automatically climb the switching hierarchy in search of an idle circuit. If the most direct route is busy, the call speeds up the hierarchy to the next switching center and the next in a fraction of a second until it finds a path to complete the call. In this process the Class 1 office is the "court of last resort," the highest level to which the search can be carried.

Class 2 office   A sectional toll center; the second level toll office in North America, of which there are 55 in the U.S. and 10 in Canada.

Class 3 office   A primary outlet in the North American long distance switching hierarchy.

Class 4C office   A toll center in the North American long distance switching hierarchy.

Class 4P office   A toll point in the North American long distance switching hierarchy.

Class 5 office   An end office; a local central office or exchange in the North American hierarchy.

classical MUF   (maximum useable frequency) Highest frequency that can be propagated in a particular mode between specified terminals by ionospheric refraction alone; on an oblique ionogram, the frequency at which the high- and low-angle rays merge into a single ray. Also called junction frequency.

classmarks   Designators used to describe the service privileges and restrictions for lines accessing a switch, e.g., precedence level, conference privilege, security level, or zone restriction. Syn.: class-of-service marks

cleaning, ultrasonic   A cleaning method often used for telephone equipment, in which the items are immersed in a solvent that is agitated by a sound wave with a frequency above the human hearing range.

clear   1. In plain language, i.e., not in code or cipher. 2. To repair a fault on a circuit. 3. To give an on-hook signal and release a circuit from occupation. 4. To empty a data storage device.

clear-back signal   A signal sent in the backward direction indicating that the called party has cleared.

clear channel   Transmission path wherein the full bandwidth is available to the user, e.g., with no portions of the channel used for control, framing or signaling.

clear collision   In data transmission, the contention that occurs when a DTE and a DCE simultaneously transfer a clear request packet and a clear indication packet specifying the same logical channel.

clear confirmation   In data transmission, a call control signal to acknowledge receipt of the data terminal equipment (DTE) clear request by the data circuit terminating equipment (DCE) or the reception of the DCE clear indication by the DTE.

clear confirmation signal In data transmission, a call control signal to acknowledge reception of the DTE clear request by the DCE or the reception of the DCE clear indication by the DTE.

clear-forward signal A signal sent in the forward direction to terminate the call or call attempt and release the circuit concerned. This signal is normally sent when the calling party clears.

clear message In data common channel signaling, a message sent in the forward and the backward direction containing a circuit-released or circuit-released-acknowledgement signal. The clear message will contain an indication whether the message is in the forward or the backward direction.

clear request In data services signaling, a condition, appearing in the forward and backward directions in the interexchange data channels, sent by the customer terminals when clearing.

clear text A message written in plain language.

clearing In data transmission, a sequence of events to disconnect a call and return to the ready state.

cleat The process of securing wires or cables to a pole or wall using an insulating clamp (for open wires) or a soft metal strap (for cables).

cleaved coupled cavity laser Type of laser able to switch between many different wavelengths so as to encode several communications systems on a single beam of light.

clevis, insulator Metal bracket device used to support an open wire insulator on a pole.

click, key Distinctive noise introduced into a circuit by the operation of a key.

click reducer A shock absorbing circuit connected across an operator's headset to minimize inconvenience caused by clicks.

climbers, lineman's Climbing irons which are strapped to lineman's legs so that they may climb telephone poles. Some utilize pointed gaffs, others utilize nearly circular grips.

climbers, tree A type of lineman's climbing irons with extra long gaffs to penetrate into the wood of a tree and make climbing safe.

clip, adhesive cable Plastic clip with strong adhesive used to hold cables on clean walls.

clip, alligator Small spring loaded test clip with long jaws.

clip, battery Metal clip with strong spring jaws to make contact with the terminal post of a secondary battery.

clip, busy Clip inserted in equipment in a central office to busy it out so that traffic will not be routed to it.

clip, drop wire Small metal clip used to support a drop wire.

clip, fuse The spring contacts which hold a cartridge fuse.

clip, test Clip on the end of a test lead.

clipper A limiting circuit which ensures that a specified output level is not exceeded.

clipping 1. Deforming of speech signals by the cutting-off of initial or final syllables (due to the action of, for example, an echo suppressor or TASI equipment) or the cutting-off of peaks of sound waveforms which would have otherwise overloaded amplifiers or modulation stages. 2. The shearing-off of the peaks of a TV signal. For a picture signal this may affect the positive (white) or negative (black) peaks. The sync signal may also be affected.

clock 1. A reference source of timing information for equipment, machines, or systems. 2. Equipment providing a time base used in a transmission system to control the timing of certain functions such as the control of the duration of signal elements or the sampling rate. 3. A source of timing signals used for the synchronization of isochronous data channels. 4. A device for time measurement and time display, generally using periodic phenomena. 5. An ultra-stable square wave oscillator capable of maintaining a precise output frequency for long periods of time. Note that an assembly of timing devices replicated for security reasons is regarded as a single clock.

clock, switchboard A clock mounted on or near a manual telephone operator position.

clock circuit A time pulse circuit used for synchronization.

clock difference The time difference between two clocks, i.e., a measure of the separation between their respective time marks.

clock interrupt A type of interrupt which occurs at regular intervals, e.g., to initiate processes such as polling.

clock phase slew The changing in relative phase between a given clock signal and a phase-stable reference signal.

clock pulse See clock signal.

clock signal A synchronization signal provided by a clock. Syn.: Clock pulse

clock time difference See time scale difference.

closed circuit 1. In radio and television transmission, used to indicate that the programs are transmitted directly to specific users and not broadcast for general consumption. 2. A completed electrical circuit.

closed circuit television A nonbroadcast TV system, typically a security system with cameras in vulnerable points and remote display at a manned console, or an educational system, for example, with cameras focused on the surgeon's hands so students may learn new operating techniques.

closed end The end of an FX private line (out-of-area line, in Britain), terminating on a PBX or telephone instrument in such a way that a call cannot advance further.

closed user group Telecommunication facilities available on a switched basis for limited groups of customers. (Videotex) A group of users that restricts a specified part of the videotex database for use only by its own members. (Data or telephony) A group of users able to use a switched telecommunications network to obtain access only to other members of the same user group.

closed-user-group indicator In data common channel signaling, information sent in the forward direction and, in some circumstances, in the backward direction indicating whether or not the calling customer belongs to a closed user group. In the positive case it also indicates that: a) an interlock code is included and outgoing access is not allowed; or b) an interlock code is included and outgoing access is allowed; or c) no interlock code is included because the call is a direct call.

closed user group with outgoing access In telegraphy and data, a facility assigned to a user in a closed user group to enable that user to communicate with other users of a public data network transmission service; a facility assigned to users having a data terminal equipment connected to any other public switched network to which interworking facilities are available.

closure, buried ready-access Metal or plastic case that opens readily for easy access to pairs in buried cables; it may be direct buried, but more often is placed in a concrete handhole accessible at ground surface level to craft people.

closure, cable Any device which in effect replaces the sheath around a cable at a splice or joint and provides a waterproof covering plus mechanical security.

closure, ready access Various devices used for enclosing cable splices in ways which can readily be reopened; most utilize plastic components, mechanically designed to give an acceptable measure of security.

closure, splice Various devices for enclosing cable splices, providing waterproof protection for the spliced conductors and mechanical protection for the splice itself.

closure, terminal Method of splicing cable conductors and enclosing a cable splice by providing terminals for the pairs concerned on the split cases of the splice closure itself.

closure, vault cable Large cable sleeve used in vaults beneath main distributing frames; each large external cable is jointed onto several smaller cables designed to fill one vertical of the MDF.

cloth, catch A cloth with which a cable splicer catches molten solder as he pours it on the lead cable joint he is wiping. The hot solder held to the bottom of the joint helps to bring it up to wiping temperature.

cloth, wiping Cloth used for wiping soldered and plumbed joints in lead-sheathed cables. (UK: moleskin).

cluster Type of data network or processing hierarchy in which two or more terminals or processors are connected to a line, highway, bus or data channel, usually at a single point.

C-message weighting A noise weighting used to measure noise on a line that would be terminated by a 500-type telephone or similar instrument. The meter scale readings are in dBrn (C-message) or in dBrnC.

CMOS A logic circuit using complementary metal-oxide semi-conductor transistors as logic functions. Complementary means that pairs of opposite types of transistors are used together on a push-pull basis, thus eliminating resistors and greatly reducing power consumption.

CNET French national telecommunications research organization (Centre National d'Etude des Telecommunications).

coast station A radio station on land in the maritime mobile (terrestrial) service.

coat, splice A rapid-action manufactured plastic sheath closure device for use on plastic sheathed cable joints/spices.

coating Protective layer applied to the cladding surface of an optical fiber.

coaxial On a common axis. A coaxial pair is one with a central conductor surrounded by insulant which in turn is surrounded by a tubular outer conductor, which is covered by more insulant.

coaxial cable A cable with one or more coaxial pairs under one outer sheath.

coaxial carrier A wideband carrier telephone system routed over coaxial pairs.

coaxial line Transmission system using coaxial cable.

cob, insulator Plastic or wooden bushing used in some countries to cover a metal insulator pen or spindle and onto which is threaded an insulator to carry an open wire conductor.

COBOL Common business oriented language. An internationally standardized high level programming language developed for commercial use.

co-channel Any two or more TV signals are considered co-channel when their video carriers, either off-air or after conversion by CATV equipment, occupy the same TV channel.

co-channel interference Interference resulting from two or more transmissions in the same channel.

code 1. A set of unambiguous rules specifying the manner in which data may be represented in a discrete form. 2. Any system of communication in which arbitrary groups of symbols represent units of plain text. 3. A cryptosystem in which the cryptographic equivalents (usually called code groups) typically consisting of letters and/or digits in otherwise meaningless combinations are substituted for plain text elements which are primarily words, phrases or sentences. Codes may be used for brevity or for security.

code, access Digits that must be dialed in order to gain access to a particular circuit or facility. A PABX extension user, for example, often has to dial "9" to access the public network.

code, accounting Numerical accounting codes used in the U.S. to identify plant accounts; these are FCC codes used by all telephone companies.

code, area See area code.

code, ARQ A code providing automatic repeat request for teletypewriters.

code, ASCII American Standard Code for Information Interchange. The standard code, using a coded character set consisting of seven-bit coded characters (eight bits including parity check), used for information interchange among data processing systems, data communication systems, and associated equipment.

code, Baudot See Baudot code.

code, BCH See BCH code.

code, cable color A code which identifies cable pairs by color. Black, brown, red, orange, yellow, green, blue, violet, gray and white are usually used for the ten decimal numbers. (Some administrations use color combinations giving 25 different pair identifications.)

code, central office In North America, any of the NNX codes used as an address for a 10,000-line central office unit (N= 2 thru 9, X= 0 thru 9). Codes in the N11 series are not normally available for central offices but are used for services.

code, Codel A four- or five-in-parallel code sometimes used for transfer of digital information between stages within a central office. In new exchanges, multifrequency code signaling and common channel signaling have replaced this parallel code system but similar principles are often encountered. The Codel 4 relay system, using W X Y and Z relays, was: $1 - W + X; 2 - W + Y; 3 - W + Z; 4 - X + Y; 5 - X + Z; 6 - Y + Z; 7 - W; 8 - X; 9 - Y; 0 - Z$

code, color A system of colors which indicates the electrical value, voltage rating, and tolerance of a component, or identifies the pair or quad number of conductors in a multiwire cable.

code, constant ratio, data A data transmission code in which the number of 0s and the number of 1s in a binary word is fixed; if the signal as received does not indicate the same ratio of 0s to 1s, a repeat transmission of the doubtful word is requested.

code, data interchange A variant of ASCII code, including various nonprinting control characters.

code, destination In North America, the complete ten-digit number which constitutes a telephone address, comprised of a three-digit area code plus a three-digit office code plus a four-digit station number.

code, directing Routing digits or a special code dialed before the normal directory number.

code, extended binary coded decimal interchange An eight-bit extension of the standard binary coded decimal enabling it to deal with various control codes.

code, field accounting Basic plant codes used by many telephone companies on current work documentation. Full codes are usually allocated later by clerical staff.

code, Fieldata A U.S. military field data code.

code, five-unit The internationally used start-stop code for teletypewriters, International Telegraph Alphabet No. 2.

code, four-out-of-eight A data code which makes every character four 1s plus four 0s in different order arrangements.

code, geometric parity check An error control code using a matrix type of parity check.

code, Golay An error-correcting data code using 23 bits per character.

code, Gray A digital code especially suited for advising angular position.

code, Hollerith Code used for punched holes on an 80-column data or computer card.

code, instruction Code used in programming switching systems.

code, ITA #2. International Telegraph Alphabet No. 2. The standard code for start-stop teletypewriter operation, commonly known as the Baudot code. It consists of a one-unit open start pulse, five one-unit information bits, and (in the U.S.) a 1.42-unit closed stop pulse. (1.5 unit in Europe) It has both a letters and figures shift, and each provides 28 characters.

code, Morse The dot-dash code used for early telegraphy systems, and still used in some radio telegraphy applications.

code, multiple address Code which lists the addresses of a multiple-addressed message.

code, NNO (Where N=any digit 2 thru 9). In North America used as central office codes, but initially used for some NPA codes.

code, NNX (Where N=any digit 2 thru 9, X=0 thru 9) In North America used as a central office code.

code, NXX In North America, code group reserved for later use as NPA codes.

code, N 0/1 X In North America, the original group of NPA codes, with second digit 1 or 0.

code, plant accounting A two- or three-unit code used for control and recording of plant costs: C = construction; X = removals; R = repairs & maintenance; M = rearrangements; T = toll plant.

code, redundant A code which utilizes more elements than are needed to represent the information concerned.

code, regional identity A code used in international telephone dialing to route calls to a particular continent. This first digit has to be followed by a country code, then the required national number. Zone: 1 — North America, Caribbean and Hawaii; 2 — Africa; 3 — Europe; 4 — Europe; 5 — South America, Cuba; 6 — Australasia; 7 — Russia; 8 — East Asia; 9 — Far East and Middle East.

code, ringing A coded ringing signal sent out on a party line to alert a subscriber on the line.

code, route In North America, a two-digit code dialed before the number of the called line when a special grade of circuit is required, e.g., for a data call.

code, simple parity check The addition of one parity or check bit to code words of any length to make the total number of 1s always odd or always even.

code, telegraph   Any code system used to transmit letter characters by a digital scheme, e.g., the Morse code or teletypewriter code.

code, teletypewriter   The five-unit code used in the International Telegraph Alphabet No. 2, the Baudot start-stop code.

code, terminating toll center   A three-digit code used by operators at distant switching centers to reach long distance information or test panel operators at a terminating toll center. First digit of a TTC code is 1 or 0.

code, ternary   Code with three possible values for each element.

code, toll authorization   A special code a customer may dial to prevent his line being used for long distance calls, until he removes the ban by dialing in a special release or unlatch code.

code, two-out-of-five   Code used for parallel transmission of digital information, using five basic stores, each of which receives a specific value, e.g., 0-1-2-4-7. Every decimal digit may then be represented by pulses into two of these stores.

code, UNIVAC   A data code used in some computer applications.

code 11   Employed to route a call direct from an operator in one country to an operator in another country for assistance purposes; also used when the originating operator cannot dial or obtain connection to the destination in the remote country.

code 12   Signal employed to route a call direct from an operator in one country to an operator in another country for the inward operator to handle transit calls on delay routes, calls which the outward operator has difficulty in connecting semiautomatically, and special categories of call.

code 911   An industry universal emergency code in the U.S. which can be dialed from any phone without charge to report any kind of emergency. Outside the U.S. 999 often is used for this same purpose.

code BCD   Eight-bit code which includes a parity check; used by IBM; stands for binary coded decimal.

code call   Calling systems by which a coded dialed input is translated to a lamp indication or musical tone, inviting the called party to go to the nearest phone and make contact with the caller. The more modern equivalent are radio paging systems which are more flexible.

code character   The representation of a discrete value or symbol in accordance with a code.

code conversion   1. Conversion of character signals or groups of character signals in one code into corresponding signals or groups of signals in another code. 2. A process for converting a code of some predetermined bit structure (for example, 5, 7, or 14 bits per character interval) to a second code with approximately the same number of bits per character interval. No alphabetical significance is assumed in this process.

code division   The separation of a plurality of transmission channels by using specific values of codes belonging to the same set.

code-division multiple access   A form of modulation whereby digital information is encoded in an expanded bandwidth format. Several transmissions can occur simultaneously within the same bandwidth with the mutual interference reduced by the degree of orthogonality of the unique codes used in each transmission. It permits a high degree of energy dispersion in the emitted bandwidth.

code element   One of a finite set of parts of which the characters in a given code may be composed.

code group   A group of letters or numbers or a combination of both assigned in a code system to represent a plain text element.

code-independent data communication   A mode of data communication that uses a code-independent protocol and does not depend on the character set or code used for its correct functioning. Syn.: code-transparent data communication

code number   Code numbers used for routing or other purposes.

code set   The complete set of representations defined by a code or by a coded character set, e.g., all of the three-letter international identifications for airports.

code-transparent data communication   See code-independent data communication.

code word   1. A word which has been assigned a classified meaning to safeguard information. 2. A cryptonym used to identify sensitive intelligence data.

CODEC   Coder plus decoder. An assembly comprising an encoder and a decoder in the same equipment; a device which produces a coded output from an analog input, and vice versa.

coded character set   A set of unambiguous rules that establishes a character set and the one-to-one relationships between the characters of the set and their coded representations.

coded ringing   Selective calling on a multi-party line, with a combination of long and short rings.

coded speech   A speech signal that has been encrypted or changed into a standard digital code form.

codelreed   A reed switch used as a binary digit store.

codes, error control   An error detection and correction code used in data transmission. Most of these codes depend on extra parity bits or on the use of codes with redundancy.

codes, hotel service   One-digit calling used on most hotel PABXs to obtain services such as room service, laundry, or porter.

codes, operator   Codes used in North America to reach special or auxiliary service operators.

codes, toll center operator   Code used after a toll center code to route a call to a particular operating service, for example: 121 inward; 131 information; 141 route; 11X leave word operator

coding, duobinary   Signal technique which produces a three-voltage level signal from a binary input, thereby giving, in effect, twice the data capacity for a given bandwidth. This technique also detects errors without having to add check-bits to character signals.

coefficient, coupling   A decimal fraction of less than 1 representing the degree of coupling between two circuits.

coefficient, reflection   A decimal fraction of less than 1 which is the ratio between the amplitudes of reflected and incident waves.

coefficient, telegraph transmission   A figure from 1 to 10 representing the transmission degradation and distortion contributed by a particular link in a system. Coefficients for all sections concerned are added together numerically and the total should not exceed 10 (representing one error in 44,000 characters) for satisfactory operation.

coefficient, traffic   Work unit coefficients designed to measure the work done by telephone operators based on the average time to handle each class of call.

coefficient, transmission   A system (not generally used) by which performance of a circuit is given a number inversely related to the subjective transmission quality of the circuit.

coherence area   The area in a plane perpendicular to the direction of propagation over which light may be considered highly coherent. Commonly the coherence area is the area over which the degree of coherence exceeds 0.88.

coherence length   The propagation distance over which a light beam may be considered coherent. If the spectral linewidth of the source is $\Delta\lambda$ and the central wavelength is λ_o, the coherence length in a medium of refractive index n is approximately $\lambda_o^2/n\Delta\lambda$.

coherence of frequency   Same as coherence of phase.

coherence of phase   The condition of two frequencies M and N to resume the same phase difference after M cycles of the first and N cycles of the second, M/N being a rational number, obtained through multiplication and/or division from the same fundamental.

coherence time   Coherence length divided by the phase velocity of light in a medium.

coherent   Characterized by a fixed phase relationship among points on an electromagnetic wave.

coherent bundle   Same as aligned bundle.

coherent detection   Detection when a replica of the transmitted carrier is available at the demodulator to be used in the demodulation process.

coherent network   A network in which inputs, outputs, signal levels, bit rates, digital bit stream structures and signaling information are all inherently compatible, throughout the network. If precise matching or coherence is missing, conversion will be needed, which may involve extra hardware and/or revised software.

coherent pulse   Condition in which a fixed phase relationship is maintained between consecutive pulses during pulse transmission.

coherent transmission   Type of optic fiber system using lasers with high accuracy and stability; ordinary lasers produce broad spectrum light signals with line width of up to 700 GHz whereas a coherent system usually has an optical cavity or filter external to the laser giving a very narrow line width of 10 — 100 kHz.

coil   One or more helical turns of wire, often forms a doughnut shape; provides inductance.

coil, antisidetone induction   Coil in a telephone instrument which balances out most of the sidetone resulting from locally generated speech power.

coil, drainage   High impedance coil with center tap grounded; drains static electric charges to ground.

coil, dummy heat   Small metal stamping which fits into holder on a main distributing frame normally occupied by a heat coil but which is itself an inactive item.

coil, exploring   Large diameter (about 1 meter) coil of wire, mounted on a frame, used as a search coil in conjunction with a tone receiver to locate underground cables.

coil, heat   A device used to protect central office equipment from steady excess currents. Common types will carry 350 mA for three hours but will operate within three minutes if the current exceeds 540 mA.

coil, hybrid   Balance coil which joins together the four-wire portion and the two-wire portion of a transmission circuit.

coil, induction   Coil in telephone instruments which produces the voice frequency ac signal which is sent out to line from the modulated dc signal produced by the microphone.

coil, loading   Small high inductance (but low resistance) coils inserted at regular intervals in toll circuits or long subscriber lines in order to reduce line loss at audio frequencies. Loading coils usually have to be removed for new-generation, high-speed services because they severely attenuate signals at frequencies above the voice band.

coil, phantom   Coil providing center tap on a physical circuit. One leg of a phantom circuit.

coil, repeating   A transformer which separates one section of a telephone circuit from another, with no dc link between the sections.

coil, retard   A coil with high inductance which resists rapid changes of current; a choke.

coil, shading   A single turn of wire in an induction motor which is sufficient to produce a field to start the motor.

coil, side circuit   A phantom coil or repeating coil in a physical circuit.

coil, simplex   A repeating coil with center tap used to derive a simplex circuit.

coil-loaded   A circuit with inductance increased by the insertion of loading coils at regular intervals.

coil-loading   The use of series or lumped loading coils at regular intervals in order to improve transmission. Sometimes called pupin coil loading.

coiling   The process of stowing submarine cable into a cable tank.

coin box   In North America, the receptacle into which coins drop in a pay phone. In Britain, a pay phone itself is often called a coin box or a coin box phone.

coin collecting box   A payphone (in the U.K.)

coin collector   A pay phone or coin collecting box. More often, this means the person who collects full coin boxes or cash boxes from pay phones and replaces them with empty ones.

coin control   The mechanism and signals by which the collection of coins in a coin box is controlled by either an operator, a relay set, or a local microprocessor.

coin free services   Services which can be obtained from a paystation/coin phone without inserting coins, or if coins have been inserted they will be refunded. These services differ from one administration to another, but generally include: operator assistance, information, emergency calls, police and fire calls, and repair services.

coin pilot lamp   A supervisory light on an operating position which lights when a collect key is operated on the position.

coin service   A coin collecting telephone station.

coin supervisory circuit   Relay group connected to a trunk that is serving a coin call. Provides coin collect battery at the end of a completed charged call, or coin refund battery if the call is not completed or if there is no charge.

coin telephone   A public telephone with a coin box which accepts coins for the payment of call charges.

coin timing circuit   Circuit at a coin box (or in the coin box line equipment in the central office) which calculates the duration of the call which has been paid for and advises the caller to insert more money when his paid time limit has been reached to continue the call.

coinbox (payphone) telephone service   A service which permits outgoing telephone calls after insertion of adequate coins or tokens and, without payment, incoming calls.

coincidence circuit   A circuit which has two or more inputs and one output. An output signal is produced only when all the input circuits receive inputs of specified level and within a specified time interval.

cold joint   A soldered connection which was inadequately heated, with the result that the wire is held in place by rosin flux, not solder. Sometimes called dry joint.

cold restart   An initialization phase during which temporary storage is deallocated and cleared. All calls are dropped and peripheral processors clear all channel assignments.

cold work   Repeated flexing or bending of an unannealled conductor, likely to lead to breakage.

collar, manhole   A small manhole chimney; a brick or concrete tube section between the manhole roof and the road surface.

collate   To compare or merge two things into one thing. May refer to files in the computer.

collator   A punched card processing machine which combines two decks of punched cards into a single deck, all in sequence with respect to one item of data on the cards.

collect   The process of transferring cash on a coin telephone from a temporary hold position (from which it could be returned to the caller) into the coin cash box.

collect call   A call paid for by the called party, who must be asked if he will accept the call before it is connected.

collect key   The key which, when depressed by an operator, activates the coin collection mechanism in a coin telephone.

collection charge   The charge in its national currency collected by a telephone administration from its customers for the use of international telephone service.

collector   Transistor terminal which acts similarly to the plate, or anode, in an electron tube.

collett   Small tube, usually plastic, on which a cable pair number is printed or written. The collett is slipped over the wires at a cable joint so that pairs may readily be identified. Sometimes collett, pair.

collimation   The process by which a divergent or convergent beam of radiation is converted into a beam with the minimum divergence possible for that system (ideally, a parallel bundle of rays).

collision   1. In a data transmission system, the situation that occurs when two or more demands are made simultaneously on equipment that can handle only one at any given instant. 2. In a computer, the situation that occurs when the same address is obtained for two different data items that are to be stored at that address.

co-located   1. In the same building. 2. Separated by not more than 25 ohms of cable.

co-located exchange concentrator   A concentrator in the same location as the exchange that controls it and to which its higher traffic volume circuits are connected.

color burst   That portion of a color TV signal, usually a few cycles of sinusoidal chrominance subcarrier, used to establish a reference for demodulating the chrominance signal. In NTSC color this is a burst of about nine cycles of 3.6 MHz subcarrier on the back porch of the composite video signal; it establishes frequency and phase reference synchronization for the chrominance signal.

color code   See code, color.

color subcarrier   In NTSC color, the carrier whose modulation sidebands are added to the monochrome signal to convey color information, i.e., 3.6 MHz (3.579545 MHz). (PAL color as used in Britain utilizes a color subcarrier of 4.43 MHz).

column, card   One of the 80 columns of a data card into which holes are punched to record information.

combat-net radio   A radio operating in a network, providing a half-duplex circuit employing a single radio frequency. These radios are primarily used for command and control of combat, combat support, and combat service support operations between and among ground naval, and airborne forces.

combination switch A switch which is able to carry out two different tasks, depending on the way it has been accessed.

combination trunk A trunk available for both incoming and outgoing traffic.

combinational logic element A device having at least one output channel and one or more input channels, all characterized by discrete states, such that at any instant the state of each output channel is completely determined by the states of the input channels.

combined distribution frame A distribution frame which combines the functions of main and intermediate distribution frames. The frame contains both vertical and horizontal terminating blocks. The vertical blocks are used to terminate the permanent outside lines entering the station. Horizontal blocks are used to terminate inside plant equipment. This arrangement permits the association of any outside line with any desired terminal equipment. These connections are made with twisted pair wire normally referred to as jumper wire.

combined engineering plant exchange record (CEPER) A property and plant record system used by some U.S. independent telephone companies.

combined exchange An exchange where some of the equipment serves more than one application, e.g., local and long distance or transit switching.

combined local/transit exchange An exchange where subscribers' lines terminate that also is used as a switching point for traffic between other exchanges.

combined loss and delay system A system in which bids that cannot be served immediately are permitted to wait until service can begin, provided a waiting place is free or until a time-out occurs.

combined station A data terminal using high-level data link control (HDLC) protocol and capable of assuming the role of either a primary or a secondary station.

combined switch train Switches which handle both local and toll traffic.

combined toll/DSA board Manual operating board where both dial service assistance (DSA) functions and outward toll functions are performed.

combiner Circuit used in diversity reception of radio signals to produce a fade-free output.

combiner, channel Device for combining several radio signals for delivery to a CATV system.

combiner, diversity Device to combine two signals received via different routes or frequencies to produce a good usable and fade-free output.

combiner, linear A diversity combiner which adds the two received signals together on the principle that the sum (or mean) is unlikely to fade as violently as either one of the signals.

combiner, post-detection A diversity combiner operating on received basebands or channels.

combiner, pre-detection A diversity combiner operating on radio frequency signals, usually at the intermediate frequency stage.

combiner, ratio-squared A diversity combiner which weights the two received signals, the one with less noise being given more weight than a signal accompanied by much noise.

combiner, switching A diversity combiner which selects the received signal which is of higher pilot level or lower noise level, changing over when justified.

combining network (cable TV) A passive network which permits the addition of several signals into one combined output with a high degree of isolation between individual inputs. It may be a power or frequency combiner.

come along Slang name for clamp used in pulling a messenger strand up to the required sag in a section.

command 1. In data transmission, an instruction sent by the primary station instructing the secondary station to perform some specific function. 2. That part of a computer instruction word which specifies the operation to be performed. 3. In telephone services, a single specific manipulation at the subscriber set causing transmission of a signal which specifically indicates the manipulation to the exchange. For certain control procedures either one single command or a succession of commands are required. 4. In man-machine language, a specification of an expected action or function by the system.

command and control system The facilities, equipment, communications, procedures, and personnel essential to a commander for planning, directing, and controlling operations of assigned forces pursuant to the missions assigned.

command frame In data transmission, a frame containing a command transmitted by a primary station.

command language A source language consisting primarily of procedural operators that indicate the functions to be performed by an operating system.

command net A communications network which connects an echelon of command with some or all of its subordinate echelons for the purpose of command control.

comment 1. In man-machine language (MML), a character string enclosed between the separator strings/* (solidus asterisk) and */ (asterisk solidus). Has no MML syntactical or semantical meaning. (Asterisk solidus sometimes is called star, slash.) 2. Information which is in addition to or clarifies a specification description language program. Comments may be attached by a single square bracket connected by a dashed line to a symbol or flow line.

Commission Internationale de l'Eclairage (CIE) International body which sets standards for color television services.

Commitment, Concurrency and Recovery (CCR) Services provided by a data network to give reliable operation of distributed applications in the presence of delays and crashes.

common base Type of transistor circuit in which the base is the common or grounded electrode.

common-base connection Method of using a transistor with base common to both the input and output circuit, and usually this is grounded. Also called grounded-base connection.

common battery A battery that serves as a central source of energy for many similar circuits. In many telecommunications applications, a common battery is 48 volts.

common battery signaling A system in which the signaling power of a telephone is supplied by the battery at the servicing switchboard.

common carrier A company which furnishes public telecommunications facilities, at non-discriminatory rates, e.g., a telephone or telegraph company. The carrier cannot control message content.

Common Carrier Bureau (CCB) Department which implements FCC's policies regarding common carriers and their regulation.

common channel exchange An exchange utilizing a common channel signaling system.

common channel exchange, first The exchange closest to the calling party in each common channel section of a connection where, unless it is the calling party's exchange, interworking takes place with other signaling systems.

common channel exchange, intermediate A transit exchange where interworking takes place between common channel signaling systems.

common channel exchange, last The exchange closest to the called party in each common channel of a connection where, unless it is the called party's exchange, interworking takes place with other signaling systems.

common-channel interoffice signaling A method of transmitting all signaling information for a group of trunks by encoding it and transmitting it over a separate voice channel using time-division digital techniques.

common channel signaling A signaling method in which a single channel conveys, by the means of labeled messages, signaling information relating to a multiplicity of circuits and other information such as that used for network management. No supervision or addressing appears on individual trunks, they are used for voice only. Examples of common channel signaling are CCITT Signaling System No. 7 and North American CCIS, common channel inter-office signaling.

common channel transit exchange An intermediate exchange where interworking of common channel signaling systems takes place.

common collector Type of transistor circuit in which the collector is the common or grounded electrode.

common-collector connection Method of using a transistor in which the collector is common to both the input and output circuits, and it is usually grounded. Also called a grounded connection.

common control Control of an automatic switching arrangement in which the functional units necessary for the establishment of connections are shared, being associated with a given call only during the period required to accomplish the control of the setting-up function.

common control office A telephone central office where control of switching functions is centralized, or in which each control unit is able to establish and supervise more than one connection. This compares with Strowger type step-by-step exchanges in which a control function is directly associated with each switch at each selection stage.

common control switching arrangement A switch service offering by AT&T for private networks that allows any station in the network to call any other using a seven-digit number.

common direct control Digital control of switches on a step-by-step basis, but in which the digit receiving elements are common to a number of selectors.

common emitter Type of transistor circuit in which the emitter is the common or grounded electrode.

common-emitter operation Method of using a transistor in which the emitter is common both to the input and output circuits, and is usually grounded. Also called grounded-emitter operation.

common equipment Items used by more than one channel or equipment function.

.tfh2 **common indirect control** Control of switches by digital signals from line which are stored in a register, and fed to a marker for the purpose of setting the switches.

common language location identifier A standardized identification scheme for trunk groups.

common logic Logic systems in which the logical functions for a number of registers or markers are provided on a common basis and shared by time division between each of the registers or markers.

common mode For central offices: a method of pooling equipment, such as loop extenders, so that these are brought into use when required and not provided on a permanent per-line basis for all those lines which require them.

common-mode interference Interference that appears between signal leads, or the terminals of a measuring circuit, and ground.

common-mode rejection ratio The ratio of the common-mode interference voltage at the input of a circuit to the interference voltage at the output.

common-mode voltage 1. Any uncompensated combination of generator-receiver ground potential difference, the generator offset voltage, and the longitudinally coupled peak random noise voltage measured between the receiver circuit ground and cable with the generator ends of the cable short-circuited to ground. 2. The algebraic mean of the two voltages appearing at the receiver input terminals with respect to the receiver circuit ground.

common return A return path that is common to two or more circuits and returns currents to their source or to ground.

common return offset The dc common return potential difference of a line.

common strap A wire or bus connection linking together like terminals on adjacent circuit units to provide ground or a ringing lead.

common-user Facilities available to be used by many customers on a switched basis.

common user circuit A circuit designated to furnish communications service to a number of users.

common user network A system of circuits or channels allocated for communication paths between switching centers to provide communication service on a common basis to all connected stations or subscribers. It is sometimes called a general purpose network.

common user service A type of communications service provided by a common user network.

common wire Wire connecting a particular signal or power source to several items or components.

commonality A term applied to equipment or systems which have the quality of one entity possessing like and interchangeable parts with other equipment.

commons Wires used to make connections in several similar circuit units.

COMMS Central office maintenance management system. An AT&T software routine for electronic offices/exchanges.

communicating sequential processes A concurrent programming procedure.

communication 1. The transmitting and/or receiving of information, signals, or messages. 2. The information thus received. 3. Information transfer according to agreed conventions.

communication center 1. Building or room in which communications equipment has been centralized. 2. A facility responsible for the reception, transmission, and delivery of messages. Its normal elements are a message center section, a cryptographic section, and a sending and receiving section, using electronic communications devices.

communications 1 The aggregate of several modes of communication used to convey information, signals, or messages. 2. The art and science of communicating, as a study. 3. A method or means of conveying information of any kind from one person or place to another, except by direct unassisted conversation or correspondence through postal agencies.

Communications Act of 1934 U.S. legislation which established the FCC and set a goal of high quality universally available telephone service at reasonable cost.

communications common carrier A company that offers its facilities to the public for communications services.

communications-electronics The specialized field concerned with the use of electronic devices and systems for the acquisition or acceptance, processing, storage, display, analysis, protection, and transfer of information.

communications-electronics EMC analysis An investigation of the electromagnetic compatibility of C-E material and concepts within their intended C-E environment.

communications facilities Installations and equipment that provide telecommunications.

communications front-end equipment Interface equipment between a computer and a communications line.

communications net An organization of stations capable of direct communications on a common channel or frequency.

communications network An organization of stations capable of intercommunications but not necessarily on the same channel.

communications processor unit The message data processor of an AUTODIN switching center.

communications satellite An orbiting vehicle which relays signals between communications stations. There are two types: (a) active communications satellite — a satellite which receives, regenerates, and retransmits signals between stations; and (b) passive communications satellite — a satellite which reflects communications signals between stations.

Communications Satellite Corporation (COMSAT) A U.S. company created by act of Congress in 1962 to provide communications via satellites. COMSAT leases satellite circuits to many U.S. companies and is active in international communications through partial ownership in the International Telecommunications Satellite Organization (INTELSAT) and the International Maritime Satellite Organization (INMARSAT).

communications satellite system System of orbiting satellites and associated ground stations transmitting telephone, television and data signals between terrestrial points.

communications security(COMSEC) The protection resulting from the application of cryptosecurity, transmission security, and emission security measures to telecommunications and from the application of physical security measures to COMSEC information.

communications security equipment Equipment designed to provide security to telecommunications by converting information to a form unintelligible to an unauthorized interceptor and by reconverting such information to its original form for authorized recipients, as well as equipment designed specifically to aid in, or as an essential element of, the conversion process. COMSEC equipment includes cryptoequipment, cryptoancillary equipment, cryptoproduction equipment, and authentication equipment.

communications security material All documents, devices, equipment, or apparatus, including cryptomaterial, used in establishing or maintaining secure communications.

Communications Service Authorization (CSA) U.S. Defense Communications System order to a common carrier for the provision of a communications service.

communications sink A device which receives information, control, or other signals from communications source(s).

communications source A device which generates information, control, or other signals destined for communications sink(s).

communications subsystem A major functional part of a communications system, usually consisting of facilities and equipment essential to the operational completeness of a system.

communications survivability The capability of a system to continue to operate effectively even though portions may suffer physical damage or destruction due to enemy attack or other causes. Methods include dispersing routing facilities, utilizing different transmission methods, having equipment redundancy, provision of protection against electromagnetic pulse (EMP) damage and site hardening.

communications system A collection of individual communications networks, transmission systems, relay stations, tributary stations, and terminal equipment capable of interconnection and interoperation to form an integral whole. These individual components must serve a common purpose, be technically compatible, employ common procedures, respond to some form of control, and, in general, operate in unison.

communications zone The rear part of a theater of operations (behind but contiguous to the combat zone) which contains the lines of communications, establishments for supply and evacuation, and other agencies required for the immediate support and maintenance of the field forces. (UK: lines of communications [L of C] area).

community antenna relay service (CARS) FCC-authorized system for relaying television programs to CATV systems.

community antenna television (CATV) CATV, in most countries, originated in areas where good reception of broadcast TV was not possible. Signals from distant broadcasting stations are picked up with antennae and amplified, then sent by cable or microwave links to subscribers. Now CATV also provides a highly sophisticated cable distribution system in large metropolitan areas in direct and successful competition with direct broadcasting.

community automatic exchange Small rural automatic telephone office.

community dial office Small rural automatic telephone office.

community of interest A grouping of telephone subscribers who generate a significant proportion of their traffic in calls to each other.

community reception The reception of emissions from space stations in the broadcasting-satellite service by receiving installations, which in some cases may be complex and have antennae larger than those used for individual reception, and intended for use either by a group of the general public at one location, or through a distribution system covering a limited area.

commutation Successive switching process carried out by a commutator.

commutator A circular assembly of contacts, insulated one from another, each leading to a different coil from its neighbors.

compact towed sonar systems Towed array sound detection system for submarine location.

compaction See compression, bit.

companding, instantaneous Process used in some PCM systems which compresses analog signals before they are sampled for quantizing; at the receiving end the signals are expanded after decoding, which reduces the effect of quantizing noise.

companding, syllabic A rapid action compression/expansion system, operating on individual speech syllables.

compandor Contraction made up from compressor and expander. Compandors are often used on radio circuits to improve the signal-noise ratio, with the compressor to reduce the volume range at the transmit end and the expandor to compensate the signal back to its original range at the receiving end.

comparator Circuit which is designed to compare two variable signals, or a variable signal against a constant signal.

compatibility 1. The facility of diverse systems to exchange necessary information at appropriate levels of command directly and in usable form. Communications equipment items are compatible if signals can be exchanged between them without the addition of buffering, translative, or similar devices for the specific purpose of achieving workable interface connections and if the equipment or systems being interconnected possess comparable performance characteristics, including suppression of undesired radiation. 2. Capability of two or more items or components of equipment or material to exist or function in the same system or environment without mutual interference. 3. In data processing, two computers are compatible if programs can be run on both without modification. Sometimes this is not a two-way possibility: a computer is upwards compatible with a second computer if a program written on and for the first can be run on the second, but not the reverse.

compatible sideband transmission That method of independent sideband transmission wherein the carrier is deliberately reinserted at a lower level after its normal suppression to permit reception by conventional AM receivers. Syn.: amplitude modulation equivalent (AME); compatible SSB.

compelled signaling A signaling method in which, after one signal or group of signals has been sent, the sending of any further signals in the same direction is inhibited until the receiving terminal has acknowledged the send signal and the acknowledgement received.

compensation Passing a signal through an element with characteristics which are the reverse of those in the transmission line so that the net effect is a received signal with an acceptable level/frequency characteristic.

compensation, earth potential A special arrangement needed when duplex signaling is used between locations where the earth has different potentials. This usually involves providing a physical path between the two locations, and not using a ground return path.

compensation, high frequency A modification to the output circuit of an amplifier to improve its high frequency response.

compensation, low frequency A modification to the output circuit of an amplifier to improve its low frequency response.

compensator, impedance A return/loss (R/L) network connected across a loaded circuit in order to improve R/L characteristics at higher frequencies.

compile 1. To translate a computer program expressed in problem-oriented language into a computer-oriented language. 2. To prepare a machine language program from a computer program written in another programming language by making use of the overall logic structure of the program, or generating more than one computer instruction for each symbolic statement, or both, as well as performing the function of an assembler.

compiler A program which converts instructions written in a source language (COBOL, ALGOL, etc.) into machine code which can be read and acted upon by the computer.

compiler, data Tool for collecting, vetting and providing input tapes for exchange data, to be used on host procesors.

complement In the computer world, the negative of a number. The usual way of deriving a complement is to subtract each digit in turn from one less than the number base (i.e., take away from 9 if the number is decimal) then add 1 to the total: this is called the true complement or the radix complement.

complement, cable Group of pairs in a cable which terminate at the same point or have some other feature in common.

complementary channels, pair of Two channels, one in each direction, which provide a bidirectional communication.

complementary metal oxide semiconductor (CMOS) A technology used in the manufacture of integrated circuits by combining N-channel and P-channel MOS transistors.

complementary operation When one Boolean operation has a result which is the negative of another, the two operations are said to be complementary.

completed call A call which has been switched through on an established path and conversation has begun. In telephone traffic usage completed does not mean the end of a call.

completion of calls to busy operators in public exchanges A call to the operator which encounters the busy condition may be completed when both the caller and an operator are free and without the caller generating a second call or waiting on the line.

completion of calls to busy subscribers A call to a busy number may be completed when both called number and caller are free and without the caller generating a second call or waiting on the line.

completion ratio The ratio of the number of completed call attempts to the total number of call attempts, as measured at a given point of a network.

complex wave A wave which is not a simple sinusoidal signal; it can be broken down into sine wave components.

compliance For mechanical systems: a property which is the reciprocal of stiffness.

compliance, statement of A detailed list which has to be submitted with most international tenders indicating whether or not the equipment offered is fully compliant with the requirements in the specifications.

component An assembly, or part thereof, which is essential to the operation of some larger assembly. It is an immediate subdivision of the assembly to which it belongs.

component, circuit An essential item that becomes an integral part of a circuit.

component, discrete A separately packaged circuit element with its own external connections.

composite signaling A signaling arrangement to provide means for direct current signaling and dial pulsing beyond the range of loop signaling methods. Sometimes alternating current at sub-audio frequencies (e.g., below 100 Hz) is used, separated from voice signals by filters.

composite two-tone test signal A test signal used for intermodulation distortion measurements.

composite video signal The complete video signal. For monochrome, it consists of the picture or luminance signal and the blanking and synchronizing signals. For color, additional color synchronizing signals and color picture information (the chrominance signal) are added.

composited circuit A circuit which can be used simultaneously for telephony and dc telegraphy or signaling; separation between the two being accomplished by frequency discrimination. Syn.: voice-plus circuit

compression An undesired decrease in amplitude of a portion of the composite video signal relative to that of another portion. Also, a less than proportional change in output of a circuit for a change in input level. For example, compression of the sync pulse means a decrease in the percentage of sync during transmission.

compression, bit Application of a technique to reduce the number of bits required to represent the information in data transmission or storage, and so conserving bandwidth or memory. Syn.: Compaction

compressor A device with a non-linear gain characteristic that acts to reduce the gain more on larger input signals than it does on smaller input signals. Allows signals with a larger dynamic amplitude range to be sent through devices and circuits with a more limited range.

compulsory license Legislation requiring copyright holders to license users of copyrighted material (CATV operators) on a uniform basis and for a stipulated fee.

compunications A made-up word to describe the merging of telephones, computers, television and data systems. Syn.: telematics

computer A device capable of accepting and processing information and supplying the results. It usually consists of input, output, storage, arithmetic, logic, and control units.

computer, analog Computer using electrical analog value for continuous variable input and producing analog output.

computer, digital Computer processing information on a discrete or discontinuous basis.

computer-aided manufacturing (CAM) The use of computers to control manufacturing processes.

Computer and Business Equipment Manufacturers Association (CBEMA) Trade association for manufacturers of office automation and business equipment.

Computer and Communications Industry Association (CCIA) Trade association for computer manufacturers and related organizations.

computer controlled message switching A message switching system in which messages from outlying stations are automatically accepted, stored and forwarded to the destination station given in the header of the message.

computer-dependent language See assembly language.

computer floor See floor, raised.

Computer Inquiry Inquiries I and II were the main FCC Dockets dealing with regulating differences between communications and information processing services. CI I (1971) and CI II (1976/1980) led eventually to AT&T (and its successor companies) setting up subsidiary companies to offer nonregulated services.

computer language 1. A computer-oriented language whose instructions consist only of computer instructions. 2. A language that is used directly by a computer. Syn.: machine language.

computer-oriented language 1. A programming language that reflects the structure of a given computer or that of a given class of computers. 2. A programming language whose words and syntax are designed for use on a specific class of computers.

computer peripherals The auxiliary devices under control of a central computer, such as card punches and readers, high-speed printers, magnetic tape units, and optical character readers.

computer routine An ordered set of computer instructions that may have some general or frequent use.

computer utility A service that provides computational ability; a time-shared computing system. Programs as well as data may be made available to the user, or he may store his own data and programs with the utility. In some countries called a computer bureau or a service center.

computer word In computing, a sequence of bits or characters which occupy one storage location and are treated by the computer circuits as a unit and transferred as such.

computerized branch exchange Name used by some manufacturers for their processor-controlled (SPC) PBXs.

COMSTAR U.S. domestic satellite system, owned by a COMSAT subsidiary.

concatenate To connect together as in a chain; linking data in a computer program.

concentrated range extension with gain (CREG) A U.S. technique for installing loop extenders in a central office/exchange after the first concentration stage, i.e., so that they may be switched in or out depending on the characteristics of the line circuit concerned. The nearest equivalent in Europe is usually called pooled or common mode loop extenders, but these are usually directly associated with the line circuit stage of an exchange, as was done in AT&T's unigauge system, now being replaced by CREG.

concentration 1. A configuration wherein the number of inlets into a switching stage is larger than the number of outlets. 2. The process of allowing a large group of users to share a smaller group of facilities. This is generally achieved by sequentially switching each user onto a facility one at a time as they become available.

concentrator 1. A switching system in which a large number of inlets is connected to a smaller number of outlets. 2. In data transmission, a functional unit that permits a common path to handle more data sources than there are channels currently available within the path. A concentrator usually provides communication capability between many low-speed, usually asynchronous channels and one or more high-speed, usually synchronous channels. Usually different speeds, codes and protocols can be accommodated on the low-speed side. The low-speed channels usually operate in contention and require buffering, with the concentrator dynamically allocating channel space according to demand in order to maximize data throughput at all times.

concentrator, line A pair-saving device which consists of two units of equipment—one in the central office, the other near a remote group of subscribers. The subscribers' lines are switched to access a small number of trunks from the concentrator back to the central office, where the group is expanded again to one line per subscriber basis. With modern digital equipment however, a concentrator is sometimes the first stage in a central office and can be colocated with the office or remote, working in both cases by PCM to the group switching stage. For these remote concentrators there is no expansion stage needed in the central office

concentrator, station line A concentrator providing pair saving between multi-key telephones (keysets) and keyset line circuits.

concentrator, telegraph An automatic traffic concentration switch for telegraph instruments.

concentrator-identifier Concentration of calls, together with the provision of an automatic line identification signal so an operator knows which line is calling in.

concentric-lay A method of making-up a multipair cable with pairs of conductors in concentric layers.

Usually alternates with clockwise and anticlockwise layers.

concentric line   A coaxial line.

concentricity   A measure of the deviation of the center conductor position relative to its ideal location in the exact center of the dielectric cross-section. This is a very important mechanical characteristic of the core of a submarine cable.

conceptual modeling   A method of problem-solving when some element of forecasting enters into the problem. A mathematical model is constructed to suit the results of one or a series of experiments and the model is then used to carry out further calculations to see how close it comes to the actual answers, i.e. to see if the mathematical assumptions and calculations are reasonably well justified.

concrete-encased reinforcing bar ring ground (RBRG)   Similar to a ring ground, but employs welded reinforcing bars to complete the circle instead of a copper conductor.

concrete-encased ring ground (CERG)   Similar to a ring ground, but located in the building foundation.

conditioned circuit   A circuit that has conditioning equipment to obtain the desired characteristics for voice or data transmission.

conditioned diphase modulation   A method of modulation employing both diphase modulation and signal conditioning to eliminate the dc component of a signal, to enhance timing recovery, and to facilitate transmission over VF circuits or coaxial cable facilities.

conditioned loop   A loop which has conditioning equipment to obtain the desired line characteristics for voice or data transmission.

conditioned voice grade circuit   A voice grade circuit with conditioning equipment to equalize envelope or phase delay response to improve data transmission through the circuit. Syn.: data grade circuit

conditioning   To bring to a standard. Line conditioning, for example, will bring attenuation, impedance, and delay characteristics to within set limits.

conditioning, circuit   Procedures followed in bringing circuits to acceptable standards.

conditioning equipment   1. At circuit junctions, equipment used to match transmission levels and impedances, and to provide equalization between facilities. 2. Corrective networks used to equalize the insertion loss versus frequency characteristic and the envelope delay distortion over a desired frequency range in order to improve data transmission.

conditioning schedules   (U.S. Dept. of Defense)
Type D = for data transmission circuits.
Type N = for teletypewriter speed circuits.
Type S = for secure voice circuits.
Type V = for facsimile, slow data or nonsecure voice circuits.
Type Z = for wideband (up to 50 kHz) secure voice circuits.

conductance   1. A measure of the ability of a substance to conduct electricity. It is the reciprocal of resist-

ance, and is expressed in siemens. 2. The real part of admittance.

conductance, mutual   See transconductance.

conducted interference   1. Interference resulting from radio noise or unwanted signals entering a device by direct coupling. 2. An undesired voltage or current generated within a receiver, transmitter, or associated equipment, and appearing at the antenna terminals.

conduction   Transfer through a medium, such as the conduction of electricity by a wire, or of heat by a metallic frame, or of sound by air.

conduction band   A partially filled or empty atomic energy band in which electrons are free to move easily, allowing the material to carry an electric current.

conductivity   Specific conductance, i.e., conductance per unit length. The reciprocal of resistivity.

conductivity, n-type   In a semiconductor, conductivity due to electron movement.

conductivity, p-type   In a semiconductor, conductivity due to the movement of holes.

conductivity, percentage   The conductivity of the material as a percentage of the conductivity of copper.

conductor   Wire which carries electric current.

conductor, adjacent   Wire immediately next to another wire, i.e., in a multi-pair cable.

conductor, aluminum   Wire, either solid or stranded, made of pure aluminum.

conductor, flat   Wire manufactured in tape form so that multi-wire tapes for telephones may be laid easily under carpets or on walls.

conductor, solid   Solid wires, rather than stranded ones.

conductor, steel/aluminum   (or aluminum conductors, steel-reinforced) (ACSR) High tensile steel core (for strength in long spans) surrounded by aluminum wires (for low resistance).

conductor, stranded   Conductor made up of many small soft strands twisted together.

conduit   Pipes installed underground to carry telephone distribution network cables; ducts.

conduit, asbestos-cement   Asbestos-cement pipe used as cable conduits or ducts. Joined by collars or by tapered couplings.

conduit, clay tile   A multiple duct sectional conduit, made of vitrified clay tile.

conduit, corrugated   Plastic conduit with corrugated walls to give flexibility.

conduit, fiber   Fiber pipe used as telephone cable ducts.

conduit, lateral   Conduits laid out from a main or primary duct route towards flexibility or distribution points.

conduit, main   Principal routes followed by conduit leaving central office; a route used for primary distribution cables.

conduit, monolithic   Type of conduit formed in the trench by laying concrete around inflated flexible tubes. After the concrete has set the tubes are deflated, withdrawn, and used in the next section.

conduit, multiple duct   Duct units with more than one duct way, typically 2, 4, 6, 8, 9, 12 tubes.

conduit, plastic   Plastic tubing used as ducts. Thin-wall tubes require protection by concrete, thick-wall may be laid directly in the ground. Probably the most popular material for telephone cable ducts today is polyvinyl chloride (PVC) because it is easy to pull long lengths of plastic-sheathed cables through PVC ducts.

conduit, protected   A nest of ducts protected by concrete, either poured and set around the ducts as they are laid or in a slab above the ducts, or in box form surrounding the duct-nest.

conduit, rigid metal   Iron pipes were once common in duct systems but are now rarely encountered unless the ducts need substantial rigidity.

conduit, riser   A vertical duct, typically from floor to floor in a multi-story building.

conduit, steel   Galvanized or enameled steel pipe used where rigidity is essential.

conduit, subsidiary   A minor or spur duct route leading from a manhole or jointing chamber to a building or distribution point.

conduit, thin wall asbestos-cement   Asbestos-cement pipe having a ¼-inch wall thickness. Used for underground electrical conduit where concrete encasement is required.

conduit, thin wall metallic   Electrical metallic tubing conduit used for internal cabling in buildings.

conduit, wrought iron   Ducts of iron pipes, which resist corrosion but are expensive to install.

conduit run   The route to be followed by a nest of ducts.

cone   The structure rising in the center of a submarine cable ship's tank, in the form of a truncated cone (conic frustum), around which the cable is coiled at the inside of a flake. The diameter of the cone at the top is commensurate with the allowable bending radius of the cable.

conference, broadcast   Multi-party telephone call or signal from one station on a system to several other stations. Some conference type features permit only one-way transmission of information, others permit talk-back.

conference, meet-me   A type of call in which all concerned key out a special number which is connected to a multi-party conference bridge.

conference, preset   Procedure used for regularly established conference calls: the originator keys out a code number, on receipt of this the central office calls all those parties required to be connected to the conference call.

conference, progressive   An operator-controlled conference call procedure, the operator calls each party in turn and adds him to the conference circuit.

conference, teletypewriter   A type of conference call between teletypewriters in which an enlarged view of the typed copy is projected onto a screen or shown on visual display units.

conference call   Telephone call between three or more parties.

conference call, add-on   A conference call established by the caller, successively connecting different subscribers to a common speech path; further parties may be added at any time at the discretion of the caller.

conference call, meet-me   A number of subscribers each can, by prior arrangement, make a telephone call at an agreed time to a common point within the telephone network.

conference call, operator controlled   A conference call set up by an operator at a customer's request.

conference call, subscriber controlled booked   A conference call, which is established by means of an automatic device, which is programmed in advance and can be activated at any moment by the caller.

conference call, subscriber controlled occasional   A conference call which is established by means of an automatic device which is programmed by the caller for each occasion.

conference operation   1. In a telephone system, that type of operation in which more than two stations can carry on a conversation. 2. In telegraph or data transmission, that form of simplex or half-duplex operation in which more than two stations may simultaneously exchange information, carry on conversations or pass messages.

conference repeater   A repeater, connecting several circuits, which receives telephone or telegraph signals from any one of the circuits and automatically retransmits them over all the others.

confidence test   For submarine cable systems, a test of several weeks' duration to which repeaters are subjected before shipping in order to discover any flaws.

configuration, conductor   Physical geometrical arrangement of conductors within a cable.

configuration management   Procedures aimed at ensuring that the build state of systems is ascertainable and recorded.

confirmation signaling   A compelled signaling procedure in which the receipt of each digit is acknowledged in turn by sending a backwards signal representing the same digit.

conflict, structure   The location of pole routes in such a way that if poles overturn, the wires on one route will contact the wires on the other. Conflict can be avoided by separation, e.g., using opposite sides of a road.

Confravision   A conference call using videophones or large screen projection of persons or objects at the distant end.

confusion reflector   Devices used to produce false images on military radar.

congestion   Condition which exists when the demands for service upon a communications network exceed its capacity, thus necessitating a restriction of traffic.

congestion, time   The probability that a system is congested over any time period.

70

congruency The ability of a facsimile transmitter or receiver to perform in an identical manner as another facsimile system.

conjugate selection The simultaneous operation of several switches establishing a speech path through an office.

conjunction, satellite When two bodies in space are directly in line with each other as viewed from a point on earth.

connect time The amount of time that a circuit is in use. Syn.: holding time

connecting circuit Facilities used to connect specific input and output points either on a temporary, semipermanent, or permanent basis.

connecting company An Independent telephone company in the U.S. with arrangements with a Bell company for the interchange of telephone traffic.

connection 1. A direct wire path for current between two points in the circuit. 2. The current-carrying junction which results when two conductors are clamped, wrapped, or wrapped and soldered into electrical contact. 3. An association of channels and other functional units providing means for the transfer of information between two or more terminal points.

connection, compression A low resistance electrical joint made by crimping two conductor wires together inside a sleeve.

connection, delta A method of providing a three-phase electric power supply over three conductors, with each of the three phases connected across a different pair of the three wires.

connection, drainage A connection to ground so that stray currents do not interfere with the operation of the component.

connection, loose A bad contact between wires or a wire and a terminal. Produces noise and heat.

connection, series Connected so that the same current flows through all components concerned.

connection, telephone A bidirectional facility between two points permitting transmission and reception of voice signals together with all necessary signaling and control signals.

connection, through A telephone connection which does not terminate in a particular office but is transmitted through it on either a four-wire or two-wire basis.

connection, wye Method of providing a three-phase electric power supply over four conductors, with each of three phases connected across a phase wire and a common ground wire; a star-type network.

connection in progress A call control signal at the data circuit-terminating equipment (DCE)/data terminal equipment (DTE) interface which indicates to the DTE that the establishment of the data connection is in progress and that the ready-for-data signal will follow.

connection time Total time in seconds between receipt of the last digit dialed and the first ring sent to the called line.

connections per circuit hour A unit of traffic measurement, i.e., the number of connections established at a switching point per hour.

connector 1. Any device for making a temporary or semipermanent connection. 2. A Strowger-type switch used as final selector in a step-by-step office. 3. A protective device mounted on main distributing frame. 4. In CCITT specification description language, a connector is either an in-connector or an out-connector. Flow is always assumed to be from the out-connector to its associated in-connector.

connector, bridging A screw-type unit used to connect drop wires to open-wire conductors or to provide a test point on an open-wire route.

connector, cable Multi-conductor receptacle used to terminate the end of equipment cables. Connectors mate with similar multi-conductor plugs.

connector, cell Lead or copper strapping between adjoining cells in a central office storage battery.

connector, coil spring Small coil made of square-cut springy wire; plastic insulated conductors may be pulled between adjacent coils in the spring, the spring wire then penetrates the insulation and makes good contact with the conductor. Sometimes used instead of screw terminal binding posts.

connector, crossbar A rectangular matrix switch with horizontal and vertical bars operated by electromagnetic relays.

connector, hermaphrodite Connector with both units of identical shape, e.g., a two-pronged fork.

connector, marker A multi-contact relay which connects a decoder to a marker upon command. Each connector serves one decoder and has access to all markers.

connector, PBX A Strowger or step-by-step final switch able to search a group of lines for a free one to the subscriber. A trunk-hunting connector.

connector, pole strand A galvanized steel fastener used to terminate messenger wires on a pole.

connector, pretranslator Circuit sometimes used to connect a route/digit translation circuit to a register.

connector, recorder Method of connecting warning tone or beeps to a line to indicate that the conversation is being recorded.

connector, solderless Heavy bolt-activated clamps sometimes used for joining power conductors.

connector, step-by-step A two-motion switch which receives the last two of the dialed digits and steps wipers up and around to make contacts with the required telephone line. Each connector has relays which enable testing, ringing and supervisory functions to be carried out. In Britain, called a final selector.

connector, test A connector to which only operators or test staff have access, it switches through to the required line even if the line is busy. Used for testing purposes.

connector, trunk and level hunting Connector used in a step-by-step office to connect calls to subscribers who have a large number of exchange lines. The switch is able to step up to a level where there is an idle line, then around to the idle line. In Britain, called an 11-and-over PBX final selector.

connector, trunk hunting Connector used in a step-by-step office to connect calls to subscribers who have more than one (but less than 11) exchange lines. The switch is driven up and around to the directory number of the subscriber, then hunts automatically for an idle line. In Britain, called a 2-10 PBX final selector.

connector, wire Device used to join together conductors in cables. Modern connectors accept insulated conductors which are crimped together to provide a low resistance splice.

consent decrees The 1956 Decree arose from a lawsuit between the U.S. federal government and the Bell System, resulting in Bell being restricted (in general) to the provision of regulated services and manufacturing equipment used for such services. The 1982 Decree, together with the Modified Final Judgment, divested Bell Operating Companies from AT&T. See entry: Bell System.

conservation of radiance A basic principle stating that no passive optical system can increase the quantity Ln^{-2} where L is the radiance of a beam and n is the local refractive index. Syn.: radiance theorem

console A small manually-operated attendant's unit mounted on a desk or table.

console, telephone PABX attendant's control unit where incoming calls may be answered.

constant-current source Source with infinitely high output impedance so that output current is independent of voltage, for a small specified range of output voltages.

constant-current working Design of line terminating unit in a central office/exchange which applies sufficient voltage to the subscriber's line to produce a current of a specified value.

constant-voltage source Source with low, ideally zero, internal impedance, so that voltage will remain constant independent of current supplied.

constant-voltage working Design of line terminating unit in a central office/exchange which always applies the same voltage to the subscriber's line (usually 50 volts) with the result that the current varies with the length and resistance of the line.

constantan Alloy of nickel and copper with low temperature coefficient of resistance, used for high quality precision-wound resistors.

constrictor, cable sheath Tool used to compress a cable sheath to increase resistance to gas flow past the point.

consultation call PABX facility enabling an extension to hold an incoming call while connected to another local extension.

contact 1. The strip or disc of metal which makes electrical contact when a relay or other electromechanical device operates. Contacts are often made of precious metal to eliminate resistance faults due to oxidation. 2. (as verb) To communicate with any person or organization.

contact, back Stationary contact of a pair of contacts which open when a relay is operated.

contact, break Contacts which open when a relay is operated.

contact, dry Contacts through which no direct current flows at any time.

contact, front Stationary contact of a pair of contacts which close when a relay is operated.

contact, make Contact which closes when a relay is operated.

contact, moving A contact which moves to or from a fixed contact.

contact, normal Stationary contact in a jack which closes when there is no plug in the jack but opens when a plug is inserted.

contact, sealed Contact enclosed in a glass, metal or plastic tube or box.

contact, wiping Contact which moves over a stationary bank of contacts, keeping surfaces clean.

contact bounce Rebound of a contact, which temporarily opens the circuit after its initial make.

contact common A continuous or common segment in a bank assembly, which wipers are able to ride along.

contact follow The small distance two contacts move together after the moving one has first made contact with the stationary one.

contact form The configuration of a contact assembly on a relay. Many different configurations are possible from simple single make contacts to more complex ones involving breaks and makes.

contact potential Potential difference which arises when two different conducting materials are brought into contact. Results from the difference between the work functions of the two materials.

contact resistance Resistance at the surface when two conductors make contact.

contactor An alarm unit used in cable pressurization schemes which activates when there is a fall in pressure.

contactor, gas pressure Alarm-giving contactor, may be located in manholes remote from the central office. When operated, the resistance of an alarm pair, as measured from the central office, indicates the location of the contactor giving the alarm.

contactor, magnetic Heavy duty relay used for switching on and off power circuits.

contactor terminal A gas-pressure contactor which is also the terminal of an alarm circuit.

contacts Points which are brought together or separated to complete or break an electric circuit.

contacts, break Contacts which open when a relay is operated or a key thrown.

contacts, dissimilar Two contacts, which make or break circuits, made of different metals to reduce contact resistance and minimize the effects of possible corrosion.

contacts, make-before-break Three contacts in a springset so that A makes contact with C before it breaks from B.

contacts, mercury Contacts wetted by mercury, giving very low resistance.

contacts, twin Doubling-up of contacts by placing contact points in parallel, to reduce likelihood of dirt or dust interfering with the operation of the circuit.

contacts, wetted Contacts through which a small direct current is passed in addition to a voice or other signal.

content In a message handling system (MHS) the piece of information that the originating User Agent wishes delivered to the recipient User Agent.

content addressable memory A memory from which data can be retrieved without supplying the location in store, e.g., a directory enquiry system programmed to work on an input of a name, possibly misspelt, or with only part of a street address.

contention A condition arising when two or more data stations attempt to transmit at the same time using a shared channel.

Continental Telecom, Inc. Major U.S. telecommunications organization, which owns many local telephone companies and manufacturing firms.

contingency planning Planning of a course of action such that the plan will only be invoked if the contingency materializes.

continuity An uninterrupted electrical path.

continuity check A check made to a circuit in a connection to verify that an acceptable path for transmission of data or speech exists. With the increased use of separate or common channel signaling systems, a continuity check on the actual voice path itself is normally carried out before a circuit is put through to the customer.

continuity check tone A tone (usually 2000 Hz) used for transmission path checking of circuits established using common channel signaling. Syn.: continuity signal

continuity check transceiver A combination of check-tone transmitter and receiver, used to confirm that a speech path exists.

continuity check transponder A device which is used to interconnect the paths of a circuit at the incoming end, which, on detection of a check tone, transmits another check tone to permit continuity checking of the circuit.

continuity-failure signal In data common channel signaling, a signal sent in the backward direction indicating that the call cannot be completed due to failure of the forward continuity check.

continuity signal See continuity check tone.

continuous operation In data transmission, a type of operation in which the master station need not stop for a reply after transmitting each message or transmission block.

continuous wave laser A laser in which the coherent light beam is generated continuously, as is normally needed for optical fiber communications systems.

With present technology "continuous" usually means for a period greater than 0.25 seconds.

continuous wave radio telegraphy Radio system in which unmodulated carrier is transmitted (for a mark signal) and nothing transmitted (for a space signal). At the receiving end, a beat frequency oscillator is used to make the CW signal audible.

continuously variable slope delta modulation A type of delta modulation in which the size of the steps of the approximated signal is progressively increased or decreased as required to make the approximated signal closely match the input analog wave.

contrahelical Cable manufacturing procedure by which successive layers are helically assembled in opposite directions.

contrast The range of light and dark values in a picture, or the ratio between the maximum and minimum brightness values. A high-contrast picture would contain intense blacks and chalky whites, a lower contrast picture would contain only shades of gray.

control, automatic frequency Device which maintains frequency of an oscillator within predetermined limits.

control, automatic gain Maintenance of intermediate frequency output constant in a radio receiver despite variations in received signal strengths.

control, automatic volume An automatic gain control operating either on the intermediate frequency or audio stage of a radio receiver.

control, circuit The organization which controls testing and maintenance of a toll or long distance circuit.

control, line load Selective denial of call origination capability to specified subscribers' lines when excessive demands for service are offered to a switching center. It does not affect the capability to receive calls. Normally, in North America, a central office is divided into three groups of lines and control is rotated, with one or two groups being denied outward service at a time.

control, remote A system for the control of equipment from a distance.

control, technical Authority with technical control of circuit engineering features.

control character In man-machine language, a character whose occurrence in a particular context initiates, modifies, or stops an action that affects the recording, processing or interpretation of data.

control electrode An electrode to which a signal is applied in order to produce changes in currents in one or more of the other electrodes.

control grid The grid in an electron tube that controls the flow of current from cathode to anode.

control of electromagnetic radiation An operational plan to minimize the use of electromagnetic radiation in the event of attack or imminent threat thereof, as an aid to the navigation of hostile aircraft, guided missiles, or other devices.

control office An office, testboard, or technical control facility which establishes, maintains and rearranges circuits, trunks and lines.

control procedure A method in which information is exchanged in a predetermined forward and backward order between telephone subscriber and exchange to effect control of a service.

control station That point within the general maintenance organization that fulfills the control responsibilities for the circuit, group, supergroup or line section assigned to it. In a data network, the station that selects the master station and supervises operational procedures such as polling, selecting, and recovery. The control station has the overall responsibility for the orderly operation of the entire network.

control switching point A toll switching center in the North American telephone system, which may be a Class 1, 2, or 3 office, where intertoll trunks are connected to other intertoll trunks. At least one other office of lower rank homes on it.

control word A word whose appearance initiates a predetermined control function.

controlled area An area to which security controls have been applied to provide protection to an information-processing system's equipment and wirelines equivalent to that required for the information transmitted through the system.

controlled-not-ready signal In data common channel signaling, a signal sent in the backward direction indicating that the call cannot be completed because the called number is in a controlled-not-ready condition.

controlled out-pulsing Method of interworking between central offices which delays the out-pulsing of digits until a signal has been received indicating that a register is available at the distant end to receive these digits.

controlled re-ring Feature of some operator controlled switching equipments which permits operators to ring back to a telephone without having to redial the number.

controlled rerouting A procedure of transferring, in a controlled way, signaling traffic from an alternative signaling route to the normal signaling route, when it is available.

controlled start of ringing Feature of some operator controlled switching equipment which permits operators to dial a number but to delay ringing the telephone until a key is operated.

controlling exchange In international telephone services: The exchange responsible for setting up calls and deciding the order in which they are to be connected. The Administrations concerned agree among themselves to designate the controlling exchange.

controlling operator The outgoing operator in the controlling exchange who operates the international circuit.

controls, dynamic overload Automatic telephone traffic control and network management equipment which keeps regular check on traffic patterns and possible congestion paths and instructs offices on switching possibilities and alternative routing procedures.

convention A generally acceptable symbol, sign or practice.

convergence 1. Intersection of beams in a multi-beam electron tube, such as a color TV tube. 2. The bringing together and unification of technologies emerging from computing and telecommunications.

convergence electrode Electrode used for electrostatic control of convergence in a multi-beam electron tube.

convergence magnet Used for electromagnetic control of convergence in a multi-beam electron tube.

conversation time Normally equal to call duration in fully automatic service.

conversational mode A mode of operation of a data processing system in which a sequence of alternating entries and responses between a user and the system takes place in a manner similar to a dialogue between two persons.

conversion, code The translation of one dialed code into another, e.g., a toll directing code into a code used for route selection.

conversion, frequency Any change in frequency, particularly the change from radio frequency to intermediate frequency in a radio receiver.

converter 1. Electron tube or transistor used in a frequency conversion stage. 2. A device used to convert non-VHF television signals into standard VHF channels. Cable systems often install converters where more than 12 channels are relayed on a signal cable. Converters also protect signals from the interference of strong local signals.

converter, analog-to-digital A coder; a sampling device which produces a digital output from an analog input conveying the same information.

converter, code Circuit which changes an information-carrying signal from one code system to another.

converter, dc-dc Device which converts dc at one voltage to dc at another voltage. These are common in modern telephone switching systems where low voltages (typically 5v) are needed to power transistorized circuitry.

converter, digital-to-analog A decoder; produces an analog output from a coded digit input signal. Usually combined with a coder as a codec.

converter, frequency Device which changes a signal from one frequency to another.

converter, inverted rotary Rotating machine used to convert dc to ac.

converter, language and mode Interface unit which enables teletypewriters using different codes and speeds to work with each other.

converter, ringdown Relay set used to convert receive and transmit signals into ringdown signals and vice versa.

converter, ringing A signal conversion set to interface between a switchboard ringing signal (typically 20 Hz AC) and tone signaling (eg 2600 Hz, 1000/20 Hz).

converter, rotary Rotating machine used to convert ac to dc.

converter, signaling A signaling conversion relay set, typically one used to interface between signaling used on a circuit and a switchboard.

converter, static A converter with no moving parts.

converter, tape-to-card Device that reads magnetic or punched tape and produces a punched data card with the same information.

converter, touch-tone A device which receives dual-tone multi-frequency signals and converts them into dial pulses. Required when converting a common control dial office, with no in-built ability to receive MF push-button tone calling, to accept address signals from push-button telephones.

cooperating User Agent A user agent in a message handling system that cooperates with another recipient's User Agent in order to facilitate communication between originator and recipient.

coordinate clock A clock in a set of clocks distributed over a spatial region producing time scales which are synchronized to the time scale of a reference clock at a specified location.

coordinate time The concept of time in a specific coordinate frame, valid over a spatial region with varying gravitational potential. If a time scale is realized according to the coordinate time concept, it is called a coordinate time scale. TAI (International Atomic Time) is a coordinate time scale. Its reference is the earth's surface at sea level.

coordinated time scale A time scale synchronized within given limits to a reference time scale.

Coordinated Universal Time (UTC) The time scale, maintained by the BIH (Bureau International de l'Heure) which forms the basis of a coordinated dissemination of standard frequencies and time signals. It corresponds exactly in rate with TAI (International Atomic Time) but differs from it by an integral number of seconds, having been adjusted by the insertion or deletion of seconds (positive or negative leap seconds).

coordination, frequency The cooperative selection and allocation of radio frequencies such that all systems can operate without giving or receiving interference. Also called frequency management or frequency planning.

coordination, inductive 1. Reducing inductive interference on communication or power circuits by the application of corrective measures. 2. Cooperation by communication and power engineers to reduce inductive interference.

coordination, level Assigning and maintaining specific power levels on a multi-channel system so as to eliminate crosstalk on all channels.

coordination, structural 1. Taking preventive measures to reduce the danger that conductors on power and communication pole lines will make accidental contact. These measures deal with pole strength, proper guying, and adequate clearances both horizontal and vertical. 2. Cooperation by communication and power engineers to insure that such measures are taken.

coplanar process Complex process used during the manufacture of LSI and MOS integrated circuits.

copper, soldering A wedge-shaped bar of copper fitted with a steel shank and heat-insulating handle which is used for heavy soldering jobs. In Britain, called a soldering iron even though it is made of copper.

copper-clad aluminum Aluminum wire with welded copper cladding, somewhat cheaper than pure hard drawn copper but very similar in electrical characteristics for telephone purposes.

copper-clad steel Steel wire (for strength) with welded copper cladding. Can be used in areas where long spans are economically necessary.

copper loss Loss due to heating effect of current, sometimes called the I^2R loss.

copper wire counterpoise ground (CWGC) A length of bare copper wire buried at least 18 inches below ground to provide an earth interface. Driven rods are not used.

copy, hard Printed copy.

copy, page Printed copy of a message as delivered by a teletypewriter.

copy, tape Punched tape version of teletypewriter message, as received or transmitted.

CORAL A high-level programming language.

cord A flexible conductor for making temporary connections to jacks on switchboards.

cord, closure sealing Sealing compound used between halves of a split closure for cable splices.

cord, handset The three- or four-conductor cord, usually helical, used between a telephone instrument and its handset. Plug and socket arrangements are now becoming common.

cord, line The two-, three-, or four-conductor flexible cord used to connect a hand telephone set to the incoming line at a connecting block.

cord, nylon lacing Nylon filament cord used for lacing cable forms.

cord, patching Cord with plugs at each end, used for temporary patching of circuits to equipment.

cord, plugging-up Cord used on a switchboard to plug-up faulty circuit.

cord, power A three-wire (sometimes two-wire) cord used to make connection with an ac public power supply.

cord, retractile A helically wound cord used between telephone instruments and handsets; it may be extended to full length, but when not under tension it shortens to a neat length.

cord, sealing Closure sealing cord used to join the two halves of a metal splice case, which provides protection for a cable splice.

cord, switchboard The two- or three-conductor cord used by telephone operators at manual operating positions to connect subscribers.

cord, telegraph facility patching Cords used on a telegraph test position. The circuit can be operated through this test position and traffic in both directions observed to confirm that faults have been cleared.

cord, telegraph test signal A cord used on a telegraph test board and its associated relays which together send signals automatically to a telegraph line under test.

cord, test Cord used in making connections from a line to a test set or test panel.

cord, tinsel The flexible electrical conductors used in switchboard cords. They consist of many thin copper tapes wound around textile bases.

cord circuit A pair of cords on a manual operating position along with keys, lamps, supervisory relays, etc.

cord lamps Lamps associated with a cord circuit that indicate supervisory conditions for their respective part of the connection.

cord pair The two halves of the cord circuit (answer cord and calling cord) which, together with keys, lamps etc., constitute a complete cord circuit.

cord shelf The table section of an operating position, with plugs and cords used by the operator.

cord-splitting Throwing a key on the operator's position to separate the two halves of a coin circuit, the operator may talk to either side but not both.

cord test circuit A test jack on manual operating positions so that operators may test their own cords.

cordless switchboard A telephone switchboard in which manually operated keys are used to make connections.

core 1. The center region of an optical waveguide or fiber through which light is transmitted. 2. For a coil, the magnetic material at the center of the winding. 3. In submarine cable, the center conductor and the dielectric. 4. For telephone cable, the bundle of insulated conductors within the cable sheath.

core, closed A closed loop core of ferromagnetic material used in a transformer.

core, laminated Ferromagnetic transformer core built up from thin iron laminations, separated by varnish or other insulant. Laminated cores have low eddy current losses.

core, magnetic 1. Ferromagnetic material placed within a coil of wire to provide a low reluctance path for magnetic flux and increase the inductance of coil and the coupling between coils. 2. Devices used as stores for binary information; magnetic material cores arranged in a matrix formation, threaded with 'write' wires and a 'read' wire. Ferrite stores of this nature were at one time common in computers and in the control circuitry of telephone central offices.

core, open A magnetic core which is within a coil winding but does not have a closed magnetic loop. Low magnetic flux results.

core, relay The metallic base on which a relay coil is mounted.

core, tuning A core which can be moved in relation to its coil winding to vary its inductance.

core diameter The diameter of an optical waveguide at the point where the refractive index of the core exceeds that of the cladding by k times the difference between the maximum refractive index in the core and the minimum refractive index in the cladding, where k is a specified constant ($0 < k < 1$). Specified through the use of a tolerance field.

core hitch Attachment of a pulling line directly to a cable core rather than to the cable sheath, to permit pulling it into a duct without damaging the cable.

core mode Synonym for bound mode.

core packing fraction The fractional area occupied by core material in a group of optic fibers brought together to form a bundle or area.

core-store Main store of a computer. This dates back to the days of ferrite core stores but the term is still in use even if the main store is in integrated circuits.

core tuning Moving a powdered iron or ferrite core in or out of a coil and thereby adjusting the frequency or resonance of the coil.

corner pole A pole located at a point where an open wire route changes direction. The unbalanced pull on the pole means that the pole must be guyed or braced or strutted.

corona A blueish luminous discharge due to ionization of the air near a conductor caused by a voltage gradient above a critical level.

corrective maintenance Tests, measurements, and adjustments made to remove or correct a fault.

corrector, impedance A network designed to improve impedance matching of a basic balancing network, particularly at low audio frequencies.

corrector, pulse Circuit incorporated in a pulse repeater which transmits cleaned-up pulses even if distorted pulses have been received. This is particularly relevant to dial pulses at 10—20 pps.

correed A reed relay. A bobbin and operating coil within which are inserted one or more sealed glass tubes containing flat metal reeds, overlapping at the center of the tube. Current through the coil causes the reeds to make contact. Reed relays have been widely used in central offices to provide speech paths.

correlation, reverse path A check made to determine if fading of a radio path is the same in both directions.

correlation, signal A check made on received signals of different frequencies; to obtain reliable diversity reception the two received signals must not be likely to fade at the same time.

correlator Logic device which creates a canceling-out noise signal to enable the recovery of information signals which would otherwise be masked by noise.

corrosion Chemical or acidic action that wastes away material.

cosine emission law Energy emitted in any direction is proportional to the cosine of the angle which that direction makes with the normal to the emitting surface. Also, Lambert's Emission Law.

cosmic noise Random noise originating outside the earth's atmosphere.

Cosmos Russian series of low-altitude satellites: more than 1,100 of these have been launched.

cost of removal In telephone company accounting this includes the cost of removing or otherwise disposing of plant and of recovering salvage.

cost of service pricing Telephone company accounting rule which prices services strictly in accordance with attributable costs.

coulomb The SI unit of electric quantity or charge; the quantity of electricity transported in 1 second by a current of 1 ampere.

count, pair Pair numbers in cables, as shown by pair positions on an MDF or terminal; the actual count.

count, theoretical The cable pair designation of the pair in the cable itself, not its MDF designation.

count, usage A traffic count of the number of times the circuit is busy. If the scanning rate is 100 seconds the count will be approximately the CCS figure for the route.

count rate Average rate of occurrence of events counted by a counter circuit or device.

counter Device that detects and counts particle or photons, e.g. geiger counter. Also, any form of device which counts electric pulses, e.g. a digital counter, which is used in telephone exchanges to count received dial pulses.

counter, peg-count A small mechanical counter used to count occurrences, such as numbers of circuits busy at an instant in time.

counter, 3-type A electromechanical timing device used on a traffic service position to indicate the completion of the initial talking interval on coin calls or calls where the customer requests notice of completion. Adjustable to any of three initial periods from one to five minutes. Provides a flashing light six seconds before end of the initial period, changing to a steady light eighteen seconds later. In Britain a similar device is called a timing clock, or Clock No. 44.

counter lag time Delay between the actual event and the indication of a count, in a counter.

counterpoise Special wire network arranged just above the earth in order to capacitively couple a radio transmitter to the ground when the ground resistance is high.

counting circuit Circuit actuated by received pulses.

country code The combination of one, two or three digits characterizing the called country, or integrated numbering plan, e.g., 1 = North America, 33 = France, 44 = U.K.

country-code indicator Information sent in the forward direction indicating whether or not the country code is included in the address information.

country-or-network identity Information sent in the backward direction consisting of a number of address signals indicating the identity of a country or network in which the call has been internationally transit switched.

couple 1. A thermocouple; a galvanic pair of two dissimilar metals. 2. Rotating forces. 3. To link together, e.g., by inductance, so that energy is transferred from one circuit to another.

coupler, antenna An impedance matching transformer, matching antenna feed with transmitter or receiver.

coupler, directional 1. A matched tapping device for insertion in a CATV cable, which introduces a loss of only 1 dB in the through transmission and provides a spur outlet. 2. A transmission coupling device for separately sampling either the forward (incident) or the backward (reflected) wave in a transmission line.

coupler, multioutlet TV A bridge device which enables several TV receivers to be served from one CATV outlet.

coupling The means by which signals are transferred from one conductor (including a fortuitous conductor) to another. Types of coupling include: (a) capacitive coupling: the linking of one conductor with another by means of capacitance (also known as electrostatic coupling); (b) inductive coupling: the linking of one conductor with another by means of inductance; (c) direct coupling: hard-wire connection of one conductor to another.

coupling, capacitive The linking of one stage or circuit with another by means of capacitance.

coupling, close Coupling between circuits so that most of the power in one is transferred to the other.

coupling, conduit Jointing practices and accessories for ducts and conduits.

coupling, critical Coupling for the maximum transfer of energy between circuits.

coupling, crosstalk Coupling between circuits which produces audible crosstalk.

coupling, direct Coupling of stages or circuits by a direct metallic connection.

coupling, ground-rod Coupling pipe used to join together two sections of rod being driven into the ground to provide a ground connection.

coupling, impedance Linking of one stage or circuit with another by a mutual impedance.

coupling, inductive Linking of one stage or circuit with another by the mutual inductance of a transformer.

coupling, link Linking of one stage or circuit with another via the interaction of an intermediate coupling circuit.

coupling, loose A loose linkage between one stage or circuit and another so that very little power is transferred from one to the other.

coupling, magnetic Linking of one stage or circuit with another by a magnetic flux which is common to both coils.

coupling, power/communications circuit The amount of noise induced into a communications circuit from a power circuit. This is governed by the voltages involved, physical separation, height above ground, physical shielding, whether or not the power circuits are well balanced between phases, and by the general standards of construction of both routes.

coupling, reference crosstalk Coupling which gives 90 dB loss between disturbing and disturbed circuits.

coupling, resistance Linking of one stage or circuit with another by transfer of power through a resistor.

coupling, variable Inductive coupling with provision for the movement of coils with respect to each other, giving control over the degree of coupling.

coupling coefficient A measure of the electrical coupling that exists between two circuits; it is equal to the ratio of the mutual impedance to the square root of the product of the self impedances of the coupled circuits, all impedances being of the same kind.

coupling factor A figure representing the combined effect of the various couplings between a power circuit and a communications circuit in producing noise in the communications circuit.

coupling loss The power loss suffered when coupling light from one optical device to another.

cover Depth of buried plant, i.e., distance between tops of ducts and cables and the road surface. Cover is also the material used as backfill on top of ducts and cables.

cover, manhole Heavy cast iron circular, rectangular or triangular plate which gives access from road level into a manhole.

cover, wall rack Light steel or plastic cover which covers relay sets mounted on walls.

coverage Area within which a radio or television broadcast transmitter gives an acceptable signal.

cradle The hook switch operated by a telephone instrument handset.

crankback When a call switched out on one route encounters congestion, crankback is the ability for a call to go back to an earlier switching stage and choose an alternative route avoiding the congested section.

crash A complete failure of a hardware device or a software operation.

cream skimming Ability of service companies, e.g., other common carriers, to choose to serve only the more profitable market sectors.

credit and load management system (CALMS) Telecontrol system which enables power supply meters to be read remotely using telephone lines.

credit card call A toll call charged to an acceptable credit card. In most systems requires action by an operator.

creepage 1. In a storage battery, the lifting of the level of electrolyte by capillary action. 2. In an underground system, the slow movement of cables in ducts due to traffic vibrations and gradients in duct routes.

crimp termination A terminal connected to a conductor by high pressure crimping of wire into a lug.

crimping sleeve A sleeve used to joint conductors in a cable joint. A common type of such sleeve is called a B-connect. Tests have shown that machine-made joints are usually more satisfactory than any form of hand twisted conductor joints, both in the short term and after aging.

crinoline Ring-shaped structure on a submarine-cable ship which surrounds the cone and can be vertically adjusted so its distance from the top of the cable coil can be kept virtually constant; with the cone it forms a relatively narrow opening through which the cable is paid out smoothly.

critical angle 1. In optics, when light propagates in a homogeneous medium of relatively high refractive index (n_{high}) toward a planar interface with a homogeneous material of lower index (n_{low}), the critical angle of incidence is defined by:

$$\theta_c = \sin^{-1}(n_{low}/n_{high}).$$

2. In fiber optics, the maximum angle at which a light wave striking the cladding will be reflected back into and along the core.

critical dates In Bell System Universal System Service Orders, specific dates in the life of an order established for internal order control and progress monitoring purposes based upon the type, complexity and geographic extent of service.

critical distance or cable length (cable TV) The length of a particular cable which causes a worst-case reflection if mismatched; depends on velocity of propagation, attenuation of cable, and frequency.

critical frequency 1. In radio propagation, the limiting frequency below which a wave component is reflected by, and above which it penetrates through, an ionospheric layer. 2. The limiting frequency below which a wave component is reflected by, and above which it penetrates through, an ionospheric layer at vertical incidence.

critical path method A project management system: all those activities which build up into the complete project are drawn out on a network diagram showing which items must be completed before others can start. The particular chain of events which is the longest one is the one which determines completion of the project on time and is called the critical path, any actions or delays which affect this longest chain of events will directly affect completion of the project.

critical technical load That part of the total technical power requirement which is needed for synchronous communications and automatic switching equipment.

CRO (Complete with related order) Abbreviation used in Bell System Universal Service Orders: an entry that indicates that the order is to be completed simultaneously with a related order.

cross Accidental contacts between wires of different pairs in a cable or on a pole route.

cross-border traffic Traffic between border towns in adjacent countries.

cross connect Connections between terminal blocks on the two sides of a distribution frame or between terminals on a terminal block, sometimes called jumpers. Connections between terminals on the same block are called straps.

cross coupling The coupling of a signal from one channel, circuit, or conductor to another, where it becomes an undesired signal.

cross elasticity   Changes in price or service made to an existing facility because of competition or threat of competition from a new facility or system.

cross-exchange check   A check made across the exchange to verify that a speech path exists.

cross modulation   Interference when a carrier signal becomes modulated by an interfering and unwanted signal as well as being modulated by its wanted signal.

cross-office check   A check made across the central office to verify that a speech path exists.

cross-ownership   Ownership of two or more kinds of communications outlets by the same individual or business. The FCC prohibits television stations and telephone companies from owning cable systems in their service areas. Television networks are prohibited from owning cable systems anywhere in the U.S.

cross-polarized operation   The use of two transmitters operating on the same frequency, with one transmitter-receiver pair being vertically polarized and the other pair horizontally polarized.

cross-ring   On a party line, the ringing of a second, undesired party after one party has already been rung.

cross-subsidization   Practice of using revenues generated by one service (e.g., long distance traffic) to support another (e.g., local telephone services).

crossarm   Wooden arm bolted to a pole and used to support insulators to which overhead wires are attached. U.S. standard is a 10-pin crossarm, 10 feet long.

crossbar switch   A switch having a plurality of vertical and horizontal paths, and electromagnetically-operated mechanical means for interconnecting any one of the vertical paths with any of the horizontal paths.

crossbar system   An automatic switching system in which the selecting mechanisms are crossbar switches.

crossbar system, No. 4A   A toll-switching system using four-wire crossbar switches with common-control for interconnecting intertoll trunks, tandem trunks, toll-switching trunks, and miscellaneous terminating trunks. Provides up to six-digit translation and automatic alternate routing. Being replaced in the Bell System by AT&T's digital tandem, No. 4 ESS.

crossbar system, No. 5   A local switching system using two-wire crossbar switches with common-control for interconnecting subscriber's lines with other lines or with interoffice or tandem trunks. Being replaced in the Bell System by AT&T's No. 1A and No 2. ESS and by digital offices, e.g., AT&T's No. 5 ESS.

crossfire   Interference from one telegraph circuit to another.

crossing, power   Place at which an overhead power line crosses over a pole route communication line. Separation and safety factors which must be observed are defined in local and national safety codes.

crossing, railroad   Place where a pole route communication line passes over a railroad track, requiring special clearance and strength. Factors to be observed are defined in local and national safety codes.

crossover frequency   Frequency at which output signals pass from one channel to the other in a crossover network. At the crossover frequency itself the outputs to each side are equal.

crossover network   A type of filter which divides an incoming signal into two or sometimes more outputs, with higher frequencies to one output, lower ones to another. Widely used in domestic radio loudspeaker systems, where treble components are dealt with by special high frequency speakers (tweeters), and low frequencies by woofers.

crosspoint   A single element in an array of elements that comprises a switch. It is a set of physical or logical contacts that operate together to extend the speech and signal channels in a switching network.

crosstalk   The phenomenon in which a signal transmitted on one circuit or channel of a transmission system creates an undesired effect in another circuit or channel.

crosstalk, direct   Crosstalk induced from the disturbing circuit directly into the disturbed circuit.

crosstalk, far end   Crosstalk measured by applying a tone to the far end of the disturbing circuit and then measuring it at the near end of the disturbed circuit. It measures the crosstalk coupling which exists at the far end of the circuit.

crosstalk, indirect   Crosstalk induced by a disturbing circuit into a second circuit and from this, in turn, into the disturbed circuit.

crosstalk, interaction   Crosstalk from one carrier system to another via an intermediate circuit which need not be a carrier circuit.

crosstalk, near end   Crosstalk due to induction from a circuit carrying a comparatively high level output signal leaving a repeater in one direction into a circuit leading into the input of a repeater at the same location, but dealing with the other direction of transmission of a different circuit. Both the levels of applied tone and the level in the disturbed circuit are measured at the near end.

crosstalk, runaround   Indirect crosstalk in which a high level signal leaving a repeater couples into a second but unrepeatered circuit which in turn couples the interfering signal into a repeated circuit at a low level point near a repeater input.

crosstalk coupling   The ratio of the power in a disturbing circuit to the induced power in the disturbed circuit observed at definite points of the circuits under specified terminal conditions, expressed in dB. Syn.: crosstalk coupling loss

crosstalk level   A crosstalk signal's effective power, expressed in decibels below one milliwatt.

crowbar   A short-circuit or low resistance path placed across the input to a circuit.

crown, cabinet   Space at the top of an equipment cabinet, normally reserved for terminal blocks.

cryogenics The science of physical phenomena at very low temperatures, approaching absolute zero, 0 K or —273°C.

cryotron Switching device which is cryogenic, i.e., demands very low temperatures and utilizes the phenomenon of superconductivity.

cryptanalysis The steps and operations performed in converting encrypted messages into plain text without initial knowledge of the key employed in the encryption.

cryptochannel A complete system of crypto-communications between two or more holders. The basic unit for cryptographic communication. It includes: (a) the cryptographic aids prescribed; (b) the holders thereof; (c) the indicators or other means of identification; (d) the area or areas in which effective; (e) the special purpose, if any, for which provided; and (f) pertinent notes as to distribution, usage, etc. A cryptochannel is analogous to a radio circuit.

cryptographic Pertaining to equipment or systems that encode data to conceal real meanings.

crypto-information Information which would make a significant contribution to the cryptanalytic solution of encrypted text or a cryptosystem.

cryptologic Of or pertaining to cryptology.

cryptology The science of hidden, disguised, or encrypted communications. It embraces communications security and communications intelligence.

cryptomaterial All material, including documents, devices, or equipment, that contains crypto-information and is essential to the encryption, decryption or authentication of telecommunications.

cryptosecurity See communications security.

cryptosystem The associated items of cryptomaterial that are used as a unit and provide a single means of encryption and decryption.

crystal A solidified form of a substance which has atoms and molecules arranged in a symmetrical pattern.

crystal, piezoelectric A crystal, such as quartz, which will generate voltage. Used for precision frequency control of oscillators.

crystal, quartz A piezoelectrial crystal cut from natural quartz.

crystal, X-cut A crystal with its major flat surfaces cut so that they are perpendicular to an electrical (X) axis of the original quartz crystal.

crystal, XY-cut A cut crystal which has characteristics similar to those of the X-cut and the Y-cut crystals.

crystal, Y-cut A crystal with its major flat surfaces cut so that they are perpendicular to a mechanical (Y) axis of the original quartz crystal.

crystal-controlled oscillator Oscillator with a piezoelectric effect crystal coupled to a tuned oscillator circuit in such a way that the crystal pulls the oscillator frequency to its own natural frequency and does not allow frequency drift.

crystal filter A filter which uses piezoelectric crystals to provide its resonant or anti-resonant circuits.

crystal oscillator Oscillator with a piezoelectric crystal as the tuned circuit which provides the resonant frequency.

crystallization In lead cable sheaths, especially those of pure lead laid in roads carrying heavy vehicular traffic, the formation of small crystals which in time produce holes in the sheath and result in moisture entering the cable.

CSELT Italian Telecoms research organization (Centro Studi e Laboratorie Telecomunicazioni).

cupeth Cable sheath made of copper tape under a polyethelene outer sheath.

curb, manhole Steel chimney placed over a manhole opening during adverse weather conditions to stop rain or snow from getting into a manhole while cables are being spliced.

curie point Temperature at which piezoelectric properties cease.

current General word for the transfer of electricity, or the movement of electrons or holes.

current, alternating An electric current which is constantly varying in amplitude and periodically reversing direction; ac is usually sinusoidal.

current, average Arithmetic mean of instantaneous values of current, averaged over one complete half-cycle.

current, charging 1. Current which flows in to charge a capacitor when it is first connected to a source of electric potential. 2. Current flowing through a secondary cell or accumulator which results in chemical changes which enable the cell to operate as a source of electric power.

current, direct Electric current which flows in one direction.

current, eddy A wasteful current which flows in cores of transformers and produces heat. Largely eliminated by use of laminations and of dust-type cores.

current, effective The ac which will produce the same effective heat in a resistor as is produced by dc. If the ac is sinusoidal its effective current value is 0.707 times the peak ac value.

current, fault Currents which flow between conductors or to ground during fault conditions.

current, ground fault A fault current which flows to ground.

current, ground return Current returning through the earth. Examples are trolley bus/tramway currents, and unbalanced power route currents.

current, lagging In an inductive circuit an alternating current lags behind the voltage which produces it.

current, leading In a capacitive circuit an alternating current leads the voltage which produces it.

current, longitudinal Current flowing in the same direction along both wires of a two-wire circuit.

current, magnetizing Current in a transformer primary winding which is just sufficient to magnetize the core and offset iron losses.

current, neutral Current which flows in the neutral conductor of an unbalanced polyphase power circuit, which if correctly balanced would carry no net current at all.

current, peak The maximum value reached by a varying current during one cycle.

current, pick-up Minimum current at which a relay just begins to operate.

current, plate The anode current of an electron tube.

current, pulsating Current of constant direction but varying amplitude.

current, residual Vector sum of currents in the phase wires of an unbalanced polyphase power circuit.

current, sneak An unwanted, small but steady current which has found its way into a telephone conductor system. MDFs are sometimes equipped with protective devices to guard against sneak currents.

current, space Total current flowing through an electron tube.

current amplifier Low output impedance amplifier capable of providing high current output.

current transformer A transformer type of instrument in which the primary carries the current to be measured and the secondary is in series with a low-current ammeter. Measures high values of alternating current.

cursor Transparent slide or sheet used with radar displays or with calculating slide rules to give base position or location; also used with computer or word processor input devices on a video display unit to indicate where the next keyboard character will be entered.

curtain, antenna Antenna whose radiating elements are mounted in the same vertical plane.

curvature loss See macrobend loss.

curve, gas pressure Diagrammatic plot of pressure readings at different points in a pressurized cable system. Points of low pressure indicate possible cable sheath faults.

custom calling service Provision of enhanced services to small business users and residential telephone lines.

custom card services Special AT&T service which provides services on production of credit card, including ability to charge calls to a third telephone number (e.g., a customer's office line).

custom chip An integrated circuit chip designed for a specific task.

customer access line charge (CALC) Fee to be paid by customers for access to a telephone company system.

Customer Information Control System (CICS) IBM facility enabling transactions at remote terminals to be processed centrally.

customer premises equipment (CPE) All telecom equipment located at a customer's premises (except pay phones).

customer-recorded information service This service gives customers the ability to distribute information transmitted from recording equipment to calling subscribers.

customer service unit Terminal which performs signal shaping and loop-back testing for AT&T's Dataphone Digital Service (DDS).

cut Orientation of a crystal with respect to its electrical and mechanical axes. Of a circuit: to open or disconnect.

cut, ring Abrasion fault on an aerial cable due to ring supporting the cable from a messenger strand.

cutback technique Method of measuring the attenuation in an optic fiber by cutting it about one meter from the transmitter, measuring the power at the cut-point without changing the launching conditions, rejointing the fiber, then measuring the received power at the output end.

cut-in Of a switch: the way in which the wipers of an electromechanical switch enter a bank of contacts. Of operating: to throw a key to monitor a call and assist the parties concerned.

cut-off Of an audio transmission circuit: frequency at which the loss is 10db worse than that at 1000 Hz. Of a connection: the premature breaking-down of a call.

cutoff, waveguide Lowest radio frequency that can effectively be carried by a waveguide.

cut-off frequency 1. The frequency above or below which the output current in a circuit, such as a line or a filter, is reduced to a specified level. 2. The frequency below which a radio wave fails to penetrate a layer of the ionosphere at the angle of incidence required for transmission between two specified points by reflection from the layer.

cutoff relay A relay used to disconnect a subscriber's loop from a line circuit.

cutoff wavelength The shortest wavelength at which only the fundamental mode of an optical wavelength is capable of propagation.

cutout Key on a manual operating position which disconnects the operator's circuit from all cord circuits.

cut-out, open space An over-voltage protector, typically two carbon blocks separated by a small air gap, as used on MDFs to protect central office equipment against lightning.

cut-over To transfer a facility from one circuit or system to another; to bring a new central office into service.

cuts off Fails to provide satisfactory transmission at a particular cut-off frequency.

cuts out Intermittent fault condition.

cutter, cable Strengthened shears for cutting cables.

cutter, strand A long-handled bolt cutter, used for cutting steel messenger strand.

CV curve Graph of total capacitance against gate voltage for a semiconductor.

cybernetics Relationship between animal and machine behavior and the study of control system theory.

cycle One complete sequence of an alternating characteristic.

cycles per second Frequency; SI unit is the Hertz, one cycle per second.

cyclic distortion   In telegraphy, distortion which is neither characteristic, biased, nor fortuitous, and which generally has a periodic character.

cyclic redundancy check (CRC)   Method of error detection. Typically, a 16-bit check character is added to each data block based on a several-times-repeated examination of each information bit. This provides efficient error detection even on noisy circuits. Syn.: cyclic sum check

cyclic sum check   Syn.: Cyclic redundancy check

cycling   Not steady on a specified level, but searching between limits in an automatically controlled system.

cylinder, propane   Steel cylinder, available in various sizes, for storage of propane or liquid petroleum (LP) gas (called LPG in Britain). Used by cable splicers.

cymomotive force   The product formed by multiplying the electric field-strength at a given point in space, due to a transmitting station, by the distance of that point from the antenna. This distance must be sufficient for the reactive components of the field to be negligible; moreover, the finite conductivity of the ground is supposed to have no effect on propagation. The cymomotive force (cmf) is a vector; when necessary it may be expressed in terms of components along axes perpendicular to the direction of propagation. The cmf is expressed in volts; it corresponds numerically to the field-strength in mV/m at a distance of 1 km.

D clip   Soft aluminum clip used with drop wires.

D flip-flop   A type of bistable with a single trigger input terminal, the D input.

D layer   A layer in the ionosphere, about 50-90 km high, which reflects low frequency radio communication signals.

D*   (Pronounced D-star.) In optoelectronics: A figure of merit often used to characterize detector performance, defined as the reciprocal of noise equivalent power, normalized to unit area and unit bandwidth.

$$D^* = \sqrt{A(\Delta f)} \, / \, NEP,$$

where A is the area of the photosensitive region of the detector and Δf is the bandwidth of the modulation frequency of the incident radiation. Syn.: specific detectivity

daisy chain   A method of sending data signals along a bus. Any devices which do not need the signal pass it on until it reaches the device which does want it, this device then breaks the daisy-chained signal continuity.

daisy wheel   Type of wheel used in fast printers; alphabetic type is on radial petals, the wheel spins and the petals impact paper to print the message.

dam   A blockage inside a cable, usually a plastic or wax compound which fills the gaps between insulated wires so that gas cannot pass the blocked point.

damped oscillation   Oscillation with progressive diminution of amplitude with time.

damper, vibration   Device attached to an overhead line wire or messenger strand to prevent vibration and "dancing."

damping   1. The progressive diminution with time of certain quantities characteristic of a phenomenon. 2. The progressive decay with time in the amplitude of the free oscillations in a circuit.

damping, magnetic   The slowing down of movement by the interaction of magnetic fields.

dancing, cable   The vibration of an aerial cable in windy locations; slow oscillation at resonant frequency, which usually depends on span length.

daraf   The unit of electrical elastance which is equivalent to the reciprocal of the capacitance in farads. Daraf is farad spelled backwards.

dark current   In optoelectronics: the external current, under specified biasing conditions, that flows in photosensitive detectors when there is no incident radiation.

dark current noise   Noise component in a photodiode resulting from the shot noise of the current flowing in the unilluminated photodiode.

dart leader stroke   The initial discharge which largely determines the path to be taken by a lightning flash. A dart leader develops continuously, a stepped leader develops in short steps. Both are followed by the high-current discharge flash, usually in the reverse direction to the leader stroke.

dashpot   Device with a cylinder and loose-fitting piston used to slow down movement of an armature or relay.

data   Any representation, such as characters or analog quantities, to which meaning is or might be assigned. Data is the plural of datum, the Latin word which means given information or basis for calculations.

data, high-speed   Data transmission which cannot normally be carried by a voice-frequency analog circuit, typically above 10,000 bits per second.

data, low-speed   Data transmission at the bottom end of the voice-frequency band, typically 300 bits per second or less.

data, medium-speed   Data transmission at rates which may be carried (using a modem) by voice-frequency analog circuit, typically between 300 and 2400 or 4800 bits per second.

data, raw Data as received, which may not be in suitable format for further processing.

data bank A comprehensive collection of data.

data block A group of bits transmitted as a unit over which a parity check procedure is applied for error control purposes.

data bridging capability Service which enables digital data to be transmitted from one location to several others.

data carrier failure detector A monitoring unit designed to indicate that the level of the data carrier on a voice-frequency channel is below the minimum sensitivity of the receiver.

data channel A unidirectional transmission path for data, with transmission terminal equipment at both ends.

data channel, analog A one-way path for data signals which includes a voice-frequency channel and an associated data modulator and demodulator.

data channel, digital A one-way path for data signals which includes a digital channel and associated interface adaptors at each end.

data channel failure detector A data carrier failure detector or loss of frame alignment detector.

data channel signaling conditions Interexchange data channel conditions employed in the call set-up and clear-down procedures.

data circuit A means of two-way data transmission between two points comprising associated transmit and receive channels. (a) Between data switching exchanges, the data circuit may or may not include data circuit-terminating equipment, depending on the type of interface used at the data switching exchange. (b) Between the data terminal installation and a data switching exchange and/or concentrator, the data circuit includes the data circuit-terminating equipment at the data terminal installation end and may also include equipment similar to a data circuit-terminating equipment at the data exchange or concentrator location. (c) Either physical or virtual data circuits may be established. (In North America, a two-way transmission is sometimes called a channel, rather than a circuit).

data circuit connection The interconnection of a number of links or trunks, on a tandem basis, by means of switching equipment to enable data transmission to take place among data terminal equipment.

data circuit-terminating equipment (DCE) The interfacing equipment sometimes required to couple the data terminal equipment (DTE) into a transmission circuit or channel and from a transmission circuit or channel into the DTE. In a CCITT X.25 (packet switching) connection the network access and packet switching node is viewed as the DCE. Syn.: data communications equipment; data set

data collection 1. A facility for gathering small quantities of data from a nominated group of addresses, assembling them within the network into a single message for delivery to another nominated address. 2. The act of bringing data from one or more points to a central point.

data common channel signaling units Each signal unit carrying signaling information contains, apart from transfer control information, a signaling information field, a signal unit indicator and a service indication. These are the elements of a signal message that are common to all user parts. A signal message that is carried by one signal unit only is called a one-unit message and the corresponding signal unit is called a lone signal unit (LSU). A signal message that is carried by two or more signal units is called a multiunit message (MUM).

data communication The movement of encoded information by means of electric transmission systems via one or more data links according to a protocol.

data communication control character A functional character intended to control or facilitate transmission of information over communication networks.

data communication control procedure A means to control the orderly communication of information among stations in a data communication network.

data communications exchange Hardware which enables data to be received, switched and transmitted in real time simultaneously to several different destinations over several different routes and channels (even when some of these are operating in different modes). Also called a data switch.

data compaction Pertaining to the reduction of space, bandwidth, cost, and time for the generation, transmission, and storage of data by employing techniques designed to eliminate repetition, remove irrelevancy, and employ special coding.

data compression 1. A method of increasing the amount of data that can be stored in a given space or contained in a given message length. 2. A method of reducing the amount of storage space required to store a given amount of data or reducing the length of message required to transfer a given amount of information.

data concentrator A unit that permits a common transmission medium to serve more data sources than there are channels currently available within the medium.

data conferencing repeater A device that enables a group of users to operate such that if any one user transmits a message it will be received by all others in the group. Syn.: technical control hubbing repeater

data connection The interconnection of a number of data circuits on a tandem basis by means of switching equipment to enable data transmission to take place between data terminal equipments. (a) Where one or more of the data circuits which are interconnected is a virtual data circuit, the overall connection is known as a virtual data connection. (b) The overall connection includes the data circuit terminating equipment at the respective data terminal installation locations.

data coupler Device which interfaces data terminal with communications network. A coupler includes any necessary protective devices.

data-grade media Special type of twisted-pair, copper-wire cable designed to carry high frequency signals satisfactorily — typically 4 Mbit/s for distances up to 1 km between repeaters. DGM-1 is usually 22 AWG with attenuation 21 dB/km at 4 MHz, DGM-2 is usually 26 AWG, 33 dB/km at 4 MHz.

data inquiry-voice answer (DIVA) System by which a computer, using pre-recorded or synthesized voice signals, provides verbal answers to inquiries.

data link An ensemble of terminal installations and the interconnecting network operating in a particular mode that permits information to be exchanged between terminal installations. A bidirectional transmission path for data comprising two data channels in opposite directions which operate together at the same data rate.

data link escape character A transmission control character that changes the meaning of a limited number of contiguously following characters or coded representations, and is used exclusively to provide supplementary transmission control characters.

data link layer Layer in the Open Systems Interconnection (OSI) protocol that establishes, maintains and releases data connections between adjoining nodes in a network.

data logger Digital circuit which converts an analog output (e.g., from a transducer such as a strain gauge or potentiometer) into digital form.

data logging Recording of data about events that occur in time sequence.

data mode The state of data circuit-terminating equipment when connected to a communication channel but not in a talk or dial mode.

data network 1. The assembly of units that establishes data circuits between data terminal equipment units. 2. The interconnection of a number of locations using telecommunications links for the purpose of transmitting or receiving data.

data origination The earliest stage at which source material is first put into machine-readable form or directly into electric signals.

data over voice Procedure permitting data services to share local cabling with telephones: filters are used to separate frequency bands, frequencies above normal commercial speech being used for tdm (analog) data channels.

data phase That phase of a data call when data signals may be transferred between data terminal equipments which are interconnected via the network.

Data-Phone Bell System trade name for a system which permits calls to be established over normal switched telephone circuits using modems at both ends to enable data terminal equipment to work over the circuit, at speeds up to 2 kb/s.

Data-Plus A programming language which uses English words and syntax.

data processing The automatic organization of data; information is received, stored, acted upon mathematically, and transmitted as in a computer system; the form, location, meaning, appearance or any other aspect of the data may be changed.

data rate, high speed Data transmission at speeds faster than 2400 bauds.

data rate, low speed Data transmission at speeds up to 150 bauds.

data rate, medium speed Data transmission at speeds between 150 and 2400 bauds.

data security The protection of data against unauthorized disclosure, transfer, modifications, or destruction, whether accidental or intentional.

data set A device that converts the signals of a business machine to signals that are suitable for transmission over telecommunication circuits and vice versa. It may also perform other related functions. The Bell System name for modem.

data signaling rate A measure of signaling speed given by:

$$DSR = \sum_{i=1}^{m} (1/T_i) \log_2 n_i$$

where DSR is the data signaling rate, m is the number of parallel channels, T is the minimum interval for the i-th channel expressed in seconds, n is the number of significant conditions of the modulation in the i-th channel. Data signaling rate is expressed in bits per second (b/s).

data signaling rate transparency A network parameter which enables the transfer of data between users without placing any restrictions, within certain limits, on the data signaling rate used.

data sink A store in which data may be temporarily retained.

data skew Distortion of a signal due to lack of synchronization between the source of the signal and the receiver.

data source A device which generates data signals destined for data sinks.

data stream A sequence of binary digits used to represent information for transmission.

data subscriber loop carrier A pair-gain system used for local distribution which enables local loops to be used for both voice and data.

data subscriber terminal equipment A general purpose AUTODIN terminal device consisting of all necessary equipment: (a) to provide AUTODIN interface functions; (b) to perform code conversions; and (c) to transform punched card messages, punched paper tape, or magnetic tape to electrical signals for transmission, and the reverse of this process.

data switching exchange Equipment installed at a single location to switch data traffic. A data switching exchange may provide only circuit switching, only packet switching, or both.

data terminal equipment 1. Equipment consisting of digital end instruments that convert user information into data signals for transmission, or reconvert the received data signals into user information. 2. The functional unit of a data station that serves as a data source or a data sink and provides for the data communication control function to be performed in accordance with link protocol. The DTE may consist of a single piece of equipment which provides all the required functions necessary, or it

may be an interconnected subsystem of multiple pieces of equipment, which perform all the required functions.

data terminal installation Installation comprising: the data terminal equipment, the data circuit-terminating equipment, and any intermediate equipment.

data transfer rate The number of bits, characters or blocks per unit time passing between corresponding equipments in a data transmission system.

data transfer requested A call control signal transmitted by the data circuit terminating equipment to the data terminal equipment (DTE) in leased circuit service to indicate that the distant DTE is wishing to exchange data.

data transfer time The time that elapses between the initial offering of a unit of user data to a network by transmitting data terminal equipment and the complete delivery of that unit to receiving data terminal equipment.

data transmission The sending of data from one place to another by means of signals over a channel.

data transmission circuit The transmission media and intervening equipment involved in the transfer of data between data terminal equipment.

data under voice Bell System service providing wideband digital signals (up to 56 kb/s) to be carried on existing microwave radio systems, in addition to the usual multiplexed voice signals.

database A collection of information of technical or commercial importance to an organization. Database information held in a computer-accessed memory is usually subdivided into pages, with each page accessable to all users unless it belongs to a closed user group. The word does, however, cover catalogs on a library shelf as well as facts stored in computer memories.

database access service A service which provides remote data terminal equipment in one country with access to host computers in another country for the purpose of accessing filed information housed in a database or of performing remote computing using facilities provided by that host.

datafax Public facsimile service between subscribers' stations via a data network.

datafile A collection of related records held in a database.

datagram A message in a packet-switching system which is handled as a self-contained entity, using a single packet.

Dataphone Digital Service (DDS) AT&T service with signals transmitted from the customer's premises in digital form, i.e., without needing a modem.

date-time group Date and time of origin of a message, in a six figure plus one letter group. The first two figures are the date in the month, the next four are the time (by the 24-hour clock), and the letter is time zone. In some organizations, two figures for the month are included, but care must be taken since in Britain the 3rd of February would usually be 3/2 or 0302, while in North America it would be ⅔ or 0203.

datel General title for data transmission facilities offered by British Telecom.

dating format The time of an event on the Universal Time System given in the following sequence: hours, day, month, year; e.g., 1445 UT, 23 August 1971. The hour is designated for a 24-hour system.

datum line For a cableship, the axis of the cable from the point it enters the linear cable engine, through the engine, and tangent to the stern sheave or chute.

daughter board Small printed circuit board mounted on a standard-sized (or mother) board. This often provides a special service or facility not always required.

day to busy hour ratio The ratio of the 24-hour day traffic volume to the busy hour traffic volume, sometimes abbreviated Day/BH; the reciprocal busy hour to day ratio, or BH/day, also is used.

dB Abbreviation for decibel. The standard unit for expressing transmission gain or loss and relative power ratios. The decibel is one tenth the size of a Bel, which is too large a unit for convenient use. Both units are expressed in terms of the logarithm to the base 10 of a power ratio, the decibel formula being:
$$dB = 10 \log_{10}(P_1/P_2)$$

dBa, dBrn adjusted Weighted noise power, in dB referred to 3.16 picowatts (—85 dBm), which is 0 dBa. Use of F1A-line or HA1-receiver weighting shall be indicated in parentheses as required.

dBa (F1A) Weighted noise power in dBa, measured by a noise measuring set with F1A-line weighting.

dBa (HA1) Weighted noise power in dBa, measured across the receiver of a Western Electric 302-type or similar subset, by a noise measuring set with HA1-receiver weighting.

dBa0 Noise power in dBa referred to or measured at a zero transmission level point; also called a point of zero relative transmission level.

dBm A dB referred to one milliwatt; employed in communication work as a measure of absolute power values. Zero dBm equals one milliwatt.

dBm0 Noise power in dBm referred to or measured at a zero transmission level point (0TLP). The 0TLP is also called a point of zero relative transmission level (0 dBr0).

dBm0p Noise power in dBm0, measured by a psophometer or noise measuring set having psophometric weighting. See also: noise

dBm(psoph) A unit of noise power in dBm, measured with psophometric weighting. For conversion to other weighted units:
$$dBm(psoph) = [10 \log_{10}pWp] - 90 = dBa - 84$$

dBr The power difference expressed in dB between any point and a reference point selected as the zero relative transmission level point. Any power expressed in dBr does not specify the absolute power. It is a relative measurement only.

dBrn Decibels above reference noise. Weighted noise power in dB referred to 1.0 picowatt. Thus, 0dBrn = —90 dBm.

dBrn(144 line)   Weighted noise power in dBrn, measured by a noise measuring set with 144-line weighting.

dBrnC   Weighted noise power in dBrn, measured by a noise measuring set with C-message weighting.

dBrnC0   Noise power in dBrnC referred to or measured at a zero transmission level point.

dBrn(f$_1$-f$_2$)   Flat noise power in dBrn, measured over the frequency band between frequencies f$_1$ and f$_2$.

dBW   Decibels referred to one watt.

direct current (dc)   Electric current flowing in one direction.

dc amplifier   Amplifier capable of amplifying dc and slowly varying alternating signals.

dc component   The average value of a varying signal such as the luminance of a TV picture.

dc loop signaling   Signaling on a physical trunk by opening and closing a loop across a metallic pair or by reversing polarity of voltage on the pair.

dc patch bay   A patch bay in which dc circuits are grouped.

dc spark-over voltage   Voltage at which a gas discharge protector sparks over with slowly increasing dc voltage. Indicates the circumstances in which a protector is suitable for use.

dc/dc converter   Unit which converts one dc voltage value to a different one. Much used in transmission equipment, where power is fed to each rack and converted to the different dc voltages needed.

DCE clear indication   For data: a call control signal transmitted by the data circuit terminating equipment (DCE) to indicate that it is clearing the call.

DCE waiting   For data: a call control signal at the data circuit terminating equipment (DCE)/data terminal equipment interface which indicates the DCE is waiting for another event in the call establishment procedure.

DD   Due date. Abbreviation used in Bell System Universal System Service Orders for special services: the date that turn-up, change or denial (in the case of disconnects) of service is required to be provided to the customer.

DDD   See direct distance dialing.

dead   Not connected to any live electrical source.

deadbeat   When the pointer of a meter goes to a position without overshooting.

dead-end   The end of a cable or route at a pole.

dead-end, false   Securing of a cable in both directions as if it were terminating at a pole.

dead-end, preformed   A helically wound steel wire which can be easily put over an aerial cable's messenger strand, but which when tension is applied, grips the strand very firmly, enabling it to be pulled in.

dead front   Equipment manufactured with no exposed high voltage points on its front.

dead man   1. Large log used as anchor for a beach anchor or pole stay. 2. Short pole used as support while erecting a pole before dropping it into its hole.

dead pair   Cable pair which is not connected through to the main distributing frame.

dead sector   In facsimile, the elapsed time between the end of scanning of one line and the start of scanning of the following line.

dead short   A short circuit.

deaf set   An amplifying handset, available for the hard-of-hearing.

debarnacle   To redesign a component to suit normal manufacturing processes by the elimination of manually added-on modifications.

deblocking   The action of making the first and each subsequent logical record of a block of information available for processing.

debugging   The technique of detecting, diagnosing and correcting errors (bugs) which may occur. Bugs can exist both in programs, e.g., failure to understand and describe a process, or in hardware, e.g., in the way a hardwired logic unit has been strapped.

deca-   A prefix to a unit, indicating a multiple of 10. Not now widely used because SI units concentrate on multiples of 10^3.

decay   Reduction in amplitude of a signal on an exponential basis.

deci-   One-tenth. Not an SI recommended prefix.

decibel   See entries under dB.

decimal notation   Normal to-the-base-10 figures. Although computers operate on a binary basis (to the base 2, i.e. either 0 or 1) it is common when normal decimal numbers are to be operated upon (i.e. in most telephone exchanges) for computers to use binary numbers in groups of four, each group of four representing one decimal digit. This system is called binary coded decimal notation.

decimonic frequencies   Party line ringing frequencies used where selective ringing is available; frequencies used are from 20Hz to 60Hz in 10Hz steps.

decision   In specification description language, a decision is an action within a transition which is a question to which the answer can be obtained at that instant and chooses one of several paths to continue the transition.

decision circuit   A circuit which decides the probable value of a signal element.

decision element   A logic circuit which accepts inputs and gives an output, all in binary form.

decision instant of a digital signal   The instant at which a decision is taken by a receiving device as to the probable value of a signal element.

decision table   A chart showing all the possibilities to be considered, and the action to be taken in all circumstances, typically on a yes or no basis.

decision value   For PCM: a reference value defining the boundary between adjacent intervals in quantizing or encoding.

deck   Tape recording or playback device; also a set of data cards as presented to or received from a reader, card puncher or sorter.

decode To recover an original message from a coded form. In telephony the most commonly-encountered decoders receive a PCM input and produce an analog output.

decoder A device for decoding signals. In electromechanical central offices routing information is sometimes advised to marker stages by a decoder stage which analyzes and decodes the digits dialed to obtain the required routing information.

decoder, mobile radio A selective calling decoder which reacts to only one particular incoming call signal.

decoding For PCM: a process in which one of a set of reconstructed samples is generated from the character signal representing a sample.

decompiler, data Tool for dumping the current data from a working exchange, out-putting it in man-readable form and also in input format for a data compile.

decoupling The breaking down, reduction or removal of undesired coupling between two circuits or stages.

decrement The opposite of increment. In a damped oscillation each oscillation is of a lower amplitude than the one before; this decrement is expressed as the ratio of two successive peak values. Logarithmic decrement is the natural logarithm of the ratio of peak values (in the same direction).

decryption Returning an encrypted message to its original clear text.

dedicated access lines A group of leased lines which interconnect a switching system to a dedicated customer. Dedicated access lines may be connected to a customer's telephone, key telephone system, or PBX.

dedicated channel A channel leased by a private concern from a common carrier which is at the exclusive disposal of the leasor. Also called leased line and private line.

dedicated outside plant System by which external plant facilities (cable pairs in the central office distribution network) are semi-permanently allotted and connected through from the main distributing frame to residential or office locations.

dedicated service Service allocated to and used only by a particular user.

de-emphasis About 6 db per octave of extra loss added at the receiving end of a radio communication link to offset pre-emphasis of high frequencies by the transmitter.

de-energize To switch off.

de-energized A system from which sources of power have been disconnected.

deep-sea cable Armorless or lightweight cable used in deep water where physical danger to the cable is considered to be almost non-existent.

deep space Space at distances from the Earth greater than or approximately equal to the distance between the Earth and the Moon.

defect An error made during initial planning which is normally detected and corrected during the development phase. Note that a "fault" is an error which occurs in an in-service service system. A system may be designed to overcome or recover from faults; it cannot at present be designed to overcome defects.

deflector, coin Mechanical shield in a type of coin telephone which stops coins from sticking in the mechanism.

degenerate waveguide modes In optoelectronics: waveguide modes that have either the same phase or group velocity.

degeneration Negative feedback.

deglitcher A special circuit used to limit the duration of switching transients in digital converters.

degradation, ultraviolet The breakdown of a material's insulating qualities by exposure to bright sunlight.

degraded service state A telecommunication service condition defined to exist during any period over which the established limiting value for at least one of the performance parameters is worse than its specified threshold value.

degree, electrical One complete ac cycle is divided into 360 electrical degrees.

degree of coherence A measure of the coherence of a light source; the magnitude of the degree of coherence may be shown to be equal to the visibility of the fringes of a two-beam interference experiment, where:

$$V = \frac{I_{max} - I_{min}}{I_{max} + I_{min}}$$

I_{max} is the intensity at a maximum of the interference pattern, and I_{min} is the intensity at a minimum. Light is considered highly coherent when the degree of coherence exceeds 0.88, partially coherent for values less than 0.88, and incoherent for very small values.

degree of individual distortion of a particular significant instant As applied to a modulation or a demodulation, the ratio to the unit interval of the maximum displacement, expressed algebraically, of this significant instant from an ideal instant. This displacement is considered positive when a significant instant occurs after the ideal instant. The degree of individual distortion is usually expressed as a percentage.

degree of isochronous distortion The ratio to the unit interval of the maximum measured difference, irrespective of sign, between the actual and the theoretical intervals separating any two significant instants of modulation (or demodulation), these instants being not necessarily consecutive (usually expressed as a percentage).

degree of start-stop distortion 1. The ratio to the unit interval of the maximum measured difference, irrespective of sign, between the actual and theoretical intervals separating any significant instant of modulation (or of demodulation) from the significant instant of the start element immediately preceding it. 2. The highest absolute value of individual distortion affecting the significant instants of a start-stop modulation. The degree of distortion of a start-stop modulation (or demodulation) is usually expressed as a percentage.

deka Ten times. Not an SI recommended prefix.

delay The amount of time by which a signal is delayed or an event retarded.

delay, absolute Total time for a signal to pass through a network or channel. Where delay varies with frequency, absolute delay usually means the least delay found in the channel passband.

delay, answering Time interval between the setting up of an end-to-end connection between the calling and called stations, and the detection of an answer signal.

delay, average The total waiting time of all bids divided by the total number of bids, including those not delayed. The mean waiting time.

delay, call The delay in obtaining access to an item of common equipment in a central office. North American acceptable standard is: not more than 1 ½% of all calls should be so delayed by 3 seconds during the busy hour.

delay, differential Difference in delay of two specified frequencies of interest in a band or channel.

delay, envelope Time delay of the modulation envelope of a signal in passing through a channel or network.

delay, exchange call set-up The interval from the instant when the address information for a call is received at the incoming side of the exchange, to the instant when the seizing signal or the corresponding address information is sent to the subsequent exchange, or the calling signal is sent to the required terminating point.

delay, group Same as envelope delay.

delay, hangover In an echo suppressor, the time interval, proportional to circuit delay, that a directional transmission is suppressed after the control signal ends.

delay, incoming response Interval from the instant an incoming circuit seizure signal is received until the exchange is able to receive signaling information or a proceed to send signal is sent backwards by the exchange.

delay, phase Delay encountered by a single unmodulated frequency in passing through a network or channel.

delay, post dialing Time interval between the end of dialing by the subscriber and the reception of the appropriate tone or recorded announcement, or the abandonment of the call without tone.

delay, relative Difference between maximum and minimum delays occurring in a channel or band of frequencies.

delay, through connection The interval from the instant when the information required to set up a through connection in an exchange is avaiiable for processing, to the instant that the switching network through connection is established between the inlet and outlet.

delay call Call set up by an operator who establishes links to both parties concerned. Once the call is established, supervision follows normal subscriber-originated call procedures.

delay circuit A circuit designed to delay a signal passing through it.

delay dial A signaling protocol which responds to a request for service and indicates a "start dial" to the sending switch. The receiving switch returns an off-hook signal immediately in response to the incoming request for service and returns an on-hook signal when it is prepared to receive the incoming digits.

delay distortion Distortion which results when some portions of a wave train are delayed more than others, i.e., because of non-uniform speed of transmission of the various frequency components of the signal. Envelope delay distortion is usually considered to be the most important form of delay distortion.

delay equalizer Network which makes phase delay or envelope delay substantially constant over a desired frequency range.

delay line Cable with low velocity of transmission.

delay system A system in which bids that cannot be served immediately are permitted to wait until service can begin.

delay time The delay to which an international call is subject at the controlling exchange.

delay traffic table Traffic tables used by telephone companies; given the acceptable delay in seizing an idle trunk, the holding time, and the load in CCS, these tables give the number of trunks which should be in the group in order to meet required standards.

delete character A control character usually used to delete errors in a punched tape.

deletion, digit A procedure through which a central office deletes unneeded digits dialed in to it. A digit deletion stage is able to delete up to six digits (e.g. the NPA code and the central office code) if for example there are direct tie lines to the wanted central office.

delimiter Character or word which indicates the beginning or end of a portion of a program in software.

delivery (Message Handling Systems) The interaction by which the Message Transfer Agent transfers to a recipient User Agent the content of a message plus the delivery envelope.

delivery confirmation In data: a user facility which provides information to the sending data terminating equipment that a given packet has been delivered to the nominated address.

delivery envelope (Message Handling Systems) The envelope which contains the information related to the delivery of the message.

Delphi technique A forecasting method for use when there is no directly relevant historical data available. It is an intuitive forecast made after combining the views of many experts in the particular field.

delta connection A method to join together a three-phase power supply, with each phase across a different pair of the three wires used.

delta matched A radio-frequency transmission line fed with transmitter power through a delta-matched transformer.

delta modulation A digital transmission method. An analog signal is scanned, typically 32,000 times per second, to see if its instantaneous value is less or greater than it was at the previous scan, and a digital pulse sent to line to indicate this. Although delta mod samples speech waveforms more rapidly than PCM (PCM only 8000 times per second), each DM sample is represented by a single bit, 0 or 1, whereas each PCM sample needs eight bits to indicate the sample level. DM's advantages over PCM are: (a) it provides greater channel capacity for a given bit rate, resulting in higher pair gain and lower per-channel cost; and (b) It does not inherently require synchronization as PCM does, and is more tolerant of system noise. Against Delta Mod is the fact that it took very many years to reach even the limited international agreement which has been reached over PCM, (and the American 24 channel PCM will not interface directly with the European 30 channel PCM) and there is no sign of any agreement at all yet on Delta Modulation although one form or other, possibly Continuously Variable Slope Delta Modulation, may one day be specified internationally.

demand, power system Average total load of a power system, in KW or KVA.

demand assignment An operational technique whereby various users share a satellite capacity on a real-time demand basis. This service is analogous in many ways to an ordinary telephone switching network that provides common trunking for many subscribers through a limited size trunk group on a demand basis. With modern digital telephone networks, facilities are sometimes assigned on a dynamic basis. See dynamic non-hierarchic routing.

demand assignment multiple access (DAMA) When traffic is not sufficient to justify circuits being put permanently through from one exchange to another, satellite channels are assigned for use on a "demand" basis. This means that the two ground stations concerned are instructed to use particular channels for transmit and receive while other ground stations using the same satellite transponder are kept off these channels while the call is in progress.

demand assignment signaling and switching unit (DASS) Part of the equipment in a ground station working to an Intelsat satellite and using SPADE (single channel per carrier PCM multiple access demand assignment equipment).

demand factor The ratio of the maximum demand on a power system to the total connected load of the system.

demand load 1. General: the total power required by a facility. 2. Communications center: the power required by all automatic switching, synchronous and terminal equipment, operated simultaneously online or standby, control and keying equipment, plus lighting, ventilation and air conditioning equipment required to maintain full continuity of communications. 3. Non-technical: the power required for ventilating equipment, shop lighting and other support items which may be operated simultaneously with the technical load. 4. Operational: the sum of the technical demand and non-technical demand loads of an operating facility. 5. Receiver facility: The power required for all receivers and auxiliary equipment which may be operated on prime or spare antennas simultaneously, those in standby condition, multi-couplers, control and keying equipment, plus lighting, ventilation and air conditioning equipment required for full continuity of communications. 6. Transmitter facility: the power required for all transmitters and auxiliary equipment which may be operated on prime or spare antennas or dummy loads simultaneously, those in standby condition, control and keying equipment plus lighting, ventilation and air conditioning equipment required for full continuity of communications.

demand operating The action taken by an operator after a request has been recorded in an outgoing international exchange to immediately attempt to set up the call.

demand operating, manual There are two operating methods: (a) indirect manual demand operating, where the operator at the incoming international exchange always acts as an interpreter between the operator in the outgoing international exchange and the called party; and (b) direct manual demand operating. In this method, the operator in the outgoing international exchange speaks with the called party direct.

demand operating, semi-automatic The operator in the outgoing international exchange controls the automatic switching operations to obtain either the called station, or an operator in the incoming or transit international exchange (or an operator in a manual exchange in the country of destination).

demarcation strip A terminal acting as a physical interface between equipment which are the responsibilities of different carriers.

democratic (mutually synchronized) network All clocks in this network are of equal status and exert equal amounts of control on the others, the network operating frequency being the mean of the natural (uncontrolled) frequencies of the whole population of clocks.

demodulation Process by which a modulated wave input produces the original modulating wave signal as output, i.e., the extraction of transmitted information from a modulated carrier signal.

demodulation, phase-lock loop Demodulation process used in ground stations working to communications satellites, capable of detecting and demodulating very weak received signals.

demodulator A device that demodulates.

demodulator, SCA Subsidiary carrier authorization demodulator. Used to demodulate the special extra-charge subcarrier program broadcast by some FM radio stations in addition to their normal programs.

demodulator, television Circuitry of a TV receiver which demodulates a received signal and produces a video signal and a sound signal.

demonstration jacks Jacks on a manual operating position into which calling lines may be connected to enable customers to hear the various information tones used in the telephone network.

demonstration test Test to show a particular audience, normally including customer representatives, that a particular system or subsystem is performing satisfactorily.

De Morgan's Rules Rules used in Boolean algebra: NORing two variables is equivalent to ANDing their inverses, and NANDing two variables is equivalent to ORing their inverses.

demultiplexer Device which accepts as input a broadband carrier signal and produces as output a group of separate audio frequency channels.

denied usage channel evaluator (DUCE) DUCE procedure can construct an interference-free frequency list, given appropriate input constraints. Clearly the more radiocommunications systems already working, the greater the number of "denied channels." If you need to reduce the number of denied channels (e.g. in order to open new links) one way would be to use better receivers. (This is a grossly oversimplified description of one feature only of a highly complex procedure, involving adjacent signals, spurious responses, transmitter noise, spurious emissions, receiver intermodulation, transmitter intermodulation, etc.)

density In a facsimile system, a measure of the light transmission or reflection properties of an area. It is expressed by the common logarithm of the ratio of incident to transmitted or reflected light flux.

density, power Value of the Poynting vector (the local surface density of energy flow per unit time) at a point in space.

density packing Number of units of useful information within given linear dimension, usually in units per inch.

departure, frequency Variation from its assigned or center frequency of a carrier.

departure angle The angle between the axis of the main lobe of an antenna pattern and the horizontal plane at the transmitting antenna.

depletion layer Sensitive zone at the junctions on n-type and p-type semiconductors in which there are no current carriers until the device is biased or operated. Also called depletion region.

depolarizer Chemical included in primary cells to react with any products produced by the main action of the cell which might build up into an insulating or polarizing layer around an electrode.

deposit 1. To preserve information in memory. 2. Action of placing coins or tokens into a coin telephone.

depreciation 1. A method of spreading the cost of assets over their useful life so that stable and realistic accounting and pricing policies may be followed. 2. A method of providing a sum that needs to be set aside by systematic allocation to replace assets at the end of their operating life. 3. The loss in service value, not replaced by maintenance, due to the consumption or prospective retirement of the plant.

depth profile The graphic description of the water depth along a sea cable route where the abscissa represents the distance along the route and the ordinate the water depth, usually at an exaggerated scale.

depth recorder The instrument which makes a continuous plot of water depth versus time. Also see echo-sounder.

derate To use a device or component at a lower current, voltage or power level than it can handle, in order to give longer life or reduce occurrence of stress-related failures.

derating factor, wire Factor giving the reduced current carrying capacity of a conductor when used in harsher environmental conditions than normal.

derivation equipment, channel Device which enables several narrow bandwidth channels to be obtained from one wider band.

derive To produce a channel by time or frequency division of a transmission facility.

derived filter Filter designed as a development of a simple prototype filter but with sharper attenuation-frequency characteristics; an m-derived filter.

derrick, gas cylinder Small, truck-mounted hoist used to lift heavy butane gas tanks into manholes for use by cable splicers.

derrick, pole setting Two- or three-pole derrick used to hoist poles into their holes.

descrambler Device or signal used to unscramble TV reception from satellite provided by subscription-supported services (scrambling prevents unauthorized reception).

description In specification description language: the implementation of the requirements of a system. Descriptions consist of general parameters of the system as implemented and the function description (FD) of its actual behavior.

descriptive name In Message Handling Systems: a name which denotes exactly one user in the system.

desensitization, receiver Situation in which a radio receiver receives a very strong interfering signal and its automatic gain control circuits react to reduce the gain not only for unwanted signals but also for wanted ones, with the result that a weak wanted signal may be lost completely.

deserializer See serial-to-parallel converter.

desiccant A drying agent; a chemical which absorbs water and is used for drying out cable splices or sensitive equipment.

design objective A desired electrical or mechanical performance characteristic for communication circuits and equipment which is based on engineering judgment but, for a number of reasons, is not considered feasible to establish as a system standard at the time the standard is written.

designated carrier An inter-exchange carrier selected by a telephone customer (in the U.S.) to be his first-choice carrier for inter-LATA calls.

designation strip Strip, usually plastic, mounted on manual operating positions and labeled to show circuit names/numbers.

desk, special service Manual operating positions with key-ended terminations, for use as emergency or intercept positions.

desk stand Early model telephone instrument now made as a decorator phone; a candlestick telephone.

despotic (synchronized) network A network in which a complete master-and-slave relationship exists: the master clock has full power of control over all other clocks.

despun antenna Antenna on a satellite which is made to rotate at a rate equal and opposite to the rate at which the satellite itself is spinning for its own stabilization, so that the directional antenna remains pointing at the same point on earth or in space.

destination address Data, common channel signaling: information sent in the forward direction consisting of a number of address signals indicating the complete address of the called customer.

destination code The complete ten-digit number (three-digit NPA code + three-digit CO code + four-digit station number) which pinpoints the location of a telephone in North America. Also called a telephone address.

destination code cancel (DCC) A network management control which permits blocking a set percentage of traffic into a congested area.

destination-code indicator Data, common channel signaling: information sent in the forward direction indicating whether or not the destination code is included in the destination address.

destination field The field in a message header (e.g., in packet switching) that contains the address of the station to which the message is addressed.

destination point A signaling point to which a message is destined.

destination point code A part of the label in a signaling message which uniquely identifies, in a signaling network, the destination point of the message.

destination user A user to whom the source user information is to be delivered during a particular information transfer transaction.

destruction characteristic Of a gas discharge protector, indicates the relationship between the value of the discharge current and its time of flow until the protector is mechanically destroyed. It is obtained from the average of measurements on several protectors.

de-stuffing The controlled deletion of digits from a stuffed digital signal to recover the original signal prior to stuffing.

DETAB A programming language based on COBOL, but using decision tables.

detached contact drawing Circuit drawing in which contacts operated by a relay are not drawn immediately next to the relay coil circuit itself but are located at positions convenient for their own circuits.

detect The rectification of a modulated wave in order to recover the modulating signal.

detection Rectification process which results in the modulating signal being separated out from a modulated wave.

detectivity The reciprocal of noise equivalent power.

detector A demodulation stage in a radio receiver.

detector, ANI Automatic number identification equipment which enables a calling line's number (in an automatic telephone exchange) to be identified for the preparation of bills.

detector, carbon monoxide Small glass ampoule which is broken open and held in a manhole to check for the presence of poisonous carbon monoxide. Also called a gas detector.

detector, coin level Mechanical detector placed in some pay station cash boxes which gives a remote alarm when the level of coins reaches a predetermined level.

detector, crystal A crystal diode capable of demodulating a received radio signal. Early radio receivers used small pieces of crystal and a spring wire contact, the catswhisker, to act as detector.

detector, first Mixer stage in a superheterodyne radio receiver, produces an intermediate frequency.

detector, gas leak Microphone on long probe which can be held close to a pressurized aerial cable while localizing faults due to sheath perforation. The escaping gas from the fault makes an ultrasonic hissing noise which is picked up by the microphone and converted to an audible tone.

detector, ratio An FM discriminator.

detector, second The detection stage in a superheterodyne radio receiver which takes the intermediate frequency as input and gives an audio frequency output.

detector, ultrasonic leak A gas leak detector.

detector guard Circuit which stops voice signals from operating signaling frequency receivers, usually by amplifying all frequencies other than signal frequencies, rectifying the totality of these, and using the result to offset and cancel any effect there might be at the signal frequency itself. Also called a voice guard circuit.

detector noise limited operation Operation of an optoelectronic system when the amplitude of the pulse, rather than its width, limits the distance between repeaters.

detent A tooth, notch or latch used to lock a mechanical device.

detune To change the tuning of a circuit so that it no longer resonates at a particular frequency, or the circuit is tuned to receive or operate only upon a particular frequency.

Deutsche Industrie Normenausschus (DIN) The German standard-setting organization; plugs, sockets, and photographic film, for example, are now made to DIN standards worldwide.

deviation, allowable Acceptable variation from a reference condition.

deviation, frequency Peak difference between the instantaneous frequency of an FM signal and the carrier frequency.

deviation, phase Peak difference between the instantaneous angle of a phase modulated wave and the angle of the carrier.

deviation, standard A statistical figure indicating how the various values (of a reading or measurement) compare and are distributed relevant to the mean value of all values in the sample.

deviation ratio Ratio of the maximum frequency deviation to the highest modulating frequency.

deviator Modulator in an FM system.

device A component, circuit or functional unit.

device, analog A simple computer-type device able to operate on analog inputs and produce analog outputs.

device, anti-singing See voice operated device, anti-singing (VODAS).

device, input-output Equipment used to put information into or take information out of a computer system.

device, semiconductor Component which uses a semiconducting material, e.g., transistor.

dewpoint Temperature at which moisture will condense out. In tropical conditions relative humidity can often exceed 95%, so very slight cooling results in moisture deposits.

diad See dibit.

diagnosis Location of errors in software or equipment faults in hardware.

diagnostic routine A program designed to trace errors in software or to locate hardware faults or the causes of breakdown.

diagnostics Syn.: diagnostic routine

diagram, block Simplified schematic diagram in which functional blocks and the links between them are shown rather than actual components and circuits.

diagram, schematic A functional diagram of a circuit using conventional symbols for components.

diagram, traffic A block diagram showing traffic, routing patterns and circuit quantities between different central offices or toll centers.

diagram, wiring A detailed diagram of a circuit showing all components, color coding of all wires, etc.

dial Rotary device on telephone instrument which makes and breaks loop to the office to indicate the digit dialed. Also used for the push-button sets which send a dual-tone multi-frequency signal to the central office to indicate the digit.

dial, automatic Unit which will automatically initiate dialed calls to one or more lines.

dial, card-reader Device attached to a telephone which, by the simple insertion of a small pre-prepared card, will dial out an often-needed telephone number.

dial, partial When not enough digits have been received to enable a central office to route a call.

dial, rotary A standard ten-hole telephone dial which makes and breaks the subscriber's loop in accordance with the number of the digit dialed.

dial, space-saver Dial with moving finger stop with no space between 1 and 0, resulting in a smaller dial with all the circumference used for digit-holes.

dial, subscriber Standard rotary dial suitable for use on customers' premises.

dial, switchboard Dial unit made with more robust components than an ordinary unit and able to stand up to the heavy usage which a dial on an operating position is likely to have to endure.

dial, touch-calling A push-button key pad with ten or more buttons. Output is dual tone, multi-frequency (DTMF) pulses.

dial 1 Dialing code now being introduced in the U.S. to enable additional Numbering Plan Area and central office codes to be brought into service, for the expansion of the national network. Prefix 1 usually differentiates between calls to the "home" NPA and a "foreign" NPA, i.e., a call to the home NPA is dialed with the standard seven-digit code, a call to a foreign NPA is dialed using 1 as a prefix followed by the full 10-digit number.

dial-answer night service Feature provided by some PABXs which permits incoming calls to be answered by any extension when the operator's console is unmanned.

dial-back trunk A circuit on which a caller in a rural area, particularly one served by a community dial office (CDO) may call the nearest operator-controlled switchboard, and the operator can then dial back on the same circuit to connect the caller with a called party on the same CDO. A form of revertive calling.

dial cord 1. Small group of insulated wires inside a telephone instrument, joining the various terminals on the dial to terminals on the mounting board or circuit of the instrument. 2. Cord on an operator's manual position used for dialing out on a circuit.

dial exchange An automatic exchange or central office, i.e., one controlled by rotary or pushbutton dialing rather than by an operator.

dial-in-handset Compact telephone instrument with the dial mounted in the handset unit.

dial key Key on a manual operating position which links the operator's dial with a cord circuit on which the operator has her talk key thrown. On some types of position it is not necessary to throw two keys to dial, the talk key automatically associates the operator's dial with the cord circuit concerned.

dial office An automatic central office or telephone exchange.

dial pilot lamp Indicator lamp on some manual operating positions which glows when the operator's dial has been moved.

dial pulse One of two methods of dialing digits from a telephone (the other being key pulse). Dial pulses are generated by alternately opening and closing a contact in the telephone through which dc current flows. Each digit is represented by a string of the appropriate number of pulses. Also called rotary dialing.

dial release key Key on some manual operating position which is used to disconnect the dial from a cord circuit after dialing has been completed.

dial service assistance switchboard A switchboard associated with switching center equipment to provide operator services such as information, intercepting, random conferencing, and precedence calling assistance.

dial signaling A type of signaling in which pulse trains are transmitted to a receiving terminal to operate automatic line selection equipment. The sequence of the dial pulses is determined by an operator, but their duration is predetermined by equipment adjustments.

dial tandem network A private network in which the caller must route himself through the network to his destination by dialing a sequence of access codes to string together the correct trunks.

dial teletypewriter exchange service Subscriber dialing of teletypewriter calls. This is now possible to most countries of the world; outside North America, called telex service.

dial through A technique applicable to access circuits which permits an outgoing routine call to be dialed by the PBX user after the PBX attendant has established the initial connection.

dial tone 1. A steady signal from a CO or PBX used to inform the caller to start dialing. 2. A signaling protocol which responds to a request for service and indicates a "start dial" to the sending switch. The dial tone is returned only after the receiving switch is prepared to collect the incoming digits.

dial-tone delay Time elapsed between the subscriber going off-hook and reception of dial tone.

dial tone first A type of pay phone/coin telephone where the dial tone is received as soon as the caller goes off-hook, coins only have to be inserted after the call has been connected through to the called line. Also called post-payment boxes. Permits caller to reach an operator or to dial 911 emergency service without depositing coins.

dial train A series of pulses or tones containing digital information from a customer's dial or pushbuttons.

dial-up The use of a rotary type dial or push-button set to initiate a station-to-station telephone call.

dialer, repertory Automatic device which dials complete telephone numbers when a key or button is pushed.

dialing, intertoll Dialing over trunks between toll centers.

dialing, nationwide Dialing of long distance calls by customers, i.e., direct distance dialing (DDD) (U.S.) or subscriber trunk dialing (STD) (U.K.).

dialing area Area within which customers may make dial calls. This term is now largely outdated. Nowadays international distance dialing or international subscriber dialing are commonplace.

dialing-time Time elapsed between the reception of dial tone until the end of dialing by the calling subscriber.

diamagnetic A material which, when placed in a nonuniform magnetic field, tends to move from the stronger to the weaker region of the field. A bar of diamagnetic material placed in a uniform magnetic field tends to swing around so that its longer axis is at right angles to the field. Copper, bismuth, and antimony are diamagnetic.

diametral index of cooperation In facsimile systems, the product of the drum diameter and the line advance in scanning lines per unit length. The unit of length must be the same as that used for expressing the drum diameter.

DIANE Direct Information Access Network for Europe: a grouping of otherwise independent computerized information services located throughout Europe.

diaphragm Thin flexible sheet that vibrates to sound waves, as in a microphone, or to electrical impulses, as in a telephone earpiece or a loudspeaker.

diaphragm, damped A mechanically loaded diaphragm which avoids resonant effects and gives reduced distortion.

diary service With prior indication from a subscriber, a call is made automatically to his telephone number at a specific date and time and, when an answer condition is detected, a recorded message is connected to his telephone termination to remind him of a particular event, e.g., the birthday of a relative.

dibit A pair of binary digits which can be represented by a single modulation condition in a data transmission system, such as four-phase modulation. The four possible states for a dibit are 00, 01, 10, and 11. Syn.: diad

dichotomizing search A search in which an ordered set of items is partitioned into two parts, one of which is rejected, the process being repeated until the search is completed.

dichroic filter An optical filter that transmits light selectively according to wavelength.

dichroic mirror A mirror that reflects light selectively according to wavelength.

dictation service Feature of many PABXs which enables extension telephones to be used to access a centralized stenographic service.

dictation service, remote Feature of some PABXs which enables extension telephones to be used to access a centralized dictating machine service and to control the recording and playback action from the telephone.

DID See direct inward dialing.

dielectric Insulating material which will store but not conduct electricity. In submarine cables: the insulation between the inner conductor and the outer conductor of the coaxial structure of a sea cable, which usually consists of special grade polyethelene with high insulation resistance and a low dissipation factor.

dielectric constant A dielectric material's ability to store electrostatic energy, compared to air. Also called specific inductive capacity.

dielectric lens A lens made of a dielectric material which is used to refract or focus radio waves in the same way as a glass lens refracts light waves. Used with some specialized microwave antennae.

dielectric phase angle Angular difference in phase between sinusoidal alternating voltage applied to a dielectric and the component of the resulting alternating current having the same period as the voltage.

dielectric strength Potential gradient at which electric breakdown occurs.

difference, phase Radians by which one electrical wave leads or lags another of the same periodicity.

difference, potential Algebraic difference between voltages at two points.

differential Any process or procedure which depends on differences between values or quantities.

differential gain Variation in output gain for different levels of signal input. In color TV, the change in gain (expressed in dB) for the 3.58 MHz color subcarrier as the level of the luminance signal is varied from blanking to white.

differential gain control Adjustment, usually automatic, to obtain desired output levels from alternately applied unequal input signals.

differential modal delay Optoelectronics: the differences in propagation delays among modes owing to their differing group velocities. Syn.: multimode group delay

differential-mode interference 1. Interference causing a change in potential of one side of a signal transmission path relative to the other side. 2. Interference resulting from an interference current path coinciding with the signal path.

differential modulation A type of modulation in which the choice of the significant condition for any signal element is dependent on the choice for the previous signal element. Delta modulation is an example.

differential phase Variation in output phase for different input signal levels. For TV, variation in the phase of the color subcarrier as the level of the luminance signal is varied from blanking to white.

differential phase-shift keying (DPSK) A method of modulation employed for digital transmission. In DPSK, each signal element is a change in the phase of the carrier with respect to its previous phase angle.

differential quantum efficiency The slope of the light-current curve, which describes devices that have nonlinear output-input characteristics.

differential sensitivity Of an echo suppressor: the difference, in dB, between the level of the test signals applied to the send path and receive path when break-in occurs.

differential windings Typically two coils on the same core so arranged that when both carry current their magnetomotive forces are in opposition so that, if the coils are relay coils, for example, the relay will only operate if only one coil carries current, or if the current through one coil is significantly greater than that through the other.

differentially coherent phase-shift keying (DCPSK) A method of modulation in which information is encoded in terms of phase changes, rather than absolute phases, and detected by comparing phases of adjacent bits.

diffraction The deviation of a wavefront from the path predicted by geometric optics when a wavefront is restricted by an opening or an edge of an object. Diffraction occurs whenever a light or radio beam is restricted in any way and may still be important when the opening is many orders of magnitude larger than the wavelength.

diffraction grating An array of fine, parallel, equally spaced reflecting or transmitting lines which mutually enhance the effects of diffraction.

diffraction limited A beam of light is diffraction limited if: (a) the far field beam divergence is equal to that predicted by diffraction theory; or (b) in focusing optics, the impulse response or resolution limit is equal to that predicted by diffraction theory.

diffraction propagation Propagation of electromagnetic waves, especially radio waves, around objects.

diffraction region The region beyond the radio horizon.

diffuse reflection The scattering effect that occurs when light, radio or sound waves strike a rough surface.

diffused-base transistor Transistor with a nonuniform base region produced by diffusion.

diffused-junction transistor A transistor in which both the emitter and collector electrodes have been formed by diffusion of an impurity into the semiconductor wafer.

diffused-mesa transistor A diffused-junction transistor in which an N-type impurity has been diffused into one side of a P-type wafer. After a second PN junction has been diffused into the new N-type surface to become the emitter the unwanted ends of the unit are etched away, giving the transistor the appearance of a flat-top mountain or mesa.

diffusion 1. The spreading or scattering of a wave, such as a sound wave. 2. The process through which the crystalline structure of semiconductor material can be "doped" so that either positive or negative carriers effectively carry electric current through the material.

digger, hydraulic Mechanical pole-hole drill or auger, powered from a truck engine by hydraulic drive.

digit 1. Any single number from 0 to 9. 2. In digital transmission, a digit may be represented by a signal element, characterized by the dynamic nature, discrete condition and discrete timing of the element. 3. In equipment used in digital transmission, a digit may be represented by a stored condition being characterized by a specified physical condition, e.g., it may be represented as a binary magnetic condition of a ferrite core.

digit, language On international telephone calls, an additional digit inserted, usually automatically, as an indication of the language desired to be spoken by an operator at the distant end.

94

digit position The position of a particular digit in a number, the lowest significant position is usually called 0, the next is 1, then 2 and so on. The position in time or space into which a representation of a digit may be placed.

digit rate The number of digits per unit time. An appropriate adjective should precede the word digit, for example, binary digit rate.

digit time slot A time slot allocated to a single digit.

digit translation The capability of the switching system to determine a traffic route choice based on a set or subset of the dialed digits.

digital Using digital or discrete signals; not a continuously variable analog type signal.

digital access and cross-connect system (DACS) AT&T facility providing for direct connection of T 1 carrier systems and cross-connection of individual 56/64 kbit/s channels.

digital alphabet A coded character set in which the characters of an alphabet have a one-to-one relationship with their coded representations.

digital block A set of multiplexed equipment that includes one or more data channels and associated circuitry.

digital circuit A circuit which transmits information signals in digital form between two exchanges. It includes termination equipment but not switching stages.

digital circuit switch A time-switch established to interconnect digital signals.

digital combining A method of interlacing digital data signals, in either synchronous or asynchronous mode, without converting the data into a quasianalog signal.

digital connection An association of digital circuits, digital switches and other functional units to transfer digitally-encoded information signals between two terminals.

digital data 1. Data represented by discrete values or conditions, as opposed to analog data. 2. A discrete representation of a quantized value of a variable, i.e., the representation of a number by digits, perhaps with special characters and the "space" character.

digital data bus A type of bus connection used for domestic Small Area Networks.

Digital Data Service Bell System's digital data service in the U.S.

digital echo suppressor (DES) Voice-activated device which monitors levels of digital speech signals on both directional paths and automatically applies attenuation when necessary to reduce echo effects on long-haul circuits.

digital error A single digit inconsistency between the transmitted and received signals.

digital exchange An exchange that switches information in digital form.

digital facsimile Facsimile equipment which transmits and receives digital signals.

digital filling The addition of a fixed number of digits to a digital signal to change the digit rate from its existing nominal value to a higher predetermined nominal value. The added digits are not used to transmit information.

digital frequency modulation The transmission of digital data by frequency modulation of a carrier, as in binary frequency shift keying.

digital line engineering program (DILEP) An AT&T software routine used to plan spacing of regenerative repeaters for PCM digital line transmission systems.

digital line path Two or more digital line sections interconnected in tandem in such a way that the specified rate of the digital signal transmitted and received is the same over the whole length of the line path between the two terminal digital distribution frames (or equivalents). Where appropriate, the bit rate should qualify the title.

digital line section A unit made up of two consecutive line terminal equipments together with their interconnecting transmission medium and the cabling between them, plus their adjacent digital distribution frames (or equivalents), all of which provide the means of transmitting and receiving between two consecutive digital distribution frames (or equivalents) a digital signal of specified rate.

digital loop-back Diagnostic feature providing for a loop-back of a digital signal from the remote modem, to provide a complete test of the line and remote modem.

digital microwave Transmission of voice or data in digital form on microwave links. Second or higher levels of PCM multiplexing are used to carry information.

digital modulation The process of varying one or more parameters of a carrier wave as a function of two or more finite and discrete states of a signal.

digital multiplex equipment Equipment for combining, by time division multiplexing, (multiplexer) a defined integral number of digital input signals into a single digital signal at a defined digit rate and also for carrying out the inverse function (demultiplexer). When both functions are combined in one equipment at the same location the abbreviation MULDEX may be used to describe this equipment.

digital multiplex hierarchy A series of digital multiplexes graded according to capability so that multiplexing at one level combines a defined number of digital signals, each having the digit rate prescribed for a lower order, into a digital signal having a prescribed digit rate which is then available for further combination with other digital signals of the same rate in a digital multiplex of the next higher order.

Digital Multiplex System (DMS) Northern Telecom (Canada) switching system in which all external signals are converted to digital data and stored in assigned time slots. Switching is performed by reassigning the original time slots.

digital overlay A digital network arranged to overlay an existing analog network with suitable interconnections between the two.

95

digital path   The whole of the means of transmitting and receiving a digital signal of specified rate between two digital distribution frames (or equivalent) where terminal equipment or switches will be connected.

digital radio   A microwave radio system designed for the transmission of digital signals, which are usually pulse code modulated signals.

digital radio path   Two or more digital radio sections interconnected in tandem in such a way that the specified rate of the digital signal transmitted and received is the same over the whole length of the radio path between the two terminal digital distribution frames (or equivalent).

digital radio section   Two consecutive radio terminal equipments and their interconnecting transmission medium which together provide the whole of the means of transmitting and receiving, between two consecutive digital distribution frames (or equivalent), a digital signal of specified rate.

digital section   The whole of the means of transmitting and receiving between two consecutive digital distribution frames (or equivalent) a digital signal of specified rate. A digital section forms either a part or the whole of a digital path.

digital signal   1. A nominally discontinuous electrical signal that changes from one state to another in discrete steps. 2. A signal that is time-wise discontinuous, i.e., discrete, and can assume a limited set of values.

digital speech interpolation (DSI)   Technique to improve utilization of channels by the allocation of time-slots to other users during periods of silence from the original user. See time assignment speech interpolation (TASI).

digital sum   In a multilevel pulse code, the sum of pulse amplitudes from some arbitrary time-origin to the last transmitted pulse at the time considered, the amplitude unit being chosen in such a way that adjacent levels differ by one unit.

digital switch   Solid state device that routes a signal in digital format. Some mechanical devices carry digital signals but these normally operate on a circuit switch basis.

digital switching   A process in which connections are established by operations on digital signals without converting them to analog signals.

digital-to-analog converter   A device that converts a digital input signal to an analog output signal carrying equivalent information.

digital transmission group   A number of voice or data channels or both that are combined into a digital bit stream for transmission over various communications media.

digital transmission system   A transmission system suitable for digital signals, in particular for pulse code modulated signals.

digital voice transmission   Transmission of analog voice signals that have been converted into digital signals.

digital voltmeter   A voltmeter which displays its readings in a digital format, either by LCDs or by a digital output signal into another device.

digitize   To convert information into a digital code.

digitizing   The conversion of a signal into digital format; use of analog-to-digital coder.

digitron   A cold-cathode tube with several cathodes, usually ten in the shape of the digits 0 to 9; as voltage pulses are received the power connection is switched to the appropriate cathode which glows brightly. Now displaced by LCDs and LEDs. Called Nixie tubes in the U.S..

digits, partial   When a register has not received enough signaling information to enable a call to be put through.

digits, route control   Digits received which specify digits to be outpulsed and routes to be used for the establishment of toll calls.

digroup   A basic group of PCM channels assembled by time division multiplex. In North America, usually derived from 1544 kbit/s PCM multiplex equipment.

diminished radix complement   Number obtained by subtracting each digit of the specified number from a number which is one less than the radix, i.e., for decimal figures taking each away from 9. Many computation programs use complements such as this to represent negative number values.

dinkey, pole   A small pole trailer.

diode   A two-electrode electron tube or its semiconductor equivalent.

diode, crystal   A diode rectifier which usually utilizes a silicon or germanium crystal and a point contact.

diode, germanium   Semiconductor diode with germanium as the active and rectifying element.

diode, junction   A semiconductor distinguished by the fact that rectification occurs at a junction between types of semiconducting material rather than at a point contact.

diode, tunnel   Junction diode based on the fact that electrons can in some circumstances move across potential barriers which their energy levels are insufficient to jump over.

diode, varactor   A variable capitance diode.

diode bridge   A bridge which permits polarity sensitive devices to be used across a line.

diode laser   See injection laser diode.

diode transistor logic   (DTL) A family of integrated logic circuits, with each input coming through a diode, and the output from the collector of an inverting transistor. DTL circuits are slow in operation and have largely been replaced by ttl and CMOS logic families. The basic gate in the DTL family performs the NAND function.

DIP   See dual in-line package.

diphase line codes   Special coding to avoid certain data transmission difficulties. Simple binary digital signals can be sent to line on physical pairs but there are difficulties which have led most Administrations to code these signals before transmission. The main reasons for this are that some data may contain long sequences of consecutive ones or zeros, and that some data terminals present a continuous steady condition when in the idle state. This means that the line signal would effectively become DC so could not get through transformers or line transmission equipment, and also that there would be insufficient information in the transmitted line signal to ensure satisfactory synchronization between both ends. To get around these difficulties a 'dipulse' or 'diphase' code is normally adopted: instead of using 0 and 1 as the two significant conditions it uses 01 and 10.

diplex operation Either the operation of two radio transmitters into one antenna, or the simultaneous utilization of transmission and reception facilities, with different frequencies, using a single antenna.

diplexer Combining device which permits use of one antenna by two transmitters.

dipole A straight wire antenna, one half-wave length long, center fed.

dipole, folded Antenna similar to an ordinary dipole but made from one piece of tubing, folded back upon itself so that the two ends of the tubing are both brought to the central point at which the feeder is connected.

dipole, wideband A dipole antenna with large diameter effective radiating elements, usually made up of spaced groups of wires or rods.

direct access The facility to obtain data from a storage device, or to enter data into a storage device in such a way that the process depends only on the location of that data and not on a reference to data previously accessed.

direct access (videotex) The process of retrieving information by directly calling the page in the videotex system where it occurs.

direct address An address that designates a storage location of an item of data to be treated as an operand.

direct broadcasting from satellite Satellite service, usually via spot-beam antenna with a relatively small footprint on earth, so that the received signal is sufficiently strong to be picked up using small and cheap antennae at the viewer's own premises.

direct burial See burial, direct.

direct call Telegraphy and data: a facility which avoids the use of address selection signals. The network interprets the call request signal as an instruction to establish a connection with a single destination address previously designated by the user.

direct circuit An intertoll or office trunk, in particular a high usage circuit connecting two toll centers.

direct circuit call A toll call which is routed on a direct circuit between originating and terminating toll centers.

direct control Control of an automatic switching arrangement in which the functional units necessary for the establishment of connections are associated with a given call for its duration and are set directly in response to signals from the calling device.

direct control office A central office, switching center or telephone exchange in which switching equipment at each selection level is actuated directly by signals dialed in by the calling subscriber.

direct coupling Coupling between stages which permits dc to flow.

direct current (dc) Current flowing in one direction.

direct current signaling In telephony, a method whereby the signaling circuit receive and transmit leads use the same cable pair as the voice circuit and no filter is required to separate the signaling frequency from the voice transmission. Syn.: DX signaling.

direct dialing-in Calls can be dialed from a telephone connected to the public network directly to extensions on a PABX. Called direct inward dialing in the U.S.

direct distance dialing (DDD) A telephone service which enables a user to dial directly telephones outside the user's local area without the aid of an operator.

direct distribution Use of the fixed-satellite service to relay programs from one or more points of origin directly to terrestrial broadcasting stations without any intermediate distribution stages.

direct exchange line (DEL) A line from an exchange or central office serving only a single subscriber. Syn.: exclusive exchange line

direct feed cable A distribution cable from a central office/exchange MDF feeding directly to a large building or complex, i.e., not feeding through a cabinet or other flexibility point.

direct inward dialing A feature of PBXs and Centrex systems which allows callers to dial from the public network straight to a wanted extension on a PABX without intervention by an operator. Called direct dialing-in in Britain.

direct line service A semipermanent telephone connection between two locations, typically with a telephone instrument or PBX at each end, and a circuit which rings the other end as soon as the instrument at one end goes off-hook. In Britain, a direct line is a subscriber's line to a central office.

direct memory access A device for transferring blocks of continuous data to and from memory at a high rate.

direct orbit A satellite orbit such that the projection of the center of the satellite's mass on the reference plane revolves about the axis of the primary body in the same direction as that in which the primary body rotates. If it revolves in the reverse direction it is called a retrograde orbit.

direct outward dialing (DOD) A telephone service in which outgoing calls can be placed directly by dialing an initial digit (access digit) and then the desired number without the aid of an operator.

direct ray A ray of electromagnetic radiation that follows the path of least possible propagation time between transmitting and receiving antennas.

direct recording That type of facsimile recording in which a visible record is produced, without subsequent processing, in response to the received signals.

direct route Route between two exchanges with no intermediate switching.

direct service area Geographic area near a central office/exchange fed by distribution cables (secondary cables) direct from the MDF, i.e., there are no normal primary or main cables leading to a cabinet or flexibility point (from which distribution or secondary cables normally feed out).

direct station selection A PBX feature which permits an attendant to extend an incoming call out to an extension by means of a single (usually pushbutton) switch.

direct transit country A country through which traffic is routed on direct circuits provided for the exclusive use of other countries.

direct voltage A voltage which produces a current flowing in one direction only.

direct-writing recorder A meter with a moving paper record which produces a permanent record of readings, e.g., of exchange battery voltages or current flow.

directing code A routing digit dialed before the called subscriber's telephone number in order to obtain access to another network.

direction of scanning In a facsimile transmitting apparatus, the scanning of the plane of the message surface along lines running from right to left commencing at the top. Scanning therefore commences at the top right-hand corner of the surface and finishes at the bottom left-hand corner; this is equivalent to scanning over a right-hand helix on a drum. At the receiving apparatus, scanning takes place from right to left and top to bottom (in the above sense) for positive reception and from left to right and top to bottom for negative reception.

directional control, hub Circuit feature in a telegraph hub circuit which stops an incoming signal from being looped around and sent out on the sending side of the same leg.

directional coupler (cable TV) A high-quality tapping device providing isolation between tap and output terminals.

directional filter A combination high-pass and low-pass filter with a common branching point; used to separate the higher and lower transmission bands of a bidirectional system.

directional reservation equipment (DRE) Network management control applied to trunk groups that gives priority for completing traffic (at the expense of originating traffic).

directive gain The ratio of 4π times the power delivered per unit solid angle (steradian) in a given direction to the power delivered to 4π steradians. Usually expressed in dB as $10 \log_{10}$ of the ratio obtained.

directivity The value of the directive gain of an antenna, in the direction of its maximum value.

directivity pattern A diagram relating power density (or field strength) to direction relative to the antenna, at a constant large distance from the antenna.

director 1. A centralized control device in an automatic telephone office which combines the functions of registers, translators and senders. Used in some major metropolitan areas to control the routing of junction and tandem calls, particularly between step-by-step central offices. 2. A parasitic element in an antenna on the major lobe side of the active dipole, i.e., facing the transmitter whose signal is to be received.

directory assistance Telephone company information service; operators help callers who cannot find the telephone numbers they wish to call.

directory enquiry service See directory assistance.

directory number The full complement of numbers required to designate a subscriber's station within a numbering plan area.

directory number hunt Feature which permits calls to a busy number to be rerouted automatically to other lines serving the same subscriber.

disable To act on a circuit to stop it operating.

disabler, echo suppressor Control device which stops echo suppressors from operating whenever a circuit is to be used for data or telegraphy signals in both directions of transmission at the same time.

disabling tone A selected tone transmitted over a communications path to control equipment.

disc See disk.

discharge Conversion of a storage battery's chemical energy into electrical energy, or of a capacitor's dielectric stress into an electric current.

discharge blocks Lightning protectors, such as the small carbon blocks mounted on main distributing frames, through which lightning-induced voltages are discharged to ground.

discharge tube Gas-filled tube which acts as a lightning protector by providing a very low resistance path to earth.

discharger Protective device, particularly a pair of carbon blocks or a gas-filled discharge tube.

disconnect The dissociation or release of a switched circuit between two telephones or data sets, or to switch off a power supply to deactivate a device.

disconnect command In a data communication network, an unnumbered command used to terminate the operational mode previously set.

disconnect-make busy, carrier Feature which cuts in when a carrier system fails: all the affected trunks are automatically disconnected and switches released, and the busy condition applied so that the trunks cannot be picked up again until the fault has been cleared.

disconnect signal The on-hook signals given by both ends to signify the end of a call and to instruct the offices concerned to release all the switches used to establish the connection.

disconnecting switch In a power system, a switch used for closing, opening, or changing the connections in a circuit or system or for isolating purposes. It has no interrupting rating and is intended to be operated only after the circuit has been opened by some other means.

discontinuity An abrupt nonuniform point of change in a transmission circuit where impedances suddenly alter and power is lost by reflection.

discrete address The individual address of a particular terminal.

discriminate To restrict, or detect and give different treatment to different signals.

discrimination, number The ability to treat groups of numbers served by a central office in different ways.

discriminator 1. That part of a circuit which extracts the desired signal from an incoming frequency-modulated wave by changing frequency variations into amplitude variations. 2. A device which, when the input voltage surpasses a given level, produces a voltage output. 3. A device which responds only to a pair of frequencies which share some characteristic, such as amplitude.

discriminator, FM An electron tube or transistor device which, after detecting the frequency rate of change of an FM signal, is able to change this to a signal voltage equal to the original modulating signal.

disengagement attempt The process by which one or more users interact with a telecommunication system in order to end an established access.

disengagement denial Disengagement failure due to excessive delay by the telecommunication system.

disengagement-denial probability Ratio of disengagement attempts which result in disengagement denial to total disengagement attempts counted during a measurement period.

disengagement failure Failure of a disengagement attempt to return the participating user to the idle state within a specified maximum disengagement time.

disengagement originator The functional entity responsible for initiating a particular disengagement attempt, which can be either the source user or the destination user, or in the case of systems with preemption, by the communication system.

disengagement phase In an information transfer transaction, the phase during which successful disengagement occurs.

disengagement request A control or overhead signal issued by a disengagement originator for the purpose of initiating a disengagement attempt.

disengagement time Elapsed time between the start of a disengagement attempt and successful disengagement.

dish A parabolic microwave antenna.

disk A memory system based on rotating disks coated with a magnetic recording medium.

disk, fixed head A disk memory with a multiplicity of heads, each giving access to one read/write track.

disk, floppy A disk memory based on interchangeable flexible plastic disks, usually with a moveable head able to access many read/write tracks.

disk, moving head A multi-track disk with a single head which moves to access any read/write track.

disk, rigid A disk memory in which the magnetic medium is coated onto a rigid substrate.

disk, Winchester A disk memory in which the disks and magnetic head assemblies are contained within an enclosure sealed against dust or other contamination.

diskette Floppy disk used as a magnetic store.

disparity For PCM: the digital sum of a set of n signal elements.

dispersion 1. The chromatic or wavelength dependence of a parameter. Also, the process by which an electromagnetic signal is distorted because the various frequency (i.e. wavelength) components of that signal have different propagation characteristics. The term also describes the relationship between refractive index and wavelength. Signal distortion in an optical waveguide is caused by several dispersive mechanisms: waveguide dispersion, material dispersion, and profile dispersion. In addition, the signal suffers degradation from multimode distortion, which is often (erroneously) referred to as multimode dispersion. 2. The scattering of microwave radio radiation by rain drops, or similar obstructions. 3. The separation into the components of an electromagnetic wave with different frequencies. 4. The solution of finely divided solids in a liquid, such as graphite particles in oil. 5. The allocation of circuits between two points over more than one geographic or physical route.

dispersion, chromatic Pulse distortion in an optic fiber system caused by different wavelengths of light within the pulse traveling at different speeds.

dispersion, facility Routing of circuits over geographically separated paths for additional security.

dispersion, modal Pulse distortion in an optic fiber system caused by some parts of the light pulses following longer paths (modes) than other parts.

dispersion limited operation Operation of an optoelectronic system when the dispersion of the pulse, rather than its amplitude, limits the distance between repeaters/regenerators.

display Visual presentation of information, as on a video display unit.

display, dialed number Face equipment on some manual operating or service observation positions which gives a visual display of the number actually dialed by a customer.

display device An output unit that gives a visual representation of data.

dissipation factor Ratio of energy dissipated in a dielectric material (typically lost in heat) to the energy stored for one cycle.

dissipator, heat A heat sink.

distant signal (CATV) TV signals which originate at a point too far away to be picked up by ordinary home reception equipment; also signals defined by the FCC as outside a broadcast's license area. Cable systems are limited by FCC rules in the number of distant signals they can offer subscribers.

distinctive ringing Facility on some PABXs and keyphone systems by which ringing current periodicity is coded, differentiating between internal and outside calls.

distort To change a signal wave form during transmission.

distortion 1. The difference between the wave shapes of an original signal and the signal after it has traversed the transmission circuit. 2. In an optical waveguide, signal distortion is caused by several dispersive mechanisms: waveguide dispersion, material dispersion, and profile dispersion. In addition, the signal suffers degradation from intramodal distortion and multimode distortion which is often (erroneously) referred to as multimode dispersion. 3. Any departure from a specified input-output relationship over a range of frequencies, amplitudes, or phase shifts, during a time interval.

distortion, amplitude Amplitude-frequency distortion; when different frequencies, all present in a complex wave, are attenuated differently.

distortion, attenuation Same as amplitude distortion.

distortion, bias Distortion affecting a two-condition (or binary) modulation (or restitution) in which all the significant intervals corresponding to one of the two significant conditions have different durations than the corresponding theoretical durations. Also called asymmetrical distortion.

distortion, characteristic Distortion caused by transients which, as a result of the modulation, are present in the transmission channel and depend on its transmission qualities.

distortion, degree of gross start-stop Degree of distortion determined when the unit interval and the theoretical intervals assumed are exactly those appropriate to the standardized modulation rate.

distortion, degree of individual Of a particular significant instant of a modulation or a restitution of a telegraph signal: ratio of the unit interval of the displacement, expressed algebraically, of this significant instant from an ideal instant. This displacement is considered positive when the significant instant occurs after the ideal instant. The degree of individual distortion usually is expressed as a percentage.

distortion, degree of isochronous 1. Ratio to the unit interval of the maximum measured difference, irrespective of sign, between the actual and the theoretical intervals separating any two significant instants of modulation or of restitution, these instants being not necessarily consecutive. 2. Algebraic difference between the highest and the lowest value of individual distortion affecting the significant instants of an isochronous modulation or restitution. The difference is independent of the choice of the reference ideal instant.

distortion, degree of start-stop 1. Ratio to the unit interval of the maximum measured difference, irrespective of sign, between the actual and theoretical intervals separating any significant instant of modulation or of restitution from the significant instant of the start element immediately preceding it. 2. The highest absolute value of individual distortion affecting the significant instants of a start-stop

modulation or restitution. The degree of distortion usually is expressed as a percentage.

distortion, degree of synchronous start-stop Degree of distortion determined when the unit interval and the theoretical intervals assumed are those appropriate to the actual mean rate of modulation or of restitution.

distortion, delay Distortion caused by the later arrival of higher frequency components of a complex waveform due to slower travel speed of higher frequency signals. Also called phase distortion.

distortion, envelope delay Distortion due to delay of the envelope or group of signals passing through a network. Sometimes defined as one half of the relative delay, i.e., half the difference between maximum and minimum delays occurring in a channel or band of frequencies.

distortion, fortuitous Distortion resulting from causes generally subject to random laws.

distortion, frequency Changes in relative amplitudes of different frequency components of a complex waveform.

distortion, harmonic Distortion due to the creation of harmonics of a fundamental frequency

distortion, intermodulation Distortion produced when two or more waves (or a complex wave form involving two or more frequencies) pass through a non-linear device which produces sum-and-difference modulation product frequencies.

distortion, linear Distortion which is independent of the signal amplitude.

distortion, non-linear Distortion which is dependent on signal amplitude, e.g., compression, expansion.

distortion, phase See distortion, delay.

distortion, phase-frequency Distortion due to the difference between phase delay at one frequency and at a reference frequency.

distortion, single harmonic Ratio in dB of output power at the fundamental frequency to output power at any single harmonic frequency when a single fundamental frequency signal of specified level is input to the system

distortion, spaced Distortion where there is speed difference between sending and receiving devices.

distortion, standard Phrase used to describe a commercial speech transmission path with a bandwidth of 250 Hz to 3000 Hz. (300 Hz to 3400 Hz in some circumstances)

distortion, systematic Distortion of a signal in a pattern which is regularly repeated.

distortion, telegraph Distortion of a modulation due to the fact that the significant intervals do not conform exactly to their theoretical durations. A modulation or restitution is affected by telegraph distortion when significant instants do not coincide with the corresponding theoretical instants.

distortion, teletypewriter signal Distortion of significant intervals, i.e., transition points of signal pulses, from their correct positions relative to the start signal.

distortion, total harmonic Ratio in dB of output power at fundamental frequency to that at all harmonic frequencies that appear at the output, when a single fundamental frequency signal of specified level is input to the system.

distortion, transmitter A signal transmitted by an apparatus (or a signal at the output of a local line with its termination) is affected by telegraph distortion when the significant intervals of this signal have not their exact theoretical durations.

distortion limited operation The condition prevailing when the shape of the signal, rather than its amplitude (or power), is the limiting factor in communication capability. The condition is reached when the system distorts the shape of the waveform beyond tolerable limits. For linear systems, distortion limited operation is equivalent to bandwidth limited operation.

distortion transmission impairment (DTI) A subjectively measured amount of attenuation which has to be included in a distortionless voice circuit to produce the same effective transmission degradation as a particular bandwidth reduction below 3 kHz.

Top frequency reduced to	DTI
3 kHz	0 dB
2.9 kHz	1 dB
2.7 kHz	2 dB
2.5 kHz	3 dB
2.35 kHz	4 dB
2.2 kHz	5 dB

distributed Of a transmission line: the regular distribution along the whole length of a line of its electrical parameters (inductance, capacitance, and resistance), as opposed to lumped loading, or the increase of circuit inductance by insertion of coils at regular intervals.

distributed common control Systems which physically separate the various common equipment items which control the different modules of a switching system. Centralized common control began in the 1940s, replacing step-by-step exchanges in which every switch had its own control and thus the controls were distributed throughout the exchange. With the move from SXS to crossbar there was a strong economic case for centralizing controls for greater efficiency. However, some manufacturers designed crossbar exchanges with distributed electromechanical control units for the various modules. These were largely for small town exchanges, not for the big city exchanges. The first computer-controlled or SPC (stored program controlled) exchanges were in effect centralized common control exchanges using computers to perform most of the functions of the centralized electromechanical registers, markers, senders etc. With technology changing the economics of engineering design it is now becoming practical to go back to distributed control with microprocessors incorporated in circuits wherever their use is economical or when such distribution is more efficient than either having the task performed centrally (i.e. by the exchanges's central processing unit) or having the task performed on a distributed basis by hard-wired logic.

The microprocessor does not always win out, but as time goes on it will soon be very rare to find any type of exchange, being made by a major manufacturer anywhere in the world, which does not use distributed microprocessors performing functions more cheaply and efficiently than was possible before distribution of control again became economic.

distributed data processing (DDP) A network of geographically dispersed, though logically interconnected, data processing nodes, generally configured so that nodes may share common resources.

distributed frame alignment signal A frame alignment signal in which the signal elements occupy non-consecutive digit time slots.

distributing terminal assembly A grading frame in a space-division electromechanical central office.

distribution A process performed by a switching stage in which the number of inlets is approximately the same as the number of outlets.

distribution, automatic call The distribution of incoming calls to different attendants to increase traffic handling efficiency.

distribution frame (DF) 1. A hardware entity which on one side provides metallic terminations for cables carrying incoming and outgoing voice paths to telephone switching equipment and, on the other, terminations for the outside distribution cable network. 2. A structure with terminations for connecting the permanent wiring in such a manner that interconnection by cross-connections may be readily made.

distribution network 1. Part of the local exchange cable network, comprising small cables between subscribers' distribution points (DPs) and cabinets, remote line units (RLUs) or other flexibility points. 2. That part of a cable TV (CATV) system used to carry signals from the head end to subscribers' receivers. Sometimes used to mean only that part of the system between bridger amplifiers and subscribers' receivers.

distribution plant (CATV) The hardware of a cable system — amplifiers, trunk cable and feeder lines, attached to utility poles or fed through underground ducts or conduits.

distribution substation A substation that modifies electric energy for service to utilization equipment.

distribution voltage drop The voltage drop between any two defined points of interest in a power distribution system.

distributor Switch in an electromechanical central office which allots or pre-selects the first available line-finder to deal with a calling line.

distributor, automatic call Device which distributes incoming calls to different operating positions to spread traffic load and increase efficiency.

distributor, CATV transmission An amplifier and coupler used to transmit CATV signals to several branching routes.

distributor, position An automatic call distributor.

distributor, telegraph A time-division device which switches both ends of a single telegraph channel in sequence to different telegraph transmitters and receivers.

disturbance Interference with normal conditions and communications.

disturbance, sudden ionospheric Interference with radio communications caused by sudden high ionization in the earth's ionosphere due to solar flares.

disturbance current The unwanted current of any irregular phenomenon associated with transmission which tends to limit or interfere with the interchange of information.

disturbance power The unwanted power of any irregular phenomenon associated with transmission which tends to limit or interfere with the interchange of information.

disturbance voltage The unwanted voltage of any irregular phenomenon associated with transmission which tends to limit or interfere with the interchange of information.

dither signal A noise signal with regular repetition rate and constant energy per cycle. Dither signals of small amplitude are sometimes used in feedback control circuits to offset hysteresis or friction effects.

diurnal variation Daily variation in radio propagation characteristics.

DIVA See data inquiry-voice answer.

divergence See beam divergence angle.

diversion (of signaling traffic). See controlled rerouting.

diversion, call System by which calls to PABX extensions are diverted to the attendant if the extension does not answer or if it is repeatedly busy.

diversion, slumber time Diversion feature in hospital PABXs which diverts to an operator any calls to patients during the night.

diversity Method of combatting the effects of path fading in a radio communications system by combining two or more received signals.

diversity, angle Diversity by combining signals with different polarization angles.

diversity, cross-band Diversity by combining two signals in different frequency bands.

diversity, dual Diversity by combining two received signals.

diversity, frequency Diversity by combining two signals at slightly different frequencies in the same band.

diversity, polarization Diversity by combining two signals with different polarization angles. This can be done by using two different feed horns, one for each angle of polarization, and one shared parabolic antenna.

diversity, quadruple Diversity by combining together four signals received over different paths, or using different frequencies or polarizations.

diversity, space Path diversity obtained by positioning antennas at different locations or at different heights on the same tower.

diversity combiner A circuit or device for combining two or more signals carrying the same information received via separate paths or channels with the objective of providing a single signal which is superior in quality to the contributing signals.

diversity factor The ratio of the sum of the individual maximum demands of the various parts of a power distribution system to the maximum demand of the whole system. The diversity factor is always greater than unity.

diversity reception That method of radio reception whereby, in order to minimize the effects of fading, a resultant signal is obtained by combination or selection, or both, of two or more independent sources of received-signal energy which carry the same modulation or information, but which may vary in their fading characteristics at any given instant.

diverter, call Device which diverts calls from a barred code to an operator.

divestiture The dividing-up of AT&T, 1984. See: Bell System.

divider, frequency Circuit device whose output is a sub-multiple of the input frequency.

divider, voltage A resistance across a potential difference, tapped so that different proportions of the original voltage may be obtained as output.

division code A method of designating a facility, usually a PBX, which is directly connected to the switching network via leased lines. Also, a method of dialing where a division code (comprised of two or three digits) and extension digits (comprised of four or five digits) are dialed and the switching system routes the call to the specific facility and outpulses the extension digits to direct the call to the specific extension.

DLRD Design layout report date: a Bell System Universal System service order issued to other common carriers, (OCC), the date by which the design layout report (DLR) or termination layout report (TLR) must be forwarded to the design OCC representative.

D/MOS Diffusion/metal-oxide semiconductor; MOS circuits or transistors made by a double diffusion process.

do-not-disturb service A call to a subscriber's number is automatically transferred to an answering machine in the public exchange, when the subscriber has requested the service by dialing a special code.

docket 1. Papers giving details of FCC investigations and determinations, e.g., Docket 80-54, which authorized other common carriers to provide long distance services, and Docket 81-893, dealing with costs and payments for wiring in customers' premises. 2. A trouble report or ticket or fault record.

documentation A written description of a program; any record that has permanence and can be read by humans or machines.

DoD master clock The U.S. Naval Observatory master clock has been designated as the DoD (Dept. of Defense) master clock to which DoD time and frequency measurements are referenced (traceable).

domain 1. Region in a crystalline structure with parallel electric fields for all molecules. 2. The range of values assumed, e.g., the domain of a trunk may be incoming, outgoing or bothway.

domain, magnetic Region in a magnetic material where direction of magnetization is uniform.

domestic satellite (DOMSAT) A satellite system used purely for national telecommunications services. Some DOMSAT services are provided by using transponders in INTELSAT satellites, on lease from the international body responsible; other national systems use their own private satellites.

domestic satellite carrier Common carrier which owns or leases satellite facilities to provide communications services within the U.S.

dominant carrier The major provider of service in a particular market, e.g., AT&T Communications is likely to be the dominant long distance carrier in many areas. Dominant carriers are more stringently regulated than minor carriers.

donor Material added to a semiconductor to increase the number of free electrons; this produces an n-type semiconductor.

dopant An impurity added to a super-pure semiconductor in order to produce the required electrical qualities.

doping The adding of impurities to semiconductors to produce n-type or p-type material.

doppler effect The phenomenon evidenced by the change in the observed frequency of a sound or radio wave caused by a time rate of change in the magnitude of the radial component of relative velocity between the source and the point of observation.

dot-cycle One total segment of a signal with two alternating signaling conditions.

dot-matrix printer A printer which, instead of having a type wheel, has five or more carefully spaced and segmented wires, elements of which are pushed against the printing paper under the control of solenoids so that a dot-pattern is printed for each character. Variants of this technique use tiny ink-jets, sometimes with a matrix of up to 20 x 15 spots per letter enabling suitably programmed printers to print Chinese, Japanese or Arabic characters.

double armor The application of two layers of armor wires to a submarine cable to provide a high order of abrasion resistance and high-breaking strength to resist parting by dragging anchors and trawls.

double crucible technique A method of fabricating an optical waveguide by melting core and cladding glasses in two suitably joined concentric crucibles and then drawing a fiber from the combined melted glass.

double-dog Detent which holds the shaft of a two-motion switch in its operated position during a call.

double-ended control For digital networks: a synchronization control system between two exchanges is double-ended if phase error signals used to control the clock at a particular exchange are derived from comparison of the phase of the incoming digital signal and the phase of the internal clock at both exchanges.

double frequency-shift keying A multiplex system in which two telegraph signals are combined and transmitted simultaneously by a method of frequency shifting among four radio frequencies.

double heterostructure Layer sequence in an optoelectronic semiconductor device in which the active layer is bounded by two cladding layers with higher band gaps.

double-hop path Radio signal which has gone up to the ionosphere twice during its journey from transmitting station to receiver.

double-pole Contacts which open or close both sides of the same circuit simultaneously.

double precision Quantity with twice as many digits as are normally carried in a specific computer word.

double precision arithmetic Arithmetic giving more accuracy than a single word of computer storage will provide; two computer words are used to represent a single number.

double sideband reduced-carrier transmission Double sideband transmission in which the carrier's power is reduced substantially so that more of the authorized power can be utilized for intelligence-carrying sidebands. Used for some Citizens Band equipment to improve performance.

double-sideband suppressed carrier transmission That method of transmission in which the frequencies produced by the process of amplitude modulation are symmetrically spaced both above and below the carrier. The carrier level is suppressed to a predetermined value below the level of the transmitted sidebands.

double-sideband transmission That method of sideband transmission in which both sidebands are transmitted.

double-throw Switch which changes a circuit over from one output to another.

double window An optical fiber having desirable transmittance characteristics in both first and second window regions.

doubler Circuits providing frequency or voltage doubling, or special contact arrangements in a crossbar switch which increase flexibility.

doubler, frequency Circuit which provides output current of double the input frequency.

doubler, voltage Circuit which uses both halves of an alternating input to produce an output of double the D.C. voltage.

doublespace, telegraph Condition when a telegraphy receive hub receives space signals simultaneously from two half-duplex telegraph circuits. Under these circumstances, the hub will send spaces out on all legs.

dowelpin, conduit An aligning pin used when laying some types of multi-way conduit to check that ductways are accurately aligned.

down-converter A type of converter which is characterized by the frequency of the output signal being lower than the frequency of the input signal. It is the converse of up-converter.

down-lead A leading-in wire from an antenna to a receiver or transmitter.

downlink That portion of a communication link used for transmission of signals from a satellite or airborne platform to a surface terminal. It is the converse of uplink.

down-loaded A programming method in which the program to be used in a remote terminal is sent 'down the line' from a central location to that terminal, and stored there for local use.

downstream (CATV) Signals traveling from the headend to subscriber's homes.

downtime The time when a system is not operating due to a fault condition.

drag coefficient Relates drag forces on a towed cable to towing velocity and cable diameter; mostly determined by towing tests.

drain Total current drawn by a load.

drain, busy-hour battery Maximum current draw of a central office during its busy hour.

drain, current Total current drawn from a source.

drainage connection A wire linking together metal cable sheaths and a galvanic anode, which makes the sheaths negative to ground and protects them from electrolytic action; cathodic protection.

drawbar pull The effective pulling force delivered when plowing in cables.

drawing, circuit description A circuit drawing and outline of its operation.

drawing, circuit schematic Drawing giving symbols or representations of components used therein.

drawing, detached schematic Simplified schematic circuit drawing in which relay contacts are not drawn next to the relay coil which operates them but at the most convenient point on the circuit.

drawing, single line Simplified schematic drawing in which one line is used to represent two or more wires which go together.

drawing, wire Pulling out a hot metal rod and making it into wire.

drawing, wiring layout A wiring diagram showing colors, gauges, etc., of wires and detailed layouts thereof.

draw-off gear For submarine cable systems: hold-back gear becomes draw-off gear when picking up. The draw-off device rotates slightly faster than the drum, exerting a pull tending to tighten the turns around the drum.

DRCS Distress radio call system: developed for maritime services involving the use of a satellite system.

dress To make cables look neat and tidy in cable vaults and on main distributing frames.

dresser, cable A very hard wooden block used by cable splicers to beat lead sleeves into shape prior to the sleeves being plumbed and joints spliced or closed.

drift Slow changes in frequency or attenuation, or the movement of electrons or holes in semiconductors.

drift compensation For a common channel signaling system: the process of adjusting for the difference in relationship of the backward acknowledgement information contained in the acknowledgement signal unit to the forward signal units it acknowledges which occurs as a result of drift in the bit rates of the data channels.

drift plug Tapered hardwood plug used to open lead cable sleeves and restore them to their original cylindrical shape before using them for a cable splice.

drift-space Space in a klystron tube in which electrons drift at their entering velocities and form bunches.

drill, bell hanger's Long drill for making holes in wood; it is designed to be used like a large needle because wire may be threaded through an eye hole near its cutting end.

drill, cable Small drill used for making holes in metal cable sheath to affix pressure testing gauge or valve.

drill, concrete core Slowly-rotating drill used to bore annular ring holes in concrete masonry.

drill, masonry Drill with hard tip, usually of carbide steel, to make holes in masonry.

drill, star A steel chisel used with a hammer to make round holes in masonry.

drilled well ground (DWG) A well drilled for the sole purpose of providing an earth interface for the ground conductor.

drive Input signal to a circuit stage, particularly to an amplifier.

driver Circuit supplying input to a high-power amplifying circuit.

driving current Optoelectronics: the electrical input current that drives a semiconductor light source.

DRO Destructive read operation: the act of reading information from the store of a computer which usually means that the store is cleared; if the information is to be retained in the computer's store after it has been read out, it is usually necessary for the information to be written-in to the store again as a special operation.

drop 1. An indicator in a manual operating position. 2. Wire from a cable terminal to a subscriber's premises. 3. A fall in potential or voltage.

drop, CATV subscriber Flexible, self-supporting cable which feeds a CATV subscriber from the tapping point or directional coupler on the CATV cable.

drop, clearing Old-style indicator in a cord circuit which operates as an indication that the call has been completed, and that the cords, at a manual operating position, should be disconnected.

drop, magneto Calling indicator for a line on an operating position.

drop, potential The differential in potential, often due to a voltage drop across a resistance.

drop, subscriber's Wire which runs from a cable terminal or distribution point to the subscriber's premises.

drop, switchboard An indicator on a manual operating position which indicates a subscriber's line or junction termination.

drop, voltage Difference in voltages measured at both ends of a device.

drop-and-insert That process wherein a part of the information carried in a transmission system is terminated (dropped) at an intermediate point and different information is entered (inserted) for subsequent transmission in the same position.

drop-back Way in which some electromechanical step-by-step switches drop back to their rest position after receiving and discriminating upon particular digits.

drop channel operation When one or more channels of a multichannel system are terminated (dropped) at any intermediate point between the end terminals of the system.

drop-in The presence of an extra and spurious bit in an output due to faulty reading of input signals. The absence of a bit from the output which was in the original input is called drop-out.

drop-out 1. A momentary loss in signal, causing errors and loss of synchronization especially with phone-line data transmission. Defined as an unexpected drop of at least 12 dB for more than 4 milliseconds. Bell standard allows no more than two such drop-outs per 15-minute period. 2. A failure to read a binary character from magnetic storage, generally caused by defects in the magnetic media, or failure in the read mechanism. 3. In magnetic tape, a recorded signal whose amplitude is less than a predetermined percentage of a reference signal.

drop-out value Value of current or voltage at which a relay will cease to be operated.

drop side Equipment facing the central office (rather than out to line).

drop wire Paired wires, insulated and under a common cover, which connect a subscriber's line from the terminal on the pole to the protector on the house.

dropped channel Channels terminated (dropped) before reaching the terminal of a multiplex system.

dropping resistor Resistor designed to carry current which will make a required dropped voltage available.

drum, cable 1. On a submarine cable ship, the principal part of a cable engine: the cylindrical power-driven and power-braked member around which the cable passes. 2. Cable drum is the British word for a cable reel: the large steel or wooden reel on which cables are stored.

drum factor In facsimile systems, the ratio of drum length to drum diameter. Where drums are not used, the ratio of the equivalent dimensions.

drum room The location on board a submarine cable ship where the instrumentation and control of the mechanical activity of cable laying is concentrated.

drum speed The angular speed of the facsimile transmitter or recorder drum, measured in revolutions per minute.

drum storage, magnetic Continuously rotating drum, coated with magnetic material, used for backing stores: data can be read from tracks on the drum by heads mounted at different levels.

dry 1. Contacts which do not carry dc. 2. Free from moisture.

dry circuit Circuit with ac voice currents, and no dc flow.

dry contacts Contacts which do not carry any direct current flow.

dry run To check a program on paper before trying it out on the computer itself.

dryer, air Dry air unit used to provide air for pressurizing telephone cables.

DSA board Dial system A board in an automatic network which provides operator assistance in a dial office.

DSCS The U.S. Air Force Defense Satellite Communications System's geostationary telecommunications satellite system.

D-side Secondary cables in a local distribution network, i.e., cables between a cabinet or flexibility point and a distribution point serving subscribers.

DTE clear request For data: A call control signal sent by the data terminal equipment (DTE) to initiate clearing.

DTE waiting For data: A call control signal condition at the data circuit terminating equipment (DCE)/data terminal equipment (DTE) interface which indicates that the DTE is waiting for a call control signal from the DCE.

DTF (date to follow): abbreviation used in a Bell System Universal System Service Order (USO) on an Intercompany Service Coordination order involving an independent company, the code which follows the due date (DD) to convey that a firm DD is being negotiated and will follow on a USO correction order.

DTL Diode transistor logic: a family of integrated circuits with input through a diode. The output of each diode gate is connected to a transistor which inverts and amplifies the signal thus compensating for losses in the diode gate. DTLs have been largely replaced by transistor-transistor logic.

DTMF Dual tone, multi-frequency: push button dialing.

dual access 1. The connection of a user to two switching centers by separate access lines using a single message routing indicator or telephone number. 2. In satellite communications, the transmission of two carriers simultaneously through a single communications satellite repeater.

dual cable (CATV) A method of doubling channel capacity by using two cables installed side by side to carry different signals.

dual channel office Telephone office in which code O is used to reach assistance operators and codes 110 or 211 to reach operators for toll services.

dual diversity The simultaneous combining of, or selection from, two independently fading signals and their detection through the use of space, frequency, angle, time, or polarization characteristics.

dual element charge plan Pricing plan for installation-related costs up to and beyond the standard network interface (SNI).

dual homing   The connection of a terminal so that it is served by either of two switching centers. This service uses a single directory number.

dual in-line package   A standard method of packaging integrated circuits with input/output pins bent at right angles and in lines along the two long sides of the unit so that they go straight into holes in a printed circuit board.

dual seizure   The condition which occurs when two toll centers attempt to seize the same circuit at the same time.

dual synchronous operation   A method of improving the reliability of a processing system by providing two processors which simultaneously perform the same workload in step synchronism with comparison of their outputs. If one processor fails, an immediate indication is obtained.

dual-tone multi-frequency signaling (DTMF)   A signaling method employing set combinations of two specific voice-band frequencies, one of which is selected from a group of four low frequencies, and the other from a group of either three or four relatively high frequencies.

dual-use access line   A subscriber-access line normally used for voice communications but which has special conditioning for use as a digital transmission circuit.

duct   1. A pipe or conduit installed underground and used for telephone distribution cables. 2. Layer of cold air under warm air experienced in some semitropical areas which takes microwave signals (including domestic TV) much further than nonanomalous propagation would lead one to believe possible.

duct, metal floor wiring   Various proprietary schemes of ducting to provide flexibility in telephone installation in offices.

duct, multiple tile   A factory made multiple clay tile duct in units of up to 12 ducts in a nest, which are used to construct a large duct route.

duct, rubber floor wiring   Various proprietary schemes of ducting to provide flexibility in telephone installation in offices.

duct, surface   A radio duct whose lower boundary is the Earth's surface.

duct height   For radio: the height above the Earth's surface of the lower boundary of an elevated duct.

duct-kilometer   Product of the number of ducts in a nest and the length of the duct system.

duct nest   A number of cable ducts provided and laid in one trench.

duct run   A duct system for distribution of telephone cables in a city.

duct thickness   For radio: the difference in height between the upper and lower boundaries of a tropospheric radio duct.

dummy   Item made to the same shape as a circuit element but with no operational role, e.g., dummy fuse or dummy heat coils.

dummy antenna   Network which can be connected to a radio transmitter and creates the same impedance and power dissipation characteristics as an antenna but does not radiate any power.

dummy load   Network with impedance matched to the transmission circuit under test, enables the circuit to be tested without reflected waves causing complications.

dump, to   To write all or part of the contents of a storage, usually from an internal storage, onto an external medium, e.g., to be displayed on an output device. This would be for a specific purpose, such as to allow other use of the storage, to provide a safeguard against faults or errors, or in connection with debugging.

dunnage   Wooden slats or boards that are placed between layers (flakes) of cable as it is coiled into a cable tank on a cable ship.

duodecimal   Coding to a base 12.

duplex   1. Two units in one. 2. Simultaneous two-way and independent transmission in both directions. Sometimes called full duplex.

duplex, half   Communication between terminals one direction at a time.

duplex control (telegraph)   Control conditions which inhibit space signals received on a half-duplex telegraph leg from being sent out again on the same leg.

duplex operation   Simultaneous transmission and reception.

duplex (DX) signaling   An extended range AC signaling system which involves the careful balancing of circuits.

duplexer   1. Microwave device utilizing tuned filters which permits transmitter and receiver to operate on the same antenna while stopping transmitter power feeding straight in to the receiver. 2. Device which combines audio and video signals on a single transmission path.

duration of a call   See call duration.

duration, pulse   Time between leading and trailing edges of the pulse.

dust-core   A magnetic core made of a powdered material, e.g., ferrite, giving low eddy-current loss at high frequencies.

duty cycle   Ratio of operating time to total elapsed time, as a percentage.

duty factor, pulse   The ratio of average pulse duration to average pulse spacing.

DVA   Designed, verified and assigned date: abbreviation used in Bell System Universal System Service Orders. The date on which central office and station installation forces are to verify and report whether everything required for the installation is available as prescribed by the service order. DVA is a positive report date to the control system.

dx   Distant; in amateur radiocommunication jargon, refers to the reception of a distant station.

dye laser   A laser whose active medium is an organic dye, generally in solution, with the liquid often encapsulated within a cell.

dynamic   Equipment in which the operating parameters are continually changing.

dynamic hazard   See glitch.

dynamic margin   Step-by-step attenuation is applied to a data test signal; the amount of attenuation added when the receive end gives total errors or "wipe-out" is noted, this becomes the benchmark for future trouble isolation.

dynamic memory   Memory devices in which the stored information decays over a period of time, only milliseconds in some circumstances. The design intention is that information would normally be read out only nanoseconds after it has been written-in.

dynamic multiplexing   Form of time-division multiplexing in which the allocation of time slots to constituent channels is made according to the actual demands of these channels. See statistical multiplexing.

dynamic non-hierarchial routing (DNHR)   Computer-controlled routing of long distance telephone calls dependent on actual traffic flow at the time, i.e., calls are not automatically passed right up the hierarchic structure to a Class I office for routing on a final choice circuit if high usage circuits are busy. The availability of relatively inexpensive digital switching centers enables through-circuits to be established on routes passing through several such centers in tandem without degrading the transmission quality of the overall circuit.

dynamic overload control (DOC)   1. A regional toll center control system which follows the growth of traffic congestion and automatically cancels or reroutes traffic when the toll office becomes too congested. 2. A network management control applied automatically in response to an external signal or an overload condition detected in the switch itself.

dynamic ram   A memory which stores data on capacitances between gate and source of a MOFSET. A dynamic memory.

dynamic range   1. In a transmission system, the difference in decibels between the noise level of the system and its overload level. 2. The difference, in decibels, between the overload level and the minimum acceptable signal level in a system or transducer.

dynamic variation (transient)   Short time variations outside of steady state conditions in the characteristics of power delivered to the communications equipment.

dynamo   A rotating machine, normally a dc generator.

dynamometer   1. An indicating instrument placed between the chain hoist and strand puller when tensioning guys or suspension strand. Reads tension 0-10,000 pounds. 2. Device to measure cable tension on board a cable ship during pick-up or pay-out.

dynamotor   A rotating machine which converts dc to ac.

dyne   A cgs unit of force, now rarely used. Equal to 10^{-5} Newton.

E layer   Layer in ionosphere about 65 miles above Earth's surface which reflects radiocommunication waves. The E layer tends to disappear during darkness.

E layer, sporadic   An occasional variation in ionization levels in the E layer, sometimes produces propagation/reception problems.

E & M signaling   A method of signaling between junction equipment in an exchange and a signaling unit associated with the transmission equipment using two leads — a receive (E) lead and a transmit (M) lead. The system is used where signaling is transmitted on a circuit separate from the speech path but associated with it, e.g., over carrier or radio systems. E&M type 1 is used between central offices with electromechanical switches. E&M type 2 is used between central offices with electronic switches. E&M type 3 is a partially looped system sometimes used between electronic offices.

Early Bird   The first geostationary telecommunications satellite operated by COMSAT. It was launched 6 April 1965 and used for communications between Europe and the U.S.

early contacts   Relay contacts which operate before other contacts operated by the same relay.

earphone   Transducer in telephone handset which accepts an ac signal and makes a diaphragm vibrate to produce sound waves.

earth   Ground; a large conducting body which is the zero level in the scale of electrical potential. An earth is a connection made either accidentally or by design between a conductor and earth. Common ways of obtaining good ground connection are to connect to main water pipes, taking care to ensure that these are not plastic. To earth (to ground) is to connect to earth, usually for safety reasons.

earth capacitance   Capacitance between any circuit or component and a point at ground potential.

earth coverage   In satellite communications, the condition obtained when a beam is sufficiently wide to cover the surface of the earth exposed to the satellite.

earth current   A current that flows to earth/ground, especially one that follows from a fault in the system. Also, currents which flow in the earth, due either to ionospheric disturbances or lightning, or to faults on power lines.

earth fault   A fault which occurs when a conductor is accidentally grounded/earthed, or when the resistance to earth of an insulator falls below a specified value.

earth potential   The potential taken to be the arbitrary zero in a scale of electric potential.

earth return circuit   A telecommunications circuit that has one metallic path and one path which depends on currents flowing back through the earth; early telephone and telegraph circuits were of this type.

earth stations   Ground terminals that use antennas and associated electronic equipment to transmit, receive and process communications via satellite. Syn.: ground station

EAS, optional   Optional extended area service; option provided to telephone subscribers either to retain previous tariffs with payment for every toll call, or to pay a higher tariff and be able to call nearby offices on a toll-free basis.

easement   Authority to enter land owned by another for the purpose of installing or maintaining telecommunications plant; a wayleave.

East terminal   Transmission systems in North America are usually defined as east-west or west-east, so one end or the other (usually east or north) is designated the East terminal, on an arbitrary basis.

EC lead   Extra control lead: an additional wire in some step-by-step central offices which enables class of service priorities to be given to certain lines.

eccentricity   Measure of the concentricity of conductors in a cable. This is especially important for coaxial cables to be used in wideband transmission systems.

echo   A wave which has been reflected or otherwise returned with sufficient magnitude and delay to be perceived. An effect which can be experienced on long distance calls. It is usually associated with relatively long round-trip delay in the four-wire portion of the circuit so that the disturbing sound is perceived as being separate in time from the wanted sound. The phenomenon of echo primarily affects the talker and not the listener. In television echoes appear as "ghosts" displaced to one side or another of the primary picture.

echo, talker   The echo heard by a talker, reflected from a mismatch point, sometimes at the distant end of a connection.

Echo I   The large (100 ft. diameter) plastic balloon, coated with aluminum, which was sent into low orbit around the Earth in August 1960 and used to reflect radio signals between different locations.

echo attenuation   In a two- or four-wire circuit, in which the two directions of transmission can be separated from each other, the attenuation of the echo signals is determined by the ratio of the transmitted power to the echo power received; expressed in dB.

echo canceller   The traditional form of echo suppressor (used on circuits with long transmission time) attenuates the direction of transmission which is not active so that any echoes which are fed into a circuit do not return to the starting point and cause confusion. Echo cancellation stops a received signal from being transmitted back to its origin by con-

structing a signal closely approximating the echo component and subtracting this from the locally transmitted signal. This avoids the difficulties caused by double-talking.

echo check   A method of checking the accuracy of transmission of data: the received data are returned to the sending end for comparison with the original data.

echo path loss   Total path loss that an echo encounters during a round trip.

echo plexing   A checking system: characters received at a central processor are echoed back to the terminal which originated them, and this terminal prints the received echo so the operator may check for any errors.

echo return loss   The difference, in dB, between the level of a composite-frequency signal sent into a circuit and the level of the echo of that signal as it reaches the point of application. The signal normally contains all frequencies between 500 Hz and 2500 Hz at equal amplitudes.

echo suppressor   A voice-operated device placed in the four-wire portion of a long telephone trunk that blocks or attenuates transmission in the return direction when a speaker is talking. This switching action prevents a speaker from hearing echoes.

echo suppressor, controlled   Echo suppressor associated with a circuit but only activated when the toll office or gateway office concerned switches it on, typically used only on transit calls.

echo suppressor, differential   Echo suppressor in which the action is controlled by the difference in level between the signals on the two speech paths.

echo suppressor, fixed   Echo suppressor associated with a circuit which is permanently switched on.

echo suppressor, full   Echo suppressor in which the speech signals on either path control the suppression loss in the other path.

echo suppressor, half   Echo suppressor in which the speech signals of one path control the suppression loss in the other path but this action is not reciprocal.

echo suppressor, terminal   Echo suppressor designed for operation at one or both terminals of a circuit.

echo suppressor disable   A tone signal, often 2025 Hz, sent to disable echo suppressors so that circuits may be used for non-voice signals such as data.

echo suppressor testing system   A set designed to test all the operational characteristics of echo suppressors assigned to all categories of international circuits.

echo tolerance   Subjectively obtained figures for the loudness of an echo which can be tolerated and how this varies with the delay before the echo is heard. If the echo is heard only one or two milliseconds after the words have been spoken a loss in the echo path of only 3dB is sufficient, but if the delay is longer the echo signal has to be attenuated much more before it is considered tolerable, e.g., 30dB for a 100ms delay.

echoplex An echo check applied to network terminals operating in the two-way simultaneous mode.

echosounder Shipboard equipment which measures the distance between the ship's bottom and the ocean floor by sending out ultrasonic pulses via a transducer which are reflected at the sea bottom and received with the same transducer; the elapsed time between sending the pulse and receiving the echo is a measure of the depth. A PDR (precision depth recorder) is often used with the echosounder to provide an increased degree of accuracy.

eclipse, satellite When one orbiting body passes between two others, e.g., when a geostationary satellite is hidden from the sun by the shadow of the Earth.

ecliptic The apparent path of the sun around the earth.

economic standard antenna The directional antenna with specified characteristics, such as directive gain and service sector at its operating frequencies, which are justifiable on economic grounds.

eddy current Locally circulating current, in the cores of transformers, which produces heat and loss and is eliminated whenever possible.

edge connector A track on a printed circuit board or printed wiring board taken to one edge of the board where it may be plugged into a suitable socket.

Edison base The screw thread used on domestic lighting fittings in the U.S.

edit To rearrange information for machine output or input.

editor A special computer program that allows changing, moving and general editing of statements.

education channel FCC rules require cable systems in the top 100 markets to set aside one channel for educational uses, to be available without cost for the "developmental period." The developmental period of a CATV system in the U.S. runs for five years from the time that subscriber service began, or five years after the completion of the basic trunk line.

Educational Television Noncommercial service providing transmission of TV signals to schools.

effect, end The effect of capacitance at the ends of a dipole rod antenna, such that the half-wave dipole is about 5% less than exactly half the wavelength.

effective antenna length The ratio of the open-circuit voltage of an antenna to the electric field intensity.

effective cable pairs Total number of pairs of wires available for use in a local distribution network (i.e., terminated at a distant distribution point and at the central office or exchange).

effective call A telephone call which has been set up and answered.

effective data transfer rate The average number of bits, characters, or blocks per unit time transferred from a data source and accepted as valid by a data sink. It is expressed in bits, characters, or blocks per second, minute, or hour.

effective ground Connection to ground through a medium of sufficiently low impedance and adequate current-carrying capacity to prevent the building up of voltages which may be hazardous to equipment or personnel.

effective height 1. The height of the center of radiation of an antenna above the effective ground level. 2. In low frequency applications involving loaded or nonloaded vertical antennas, the moment of the current distribution in the vertical section divided by the input current.

effective input noise temperature The source noise temperature in a two-port network or amplifier that will result in the same output noise power, when connected to a noise-free network or amplifier, as that of the actual network or amplifier connected to a noise-free source.

effective monopole-radiated power The power supplied to an antenna, multiplied by its gain in a given direction, referred to that of a short vertical antenna in the horizontal direction.

effective radiated power The power supplied to the antenna multiplied by the power gain of the antenna in a given direction.

effective radius of the Earth Radius of a hypothetical spherical Earth having no atmosphere, for which propagation is rectilinear, and for which the distance to the horizon is the same as that for the actual Earth enveloped in an atmosphere having a constant vertical gradient of refractivity. (For the standard radio atmosphere the effective radius is four-thirds that of the true radius.)

effective resistance The increased resistance of a wire to ac compared with dc. This is because higher frequencies tend to travel only on the outer skin of the conductor whereas dc flows uniformly through the entire area.

effective speed of transmission The rate at which information is processed by a transmission facility expressed as the average rate over some significant time interval. This quantity is usually expressed as the average number of characters or bits per unit time. Syn.: average rate of transmission

effective traffic The traffic intensity corresponding to the call durations.

effective transmission A system of rating transmission performance based upon subjective tests of repetition rates.

effective value Of an alternating voltage or current: the root mean square value. (Effective value is a deprecated term.)

effectively transmitted signals in sound-program transmission For sound-program transmission, a signal at a particular frequency is effectively transmitted if the nominal overall loss at that frequency does not exceed the nominal overall loss at 800 Hz by more than 4.3 dB. For sound-program circuits, the overall loss (relative to that at 800 Hz) defining effectively transmitted frequency is 1.4 dB, i.e., about one-third of the allowance.

efficiency The useful power output of an electrical device divided by the total power input, expressed in percent.

efficiency, ampere-hour Number of ampere-hours which result from a test discharge of a storage battery, divided by the number of ampere-hours needed to recharge the battery.

efficiency, data transmission The number of data bits correctly received divided by the total number transmitted; the through-put rate.

efficiency, radiation For an antenna: ratio of power radiated to total power supplied to the antenna.

efficiency, watthour For a storage battery: ratio of watthours output to watthours needed for recharge.

efficiency factor In telegraph communications, the ratio of the time to transmit a text automatically and at a specified modulation rate, to the time actually required to receive the same text with a specified error rate.

efficiency rate The ratio of the number of effective or completed call attempts to the total number of call attempts, as measured at a given point of a network.

eight-party line A common telephone line arranged to serve eight main stations. Multiparty ringing procedures give selective ringing.

eighty-column card A punched card with 80 vertical columns; used in conjunction with the Hollerith punch code.

EIR Engineering information report: abbreviation used in Bell System Universal System Service Orders. A report from a local control design group to the design control group for the purpose of design coordination, which includes minimum design data pertaining to equipment and facility assignments, requisition numbers, and station termination equipment.

EIRD Engineering information report date: abbreviation.used in Bell System Universal System Service Orders. The date on which the design control group should have received an EIR from local design groups.

Ekran Russian geostationary TV distribution satellite.

elapsed time Total time during which a circuit is occupied by a connected call.

elastance The reciprocal of capacitance, expressed in darafs.

elasticity of demand Term used in tariff discussions to indicate the relationship between demand for a service and the price charged.

electret A substance that is permanently charged electrically; the electric equivalent of a permanent magnet.

electric Operated by, produced by, producing, transporting, or using electricity.

electric field strength Strength of the electric field at a point, measured in volts per meter.

electric flux Quantity of electricity displaced across a given area in a dielectric, measured in coulombs.

electric vector The electric field vector associated with a light wave. The electric field vector specifies the polarization and amplitude of the electric field.

electrical degree One complete cycle of alternating voltage (or current) may be divided into 360 degrees.

electrical length The length expressed in wavelengths, radians, or degrees. When expressed in angular units, it is the distance in wavelengths multiplied by 2π to give radians, or by 360 to give degrees.

electrically alterable read-only memory (EAROM) A memory unit made up in such a way that electrical pulses on appropriate pins can erase some or all of the stored data so that new information can then be written in.

electrically erasable read-only memory (EEROM) A memory circuit in which an applied electric pulse erases all the stored memory, so that it must be rewritten.

electro-acoustic transmission measuring system A system used to judge a transmission system's performance by giving the loss in db from the acoustical power (in millibars) into a transmitter to the acoustical power out of the receiver at the end of the circuit. Noise and bandwidth impairments are assumed to be negligible.

electrocardiogram data set Modem and coupling unit which permits a doctor to check on a patient's heart condition and send the information by any telephone line to a hospital for study, without having to move the patient to a hospital or clinic.

electrochemical recording Facsimile recording by means of a chemical reaction brought about by the passage of a signal-controlled current through the sensitized portion of the record sheet.

electrode A terminal or an element which controls the flow of electricity through a medium.

electrode potential Potential of an electrode when immersed in an electrolyte compared with the potential of a hydrogen electrode.

electrode-potential series Chemical elements arranged in a series in order of their electrode potentials.

electroluminescence Nonthermal conversion of electrical energy into light in a liquid or solid substance. An example is the photon emission resulting from electron-hole recombination in a pn junction (the mechanism involved in the injection laser).

electrolysis Chemical changes caused by the passage of electric current through an electrolyte.

electrolysis, solid state The migration of ions from one contact to another even if there is no liquid electrolyte.

electrolyte Non-metallic conductor of electricity in which current is carried by the physical movement of ions.

electrolytic Relevant to the magnetic and electric fields which are produced by electric currents.

electrolytic recording That type of electrochemical facsimile recording in which the chemical change is made possible by the presence of an electrolyte.

electromagnet A device which becomes magnetized only when an electric current flows through windings.

electromagnetic compatibility The condition which prevails when telecommunications equipment is collectively performing each of its individual functions in a common electromagnetic environment without causing or suffering unacceptable degradation due to electromagnetic interference to or from other equipments/systems in the same environment.

Electromagnetic Compatibility Figure of Merit (EMC FOM) A tentative universal equipment rating method for single channel voice radiocommunication systems; set up by a committee of experts who reviewed parameters and selected three most significant parameters for receivers and three for transmitters. Significant in that it points to possible future acceptance for a wide variety of telephone equipment.

electromagnetic emission control The control of friendly electromagnetic emissions, e.g., radio, radar, and sonar transmissions, for the purpose of preventing or minimizing their use by unintended recipients.

electromagnetic environment The power and time distribution, in various frequency ranges, of the radiated or conducted electromagnetic emission levels which may be encountered by the equipment, subsystem, or system when performing its assigned mission. The electromagnetic environment may also be expressed in terms of field strength.

electromagnetic field Electric and magnetic fields associated with radio and light waves.

electromagnetic induction The production of an electromotive force in a conductor caused by a change in the magnetic flux through the conductor.

electromagnetic interference The phenomenon resulting when electromagnetic energy causes an unacceptable or undesirable response, malfunction, degradation, or interruption of the intended operation of the electronic equipment, subsystem, or system. Syn.: radio frequency interference

electromagnetic interference control The control of radiated and conducted energy such that the emissions unnecessary for system, subsystem, or equipment operation are minimized or reduced. Electromagnetic radiated and conducted emissions, regardless of their origin within the equipment, subsystem, or system are therefore controlled. Successful EMI control, along with susceptibility control, leads to electromagnetic compatibility.

electromagnetic lens An arrangement of coils used to focus an electron beam, e.g., in a TV tube.

electromagnetic pulse A broadband, high-intensity, short-duration burst of electromagnetic energy. A high-altitude nuclear explosion could produce damaging fields several thousand kilometers away from the explosion itself, making unprotected radio and line transmission and switching systems inoperative.

electromagnetic radiation Radiation made up of oscillating electric and magnetic fields and propagated with the speed of light. Includes gamma radiation, X-rays, ultraviolet, visible and infrared radiation, and radar and radio waves.

electromagnetic spectrum The frequencies (or wave lengths) present in a given electromagnetic radiation. A particular spectrum could include a single frequency or a wide range of frequencies. See Appendix B.

electromagnetic survivability The ability of the equipment, subsystem, or system to resume functioning without evidence of degradation following temporary exposure to an adverse electromagnetic environment. This implies that system performance will be degraded during exposure to the adverse electromagnetic environment but the system will not experience any damage.

electromagnetism The study of phenomena associated with varying magnetic fields, electromagnetic radiation, and moving electric charges. In the telecommunications field, electromagnetic often describes electro-mechanical apparatus which depends on magnetic phenomena, e.g., relays, two-motion switches, or crossbar switches, as contrasted with electronic apparatus which, in general, has no moving parts and requires no adjustments during its service life.

electromechanical exchange A telephone exchange in which speech paths are switched by metallic contacts actuated by electromagnetic devices.

electromechanical recording Recording by means of a signal-actuated mechanical device.

electromotive force The force that produces movement of electric charges, measured in volts.

electron A stable elementary particle with a negative charge which is mainly responsible for electrical conduction. Electrons move when under the influence of an electric field. This movement constitutes an electric current.

electron, valence An outer electron in an atomic structure. Valence electrons play important roles in chemical actions and in the flow of electric current. In n-type semiconductors, valence electrons are the majority carrier whose movement determines transistor performance.

electron beam A beam of electrons emitted from a source, e.g., a cathode ray tube has a thermionic cathode electron gun which produces the beam.

electron gun A hot cathode that produces a finely focused stream of fast electrons, which are necessary for a cathode ray or television tube. The gun is made up of a hot cathode electron source, a control grid, accelerating anodes, and usually focusing electrodes. This device makes up an essential part of instruments such as cathode ray tubes, TV receivers, video display units, and electron microscopes.

electron lens A device for focusing an electron beam. These can be either magnetic (by external coils creating magnetic fields within the tube) or electrostatic (by metallic plates within the tube which are charged electrically in such a way as to divert the moving electrons in the beam).

electron volt Energy acquired by an electron in passing through a potential difference of one volt, in vacuum.

electronic Description of devices (or systems) which are dependent on the flow of electrons in electron tubes, semiconductors, etc, and not solely on electron flow in ordinary wires, inductances, capacitors, etc.

electronic central office Modern telephone central office using solid state devices (not electromechanical switches).

electronic computer originated mail (ECOM) The U.S. Postal Service's version of electronic mail.

electronic counter-countermeasures That division of electronic warfare involving actions taken to insure friendly effective use of the electromagnetic spectrum despite the enemy's use of electronic warfare.

electronic countermeasures That division of electronic warfare involving actions taken to prevent or reduce an enemy's effective use of the electromagnetic spectrum.

electronic data processing Data processing performed by electronic machines, as compared to automatic data processing which does not involve electronically controlled or operated machines.

electronic data processing machine Any machine or device used in data processing which uses electronic circuitry in order to perform logical and arithmetic operations.

electronic deception The deliberate radiation, reradiation, alteration, absorption, or reflection of electromagnetic energy in a manner intended to mislead an enemy in the interpretation or use of information received.

electronic emission security Those measures taken to protect all transmissions from interception and electronic analysis.

electronic funds transfer system (EFTS) A data and communications system used by many banks to speed up financial transactions.

Electronic Industries Association (EIA) A U.S. organization made up of manufacturers of a wide variety of electronic products, including telecommunications equipment; active in industry standard-setting.

electronic intelligence (ELINT) The intelligence information product of activities engaged in the collection and processing, for subsequent intelligence purposes, of foreign, noncommunications, electromagnetic radiations emanating from sources other than nuclear detonations and radioactive sources.

electronic jamming The deliberate radiation, reradiation, or reflection of electromagnetic energy with the object of impairing the use of electronic devices, equipment or systems being used by an enemy.

electronic line scanning A method of scanning that provides motion of the scanning spot along the scanning line by electronic means.

electronic mail System for transmitting documents via telecommunications facilities: includes facsimile, teletex, and mailbox services.

electronic memory A memory in which read and write operations are entirely electronic, includes ferrite cores, magnetic bubbles, and MOS devices.

electronic message service (EMS) Broadband digital service which uses satellites and is able to bypass the distribution network of local telephone companies.

electronic message-service system (EMSS) The U.S. Postal Service's nationwide electronic mail service.

electronic PABX Private automatic branch exchange in which switching is accomplished by the flow of electrons through solid state devices rather than by electromechanical means.

electronic scratch pad A memory area used as a temporary working area for intermediate results.

electronic switching system Any switching system whose major components utilize semiconductor devices.

electronic tandem switching A feature offering on some PBXs which allows the PBX to perform many of the functions formerly provided only by switches designed specifically for tandem applications.

electronic warfare Military action involving the use of electromagnetic energy to determine, exploit, reduce, or prevent hostile use of the electromagnetic spectrum and action which retains friendly use of the electromagnetic spectrum.

electronics Pertaining to that field of science and engineering that deals with electron devices and their utilization.

electro-optic effect A change in a material's refractive index under the influence of an electric field. Pockels and Kerr effects are electro-optic effects that are respectively linear and quadratic in the electric field strength.

electroplate To coat with a deposit of metal by electrolytic action.

electropositive Refers to an electrode which is more positive than the reference electrode, or to a chemical element more positive than hydrogen, which is zero in the electrode-potential series.

electrostatic Pertaining to electric charges at rest, with no electric current flowing.

electrostatic deflection The use of electrostatic fields produced between metallic deflection plates within electron-beam devices, such as TV tubes.

electrostatic field Space in which there is electric stress produced by static electric charges.

electrostatic headphone A sound receiver in which the movements of the diaphragm are controlled by varying dc voltages between two acoustically transparent metal plates; the diaphragm itself is also conducting and is placed between the two metal plates.

electrostatic induction Inducing static electric charges on bodies by bringing them near other bodies which carry high electrostatic charges.

electrostatic printing A type of printing, now common in photocopying, in which a pattern of electrostatic charges is produced on the surface of paper, a very fine powder is then applied to the paper, which sticks to the charged lines and letters and is made permanent by applying heat.

electrostatic recording Recording by means of a signal-controlled electrostatic field.

electrothermal recording Recording produced principally by signal-controlled thermal action.

element   1. In chemical sense: substance that consists of atoms of the same atomic number. Elements are the basic units in all chemical changes other than those in which atomic changes, such as fusion and fission, are involved. 2. In general sense: component or basic part of a whole.

element, circuit   Components which, when interconnected, form an electrical circuit.

element, code   One of the component parts of a code signal.

element, driven   Part of an antenna which is energized directly from the transmitter via the antenna feed.

element, parasitic   Element of an antenna which is not fed directly by the transmitters (or connected electrically with a receiver) but reflects or directs radio waves to provide directivity to the antenna.

element, radiating   An active section of an antenna.

element, signal   A basic unit of a data signal.

element error rate   In telegraph distortion: the ratio of the number of incorrectly received elements to the number of emitted elements.

elemental area   In facsimile transmission systems, any segment of a scanning line of the subject copy the dimension of which along the line is exactly equal to the nominal line width.

elementary particle   Any of the particles of matter which cannot be further subdivided.

elevated duct   A tropospheric radio-duct in which the lower boundary is above the surface of the Earth.

eleventh rotary step   In step-by-step electromechanical central offices, when a switch hunts over a level of 10 outlets and cannot find a free one, it moves onto the eleventh step and returns a busy tone to the caller.

eliminator, battery   A rectifier with low impedance and well-filtered output capable of providing quiet speech battery power for a very small central office or a PABX without needing a storage battery.

ellipsometer   Instrument which measures the ellipticity of the polarization of light, usually to measure thickness of very thin transparent films.

elliptical orbit   A closed satellite orbit in which the distance between the centers of mass of the satellite and of the primary body is not constant.

elliptically polarized   Electromagnetic wave whose plane of polarization rotates continually as the wave propagates forward. The magnitude of the wave varies, depending on the instantaneous value of the angle of polarization.

emanations security   The protection resulting from all measures designed to deny unauthorized persons information of value which might be derived from intercept and analysis of compromising emanations from other than crypto-equipment and telecommunications systems.

embedded base   Customer premises equipment (CPE) which was in service for a Bell Operating Company before January 1, 1984.

embedded direct analysis (EDA)   Method of calculating costs and revenues of Bell Operating Companies.

embedment   A process of protecting an electronic part or assembly with an insulating material which fills in all voids and needs a mold or container while it is hardening.

emergency call service   A fast, easy means of giving information about an emergency situation to the appropriate emergency organization (e.g., fire brigade, police, ambulance).

emergency changeover   A modified common channel signaling changeover procedure to be used whenever the normal one cannot be accomplished, i.e., in case of a failure in the signaling terminal equipment or of inaccessibility between the two involved signaling points.

emergency manual line   Temporary telephone service provided under emergency conditions giving connection to an operator.

emergency position indicating radio beacon (EPIRB)   Used for maritime services, involving the use of a satellite system. This has been developed by IMCO, the Inter-Governmental Maritime Consultative Organization.

Emergency Reporting System (ERS)   A network of manual street telephones in many U.S. cities which are connected to a central ERS PBX, where any emergency may be reported.

emergency restart   The procedure of reestablishing signaling communication, when all other signaling links fail, on a common channel signaling system.

emergency ring-back (ERB)   System feature used to recall a subscriber station immediately after disconnection of a call to an emergency bureau or operator.

emergency route   The circuits to be used in case of complete interruption or major breakdown of primary and secondary routes. Such an emergency route (for international calls) may pass through any country.

emission   1. Radiation produced, or the production of radiation, by a radio transmitting system. The emission is considered to be a single emission if the modulating signal and the other characteristics are the same for every transmitter of the radio transmitting system and the spacing between antennae is not more than a few wavelengths. 2. Release of electrons from the cathode of a vacuum tube.

emission, parasitic   Spurious emission, accidentally generated at frequencies which are independent both of the carrier of characteristic frequency of an emission and also of frequencies of oscillations resulting from the generation of the carrier or characteristic frequency.

emission, primary   Emission of electrons because of high temperature or high negative potential of the item.

emission, radio   Electromagnetic radiation in a radio frequency band. A classification and designation scheme for radio emissions is given in CCIR Radio Regulations.

emission, secondary   In an electron tube, emission of electrons by a plate or grid because of bombardment by primary emission electrons from the cathode of the tube.

emission, spurious   Emission outside the radio frequency band authorized for a transmitter.

emission, thermionic   Emission from a cathode due purely to its high temperature.

emission beam angle between half-power points   Angle centered on the optic axis of a light emitter within which the radiant power output is not less than half the maximum.

emissivity   The ratio of power radiated by a substance to the power radiated by a blackbody at the same temperature. Emissivity is a function of wavelength and temperature.

emitter   The region of a transistor from which the carriers flow into the base.

emitter-coupled logic   (ECL or CML, current mode logic) A family of logic circuits in which transistor pairs are coupled together by their emitters. Such logic circuits can be made to operate extremely rapidly, in 10^{-6} second.

emitter follower   Transistor using a grounded collector.

emitter-follower logic   A form of logic in which pnp and npn transistors are combined together in an integrated circuit.

emphasis   The intentional alteration of the frequency-amplitude characteristics of a signal to reduce adverse effects of noise in a communication system. An example is the pre-emphasis used in the transmission of a frequency modulated wave.

empirical   Not based on pure theory, but on practical and experimental work.

empty slot   Transmission method used in some ring-type local area networks: packets circulate around the ring, either containing data traffic or empty. A sending station waits for (and fills) an empty slot packet.

emulation   Using one system to imitate the capabilities of another system.

emulator   A program or hardware device which duplicates the instruction set of one computer on a different computer. This allows the development of programs for the emulated computer without that computer being directly available at the same site.

enable   To prepare a circuit for operation or to allow an item to function.

enabling signal   A signal that permits the occurrence of an event.

encapsulation   Providing a protective outer casing, usually of transparent plastic.

encode   1. To apply a code, frequently one consisting of binary numbers, to represent individual characters or groups of characters in a message. 2. To substitute letters, numbers, or characters for other numbers, letters, or characters, usually intentionally to hide the meaning of the message. 3. To convert plain text into unintelligible form by means of a code system.

encoded information type   The code and format of information that appears in the body of an

interpersonal message in a Message Handling System. Examples are teletex, facsimile, voice.

encoder   In computation, a network with one input and a combination of outputs.

encoding   The generation of character signals to represent quantized samples.

encoding law   The law defining the relative values of the quantum steps used in quantizing and encoding.

encrypt   To convert plain text into unintelligible form by means of a cryptosystem.

encrypted voice   Telephone communications which are protected against compromise through use of an approved ciphony system. Syn.: secure voice

encryption   Conversion of plain language (clear text) into a coded signal.

encryption, bulk   Feeding of several information channels to a single encryption device.

encryption, end-to-end   Encryption with the same keys and devices available at both send and receive terminals, so the message is transmitted entirely as an encrypted message.

encryption, link   Encryption of various segments of a multi-link circuit separately.

encryption, super   Higher grade of encryption sometimes used, with end-to-end encryption supplemented by link-to-link encryption of the already encrypted message.

end cells   Three extra storage cells in a central office which are switched into service to aid the usual 23 cells if the rectifier fails temporarily, in order to maintain the required voltage on the switching equipment.

end delay   Round trip time delay between a point of echo reflection and an echo suppressor.

end distortion   In start-stop teletypewriter operation, the shifting of the end of all marking pulses from their proper positions in relation to the beginning of the start pulse.

end instrument   Terminal set on the end of a line, e.g., telephone set, data terminal, etc.

end of block   A set of characters which indicates that the block of information being transmitted has been completed.

end-of-dialing signal   Signal sent after sending an abbreviated dialing code signal, so the office knows it is an abbreviated code and not an incompletely dialed normal code.

end of file (EOF)   In data processing, the end of a particular set of records.

end of header   A set of characters which separates the address to which a message is to be sent from the message itself.

end-of-message functions   In tape relay procedure, the letter and key functions, including the end of message indicator, which comprise the last format line.

end-of-pulsing signal   A telephone signaling address signal sent in the forward direction indicating that there are no more address signals to follow.

end of run (EOR)   In data processing, the completion of a single program or programs forming a run, usually of similar or directly related routines, actions and programs.

end of selection The character in telegraphy and data which indicates the end of selection signals.

end of tape (EOT) A marking strip near the end of the permissible recording area on magnetic tapes. This strip generates an EOT signal indicating that a reel change is necessary.

end-of-text character A transmission control character used to terminate a text.

end of transmission A standardized, uninterrupted sequence of characters and machine functions used to terminate a transmission and disconnect the circuit and transmitting equipment.

end-of-transmission-block character A transmission control character used to indicate the end of a transmission block of data.

end-of-transmission character A transmission control character used to indicate the conclusion of a transmission that may have included one or more texts and any associated message headings.

end office A Class 5 office in the North American hierarchic routing plan; a switching center where subscribers' loops are terminated and where toll calls are switched through to called lines.

end section 1. An operating position added to the end of a suite so that all operators may readily access all terminations. 2. Distance between the end of a loaded cable and the first loading coil.

end-to-end 1. A function which is carried out from one end to another of a circuit. 2. Connection of circuit elements in series.

end-to-end responsibility Assignment to a common carrier of responsibility for equipment design, operation and maintenance.

end-user Subscriber who uses (rather than provides) telecommunications services.

ends, clearing Preparing the end of a cable for splicing to another cable.

energize To connect to a source of power.

energized Switched on; alive; powered-up; electrically connected to a source of potential difference.

energy communication services (ECS) An AT&T enhancement to its Dimension® PABX, which provides centralized control for up to 64 individual energy-consuming devices at each of up to 14 remote locations.

energy density The energy per unit area of a light beam, expressed in Joules per square meter. Syn.: irradiance

engine, cable A submarine cable term for the facilities that pick up and pay out cable from a cableship.

engine, drum cable Alternative name for type of cable engine on submarine cable-laying ships.

engine, linear cable Newer type of cable engine which utilizes flexible tracks which grip the cable and feed it overboard at a carefully controlled rate.

engine, pay-out Submarine cable term for the mechanism for controlling the overboarding of cable and repeaters during a normal cable-laying operation.

engine, propane gas Small gas-powered engine used by cable splicers.

engineer, communications Engineer who specializes in design, maintenance, and operation of telecommunications facilities and equipment.

engineer, professional An engineer with the education and experience to qualify for the responsibility of important engineering work, and who is, in many countries, registered as a professional engineer by a state authority.

engineered circuit Circuit which has been designed and made ready to meet a special requirement.

engineering, circuit record The design of a complete circuit from one terminal to the other and taking into account all relevant factors, including economics, transmission, and maintenance.

engineering, electrical Branch of engineering dealing with the principles and practices of electricity in all its ramifications.

engineering, electronic Branch of electrical engineering specializing in light current, radio communication, transmission and switching procedures and practices.

engineering, television Branch of electronic engineering specializing in television procedures and practices.

engineering, traffic Branch of electrical engineering with special responsibilities for mathematical, statistical and economic matters concerning the design of switching plant.

engineering orderwire (EOW) A communication path for voice and/or data that is provided to facilitate the installation, maintenance, restoral, or deactivation of segments of a communication system by equipment operators, attendants, and controllers.

engineering route A set of one or more traffic routes, entering and leaving the transmission network at the same terminal transmission nodes, and treated as a unit for transmission routing purposes.

enhanced private switched communications service A tariffed offering by AT&T that provides the same basic function as CCSA, but with the addition of several sophisticated features, including the capability of giving the customer his own network control center.

enhanced services Services offered over transmission facilities that utilize computer-based processing applications to provide the subscriber with additional or restructured information.

enquiry character A transmission control character used as a request for a response from the station with which a connection has been set up.

Enterphone Proprietary brand of security telephone used in apartment blocks.

Enterprise number A "toll-free" number; i.e., one which permits the charge for a call to be billed to the called number instead of to the calling number.

entrance cable protector ground bar (ECPGB) The copper ground bar located on the main or protector distributing frame (MDF or PDF).

entropy A measure of the amount of information in a message, based upon the number of possible equivalent messages.

envelope 1. The boundary of the family of curves obtained by varying a parameter of a wave. 2. A group of binary digits formed by a byte augmented by a number of additional bits which are required for the operation of the data network. 3. In a message handling system, the place in which the information to be used in the submission, delivery and relaying of a message is contained.

envelope delay distortion In a given passband of a device or a transmission facility, the maximum difference of the group delay time between any two specified frequencies.

ephemerides The computed positions of satellites.

ephemeris An astronomical tabulation giving the location of many celestial bodies at specific intervals. The term is also applied to artificial satellites.

ephemeris, satellite Tables giving the location of satellites at particular times and dates.

epitaxy A technique used in the manufacture of semiconductors in which thin layers of one material are grown onto a base such as a silicon crystal, maintaining the same crystal lattice structure.

epoxy resin Two-part resin used as an adhesive.

equal access Coding method enabling telephone customers to choose for themselves the long distance network in which their calls are to be routed, e.g., via AT&T or MCI in the U.S., via British Telecom or Mercury in the U.K., using the same number of digits whichever carrier is chosen.

equal gain combiner A diversity combiner in which each channel's signals are added together; the channel gains are all equal and can be made to vary equally so that the resultant signal is approximately constant.

equalization The reduction of frequency distortion and/or phase distortion of a circuit by the introduction of networks to compensate for the difference in attenuation, time delay, or both, at the various frequencies in the transmission band.

equalization program The sequence of events while laying a submarine cable to determine the required ocean block equalizer characteristics during the laying operation and the assembly or adjustment or selection of the equalizer networks; related to system tests during laying.

equalize To insert in a line a network with complementary transmission characteristics to those of the line, so that when the loss or delay in the line and that in the equalizer are combined, the overall loss or delay is approximately equal at all frequencies.

equalizer A network which corrects a circuit's transmission-frequency characteristics to allow it to transmit selected frequencies in a uniform manner.

equalizer, absolute delay Network(s) which are capable of equalizing the absolute time delay of two or more circuits carrying identical signals to a single receiving terminal.

equalizer, adaptive An equalizer which automatically and constantly readjusts itself in order to produce an equalized output signal from a distorted input.

equalizer, attenuation An equalizer which corrects the attentuation-frequency characteristic of a circuit.

equalizer, delay An equalizer inserted in a line which delays the higher frequencies, so that all useful frequencies will have the same delay, and the minimum delay distortion.

equalizer, mop-up A manually-adjusted equalizer which allows for imperfect equalization which has occurred previously in the system.

equalizer, ocean block Assembly of networks placed in a water-tight enclosure and inserted in a submarine cable which reduce the level deviations accumulated over an ocean block; typically they are assembled or adjusted on board the ship during the laying operation. Mechanically they are practically identical to repeaters.

equalizer, program A wide-band (10 kHz or 15 kHz) attenuation equalizer.

equalizer, relative delay A network with a delay characteristic which is the inverse of the line or equipment to which it is connected, and which can achieve a relatively constant delay over a given frequency band.

equalizer, terminal station Adjustable network for equalization and control of transmission parameters properly controllable at the ends of a submarine cable system.

equalizer assembly Operation on board a cable-laying ship where equalizers are assembled during the laying operation after their characteristic requirements have been determined by computation.

equalizer section A submarine cable repeater section containing an ocean block equalizer.

equalizing charge An overcharge applied to a lead-acid storage battery in order to correct variations in an individual cell's specific gravity. The equalizing charge should be maintained until all cells have been completely charged.

equalizing pulses (CATV) Pulses of one-half the width of the horizontal sync pulses which are transmitted at twice the rate of the horizontal sync pulses during the blanking intervals immediately preceding and following the vertical sync pulses. The action of these pulses causes the vertical deflection to start at the same time in each interval, and also serves to keep the horizontal sweep circuits in step during the vertical blanking intervals immediately preceding and following the vertical sync pulse.

equatorial orbit The plane of a satellite orbit which coincides with that of the equator of the primary body.

equilibrium condition See equilibrium mode distribution.

equilibrium coupling length See equilibrium length.

equilibrium length The length of multimode optical waveguide, excited in a specified manner, necessary to attain the equilibrium mode distribution. Syn.: equilibrium coupling length; equilibrium mode distribution length.

equilibrium mode distribution The distribution of power among the modes after transmission through a requisite length of multimode optical waveguide such that thereafter (in distance) the relative power distribution among the various modes remains constant. The requisite length, which typically varies from several hundred meters to a few kilometers, is dependent upon various parameters of the waveguide, wavelength, and initial launching conditions. Syn.: equilibrium condition; steady state condition

equilibrium mode distribution length See equilibrium length.

equilibrium mode simulator A device used to create an approximation of the equilibrium mode distribution. This distribution may be achieved by using selective mode excitation or mode filters either with or without mode scramblers.

equilibrium radiation angle The radiation angle of an optical waveguide having an equilibrium mode distribution.

equilibrium radiation pattern The output radiation pattern of an optical waveguide having an equilibrium mode distribution, measured as a function of angle or distance from the waveguide axis. Far field equilibrium radiation pattern is measured as a function of angle. Near field equilibrium radiation pattern is measured as a function of distance from the waveguide axis. The equilibrium radiation pattern is independent of waveguide length (beyond the equilibrium length) and excitation conditions but may be a function of wavelength.

equipment Telephone equipment located indoors.

equipment, central office Equipment used in a telephone central office to provide communication services.

equipment, digital subscriber terminal Equipment used in the AUTODIN network.

equipment, diversion Equipment which analyzes digits dialed and directs calls for nonauthorized services to a busy tone circuit or to a recorded announcement.

equipment, key telephone A key set; a multi-button instrument which provides intercom and public network service and many of the same features as PABX extensions.

equipment, line Circuit in a central office which is directly associated with a particular line.

equipment, off-line Equipment which is not connected to an active line but is used for a local purpose, e.g., a teletypewriter used to prepare a tape or message for later transmission.

equipment, on-line Equipment which is in direct communication with a switching center or distant terminal.

equipment, peripheral Equipment which itself has no on-line role but works closely with on-line equipment, e.g., a card punch or sorter.

equipment, station Communications equipment used at a subscriber's premises.

equipment, subscriber's line The line circuit in a central office directly associated with a particular subscriber's line.

equipment, terminal The man-machine interface devices on the ends of a circuit, or the relays and components which terminate a circuit on a switch.

equipment, unitized power A modular power system designed for central office use, includes rectifiers, tone generators, and supervisory alarms.

equipment clock A clock which satisfies the needs of equipment and, in some cases, may control the flow of data at the equipment interface.

equipment compatibility When programs and data are interchangeable between different types of equipment.

equipment failure When a hardware fault stops the successful completion of a task.

equipment ground (EG) A protection ground consisting of a conducting path to ground of non-current carrying metal parts.

equipment intermodulation noise Intermodulation noise introduced into a system by a specific piece of equipment.

equipment side That portion of a device that faces the in-station equipment.

equipment supervisory rack A central office rack where the power supervisory equipment is mounted.

equipotential At the same voltage level.

equivalent For any circuit element there is an energy loss in transmission. When this is expressed in dB it is called its equivalent. The term is a historic one, dating back to comparison with equivalent miles of standard cable.

equivalent, circuit The overall loss at 1000 Hz of a circuit, two-wire to two-wire, in dB.

equivalent binary content The content of a signal generated by a digital source, expressed in binary terms.

equivalent binary digits The number of bits necessary to represent each of a group by a unique binary number, e.g., if all numbers 0 to 9 are the group then four bits are needed.

equivalent bit rate In a line coded signal, the number of binary digits that can be transmitted in a unit of time.

equivalent circuit Simplified circuit with similar characteristics to a real and complex circuit, used in the mathematic treatment of problems.

equivalent circuits Circuits are equivalent if the same applied voltages result in the same currents flowing.

equivalent four-wire Circuit using different frequency bands for the two directions of transmission, permitting four-wire operation over only two physical conductors.

equivalent isotropically radiated power The product of the power supplied to the antenna and the antenna gain in a given direction relative to an isotropic antenna.

equivalent line Term used to equate four-wire toll (transit) and international telephone switching centers (which incorporate complex signaling systems) with much simpler two-wire end offices/local exchanges by saying that every long distance circuit terminated at the four-wire office was equivalent to say six local lines, i.e. a 1000-circuit gateway exchange would take up as much rack space and cost about as much as a 6000-line local exchange/end office. This was a convenience for users whose procurement policies were based on price per line or equivalent. With the advent of digital techniques the costs are much closer to equal.

equivalent network 1. A network that may replace another network without altering the performance of that portion of the system external to the network. 2. A theoretical representation of an actual network.

equivalent noise resistance A quantitative representation in resistance units of the spectral density of a noise voltage generator at a specified frequency.

equivalent PCM noise Through comparative tests, the amount of thermal noise power on an FDM or wire channel necessary to approximate the same judgment of speech quality created by quantizing noise in a PCM channel. Generally, 33.5 dBrnC $\pm$ 2.5 dB is considered the approximate equivalent PCM noise of a seven-bit PCM system.

equivalent random circuit group A number of theoretical circuits used in conjunction with an equivalent random traffic intensity to permit traffic theories that do not explicitly recognize peakedness to be used in peadkedness engineering.

equivalent random traffic intensity The theoretical pure chance traffic intensity that, when offered to a number of theoretical circuits (equivalent random circuits), produces an overflow traffic with a mean and variance equal to that of a given offered traffic. The equivalent random concept permits traffic theories that do not explicitly recognize peakedness to be used in peakedness engineering.

equivalent step-index profile (ESI) Method of describing a single-mode fiber by referring to a step-index fiber (with specified core diameter and refractive index difference) having almost identical field distribution characteristics.

erasable and programmable read-only memory (EPROM) A general name used for ultraviolet light-erasable PROMS, sometimes called MOS PROMS. An advantage of using these, though they are likely to be more expensive, is that reprogramming can be done in the field to keep up with software update.

erasable storage A store which is used repeatedly; when new information is written-in, the earlier information is erased before the new is stored.

erase 1. To obliterate information from any storage medium, e.g., to clear or to overwrite. 2. To remove all previous data from magnetic storage by changing it to a specified condition that may be an unmagnetized state or predetermined magnetized state.

erect position In frequency division multiplexing, a position of a translated channel in which an increasing signal frequency in the untranslated channel causes an increasing signal frequency in the translated channel.

ergonomic Relating to biotechnology: the design of equipment to suit human habits and dimensions.

erlang An international (dimensionless) unit of the average traffic intensity (occupancy) of a facility during a period of time, normally a busy hour. The number of erlangs is the ratio of the time during which a facility is occupied (continuously or cumulatively) to the time this facility is available for occupancy. One erlang is equivalent to 36 ccs. Syn.: traffic unit

erroneous block A block in which there are one or more erroneous bits.

error 1. The difference between a computed, estimated, or measured value and the true, specified, or theoretically correct value. 2. A malfunction that is not reproducible. 3. A deviation occurring during transmission so that a mark signal is received instead of a space signal, or vice versa. 4. Bipolar violation: the appearance of two successive pulses of the same polarity in a bipolar pulse stream.

error, aliasing Badly reconstituted analog signals resulting from attempts to sample and quantize for PCM at too low a sampling frequency.

error, probable Error with 50% or more chance of occurrence.

error, random An error that varies in a random fashion, e.g., an error resulting from radio noise or interference.

error, systematic A constant error or one that varies in a systematic manner, e.g., equipment misalignment.

error budget The allocation of a bit error rate requirement to the segments of a circuit, e.g., trunking, switching, access lines, or terminal devices, in a manner which permits the specified system end-to-end error rate requirements to be satisfied for traffic transmitted over a postulated reference circuit.

error burst A group of bits in which two successive erroneous bits are always separated by less than a given number (x) of correct bits. The number x should be specified when describing an error burst.

error checksum Data group added to a message (e.g., in packet switching systems) to enable received message to be checked for transmission errors.

error code The identification of a particular error by a code in order to draw attention to it during the next stage of operation.

error control The improvement of digital communications through use of certain techniques such as error detection, forward-acting error correction, and use of block codes.

error control loop The number of common channel signaling system signal units transmitted on the signaling link between the time a particular signal unit is sent and the time that the acknowledgement of that signal unit is recognized.

error control procedure That part of the link protocol (in data communication) controlling the detection and sometimes also the correction of transmission errors.

error-correcting code A code in which each telegraph or data signal conforms to specific rules of construction so that departures from this construction in the received signals can be automatically detected. This permits the automatic correction, at the receiving terminal, of some or all of the errors. Such codes require more signal elements than are necessary to convey the basic information.

error-correcting system A system employing an error-correcting code and so arranged that some or all signals detected as being in error are automatically corrected at the receiving terminal before delivery to the data sink. In a packet switched data service, the error correcting system might result in the re-transmission of one or more complete packets should an error be detected.

error correction Automatic process in some data transmission systems, when an error has been detected the receiving end instructs the sending end to re-transmit poorly received signals.

error-detecting code A code in which each telegraph or data signal conforms to specific rules of construction, so that departures from this construction in the received signals can be automatically detected. Such codes require more signal elements than are necessary to convey the basic information.

error-detecting system A system employing an error-detecting code and so arranged that any signal detected as being in error is either deleted from the data delivered to the data sink, in some cases with an indication that such deletion has taken place, or delivered to the data sink together with an indication that the signal is in error.

error rate The ratio of the number of signal elements incorrectly received to the total number transmitted. This also may be considered as the probability of an error occurring during the transmission of a message.

error rate monitor A common channel signaling system device which receives an indication for each signal unit found in error and which measures the rate of occurrence of errors according to a pre-scribed rule.

escape character A predetermined character which indicates to the receiving station that the subsequent characters are to be interpreted in a nonstandard way.

escape velocity Minimum velocity required to enable a man-made satellite to escape and maintain itself in orbit around the Earth.

E-side Main (primary) distribution cables between an exchange or central office and a cabinet or flexibility point.

established connection A connection which has been put through at all necessary switching points. However, the instruments at sending and receiving ends are not necessarily actively connected.

estimate 1. An estimation of the cost of doing a given piece of work. 2. A written statement of proposed costs. 3. A formal plan which includes a cost esti-mate for adding a large amount of telephone plant, where cost exceeds a specified limit.

estimated junction frequency (EJF) Standard maximum useable frequency. An approximation to the classical MUF or junction frequency, usually obtained by application of the conventional transmission curve to vertical-incidence ionograms, together with the use of a distance factor.

Ethernet A packet-switched local network design (by Xerox Corp.) employing CSMA-CD as access control mechanism.

Eureka European co-operative research and development project for data processing.

Euroboard European standard sizes of printed circuit board: 100mm, 160mm, 220mm or 280mm in depth.

Euronet A European packet switching network with master packet switching exchanges (PSES) in London, Paris, Rome, and Frankfurt, and slave remote access points (RAPS) in Dublin, Brussels, Luxembourg, Amsterdam and Copenhagen. The four PSES are linked by a ring of 48 kbit/s digital links and the RAPS feed in to adjacent PSES by 48 kbit/s or 9.6 kbit/s links. Euronet is sponsored by the European Economic Community; its network management center is in London.

European Space Agency (ESA) The organization responsible for launching rockets carrying spacecraft for various operations.

evanescent field A time-varying electromagnetic field whose amplitude decreases monotonically but without an accompanying phase shift in a particular direction is said to be evanescent in that direction.

even parity check A parity check in which the number of zeros (or of ones) in a group of binary digits is expected to be even. The other common parity check is an odd parity check.

event A definable action, particularly in a critical path or PERT control procedure.

exalted-carrier reception A method of receiving either amplitude or phase modulated signals in which the carrier is separated from the sidebands, filtered and amplified, and then combined with the sidebands again at a higher level prior to demodulation.

exception condition In data transmission, the condition assumed by a secondary station when it receives a command which it cannot execute.

excess-fifty A binary code in which a number x is represented by the binary version of $50 + x$.

excess-three A binary code in which a number x is represented by the binary version of $3 + x$.

exchange 1. A room or building equipped so that telephone lines terminating there may be interconnected as required. The equipment may include manual or automatic switching equipment. 2. Switching exchange: an aggregate of traffic-carrying devices, switching stages, controlling and signaling means at a network node that enables subscriber lines and/or other telecommunication circuits to be interconnected as required by individual callers.

exchange, multioffice A telephone exchange area in which there are several central offices. Called a multiexchange area (MEA) in the U.K.

exchange, private automatic (PAX) Small local automatic telephone office, normally with no external lines to the public network but serving extensions in a business complex.

exchange, private automatic branch (PABX) Small local automatic telephone office serving extensions in a business complex and providing access to the public network.

exchange, private branch (PBX) Small local telephone office, either automatic or manually operated, serving extensions in a business complex and providing access to the public network.

exchange, single-office An exchange area served by a single central office.

exchange, telephone In the U.S., an area within which telephone service is provided without toll charges. In Europe, a telephone central office.

exchange, teletypewriter A switching device which interconnects teletypewriters for local, domestic long distance and international service.

exchange area In the U.S., an area within which there is a uniform tariff for telephone service. There may be more than one central office in an exchange area. In Europe, the area of service of a single exchange or central office. On both continents a call between any two points within an exchange area is a local call.

exchange call release delay For common channel signaling: an exchange call release delay is the interval from the instant at which the last information required for releasing a call in an exchange is available for processing to the instant that the switching network through connection is no longer available between the incoming and outgoing circuits and the disconnection signal is sent to the subsequent exchange. This interval does not include the time taken to detect the release signal, which might become significant during certain failure conditions, e.g. transmission system failures.

exchange call set-up delay The interval from the instant when the address information required for setting up a call is received at the incoming side of the exchange to the instant when the seizing signal or the corresponding address information is sent to the subsequent exchange, or to the instant when the ringing signal is sent to the appropriate user.

Exchange Carrier Association (ECA) Organization of local telephone companies (U.S.) which deals with tariff and access charge matters relevant to long-haul carriers having access to local networks and local loops.

exchange code A digit, or group of digits, which identifies the central office or exchange to which a subscriber belongs.

exchange concentrator A switching stage wherein a number of subscriber lines or interexchange circuits carrying relatively low traffic volumes can be through-connected to a smaller number of circuits carrying higher traffic volumes.

exchange control system The central control system of a stored program controlled switching system. It may consist of one or more processors.

exchange feeder route analysis program (EFRAP) An AT&T software routine. A program used to develop long-range plans for cable-supporting conduit structure augmentation over an entire feeder cable network.

exchange line capacity (ELC) Total number of customers' lines which may be connected to an exchange/central office, i.e., the equipped multiple of the office.

exchange network facilities for interstate access (ENFIA) Access circuits and procedures enabling telephone subscribers in the U.S. to route long distance calls over the AT&T System or over one of the competing OCC (other common carrier) networks. ENFIA A (Feature Group A, or FGA) uses off-network access lines (ONAL). Subscribers have to dial a full local subscriber number to connect to the OCC, then have to key out (using DTMF) their personal authorization code followed by the distant number required. ENFIA A (Feature Group B, or FGB) gives no hardware answer code so call timing was necessarily inaccurate. ENFIA B (for local offices) and ENFIA C (Feature Group C, or FGC) (for local tandem offices) uses off-network access trunks which eliminate most ONAL difficulties and greatly improve transmission quality.

exchange service Furnishing of ordinary telephone service. In the U.S., this is a regulated service within specific geographic areas.

excitation Current which energizes field coils in a generator.

excitation, series Field coils in a generator or motor in series with the armature.

excitation, shock Feeding into a resonant circuit a sudden steeply-fronted signal, making the circuit oscillate; such oscillations usually are damped.

exciter 1. Driven element in an antenna which includes parasitic elements. 2. Generator producing field current for an alternation.

exclusion, attendant PABX feature which stops the operator from being able to cut in on an established call.

exclusivity The provision in a commercial television film contract that grants exclusive playback rights for the film or episode to a broadcast station in the market it serves. Under the FCC's rules, U.S. cable operators cannot carry distant signals which violate local television stations' exclusivity agreements.

Execunet Interstate public switched voice service provided by MCI.

execute To perform a specified computer instruction; to run a program on a computer.

executive program A program, usually part of an operating system, that controls the execution of other programs and regulates the flow of work in a data processing system. Sometimes called a supervisory program or supervisor.

executive right-of-way Over-ride feature on some PABXs which permits privileged priority extensions to cut in on existing calls.

exempt items Small stores which, in most telephone companies, do not have to be allocated exactly to the particular project on which they were used but are treated as an overall cost.

expandor A device with a nonlinear gain characteristic that acts to increase the gain more on larger input signals than it does on smaller input signals.

expansion 1. A process in which a compressed voice signal travels through an amplifier whose gain varies with the signal magnitude, amplifying large signals more than weaker signals. 2. A configuration wherein the number of inlets into the switching stage is smaller than the number of outlets.

expansion matrix A submarine cable system computation which reflects the nonlinear gain behavior of bidirectional repeaters with a common amplifier for both directions of transmission; permits the establishment of the margin of a system against nonlinear singing.

expected measured loss (eml) The expected reading, in dB, at a test point at one end of a trunk when a sending power of a specified value is applied to a prescribed test point at the other end of the trunk.

expediting of a call in progress Intervention by an operator, interrupting a call in progress, in order to allow another incoming call to be offered.

exploring coil A coil which can be moved manually in order to detect a radiation or stray field.

explosion-proof sets Mine telephones designed so they can be used safely in potentially explosive atmospheres.

exponential function A function with the property that its rate of change with respect to the independent variable is proportional to the function, or: $dy/dx = b$ times y

exposure 1. Normally used to indicate the situation when a communications circuit is located so close to a power circuit that it could cause noise or danger. 2. For telegraph & data circuits: the vulnerability multiplied by the probability of occurrence within a given time.

extended area service (EAS) Extension of toll-free area to include nearby exchange areas in return for accepting a higher tariff.

extended binary-coded decimal interchange code (EBCDIC) An extension of ordinary binary-coded decimal (four binary digits for each decimal number) into an eight binary position code for each character, giving a possible maximum of 256 characters. Used as a system code for many commercial computers.

extender, CATV line Amplifier used in line to provide CATV service to remote subscribers.

extender, loop Device in central office which supplies augmented voltage out to line for subscribers who are at considerable distances from the office, and provides satisfactory signaling and speech for such subscribers.

extender, loop signaling Boosting device added to a subscriber's line in a central office which provides satisfactory loop-disconnect dialing and supervisory signals on longer-than-normal lines.

extender, range Device added to a subscriber's line in a central office which supplies both extra voltage to ensure satisfactory signaling on a long line and a standard-gain amplifier, usually 5dB.

extension An additional telephone instrument in parallel with the main station, or a telephone station served by a PBX.

extension, cable rack Steel spacing bars used in manholes to ensure that cable racking is not mounted flat up against walls.

extension, off-premises Extension served by a PBX which is located at a distant site so that the line to the extension has to run through other property.

extension, pole top Crossarm mounted vertically on the top of a pole to support wires which cross the route.

extension, telephone An extra telephone on the same line as the main instrument.

extension circuit, E & M Signaling circuit which repeats receive (E) and transmit (M) signals so that a received E signal is sent out as an M signal.

external blocking When referring to a switching stage, the condition in which no suitable resource, connected to that switching stage, is accessible.

external memory A backing store which is under the control of the central processor but is not necessarily connected permanently to it. It holds data in a form directly acceptable to it, to minimize access time delays.

external photo-effect The emission of photon-excited electrons after overcoming the energy barrier at a photo-emissive surface.

extinction ratio Ratio of light transmitted when a modulation is "on" to that transmitted when it is "off."

extra large scale integration (ELSI) More than one million logic gates or bits of memory in one device.

extractor, lamp Small tubular hand tool used to pull switchboard lamps out of their sockets.

extractor, lamp cap Small hooked pliers used to pull out switchboard lamp caps.

extremely high frequency Radio signals with frequencies between 30 GHz and 300 GHz, i.e., wavelengths between 1 cm and 1mm.

extremely low frequency Radio signals with frequencies below 300 Hz, i.e., wavelengths longer than 1000 km. Used for radio communications with submerged submarines.

extrinsic junction loss Those junction losses that are caused by different geometric and optical parameter mismatches when two nonidentical optical waveguides are joined.

extrusion The process in cable manufacturing of applying the insulant in hot plastic form to the conductor as it passes through the extrusion machine; also the process of applying the outer jacket.

eye, pulling　Metal loop securely fixed to the end of a cable, used to pull the cable into a duct.

eye diagram　Technique which enables the performance of digital equipment, such as regenerators, to be assessed. Oscilloscope is triggered at the digit rate and random digital signals are inputted, successive sweeps of the output trace build up a picture of all possible transitions.

eye pattern　An oscilloscope pattern used for examination of digital signal distortion; an eye diagram.

eyelet　Metal ring crimped into a hole in a printed wiring board to provide a front-to-rear connection.

F layers　Ionized layers in the upper ionosphere. The lower of these, the F1 layer, is at a height of about 200km, but it tends to disappear at night. The upper or F2 layer is located at about 400km during the day but falls to about 250km at night.

face　The side of a pole on which crossarms are secured.

face equipment　Lamp strips and jack strips on vertical or near-vertical panels of operating positions/switchboards.

facilities　General word describing elements of equipment which provide a service. Also used to mean the operating or maintenance service features made available by a system.

facilities, telecommunications　Telecommunications equipment and people maintaining it to provide telecommunications services.

facilities, transmission　General term for equipment which acts as a bearer of information signals: physical pairs in a cable, line carrier systems on open wires, on pair-type cables, on coaxial cable systems or on optic fibers, narrow and broadband radiocommunication systems.

facility　A service provided by the telecommunications network or equipment for the benefit of the subscribers or the operating administration. A general term for the communication transmission pathway and associated equipment.

facility request　For telegraphy & data: the part of the selection signals which indicates the required facility.

facility request separator　For telegraphy & data: the character which separates the different facility requests in the selection signals.

facsimile　A system for the transmission, usually over a voice band, of a picture, drawing, or document and having it reproduced at the other end. Some specialized facsimile equipment requires broader bandwidths, e.g., 48 kHz. Early systems used amplitude modulated voice-band tones, with synchronized mechanical rotation at transmit and receive ends, however, digital techniques are now beginning to be used and blank spaces ignored thereby greatly speeding up transmission time. Reproduction is normally in black and white only, but some broadband equipment is capable of producing color copies. The CCITT has produced internationally-agreed recommendations for facsimile equipment: group 1 transmits an A4 page (210 mm x 297 mm) in six minutes, group 2 in three minutes, group 3 (digital) in less than one minute. Group 4 facsimile has also been defined for operation in conjunction with teletex.

facsimile, analog　Facsimile transmission method utilizing continuously varying signals, e.g., amplitude or frequency modulation. These systems normally enable different shades of grey to be reproduced.

facsimile, digital　Facsimile transmission method utilizing digital signals. These systems usually provide black or white reproduction only and are able to operate more rapidly than analog-type systems.

facsimile converter　1. Receiving: a facsimile device which changes the type of modulation from frequency shift to amplitude. 2. Transmitting: a facsimile device which changes the type of modulation from amplitude to frequency shift.

facsimile picture signal　A signal resulting from the scanning process.

facsimile receiver　A facsimile device that converts the facsimile picture signal from the communications channel into a facsimile record of the subject copy.

facsimile recorder　That part of the facsimile receiver which performs the final conversion of the facsimile picture signal to an image of the subject copy on the record medium.

facsimile scanner　Part of a facsimile terminal with a rotating drum around which the original material is placed. As the drum rotates the copy is scanned by a moving light beam and the facsimile signal is sent out to line.

facsimile service　A service for the transmission of documents between facsimile machines.

facsimile signal level　The facsimile signal power or voltage measured at any point in a facsimile system.

facsimile transmitter　A facsimile device that converts the subject copy into signals suitable for delivery to the communication system.

factor, amplification　The voltage gain of an amplifier.

factor, energy dissipation　Ratio of energy dissipated to energy stored in one cycle.

factor, energy storage　Ratio of energy stored to energy dissipated in one cycle.

factor, loss Of a dielectric: the product of dielectric constant and the tangent of its dielectric loss angle; a measure of the heating effect which occurs in dielectrics.

factor, power Ratio of total watts to total root-mean-square volt-amperes, i.e., the active power to the apparent power.

factor, power diversity Ratio of the total of all the individual power loads of different elements to the demand power load for the complete system composed of all these elements.

factor, pulse crest Ratio of peak amplitude to root-mean-square value of a pulse.

factor, shield Ratio of noise on a circuit when shielded to that on the same circuit when shielding has been removed.

factor, space Ratio of volume occupied by the conductors in the winding of a coil to the total volume of the entire coil.

fade A change of signal strength. Also see fading.

fade margin 1. An allowance provided in system gain (or sensitivity) to accommodate expected fading to insure that the required grade of service will be maintained for the specified percentage of time. 2. The amount by which a received signal level may be reduced without causing the system (or channel) output to fall below a specified threshold.

fading The variation in time of the intensity and/or relative phase of any or all frequency components of a received signal due to changes in the characteristics of the propagation path. This can be caused by several different effects, such as:
>   interference between skywave and ground wave.
>   movement of the ionosphere, and multipath changes causing interference fading.
>   interference between multiple reflected sky waves.
>   rotation of the axes of the polarization ellipses.
>   variation of the ionospheric absorption with time.
>   focusing and temporary disappearance of the signal due to MUF failure.

There are a variety of ways for obtaining signals for which the periods of fading occur independently of one another:
>   space (spaced antenna) diversity.
>   frequency diversity.
>   angle-of-arrival diversity.
>   polarization diversity.
>   time (signal repetition) diversity.
>   multipath diversity.

fading, amplitude Fading which affects all frequencies more or less equally, so the signal is reduced in amplitude but not otherwise distorted.

fading, Dellinger A complete loss of received signal, due to the effect on the ionosphere of sudden sun spot eruptions.

fading, flat Same as amplitude fading.

fading, selective Fading which affects different radio frequency signals in different ways.

fading distribution The probability that signal level fading will exceed a certain value relative to a certain reference level.

Fahnestock clip Type of spring clip used as a simple wire termination.

fail-safe operation A type of control that prevents improper functioning in the event of circuit or operator failure.

failure, catastrophic Complete and sudden failure of a system.

failure, degradation A gradual and partial failure of a component or device.

failure, misuse Failure because an item has been exposed to conditions outside its proper working range.

failure, partial Fault condition which degrades service but does not interrupt it completely.

failure, secondary Failure caused by the failure of another item, causing the item concerned to be exposed to stresses or conditions for which it was not designed.

failure, total Fault condition which has completely interrupted a service.

failure effect The result of the malfunction or failure of a device or component.

failure in time (FIT) A unit used in some disciplines to indicate the reliability of a component or device; one FIT corresponds to a failure rate of 10^{-9}/hour.

failure rate, observed The ratio of the number of actual failures to the number of times each item has been subjected to stress conditions.

fairlead A submarine cable ship device for guiding the movement of cable to or from the tank, the deck, cable engines, and sheaves, configured to restrain the bending of the cable to the allowable bending radius.

FAMOS Floating gate avalanche metal oxide semiconductor.

fan To spread out wires in a cable or form so that they may be identified, tested, and terminated.

fan-in The greatest number of separate inputs acceptable to a single specified logic circuit without adversely affecting performance.

fan-out 1. The maximum number of inputs to other circuits which can be fed by the output of a specified logic circuit without the output voltage falling outside the limits at which the logic levels 1 and 0 are specific. (2) Same as fan.

far-end The distant end to that being considered, i.e., not the one where testing is being carried out.

far-end crosstalk Crosstalk which is propagated in a disturbed channel in the same direction as the propagation of a signal in the disturbing channel. The terminals of the disturbed channel where the far-end crosstalk is present and the energized terminals of the disturbing channel are usually remote from each other.

far field diffraction pattern The diffraction pattern of a source (including the output end of an optical waveguide) observed at an infinite distance from the source. Theoretically, a far field pattern exists at distances that are large compared with S^2/λ, where S is a characteristic dimension of the source and λ is the wavelength. Syn.: Fraunhofer diffraction pattern

far field radiation pattern See radiation pattern.

far field region 1. The region, far from a source, where the diffraction pattern is equal to that observed at infinity. 2. The region of the field of an antenna where the angular field distribution is essentially independent of the distance from the antenna.

farad The derived SI unit of capacitance: the capacitance of a capacitor between the plates where there is a potential difference of 1 volt when charged by a quantity of electricity of 1 coulomb. This is a very large unit; microfarads (10^{-6}) and picofarads (10^{-12}) are most commonly used.

farm, antenna Area of land occupied by or reserved for transmitting and receiving antennae.

fastener, jack Metal fixing used to hold a jack-strip in its correct place on the face of a switchboard.

fastener, panel Metal fixing, often a butterfly nut, used to hold a panel in place on the face of a switchboard.

fathom A length of six feet. One thousand fathoms approximates a cable mile or a nautical mile.

fatigue Reduction in strength of a metal caused by formation of crystals due to repeated flexing of the part in question.

fault Any defect which impairs the functioning of a device or circuit or system; a reproducible malfunction.

fault, intermittent A defect in a circuit or medium which is not continually apparent.

fault finder Test set or other device which enables faults to be identified and localized.

fault location Procedure of electrical tests made from a terminal station to determine the location (and sometimes the cause) of system malfunction.

fault to ground Fault due to the failure of insulation and the consequent establishment of a direct path to ground from a part of the circuit which should not normally be grounded.

faulty link information Common channel signaling: information sent on a signaling link to indicate its failure. The information consists of alternate blocks of changeover signals and synchronization signal units.

faure plate Type of plate used in some lead-acid storage batteries: consists of a grid of a lead-antimony alloy with all the grid holes filled with a lead oxide paste.

FAX Short for facsimile, a name which refers either to the service or the actual machines by which a copy of a document or picture may be transmitted over a transmission medium.

feather edging Submarine cable ship term for boards of triangular cross-section that are placed alongside cable crossing the turns of a flake in a cable tank to avoid undesirable concentrated compressive stresses caused by cables crossing at right angles under heavy weight from successive flakes above.

feature access code Code *XX or XX proposed to be used to obtain access to special functions and services provided on an inter-LATA basis. Syn.: function access code

feature groups The different types and qualities of connection services between local telephone companies and long distance companies. See ENFIA.

features Aspects of a system which are visible to the subscriber or customer (e.g., abbreviated dialing).

Federal Communications Commission (FCC) In the U.S., a board of seven commissioners, appointed by the President, with the power to regulate all interstate and foreign electrical communications systems originating in the U.S., including radio, television, facsimile, telegraph, telephone, and cable systems; the federal regulatory body.

federal-state joint board Body established by the U.S. FCC to deal with matters (especially financial and regulatory) which affect state interests.

Federal Telecommunications System (FTS) A U.S. government communications system administered by GSA covering the 50 states, Puerto Rico, and the Virgin Islands, providing services for voice, teletypewriter, facsimile, and data transmission.

feed Wires, cable or waveguide connecting an antenna with its radio transmitters or receivers.

feed, antenna The transmission line which delivers power to an antenna.

feed, current Antenna with the feed at the point of maximum current.

feed, series Feeding the input signal into an electron tube circuit through the same impedance as is used to feed in its plate voltage.

feed, shunt Separation between input signal to an electron tube and its plate voltage by applying these in parallel, with a choke in the plate circuit.

feed, voltage Connection of an antenna feed to the antenna at the point of maximum potential variation.

feed holes Holes prepunched in paper tape to permit it to be driven by a sprocket.

feedback 1. The return of a portion of the output of a device to the input; positive feedback adds to the input, negative feedback subtracts from the input. 2. Information returned as a response to an originating source.

feedback, acoustic Howling of a public address system; feeding back into a microphone some of the output from a loudspeaker in the same audio system

feedback, negative The return of some of the output of a device, such as an amplifier, to its input, but 180 degrees out of phase with the normal input voltage. When adjusted correctly, negative feedback will reduce any distortion in the device, but also reduce the gain.

feedback, positive Returning of a portion of the output of a device, such as an amplifier, to its input, in phase with the original. This is the basis of operation of most oscillators.

feeder Transmission line linking radio communication transmitter or receiver with its antenna.

feeder, cable Flexible tubular guide through which a cable is passed down into a manhole while being pulled into a duct.

feeder, negative Low resistance cable that runs parallel with the steel rails in city tramway systems: the cable is jointed to the rails at frequent intervals so that current will tend to flow through the cable rather than in the rails or through the earth.

feeder, non-resonant A matched feeder, terminated in its own characteristic impedance so that it does not generate a standing wave.

feeder, resonant A feeder which is not terminated by its characteristic impedance and so generates a standing wave pattern.

feeder echo noise Distortion of a signal as a result of reflected waves in a transmission line that is many wavelengths long and mismatched at both generator and load ends.

feeder line (CATV) Intermediate cable distribution lines that connect the main trunk line to the smaller house drops that lead into residences.

feedhorn A waveguide opened out in a carefully calculated shape so as to direct maximum energy to a reflector (typically a parabolic microwave dish antenna).

feeding bridge A circuit in a telephone exchange which provides the dc line current needed to activate the microphone in conventional telephone sets.

feedthrough Method of connecting a circuit on one layer on a printed circuit board with one in the next layer. Complex boards often involve ten or more layers so alignment sometimes becomes critical.

femtosecond One millionth of a billionth of a second (10^{-15}).

fermal principle A ray of light from one point to another follows that path which requires the least time. Syn.: principle of least time.

ferreed A reed relay: two ferrous metal reeds sealed into opposite ends of a glass tube so that the ends overlap. When current flows through an external coil the magnetic field causes the reeds to move together and close contacts at their ends. Reeds of this type have been widely used in space-division switching systems.

ferrimagnetism At a critical temperature the magnetic properties of some materials such as garnet change significantly: above this temperature (the Neel temperature) the material is paramagnetic while below it it behaves like a ferromagnetic material.

ferrite A ceramic material made up from powdered and compressed ferric oxide plus other oxides (mainly cobalt, nickel, zinc, yttrium- iron, manganese). These materials have a very low eddy-current loss at high frequencies so are much used as dust cores in computer memories.

ferrite core Small (typically 0.5 mm diameter) doughnut-shaped ring of ferrite material used as a binary digit store. When magnetized in one direction it remains in that state until a sufficiently large magnetizing force is applied to reverse the direction of magnetization.

ferroelectric material A crystalline material that exhibits, over a particular range of temperature, a remanent polarization which can be reoriented by application of an external electric field.

ferrofluid A fluid in which fine particles of iron, cobalt or magnetite are suspended. Such fluids are super-paramagnetic, i.e. they exhibit no hysteresis effects, and can be used to provide liquid seals, held in position by a magnetic field, for example, on drive shafts of magnetic-disk drive units, keeping them dust-free.

ferromagnetic material Material with permeability significantly greater than unity and dependent on a magnetizing force.

ferromagnetic material, hard Materials with low relative permeability and high coercive force so that they are difficult to magnetize and demagnetize. They retain magnetism well and are used in permanent magnets.

ferromagnetic material, soft Materials with high relative permeability and low coercive force so that they are easy to magnetize and demagnetize. Suitable for uses where rapid changes of flux are encountered, e.g., in transformers.

ferrule 1. The metal cap on the end of a cartridge fuse tube which is used as a contact for the fuse. 2. A mechanical fixture, generally a rigid tube, used to confine the stripped end of a fiber bundle or a fiber. Typically, individual fibers of a bundle are cemented together within a ferrule of a diameter designed to yield a maximum packing fraction. Nonrigid materials, such as shrink tubing, also may be used for ferrules for special applications.

fetron A junction field effect transistor of a special type.

fiber Fine glass strands, the thickness of a human hair. Coherent light systems which now can be modulated to carry broadband telecommunications multiplexing can be routed on these glass fibers. Lasers sometimes are used as light sources. Systems now in common use are working at bit rates up to 560 Mbit/s.

fiber bandwidth Frequency at which the magnitude of the transfer function of an optical waveguide has fallen to half the zero frequency value (i.e., signal attenuation has increased by 3 dB.)

fiber buffer A material used to protect an individual optical fiber waveguide from physical damage, providing mechanical isolation and/or protection.

fiber bundle An assembly of unbuffered optical fibers. A bundle is usually used as a single transmission channel, as opposed to multifiber cables, which contain optically and mechanically isolated fibers, each of which provides a separate channel.

fiber harness In equipment interface applications, an assembly of a number of multiple fiber bundles or cables fabricated to facilitate installation into a system.

fiber optics (FO) The branch of optical technology concerned with the transmission of radiant power through fibers made of transparent materials, such

125

as glass, fused silica or plastic. Communications applications of fiber optics employ flexible fibers. Either a single discrete fiber or a nonspatially aligned fiber bundle may be used for each information channel. Such fibers are often referred to as optical waveguides to differentiate them from fibers employed in noncommunications applications. Various industrial and medical applications employ typically high-loss, flexible fiber bundles in which individual fibers are spatially aligned, permitting optical relay of an image, an example is the endoscope. Some specialized industrial applications employ rigid (fused) aligned fiber bundles for image transfer, an example is the fiber optics CRT faceplate used on some high-speed oscilloscopes.

fidelity   The degree to which a system, or a portion of a system, accurately reproduces at its output the essential characteristics of the signal impressed upon its input.

field   1. In a record, a specified area used for a particular category of data. 2. A subdivision of a common channel signaling system signal unit which carries a certain type or classification of information, e.g., label field, signal information field, etc. 3. One-half of a complete picture (or frame) interval, containing all of the odd or even scanning lines of the picture.

field, card   See card field.

field, electric   1. Region in which a vector field of electric field strength or of electric flux density has a significant magnitude. 2. Condition in which stationary charged bodies are subject to forces resulting from their charges. 3. Condition characterized by spatial potential gradients or electric field vectors.

field, electromagnetic   Field associated with electromagnetic waves (radio, light, or heat) with magnetic and electric fields at right angles to each other.

field, electrostatic   Region in which there is electric stress resulting from electrostatically charged bodies.

field, jack   Face of an operating position or test panel where jack strips are mounted.

field, magnetic   State in a medium, produced either by current flowing in a conductor or by a permanent magnet, that can induce voltage in a second conductor in the medium when the state changes or when the second conductor moves in prescribed ways.

field, radiation   Electromagnetic radiation produced by a transmitting radio antenna.

field effect transistor   (FET) A unipolar semiconductor device in which current is carried by the majority carriers only, unlike junction transistors in which both the majority and minority carriers contribute to conduction, i.e. both the electrons and the holes. In a FET, the current is modulated by the voltage applied at a third electrode.

field emission   Emission of electrons from a surface because of a large external field rather than a heated filament effect.

field frequency   The rate at which a complete field is scanned, nominally 60 times a second. (in the U.S.)

field programmable logic family   (FPLF) A family of transistor-transistor logic (TTL) compatible programmable chips that provide the links between large scale integration(LSI) circuit blocks. The FPLF programming ability means that the logic can be altered to meet specific requirements. Devices of varying complexity are available: field programmable gate arrays (FPGA), field programmable logic array (FPLA), and field programmable logic sequences (FPLS).

field side   Wires on a telephone pole route on the side of the pole away from the street.

field stock   Units of plant held in local stocks to meet expected local service requirements.

field strength   The intensity of an electric, magnetic, or electromagnetic field at a given point.

fifth generation computers   "State of the Art" computers being designed in the late 1980s using knowledge-based systems or artificial intelligence and a heuristic (informed trial and error) approach. The four earlier generations of computers are generally considered all to have used Von Neumann principles of primarily sequential operation, based on: 1) vacuum tubes, 2) transistors, 3) integrated circuits and 4) very large scale integrated circuits.

filament   A wire which becomes hot when current is passed through it and is used either to emit light (as in domestic light fittings) or to heat a cathode to enable it to emit electrons (as in an electron tube).

filament, mono   Single strand filament (not braided).

filament, thoriated   Filament coated with thorium oxide which emits electrons efficiently even when heated only to a dull red temperature.

file   1. Equipment: name often given to vertical units of equipment in a central office, e.g., a file of crossbar switches. 2. A set of related records treated as a unit.

file transfer, access and management   (FTAM) Protocols used in a data network which allow for remote access to files and transfer of files between machines.

fill   A measure of the number of pairs in use in a cable or central office compared with its total capacity.

fill, cable   Percentage of working cable pairs to the total number of pairs in the cable.

fill, line   Percentage of working telephone lines in a central office to the maximum capacity of the office.

fill, percentage   Fill of cable or office expressed as a percentage.

filler   Material used to fill in gaps or voids, e.g., spaces between wires in a multi-pair cable.

filler, fibrous   A fiber-type filler used in the intersticial gaps in cables, helps maintain desired relative positions between wires and so regulates mutual capacity.

fill-in signal units   A signal unit in a common channel signaling system containing only error control and delimitation information, which is transmitted when there are no message signal units or link status signal units to be transmitted.

film resistor Resistor type made up by depositing a thin layer of resistive material on an insulating core.

filter 1. A network that passes desired frequencies but greatly attenuates other frequencies. 2. A device for use on power or signal lines, specifically designed to pass only selected frequencies and to attenuate substantially all other frequencies. There are two basic types of filters: (a) active filters: those which require the application of power for the utilization of their filtering properties; and (b) passive filters: those which use inductance-capacitance components and do not require the application of power for the utilization of their filtering properties.

filter, active RC Filter which uses solid state amplifiers in order to produce a desired frequency-attenuation characteristic.

filter, band elimination A band-stop filter which passes with negligible loss all signals except those in a specified band.

filter, bandpass A filter which greatly attenuates signals of all frequencies above and below those in a specified band

filter, band-stop A band elimination filter.

filter, capacitor-input Common type of smoothing filter: output from a rectifier is shunted by a large capacitor as the first element in a smoothing circuit.

filter, carrier line Type of highpass/lowpass filter used when physical pairs are utilized simultaneously for audio circuits and carrier systems: a carrier line filter separates out the low audio band from the high carrier frequencies.

filter, cavity A filter with very precise characteristics for separating microwave frequencies, using cavity resonance.

filter, choke input Low pass power smoothing filter with inductance as its first element.

filter, comb Filter with several sharp band-stop sections for different frequencies.

filter, composite An m-derived filter made up of several filter sections, calculated to give required impedance and sharp frequency changeover characteristics

filter, constant-k Filter in which the product of impedances of shunt components and of series components is a constant, independent of frequency.

filter, crystal Filter with sharp cut-off or changeover characteristic, obtained by the use of quartz crystal components in resonant circuits.

filter, enhancement Filter used in conjunction with an amplifier to produce an output which enhances a particular band of frequencies at the expense of other signals originally present in the circuit.

filter, high-pass Filter which attenuates signals below a specified frequency but passes with minimal attenuation all signals above that frequency.

filter, key click Filter which eliminates the power surge clicks which a high-power radio telegraphy transmitter causes.

filter, LC Filter with inductance (L) and capacitance (C) circuit elements.

filter, longitudinal suppression Filter designed to suppress unwanted noise signals flowing in the same direction on the two wires of a pair, i.e., with return path via another pair of conductors or via the earth.

filter, low pass Filter which greatly attenuates signals of higher than a specified frequency, but passes with minimal attenuation all signals lower in frequency.

filter, m-derived Filter designed from a basic prototype section which has a steeper attenuation frequency characteristic while maintaining the same characteristic impedance.

filter, mechanical Filter which utilizes magnetostriction effects to provide its resonant circuit elements.

filter, notch Bandpass filter with upper cut-off frequency twice its lower cut-off frequency.

filter, power interference A filter in series with the public power input to a rectifier which passes the fundamental frequency of the power supply but greatly attenuates higher interfering frequencies.

filter, prototype A basic filter section giving a particular characteristic impedance and providing a specified cut-off frequency, but with only a gradual slope of the attenuation versus frequency line.

filter, RFI A radio frequency induction filter often mounted in telephone instruments to eliminate interference into nearby radio receivers from the pulses of a dial circuit.

filter, roof A low pass filter, i.e., the roof sets the limit.

filtered attitude determination system System used in determining and adjusting the orientation of a satellite in orbit.

filters, directional A pair of filters used when carrier systems utilize different frequency bands in go and return directions so that one physical pair may be used for both directions of transmission.

final circuit-group A circuit-group which receives overflow traffic and for which there is no possible overflow. It also may carry first choice parcels of traffic, for which it is said to be fully provided.

final group Same as final circuit-group.

final route With respect to a particular parcel of traffic, a route which is the final choice circuit group for this traffic, i.e. from which calls do not overflow, and which is not traffic engineered as a high usage group.

final splice For submarine cable systems: cable junction between the seaward end of the shore-end cable and the end of the cable in the cable laying ship. The final splice concludes the cable laying operation.

final trunk group Same as final circuit-group.

finder Switch in a dial office which seeks and finds a particular input circuit and connects it to the next switching stage.

finder, pull Device used when constructing pole routes, it helps with determination of the angle to the route at which supporting guy lines should be installed.

finder, trunk Switch in a dial office which seeks and finds a trunk circuit incoming to the office on which there is an incoming call.

finger wheel The front plate of an ordinary telephone dial, with ten finger holes.

firmware 1. Hardwired logic, e.g., information stored by threading wires through ferrite cores. 2. Software prepared in the factory and permanently stored within a read-only memory (ROM). The implication is that firmware is almost the same as software, but that data in firmware may be changed only if hardware is itself changed, e.g., a different ROM is plugged in or wires physically changed.

first choice circuit group At a switching system, the circuit group to which a traffic item is initially offered.

first in, first out (FIFO) 1. A queuing discipline in which arriving entities assemble in the time order in which they arrive and leave in the same order in which they arrived. 2. In message switching, an order of precedence for processing message traffic based on queue position of messages such that messages are dispatched from a point in the same order in which they are received at the point, the message waiting the longest being the first to be dispatched.

first splice For submarine cable systems: cable junction between the seaward end of previously-installed shore-end cable and the first end of the cable in the cable laying ship. The first splice commences the cable laying operation.

first window Characteristic of an optical fiber having a region of relatively high transmittance surrounded by regions of low transmittance in the wavelength range 800 to 900 nanometers.

fish tape When using an empty duct for a new cable and there is no draw wire to pull the cable in, a stiff wire, with a smooth head attachment, is pushed through the duct. When it reaches the next manhole it is pulled straight out, bringing a draw-wire or draw-rope with it.

fitting A coupling or other mechanical device.

five unit code The code used between teletypewriters: International Telegraph Alphabet No. 2. See Code, ITA No. 2.

five-by-five Slang radio expression meaning good signals in both directions.

five-wheel gear The pay-out engine developed in England in 1950 for cable laying ships.

fixed Not changeable or moveable.

fixed format System in which all record communications follow a predetermined sequence giving predictability of message length, placement of control elements, etc.

fixed length numbering scheme A numbering scheme in which the length of subscribers' numbers is the same throughout a given numbering area.

fixed length working A method of organizing data in which all the records contain the same number of digits.

fixed-reference modulation A type of modulation in which the choice of the significant condition for any signal element is based on a fixed reference.

fixed station A main radiocommunication station, not mobile.

fixed storage Storage whose contents during normal operation are unalterable by a particular user.

fixture, A A double-strength telephone pole made up of two poles bolted together at the top but both slanted out so that they are separated by several feet at their bases (shaped like the letter A).

fixture, H A double strength telephone pole made up of two vertical parallel poles joined together by long crossarms at their tops. H fixtures often are used to support loading coil cases for loaded circuits in aerial cables.

fixture, loading An H-fixture pole assembly with steel angle beams instead of ordinary wooden crossarms, used to support heavy loading coil cases.

fixture, pole-extension A light steel angle bolted onto the top of a pole to provide additional height, e.g., for extra drop wires.

flag A data item or word which is set to a specific value or values to indicate the beginning or end of a frame. E.g., the unique pattern on a common channel signaling system signaling data link used to delimit a signal unit.

flag sequence In data transmission, the sequence of bits employed to delimit the beginning and end of a frame.

flake One spirally-laid-down layer of cable in a cable tank in a cable ship.

flange, ceiling Metal bracket used to suspend cable racking or runways from a ceiling in a central office or transmission equipment room.

flange, pressure testing Brass disk tinned so it may be soldered onto a lead cable sheath and with a central threaded hole which may be used for cable pressurization attachments or for testing.

flanking filter One of several filters connected in parallel (as in an FDM multiplexing terminal); each filter has an effect on the others, particularly those concerned with adjacent frequency bands. If one channel and its associated filters are removed, they must be replaced by a special network in order to maintain the specified characteristics of the other channels.

flash A succession of off- and on-hook signals; in some circumstances this can call an operator to provide assistance.

Flash (F) A high precedence signal as used on AUTOVON and similar command systems.

flash, hookswitch Repeated flash signal originated by a subscriber, the break periods (on-hook) are not long enough to be considered as calling for the circuit to be released.

Flash Override (FO) Highest preference signal as used on AUTOVON and similar command systems.

flashing jack Jack termination on some manual operating positions which enables an operator to send flashes out on a trunk circuit using a signal appropriate for that particular trunk signaling format.

flashover An arc or spark between two conductors, e.g., around or over the face of an insulator.

flashover voltage Voltage between conductors at which flashover just occurs.

flat face tube Design of TV tube with almost flat face, giving improved legibility of text and reduced reflection of ambient light.

flat fading Fading in which all frequency components of the received radio signal vary in the same proportion simultaneously.

flat loss Equal loss at all frequencies, such as caused by attenuators.

flat outputs Operation of a CATV system with equal levels of all TV signals at the output of amplifier.

flat-rate Telephone tariff in which no charges are levied for local calls, the rental charged covers unlimited local calls.

flat response Response which is generally similar for all relevant frequencies.

flat weighting In a noise measuring set, an amplitude-frequency characteristic which is flat over a specified frequency range which must be stated. Flat noise power may be expressed in dBrn (f_1-f_2), or in dBm (f_1-f_2). The terms 3 kHz flat weighting and 15 kHz flat weighting also are used for characteristics which are flat from 30 Hz to the upper frequency indicated.

fleeting The process of turning cable which is around the cable drum of a cable ship in an axial direction so that cable leading onto the drum may meet the surface of the drum perpendicularly and not pile up toward the flange.

fleeting knife Mechanical controlling device applied to the cable drum of a cable ship; it moves over the cable turns on the drum by the amount of one cable diameter per revolution of the drum, thereby insuring that the incoming cable is encountering the drum perpendicularly.

Fleming's Rule Mnemonics which help technicians remember the relationship between motion, current and magnetic field directions. ThuMb is always Motion, Forefinger is Field, and seCond finger is Current. With all three at right angles the right hand rule gives the generator or dynamo rule, and the left hand the electric motor rule. In Britain students remember that motors keep to the left.

flexible Term applied to submarine cable repeaters designed to behave like enlarged segments of cable, to permit their handling, storage, laying, and recovery with cable machinery not initially designed for repeatered cable systems, typified by the multicontainer articulated repeaters produced in the 1950—60 period; term also justifiably applied to those relatively short (about 3 to 1 length-diameter ratio) monocontainer repeaters to which the cable is attached by flexible couplings (gimbals, universal joints) permitting the repeater to safely pass around drums and sheaves.

flexible numbering Facility on some PABX systems by which individual users may move to a different location but retain the same extension number without the need for wiring changes.

flexible pricing/tariffs Regulatory procedure which permits rapid alteration of prices for services.

Flexiduct Proprietary type of rubber duct (about ½ inch high) that is laid on floors to protect cables.

flexitime Office working hours procedure encountered by telephone companies in many large cities by which actual arrival and departure times for staff are not fixed so long as all are there during core periods (e.g., 10am to 4pm) and the appropriate total time is worked by each clerk during each week.

flicker Brightness fluctuations slower than the persistence of vision, i.e., less than about 30 fluctuations per second.

flink Combination of a flash signal and a wink signal.

flip-chip Chip which is modified to enable it to be mounted upside down on a suitable substrate.

flip-flop A bistable circuit with two inputs. An input pulse on one circuit flips it into one output state in which it remains until a pulse on the other input flops it over to the other state.

float To operate a power load such as a telephone central office on a mains-driven rectifier in parallel with a low impedance storage battery which is kept fully charged by the rectifier and is itself only called upon to provide power during temporary and short-duration peaks for which the rectifier output is insufficient.

floating A circuit or device that is not connected to any source of potential or to ground.

floating battery See battery, floating.

floating in The process of drawing submarine cable from a cable ship to shore by a pulling line, the cable having floats progressively attached so that it remains near the surface until the end is made fast ashore, after which the floats are removed allowing the cable to sink to its location on the sea-floor.

floats, cable See cable floats.

flooding, tar The application of an asphaltic tar in hot fluid state to a traditional submarine cable at the point of application of the armor wires, or just afterward; also applied to jute serving.

floor, raised A floor which is made up of individually removeable 1m square panels supported above a sub-floor by leveling jacks at the panel corners. It is used in computer rooms and some communication equipment rooms, and permits rapid access to the sub-floor cabling space, allowing that space to be used as a plenum for cold air to cool the equipment racks. Also called computer flooring.

floor ground window (FGW) A copper ground bar located on each floor above the lowest floor and connected to the vertical riser (VR).

floor-wall distribution frame A single-sided distribution frame which is wall mounted but designed to come down to floor level to provide easy access for cable.

floppy disk   See disk, floppy.

Florduct   Proprietary type of metal cable duct used to protect cables laid on floors.

flow chart   1. A diagram to show procedures which are followed and actions taken. 2. A graphical representation of the logic of a program.

flow control   The procedure for controlling the rate of transfer of packets between two nominated points in a data network.

flow line   1. In man-machine language, a line representing a connection path between symbols in a syntax diagram. 2. A flow line in specification description language which connects every symbol to the symbol(s) it follows.

flow system, continuous   Method of pressurizing cables by supplying dry air on a continuously flowing basis.

fluorescence   When some elements are illuminated by light or by ultraviolet (UV) radiation they emit light of a longer, highly visible wavelength. In fluorescent lamp tubes the UV radiation produced strikes the chemical phosphors on the inner surface of the tube, and they fluoresce to give out light. In high-power mercury or sodium street lighting, the atoms of the vaporized elements themselves fluoresce as they decay from excited states to their normal states.

flutter   1. The distortion due to variation in loss resulting from the simultaneous transmission of a signal at another frequency. 2. The distortion due to variation in loss resulting from phase distortion. 3. In recording and reproducing, deviation of frequency which results, in general, from irregular motion during recording, duplication or reproduction. 4. In radio transmission, rapidly changing signal amplitude levels together with variable multipath time delays, caused by the reflection and possible partial absorption of the radio signal from aircraft flying through the beam or common scatter volume. 5. The effect of the variation in the transmission characteristics of a loaded telephone circuit caused by the action of telegraph direct currents on the loading coils. 6. Fast changing variation in received signal strength, such as may be caused by atmospheric variations, antenna movements in a high wind, or interaction with another signal.

flux   1. Electric or magnetic lines of force. 2. Chemical used during soldering which ensures that surfaces are clean enough to be able to make good electrical contact with each other.

flux, noncorrosive   Acid-free soldering flux: rosin is the most commonly used.

flux, residual   Magnetic flux left even when the magneto motive force concerned has been removed.

flux, soldering   Chemical used to remove surface films of oxide before wires or tags are soldered together.

fly wheel no-break set   Prime mover (normally a diesel engine) coupled via a common shaft and a heavy flywheel to an electric motor and a generator. The public ac mains power supply normally powers the motor; this turns the generator which feeds the load. If the public supply fails the diesel generator is automatically started, the inertia of the heavy flywheel is sufficient to get the diesel going so the generator, on the same shaft, has very little chance of slowing down.

flyback   Voltages must be applied in a special way in cathode ray tubes in order to make the electron sweep across the screen as it scans the picture. In most applications it is not desirable to see the electron spot return to its starting point; this is its flyback, so it is suppressed.

flywheel effect   That characteristic of an oscillator that enables it to sustain oscillations after removal of the control stimulus. This characteristic may be desirable, as in the case of phase-locked loops employed in synchronous system equipment or undesirable, as in the case of a voltage-controlled oscillator.

FM improvement factor   The signal-to-noise ratio at the output of the receiver divided by the carrier-to-noise ratio at the input of the receiver. This improvement is always obtained at the price of an increased bandwidth in the receiver and the transmission path.

FM improvement threshold   The point in an FM receiver at which the peaks in the radio frequency signal equal the peaks of the thermal noise generated in the receiver.

focusing   Method of making beams of radiation converge on a target such as the face of a TV tube.

foil   Very thin metal sheet, typically aluminum, used in fixed capacitors.

foil capacitor   Capacitors made with thin metallic foil coatings on insulating surfaces; sometimes plastic film or thin paper sheets are used rather than foil.

foil electret   Diaphragm used in one type of noncarbon condenser microphone, utilizes a thin film of plastic with a layer of metal deposited on one surface.

F1A-line weighting   A noise weighting used in a noise measuring set to measure noise on a line that would be terminated by a 302-type or similar instrument. The meter scale readings are in dBa (F1A).

Fonofax   A public facsimile service operated by British Telecom.

Fonofax SF   Fonofax store and forward. Proposed public facsimile service using high-speed scanning, storage and auto-calling equipment on the UK switched telephone network.

footprint   1. That portion of the Earth's surface illuminated by a narrow beam from a satellite. 2. Floor area occupied by a given unit of equipment.

force, back electromotive   Back emf: voltage developed inside an inductive component whenever the current it is carrying changes, the voltage developed always opposes change. Counter emf.

force, electromotive   Force that produces or tries to produce a movement of electric charges, an electric current.

forced backward clearing of subscriber's line   A subscriber engaged on an incoming call can clear his line independently of the caller, thus avoiding possible malicious blocking. After such clearing the line is free for new incoming as well as outgoing calls.

forced dialing Bad habit sometimes encountered: forcing the returning dial to go faster than it should, thereby increasing the likelihood of getting wrong numbers.

forced disconnect Action by a dial office which disconnects the calling subscriber if he does not go back on-hook within a specified period of time.

forced oscillations Oscillations forced into a circuit by an external drive.

forced rerouting A procedure of transferring common channel signaling system signaling traffic from one signaling route to another, when the signaling route in use fails or is required to be cleared of traffic.

foreign attachment Equipment or circuits not furnished by the telephone company or local telecommunications administration which is connected to local public telephone facilities. In the U.S. the Carterfone decision resulted in the interconnect market being opened up substantially, but in many other countries it is extremely difficult for manufacturers other than local firms to break in to this market.

foreign exchange (FX) A service by which a telephone or PBX in one city, instead of being connected directly to a central office in that city, is directly connected to a CO in a distant city via a private line. To callers it appears that the telephone or PBX is actually located in the distant city.

foreign mobile A mobile radio telephone station temporarily located in an area other than the one served by its home cellular radio system.

form, cable Insulated wires from which the outside cable sheath has been removed and which are fanned out in numerical order and in groups of wires to make connection with tags or terminations on a frame

form, contact Coding letter to indicate the type of contact assembly to be found on a relay, e.g., A=make, B=break, D=make before break, X=double make, etc.

form, fanned Cable form in which the fanning out of wires begins close to the point where the cable sheath has been removed.

form, loose wire Wiring held in formation like a loosely made-up cable by the use of fanning rings, i.e., with no lacing-up into tight, neat bundles.

form, manhole Wooden or metal templates or guides used to construct concrete manholes or cable jointing chambers. These forms are reusable.

form, sewed A cable form sewn up or laced into a neat bundle of wires with individual wires leaving the form exactly where they are to be terminated.

form factor Ratio of the root mean square value of a periodic function to the average absolute value, averaged over a full period of the function.

format 1. Arrangement of bits or characters within a group, such as a word, message, or language. 2. Shape, size, and general makeup of a document.

FORTRAN A high level computer language often used for scientific work: FORmula TRANslation.

fortuitous By chance, random. Most telephone traffic theory is based on the assumption that traffic demand originates in a random manner.

fortuitous conductor Any conductor which may provide an unintended path for intelligible signals, e.g., water pipes, wire or cable, or metal structural members.

fortuitous distortion Distortion of telegraph signals resulting from causes generally subject to random laws, e.g., accidental irregularities in the operation of the apparatus and its moving parts, or disturbances affecting the transmission channel.

forward In the direction from the calling to the called station.

forward acting code A logically constructed code with enough redundancy to correct many errors without requiring retransmission.

forward channel The channel of a data circuit that transmits data from the data source to the data sink.

forward direction A forward bias produces a larger current, i.e., forward means the direction in which a component or device has a smaller resistance or impedance.

forward error correction (FEC) A system of error control for data transmission wherein the receiving device has the capability to detect and correct any character or code block which contains a predetermined number of bits in error. FEC is accomplished by adding bits to each transmitted character or code block using a predetermined algorithm. It does not need a feedback channel, and therefore may be used with a one-way transmission system.

forward indicator bit A bit in a common channel signaling signal unit which indicates the start of a retransmission cycle.

forward routing Routing of a long distance call in a logical way towards its destination.

forward scatter 1. The deflection by reflection or refraction of an electromagnetic wave or signal in such a manner that a component of the wave is deflected in the direction of propagation of the incident wave or signal. 2. The component of an electromagnetic wave or signal that is deflected by reflection or refraction in the direction of propagation of the incident wave or signal. 3. To deflect, by reflection or refraction, an electromagnetic wave or signal in such a manner that a component of the wave or signal is deflected in the direction of propagation of the incident wave or signal.

forward sequence number A field in a common channel signaling signal unit which identifies the last message signal unit transmitted.

forward signal A signal sent in the direction from the calling to the called station, or from a data source to a data sink.

forwarding, call Service provided by some dial offices in which a subscriber may ask for incoming calls to be diverted to another number, until the request is countermanded. A similar practice may be followed on Centrex lines: if there is no reply within a specified time the call may be routed automatically to the company's Centrex attendant.

forwarding, customer-selective call An automatic call forwarding or follow-me service by which a customer may dial into the office and instruct the processor to divert calls from his line to other lines.

Fourier analysis Complex waveforms may be broken down into components which are themselves all simple sine or cosine waves, possibly of different frequencies and maximum amplitudes and differing in phase one with another.

four-row keyboard A typewriter-like keyboard with a full set of figures, letters and symbols.

four-thirds earth radius Atmospheric refraction makes the earth behave for radio communication purposes as if its radius were 1 ⅓ times its true value. Charts showing heights above sea level along the path between proposed radio terminals are therefore drawn to a scale appropriate to this four-thirds factor when planning line-of-sight links.

four-wire circuit See circuit, four-wire.

four-wire line A two-way circuit using two pairs of conductors so arranged that signals are transmitted in one direction only by one pair and in the other direction only by the other pair.

four-wire repeater A repeater for use in a four-wire circuit in which there are two amplifiers, one to amplify the signals in one direction and the other to amplify the signals in the other direction.

four-wire switching Switching using a separate path, frequency band or time interval for each direction of transmission.

four-wire terminating set A set used to terminate the transmit and receive channels of a four-wire circuit and to interconnect four-wire and two-wire circuits.

fox The most famous fox in the telecommunications world is the quick brown fox who stars in a standard test message, which includes all the alphanumerics on a teletypewriter and also the function characteristics (space, figures shift, letters shift): "The quick brown fox jumped over the lazy dog's back 1234567890."

frame 1. In data transmission, the sequence of contiguous bits bracketed by and including beginning and ending flag sequences. 2. In the multiplex structure of PCM systems, a set of consecutive digit time slots in which the position of each digit time slot can be identified by reference to a frame alignment signal. 3. In a TDM system, a repetitive group of signals resulting from a single sampling of all channels, including any additional signals for synchronizing and other required system information. 4. In facsimile systems, a rectangular area, the width of which is the available line and the length of which is determined by the service requirements. 5. A distributing frame. 6. A switch frame. 7. A trunk board. 8. One complete television picture, transmitted 30 times per second in the U.S. 9. A block of information from a videotex system's database that can be displayed as a single screen image.

frame, combined distribution frame A frame which combines a main distribution frame and an intermediate distribution frame.

frame, crossbar Most crossbar offices have their switches mounted on various types of racks, called frames, such as: junctor frames, link frames, line link frames, trunk link frames, sender frames, or number group frames.

frame, distribution A steel framework in a central office on which external cable terminations are mounted together with cables connecting to the appropriate line units in the office. Cross-connections are made between terminations by running jumper wires.

frame, floor-type distribution A floor-supported double-sided main distribution frame with access to both the vertical and horizontal sides.

frame, intermediate distribution (IDF) Frame which provides flexibility in allocation of the subscriber's number to the line unit or equipment in the office which is to be associated with the particular line.

frame, main To a computer engineer, a main frame is a large computer.

frame, main distribution (MDF) Frame on which external distribution cables terminate, together with their associated protective devices (on the vertical side) and with internal cables to the central office line units (on the horizontal side). Interconnection is made by running jumper wires between the termination blocks.

frame, manhole The heavy iron structure which acts as a road-level entry to a manhole.

frame, single-sided distribution A distribution frame designed for mounting against a wall or in a location where only one-sided access is possible.

frame, type A main distributing Obsolete type of MDF with external distribution cables terminated on horizontal blocks and equipment cables terminated on vertical blocks.

frame, type B main distributing The standard main distributing frame where outside plant cables terminate on vertical protectors, from which they cross-connect to equipment cables terminated on horizontal blocks. This frame supplies a protector on every exposed outside plant pair. Protection policy differs from country to country, depending on the incidence of lightning, the existence of overhead external plant, and the type of switching equipment in use.

frame, wall-type distributing A single-sided distribution frame, wall-mounted. Normally utilizes vertical terminating blocks both for external cables and for internal cables to switching equipment.

frame alignment For PCM: the state in which the frame of the receiving equipment is correctly phased with respect to that of the received signal.

frame alignment recovery time For PCM: the time that elapses between a valid frame alignment signal being available at the receive terminal equipment and frame alignment being established.

frame alignment signal For PCM: the distinctive signal used to secure frame alignment.

frame alignment time slot For PCM: a time slot starting at a particular phase in each frame and allocated to the transmission of a frame alignment signal.

frame check sequence (FCS) In bit-oriented protocols, a 16-bit-field usually appended at the end of a frame that contains transmission error-checking information.

frame frequency The number of times per second that a frame of information is transmitted or received. Nominally 30 frames per second in the U.S.

frame synchronization A technique for synchronizing the receive circuits with the incoming bit stream on a link.

frame synchronization pattern In digital communications, a prescribed recurring pattern of bits transmitted to enable the receiver to achieve frame synchronization.

framework ground (FG) A protection ground consisting of a conducting path to ground of non-current-carrying framework metal.

framing 1. In facsimile, the adjustment of the facsimile picture to a desired position in the direction of line progression. 2. Control procedure used with multiplexed digital channels (e.g., T1 carrier) where bits are inserted so that the receiver may identify the timeslots allocated to each subchannel. Framing bits may also carry alarm signals.

framing bit 1. A bit used for frame synchronization purposes. 2. A bit at a specific interval in a bit stream used in determining the beginning or end of a frame.

framing signal In facsimile transmission, a signal used for adjustment of the picture to a desired position in the direct line of progression.

franchise Contractual agreement between a CATV operator in the U.S. and the governing municipal authority. Under federal regulation a franchise, certificate, contract or any other agreement amounts to a license to operate.

Fraunhofer diffraction pattern See far field diffraction pattern.

Fraunhofer region See far field diffraction pattern and far field region.

free electron An electron not bound to a specific atom and free to move when an electric field is applied.

free oscillations Oscillations arising from internal sources; these are like a pendulum given one push but slowly thereafter reducing its amplitude of swing, or like a capacitor discharging its charge through an external circuit in a damped train of decaying oscillations.

free running An oscillator which is not controlled by an external synchronizing signal.

free space Space characterized by the absence of gravitational and electromagnetic fields. A theoretical concept of space devoid of all matter.

free space loss The signal attenuation that would result if all obstructing, scattering, or reflecting influences were sufficiently removed so as to have no effect on propagation.

free space transmission Theoretical transmission of a radio communication signal through a vacuum with no absorption, no reflection, and no diffraction.

free space wave That section of a radio wave which travels direct between antennae with no reflections or refractions.

freefone service A UK service whereby a subscriber can be allocated a special (Freefone) number and all charges to this number are paid by that subscriber instead of by the callers. (Generally similar to INWATS/800 series numbers in the US).

freeze frame Digital television transmission system where screen images are erased and renewed every few seconds at the receiver. Images are sent in real time but there is no motion effect. In videoconferencing these techniques permit narrower bandwidths to be used than would be needed for full motion TV. Syn. (U.K.): slowscan TV

frequency The number of complete cycles of a periodic activity which occur in a unit time, i.e., the number of times the quantity passes through its zero value in the same sense in unit time. If T is the period of a repetitive phenomenon, then the frequency f is $1/T$. In SI units the period is expressed in seconds, and the frequency is expressed in hertz.

frequency, audio Any frequency in the human audio band, roughly 15 Hz to 15 kHz.

frequency, critical singing Frequency at which the singing margin is lowest, this is the frequency at which singing first commences, when overall loop gain exceeds loop loss. Critical frequencies are usually at the low end (about 300 Hz) or at the top end (about 3000 Hz) of the spectrum.

frequency, crossover Frequency at which equal powers are delivered to two networks.

frequency, effective A frequency in a broadband system such that the total energy at frequencies lower than this effective frequency equals the total energy in the band at higher frequencies.

frequency, extremely high (EHF) Any · of the radio frequencies in the band 30 GHz to 300 GHz.

frequency, extremely low (ELF) A radio frequency below 300 hertz.

frequency, facsimile frame The repetition rate, or number of facsimile frames transmitted in one second.

frequency, high (HF) Any of the radio frequencies in the band between 3 and 30 MHz.

frequency, intermediate Frequency produced in a super-heterodyne receiver after combining the incoming signal with a signal from a local oscillator. This intermediate frequency is used for stages giving selectivity and amplification. Commonly-used intermediate frequencies are 455 kHz for broadcast receivers, 45.75 MHz for TV video signals, and 41.25 MHz for TV sound.

frequency, low (LF) Any of the radio frequencies in the band between 30 kHz and 300 kHz

frequency, medium (MF) Any of the radio frequencies in the band between 300 kHz and 3 MHz.

frequency, natural 1. A frequency at which a body is able to oscillate freely. 2. Of an antenna: the lowest frequency for which a standing wave may be established.

frequency, optimum working Frequency at which radio transmission between two points via the Earth's ionosphere can be most efficiently carried out at a particular time.

frequency, pulse repetition (PRF) Pulses per second in a pulse train.

frequency, resting The frequency of an FM transmission in the absence of a modulating signal.

frequency, standard A frequency with a known relationship to a frequency standard. This term often is used for the signal whose frequency is a standard frequency.

frequency, super high (SHF) Any radio frequency in the band 3-30 GHz.

frequency, tremendously high (THF) Any radio frequency in the band 300 to 3000 GHz.

frequency, ultra-high (UHF) Any radio frequency in the band between 300 MHz and 3 GHz. Also, television channels 14 through 83.

frequency, very low (VLF) Any radio frequency between 3 and 30 KHz.

frequency, voice For normal telephony purposes, the band 300 Hz to 3400 Hz is transmitted. This is called the commercial speech band.

frequency accuracy The degree of conformity to a specified value of a frequency.

frequency agility Ability of a device to tune to a number of different frequency bands or channels. See frequency hopping.

frequency allocation See allocation, frequency.

frequency assignment See assignment, frequency.

frequency averaging A process by which network synchronization is achieved by use of oscillators at all nodes which adjust their frequencies to the average frequency of the digital bit streams received from connected nodes. All oscillators are assigned equal weight in determining the ultimate network frequency since there is no reference oscillator.

frequency band, assigned The frequency band the center of which coincides with the frequency assigned to the station and the width of which equals the necessary bandwidth plus twice the absolute value of the frequency tolerance.

frequency band, high group or low group Carrier telephony systems working on a two-wire physical bearer use a high group of frequencies in one direction and a low group in the other. A common U.S. system uses 36—132 kHz as a low group and 172—268 kHz as a high group frequency band.

frequency-change signaling A signaling method in which one or more particular frequencies correspond to each desired signaling condition of a code. The transition from one set of frequencies to the

other may be either a continuous or a discontinuous change in frequency or in phase.

frequency changer Device for changing the frequency of a power supply. It is normally necessary to go through two transition stages, e.g. a motor driven by power supply at one frequency drives a generator which produces a power supply at the required frequency

frequency departure An unintentional deviation from the nominal frequency value.

frequency dependent rejection (FDR) A measure of the rejection produced by a receiver selectivity curve on an unwanted transmitter emission spectra. Also see frequency distance.

frequency deviation See deviation, frequency.

frequency difference The algebraic difference between two frequencies. These two frequencies can be of identical or different nominal values.

frequency displacement The end-to-end shift in frequency that may result from independent frequency translation errors in a circuit.

frequency distance (FD) A measure of the minimum distance separation that is required between a victim receiver and an interferer as a function of the difference between their tuned frequencies. FD and frequency dependent rejection (FDR) are measures of the interfering coupling mechanism between interferer and receiver and are basic solutions required for many interference evaluations.

frequency diversity Any method of diversity transmission and reception wherein the same information signal is transmitted and received simultaneously on two or more independently fading carrier frequencies.

frequency division multiple access (FDMA) In satellite communications, the use of frequency division to provide multiple and simultaneous transmissions to a single transponder.

frequency division multiplex (FDM) A method of deriving two or more simultaneous, continuous channels from a transmission medium connecting two points by assigning separate portions of the available frequency spectrum to each of the individual channels. Each information channel or group of channels is allotted a different frequency band, the information or signal channel is used to modulate a carrier frequency in order to produce a sideband in the required frequency band. At the incoming end separation is done by filtration. This is traditional carrier working, but now more and more carrier systems are made on a time-division basis.

frequency division switching The switching of inlets to outlets using frequency division multiplexing techniques.

frequency doubler See doubler, frequency.

frequency drift A slow, undesired change in the frequency of an oscillator, transmitter, receiver, or other equipment.

frequency exchange signaling A frequency change signaling method in which the change from one signaling condition to another is accompanied by decay in amplitude of one or more frequencies and by build-up in amplitude of one or more other frequencies.

frequency frogging 1. The interchanging of the frequency allocations of carrier channels to prevent singing, reduce crosstalk, and to correct for line slope. This is accomplished by having the modulators in a repeater translate a low-frequency group to a high-frequency group, and vice versa. 2. Alternate use of two frequencies at repeater sites of line-of-sight microwave systems.

frequency guard band A frequency band left vacant between two channels to provide a margin of safety against mutual interference.

frequency hopping An electronic counter-countermeasure technique in which the instantaneous carrier frequency of a signal is periodically relocated, according to a predetermined code, to other positions within a frequency spectrum much wider than that required for normal message transmission. The receiver uses the same code to keep itself in synchronism with the hopping pattern.

frequency instability Expressed by the frequency change within a given time interval. Generally one distinguishes between frequency drift effects and stochastic frequency fluctuations. Special variances have been developed for the characterization of these fluctuations.

frequency meter Instrument or test set used to measure the frequency, usually of a radio signal, but also of any alternating signal.

frequency meter, vibrating reed Series of mechanically tuned reeds with natural frequencies above, at and below a standard frequency. The reeds are all located in the field of an electromagnet carrying the current concerned, the reed showing the most vibration indicates the actual frequency of the current, thereby providing a visual check of operating frequency.

frequency modulation (FM) The form of angle modulation in which the instantaneous frequency of a sine wave carrier is caused to depart from the carrier frequency by an amount proportional to the instantaneous value of the modulating signal. Combinations of phase and frequency modulation also are commonly referred to as frequency modulation.

frequency multiplier Device with an output of an exact multiple of its input frequency.

frequency offset The fractional frequency deviation of a frequency with respect to another frequency:

$$\Delta f/f = (f_1 - f_2)/f_2$$

where Δf is the difference between the two frequencies: f_1 and f_2, and f is the reference frequency, f_2, with respect to which the offset is taken.

frequency response characteristic Variation in the transmission performance (gain or loss) of equipment or system with respect to variations in frequency.

frequency selectivity Ability of equipment to separate or differentiate between signals at different frequencies.

frequency shift 1. A type of telegraph operation in which the mark and space signals are different frequencies. 2. Any change in the frequency of a radio transmitter or oscillator. Also called rf shift, i.e., an intentional frequency change used for modulation purposes. 3. In facsimile, a frequency modulation system where one frequency represents picture black and another frequency represents picture white. Frequencies between these two limits may represent shades of gray. 4. The number of hertz difference in a frequency modulation system.

frequency shift carrier Form of multi-channel FM voice-frequency telegraphy in which a mark signal uses an audio frequency 70 Hz higher than a space signal (channels are usually spaced at 170 Hz intervals).

frequency shift keying (FSK) A form of frequency modulation in which the modulating signal shifts the output frequency between predetermined values and the output signal has no phase discontinuity.

frequency spectrum Division of electromagnetic radiation frequencies according to characteristics and use, e.g., radio spectrum, light spectrum. See Appendix B.

frequency stability A measure of the variations of the frequency of an oscillator from its mean frequency over a specified period of time.

frequency standard A stable oscillator used for frequency calibration. It usually generates a fundamental frequency with a high degree of accuracy, and harmonics of this fundamental are used to provide reference points.

frequency standard, secondary A frequency standard which is calibrated with respect to a primary frequency standard. The term secondary describes the position of the standard in a hierarchy, not the quality of its performance.

frequency tolerance The maximum permissible departure of the center frequency of the frequency band occupied by an emission from the assigned frequency or of the characteristic frequency of an emission from the reference frequency. This includes both the initial setting tolerance and excursions related to short- and long-term instability and aging. The frequency tolerance is expressel in parts in 10n, in hertz, or in percentages.

frequency translation The transfer en bloc of signals occupying a definite frequency band (such as a channel or group of channels) from one position in the frequency spectrum to another, so that the arithmetic frequency difference of signals within the band is unaltered.

fresnel An obsolete unit of frequency equal to one million megahertz.

Fresnel diffraction pattern See near field diffraction pattern.

fresnel lens A lens whose surface is broken up into many concentric annular rings.

Fresnel reflection In optics: the reflection of a portion of incident light at a planar interface between two homogeneous media having different refractive indices. Fresnel reflection occurs at the air-glass interfaces at entrance and exit ends of an optical waveguide. Resultant transmission losses (on the order of 4 percent per interface) can be virtually eliminated by use of antireflection coatings or index matching materials.

Fresnel region   The region close to a radio communication antenna, between the antenna itself and the Fraunhofer region in which the antenna's field is focused.

Fresnel zone   The unit with which a line-of-sight radio path's clearance over the earth is expressed. A cigar-shaped shell of circular cross section surrounding the direct path between a transmitter and a receiver. For the first Fresnel zone, the distance from the transmitter to any point on this shell and on to the receiver is one half-wavelength longer than the direct path; for the second Fresnel zone, two half-wavelengths, etc. A good line-of-sight radio path will have a clearance over earth of not less than 0.6 of the first Fresnel zone radius.

fringe area   Area outside the boundary of the zone for which satisfactory radio, TV or telephone service is provided.

frogging   See frequency frogging.

front contacts   The make contacts of a relay.

front cord   On an operating position, the cord of a pair nearer the operator.

front-end noise temperature   A measure of the thermal noise in the first stage of a receiver.

front-end processing   The transformation of information prior to a processing operation.

front-end processor   A programmed-logic or stored-program device which interfaces data communications equipment with an input/output bus or memory of a data processing computer.

front porch   That portion of the composite picture signal which lies between the leading edge of the horizontal blanking pulse, and the leading edge of the corresponding sync pulse.

front-to-back ratio   A ratio of parameters used in connection with antennas, rectifiers, or other devices in which signal strength or resistance, or other parameters, in one direction is compared with that in the opposite direction.

frying   Noise heard in telephones due to excessively large currents passing through carbon granule microphone capsules.

fuel cell   Chemical cell which produces electrical energy from a chemical fuel, e.g., from methane.

full carrier   Carrier emitted at a power level of 6 dB or less, below the peak envelope power. (Double-sideband amplitude-modulated emissions normally comprise a full carrier with a power level exactly 6 dB below the peak envelope power at 100% modulation; in single-sideband full-carrier emissions, a carrier at a power level of 6 dB below the peak envelope power is generally emitted, to enable the use of a receiver designed for double-sideband full-carrier operation).

full duplex   Simultaneous communication in both directions between two points.

full float   A storage battery floating across a rectifier and a load to provide power to the load during temporary peaks.

full network station   A commercial television broadcast station in the U.S. that generally carries 85 percent of the hours of programming offered by one of the three major national networks during its weekly prime time hours.

full speed (circuit)   An international leased teletype circuit capable of working at 66 ⅔ words per minute.

full trunk group   A direct group of circuits from which there is no overflow to a later choice route, so the group must be dimensioned to provide the required grade of service; a fully-provided circuit group.

full-wave rectifier   Both positive and negative half-cycles of the incoming ac signal are rectified to produce a unidirectional (dc) current through the load.

fully connected network   Network topology in which each node is directly connected to all other nodes; clearly impracticable as the number of nodes in the network increases.

fully dissociated signaling   A form of non-associated common channel signaling in which the path that signals may take through the network is restricted only by the rules and configuration of the signaling network.

fully perforated   Paper tape used to carry encoded information by having holes punched completely through.

fully provided circuit group   With respect to a particular parcel of traffic, a circuit group which is the first choice for this traffic and which is not traffic engineered as a high usage group. A route between two exchanges having sufficient circuits to carry the total busy hour traffic.

function   1. A teletypewriter signal which does not produce a printed output, e.g., line feed, carriage return, or letter/figure shift. 2. For man-machine language: an action which various staff groups wish to carry out, e.g., add subscriber's line, initiate a testing routine, read a subscriber's class of service.

function access code   See feature access code.

function code   A code indicating the type(s) of process to be applied to the service.

function identification   Information indicating the type(s) of process to be applied to the service.

function key   Key on a keyboard device that results in action such as line feed, carriage return, etc., and does not result in a character being printed.

functional block   In specification description language: an object of manageable size and relevant internal relationship, containing one or more processes.

functional block diagram   Diagram illustrating the definition of a problem on a logical and functional basis.

functional description   In specification description language: the actual behavior of the implementation of the functional requirements of the system in terms of the internal structure and logic processes within the system.

functional entity Partitioning of the design subordinate to subsystem, or to another functional entity. It may be in hardware, firmware or software.

functional progression charts Charts which give the behavior of a functional block when subjected to external stimuli.

functional signaling link A concept that describes a communications link together with the transfer control functions associated therewith.

functional specification In specification description language: a specification of the total functional requirements of that system from all significant points of view.

functional unit An entity of hardware and/or software capable of accomplishing a special purpose.

fundamental frequency The lowest frequency of a complex signal: higher components are often harmonics or multiples of the fundamental.

fundamental mode 1. The lowest order mode of a dielectric cylindrical wave-guide, in most cases the mode designated LP_{01} or HE_{11}. 2. Lowest order mode of an optical waveguide. It is the only mode capable of propagation in single-mode (monomode) fibers.

fundamental plans For a telephone company: the short and long term plans for development works to be carried out in order to improve service to customers and maximize opportunities for the company. Fundamental plans for local distribution networks include plans for future development of duct systems and main cable systems and proposed division of the whole area into cabinet areas to be served from individual flexibility points.

furnace, cable splicer's A small portable stove for melting solder, etc. Butane gas commonly is used as fuel.

fuse An overcurrent protective device with a circuit-opening fusible part which is heated and severed by the passage of overcurrent through it.

fuse, alarm and indicator Type of fuse used on distribution and equipment frames: when a fuse blows the fuse wire is no longer able to hold spring-loaded contacts apart, so these contacts are made and an indication given of the location of the blown fuse.

fuse, cartridge Type of fuse with the fusible wire enclosed within a short tube of glass or fiber with metal caps on both ends.

fuse, dummy Device the same physical shape as a fuse but plays no active role. If the circuit normally carries a current a dummy fuse will be a low resistance metal strap. Insulating dummies are sometimes used in inactive circuits to stop these from being used without proper authority.

fuse, grasshopper A light unenclosed fuse sometimes used on transmission equipment; it is spring loaded and jumps open when the fuse blows giving an alarm and a visual indication of location.

fuse, heat coil A delayed action fuse, the same external shape as a fuse cartridge but will be blown not only by short-duration high currents but also by comparatively low fault currents which last a long time.

fuse, plug A fuse mounted in a holder with a screw-thread base.

fuse, power rated Fuse rated to carry a specified current indefinitely. Currents 50% greater than the specified value will lead to the fuse blowing within five minutes.

fuse, sneak current See sneak current fuse.

fuse disconnecting switch A disconnecting switch in which a fuse unit is in series with or forms a part of the blade assembly.

fused fiber splice A splice accomplished by the application of localized heat sufficient to fuse or melt the ends of two lengths of optical fiber, forming, in effect, a continous, single fiber.

fused silica Amorphous silicon dioxide. Highly refined fused silica formed by a vapor deposition process or by other means is employed in the fabrication of low loss optical waveguides. Dopants may be added via the same process to obtain suitable index variations in the optical waveguide core and cladding regions.

fuseholder, extractor Type of fuse fitting in which the fuse pops up and may be removed as soon as a holder cap has been unscrewed.

fusetron A time delay fuse.

fusion splice The splicing together of optic fibers by the application of localized heat sufficient to fuse or melt the two ends concerned, forming a continuous fiber.

fusion splicer An instrument used for joining optic fibers by welding their cores together using a brief electric arc.

future proof Equipment which is made on a modular basis so that as new devices and technologies are developed the outdated units can be directly replaced by new ones, continuing to use the same interfaces.

fuze See fuse.

fuzzy logic Branch of mathematics used in some artificial intelligence computers — human decisions are often based on ideas and approximations rather than on mathematically rigid calculations.

gaff Sharp spur on climbing irons used by linemen.

gain 1. A notch cut into a telephone pole into which a crossarm is bolted. 2. Ratio of output to input potential, current or power, usually expressed in dB; measurements are usually in volts, amps or milliwatts.

gain, antenna Ratio in dB of signal level transmitted by an actual antenna to the signal level which would be transmitted by an isotropic antenna at the same location and fed with the same power. Gain by a receiving antenna is similarly defined.

gain, net Sum of all gains minus the sum of losses in a particular circuit or loop.

gain, obstacle Effect on received signal strength of an obstacle along a line-of-sight path. A sharply-ridged mountain can, for example, diffract a significant proportion of energy down to a receiver in circumstances where a map study indicates that the direct line of sight path is completely interrupted.

gain, pole Notch or flattened section on one side of a pole to enable crossarms to be securely attached.

gain, power Ratio of output power to input power of a device, expressed in dB.

gain, reflection When impedances at a reflection point are opposite in phase it is possible for reflection gain (not loss) to be introduced.

gain, slab Flattened section several feet long at the top of a pole which enables more than one crossarm to be secured to the pole; a special type of pole gaining.

gain, space diversity Improvement in reception of radio signals by combining outputs from two separate receiving antennae separated from each other by at least five wavelengths at the operating frequency.

gain, steel pole Steel fitting giving the effect of a flat side on which a crossarm may be bolted even though the pole itself may be of a material (such as concrete) which cannot be notched.

gain, transmission Increase in a signal's power from one point to another, expressed in dB.

gain, voltage Ratio of input voltage to a device to its output voltage, expressed in dB.

gain-bandwidth product Gain times the frequency of measurement of an avalanche photodiode when the device is biased for maximum gain.

gain compression Small reduction in repeater gain under traffic load conditions; significant in systems with a large number of repeaters in tandem, such as submarine cable systems.

gain hits Short duration surges in power, causing errors in data transmission. Bell standard for phone-line data transmission calls for eight or fewer gain hits in a 15-minute period. A gain hit is a surge of more than three dB, lasting more than 4 milliseconds.

gain margin Additional gain needed to make a circuit unstable and start to sing.

galactic radio noise Radio noise reaching the earth from outer space, in particular, from stars in our own galaxy.

galaxy Range of satellites manufactured by Hughes Space and Communications Corp.

gallium arsenide A semiconductor material used in logic circuits and lasers. Electrons move more than six times faster in Ga As than in silicon.

galvanic Device which produces direct current by chemical action.

galvanometer An instrument used to measure small currents.

galvanometer shunt A resistor connected in parallel with a galvanometer to reduce its sensitivity.

gamma rays Very high frequency electromagnetic radiation, emitted by some radioactive elements.

gang capacitor Variable capacitor with more than one set of moving plates ganged together, i.e., mounted on the same shaft.

gang tuning Simultaneous tuning of several different circuits by turning a single shaft on which ganged capacitors are mounted.

ganged Mechanically coupled, normally by using a shared shaft.

gap For electromechanical relays: distance between contacts in non-operated or circuit broken condition; also, the space between a pole piece and moving armature.

gap loss An optical power loss caused by a space between a source and an optical waveguide, between axially aligned waveguides, or between waveguide and detector.

garbage Meaningless or false data. "Garbage in, garbage out" or GIGO is a well-known adage in the computer world.

garble An error in transmission, reception, encryption, or decryption which changes the text of a message or any portion thereof in such a manner that it is incorrect or undecryptable.

gas breakdown When the voltage across a gas-filled tube reaches a critical value, ions are produced by collisions leading to a rapid breakdown discharge, similar to an avalanche process in a semiconductor.

gas-discharge tube Gas-filled tube designed to carry current during gas breakdown. Some tubes are designed to breakdown as protective devices, i.e, to protect other apparatus from high voltages, and some types are designed to carry current continuously, e.g., glow-discharge tubes (used in illuminated signs) and X-ray tubes.

gas-filled Tubes or glass envelopes containing apparatus and filled with gas either to improve the functioning of the apparatus or as an essential feature of the circuit itself.

gas-filled radiation tubes Gas-filled tubes designed to breakdown when a special type of radiation or charged particles pass through them, e.g., geiger counters, which can flash over when a single charged particle is detected.

gassing The production of gas at electrodes during electrolysis, e.g., when a lead-acid storage battery or accumulator is being given an equalizing charge and is nearing a fully-charged state.

gate 1. A device with one output channel and one or more input channels, such that the output channel state is completely determined by the input channel states except during switching transients. 2. A combinational logic element having at least one input channel.

gate, time Logic gate which provides output at predetermined time intervals.

gate electrode Electrodes in various types of solid state units which are considered to be electronic switches or logic devices.

gateway A facility which adapts the conventions and protocols of one network to those of another; performs a protocol-conversion operation across a wide spectrum of communications functions.

gateway (videotex) A facility allowing a videotex system to access computers that are outside the system, linked by data networks.

gateway office A toll office through which calls pass through to and from another country. White Plains N.Y. and Oakland are typical U.S. gateway offices. A gateway office or switching center must analyze digits dialed by the customer, decide on the route to the country concerned, ensure all necessary call accounting procedures are followed for correct billing, interwork with signaling systems on international circuits and supervise calls once established. A gateway office dealing with incoming calls must interface with international signaling systems and also have the ability to call in an operator able to speak in a specified language.

gating 1. The process of selecting only those portions of a wave between specified time intervals or between specified amplitude limits. 2. The controlling of signals by means of combinational logic elements. 3. A process in which a predetermined set of conditions, when established, permits a second process to occur.

gating pulse Pulse which operates a logic gate.

gauge A measuring instrument, particularly for wire diameters; a scale used to determine wire diameter or metal sheet thickness.

Gauge, American Wire (AWG) Standard scale for specifying diameter of non-ferrous wires (particularly copper wires); originally known as the Brown & Sharpe (B&S) Gauge.

Gauge, Birmingham Wire (BWG) Standard scale for specifying diameter of ferrous wires; originally known as Stub's Iron Wire Gauge.

gauge, feeler Sets of carefully made hard steel strips, with thickness marked in one-thousandth parts of an inch, used for checking adjustment of relay armatures.

gauge, metric Wire measurement system scaled on a metric basis in accordance with International Standards Organization recommendations. Commonly used in Europe

gauge, pressure A device used to measure gas pressure with a dial calibrated in pounds per square inch, Newtons per square meter, or mm of mercury. A manometer.

gauge, spring tension Small hand-held device used to check the tension of relay spring-sets.

gauge, thickness A feeler gauge.

gauge, wire Disc with marked slots of varying sizes around its periphery to enable a rapid check to be made of the diameter of any wire.

gauges, wire Systems or devices for measuring the thickness of wire. Examples of such systems include: Metric (ISO = International Standards Organization), used all over Europe; American Wire Gauge (AWG) (for non-ferrous wires), used in North America, where telecommunications line wires also are sometimes classified under the diameter of the wire in decimals of an inch or the weight of conductor per mile; Standard Wire Gauge (SWG), Birmingham Gauge (BG), the Lancashire Pinion Wire Gauge (LPG) and Stubs Steel Wire Gauge are sometimes encountered in Britain.

gauss The cgs electromagnetic unit of magnetic induction.

Gaussian beam A beam of light whose radial intensity distribution is Gaussian.

Gaussian distribution A statistically random distribution: normal probability distribution.

Gaussian noise Noise with random distribution of frequency components centered on a specified frequency.

Gaussian pulse A pulse that has the shape of a Gaussian distribution.

gear, cable Term for cable machinery on a cable ship.

geiger counter A gas-filled tube used to detect ionizing radiation and count particles.

general parameters In specification description language: the general parameters in both a specification and a description of a system relate to such matters as temperature limits, construction, exchange capacity, and grade of service.

general planning forecast (GPF) Bell term for forecast of requirements for telephone facilities over a specified period.

general register An internal addressable register in a central processing unit which can be used for temporary storage, as an accumulator, or for any other general purpose function.

General Telephone & Electronics See: GTE Corporation.

generator A machine which converts mechanical energy into electrical energy.

generator, alternating current An alternator; a rotating machine which produces an alternating current output.

generator, direct current Rotating machine which produces a direct current output.

generator, electrical Rotating machine which converts mechanical energy into electric energy.

generator, emergency engine A standby set comprising a prime mover (usually a diesel engine) and a coupled ac generator. When the public commercial power supply fails a standby set takes over at many main central offices.

generator, magneto Small hand powered ac generator used to generate ringing current in old-style operating positions such as those used in pre-dial days.

generator, milliwatt Test set oscillator with output of test tone at 1000 Hz and a power level of 1 mw.

generator, noise Device which produces wideband white noise for use in transmission system testing.

generator, receiver-off-hook A refined version of the howler, used to draw a subscriber's attention to the fact that his phone has been left off-hook. This new generator gives a burst of 50 ms each of four tones in sequence, repeated five times per second.

generator, ringing A machine or solid state device which generates 20 Hz ac which is sent out to line to ring the bells of called subscribers.

generator, signal An oscillator able to produce test tones at any frequency and power level.

generator, spectrum A special test tone generator which produces a series of test tones normally covering the complete audio bandwidth. These are all measured simultaneously and the results in the form of a chart showing frequency/gain characteristics of the circuit can be produced automatically.

generator, square wave Signal generator with square wave output, used when testing wideband equipment.

generator, standby power Prime mover, usually a diesel engine, and coupled ac generator; can be used when the regular power supply fails.

generator, sub-harmonic Transformer with secondary tuned to a subharmonic of the commercial power supply, e.g., a 20 Hz ringing supply may be obtained from a 60 Hz mains power supply.

generator, sweep-frequency A test set oscillator which generates a repeated sweep of tones from one end of a frequency band to the other.

generator, thermoelectric Device which converts heat into electricity by using the thermopile effect: if one junction between different metals is heated and the other cooled, a potential difference is established between hot and cold junctions. Thermoelectric generators using a wide variety of fuels have been used to power microwave repeater stations.

generator, tone Machine or solid state device which produces the information tones needed in a telephone system, e.g., dial tone, busy tone, or ringing tone.

generator, transistor ringing Ringing current generator using solid state devices available in sizes to suit all PBX and central office requirements.

generic program A set of instructions for an SPC central office that is the same for all offices using that type of switching system. Detailed differences for each individual office are listed in a separate parameter program.

geodesic A great circle path between points on the Earth's surface.

geographically distributed exchange An exchange where not all subsystems, such as switching stages and control means, are at the same location.

geomagnetism The study of the Earth's magnetic field.

geometric optics The science that treats the propagation of light as rays. Rays are bent at the interface between two dissimilar media, or may be curved in a medium whose refractive index is a function of position.

geostationary satellite A stationary satellite having the Earth as its parent body. The satellite is not really standing still, but is in orbit around the Earth so that it takes almost 24 hours for a complete revolution and seems to us on Earth to stay some 35,780 Km vertically above the same point on the Earth's surface.

geosynchronous satellite A synchronous earth satellite.

germanium A metallic semiconductor used in making transistors.

germanium diode Rectifier which utilizes the semiconducting property of germanium.

getter Metal used in vaporized form to remove residual gases from inside electron tubes during manufacture. Other getters are utilized to remove impurities from silicon used in the manufacture of transistors.

ghost A shadowy or weak image in the received picture, offset either to the left or right of the primary image; the result of transmission conditions which create secondary signals that are received earlier or later than the main or primary signal. A ghost displaced to the left of the primary image is designated as "leading" and one displaced to the right is designated as "following" (lagging). When the tonal variations of the ghost are the same as the primary image, it is designated as "positive" and when it is the reverse, it is designated as "negative."

giga Prefix to SI units, meaning 10^9 times multiple of the base unit.

gigabyte One billion (10^9) bytes.

gigahertz One American billion (10^9) cycles per second.

gilbert The unit of magnetomotive force in the now obsolete centimeter-gram-second electromagnetic system.

gimmick Capacitor made by twisting two insulated wires together. Also, a marketing ploy or a special feature of a particular type of equipment.

gin pole A temporary derrick using a telephone pole to assist with the erection of a tower or mast.

glare The simultaneous seizure of a trunk at both ends; double seizure.

glare resolution Ability of a trunk termination to ensure that if the trunk has been seized by both ends simultaneously, then one of the callers is given priority and the other is switched to another trunk.

glitch 1. A snag or minor problem. 2. A false signal caused by timing differences between circuit elements. Syn.: dynamic hazard

glodom Global approach to realize accelerated domestic telecommunications development using modern technology.

glow discharge Glow inside an electron tube due to the ionization of gas.

glow lamp Small gas-discharge tube with two electrodes; when energized the tube glows, and can be used as an indicator lamp.

go cipher Instruction to switch from plain language to encryption.

go plain Instruction to go back from encryption to plain language.

gobbler Software defect which results in the allocation of memory for illegitimate purposes, depriving other processes of their share.

goodnight plug Plug inserted in jack at a telegraph test-board which sends space signals out on all relevant circuits to indicate that the test-board is not operational during the night.

Gorizont Russian geostationary telecommunications satellite.

government channel (CATV) The FCC requires cable systems in the top 100 markets in the U.S. to set aside one channel for local government use to be available without cost for the "developmental period." That period runs for five years from the time that subscriber service began, or until five years after the completion of the basic trunk line.

governor A speed controller which is of particular importance with data services.

graceful degradation Failure which permits operations to continue, but not as fast or efficiently as normal, i.e., not a catastrophic failure.

grade To arrange connections between groups of switching equipment in an electromechanical dial office to maximize the traffic handling capacity of the office.

grade, data A circuit whose transmission characteristics, low distortion and low noise allow the circuit to be used safely for data services.

grade, special A circuit with special transmission characteristics making it suitable for special purposes, such as wideband program channels, or secure voice circuits.

grade, voice A circuit which is adequate for the transmission of voice signals.

grade of service 1. A number of engineering parameters used to provide a measure of adequacy under specified conditions. 2. For central offices/telephone exchanges, the probability of a call being blocked, during the busy hour, because of insufficient equipment or trunks. 3. A subjective rating of telephone communications quality in which listeners judge a transmission as excellent, good, fair, poor, or unsatisfactory.

graded index fiber See graded index optical waveguide.

graded index optical waveguide A waveguide having a graded index profile. The refractive index is highest at the center of the core so light pulses which travel straight down the center do not move as fast as those which take long paths, zig-zagging along; the effect is to reduce modal dispersion and increase permitted operating frequency.

graded index profile Any index profile that varies smoothly with radius. Distinguished from a step index profile.

graded multiple Grading together of outlets from several selectors in a dial office in order to maximize the traffic handling capacity of the office.

gradient, voltage Potential drop per unit length of a conductor, or per unit thickness of a dielectric.

grading A grading or graded multiple is obtained by partial commoning or multiplying of the outlets of connecting networks when each network only provides limited accessibility to the outgoing group of circuits. A grading must be characterized by the geometrical disposition of this multiple among networks and by the sequence of selection of the outlets available to each network.

grading, homogeneous A grading in which the outlets of an identical number of grading groups are connected to each outgoing trunk. Such a grading is formed by skipping and slipping.

grading, O'Dell A type of progressive grading applicable to the outlets of selecting switches which search in an invariable order from a fixed starting position, and formed by commoning together only identically numbered adjacent outlets to give a smooth progression from circuits connected to individual outlets, to circuits connected to pairs of outlets, partially commoned outlets and fully commoned outlets.

grading, progressive A grading in which the outlets of different grading groups are connected together in such a way that the number of grading groups connected to each outgoing trunk is larger for later choice outlets.

grading, symmetrical A grading in which all groups are treated alike, i.e., every group has access to the same number of individuals and partial commons of the same interconnecting number.

grading group A unit within a grading in which all inlets have access to the same outlets. Syn.: subgroup.

grandfathered (Tariff) Tariff for an obsolete or obsolescent service that permits the service to be discontinued without discontinuing the privileges enjoyed by existing customers.

grandfathered (Telephone Equipment) Terminal equipment legally installed before July 1, 1979. After that date all new installations have to be made with registered equipment.

grandfathering (CATV) Exempting U.S. cable systems from the federal rules because 1) they were in existence or operation before the rules, or 2) substantial investments were made in system construction before the rules. Grandfathering applies to signal carriage, access channel and the certification process.

graphic characters A collection of characters within the character set used to improve readability of output.

graphics The art of conveying information through the use of graphs, letters, lines, drawings, and pictures.

grapnel Device to grapple (hook) a submarine cable from a cable repair ship in order to bring it to the surface; grapnels come in different configurations, their selection for use depending upon the nature of the bottom and other considerations.

grapnel rope Special rope connecting the cable repair ship with the grapnel during a grappling operation.

gray code   A special binary code designed to minimize distortion by giving sequential numbers codes that differ in only one bit. Sometimes encountered in computing but rarely used in the telecom field because the requirements for processors in telecommunications are somewhat different from those in business machines.

Decimal	0	1	2	3	4	5	6	7	8	9
Binary	0000	0001	0010	0011	0100	0101	0110	0111	1000	1001
Gray	0000	0001	0011	0010	0110	0111	0101	0100	1100	1101

gray scale   In facsimile, an optical pattern in discrete steps of gray between black and white.

grazing path   Radio path which does not have a clean line of sight between transmitting and receiving antennae but which grazes ground level on this path.

great circle   A circle on the surface of the Earth, the plane of which passes through the center of the Earth.

green circuit   A shielded, protected circuit which can be used for classified information without encryption.

Greenwich Mean Time   (GMT) Worldwide reference time. The Greenwich observatory is located near London, England; GMT is five hours in front of New York winter time and nine hours behind Tokyo.

grid   A mesh electrode within a electron tube which controls the flow of electrons between the cathode and plate of the tube.

grid, control   Grid in an electron tube to which input signal is usually applied.

grid, screen   Grid in an electron tube which is held at steady potential and screens the control grid from changes in anode potential.

grid, suppression   Grid in an electron tube near the anode (plate) which suppresses the emission of secondary electrons from the plate.

grid bias   Potential applied to a grid in an electron tube to control its working point.

grip, cable   Braided tube which can be opened up and slipped over a large cable. When the tube is put under tension it closes up and holds the cable securely and enables the cable to be pulled into a duct.

grip, strand   Special steel grip which tightens over the messenger strand and enables aerial cables to be pulled up to an appropriate sag level.

grivation   Term used on some maps for the variation between grid north and magnetic north (from grid variation).

ground (GRD)   A conducting connection, whether intentional or accidental, by which an electric circuit or equipment is connected to earth or to some conducting body of relatively large extent.

ground, solid   Connection path direct to ground.

ground absorption   Loss of energy during transmission of radio waves through absorption in the ground.

ground-based duct   A tropospheric radio duct in which the lower boundary is the surface of the Earth. Sometimes called a surface duct.

ground clamp   Clamp used to connect ground wire to a ground rod or system.

ground constants   The electrical constants of the earth, such as conductivity and dielectric constants. These values vary with the chemical composition of the earth, moisture content, and frequency.

ground field-earth interface (GF)   A configuration of metallic conductors, rods and pipes used to develop a new resistive contact with the earth.

ground loop   Undesirable condition which can arise when circuit is grounded at several points: circulating ground currents can sometimes result.

ground plane   Conducting material at ground potential, physically close to other equipment so that connections may be made readily to ground the equipment at required points. An arrangement of buried wires at the base of an antenna tower.

ground potential   At zero potential.

ground reflected wave   The radio communications wave which is reflected from ground or water in its path between transmitting and receiving antennae.

ground return   1. Conductor which drains currents to ground. 2. Conductor used as one common shared path for several circuits back to the grounded side of the main storage battery.

ground return circuit   A voice circuit which uses metallic paths for only one leg, the other leg being carried through the ground.

ground rod   Rod which can be driven into the ground and connected up into a complex mesh of interconnected rods so as to provide a low resistance link to ground.

ground start   Off-hook calling condition given to a central office by many PABXs — one leg grounded (plus the usual loop across both legs through the apparatus). Syn. earth calling (UK)

ground station   An assemblage of communications equipment that receives from (and usually also transmits to) a communications satellite. Syn.: earth station

ground wave   A radio wave that is propagated over the Earth and ordinarily is affected by the presence of the ground. Ground waves include all components of waves over the Earth except ionospheric and tropospheric waves, and are affected somewhat by the change in dielectric constant of the lower atmosphere.

ground wire   Copper conductor used to extend a good low-resistance earth to protective devices in an office.

grounded   Connection to earth via a low resistance path.

grounded, resistance   Resistor inserted in grounding circuit to limit current.

grounding   Connecting a system or component to earth or ground.

grounding device   Impedance inserted in a grounding circuit which limits ground fault currents.

group   1. In frequency division multiplexing, a number of voice channels, either alone or within a supergroup, which in wideband systems is normally composed of up to 12 voice channels occupying the frequency band 60 kHz to 108 kHz. 2. In some long distance submarine cable systems, channels with only 3 kHz spacing are used (instead of the usual 4 kHz) enabling 16 voice circuits to be carried in one 48 kHz group instead of the usual 12 channels per group. 3. A supergroup is normally 60 voice channels, or five groups of 12 voice channels each, occupying the frequency band 312 kHz to 552 kHz. 4. A mastergroup is composed of 10 supergroups or 600 voice channels.

group, binder Group of wires in a plastic insulated cable which are under a single binder string.

group, line A switching unit in a dial office which provides access to and from line circuits.

group, marker A common group of markers designed to serve one or more central offices.

group, master carrier In the U.S., a 600-voice channel grouping, i.e., 10 60-channel supergroups. (In Europe, a master group is five supergroups, or 300 channels.)

group, number series Block of up to three x 10,000 telephone numbers associated with up to three central office codes.

group, phantom Two physical side circuits together with the phantom circuit derived from these.

group addressing In transmission: the use of an address that is common to two or more stations.

Group Alerting and Dispatching System A service which enables one controlling telephone to call as many as 480 phones simultaneously. If any of the called lines are busy, the equipment camps until it is free, then rings and plays the tape recorded message. This type of equipment is used for Volunteer Fire Departments and Civil Defense Units as well as wake-up calls in large hotels.

group busy hour (GBH) The busy hour offered to a given trunk group.

group call A message sent simultaneously to a number of predetermined receiving stations.

group channel Twelve voice grade channels (in an FDM carrier system).

group degenerate modes Modes that have the same group velocity.

group delay time The rate of change of the total phase shift with angular frequency through a device or transmission facility.

group distribution frame (GDF) In frequency division multiplexing, a distribution frame that provides terminating and interconnecting facilities for the modulator output and demodulator input circuits of the channel transmitting equipment and modulator input and demodulator output circuits for the group translating equipment operating in the basic spectrum of 60 kHz to 108 kHz.

group index Denoted N: velocity of light in vacuum divided by the group velocity in a medium of index n.

group link The whole of the means of transmission using a frequency band of specified width (48 kHz) connecting terminal equipment. The ends of the link are the points on group distribution frames (or their equivalent) to which the terminal equipment is connected. It can include one or more group sections.

group modulation Modulation of a 12-channel group (60—108 kHz) to one of the five frequency bands which make up a supergroup (312—552 kHz).

group section The whole of the means of transmission using a frequency band of specified width (48 kHz) connecting two consecutive group distribution frames (or their equivalent) via at least one link.

Group Switching Center (GSC) In the U.K., an exchange that connects junctions from local exchanges to and from trunk circuits.

group velocity The velocity of propagation of an envelope produced when an electromagnetic wave is modulated by, or mixed with, other waves of different frequencies.

grouping, position Coupling together of adjacent manual operating positions during periods of light traffic loading to enable one operator to deal with calls incoming to more than one position.

GTE Corporation Major U.S. telecommunications company controlling several local telephone companies and long distance carriers.

guard To hold a circuit busy for an interval of time to enable lines to be disconnected from the circuit before it is used to establish another call.

guard, guy Brightly colored fabric or plastic tube put over a guy line to make it more easily visible to pedestrians.

guard, manhole Frame which can be erected around an opened manhole entry as a safety measure.

guard, plastic tree wire Tough plastic sleeve which can be put over drop wires in an area where trees could cause abrasion faults.

guard, timed A network management feature which senses the approaching possibility of all-trunks-busy on a route and delays dealing with translations in order to maintain the route in top traffic handling efficiency.

guard, tree Tough plastic tubing placed around a messenger strand and aerial cable to protect it from rubbing against trees.

guard, trolley Wooden or plastic guard frame enclosing an aerial cable and its messenger strand to give security and protection where these pass over trolley wires.

guard, U cable Steel half-tubes placed for protection purposes over cables which feed up poles (e.g., from an underground cable to provide a distribution point).

guard, wire Helically-split plastic tube placed over multiple drop wires to give protection.

guard action Protection system needed when in-band signaling is used: all normal voice-frequency energy present offsets and opposes the effects of any energy at signaling frequency which may be present in speech. This minimizes voice operation of signaling units. Also called voice guard.

guard band Narrow frequency bands left unused between adjacent channels to minimize interference while still making economic separation filter provision possible.

guard lead Lead from each switch in a dial office, this lead is marked electrically to indicate when the particular switch is busy so that it will not be picked up by a second user.

guard period, busy The brief period (750 milliseconds) a trunk circuit between electromechanical switching offices must be held busy after its release so as to preclude wrong connections due to slow release of equipment in the connecting office.

guard time See guard period.

guide edge See reference edge.

guided mode See bound mode.

guided wave Electromagnetic wave which is transmitted from point to point inside a waveguide.

gulp A group of bytes, processed as a single unit or word.

gun, electron The component in a cathode ray tube which produces the stream of electrons needed to create a TV picture.

gun, pressure Hand tool used for injecting resin under pressure into cables to create an air block.

gun, soldering Hand tool which enables soldering of contacts to be performed speedily and accurately.

gun, stapling Hand tool used to shoot staples into a wall to hold small distribution cables in place.

Gunn effect A method of obtaining microwave frequency oscillation: a high dc voltage across a gallium arsenide crystal generates oscillation.

Gunn oscillator Oscillator using the Gunn effect, efficient at frequencies over 50 GHz.

gutta-percha Organic resin (predating rubber) with properties suitable for a cable insulant; widely used for oceanic telegraph cables in the 19th century.

guy Steel wire or rope used to hold a pole upright.

guy, anchor Steel wire or rope used to hold a pole upright, the top end is fixed to the pole, the lower end to an anchor which has been buried securely in the ground.

guy, crossarm Light guy line attached to a crossarm to balance tension if conductors on one side of the crossarm have all been terminated at that pole.

guy, dead-end An anchor guy supporting a pole at the end of an aerial route.

guy, down An anchor guy, either in line with the pole line (a head guy) or at an angle to the direction of the pole line (a side guy).

guy, pole-to-pole Guy used when local conditions make it impracticable to guy directly the pole which needs extra support: a guy is run to the next pole along the route, this pole is given the extra anchor guy support needed.

guys, storm Guys placed at regular intervals along a long exposed pole route, at right angles to the line of the route and in line with the route.

gyrator A directional phase changer used in some transmission lines.

HA1-Receiver weighting A noise weighting used in a noise measuring set to measure noise across the HA1-Receiver of a 302-type or similar instrument. The meter scale readings are in dBa (HA1).

Hakucho Japanese research satellite.

halation A halo; the central spot glow on a cathode ray tube, due in part to internal reflections within the glass.

half-circuit Term used when different administrations share the cost of a major submarine cable system.

half-duplex circuit A circuit that affords communication in either direction but only in one direction at a time.

half-duplex operation 1. That type of simplex operation which uses a half-duplex circuit. 2. Pertaining to an alternate, one way at a time, transmission mode of operation.

half-power frequency Either of the two frequencies on an amplifier or filter response curve where the output is $\sqrt{0.5}$ ($= 0.7071$) that at midband.

half speed (circuit) An international leased circuit using a teletypewriter which may operate at 33 ⅓ words per minute.

half-tap A parallel connection across a circuit.

half-wave dipole Antenna with length approximately one-half wavelength at the operating frequency.

half-wave rectifier A rectifier that delivers direct current output during alternate half-waves only of the incoming ac signal input.

halftone characteristic In facsimile systems, a relation between the density of the recorded copy and the density of the subject copy; or a relation of the amplitude of an amplitude-modulated facsimile signal to the density of the subject copy or the record copy when only a portion of the system is under consideration.

Hall effect The phenomenon by which a voltage develops between the edges of a current-carrying metal strip whose faces are perpendicular to an external magnetic field. Used to determine if current is being carried by holes or by electrons.

halo Most commonly, a dark area surrounding an unusually bright object, caused by overloading of the camera tube. Reflection of studio lights from a piece of jewelry, for example, might cause this effect. With certain camera tube operating adjustments, a white area may surround dark objects.

ham   An amateur radio operator.

hamming code   An error detecting and correcting code which uses special binary words in which the number of corresponding bit positions which differ in adjoining binary words is always the same, e.g., two of the four bits might be different every time.

hamming distance   Minimum number of bit positions in which any two valid bit sequences or codewords may differ.

handhole   A small cable-jointing chamber, giving easy access to a buried cable splice, with top cover flush with ground level. In some countries called surface, or pavement, or sidewalk jointing chamber.

handhole, pre-cast   Concrete handhole made centrally and lowered into excavated hole, i.e., not constructed on site.

handi-talkie   Small hand-held two-way radio transceiver.

handline   Length of rope carried by lineman working on poles which is used to hoist tools and materials to the working level.

handling, call   Operating procedure by which telephone calls are established and charges assessed.

hand-off   Process by which an established call on a deteriorating transmission path between a cell site (in a cellular radio system) and a mobile unit is rerouted to a different cell site on a different channel in order to obtain a stronger signal.

handset   The combined hand-held unit of microphone and earpiece of a standard telephone instrument.

handset, hard-of-hearing or deaf-aid   Handset with small line-powered amplifier on the receive side.

handset, sound booster   A deaf-aid handset.

hands-free answerback   Intercom system facility by which both-way conversation can be carried on without having to lift the phone off-hook.

hands-free telephone   A loudspeaking telephone, with no need to hold the handset.

handshaking   An exchange of predetermined characters or signals between two stations to provide control or synchronism after a connection is established.

hang up   Going back on hook at the end of a telephone call, clearing down the connection.

hanger, figure-8 cable   A special clamp which can be bolted to a pole, used to hold the messenger strand or supporting wire in a figure-8 type aerial cable.

hangover   Of an echo suppressor: the time between the end of a speech signal in one direction and the reopening of the speech channel in the other direction.

hard   1. Of a wire: high tensile strength and hence difficult to bend. 2. Of an electron tube: highly evacuated.

hard copy   A message typed on paper (compared with a message printed on a VDU screen).

hard-drawn   High tensile strength wire.

harden   To construct military telecommunications facilities so as to protect them from damage by enemy action, especially Electromagnetic Pulse (EMP) action.

hardness   Ability of installation to withstand a nearby explosion or bomb blast.

hardware   Manufactured items, i.e., physical material.

hardwire   1. To connect permanently equipment or components in contrast to using switches, plugs, or connectors. 2. The wiring-in of fixed logic or read-only storage that cannot be altered by program changes.

harmful interference   Any emission, radiation, or induction which endangers the functioning of a radio-navigation service or of other safety services, or seriously degrades, obstructs, or repeatedly interrupts a radio communication service operating in accordance with approved standards, regulations, and procedures.

harmful out-of-band components   Transferred currents arising from speech, pilots, additional measuring frequencies, and of frequencies that always lie outside the useful frequency band (corresponding to speech frequencies) of the carrier systems, but which may interfere with pilots or with additional measuring frequencies.

harmless out-of-band components   Transferred currents arising from speech or pilots which at all translation points have frequencies outside the useful frequency band corresponding to audio frequencies or pilot frequencies.

harmonic   An alternating signal whose frequency is an integral multiple of a fundamental or basic frequency.

harmonic, even   Fundamental frequency multiplied by an even number.

harmonic, odd   Fundamental frequency multiplied by an odd number.

harmonic analysis   Breaking down a complex wave into the sum of a fundamental and various harmonic frequency signals.

harmonic analyzer   Test set which can be fine-tuned to identify the frequencies of all the individual signals which make up a complex wave.

harmonic distortion   The presence of frequencies at the output of a device, caused by non-linearities within the device, which are harmonically related to a single frequency applied to the input of the device.

harmonic emission   Spurious emission at frequencies which are whole multiples of those contained in the frequency band occupied by an emission.

harmonic ringing   Method of calling individual stations on a party line by using a different calling frequency for each line. The frequencies used are usually harmonics of 16-⅔ Hz and 25 Hz.

harmonics, non-triple   Odd harmonics excluding those which are powers of three.

harmonics, triple   Odd harmonics which are multiples of three.

harness, wiring   Method of making up cable forms and groups of pre-cut and laced-up individual insulated wires prior to the assembly of units so that terminations on shelves or other units may be easily connected onto the correct wires.

Hartshorn bridge An ac bridge used to measure mutual inductance.

hash 1. Wideband high frequency audio noise such as that produced by the contacts of a vibrating ringing generator. 2. Rubbish or meaningless signals added to information in order to comply with requirements for standard block length for items in the store or program.

hash total A summation for checking purposes of one or more corresponding fields of transmission that would not usually be totaled.

haul In the U.S., the classification of toll calls. Short haul — less then 30 miles; medium haul — 30 to 1000 miles; long haul — over 1000 miles.

Hay bridge An ac bridge used for measurement of inductance.

hazard Condition which could lead to danger for plant or personnel.

hazard, fire Condition which could cause a fire or make a fire difficult to control.

hazards of electromagnetic radiation to fuel (HERF) Potential for electromagnetic radiation to cause spark ignition of volatile combustibles, such as aircraft fuels.

hazards of electromagnetic radiation to ordnance (HERO) Potential for munitions or electro-explosive devices to be adversely affected by electromagnetic radiation.

hazards of electromagnetic radiation to personnel (HERP) Potential for electromagnetic radiation to produce harmful biological effects in humans.

head Device that erases, records or reads information from a magnetic tape passed under it, as in a commercial or domestic cassette tape player/recorder.

head, pruner Pruning shears mounted on the end of a light pole and used from ground level to trim trees.

head end (CATV) Electronic control center — generally located at the antenna site of a CATV system — usually including antennas, preamplifiers, frequency converters, demodulators, modulators and other related equipment which amplify, filter and convert incoming broadcast TV signals to cable system channels.

header In a signal message such as in common channel signaling, provides general information, such as identification information, date and time, etc.

header, message Group of letters and figures at the beginning of a commercial telegram which give the office of origin, date/time of origin, message number and class.

heading The control information that characterizes an interpersonal message in a message handling system.

headphone A telephone receiver.

headset An operator's telephone set, worn on the head.

headup display An optical system which transfers operational or navigational information into a pilot's line of sight so that the pilot need not look down at instruments but may keep looking ahead through his windshield.

heat coil See coil, heat.

heat coil, indicating A heat coil or sneak current coil which gives a visual indication of having operated.

heat detector Device which is temperature sensitive and gives an indication when a specified temperature has been reached.

heat loss Loss of useful electrical energy due to its conversion into unwanted heat.

heat seal Method of providing a moisture-tight sheath closure for plastic cable splices by the use of techniques which melt the plastic sheath.

heat sink Device which conducts heat away from a heat-producing component so that it stays within its safe working temperature range.

heater In an electron tube, the filament which heats the cathode to enable it to emit electrons.

heater, tent Portable kerosene heater used in splicer's tents during cold weather.

Heaviside layer The E-layer, an ionized layer about 100 km above the Earth's surface, which reflects radio transmissions back to Earth.

heelpiece Base of an electromechanical relay onto which the core is mounted

height gain For a given propagation mode of an electromagnetic wave, the ratio of the field strength at a specified height to that at the Earth's surface.

henry The SI unit of electric inductance; the inductance of a closed circuit in which an electromotive force of one volt is produced when the electric current varies uniformly at the rate of one ampere per second. Mutual inductance is the same, but when an emf is produced in one circuit by a varying current in a second circuit.

heptode A seven-electrode electron tube.

hermaphrodite connection Connecting device using identical components on both sides.

Hertz Unit of frequency: one cycle per second.

heterochronous Two signals are heterochronous if their corresponding significant instants do not necessarily occur at the same rate.

heterodyne To mix a signal of one frequency with another, producing a signal with a much lower frequency.

heterodyne frequency The sum of or the difference between two frequencies, produced by combining the two together in a modulator or similar device.

heterodyne interference A whistle-type interference due to beating of two signals whose frequencies differ by a figure which represents an audible tone.

heterodyne reception Reception of radio signals by a beat method which results in incoming signals being combined with locally generated signals of nearly the same frequency as the incoming signal, thereby producing a beat note at audible frequency.

heterodyne repeater A repeater for a radio system in which the received signals are converted to an intermediate frequency, amplified, and reconverted to a new frequency band for transmission over the next repeater section. Sometimes called an IF repeater.

heterodyne wavemeter A test set which uses the heterodyne principle to measure frequencies of incoming signals.

heterogeneous multiplex A data multiplex structure in which all the information-bearing channels are not at the same data signaling rate.

heuristic Empirical. Trial and error method of solving problems.

hex Short for hexadecimal, the base-16 number system often used in computers.

hexadecimal numeral A numeral in the hexadecimal (base 16) numbering system, represented by the characters 0, 1, 2, 3, 4, 5, 6, 7, 8, 9, A, B, C, D, E, F, optionally preceded by H'.

hexode A six-electrode electron tube.

hierarchic (mutually synchronized) network A mutually synchronized system is hierarchic when some clocks exert more control than others, the network operating frequency being a weighted mean of the natural frequencies of the population of clocks.

hierarchic switching network A network arranged in classes of offices with defined levels of responsibility, such as the American Class 1 to Class 5 network, or the CCITT International network with its CT1s, CT2s and CT3s.

hierarchical computer network A computer network in which processing and control functions are performed at several levels by computers specially suited for the functions performed.

hi-fi Radio broadcast receiver, phonograph or cassette recorder with good quality sound reproduction.

high band TV channels 7 through 13.

high-capacity satellite digital service An AT&T point-to-point 1.544 Mbit/s data service based on the use of circuits routed via a domestic satellite service.

high-capacity terrestrial digital service An AT&T point-to-point data service based on a two-way 1.544 Mbit/s digital circuit routed on terrestrial bearers.

high day busy hour (HDBH) The hour (not necessarily a clock hour) which produces the highest load during the busy season.

high density bipolar 3 (HDB3) Line coding widely used for PCM signals: a binary 1 is transmitted alternately as a positive or negative pulse; this is a form of Alternate Mark Inversion, AMI. After three consecutive binary zeros, the fourth zero is substituted by a mark of the same polarity as the previous mark, giving a bipolar violation. If successive violations would otherwise be of the same polarity (i.e. an even number of marks between violations) the first of the four consecutive zeros is replaced by a mark of the opposite polarity to the previous mark, thereby maintaining zero disparity in the output signal.

high density carrier Carrier system able to transmit several supergroups, usually 10 supergroups or 600 channels.

high fidelity Hi-fi.

high frequency (HF) The band of radio frequencies between 3 and 30 MHz.

high frequency distribution frame (HFDF) A distribution frame that provides terminating and interconnecting facilities for those combined supergroup modulator output and combined supergroup demodulator input circuits occupying the baseband spectrum of 12 kHz up to and including 2540 kHz.

high-level control In data transmission, the conceptual level of control or processing logic existing in the hierarchical structure of a primary or secondary station that is above the link level and upon which the performance of link level functions are dependent or are controlled, e.g., device control, buffer allocation, or station management.

high-level data link control (HDLC) An international protocol standard governing exchange of data over a single communication link. HDLC is used within the CCITT's X.25 system for packet-switched network access.

high-level digital interface A station equipment interface operating at either 60 V dc, 20 ma, or 130 V dc, 20 or 60 ma.

high level language (HLL) A programming language that does not reflect the structure of any computer. The language is readily understandable to a human programmer, before being changed into a machine code that the computer can understand. Generally, one high level language statement will result in several machine code statements, while for low-level language, every instruction has a single machine code equivalent.

high loss Name sometimes given to optic fiber with attenuation of more than 50 dB/km.

high-low repeater Transmission system repeater which reverses bands, i.e., receives in a high frequency band and retransmits out in a low frequency band. Sometimes called frogging.

high-low signaling Type of DC signaling using a marginal supervisory relay to make it easier to differentiate between on-hook and off-hook states, particularly on high-resistance lines.

high-pass filter Network which passes signal of higher than a specified frequency but attenuates signals of all lower frequencies.

high-performance equipment Equipment with sufficiently exacting characteristics to permit their use in trunks or links; equipment designed primarily for use in global and tactical service where maximum performance and capabilities, and minimum electromagnetic interference are necessary for operation in a variety of nets or for fixed point-to-point circuits.

high power flux-density A power flux-density which enables signals radiated by broadcasting-satellite space stations to be received by simple receiving installations with a primary grade of reception quality.

high Q Inductance or capacitance whose ratio of reactance to resistance is high.

147

high-speed switched digital service   An AT&T switched data service using both terrestrial and satellite bearers and based on provision of a 1.544 Mbit/s circuit.

high tension   High voltage; means different voltages in different circumstances: in a domestic appliance it usually means that the voltage could cause injury or kill humans. When used on external power routes or cables, it can be interpreted as a warning, the voltage could be anything from 500 volts to 30,000 volts.

high-usage circuit-group   A group of trunks which is the primary direct route between two points, and with respect to a particular parcel of traffic, is traffic engineered to overflow to one or more other circuit-groups, ultimately to the final choice circuit group.

high usage trunks   Trunks for which an engineered alternate route has been provided. Circuits in the HU trunk group are heavily loaded with traffic in the busy hour, and the group operates efficiently with high occupancy.

high VHF band   Part of the frequency band which the FCC allocates to VHF broadcasting, including channels 7 through 13, or 174 through 216 MHZ.

high voltage alarm   Audible alarm given when voltage at a central office, normally 50v, begins to rise.

high voltage joint usage   Joint usage of pole routes by power company and telephone company. Safety procedures are given in the National Electrical Safety Code.

highway   1. A digital serial-coded bit stream with time slots allotted to each call on a sequential basis. 2. A common path or a set of parallel paths over which signals from a plurality of channels pass with separation achieved by time division.

highway, common   Circuit used in common by many users.

highway width   The greater the bandwidth, the more data can be handled simultaneously, hence the greater the throughput.

hi-lo tariff   Two-level rate structure which, for a few years, gave lower rates for calls on high-density routes than for calls on low-density routes; this was Bell's first break from nationwide price averaging.

hit   1. A transient disturbance to a communication medium. 2. A comparison of two items of data that occurs when specified conditions are satisfied.

hit indicator   Telegraph test board lamp which lights when the circuit concerned is open.

hitch, core   Method of pulling cable into a duct by attaching the towing line to the core of the cable rather than to the oversheath.

hi-valve assembly   Method of connecting pressure pipes to cables in a manhole and bringing these pipes to a fitting near the cover of the manhole so that pressure testing of all cables may be carried out without having to enter the manhole.

hoist, chain   Ratchet hoist used for many external plant tasks, such as erecting poles, tightening guys, pulling up aerial cables, etc.

hoist, Coffing   Proprietary brand of chain hoist.

hold   To maintain a telephone connection.

hold-back gear   On a submarine cable laying ship, a sheave with a jockey wheel coupled to the cable drum, turning a little slower on pay-out and a little faster on pick-up, to keep the cable tight so that it does not slip on the drum when under tension; properly called draw-off gear when picking up.

hold key   Pushbutton on key systems (or on PABXs) which holds an incoming call (often with music sent back to the caller) until the required extension becomes available.

holding bridge   Circuit with relatively low dc resistance but high inductance at voice frequencies, used across telephone circuits to operate supervisory relays without introducing unacceptable transmission losses.

holding circuit   Circuit which is not strong enough to cause a relay to operate but which is able to keep a relay operating as soon as another circuit has caused it to operate.

holding current   Current which is just strong enough to keep a relay operating.

holding jack   Jack on a manual operating position into which an operator can plug a cord to connect a circuit which is required to be held temporarily, correctly terminated.

holding time   The length of time a resource is busy for an attempt, call or message; usually measured in call-seconds.

holding time of an international circuit   The time interval during which the circuit is used. This interval includes the call duration, the operating time, and the time taken to exchange service information.

hole   A vacancy in an atom's valence band where an electron would normally be; such holes act in effect as positive charges (positrons) and are the majority carriers of current in p-type semiconductors.

hole conduction   Under the influence of an electric field an adjacent electron moves to fill the place of a hole, leaving a hole in the atom it has just left. This process is effectively the movement of the holes themselves through the material, in the direction of positive field.

hole current   The current in a semiconductor directly associated with the movement of holes.

hole site   The point on a tape or card where a hole could be punched, and where the absence or presence of a hole determines the information stored on the tape or card.

holes, feed   Row of small holes in paper tape used to control the movement of the tape.

hollerith code   A 12-level code used on punched cards to give input to computers. This code covers the alphabet and digits 0 to 9.

hollow   Subjective word used to describe an amplified circuit which is close to singing.

hollow circuit   An unstable amplified circuit which is close to singing because gain is too high or balances at hybrid coils are poor.

home   Parent office, or office of higher level (lower class number) to which final or last choice long distance circuits are connected.

home area toll call   Call to an office in the same numbering plan area.

home position   Rest position to which wipers in homing-type switches return on completion of a call and release of the switch.

homing, dual   Ability of a minor office to trunk to two different toll centers or of a PBX to be given lines on two different central offices.

homing arrangement   Plan by which local central offices are connected by direct trunks to offices of a higher level (lower class number).

homing beacon   A radio transmitter which emits a recognizable signal and, for example, guides an aircraft to land at an airport.

homing point   A "high density" rate center through which major private lines may be routed.

homing type switch   Switch which always steps back to the same normal or rest position when it has been released at the conclusion of a call.

homochronous   Two signals are homochronous if their corresponding significant instants have a constant, but uncontrolled, phase relationship.

homogeneous multiplex   A data multiplex structure in which all the information bearer channels are at the same data signaling rate.

honeycomb coil   Air-core inductance coil used for radio frequencies, wound on a lattice basis to reduce capacitance.

hook, cable suspension   Hook used in supporting an aerial cable on a separate messenger wire.

hook, cant   Steel-hooked tool used when moving telephone poles.

hook, drive   Hook which can be driven into a pole and used to support drop wires.

hook, house   A screw-based hook used to support drop wires at a subscriber's premises.

hook, manhole cover   Special tool used for lifting manhole covers.

hook, shave   Scraping tool used by cable splicers to remove the oxidized and dirty surface layer from lead-sheathed cable before plumbing.

hook, underground cable   A cable bearer, hooked into a bracket or rack in a manhole to support cables.

hooklatch   Special type of hookswitch used on some party lines. The first stage of operation permits listening only, to hear if circuit is or is not already occupied, the second stage gives normal off-hook condition.

hooks   Linemen's climbers or climbing irons.

hookswitch   The switch in a telephone instrument which operates when the handset is removed from its resting position. See on-hook, off-hook.

hookswitch, mercury   Special type of hookswitch used on some pay telephones: if attempts are made to go on and off hook at dial speed in order to make calls without payment, the mercury in this hookswitch

breaks up into globules and does not make contact, thus discouraging attempted fraudulent use.

hookup   A connection or circuit made for a special purpose.

hop   The excursion of a radio wave from the Earth to the ionosphere and back.

horizon angle   The angle, in a vertical plane, between a horizontal line extending from the center of the antenna and a line extending from the same point to the radio horizon.

horizontal (hum) bars   Relatively broad horizontal bars, alternately black and white, which extend over the entire picture. They may be stationary, or may move up or down. Caused by interference near the power line (mains) frequency or one of its harmonics.

horizontal, crossbar   The horizontal element of a crossbar matrix switch.

horizontal polarization   Sending radio waves in such a way that the electric field vector is horizontal.

horn antenna   Microwave antenna formed by flaring out the end of a waveguide.

horn feed   A horn antenna used in conjunction with a parabolic dish or reflector in order to obtain improved directional performance.

horn gap   Lightning arrester with a gap between two horns. When lightning causes a discharge between the horns the heat produced lengthens the arc and breaks it.

horn gap switch   A switch provided with arcing horns, ordinarily used for breaking the charging current of overhead transmission and distribution lines.

horn loudspeaker   Loudspeaker with drive unit fed into an exponentially-shaped horn, giving high directivity.

horsepower   Unit of mechanical power, equivalent to 745.7 watts.

hose, blower   Hose about six to ten inches in diameter used with a pump to supply fresh air in a manhole.

hose, suction   Small diameter hose (two inches at most) used to pump water out of manholes.

host   A main computer which acts as a parent node in a data network, enabling lower-level terminals to work through it to obtain access to other computers and their stores of information.

host processor   A processor used to run tools and support software and which is not a part of a working exchange.

hot   At a high potential.

hot dip galvanized   Steel galvanized by dipping it into a bath of molten zinc.

hot line   A direct circuit, voice or teletypewriter, between two points, needing no switching or patching. The most famous of these are the hot lines (via satellite) between Washington and Moscow.

hot list   Feature (provided by some switching systems) which permits the identification and blocking of calls charged to an unauthorized credit card, third number or special billing number.

hot-standby operation   Method of providing reliable radio service: two transmitters are kept fully energized, if one fails the other is readily available and immediately carries the signal. The same principle is used with central processors in stored program control central offices.

house, pole   Pole-mounted distribution point in a local telephone cable distribution network.

house drop (CATV)   The coaxial or other cable that connects each building or home to the nearest feeder line of the cable network.

house telephone   Telephone used for internal calls only, e.g., in a hotel.

housekeeping information   Signals which are added to a digital signal to enable the equipment associated with that digital signal to function correctly, and possibly to provide ancillary facilities.

housing   A removable protective cover, particularly for external plant equipment.

howl   A loud singing or wailing sound due to acoustic feedback making a circuit unstable.

howler   Device at the test desk or panel in a central office which sends a howling signal out to a subscriber's line when it has been left off-hook. Hopefully, the customer hears the noise and puts the instrument back on hook.

howler, signaling   Device like an auto horn which is used in noisy locations instead of a ringing signal on a telephone instrument.

hub   A communications center: a point at which channels are interconnected.

hub, full duplex telegraph   Bridging point providing interconnection of full duplex telegraph lines, i.e., simultaneous transmission in two directions.

hub, half-duplex telegraph   Interconnection point where all receive legs and all send legs of a half-duplex telegraph circuit may be bridged together.

hub, telegraph receive   Common bridged connection for all receive legs of a telegraph system.

hub, telegraph send   Common bridged connection for all send legs of a telegraph system.

hub operation   Method of interconnecting dc telegraph system legs.

hue (TV)   Corresponds to "color" in everyday use; i.e., red, blue, etc. Black, white and gray do not have hue.

hum   Audible interference at public power supply frequency.

hum modulation   Form of distortion where the power-line frequency modulates the TV signal, causing hum bars to appear in the picture.

humidity, relative   The percentage of saturation of the air by water vapor, i.e., the ratio of an amount of water vapor actually held by the air to the maximum it could hold at the same temperature and pressure.

humming   Sound often heard from apparatus powered by public mains supplies due to vibrations in power transformer cores.

hundred call seconds   See CCS.

hunting   1. The operation of a selector in searching terminals until an idle one is found. 2. Pertaining to the failure of a device to achieve a state of equilibrium, usually by alternately overshooting or undershooting the point of equilibrium.

hunting, random   A rule of selection among free links or outlets in which there is no preference for one free link over any other.

hunting, sequential   A rule of selection among free links or outlets based on their numbering from a starting position which may or may not change. If the starting positions are fixed they may be referred to as home positions, and each step is called a choice.

hybrid   1. In electrical, electronics, or communications engineering, a device, circuit, apparatus, or system made up of two or more components, each of which is normally used in a different application and not usually combined in a given requirement. For example, an electronic circuit having both vacuum tubes and transistors; or a mixture of thin-film and discrete integrated circuits. 2. Equipment used to convert four-wire transmission to two-wire transmission and vice versa.

hybrid, coil   Multi-winding bridge coil used with a balancing network to connect a two-wire circuit to a four-wire circuit.

hybrid, resistance   A two-wire/four-wire hybrid made up solely of a resistance network.

hybrid balance   A measure of the degree of balance between two impedances connected to two conjugate sides of a hybrid set; it is given by the formula for return loss.

hybrid coil   A single transformer which has, effectively, three windings and which is designed to be connected to four branches of a circuit so as to make them conjugate in pairs. Syn.: bridge transformer

hybrid communications network   A communication system which utilizes a combination of trunks, loops or links, some of which are capable of transmitting (and receiving) only analog or quasi-analog signals and some of which are capable of transmitting (and receiving) only digital signals.

hybrid interface   For computers, hybrid means the presence of both an analog computer and a digital computer, working together, using converters as interfaces.

hybrid loss   Transmission loss (about 3.5 dB) when a signal goes through a hybrid coil. The half dB is for coil losses, the 3 dB is lost because only half the power is used usefully, half is wasted.

hybrid mode   In optoelectronics: a mode possessing components of both electric and magnetic field vectors in the direction of propagation. Such modes correspond to skew (non-meridional) rays.

hybrid set   Two or more transformers interconnected to form a network with four pairs of accessible terminals to which may be connected four impedances so that the branches containing them may be made interchangeable.

hybrid system   A communication system that accommodates both digital and analog signals.

hybrid tee A waveguide junction which permits energy to be fed in from any arm.

hydrodynamic constant A submarine cable term related to the physical parameters of the cable such as weight in water, outer diameter and surface smoothness. This constant is used to determine the slope of the line that a cable forms when towed through water; the unit is expressed in degree-knots.

hydrometer Tester used to measure specific gravity, particularly the specific gravity of the dilute sulphuric acid in a lead-acid storage battery, to learn the battery's state of charge.

hygrometer Instrument which measures the relative humidity of the atmosphere.

hygroscopic Able to absorb moisture from the air.

hysteresis Property of an element evidenced by the dependence of the value of the output for a given excursion of the input, upon the history of prior excursions and the direction of the current traverse. Originally hysteresis was the name for the magnetic phenomenon only, the lagging of flux density behind the change in value of the magnetizing flux, but the term is used now also for other inelastic behavior.

hysteresis, dielectric Lagging of an electric field in a dielectric behind the alternating voltage which causes it.

hysteresis loop The plot of magnetizing current against magnetic flux density (or of other similarly related pairs of parameters) appears as a loop. The area within the loop is proportional to the power loss due to hysteresis.

hysteresis loss A magnetic core's loss due to hysteresis, for example, in a transformer energized by an alternating current.

identification friend or foe (IFF) A system using electronic transmission to which equipment carried by friendly forces automatically responds.

identification-not-provided signal A signal sent in a common-channel signaling system for data services in response to a request for calling or called line identification when the corresponding facility is not provided in the originating or destination network, respectively.

identification on outward dialing, automatic (AIOD) The ability of some Centrex units to provide an itemized breakdown of charges (including individual charges for toll calls) for calls made by each telephone extension.

identified outward dialing System which provides for identification for billing purposes (either automatic or by an operator) of a system placing an outward toll call.

identifier 1. A character, or group of characters, used to name an item of data and possibly indicate certain properties of that data. 2. In man-machine language, a representation of an entity, typically consisting of one or more characters; used to name a unique item of data.

idle Not engaged or busy, hence available for use.

idle-channel noise Noise present in a communications channel when no signals are applied to it. The conditions and terminations must be stated for the value to be significant.

idle character A transmitted character indicating "no information": nothing is printed at the receiving station.

idle line termination An electrical network that is switch controlled to maintain a desired impedance at a trunk or line terminal when that terminal is in the idle state.

idle search The order in which alternate route switches hunt through groups of circuits to find the first idle one on which to route an outgoing call.

idle state The telecommunication service condition that exists whenever user messages are not being transmitted but the service is immediately available for use.

idle trunk indication Small lamp on a manual operating position located against the jack end of a toll or trunk circuit which indicates the circuit is idle and available for use.

ignore A computer/data command to take no action on an instruction.

illegal state Unacceptable connection or condition in contravention of agreed protocols and procedures.

illuminated In an area to which a radiocommunications signal has been directed.

image One of the sidebands produced by amplitude modulation: the two sidebands are, in effect, mirror images of each other, one above the carrier frequency, the other below it.

image antenna See antenna, image.

image converter Electron tube that produces a visible image from an image usually produced by infrared radiation. (Also, sometimes other non-visible radiation is used.) Used to provide night vision, e.g., in "sniperscopes."

image frequency An undesired frequency, when mixed with the local oscillator frequency, can generate a signal at a super-heterodyne receiver's intermediate frequency and so produce an audible output signal. Also see image-rejection ratio.

image ratio Ratio of field strength at image frequency to field strength at the desired frequency, which can produce equal outputs if applied to the input to a superheterodyne receiver; the image-rejection ratio.

image rejection   Measure of the efficiency of a superheterodyne radio receiver in suppressing the effect of signals at the image frequency.

image-rejection ratio   Of a radio receiver, the ratio of the input signal level at the image frequency required to produce a specified output power level from the receiver to the level of the wanted signal required to produce the same power level. The image frequency is the wanted signal frequency plus or minus twice the intermediate frequency, according to whether the frequency-changer oscillator is respectively higher or lower in frequency than the wanted signal frequency.

image response   Way in which a superheterodyne receiver responds to an undesired signal at the image frequency.

imagery   Collectively, the representations of objects reproduced electronically or by optical means on film, electronic display devices or other media.

imbalance, traffic distribution   Occurs in an exchange when the traffic flow of an incoming switchblock is unevenly distributed among all the outgoing switchblocks.

imbalance, traffic load   Occurs in an exchange when the traffic load is unevenly distributed among the incoming switchblocks or among the outgoing switchblocks.

immediate   (I) The third highest precedence signal in command networks such as AUTOVON.

immediate access store   Store from which information can be read out without any significant delay.

immediate dialing   Signaling procedure used when receive registers are permanently connected to long distance circuits so that there is no need to wait for a "proceed to send" signal before sending digital information.

impact ionization   Ionization of an atom or molecule as a result of a high energy collision.

impairment, distortion transmission   (DTI) See distortion transmission impairment.

IMPATT diode   A semiconductor diode which may be used as a microwave oscillator. IMPATT means impact ionization avalanche transit time.

impedance   The total passive opposition offered to the flow of an alternating current. It consists of a combination of resistance, inductive reactance, and capacitive reactance. It is the vector sum of resistance and reactance $(R + jX)$ or the vector of magnitude Z at an angle θ.

impedance, average central office   For transmission purposes, the impedance of an average local central office is taken in North America as 900 ohms in series with a 2.16 mF capacitor; in Europe, most exchanges, toll and local, are designed with 600 ohms impedance.

impedance, blocked   Input impedance of a device when its output impedance has been made infinite.

impedance, characteristic   Impedance of a transmission line when it is infinitely long; the square root of the product of the impedances with distant end open and looped.

impedance, driving point   Input impedance of a transmission line.

impedance, free   Input impedance of a device when its output impedance has been made zero.

impedance, input   Impedance looking into the input terminals of a device.

impedance, iterative   Impedance which when used to terminate the output of a device or network makes the impedance as measured at the input of the device or network the same as that of the iterative terminating impedance.

impedance, line   Impedance measured looking into a transmission line.

impedance, loaded   Impedance measured at a device's input when its output is connected to its normal load.

impedance, mid-series   Impedance looking into a repetitive network when measured at the mid-point of one of the series elements.

impedance, mid-shunt   Impedance looking into a repetitive network when measured at a shunt element which has an individual impedance double that of the other shunt elements.

impedance, motional   Impedance of a receiver which is caused entirely by the motion of the diaphragm

impedance, mutual   Impedance between the primary and secondary of a transformer. Numerically, secondary voltage divided by primary current.

impedance, negative   For some inductive circuits, an increase in applied voltage results in a decrease in current and vice versa.

impedance, non-linear   Impedance which varies with applied voltage or with series current.

impedance, office   Nominal impedance of a switching center: 600 ohms (toll center), 900 ohms (local office) in series with a 2.16 mF capacitor. In Europe, toll and local switches usually are engineered to a characteristic impedance of 600 ohms.

impedance, open-circuit   Input impedance when the far end terminals of a four-terminal network are open.

impedance, output   Impedance looking into the output terminals of a device.

impedance, reflected   Impedance which if added to the primary of a transformer would change the primary current by the same amount as the same impedance connected as the load on the transformer secondary.

impedance, sending end   Impedance looking into a transmission line's input.

impedance, short circuit   Impedance at the input when the output is short-circuited.

impedance, surge   This is equal to characteristic impedance provided resistance and leakage of a transmission line are negligible compared with line inductance and capacitance.

impedance, terminal   The impedance looking into the terminals of a device.

152

impedance, transfer The impedance obtained by dividing the voltage at a network's input by the network's output current.

impedance balance ratio Of a two-terminal network (such as transmission test apparatus), the measure of the degree of symmetry with respect to the earth potential of the impedance presented by the network to the circuit connected to it.

impedance characteristic A graph of the impedance of a circuit showing how it varies with frequency.

impedance irregularity A discontinuity in an impedance characteristic, caused by the use of different cables, or a badly matched amplifier, etc.

impedance matching Matching impedances of adjoining circuit elements so that power transfer across the interface is maximized, in order to improve performance or to accomplish a specific effect.

impedance-matching transformer Transformer between two circuits of different impedances but with turns ratio which provides for maximum power transfer and minimum loss by reflection.

impedances, conjugate Impedances with equal resistive components and reactive components which are equal in value but opposite in sign, i.e., one is inductive, the other capacitive.

impedances, image Impedances that simultaneously terminate inputs and outputs of a device in such a way that at each of these inputs and outputs the impedances in both directions are equal.

impedances, reciprocal If the product of two impedances equals the square of a specified impedance the two impedances are said to be reciprocal.

impregnated Of a fibrous material: all voids are filled with a fluid, oil or insulating wax.

impregnated cable Cable with paper insulation around conductors which is treated with insulating oil or other material with similar properties. Not widely used now.

impulse A surge of electrical energy, usually of short duration, of a non-repetitive nature.

impulse current Current which rises rapidly to a peak then decays down to zero without oscillating.

impulse discharge current Through a gas-discharge protector: the peak value of the impulse current flowing through the protector after spark-over.

impulse excitation The production of an oscillatory current in a circuit by impressing a voltage for a relatively short period compared with the duration of the current produced.

impulse generator Device which produces a single surge-type pulse, usually by charging and discharging a capacitor.

impulse hits Voltage surges lasting up to four milliseconds that come to within six dB of normal signal level and cause errors in phone-line data transmission. Bell standard permits no more than 15 impulse hits per 15-minute period. Syn.: spikes

impulse noise Noise consisting of random occurrences of energy spikes, having random amplitude and bandwidths, whose presence in a data channel can be a prime cause of errors.

impulse noise counter Test set which registers and counts noise signals greater than a specified threshold value.

impulse response The amplitude-versus-time output of a transmission facility or device in response to an impulse.

impulse spark-over voltage Of a gas discharge protector: the highest voltage which appears across the terminals in the period between the application of an impulse of a given wave shape and the time when current begins to flow.

impulse spark-over voltage/time curve Of a gas discharge protector: the curve which relates the impulse spark-over voltage to the time to spark-over.

impulse voltage A one-way voltage which rises rapidly to a peak then falls to zero, without any appreciable oscillations.

impulses Making and breaking of a circuit by pulsing contacts to sympathetically operate remote switches.

impulses per minute Interruption rate for call progress tones, supervisory lamp signals.

impurities Atoms of a different substance which have either been introduced on purpose into a material (typically a semiconductor), or which occur naturally in the material. Impurities can affect the electrical properties of a material considerably, for example, turning a material into one in which holes are the majority carriers of electric current. Impurities in an optic fiber are one of the main causes of signal attenuation.

impurity level An atomic energy level outside the normal energy band of the material, caused by the presence of impurity atoms.

in phase Alternating signals which are at the same frequency and pass through the zero, maximum and minimum values simultaneously.

inactive signaling link A signaling link which has been deactivated and therefore cannot carry signaling traffic.

in-band noise power ratio For multi-channel equipment, the ratio of the mean noise power measured in any channel, with all channels loaded with white noise, to the mean noise power measured in the same channel, with all channels but the measured channel loaded with white noise.

in-band signaling A signaling method in which signals are sent over the same transmission channel or circuit as the user's communication and in the same frequency band as that provided for the users. Some payphones use in-band multifrequency tones to control coin collection and coin return.

inboard channel The voice channel nearest in frequency to the carrier frequency in independent sideband radio techniques, where four voice channels are often transmitted on one radio bearer.

in-call A call in progress, initial switching at a given exchange having been completed.

in-call rearrangement Reassignment of the switched path of an in-call during the call.

incidence angle Angle between the perpendicular to a surface and the direction of a signal's arrival.

inclination Of satellite orbit: the angle between the plane of the orbit of a satellite and the reference plane. By convention, the inclination of a direct orbit of a satellite is an acute angle and the inclination of a retrograde orbit is an obtuse angle.

inclined orbit A satellite orbit which is neither equatorial nor polar.

incoherent Characterized by a degree of coherence significantly less than 0.88. Also see coherent; degree of coherence.

incoming A trunk used only for calls coming into the center concerned from other offices or centers.

incoming call barring The ability of the administration or the subscriber to stop all or certain incoming calls to a telephone line.

incoming call indication A signal on manually controlled PBXs to indicate that an incoming call requires answering.

incoming response delay A characteristic that is applicable where channel associated signaling is used. It is defined as the interval from the instant an incoming circuit seizure signal is recognized until a proceed to send signal is sent backwards by the exchange.

incoming traffic Traffic entering a network generated by sources outside it, whatever its destination.

incoming trunk busy (ITB) A network management feature which restricts incoming attempts to access an overloaded switch by selectively removing from service a proportion of incoming trunks.

incorrect bit 1. A received bit whose binary value is the complement of that transmitted. 2. A bit that is not the bit intended.

incorrect block A block successfully delivered to the intended destination user, but having one or more incorrect bits, additions, or deletions, in the delivered content.

increment A small change in the value of a quantity.

indefeasible right of user (IRU) Term used in connection with the financing and circuit allocation of major submarine cable systems. An obligation on the part of the owners of a facility to furnish to the purchaser of IRU continuing access to and enjoyment of the agreed-upon circuitry.

independent-sideband transmission That method of double-sideband transmission in which the information carried by each sideband is different.

independent telephone company (ITC) A Non-Bell Operating Company (U.S.)

index, modulation The degree of modulation of a signal.

index error A scale reading error.

index matching materials Transparent materials of proper refractive index used to reduce Fresnel reflections at an optical interface.

index of cooperation A non-dimensional quantity used in specifying facsimile transmission: the diameter of the scanning drum multiplied by the number of scanning lines in the same unit of length; e.g., both in inches or mm. For wirephoto transmission the index is usually 352 or 528; for weather maps, 288 or 576; and for general purpose facsimile transmission, 264.

index of refraction The ratio of a wave's velocity in a specified medium to its velocity in a vacuum. Also, refractive index.

index page (videotex) A page that lists other pages referring to more specific subjects. It is used to route the user towards the information required.

index profile In an optical waveguide, the refractive index as a function of radius.

indexer, AMA call identity Device which gives an identifying number to identify the trunk being used by an automatic message accounting call.

in-dialing, network Dialing from the public network into a private or local network.

indicating, idle-line A toll switchboard's lamp indication system which will light only the next available outgoing trunk's lamp in each group, rather than lighting the lamps of all busy trunks. This system conserves power, keeps the switchboard cooler, and improves operating efficiency. Also called free-line signaling, or FLS.

indicating instrument Measuring set which gives a reading on a visual scale.

indicator, B flow A manometer or gas pressure meter which indicates differences in pressure between two points.

indicator, dial speed Test set which gives a read-out of the speed in impulses per second of a dialed signal.

indicator, direction of gas flow Indicator with small glass tube connected to two points of different pressure. The movement of a light plastic ball inside the glass tube indicates the direction of the flow of gas between two points.

indicator, end-of-message A standard procedural sequence which ends a teletypewriter message, stops the machines at both ends and gives a clear signal to the exchanges involved.

indicator, lamp Holder for switchboard lamps.

indicator, routing Letters used to define a particular station, and to facilitate routing of traffic, particularly telegraph traffic, to that station.

indicator, start-of-message Standard procedural sequence which begins a teletypewriter message and switches on the receiving machine to print the incoming message.

indicator, start signal A start signal of the type appropriate to the signaling convention being used on a circuit, e.g., a key pulse signal.

indicator, tuning Small voltage-conscious meter, LED or discharge tube signal used to show when a radio receiver has been tuned in accurately to an incoming signal.

indicator, volume A dynamic meter which shows the power level in a program circuit, measured in volume units.

indirect address An address that designates the storage location of an item of data to be treated as the address of an operand but not necessarily as its direct address.

indirect control 1. In digital data transmission, the use of a clock at a higher standard modulation rate (e.g., 4, 8, times the modulation rate) rather than twice the data modulation rate as is done in direct control. 2. Control of switches by digital signals from lines which are stored in a register and converted to other signals which are used to set the switches.

indirect distribution Use of the fixed-satellite service to relay broadcasting programs from one or more points of origin to various earth stations for further distribution to terrestrial broadcasting stations.

indirect wave Part of a radio emission that is reflected back down to Earth via the ionosphere.

individual line A telephone line providing individual service for one subscriber.

individual reception The reception of emissions from a space station in the broadcasting-satellite service by simple domestic installations and, in particular, those possessing small antennae.

induce To produce an electrical or magnetic effect in one conductor by changing the condition or position of another.

induced charge Electrostatic charge produced on one body when it is brought near to another charged body.

induced current Current which flows in a conductor because a voltage has been induced across two points in or connected to the conductor.

induced emission See stimulated emission.

induced voltage Voltage produced in a coil when the coil moves through a magnetic field or when the coil's surrounding magnetic field is varied.

inductance A coil of wire's property to oppose any change in a current which flows through it. SI unit is the Henry. Basically whenever a current ceases to be of steady value inductance causes an electromotive force to be generated.

inductance, distributed Inductance spread uniformly along the whole length of a route, in contrast with lumped inductance in loading coils installed at points along the route.

inductance, mutual The inductance between two circuits which determines the electromotive force induced in one circuit by changes of current in the other circuit.

inductance, self The property of an electric circuit analogous to resistance to change: self inductance produces an electromotive force (emf) in a conductor which tends to oppose a changing current in the same conductor; the emf is proportional to the rate of change.

inductance, variable A coil whose inductance can be varied.

induction The electrical and magnetic interaction process by which a changing current in one circuit produces a voltage change not only in its own circuit (self inductance) but also in other circuits to which it is linked magnetically.

induction, low frequency The induction of a power frequency hum (50 Hz or 60 Hz) into an audio frequency transmission circuit.

induction, noise The transfer by induction into a transmission line of noise signals, such as clicks, or stray earth currents.

induction, power Noise induced into a transmission line which can be identified as coming from a power source; noise at 50 Hz or 60 Hz or harmonics of these can usually be distinguished.

induction, ringing Noise induced into a transmission line which can be identified as due to ringing current going out to another line.

induction, self The generation of a counter-electromotive force in a conductor caused by changes of current flowing through that conductor.

induction field The predominant magnetic field in the zone near a radio transmitting antenna.

induction motor An alternating current motor whose primary winding (usually static) is connected to the external power supply and induced currents flow through a secondary winding (usually on the rotor).

inductive A circuit element with inductive reactance, i.e., one utilizing the phenomenon of inductance.

inductive circuit Circuit with more inductive than capacitive reactance.

inductive coupling Coupling which exists by virtue of a mutual inductance, e.g., between primary and secondary coils of a transformer.

inductive kick Voltage surge produced when a current flowing through an inductance is interrupted.

inductive load A load which possesses net inductive reactance.

inductive reactance Reactance of a circuit due to the presence of inductance and the phenomenon of induction.

inductometer An inductor with variable inductance.

inductor A coil of wire, usually wound on a special core of high permeability, which provides high inductance without necessarily being of high resistance.

inductor, saturable Device used in some party line units which reduces bridged tap loss by introducing coils with saturable cores in both lines concerned.

inert Inactive units, or units which have no power requirements.

inert cell A type of primary power cell which contains all needed chemicals in dry form, when water is added the cell begins to operate. The inert cell's advantage over ordinary dry cells is a long shelf life.

inertance The thermal or fluid equivalent of electrical inductance or mechanical moment of inertia.

inertia switch A switch which operates whenever there is a sudden change in its velocity or position.

infinite line A transmission line which appears to be of infinite length, e.g., there are no reflections back from the far end because it is terminated in its iterative (or characteristic) impedance.

inflexible   Term describing a submarine cable system repeater configuration not designed to be handled (loaded, stored, laid, recovered) in the same manner as cable: rigid, requiring special handling.

influence   In telecommunications work, the tendency of a power supply system to induce hum or noise into a voice frequency transmission line.

influence, power circuit   Those characteristics of a power circuit which indicate the extent of interference likely to be induced into nearby voice frequency transmission lines. The principal factors are voltages, type of supply and relationship between phases and conductor voltages, degree of balance between phases, harmonic content, route construction parameters, and transposition patterns.

influence factor   A weighted figure indicating the total inductive interference characteristics of a power supply circuit.

INFONET   A computerized information service in Britain, owned by a consortium of database producers.

information   1. The organizational content of a signal. 2. Collection of facts, data, numbers, letters, and symbols which have meaning.

information bearer channel   A channel provided for data transmission which is capable of carrying all the necessary information to permit communications, including user's data, synchronizing sequences, control signals, etc. It may operate at a greater signaling rate than that required solely for the user data.

information bit   A bit which is generated by the data source and delivered to the data sink and which is not used by the data transmission system.

information board   Operating positions to which requests for information are referred.

information channel   The transmission media and equipment used for transmission in a given direction between two terminals.

information content   The minimum amount of information needed in order to make the message understood by its addressee, without allowing for noise on the circuit.

information facility   A facility whereby a data service user, by sending a predetermined address from the terminal installation, may gain access to general information regarding data communication services.

information feedback   The sending of data back to a source, usually for the purpose of checking the accuracy of transmission of data; the received data being returned to the sending end for comparison with the original data.

information field   In data transmission, a field assigned to contain user information.

information network system (INS)   Japanese term for integrated services digital network (ISDN).

information operator   An operator working at an information board or position.

information position   Operating position specially equipped to enable the operator to provide answers to enquiries from the public concerning the telephone system.

information processing   Data processing.

information provider   Organization which provides information or services through a videotex system.

information retrieval   A branch of computer technology dealing with storage techniques for data.

information security   The protection of information against unauthorized disclosure, transfer, modification, or destruction, whether accidental or intentional.

information theory   Technique which shows the necessary amounts of information needed to transmit a message.

information tones   Tone signals sent back from central office to calling subscriber advising him of the progress of his call attempt (e.g., ringing tone, busy tone, reorder tone, pay tone).

information transfer   The process of moving messages containing information from an information source to an information sink.

information transfer phase   In an information transfer transaction, the phase during which user information blocks are transferred from the source user to a destination user.

information transfer transaction   A coordinated sequence of user and telecommunication system activities whose ultimate purpose is to cause user information present at a source user to become present at a destination user. An information transfer transaction is typically divided into three consecutive phases: the access phase, the information transfer phase, and the disengagement phase.

information transport   ISDN synonym for telecommunications.

Infotex   (From INFOrmation via TelEX.) A service which provides telex subscribers with access to various computer databases in the U.S., via the international telex network.

Infotext   Teletext service in the U.S.

infra low frequency (ILF)   Frequencies from 300 Hz to 3000 Hz.

infrared radiation   Electromagnetic radiation next in frequency to visible red light extending down into microwave frequencies. Wavelengths between 740 nm and 1 mm usually cover this band. The IR region is sometimes subdivided into near infrared, middle infrared and far infrared segments.

infrasonic   Sound waves at frequencies too low to be heard by a human ear, i.e., below about 15 Hz.

inhibit   To prevent a device or circuit from operating.

inhibit pulse   A pulse which cancels out the effect of a simultaneous drive pulse into a memory store.

inhibiting input   An input which stops a device from operating.

in-house videotex   A private videotex system dedicated to handling internal company information.

initial address message (IAM) A multi-unit message which is sent as the first message in a call set-up using common channel signaling system CCITT No 6, consisting of a minimum of three and a maximum of six signal units, and containing enough information to route the call through the international network.

initial alignment A procedure by which a signaling link (in a common channel signaling system) becomes able to carry signaling traffic either for the first time or after a failure has occurred.

initial period For manually set up and controlled toll calls: three minutes, the minimum charge period.

initial signal unit (ISU) The first signal unit of a multi-unit message in common channel signaling system CCITT No 6.

injection 1. The application of a signal to an electronic device. 2. The introduction of excess charge carriers into a semiconductor device in order to change the characteristics of the medium. In low level injection the number of excess carriers is small; in high level injection the quantities of excess carriers is comparable with the numbers at thermal equilibrium.

injection fiber Synonym for launching fiber.

injection laser diode (ILD) A laser employing as the active medium a forward-biased semiconductor diode. Syn.: diode laser

ink vapor recording That type of facsimile recording in which vaporized ink particles are directly deposited upon the record sheet.

inlet Point through which the incoming traffic flow enters a switching stage.

inline package (CATV) A housing for amplifiers or other CATV components designed for use without jumper cables; cable connectors on the ends of the housing are in line with the coaxial cable.

in-plant system A data system confined to one building or complex of associated buildings.

input 1. The signal fed into a circuit. 2. The terminals that receive the input signal. 3. The energizing power of an electrical device. 4. Data to be processed. 5. Sequence of states occurring on an input channel. 6. Devices used for bringing data from one area into another. 7. Channel for impressing a state on a device. 8. In man machine language (MML), the process that constitutes the introduction of data into a data processing system or any part of it. 9. In specification description language (SDL), an input is an incoming signal which is recognized by a process.

input impedance Impedance looking into the input terminals of a device.

input/output (i/o) The process of transmitting data to and from the processor and its peripherals.

input/output device A device that introduces data into or extracts data from a system. The most common of these are:

card reader input	card punch output
tape reader input	tape punch output
magnetic tape input	magnetic tape output
optical character recognition input	printer output
keyboard input	visual display unit output

magnetic ink character recognition input

input transformer Transformer at the input to a device which transforms/matches the impedance of the device to that of the preceding stage.

inquiry A request for information; directory enquiry service.

insert, concrete Device inserted in walls of concrete manholes while they are being cast in order to produce threaded holes for security fixing of racks and bearers in the manhole.

insertion gain The gain resulting from the insertion of a transducer in a transmission system, expressed as the ratio of the power delivered to that part of the system following the transducer to the power delivered to that same part before insertion. If more than one component is involved in the input or output, the particular component used must be specified. This ratio is usually expressed in decibels. If the resulting number is negative, an insertion loss is indicated.

insertion loss The difference between the power received at the load before and after the insertion of apparatus at some point in the line. If the resulting number is negative, an insertion gain is indicated.

insertion loss vs frequency characteristic The amplitude transfer function characteristic of a system or component as a function of frequency. The amplitude response may be stated as actual gain, loss, amplification or attenuation, or as a ratio of any one of these quantities, at a particular frequency, with respect to that at a specified reference frequency. Syn.: amplitude frequency response; frequency response

inside plant 1. For radio and radar, all fixed ground communications-electronics equipment that is permanently located inside buildings. 2. For wire and cable, all fixed ground cable plant extending inward from the main distribution frame (MDF), e.g., central office equipment, teletypewriters, and including the protectors and associated hardware on the telephone central office MDF.

in-slot signaling Signaling associated with a channel and transmitted in a digit time slot permanently (or periodically) allocated in the channel time slot.

inspection, pole Regular inspection of poles, including taking samples of wood, to determine their state of deterioration, if any.

inspector's ring back Facility provided by many offices/exchanges which allows calling for a ringing current on a line to enable the operation of an instrument to be adjusted or confirmed.

installer, telephone Technician who installs drop wire from the nearest distribution point (or local cabling), all necessary internal wiring and the telephone instrument itself.

instantaneous value Value at a particular instant of a quality which is varying, e.g., a sinusoidal variation in voltage or current.

Institute of Electrical and Electronic Engineers (IEEE) The U.S. organization for professional electrical engineers.

Institution of Electrical Engineers (IEE) The U.K. body for professional electrical engineers. Establishes standards for electrical power wiring and safety.

instruction Set of identifying characters designed to cause a processor to perform certain operations.

instruction set The menu of instructions a computer can execute.

instrument multiplier Measuring device which enables a high voltage to be measured using a meter with only a low voltage range; an instrument transformer.

instrument rating The range within which an instrument has been designed to operate without damage.

instrument sensitivity Response of a measuring instrument compared with the quantity being measured, e.g., in divisions per milliamp.

instrument transformer An efficient low-power transformer used to produce low voltage outputs from high voltage inputs so that meters with low effective voltage ranges may be used to measure high voltages.

insulant Insulating material. Also, the material between the inner and outer conductors of a submarine coaxial cable.

insulate To separate one conducting body from another conductor, e.g., to insulate wires by sheathing them with plastic.

insulated Covered by a dielectric or insulating material.

insulated, paper Insulated by a wrapping of dry paper, as is still standard practice for the conductors in lead-sheathed paper insulated cable.

insulated gate field-effect transistor A field effect transistor in which the conducting channel is made by the action of the applied gate voltage. Syn.: metal-oxide semiconductor field effect transistor (MOSFET)

insulation Material which does not conduct electricity and protects one body from others which may be at different electric potentials.

insulation, low Condition when the insulation between conductors or between a conductor and ground is not as good as it should be for reliable service.

insulation, primary First layer of electrical insulation around a conductor.

insulator The glass or ceramic cup attached to a cross-arm to which a telephone wire is bound.

insulator, double petticoat Type of insulator used on toll open wire circuits with two integral skirts which together give a long surface leakage path.

insulator, exchange line A small simple insulator with only one petticoat to provide a leakage path, used on subscribers' lines which are carried by open wire routes.

insulator, guy A double-hole or loop-shaped insulator used in a guyline to ensure that dangerous voltages will be out of reach.

insulator, metallic Support used for radio-frequency transmission lines, e.g., antenna feed lines: cut to exactly the correct $\lambda/4$ length for a particular frequency, and so acts like an open circuit and does not bridge the line at that particular frequency.

insulator, pin A type of insulator used on open wire routes which can be screwed onto pins on cross-arms.

insulator, pony A small glass insulator used on rural open wire lines.

insulator, stand off High voltage antenna feed insulator, used to hold a feeder clear of equipment

insulator, strain A guy insulator, used to provide separation between upper and lower sections of a guyline.

insulator, transposition Insulator with two separate grooves for wires instead of one. Point transposition can be done at poles equipped with these insulators without using transposition brackets.

integrated analog network Switched network providing voice or voice-simulated services when cost is minimized through the synergistic application of analog transmission and space division switching techniques.

integrated circuit A circuit with more than one function formed in or on a single semiconductor base.

integrated circuit, film An integrated circuit whose elements are films.

integrated circuit, hybrid An integrated circuit made up of several types of integrated circuits, or of a mixture of integrated circuits and conventional elements.

integrated circuit, monolithic An integrated circuit made up on a single semiconductor substrate.

integrated circuit, multi-chip An integrated circuit made up using semiconductor chips which then are attached to a single insulating substrate.

integrated digital network (IDN) Switched network providing voice or voice-simulated services where cost is minimized through the synergistic application of digital transmission and time-division switching technology.

integrated ground (ING) A ground circuit that connects the equipment ground with the framework.

integrated injection logic (I^2L) A family of integrated circuits devised from diode transistor logic (DTL) by combining some of the DTL components into single transistor units.

integrated optical circuit (IOC) A monolithic optical circuit, composed of both active and passive miniaturized components, employing planar waveguides for coupling to optoelectronic devices and providing signal processing functions such as modulation, multiplexing and switching.

integrated services digital network (ISDN) Switched network providing end-to-end digital transparency where voice and data services are provided over the same transmission and switching facilities.

integrated services exchange An exchange used to handle multiple services such as telephone and data using all or part of the switching, signaling and control devices in common.

integrated switching and transmission (IST) A telephone network consisting of time division exchanges and interconnecting digital paths provides IST so that conversion back to analog is not necessary within the network. (Also IDN, integrated digital network)

integrated system A telecommunication system that moves analog and digital traffic over the same switched network.

integrated voice/data terminal (IVDT) Device that incorporates voice telephone, keyboard and display unit.

integration The production of complete and complex circuits on a single chip, usually of silicon. The widely-accepted ranges of integrated circuits are:

small scale integration (SSI)-simple circuits, up to about 10 circuit elements per chip

medium scale integration (MSI)-more complex circuits, up to about 1000 elements per chip

large scale integration (LSI)-complex circuits with up to 16k bits of memory capacity or circuit elements (some makers say up to 64k)

very large scale integration (VLSI)-circuits with 32k, 64k or up to 1M of memory capacity or circuit elements

extra large scale integration (ELSI)-LSI with more than 10^6 bits of capacity or circuit elements.

integrity checking, signaling A speech-path continuity check performed in both directions with some in-band signaling systems before called party digits are transmitted.

intelligence Information, particularly in a message.

intelligence signal A signal containing information.

intelligent knowledge-based systems See artificial intelligence.

intelligent terminal Input/output device remote from its main computer. Usually has a keyboard for input, and a VDU for output, and a small amount of local storage and processing power so it can accomplish some tasks on its own. Syn.: smart terminal

intelligibility It is essential that a telephone caller be understood, that the signal be clear. Since this is a subjective matter, experts speak strings of nonsense words over the circuit under test, and find out how many were received correctly. See articulation.

intelligible crosstalk Crosstalk giving rise to intelligible sounds.

intelligible crosstalk components Transferred speech currents which can introduce intelligible crosstalk into certain channels at the point considered.

Intelpost The international facsimile service set up by the U.S. Postal Service using high speed digital facsimile equipment.

intensity In optics: the square of the electric field amplitude of a light wave. Intensity is proportional to irradiance and may be use l in place of the term irradiance when only relative values are important.

intensity, radio field Strength of an electromagnetic wave radio signal at a particular point in space.

intensity, sound Sound energy per unit area at right angles to the propagation direction, per unit of time.

inter integrated circuits A type of bus connection used for domestic small area networks.

interaction Process by which a system accepts input, processes the input requests and if necessary, returns appropriate response data to the originating terminal.

interaction crosstalk Crosstalk caused by coupling between carrier and non-carrier circuits; the crosstalk may in turn be coupled to another carrier circuit.

interactive Action in more than one direction, either simultaneously or sequentially. See conversational mode.

interactive data transaction A single (one-way) message, transmitted via a data channel, to which a reply is required in order for work to proceed logically.

interactive processing Processing in which the user can study the output results and modify the programs if necessary; a two-way process different from processing a huge batch of data and turning out large quantities of output such as a payroll.

interactive Teletext A synonym for videotex.

interaid A switching system in which outlets from a given connecting stage are connected to inlets of the same or a previous stage. In such systems calls may traverse a stage more than once. Usually these re-entering links are used as last choice paths and the resulting network is then heterogeneous. Sometimes called a reentrant link system.

interblock gap An area on a data medium used to indicate the end of a block or physical record, e.g., a space between blocks on magnetic tape.

intercept To stop and divert a telephone call to an operator or to a number other than the one dialed.

intercept tape storage In AUTODIN, a method of providing temporary storage for message traffic destined for nonoperating or backlogged channels.

intercepting, dual System by which intercepted calls can be routed either to an operator or to an announcing machine.

intercepting, machine The use of an answering/recording machine to deal with intercepted calls.

intercepting, matched pulse Method of intercepting calls by individual subscribers served by party lines: depends on matching the particular ringing code or frequency which has to be used to call the party concerned.

intercepting, operator Diversion of intercepted calls to an operator, who gives appropriate instructions to callers.

intercepting operator An operator who has access to detailed lists of changed numbers, and discontinued lines, and provides the information to callers.

intercepting position Position where changed number records and all other relevant information is available, for use by the intercepting operator.

interchange circuit A circuit between the data terminal equipment and the data circuit terminating equipment for the purposes of exchanging data and signaling information. Control signals, timing signals, common return functions, or other service features may be included in an interchange circuit.

interchangeability (TV) The ability to exchange tapes between different manufacturers' videotape recorders with no appreciable degradation of playback image.

intercharacter interval That time period between the end of the stop signal of one character and the beginning of the following character. This interval may be of any length. The signal sense of the intercharacter interval is always the same as the sense of the stop element, i.e., 1 or mark.

intercom A telephone apparatus which allows personnel to talk to each other within an aircraft, tank, ship, or activity.

intercom, home extension A revertive calling arrangement which permits calling between extension phones in a residence using an abbreviated dialed code which sends back a coded ringing signal.

intercom, telephone Common name for systems permitting two or more extension phones to have access to two or more exchange lines as well as to intercommunicate with each other; a key phone system.

intercommunication As used in PBX systems, any station in the system calling any other station in the system. In key systems, intercommunication generally occurs over a special intercom circuit.

intercommunication system An intercom or key phone system. When the word intercommunication is used it often means the instruments are hands-free or loudspeaking telephones, providing the same services.

Intercompany Service Coordination/Universal Service Order (ISC/USO) Bell System abbreviation used in Universal System Service Orders: The universal service order format to be used for Intercompany Services Coordination Plan, ISC, orders. This term is used to distinguish it from the local translated order (USO) associated with it.

Intercompany Services Coordination Plan (ISC) Bell System abbreviation used in Universal System Service Orders: a set of standard interdepartmental procedures, an interval guide, and a measurement plan for all intra- and inter-area data orders as well as all other inter-area special service orders.

interconnect A situation where a customer-provided piece of equipment is connected to, or has access to, the public switched telephone network. It also refers to the entire industry of non-telco manufacturers and distributors who supply equipment directly to customers. In cable TV systems: linking headends together, often by microwave, so that subscribers to different systems may see the same programs simultaneously.

interconnecting number The number of grading groups connected together in a homogeneous grading.

interconnecting number, mean Same as grading ratio. The quotient obtained by dividing the total number

of outlets in a grading by the number of trunks serving the grading.

interconnection The connection of telephone equipment to the network, also the connection of one carrier with another; i.e., the interface between carriers.

intercontinental telex circuit A circuit which connects two exchanges situated in different countries in different continents.

intercontinental telex connection Connection established between two different continents.

intercontinental transit telex circuit An intercontinental circuit used primarily for routing intercontinental transit traffic.

intercontinental transit telex exchange An exchange directly connected to intercontinental transit circuits and which provides facilities to interconnect intercontinental transit circuits and trunks to terminal exchanges. It also provides facilities for the interconnection of intercontinental transit circuits.

inter-digit pause The gap in time between the end of a train of disconnection pulses representing one digit and the beginning of the train of pulses representing the next digit. This gap is set by the mechanical construction of the loop-disconnect dial, and is usually about 600 ms.

interdigital time Same as inter-digit pause.

interexchange carrier (IEC) Any carrier registered with the FCC that is authorized to carry customer transmissions between LATAs interstate, or if approved by a state public utility commission, intrastate.

inter-exchange circuit A circuit traversing more than one exchange area.

interface 1. A shared boundary or point common to two or more similar or dissimilar command and control systems, subsystems, or other entities against which or at which, or across which useful information flow takes place. 2. A concept involving the definition of the interconnection between two equipments or systems. The definition includes the type, quantity, and function of the interconnecting circuits and the type and form of signals to be interchanged via those circuits. Mechanical details of plugs, sockets, and pin numbers, etc., may be included within the context of the definition. 3. The process of interrelating two or more dissimilar circuits or systems.

interface, CCITT For data equipment this requirement is similar to the EIA interface.

interface, ongoing Term used in submarine cable system practice for the point of connection between a submarine seacable system and the facilities that extend the service into the telecommunications network ashore.

interface EIA Standard RS 232 B or C Standardized method adopted by the EIA to insure uniformity of interface between data communication equipment and data processing terminal equipment.

interface equipment Conversion equipment which enables circuits designed to one set of characteristics to work to circuits designed to meet different specifications.

interface MIL STC 188B   The U.S. Dept. of Defense standard method of interface which provides the interface requirements for connection between data communication security devices, data processing equipment, or other special military terminal devices.

interface points (ISDN)   Points at which different devices interconnect with the integrated network.

interference   1. In a signal transmission system, extraneous power from natural or man-made sources that interferes with reception of desired signals. 2. The effect of unwanted energy due to one or a combination of emissions, radiations, or inductions upon reception in a radiocommunications system, manifested by any performance degradation, misrepresentation, or loss of information which could be extracted in the absence of such unwanted energy. 3. In optics, the interaction of two or more beams of coherent or partially coherent light usually derived from a single source.

interference, harmful   Any interference which endangers the functioning of a radio-navigation service or of other safety services or seriously degrades, obstructs, or repeatedly interrupts a radiocommunication system.

interference, inductive   Noise interference induced into a voice frequency transmission line, particularly from power circuits.

interference, intersymbol   A form of distortion sometimes encountered in pulse code modulation systems where energy in one pulse is spread over several time slots of adjacent pulses.

interference, permissible   Observed or predicted interference which complies with quantitative interference and sharing criteria contained in the internationally-agreed Radio Regulations or in recommendations of the CCIR or in regional agreements as provided for in the Radio Regulations.

interference, selective radio   Radio interference concentrated in a narrow frequency band aimed at disrupting a particular radio communication system or source.

interference, wave   The radio equivalent of diffraction pattern interference with light signals; changes in amplitude caused by the interaction of two incoming signals of almost exactly the same frequency.

interference emission   Emission that results in an electrical signal being propagated into and interfering with the proper operation of electrical or electronic equipment.

interference sector   Of a directional antenna, the horizontal sector outside the main beam.

interfering source   An emission, radiation, or induction which is determined to be a cause of interference in a radiocommunications system.

interframe time fill   In data transmission, the sequence of bits transmitted between frames.

Inter-Governmental Maritime Consultative Organization (IMCO)   Similar to the better-known International Civil Aviation Organization, but deals with shipping rather than with air services.

interlace   In a television (CRT) tube, the movement of the scanning beam in which for each frame the beam scans left to right 262 times from top to bottom of the picture (U.S. rate). This action is repeated, with the beam tracing between the previously scanned lines.

inter-LATA   Services which originate and terminate in different local access and transport areas.

interleaving   1. The transmission of pulses from two or more digital sources in time division sequence over a single path. 2. A technique used in conjunction with error correcting codes to lower the error rates of communication channels characterized by burst errors.

interleaving, pulse   Process by which pulses may be transmitted in a time division sequence.

interlock   1. Circuitry and mechanical restraints which prohibit access to the high-voltage sections of submarine cable system equipment until potentials are removed and cables safely terminated. 2. Method of interconnecting keys and latches, for example to stop two selection keys being activated simultaneously.

interlock, key   On a key telephone, a latch which releases one key when a second key has been depressed.

interlock code   For data and signaling, information sent in the forward direction and, in some cases, in the backward direction indicating the closed user group involved in the call.

intermachine trunk   In CCSA and EPSCS networks, the private lines which carry traffic between CCSA or EPSCS switches. Used more loosely, this term sometimes refers to any private line connecting two switching machines, i.e., two PBXs or tandem switches.

intermediate dialing center   A toll center office which completes calls initiated by subscribers on non-dial offices.

intermediate distribution frame   (IDF) See frame, intermediate distribution.

intermediate field region   The transition region between the near field region and the far field region in which the electric field strength of an electromagnetic wave developed by a transmitting antenna is dependent upon the inverse distance, inverse square of the distance, and the inverse cube of the distance from the antenna.

intermediate frequency rejection ratio   Of a radio receiver, the ratio of the level of a signal at the intermediate frequency, applied to the receiver input and which produces a specified output from the receiver, to the level of the wanted signal required to produce the same output power.

intermediate toll center   An office that assists in setting up and connecting through a toll call, which is not the originating toll center or the terminating toll center.

intermediate toll operator An operator who assists in setting up and/or connecting through a toll call, who is neither in the originating toll center nor the terminating toll center.

intermittent Not continuous, but (if a fault) frequent enough to cause degradation of service.

intermodal distortion See multimode distortion.

intermodulation The production, in a nonlinear transducer element, of frequencies corresponding to the sums and differences of the fundamentals and harmonics of two or more frequencies which are transmitted through the transducer.

intermodulation component Sinusoidal oscillation produced in an imperfectly linear amplitude-modulated radio transmitter in response to sinusoidal oscillations applied at the input to the transmitter, the frequency of which is, at the output of the transmitter, the sum or difference of the frequencies of the normal sideband components resulting from the modulation of a carrier by the exciting oscillations, or the sum or difference of integral multiples of these frequencies.

intermodulation distortion See distortion, intermodulation.

intermodulation noise In a transmission path or device, that noise which is contingent upon modulation and demodulation and results from any nonlinear characteristics in the path or device.

intermodulation products All the intermodulation components produced in an amplitude-modulated transmitter in response to given sinusoidal oscillations applied to the input.

internal bias The bias, either marking or spacing, that may occur within a start-stop teletypewriter receiving mechanism and which will have the same effect on the margin of operation as bias external to the receiver.

internal blocking The condition in which a connection cannot be made between a given inlet and any suitable free outlet owing to the impossibility of establishing a path within the switching element being considered.

internal dynamic overload control (IDOC) A network management control initiated by an office in response to internally-detected overload indicators. Signals are sent to other offices for them to implement a predefined series of routing controls.

internal photoeffect The absorption of photons and excitation of electrons which move from valence band to conduction band (intrinsic photoeffect) or to impurity levels, then to the conduction band (extrinsic photoeffect).

internal reflection The reflection of light from cladding back into the core in a stepped-index optic fiber.

internal resistance The actual resistance of the source of electric power; the total electromotive force produced by a power source is not available for external use, some of the energy is used in driving the current through the source itself.

internal traffic Traffic originating and terminating within the network considered.

international atomic time (TAI) The time scale established by the Bureau International de l'Heure (BIH) on the basis of data from atomic clocks operating in several establishments conforming to the definition of the second, the unit of time of the International System of Units (SI).

international circuit A circuit between two international exchanges situated in different countries.

International Civil Aviation Organization (ICAO) The organization responsible for coordinating the telecommunications aspects of civil aviation services.

international-congestion signal For data and signaling, a signal sent in the backward direction indicating the failure of the call set-up attempt due to congestion encountered in the international network or destination national network.

international direct dialing (IDD) The dialing or keying of calls by the subscriber between countries.

International Electrotechnical Commission (IEC) The electrical standardizing division of the International Organization for Standardization (ISO).

international exchange The exchange (at the end of an international telephone circuit) which switches a call destined to or originating from another country. Syn.: gateway exchange

International Frequency Registration Board (IFRB) One of the four permanent organs of the International Telecommunications Union (ITU). A Plenipotentiary Conference of the ITU meets about every five years and elects five independent radio experts, all from different regions of the world, to be the members of the IFRB. The IFRB's main task is to decide whether radio frequencies which countries assign to their radio stations are in accordance with the ITU Convention and the Radio Regulations and will not cause harmful interference to other stations.

International Maritime Satellite Organization (INMARSAT) Body founded by international agreement which operates communication satellite system for use by ships.

international Morse code Code widely used for radio telegraphy before the widespread use of telegraph machines such as teletypewriters. Morse characters vary in length from one dot to five dashes.

international number The number to be dialed following the international prefix: the country code of the required country followed by the national (significant) number of the called subscriber.

International Organization for Standardization (ISO) Made up of the national standard organizations of most industrialized countries.

international prefix The combination of digits to be dialed to obtain access to the automatic outgoing international equipment.

International Scientific Radio Union (URSI) A group that coordinates radio propagation all over the world.

international section The group, supergroup, etc., sections between two adjacent frontier stations in different countries.

international service carrier (ISC) Carrier providing overseas communications services other than voice.

international sound-program center (ISPC) A center at which at least one international sound-program circuit terminates and in which international sound-program connections can be made by the interconnection of international and national sound-program circuits. The ISPC is responsible for setting up and maintaining international sound-program links and for the supervision of the transmissions made on them.

international sound-program circuit The unidirectional transmission path between two International Sound Program Centers (ISPCs) and comprising one or more sound-program circuit sections (national or international), together with any necessary audio equipment (amplifiers, compandors, etc).

international sound-program connection The unidirectional path between the broadcasting organization (send) and the broadcasting organization (receive) comprising the international sound-program link extended at its two ends over national sound-program circuits to the broadcasting organizations.

international sound-program link The unidirectional path for sound-program transmissions between the International Sound Program Centers (ISPCs) of the two terminal countries involved in an international sound-program transmission. The international sound-program link comprises one or more international sound-program circuits interconnected at intermediate ISPCs. It can also include national sound-program circuits in transit countries.

International Standards Organization (ISO) International Organization for Standardization: International body concerned with setting standards in many disciplines.

international switching center (ISC) An exchange whose function is to switch telephone traffic between a national network and the networks of other countries.

International Telecommunication Union (ITU) (In French, UIT). Founded in 1865 as the International Telegraph Union, the ITU is an organization of member countries. There are at present (1985) 161 member countries. The ITU's headquarters is in Geneva, Switzerland (Place des Nations CH-1211, Geneva 20, telephone Geneva 346021, telegraphic address BURINTERNA GENEVE, Telex 23000/23000a uit ch). In its headquarters building are the offices of its four permanent organs: the General Secretariat; the International Frequency Registration Board (IFRB); the International Radio Consultative Committee (CCIR); and the International Telegraph and Telephone Consultative Committee (CCITT).

International Telecommunications Satellite Organization (INTELSAT) A consortium made up of various national satellite communications organizations. COMSAT, the U.S. representative in the consortium, acts as INTELSAT's manager.

international telex circuit Connects two exchanges in different countries, whether or not they are in different continents.

international telex connection Any connection between two stations situated in different countries.

international telex exchange A center where national and international circuits terminate.

international transit exchange An international telephone exchange which establishes telephone calls between two countries other than its own. Also may perform international exchange (CT3) functions.

International Ursigram and World Days Service (IUWDS) Recording, study and warning program regarding disturbances to ionospheric propagation. Laboratories all over the world cooperate in this work; there are regional warning centers in Germany, France, Japan, United States, USSR and Australia, with the IUWDS world warning agency at Boulder, Colo., USA.

inter-office Between two telephone switching centers or telephone offices.

inter-office trunk See trunk, inter-office.

interoperability The condition achieved among communications-electronics systems or equipment when information or services can be exchanged directly between them or their users, or both.

interpersonal messaging service The set of service elements in a message handling system which enables users to exchange interpersonal messages.

interpersonal messaging system (IPMS) The collection of User Agents and Message Transfer Agents in a message handling system which provide the interpersonal messaging service.

interpolate To estimate values based on knowledge of comparable data which fall on both sides of the point in question.

interpolation, speech The assignment of a long distance speech path to the two circuits at the local ends only when necessary to carry conversation. Also see time assignment speech interpolation.

interpole An auxiliary pole piece located between the main poles of a rotary electrical machine with a commutator. It is needed to eliminate undesirable effects at the instant of commutation.

interposition trunk 1. A connection between two positions of a large switchboard so that a line on one position can be connected to a line on another position. 2. Connections terminated at test positions for testing and patching between testboards and patch bays within a technical control facility.

interpositioning Equipment layout which sandwiches customer provided equipment between telephone company terminal equipment and telephone company network or facilities.

interpret 1. To print out a translation of the meaning of the series of punched holes in a punched card. 2. To translate non-machine language into machine language.

interpreter A computer program that converts high-level language statements to machine code for direct computer operation.

interpreter, data card Device which accepts a punched card and prints on it a translation of the meaning of the punched holes.

inter-processor signals Signals which pass between processor-controlled exchanges during the establishment of a call.

interrogation 1. The process whereby a signal or combination of signals is intended to trigger a response. 2. The process whereby a station or device requests another station or device to identify itself or give its status.

interrogator Device which demands a reply from a component or station, and automatically triggers such a reply from appropriate equipment.

interrupt A signal which causes suspension of the operation of the sequence of instructions and the initiation of another sequence of instructions, i.e., a jump out of one program to another due to an external event.

interrupted continuous wave (ICW) A modulation technique in which there is on-off keying of a continuous wave.

interrupted isochronous Pertaining to isochronous burst transmission.

interrupter springs Sets of cams on ringing and tone generation machines in older telephone offices which break up the ringing current, ringing tone or busy tone into cycles of the appropriate periodicity.

interrupting capacity The rating of a circuit breaker or fuse which tells the maximum current it is designed to interrupt at its rated voltage.

interruption control A system which monitors a pilot for interruptions on frequency division multiplex systems and which transmits an indication to the switching equipment.

interstate Crossing state lines, or between states.

interstice Gaps between insulated conductors in a multi-pair cable. Interstitial pairs or quads are often provided in a multi-tube coaxial cable.

interswitch trunk See trunk, interswitch.

intersymbol interference Extraneous energy from the signal in one or more keying intervals that interferes with the reception of the signal in another keying interval, or the disturbance that results therefrom.

inter-toll dialing See dialing inter-toll.

inter-toll trunk See trunk, inter-toll.

interval, character Total number of unit signals needed to send a signal denoting any particular character over a link.

interval, charge-delay Method of charging for measured time calls in which the time is noted when an on-hook signal indicates the end of the call but no action is taken on computing costs for up to five seconds in case the on-hook signal received was not a genuine end of call but only a transient or interference pulse.

interval, pulse The time from the start of one pulse to the start of the next pulse in a train of pulses.

interval, ringing In the U.S., ringing current is usually applied to a called line for one second, followed by three or four seconds of silence before another one second of ringing. In other countries, different periodicities are sometimes used, for example, in Britain there are two bursts of ringing followed by a short period of silence.

interval, silent The period between bursts of ringing current.

interval, transposition Any of the various length sections for which transposition schemes for open wire routes have been planned.

interval, unit For telegraphy: the time unit such that all significant intervals of a signal are whole multiples of this time unit interval.

interworking Joint functioning of two connected and interacting systems.

interworking between user classes of services The means whereby data terminal equipment belonging to one user class of service may communicate with data terminal equipment from another user class of service.

interzone call Call between zones in a large metropolitan area.

intra-exchange circuit A circuit lying wholly within a single exchange area.

intra-LATA Services which originate and terminate in the same local access and transport area.

intramodal distortion That distortion resulting from dispersion within individual propagating modes. It is the only distortion occurring in single mode waveguides.

intra-office Within the same telephone switching center or telephone office.

intrastate Within the boundaries of a state.

intrinsic conductivity Conductivity of pure semiconductor material itself, not contributed by the presence of dopants and impurities.

intrinsic junction loss In optoelectronics: the total loss resulting from joining two identical optical waveguides. Factors influencing this loss include spacing loss, alignment of the waveguides, Fresnel reflection loss, end finish, etc.

intrinsic noise See noise, intrinsic.

intrinsic semiconductor An i-type semiconductor, with equal concentrations of holes and electrons under conditions of thermal equilibrium.

intrusion Facility provided by some PABXs enabling operators to monitor established calls and speak to the parties concerned.

inventory available date (IAD) Bell System abbreviation used in Universal System Service Orders. The date on which disconnected equipment and facilities are to be made available for reuse.

inverse feedback Negative feedback.

inverse voltage Effective value of voltage across a rectifying device which conducts a current in one direction during one half cycle of the alternating input, during the half cycle when current is not flowing.

inversion 1. Production of ac from dc. 2. Reversing the voltage of a pulse, i.e., from positive to negative. 3. Changing phase by 180 degrees.

inversion, temperature Condition often encountered in semi-tropical areas in which moist warm air overlays cool dry air to produce an atmospheric duct which carries radio communication signals substantial distances.

inverted position In frequency division multiplexing, a position of a translated channel in which an increasing signal frequency in the untranslated channel causes a decreasing signal frequency in the translated channel.

inverter 1. A device for changing direct current to alternating current. 2. In computers, a device or circuit that inverts the polarity of a pulse.

invitation-to-send A signal, character or code sequence which calls for an outgoing terminal to begin transmission. This can be part of a polling or a signaling procedure.

inward restriction A Centrex feature which stops some extensions from receiving calls directly from outside the system.

INWATS Inward wide area telephone service; a service which allows other telephone users to call particular numbers without being charged. The charges are met by the called subscriber on a bulk-time basis.

ion An electrically charged atom or group of atoms.

ion exchange technique A method of fabricating a graded index optical waveguide by exchanging ions through the core-cladding interface.

ion implantation Technique used in the manufacture of integrated circuits: the surface of a semiconductor is bombarded in order to implant ions into the lattice of the crystal.

ionization Procedure by which an atom is given a net charge by adding or taking away an electron.

ionization chamber A gas-filled detection tube used to indicate the presence of radiation.

ionized Atoms which have more or fewer electrons than is usual and have a net negative or positive charge, and are attracted to or moved by other charges or by a magnetic field.

ionized layer See ionosphere.

ionosphere The region of the atmosphere, extending from roughly 50 km to 400 km altitude, in which there is appreciable ionization. The presence of charged particles in this region profoundly affects the propagation of electromagnetic radiations of long wavelength (radio and radar waves).

ionospheric disturbance A sudden increase in the ionization of the D-region of the ionosphere, caused by solar flares, which results in greatly increased radio wave absorption.

ionospheric scatter The propagation of radio waves by scattering due to irregularities in the ionosphere. Syn.: ionospheric forward scatter (IFS); forward propagation ionospheric scatter (FPIS).

IP message A message carried by a message handling system giving information generated by and transferred between Interpersonal Messaging User Agents.

IR drop Voltage drop across a resistance of R ohms when a current of I amperes flows through it.

iris A restriction of the opening into a waveguide which affects its susceptance properties.

iron, anchor Galvanized steel loop built into the wall of a manhole and used to pull cables into ducts. Also called an anchor ring or pulling-in iron.

iron, soldering Copper-tipped tool used to solder contacts together.

irradiance Radiant power incident per unit area upon a surface, expressed in watts per square meter. Syn.: power density

irradiation Exposure of material to radiation, usually at high energy, to modify its molecular structure.

irregularity, impedance Point at which the impedance of a transmission line changes abruptly, due to mismatch or fault.

ISC Team Bell System abbreviation used in Universal System Service Orders: Intercompany Services Co-ordination Plan: an interdepartmental group established for co-ordinating the ISC orders. The control ISC team is responsible for the overall co-ordination of the ISC order process. Local control ISC teams are designated for other ISC areas.

isochronous 1. Of a periodic signal, the time interval separating any two corresponding transitions which is equal to the unit interval or to a multiple of the unit interval. 2. Pertaining to data transmission in which the corresponding significant instants of two or more sequential signals have a constant phase relationship.

isochronous burst transmission A transmission process which may be used where the information bearer channel rate is higher than the input data signaling rate. The binary digits being transferred are signaled at the digit rate of the information bearer channel rate, and the transfer is interrupted at intervals in order to produce the required mean data signaling rate. The interruption is always for an integral number of digit periods.

isochronous data channel A communication channel capable of transmitting timing information in addition to data. Sometimes called a "synchronous" data channel.

isochronous distortion The difference between the measured modulation rate and the theoretical modulation rate in a digital system.

isochronous modulation Modulation (or demodulation) in which the time interval separating any two significant instants is theoretically equal to the unit interval or to a multiple of the unit interval.

isolated ground (ISG) A ground circuit that is isolated from all equipment framework and any other grounds, except for a single-point external connection.

isolation Condition which prevents interaction between circuits.

isolator Microwave device which passes power in one direction only and absorbs any power received in the reverse direction.

isolator, ringer Device used on party lines which, in effect, isolates such lines from ground during ringing to improve selectivity of such calling.

isolator, switching A magnetic switch which enables radio signals entering a waveguide to be allowed in or blocked.

isotropic With the same properties in all planes and directions.

isotropic antenna A hypothetical antenna that radiates or receives equally in all directions.

IT product Factor indicating the effect of a power supply system on a telephone system: I (the rms current in amperes,) times T, the telephone influence factor (TIF).

Italcom Italian joint marketing company for GTE, SIT, Telettra.

Italtel Italian telecommunications manufacturing company (Societa Italiana Telecommunicazioni, SIT).

I²R loss Power lost by the heating effect of current passing through resistance.

iterative Repeating an infinite number of times.

iterative impedance In a pair of terminals of a four-terminal network, the impedance that will terminate the other pair of terminals in such a way that the impedance measured at the first pair is equal to the terminating impedance. The iterative impedance of a uniform line is the same as its characteristic impedance.

iterative loop A repeated group of instructions in a routine.

J-K flip flop A type of bistable with more rigid control than an ordinary flip-flop.

jack A device into which a plug is inserted in order to make electrical contacts.

jack, answering Jack on a manual operating position on which an incoming circuit is terminated and into which a plug must be inserted to answer a call.

jack, cable drum A device capable of lifting heavy drums of telephone cable so cable may be drawn off for laying. Also called cable jacks.

jack, line Jack on a manual operating position where a subscriber's line is terminated.

jack, out-of-service Jack associated with a circuit into which a special plug may be inserted to take a particular circuit out of service.

jack, patching Jack used to make temporary circuit or equipment rearrangements.

jack, pin A small single-contact jack used for temporary circuit connection.

jack, pipe-pushing A heavy duty device, sometimes hydraulically operated, used to push pipes across roads without excavating trenches. Also called a thrust borer.

jack, pole Heavy duty lifting device which can pull a pole straight out of the ground without excavating around it.

jack, test A jack providing access to a circuit for test purposes.

jack-ended A circuit which is terminated on a jack, a switchboard or a manual operating position.

jacket The outer sheath of a cable, or the outer coating of high-density polyethylene on an armorless or light-weight submarine cable; also a coating of an appropriate elastomer on individual armor wires, such as polyvinyl chloride or neoprene, as an erosion inhibitor.

jacks, cable Heavy duty lifting devices used to lift full reels of cable by a central axle, permiting cable to be fed into duct or trench.

jacks, primary Test jacks in a toll testboard giving access to line and equipment.

jacks, secondary Test jacks in some older types of toll testboards giving monitor access to toll circuits and the ability to test phantom circuits.

jacks, telegraph turnover Pair of jacks with the tip of one connected to ring of the other which can be plugged into telegraph circuits for testing purposes.

jacks, test Jacks used for test access.

jammer A radio transmitter used to jam other emissions.

jamming Deliberate emission of electromagnetic radiation with the object of obstructing the use of radio or transmission equipment.

jar The container for a storage cell or for a Leclanche type primary cell.

jeopardy A Bell System term used in Universal System Service Orders. A condition resulting from any change in the rate of accomplishing scheduled activities which could cause the order to be completed later than the scheduled work completion date.

jet-bedding The process of fluidization of the soil of the sea-bottom by pumped water jets, to imbed submarine cable and repeaters into the sea floor.

jitter 1. Abrupt and spurious variations in a signal, such as in interval duration, amplitude of successive cycles, or in the frequency or phase of successive pulses. 2. Short-term variation of the significant instant of a digital signal from its ideal position in time (wander is long-term variation of this).

jitter, pulse Variation of pulse spacing in a normally regular pulse train.

job A set of data that collectively constitutes a unit of work to be done by a computer.

job management operations system (JMOS) AT&T's computerized system for controlling outside plant construction projects.

job transfer and management (JTAM) Protocol in a data network which allows transfer of jobs between machines.

jockey wheel On a submarine cable ship, a small wheel used to maintain the cable firmly in the groove of the sheave.

Johnson noise Electromagnetic noise emitted from dissipative bodies.

joint 1. The connection of one conductor to another, or of one conductor to a terminal. 2. The joining of one complete cable to another including the interconnection of conductors and the closure of the overall sheath; a cable splice. 3. Submarine cables: the union of the center conductor and the dielectric of two sections of core.

joint, high resistance An imperfect joint between conductors which can produce voltage drop, circuit noise, heat, or intermittent failure.

joint, insulating A joint between metal-sheathed cables in which the conductors are all jointed through but the outer sheaths are jointed together using a non-conducting material. This provides a gas-tight cover, but does not provide a path for electricity from one to the other, along the sheaths.

joint, pressure Joint between conductors made by a crimped sleeve, usually bronze.

joint, rolled A method of joining conductors on an end to end basis using a bronze sleeve which is tightly compressed on the wires by rollers.

joint, rosin A badly soldered joint, for example between a conductor and tag on a connecting block, in which there is no reliable electrical contact but the wire is held in place by the glue rosin; a dry joint.

joint, rotating A waveguide joint which can be rotated without loss of energy.

joint, soldered A joint, such as between a conductor and a connecting strip tag, in which solder has been used to hold the two firmly together and to provide a good electrical connection.

joint, twisted Joint between two light gauge conductors made by twisting the conductors together in a crankhandle twist. Syn.: hand twist

joint, Western Union The traditional method of jointing line wires by twisting conductors together and wrapping the wires around each other.

joint, wiped A union between lead-sheathed cables made by plumbing a sleeve over the joint and using molten solder to provide a gas-tight, continuous sheath.

joint access costs Costs associated with those parts of the telephone network which are used both for local and long distance calling.

joint use The shared use of poles or trenches by different utilities.

Josephson effect An effect which occurs at very low temperatures when an insulating material is introduced into a superconducting material: a current flows across the junction even though there is no applied voltage.

joule The SI unit of energy, work or quantity of heat. It is the work done when a force of one newton is applied over a displacement of one meter in the direction of the force.

Joule's law The rate of production of heat in an electric circuit which is proportional to the product of the resistance by the square of the current.

juice Slang for electric current or power, e.g., turn on the juice.

Julian date 1. A chronological date in which days of the year are numbered in sequence. For example, the first day is 001, second is 002. Last is 365 (or 366 in a leap year). 2. For some specialized international purposes Julian day numbers have an origin taken as noon on January 1, BC 4713: on this count January 1, 1982 would be day number 2,444,606.

jump A class of instruction which causes a software program to move forwards or backwards to a specific location.

jumper A semi-permanent cross-connection wire on a main or intermediate distribution frame or other cross-connecting point.

junction 1. A circuit connecting two local telephone offices/exchanges, or a local office and a tandem or trunk office/exchange. (The word junction is widely used outside North America where toll circuit is used in North America). 2. Plane of transition between regions in a semiconductor. 3. Joint between waveguide sections. 4. The intersection of two pole routes.

junction box Steel box inside which low-voltage power cable joints are made.

junction diode Semiconductor diode where rectifying action occurs at the junction between p- and n-type material.

junction field-effect transistor (JUGFET) Devices in which the current flow is modulated by an applied field.

junction frequency (JF) The highest frequency that can be propagated in a particular mode between specified terminals by ionospheric refraction alone; on an oblique ionogram it is the frequency at which the high and low angle rays merge into a single ray. Also called classical MUF (maximum useable frequency).

junction network The network whose nodes are local exchanges and junction exchanges interconnected by junction circuits (in the U.K.).

junction transistor The most common type of transistor, in which current flow depends on both the majority and minority carriers, with two junctions between n-type and p-type material, giving a sandwich form, either n-p-n or p-n-p. The three external connections go to each of the sections of the sandwich, the base to the center, collector to one end, emitter to the other end. Transistors of this type have now almost completely replaced thermionic valves in amplifiers and radio receivers.

junctor The interface equipment at the end of any inter-office circuit or intra-office trunk which provides circuit and signaling compatibility.

justifiable digit time slot A digit time slot in PCM multiplexing which may contain either an information digit or a justifying digit. Also known as a stuffable digit time slot.

justification A process of changing the rate of a digital signal in a controlled manner so that it is synchronized with a rate different from its own inherent rate, usually without loss of information. Syn.: pulse stuffing.

justification ratio The ratio of the actual justification rate to the maximum justification rate. Also known as the stuffing ratio.

justification service digits Digits which transmit information concerning the status of the justifiable digit time slots. Also known as stuffing service digits.

justify To adjust words on a printed page so that one or both margins are regular.

justifying digit A digit inserted in a justifiable digit time slot when that time slot does not contain an information digit. Also known as a stuffing digit.

jute Fiber used in ropes. When saturated with tar it is used as a protective wrapping around cables for direct burial in trenches.

jute bedding In submarine cables, one or more layers of jute yarn between the outer conductor and the armor layer to protect the coaxial during cable armoring and handling.

jute serving In submarine cables, a layer of jute yarn on top of the armor to give more corrosion protection to the armor wires, to resist any tendency of displacement of the wires during handling or laying, and to reduce slippage of cable on the cable drum.

K In the metric world, K means 1000, a Km is 1000 meters. In the computer world, K means 2^{10} or 1024 in ordinary decimal figures. Also stands for Kelvin.

K-factor 1. In tropospheric radio propagation, the ratio between the effective and the actual earth radius. 2. In ionospheric radio propagation, a correction factor applied in calculations involving a curved

layer: it is a function of distance and the real height of reflection.

Ka band Microwave frequencies approximately in the 12- to 30-GHz range.

karnaugh map Diagrammatic method of simplifying Boolean equations.

Kelvin The Kelvin is the base SI unit of thermodynamic temperature. It is the fraction 1/273.16 of the thermodynamic temperature of the triple point of water. In practical terms absolute zero (0 K) is $-273°$ Celsius, water freezes at 273 K, water boils at 373 K under standard pressure conditions. The degree Celsius is now defined as an interval of 1 Kelvin, rather than 1° Kelvin.

kelvin bridge A measuring bridge developed from the Wheatstone bridge, for the measurement of very low resistances.

Kendall effect A spurious pattern or other distortion in a facsimile record caused by unwanted modulation products resulting from the transmission of a carrier signal and appearing in the form of a rectified baseband that interferes with the lower sideband of the carrier. This occurs principally when the single sideband width is greater than half the facsimile carrier frequency.

Kennelly-Heaviside layer The ionospheric E layer.

Kerr cell A Kerr effect shutter in which the optical properties of a material are changed by the application of an electric field.

Kerr effect The direction of polarization of plane polarized light through a refractive material can be rotated by the application of an electric field: this can result in a light beam being interrupted or controlled by the applied field. A related effect is the Kerr magneto-optic effect, which occurs when plane polarized light is reflected from the poleface of a strong magnet, and the beam is elliptically polarized by the reflection.

kettle, melting A double pan with boiling water in the outer pan used to melt insulating compound.

key 1. A switch with several contacts used on switchboards, which has an operating handle protruding through the key shelf or escutcheon. 2. A key signal. 3. A manual switch used to send code signals on a telegraph system. Also, causing a transmitter signal to vary in amplitude or frequency. 4. One or more characters within or attached to a set of data that contains information about the set, including its identification. Also called a tag or label.

key, collect Key controlled by a pay station operator which deposits coins in the pay phone's cash container.

key, crypto A pseudo-random number signal which is combined with an information signal to produce a new signal which may be transmitted with minimal risk of unauthorized persons being able to break down its security.

key, function A key on a teletypewriter which controls an operation but does not cause a character to be printed.

key, lever   Common type of telephone equipment key with a lever projecting above an operating position key-shelf and contacts below the level of the shelf made or broken by the movement of the key.

key, monitoring   A key in a cord or position circuit which permits a telephone operator to hear established conversations but does not enable the operator to speak to either party.

key, pole   A steel plate placed alongside the buried base of a pole to give the pole more stability.

key, position transfer   Key on telephone operating position which interconnects adjoining positions so that one operator may use both positions.

key, push-button   A key operated by pushing in a button; locking versions need a second push to return contacts to an unoperated position.

key, refund   Key depressed by an operator controlling a pay station to return coins to the caller.

key, ringing   Non-locking key used by an operator at a manual operating position to send a current out to ring the called subscriber's bell.

key, splitting   Key on a manual operating position which enables an operator to split a cord circuit allowing separate connections with the parties on both the answer and called side of the cord circuit.

key, talk   Key which connects an operator's telephone set with cord circuits on a manual operating position.

key, talk-ring   A two-position key in cord circuits in an operating position; in the locking position the operator may talk to both parties, in the non-locking position the cord circuit rings the called subscriber.

key, telegraph   A Morse key; a hand-operated key which can send Morse code telegraph signals out to line or to a radio transmitter.

key, transfer   A position transfer key.

key, two-forward-motion   A lever key which has two operating positions on the same side of the unoperated vertical position, i.e., the key is pushed forward 30° for one set of operations and further forward to about 60° for the second set.

key change   The changing of a crypto key to a different version.

key-ended   Circuits which are terminated on a key rather than on a jack. Operation of the key connects the operator to the circuit concerned.

key map   A map providing an index to larger scale maps on which details are given of duct routes and distribution cable routes. A key map is usually required to show boundaries of areas to be served from individual cabinets or flexibility points, together with central office/exchange service area boundaries.

key-pulse   To transmit information using a keyset as opposed to dialing at 10 impulses per second.

key pulsing   A system of sending telephone calling signals in which the digits are transmitted by operation of a pushbutton keyset.

key pulsing keys   (KP) Pair of keys on an operating position (particularly in North America) which associate the operator's keyset with either the answer side or the calling side of cord circuits. In international signaling, there are two KP signals: one for terminated calls and one for transit calls.

key pulsing signal   Indication light which glows when KP key has been operated on a position.

key pulsing start key   Key that triggers transmission to line of multifrequency code pulses for all digits of a telephone number that has been keyed in by the operator.

key punch   A machine with an alphanumerical keyboard used to punch coded holes into data cards.

key signal   A pseudo-random signal used in enciphering and deciphering procedures.

key system   A local telephone system in a small office complex or home providing immediate access to all users by pressing one or two keys. All users may obtain access to lines on the public network and may communicate with each other without needing the services of an operator.

key telephone set   A telephone used in a key system.

key variable   A digital word used to configure cryptographic key generators to generate the same pseudo-random key stream for their associated communicating terminals.

key variable generator   A device which produces electronic keying variables for communications security (COMSEC) equipment.

key word search   Use of a key word or a logical combination of key words to help a user to retrieve information (e.g., from a videotex system) by narrowing down the extent of the search for the information required.

keyboard   1. Set of alphanumeric and functional keys as on a teletypewriter or a key punch. 2. A manual coding device which employs key depression as a means of generating the required code elements.

keyboard send-receive (KSR)   A teletypewriter with a keyboard for manual sending but no automatic sender.

keyer   Interface device in a radio transmitter which accepts input from a signal key to control the output of a transmitter.

keying   1. Transmitting information by operating a key. 2. The process of converting data into a machine-readable form. 3. To form signals for transmission representing the data which is to be sent as a message.

keying, frequency-shift   Method of transmitting digital signals, using one frequency for "space" and a different frequency for "mark."

keying, phase-shift   Method of transmitting digital signals, using different phases of signal elements for different code meanings.

keypad   An input device with a limited number of keys.

keyphone   User-controlled local switching system relying on manual key depression at the instrument to control the selection functions.

keys, interlocked   Mechanically linked keys or pushbuttons which operate so that when one is pressed any

169

other which had previously been pressed will return to its normal position.

keysender A type of conversion set used to convert digital signaling from fast MF to 10 ips before being passed into the registers of the switch itself in an electromechanical central office.

keyset Set of pushbutton keys used for sending routing digits and telephone address. Some instruments are equipped with only ten numerical buttons: 0 and 1—9, while others have two additional buttons, the star and square, and others have four extra buttons to indicate priority precedence: FO (flash override), F (flash), I (immediate), and R (routine). Keysets used on operating positions have additional buttons used to control fast signaling systems.

keyshelf The horizontal position of a manual operating position on which keys, supervisory lamps and dial or keyset are mounted.

killer, noise A wave-shaping network added to a communications circuit to reduce interference caused by other nearby signals.

killer, spark A quench circuit, usually a capacitor and resistor in series, connected across a pair of contacts to reduce sparking.

kilo Prefix meaning one thousand.

kilobit One thousand bits.

kilobits per second (kbit/s) A bit rate expressed in thousands of bits per second.

kilocycle/s One thousand cycles per second, a kilohertz.

kilogram The base SI unit of mass, equal to the mass of the international prototype of the kilogram.

kilosegment Practical charging unit for packet switched data: 1000 segments (a maximum of 512,000 bits).

kilostream British Telecom name for digital transmission service at 64 kbit/s.

kilovar One thousand volt-amperes.

kilovolt One thousand volts.

kilowatt One thousand watts.

kinescope recording A film recording made by a motion picture camera designed to photograph a television program directly from the front of a television tube. Often called a "kine."

kingwire Copper-clad steel wire used as a central core around which optic fibers are wound helically in a type of submarine cable used for long distance circuits.

kit, coinbox conversion A kit to make coinboxes more difficult to vandalize or to enable tariff changes to be introduced.

klystron A family of electron tubes which function as microwave amplifiers and oscillators. Simplest in form are two-cavity klystrons, in which an electron beam passes through a cavity which is excited by a microwave input, producing a velocity modulated beam which goes through a second cavity a precise distance away which is coupled to a tuned circuit, thereby producing an amplified output of the original input signal frequency. If part of the output is fed back to the input an oscillator can be the result.

klystron, reflex Klystrons with only one cavity, the action is the same as in the two-cavity unit but the beam is reflected back into the cavity in which it was first excited, after being sent out to a reflector, so the one cavity acts both as original exciter or buncher and as the catcher from which the output is taken. An efficient microwave oscillator.

Kn band Microwave frequencies approximately in the 10- to 12-GHz range.

knee In a response curve, the region of maximum curvature.

knife, cable sheath splitting Heavy knife used to cut lead-sheathed cable sheaths and sleeves.

knife, linesman's Heavy duty knife used for removing insulation or sheath from cable.

knife edge effect The transmission of radio signals into the line-of-sight shadow region caused by the diffraction over an obstacle, e.g., a sharply defined mountain top.

knife switch A type of switch often used in power control circuits, with large fixed contacts and a hinged blade which 'knifes' its way between two fixed plates to make good contact over a comparatively large surface area.

knitted armor wire For submarine cables: the covering of the armor wires with cotton or nylon fibers.

knob 1. A porcelain insulator used for attachment of wire to a pole. 2. Knob on the end of a control shaft.

knob, two piece Pair of small rectangular insulating blocks with grooves which can hold one-pair or two-pair local cabling, together with screws which fix the pair of blocks (and the cabling) to wall or skirting board.

knocked down Not assembled: shipped for local assembly.

knockout Discs partially punched out on the side or base of a terminal or junction box; one is knocked out by hammer appropriate to the size of cable being fed in to the box

KvT product Product of number of kilovolts and the telephone influence factor (TIF): this gives an indication of possible interference to an audio circuit by a power route.

L band Microwave band with frequencies approximately 1 GHz.

L group A 48 kHz bandwidth transmission path between two points (usually a standard 12 channel group), with standard terminal frequencies of 60 to 108 kHz.

L supergroup A 60-channel (five group) FDM carrier system routed on a coaxial cable path.

label 1. One or more characters within or attached to a set of data that contain information about the set, including its identification. 2. In communications, information within a message used to identify specific matters, such as the particular circuit to which the message is related. 3. In computer programming, an identifier of an instruction.

lace To arrange cables in neat bundles, carefully laced together with nylon or lacing twine.

lacing cord, nylon Narrow ribbon of nylon used to lace cables in offices.

lacing twine Waxed linen cord used to lace cables in offices.

lacquer Resin applied to ends of cable with fiber insulants to prevent these from fraying or absorbing moisture.

ladder, manhole A light-weight galvanized steel ladder intended to be left in underground manholes to provide access for cable splicers and electrolysis testers.

ladder, rolling A ladder which rolls on tracks between racks of telephone equipment, to enable staff to reach units at the tops of racks. Most have spring-controlled brakes so the ladder cannot be moved when in use.

ladder network A type of filter with components alternately across the line and in the line.

laddic A lattice type network of magnetic core units which resembles a ladder with cores in the sides and on the rungs.

lag 1. The difference in phase between a current and the voltage which produced it, expressed in electrical degrees. 2. The delay between the arrival of a signal and the response to it. 3. To follow behind, as a current wave follows a voltage wave in an inductive circuit.

lagging The outer protection for large diameter cables placed on reels for shipping, consisting of boards parallel to the reel's axis fitted between the peripheries of the flanges of the reel.

lagging current A current which lags behind the alternating electromotive force which produces it; a circuit which produces a lagging current is one containing inductance alone, or whose effective impedance is inductive.

lagging load A load whose inductive reactance exceeds its capacitive reactance; when an alternating voltage is applied the current lags behind the voltage.

Lambertian emitter An optical source which has a radiance distribution that is uniform in all directions of observation.

Lambertian radiator See Lambert's cosine law.

Lambertian reflector See Lambert's cosine law.

Lambertian source See Lambert's cosine law.

Lambert's cosine law The radiance of certain idealized surfaces, known as Lambertian radiators, Lambertian sources, or Lambertian reflectors, is independent of the angle from which the surface is viewed. The radiant intensity of such a surface is maximum normal to the surface and decreases in proportion to the cosine of the angle from the normal. Syn.: cosine emission law

laminate Material made by layers of the same or different materials bonded together and built up to the required thickness.

lamination Thin stamped-out sheets of iron, each one separated from its neighbor by a layer of varnish; used as transformer cores. Laminations minimize eddy current losses.

lamp, answer A supervisory lamp on the answer side of a cord circuit.

lamp, ballast A lamp in series with load which provides almost constant current for the load.

lamp, busy A supervisory lamp on the face equipment of a switchboard indicating that a particular circuit is busy.

lamp, call-waiting In a call distribution system, an indication given of the number of calls waiting for allocation to an operator.

lamp, dial pilot A supervisory lamp which stays lighted while an operator is dialing.

lamp, fuse alarm A lamp which shows that a fuse in the indicated panel or bay has blown.

lamp, group busy A lamp on a manual operating position or switchboard which lights to indicate when all circuits in a particular group are busy.

lamp, hold A lamp which remains alight while a connection is being held.

lamp, line A lamp on a manual operating position or switchboard which indicates an incoming call on a particular line.

lamp, message waiting A lamp on a telephone instrument or an associated control panel to indicate to a hotel guest that there is a message being held.

lamp, neon A small lamp which produces a red glow due to ionization of neon gas between two electrodes in the glass envelope of the lamp. Used as an indicator lamp and in other specialized uses.

lamp, pilot A general warning light indicating that individual equipment alarms need attention. At least one of the alarms must have been activated in order to switch on the pilot lamp.

lamp, precedence busy A special indicating lamp to show that a circuit is engaged for a priority call.

lamp, resistance A tungsten lamp used in series with load as a current-limiting resistance.

lamp, slide-base A switchboard lamp with a slide base.

lamp, supervisory Indicating lamps associated with each side of a cord pair which show if the call has been completed.

lamp, switchboard A small incandescent lamp with a slide base (push in, pull out) used for switchboard lamp signals.

lamps, cord The supervisory lamps associated with cord circuits.

land cable The cable portion of a seacable system installed on land.

land mobile service Radio communication service to and from mobile stations (e.g., in automobiles, or hand-held).

landing point The name of the place where a seacable or submarine cable comes ashore.

landline A circuit which connects two ground locations, in wire or cable.

Langley The monthly average of daily total solar and sky radiation on a horizontal surface, in calories per square centimeter, measured on the International Pyrheliometric Scale. Sometimes called 'flat Langleys' to distinguish them from 'corrected Langleys,' which are the Langleys for the location multiplied by a tilt factor appropriate to the latitude and the angle proposed to be made with the horizontal by the particular solar panel array being planned. Langley values are recorded and published for most areas of the world and are used in calculating the size and type of solar panel needed to provide enough power for the installation concerned.

language A set of symbols, characters, conventions, and rules used for conveying information.

language, machine Instructions written in a language which is directly comprehensible to a processor or computer.

language, plain Intelligible text or language, not coded or encrypted.

large-scale integration (LSI) Complex circuits with 16k bits or more of memory capacity or functions included in a single chip.

larynx, artificial An experimental oscillator-type device which enables persons without vocal cords to use a telephone.

laser A device that produces optical radiation using a population inversion to provide light amplification by stimulated emission of radiation and an optical resonant cavity to provide positive feedback. Laser radiation may be highly coherent either temporally or spatially, or both. Lasers produce light in the visible or infra-red or, less commonly, the ultra-violet region of the electromagnetic spectrum. In effect atoms are excited, either by electric discharge through a gas or by current through a solid state diode, and transitions in energy levels result in a stream of photons being emitted. Some lasers are pulsed, some are continuously operated. Laser beams can be modulated by very broadband input signals and are used in optic fiber communications links (typically at frequencies of 8 Mbit/s, 35 Mbit/s and 140 Mbit/s) as pulsed sources of light.

laser diode Semiconductor diode which emits coherent light above a threshold current (stimulated emission).

laser head Module containing the active medium, resonant cavity and other directly related components, all within one enclosure.

laser medium A synonym for active laser medium.

lash To fasten together by lashing, as with an aerial cable lashed to its supporting messenger strand.

lashed cable Aerial cable lashed to messenger strand, usually by a long pitch helical lashing wire.

lasher, cable Tool which is pulled along an aerial cable messenger strand and lashes the cable to the strand as it moves.

lashing A fastening, particularly with an aerial cable.

lasing threshold The lowest excitation level at which a laser's output is dominated by stimulated emission rather than spontaneous emission.

last choice circuit-group A circuit group which is not traffic engineered as a high usage group. It may be a final or only route circuit group.

last in, first out (LIFO) A method of storing and retrieving data in a stack, table or list.

last number redial Facility provided by modern telephones or key sets which redials a complete subscriber number on the operation of a simple code.

last trunk busy (LTB) Condition when the last of a group of circuits is busy. If the group has automatic alternative routing, an attempt is made automatically to route the call another way.

lastor Laser read/write optical storage which uses semiconductor lasers to digitally code information on plastic composite materials: typical credit card stripes could store up to 1 million characters using lastor techniques, or one optical disc about the size of a floppy disc could store about 10^9 characters.

latch, hookswitch Latch which enables the cradle switch of a party line telephone to operate in two stages: the first stage permits monitoring but does not interfere with established calls; the second stage is a normal complete off-hook.

latency Synonym for access time to a processor's memory, including time taken to obtain information from a back-up file or backing store.

lateral 1. Toward the side. 2. A lateral cable. 3. A lateral conduit.

lateral offset loss A power loss caused by transverse or lateral deviation from optimum alignment of source to optical waveguide, waveguide to waveguide, or waveguide to detector. Syn.: transverse offset loss

lateral routing In a switching plan, an acceptable routing which is not one of the normal logical forward routing patterns.

lattice-wound Method of winding a coil to reduce capacitance; also called honey-comb-wound.

launch 1. To lift from Earth into orbit. 2. To transfer energy from a feeder or waveguide to an antenna.

launch, satellite establishment Placing of a satellite in position for the establishment of a service.

launch, satellite replenishment Placing of satellites in position to replace those which are time-expired or no longer effective.

launch angle The angle between the light input propagation vector and the optical axis of an optical fiber or fiber bundle.

launch numerical aperture (LNA) The numerical aperture of an optical system used to couple (launch) power into an optical waveguide. LNA is one of the parameters that determine the initial distribution of power among the modes of an optical waveguide.

launcher Launching horn for a waveguide.

launching fiber A fiber used in conjunction with a source to excite the modes of a fiber in a particular fashion; most often used in test systems to improve the precision of measurements. Syn.: injection fiber

Law Enforcement Teletypewriter Service (LETS) A private teletypewriter service linking police stations and offices.

lay The distance along a cable which it takes for a pair of wires or a conductor to make one complete twist.

lay, direction of The direction (clockwise or counterclockwise) taken by a conductor or group of conductors in spiralling around a cable core.

layer A collection of related network processing functions that comprises one level of a hierarchy of functions; e.g., seven layers in OSI reference model for data transmission.

layer, cable A self-powered cable laying machine, which is basically a tractor pulling a vibrating plow and carrying a large reel of cable. As the tractor drives forward the cable is fed into a trench opened up by the plow.

laying effect Change of submarine cable attenuation caused by mechanical stress during the laying operation.

layout A proposed or actual arrangement or allocation of equipment.

layout, current facility Record of toll transmission facilities available for use in circuits.

layout, patched Temporary circuit brought into service using patch cords.

LC circuit A circuit with inductance and capacitance, providing tuning for particular frequencies.

LC ratio Ratio of inductance to capacitance.

lead 1. A lead-in, linking equipment. 2. The angle by which voltage leads current or vice versa. 3. The angle between the vertical plane of a submarine cable being laid or picked up and the lubber line of the cable ship.

lead, test Insulated wires with test clips used to make temporary connections while tests are being carried out.

leader The blank section at the beginning of a tape.

leader stroke In lightning, the first stroke, which usually determines the path to be followed by the following or return stroke where most of the energy is carried.

lead-in A single wire joining an antenna with its radio transmitter or receiver.

leading edge The initial portion of a pulse or wave, in which voltage or current rise rapidly from zero to a final value.

leading load A reactive load in which the reactance of capacitance is greater than that of inductance; current through such a load leads the applied voltage causing the current.

leads, E & M DC signaling control leads which activate (and are activated by) trunk signaling equipment on long distance currents. The M lead normally sends forward battery or ground, and E, the incoming lead, normally receives open or ground.

leak A low insulation condition which results in voltages falling and attenuation increasing,

leakage Shunt resistance through which a leakage current drains, or magnetic flux other than that on a useful path.

leakage, magnetic Magnetic flux which wanders from its useful working path.

leakage resistance Resistance of a path through which leakage current flows.

leaky modes In an optical waveguide, those modes that are weakly bound to the core of the waveguide and have comparatively high loss as a result of tunneling.

leap second A time step of one second used to adjust coordinated universal time to ensure approximate agreement with international universal time. An inserted second is called a positive leap second and an omitted second is called a negative leap second.

leapfrogging CATV operators' practice of skipping over one of more of the nearest TV stations to bring in a farther signal for more program diversity on the cable. FCC rules establish priority for carrying stations that lie outside the cable system's service area.

leaseback The practice by telephone companies of installing and maintaining CATV distribution systems and leasing the facilities back to separate contractors for operation of the system.

leased circuit data transmission service A service whereby a circuit of the public data network is made available to a user for exclusive use. Where only two data circuit-terminating devices are involved it is a point-to-point facility; where more than two are involved it is known as a multipoint facility.

leased line Any circuit (typically supplied by and rented from a local telecommunications administration or telephone company) designated to be at the exclusive disposal of a given subscriber: a private wire.

leased network A data network using circuits or channels leased from a PTT or major telecommunications carrier and dedicated to use solely by the lessee.

leased service The arrangement whereby a user contracts for the exclusive, and generally continuous, use of a circuit or facility.

least-cost routing Function carried out by specialized carriers (and by some PABXs) by which outgoing calls are routed over the lines which at that particular time provide the lowest cost circuits, e.g., over WATS lines or OCC lines rather than by normal DDD.

least-replaceable unit A component or assembly which is the unit replaced when a fault condition makes repair or replacement necessary.

least significant bit (LSB) In binary, refers to the bit representing the lowest power of 2.

least time principle See fermal principle.

Lecher wires Parallel wires on which radio frequency standing waves are established, used to measure signal wavelength.

Leclanche cell A primary cell consisting of a carbon rod anode and a zinc cathode. A depolarising agent surrounds the anode and is enclosed in a porous pot or fabric bag, the whole being immersed in an ammonium chloride electrolyte. If only intermittent demands are made on this cell, it will operate a long time. The Leclanche's descendant is the common dry cell or torch battery, which operates on exactly the same principles.

lecture call An established connection between one caller and two or more parties in which the speech path is used in a unidirectional way from the caller to the other connected parties. The call is set up either by an operator or by an automatic device programmed by the caller.

left hand rule See Fleming's Rule.

left-in Equipment which is left in place when a telephone is disconnected and recovered.

leg 1. A circuit of a network between the serving office and operating station or between central offices. 2. One of the two wires in a pair, e.g., A-leg and B-leg.

leg, balanced telegraph Method of connecting local lines out to teletypewriters, with the positive battery at the machine and the negative battery for a mark signal from the central office repeater, positive battery for a space signal. This combination gives efficient operation at teletypewriter speeds.

leg, butt Leg nearest the main part of a cable form.

leg, composite Terminal on a composite set connected to telegraph equipment.

leg, dial Terminal which is to be brought out to a loop-disconnect dial for signaling.

leg, telegraph The dc path out to a subscriber's telegraph equipment.

leg, tip Leg farthest from the main body of a cable form.

length, electrical The length in wavelengths of any circuit or component for which physical dimensions and their relationship with frequency are important.

length indicator A six bit field in common channel signaling which differentiates between message signal units, link status signal units and fill-in signal units and, in the case that its (binary) value is less than 63, indicates the length of a signal unit.

lens, microwave The radio equivalent of an optical lens: dielectric material is used to focus a radio frequency transmission.

lens antenna A microwave lens placed in front of an antenna in order to improve directional qualities

letter In man-machine language, a character of the character set representing the alphabet.

letters shift The key or code which means that the following characters are alphabetical.

level 1. Number of bits per character in a code. 2. The difference between measured signal strength and a standard, expressed in dB.

level, crosstalk Power of crosstalk, compared with a standard signal.

level, decision Predetermined signal amplitude which serves as a standard for determining the value of a signal pulse. If signal amplitude is above the decision level at the time of sampling a binary 1 is usually indicated, with binary 0 usually indicated when the amplitude is lower then the decision level. Ideally a signal pulse is measured at its time center. Also called decision threshold and slicing level.

level, group selector Any of the switches at the same level of selection or which serve the same groups of numbers. In a step-by-step office levels are the same as the actual bank levels to which the switches step, for example, level 2 is accessed by dialing 2; in a common control or SPC office this does not apply.

level, noise The volume of noise energy, expressed in decibels above a reference level. Also see dba, dbrn, dbrnc, and noise.

level, peak Maximum sound or signal level. Volume unit (VU) meters are used to measure signal peaks in program circuits.

level, precedence Scale of relative priority given to messages. One common scale, in descending order of importance:

FO	flash override	P	priority
F	flash	R	routine
I	immediate	D	deferred

level, slicing Signal level at the receiver of a binary signal below which the bit is considered to be a space, above which it is taken to be a mark.

level, sound The atmosphere's pressure level caused by a sound wave, specified in decibels above a 1000 Hz sound which can just be heard by a young person.

level, speech Energy of speech measured in volume units (VU). A VU meter has carefully specified dynamic characteristics to make its readings subjectively acceptable.

level, speech interference (SIL) The SIL of a noise is the average of the sound pressure levels of the noise in three octave bands, centered on 500 Hz, 1000 Hz and 2000 Hz.

level, speech power The acoustic power in human speech. This varies from —60 dBm to +10 dBm.

level, testing Normal transmission test level, in North America: one milliwatt of power at 1000 Hz.

level, transmission The ratio of a test signal's power at one point to the test signal power applied at another point in the system used as a reference point.

level, volume Energy level measured in volume units.

level, zero 1. One milliwatt of power: the zero comes from zero dB above one milliwatt. 2. Transmission power at a reference point to which all other power measurements in the system are compared.

level alignment The adjustment of transmission levels of single links and links in tandem to prevent overloading of transmission subsystems.

level diagram A graphic diagram indicating the signal level at any point in the system.

lever, armature Lever attached to a relay armature which transmits the armature movements to the springsets where contacts are mounted.

liberalization The opening of interconnect and other equipment markets to suppliers other than the telephone company or administration.

life, flex Indication of the life which may be expected from a wire or spring exposed to repeated bending.

life, service Life expectancy under normal conditions of use.

life, shelf Life when not in service of such items as batteries. Shelf life is affected greatly by environmental conditions during storage.

life, system design Of a submarine cable system, the period over which it is reasonable to continue maintenance, or during which no unrestorable deterioration or failure of cable or submerged electronics will occur; this period is usually twenty years or more for submarine telephone cable systems.

life cycle Tests carried out to determine the probable useful life of similar equipment in normal working conditions.

life test Test in which random samples of a product are tested to see how long they can continue to perform their functions satisfactorily. A special form of testing is used, usually with temperature or current or voltage or vibration effects cycled at many times the rate which would apply in normal usage.

lifeline rates Specially low telephone rates for low-income subscribers.

light An electromagnetic radiation with wavelengths between 400 nm (violet) and 740 nm (red) propagated at a velocity of roughly 300,000 km/s (186,000 miles/s), and detected by the human eye as a visual signal. In the laser and optical communication fields, the term also includes the much broader portion of the electromagnetic spectrum that can be handled by the basic optical techniques used for the visible spectrum. This is considered to extend from the near-ultraviolet region of approximately 300 nm through the visible region, and into the mid-infrared region of 3.0 μm to 30 μm.

light, coherent Light in which waves are of a single frequency and in phase so they can be modulated.

light, fuse alarm pilot Warning light on a manual operating position or switchboard to indicate that a switchboard fuse has blown.

light, pilot General warning lamp drawing attention to a failure which will be indicated in detail by equipment alarms.

light, supervisory pilot Indicating light on an operating position which shows that one or more of the supervisory lamps on that position are alight.

light-emitting diode (LED) A pn junction semiconductor device that emits noncoherent optical radiation when biased in the forward direction, as a result of a recombination effect. Gallium arsenide is one of the most common semiconductor crystals used in LEDs. LEDs are much used in fiber optic systems where long life of the light-producing element is necessary.

light energy converter (LEC) A solar cell. A photovoltaic semiconductor device which converts light energy into electric energy.

light pen A pen-like device used as input to a processor: for example, it can be used to draw circuits onto the screen of a visual display unit in an intelligent terminal to amend data in store.

light ray The path of a given point on a wavefront. The direction of a light ray is generally normal to the wavefront.

light source A generic term that includes lasers and LEDs even though these may operate outside the visible light band.

light-weight cable Armorless submarine cable; although light in weight relative to armored cable, light-weight cable is heavier than water.

lightguide Synonym for optical waveguide.

lighthouse tube A special type of electron tube designed for use at very high microwave frequencies.

lightning Electric discharge due to flow of current from part of a charged cloud to the ground.

lightwave Electromagnetic waves in the optical frequency band and in spectral regions adjacent to visible light, e.g., the near infra-red.

limited distance modem A modem designed for operation over distances of up to about 50 km., usually over privately-owned line circuits, able to transmit signals at higher frequencies than the normal commercial speed band. Sometimes called a baseband modem or a short-haul modem.

limited protection A form of short-term communications security applied to the electromagnetic or acoustic transmission of unclassified information which warrants a degree of protection against simple analysis and easy exploitation.

limited protection voice equipment Equipment which provides limited protection for unclassified voice communications.

limiter A device in which the power or some other characteristic of the output signal is automatically prevented from exceeding a specified value. There are also base limiters which cut out outputs lower than a predetermined level.

limiter, call Device which automatically cuts off callers if their calls last longer than a permitted time.

limiter, noise Device which cuts off noise peaks above a specified level (which is usually the signal peak level).

limiter circuit A circuit of nonlinear elements that restricts the electrical excursion of a variable in accordance with some specified criteria.

175

limiting A process by which some characteristic at the output of a device is prevented from exceeding a predetermined value. 1. Hard limiting is a limiting action with negligible variation in output in the range where the output is limited (controlled) when subjected to a fairly wide variation of signal input. 2. Soft limiting is a limiting action with appreciable variation in output in the range where the output is limited (controlled) when subjected to a fairly wide variation of signal input.

limits of interference In radio transmission, the maximum permissible interference as specified in recommendations of the International Special Committee on Radio Interference or other recognized authorities.

Lincompex Linked compressor for transmission and expander for reception which is used to improve signal-to-noise ratio of radio communication links.

line 1. A device for transferring electrical energy from one point to another, such as a transmission line. 2. In facsimile or television, the scanning element. 3. The path (trace) of a moving spot on a CRT. 4. A communications channel. 5. In cryptography, a horizontal sequence of symbols or groups thereof. 6. A pole line. 7. A subscriber's line. 8. A toll line. 9. Any set of conductors with supporting poles.

line, artificial A network of components made up to simulate an actual line in impedance and frequency characteristics.

line, cable A pole route which supports only aerial cables.

line, delay A transmission line constructed to introduce delay, i.e., a line with substantial propagation time.

line, dissipation An open-wire line connected to a rhombic antenna opposite the feeder and constructed so as to dissipate the power which reaches it.

line, distortionless An ideal transmission line, with attenuation at a minimum, impedance purely resistive, and both the attenuation and the impedance independent of frequency.

line, exponential transmission A transmission line whose inductance and capacitance vary exponentially with distance; this effect can be produced by varying the spacing of conductors.

line, four-wire Two-way transmission with separate paths in the two directions. This requires two pairs (four wires) if all transmission is at audio frequencies.

line, individual A direct exchange line; a public switched telephone service for the use of one subscriber.

line, keying A remote control circuit for a radio telegraph transmitter.

line, lossless An imaginary transmission line in which no attenuation occurs because there is no series resistance or leakage path across the pair of wires.

line, lossy A transmission line designed to have a very high loss per unit length at the frequencies concerned.

line, message A multiparty selective calling telephone circuit sometimes used for minor railway control facilities.

line, multiparty A line to a central office serving more than one subscriber. In some rural areas up to 10 parties are served by the same line.

line, off-premises A PBX extension in a location separated from the location of the PBX itself.

line, open-wire A pole route carrying open-wire circuits.

line, party A central office line serving more than one subscriber. In towns, four is the usual maximum; in some rural areas up to 10 subscribers use a single line.

line, PBX tie A tie line between two PBXs, permitting extensions in one PBX to be connected to extensions on the other PBX.

line, pole A pole route; poles on which open-wire pairs or aerial cables have been erected.

line, power A pole route carrying wires or aerial cables distributing electric power to customers.

line, private A point-to-point telephone line for the private use of one party.

line, radio frequency transmission A transmission line or feeder from a radio transmitter to an antenna, designed to give minimum radiation from the feeder and maximum power transfer to the antenna.

line, rural Subscriber's line in rural areas.

line, slotted A coaxial line with slots in the outer conductor through which probes can be inserted to determine standing wave patterns and calculate signal wavelengths.

line, subscriber's A telephone circuit from a subscriber's telephone to the end office or central office.

line, surface wave A type of radio frequency transmission line made up of a single wire suspended, with few bends; it acts like the central conductor of a coaxial cable.

line, telegraph Circuit for passing teletypewriter signals.

line, telephone Outside plant which provides a telephone line between subscriber and central office, either in whole or in part by cable (underground or aerial) or by open wire.

line, toll Circuit between toll switchboards which carries toll calls.

line, toll pole (TPL) Pole route which carries toll circuits either on aerial cable or on open wire, on a physical (metallic) basis or by carrier systems.

line, transmission A circuit designed to transmit electrical energy from one point to another. A transmission line for a power system must be built for minimum loss; a line for a telephone system can tolerate large losses.

line, two-wire A transmission path using two wires for both directions of transmission.

line, unbalanced A transmission system with one leg at ground potential (as with most coaxial cable systems).

line, uniform A transmission system with identical properties per unit length, over its whole length.

line adapter A modem or interface device often incorporated in equipment.

line adapter circuit A specific circuit that is for example used at the station end of a subscriber access line and connects to the four-wire telephone.

line balance The degree of electrical similarity of the two conductors of a transmission line.

line bank The multiple banks in a Strowger type central office whose two contacts represent the line itself.

line build-out (LBO) The build-out of a line, i.e., components connected to or across the line in order to bring its impedance to a desired level.

line charge Charge levied on telephone subscribers for the use of the line and access to the central office switching equipment.

line code For PCM, a code chosen to suit the transmission medium and giving the equivalence between a set of digits generated in a terminal or other processing equipment and the pulses chosen to represent that set of digits for line transmission.

line communications Telecommunications using a physical means such as wire, cable or fiber to carry the signal.

line concentrator A switching device which concentrates traffic from a number of circuits or subscriber's lines onto a smaller number of circuits to a parent local exchange, where a similar switching device deconcentrates the traffic to the original number of lines. In the case of subscriber's lines, the correspondence of the lines before concentration and after deconcentration must be maintained. Also called a stand alone concentrator. See also concentrator, line.

line conductors The wires running between poles along a pole route.

line extender (CATV) Type of amplifier used in a cable TV feeder system, also called distribution amplifier.

line filter A filter in a transmission line, for example, one which separates carrier frequencies from audio frequencies.

line filter balance A network which maintains phantom group balance when one side of the group is equipped with a carrier system. Since it must balance the phantom group for only voice frequencies, its configuration is usually quite simple compared with the filter which it balances.

line finder A switch in the concentration stage of a central office which searches out a line which wishes to initiate a telephone call, and connects this line through to the next stage of switching.

line frequency (TV) The number of horizontal scans per second, nominally 15,750 times per second in U.S. TV systems.

line hit Electrical interference causing the introduction of spurious signals into a circuit, or the incorrect signals themselves.

line hunting Procedure for searching a number of lines to find one which is idle and available for use.

line level The signal level, usually in dB, at a particular point.

line link A transmission path and all the associated equipment such that the bandwidth available, while not having any specific limits, is effectively the same throughout the length of the link.

line load The percentage of maximum traffic capacity of a channel during a specific time period.

line load control Selective denial of call origination capability to specified lines when excessive demands are offered to a switching center.

line loop A single connection, from a switching center or an individual message distribution point, to the terminals of an end instrument. Syn.: local line; local loop; subscriber line

line loss The total end-to-end loss in a transmission line, in dB.

line of force Imaginary line indicating the direction of the electric or magnetic field at all points along it.

line-of-sight (LOS) propagation Radio propagation in the atmosphere which is similar to light transmission in that intensity decreases mainly due to energy spreading according to the inverse-distance law with relatively minor effects due to the composition and structure of the atmosphere. LOS propagation is considered to be unavailable when any ray from the transmitting antenna, refracted by the atmosphere, will encounter the Earth or any other opaque object that prevents the ray from proceeding directly to the receiving antenna. To be usable, an LOS path must have additional clearance at least equal to the first Fresnel zone radius.

line printer A very high speed printer output device that can type text at speeds up to several hundred lines per minute.

line-route map A map or overlay for signal communication operations that shows the actual routes and types of construction of wire circuits in the field, and the locations of switchboards and telegraph stations.

line side That portion of a device that looks toward the transmission path, e.g., channel, loop, or trunk.

line signal A signal indicating that a particular line is carrying an incoming call, or a subscriber wishes to initiate a call.

line signaling A signaling method in which signals are transmitted between pieces of equipment that terminate and continuously monitor part or all of the traffic circuit.

line source 1. In the spectral sense, an optical source that emits one or more spectrally narrow lines as opposed to a continuous spectrum. Also see monochromatic. 2. In the geometric sense, an optical source whose active (emitting) area forms a spatially narrow line.

line spectrum An emission or absorption spectrum consisting of one or more narrow spectral lines, as opposed to a continuous spectrum.

line switching The switching technique of temporarily connecting two data lines together so that the two stations may directly exchange information. See circuit switching.

line traffic coordinator (LTC)   The processor in an AUTODIN switching center designated to coordinate the line traffic.

line turnaround   Time required to reverse direction of transmission for a half-duplex channel. Syn.: turnaround time

line-up   To adjust transmission parameters to bring a circuit up to its specified values, such as attenuation, delay, signaling, distortion, etc.

line voltage   Voltage of public power supplies, normally 117v ac in the U.S., 240v ac in the UK.

linear   1. A device with components arranged in a line. 2. A device whose output is directly proportional to its input, i.e., there is no variation of loss or gain with frequency.

linear analog control   An analog system in which the functional relationships are of simple proportionality.

linear analog synchronization   A synchronization control system in which the functional relationships used to obtain synchronization are of simple proportionality. Syn.: linear analog control

linear combiner   A diversity combiner which adds two or more receiver outputs.

linear engine   For a submarine cable ship, a cable laying engine wherein the cable forms a straight line while going through the engine.

linear optical element or system   A system in which the radiant power output is proportional to the radiant power input, and no new optical wavelengths or modulation frequencies are generated. The proportionality constant can vary with source wavelength and modulation frequency. A linear element can be described in terms of a transfer function and an impulse response function.

linear predictive coding (LPC)   A narrowband analog-to-digital conversion technique employing a one-level or multi-level sampling system in which the value of the signal at each sample time is predicted to be a particular linear function of the past values of the quantized signal. A 64 Kbit/s voice channel can be analysed by LPC and converted into a 2.4 Kbit/s stream. This would enable four voice channels to be carried on a single 9.6 Kbit/s data line.

linear receiver   A radio receiver which operates in such a manner that the signal-to-noise ratio at the output is proportional to the signal level at the input, and/or to the degree of modulation.

linearity   A constant relationship, over a designated range, between input and output characteristics of a device.

linearizer   A negative feedback device to produce a linear output from klystron modulation.

linearly polarized (LP) mode   In optoelectronics, a mode for which the field components in the direction of propagation are small compared to components perpendicular to that direction. Each LP mode consists of several phase degenerate modes. The LP description (which is an approximation for weakly guiding waveguides) becomes more accurate as the difference between the maximum and minimum values of the refractive index becomes a smaller fraction (typically less than 2%) of the mean index value across the profile.

lineman   Person who works on external telephone plant, especially on poles and pole routes.

lineswitch   A switch in a dial office which connects a subscriber's line with the first available idle trunk in the next switching stage.

lineswitch, primary   A lineswitch directly connected to the line equipment of a particular subscriber.

lineswitch, rotary   A rotary stepping switch, one per subscriber's line, which hunts for an idle trunk whenever the subscriber goes off hook to call the central office.

lineswitch, secondary   A second lineswitch used in tandem with the first when traffic requirements make it necessary for subscribers to be given the possibility of accessing 100 outlets instead of only 10.

linewidth   The frequency range over which most of a laser's beam energy is distributed. See spectral linewidth.

link   1. The communication facilities existing between adjacent nodes. 2. A portion of a circuit designed to be connected in tandem with other portions. 3. A radio path between two points, called a radio link, which may be unidirectional, half duplex, or full duplex. 4. In computer programming, the part of a computer program, in some cases a single instruction or address, that passes control and parameters between separate portions of the computer program. 5. A small metal strap used as a removable connection between two adjacent screws on a connecting block. 6. In crossbar systems, a circuit extending between the primary and secondary selectors of a selection stage.

link, cable reinforcing   Steel rod placed where a pole line turns a corner in order to avoid bending aerial cables carried on the pole line.

link, centrex data   A data link between a central office providing centrex facilities and the offices of centrex customers which provides information to the centrex console attendant.

link, connecting   Removable metal strap between terminal screws, used where a connection is needed between these particular terminals.

link, data   A circuit carrying information in digital form.

link, fast   A special link designed to provide fast access to a register so that incoming calls may dial straight in without having to wait for a second dial tone.

link, fusible   That part of a fuse which is designed to melt when the rated current is exceeded. Fusible links within an integrated circuit chip are blown when a programmable read-only memory (PROM) is being programmed.

link, incoming register   A switch which connects incoming trunks directly to incoming registers without going through a marker stage.

link, line   A switch, particularly in crossbar offices, which connects subscribers' lines to outgoing

178

junctors, and incoming junctors to subscribers' lines.

link, radio A radio system which is normally one hop only but repeaters also could be involved.

link, terminating A toll connecting trunk from toll office to a subordinate end office.

link, trunk Switches in a crossbar office which connect originating registers to junctors.

link access procedure (LAP) Internationally specified protocol (CCITT X.25) for data-link access.

link-by-link signaling A method of signaling in which signals are passed from one end to the other of a multi-link connection by repetition at intermediate switching points.

link encryption The application of on-line crypto-operation to a link of a communications system so that all information passing over the link is entirely encrypted.

link layer The logical entity in the OSI model concerned with transmission of data between adjacent network nodes.

link level In data transmission, the conceptual level of control or data processing logic existing in the hierarchical structure of a primary or secondary station that is responsible for maintaining control of the data link.

link orderwire Voice and data communications circuits between adjacent communications facilities interconnected by a transmission link; specifically used for coordination and control of link activity.

link protocol A set of rules for data communication over a data link specified in terms of a transmission code, a transmission mode, and control and recovery procedures.

link status signal unit A signal unit (in common channel signaling) which contains status information about the signaling link in which it is transmitted.

link system A switching system where one or more intermediate links, connected in tandem, are necessary to establish a connection between a given inlet and a chosen outlet, using conjugate selection. Also called conjugate selection system, or conditional selection system.

link terminal The ground station of an earth-to-satellite link.

linkage Instructions which connect one program to another, providing continuity of execution between programs.

linkage, flux The product of the magnetic flux and the number of turns in a coil linked with this flux.

linked numbering scheme An integrated local numbering scheme pertaining to more than one central office or exchange.

linker Program which combines two or more segments of code that have been independently compiled by filling in the references between them.

liquid crystals Material of low viscosity which matches the shape of the vessel in which it is contained (like a liquid) but has different refractive indices for light, depending on the path direction of the light

through the material, like a solid crystal. Under the influence of an electric field, molecules align themselves in specific directions, changing the polarization plane and enabling digits to be made visible in display panels, watches, etc. Liquid crystals are low power consumption devices (10 to 100 μW/cm^2) giving a non-emissive display system, i.e., they do not emit light themselves but modulate ambient light so have good visibility even in sunlight.

Lissajous figures The pattern traced out on a cathode ray tube when varying quantities are fed to both the horizontal and vertical plates of the tube.

listening key A monitor key on a manual operating position or switchboard which permits the operator to listen to the conversation on that cord pair.

lister Program which lists code in a readable form.

lister, combined source and object A lister which interleaves source and object code.

listing, private Telephone service where the telephone number is not listed in the directory or with information service. Also called exdirectory.

listing, reference An alternative number, often an answering service number, sometimes listed in directories as the number to be called if there is no reply at the number itself.

listing, semi-private A telephone number not listed in the printed directory but held by information and advised to callers on request.

Litz wire A Litzendraht wire. A braided wire with individually insulated fine strands, which gives low resistance at high radio frequencies.

live 1. Connected to a source of electric potential. 2. Acoustically reverberant.

live front Equipment with live components exposed.

live room A room in which sound is not greatly absorbed by the walls and fittings: the room reverberates.

liveware The people who work on computers, with hardware, firmware and software to complete the packaged discipline.

load 1. The power consumed by a device or circuit in performing its function. 2. A power-consuming device connected to a circuit. 3. To put programs or data into a register or storage. 4. To place a magnetic tape reel on a tape drive, or to place cards into the card hopper of a card reader.

load, artificial A device which can simulate a real load and absorb energy without radiating electromagnetic radiation.

load, connected power The total rated capacities of all possible loads, i.e., the maximum load if all individual loads were switched on together.

load, demand power The maximum long-term power load.

load, matched Load with an impedance such that there is maximum transfer of power from source to load.

load capacity In PCM, the level expressed in dBmO, of a sinusoidal signal the positive and negative peaks of which coincide with the positive and negative virtual decision values of the encoder. Also known as overload point.

load control, high-traffic-day Facility by which calls which would involve switching through several different switching centers are automatically given a delay recorded message when the equipment is experiencing a high percentage of all-trunks-busy conditions.

load factor The ratio of the average load over a designated period of time to the peak load occurring during that period.

load line 1. Straight line drawn across grouping of plate current/plate voltage characteristic curves showing the relationship between grid voltage and plate current for a particular plate load resistance of an electron tube. 2. Structure between the tank building of a submarine cable factory and the pier, to support and guide cable being loaded into a cable ship.

load sharing 1. Operating mode of duplicated processors whereby the two units share processing operations. In the event of a failure of one unit, the other can take over the entire load. Contrast with hot standby. 2. In common channel signaling, a process by which signaling traffic is distributed over two or more signaling or message routes, in view of traffic equalization or security.

load transfer The transfer of signaling traffic from one signaling link to another, in a common channel signaling system.

loader Program which controls peripheral device operations when reading programs for execution into system memory.

loading 1. The insertion of impedance into a circuit to change its characteristics. 2. In multichannel communications, the insertion of white noise or equivalent dummy traffic at a specified level, to simulate system traffic performance. 3. In multichannel telephony systems, the loading imposed by the busy hour traffic (the equivalent mean power and the peak power) as a function of the number of voice channels; or the equivalent power of a multichannel complex or composite signal(s) referred to zero transmission level point (0TLP). These loadings are a function of the number of channels and the specified mean voice channel power. 4. The total ice and wind pressure which is allowed for in the design of a pole line. 5. Overhead expenses, such as supervision, exempt material, and motor vehicle expense, above direct charges.

loading, antenna The addition of an inductance to enable an antenna to be tuned to a lower frequency than its natural frequency.

loading, conductor The mechanical loading per unit length of line wire due to the combined effect of iceloads, winds, and conductor weight.

loading, inductive The addition of lumped loading coils at regular intervals along a cable to reduce attenuation over the voice frequency band.

loading, lumped Loading coils inserted at regular intervals along a route to improve transmission characteristics.

loading, storm Stress on a pole route due to the combined effect of ice and wind.

loading, system The total signal power of a multichannel system, expressed as the total of the average power on all channels, or as the per-channel load which may be carried by all channels.

loading, uniform Now outdated practice of adding inductance uniformly along the whole length of a line in order to improve transmission performance.

loading characteristic In multichannel telephony systems, a plot for the busy hour of the equivalent mean power and the peak power as a function of the number of voice channels.

loading coil See coil, loading.

lobe A representation of the transmission directional efficiency of a radio antenna: the bigger the major lobe, compared with minor lobes, the more directive is the system.

lobe, back A lobe in an antenna radiation directivity pattern pointing directly away from (at 180° to) the intended direction.

lobe, front The lobe in the required direction of an antenna radiation directivity pattern; the major lobe.

lobe, major The largest of the three-dimensional lobes representing the radiation directivity pattern of an antenna.

lobe, minor Any of the lobes in the radiation directivity pattern of an antenna, except the major or front lobe.

lobe, side A minor lobe.

local An extension on a PBX.

local, long distance An extension of a PBX not located on the same premises as the PBX itself.

local access and transport area (LATA) Area served by a single local telephone company. Long distance calls are inter-LATA calls, and are handled by an interexchange carrier. Circuits with both end-points within the LATA (intra-LATA calls) are generally a purely local telco responsibility.

local action Local chemical action due to the presence of impurities which may interfere with the prime action of the unit.

local area data transport (LADT) AT&T method of using local loop to provide both data and analog voice services for subscribers. Dial-up LADT permits occasional use of telephone lines for data; direct access LADT permits simultaneous voice and data.

local area network (LAN) Network permitting the interconnection and intercommunication of a group of computers, primarily for the sharing of resources such as data storage devices and printers. LANs cover short distances (less than 1 km), usually within a single building complex. Different data transfer rates are possible and shared centralized data storage access is available. Response times are comparable with those of a single computer. Arrangements can be made for one LAN to communicate with another, which is then called a WAN (Wide

Area Network) or a MAN (Metropolitan Area Network).

local automatic message accounting (LAMA) Automatic message accounting equipment in the same office as number identification equipment, enabling all necessary accounting messages to be taken for dialed toll or long distance calls, in order to produce itemized bills.

local battery 1. In telegraphy, the battery that actuates the telegraphic station recording instruments, as distinguished from the battery furnishing current to the line. 2. In telephony, a system where each telephone set has its own individual source of power.

local call Any call within a local charging area. For traffic engineering purposes, it is a call to another number on the same central office.

local call timing An exchange facility for the measurement of the length of local calls for charging purposes.

local code A digit, or combination of digits, for obtaining access to an exchange in an adjacent numbering area without using the national number.

local distribution system (LDS) A wide-band microwave system or cable system which is capable of transporting a number of television signals simultaneously. Used to interconnect cable system headends.

local engineering control office (LECO) The office responsible for the design of the portion of the circuit within its territory.

local exchange An exchange where subscribers' lines are terminated.

local issue date (LID) The date on which the local orders are scheduled to be issued.

local line network (local distribution network) The network interconnecting subscriber terminals and local exchanges.

local loop The circuit (usually a 2-wire path) connecting a subscriber's station with the line terminating equipment in a central office. (4-wire loops are sometimes provided for leased circuits used for data).

local manual attempt recording (LOMAR) System used on some PBXs to record calling party extension number for outgoing calls.

local measured service A usage-sensitive method of pricing local telephone calls, dependent on distance and time.

local multiple connection A call established by the operator at an A board to a line in the multiple to which that operator has direct access.

local number In national telex networks, the abridged call numbers used for local or short-distance traffic.

local numbering area A group of local exchanges associated for the purpose of local numbering and national identity.

local off-net access line (LONAL) A telephone line which connects a CCSA or EPSCS switch to a local central office to permit off-net calling.

local orderwire A communications circuit between a technical control facility and selected terminal or repeater locations within the communications complex.

local origination channel A channel on a cable system (exclusive of broadcast signals) which is programmed by the cable operator and subject to his exclusive control.

local service area Area within which local call rates apply.

local side That portion of a device which faces the internal station facilities.

local transit exchange An exchange used as a switching point for traffic between local exchanges within a multi-exchange area. Syn.: tandem exchange

location A position in a store which may hold a data word or part of a word.

location life Term used in telephone economic studies for the length of time a customer uses the same equipment on the same line.

locking A relay, key or circuit which remains operated without needing a physical hold or holding current.

lock-on The automatic tracking and holding action of a ground station working to a communications satellite.

lockout 1. In a telephone circuit controlled by two voice-operated devices, the inability of one or both subscribers to get through, because of either excessive local circuit noise or continuous speech from either or both subscribers. 2. In mobile communications, an arrangement of control circuits whereby only one receiver can feed the system at one time. Syn.: receiver lock-out system 3. In telephone systems, the treatment of a subscriber's line which is in trouble or in a permanent off-hook condition by automatically disconnecting the line from the switching equipment. 4. To place terminals on a line in a mode which makes them incapable of receiving transmissions. 5. Programming technique used to prevent access to private or critical data by unauthorized terminals.

lockout, echo suppressor Condition when suppressors at both ends of a long distance circuit simultaneously block transmissions in both directions.

lockstitch Method of lacing used on cable forms.

lock-up The operation of an echo suppressor caused by high noise level rather than by received signal.

logarithm The power to which a base must be raised to produce a given number. Common logarithms are to base 10.

logarithm, natural A logarithm using the base e (=2.71828183) instead of the base 10. Sometimes called Napierian log, after its Scottish inventor.

logarithmic scale 1. Meter scale with displacement proportional to the logarithm of the quantity represented. 2. Graph paper with one or both of the grids on a logarithmic rather than an arithmetic scale.

logatom A meaningless syllable used in articulation tests to obtain a measure of the intelligibility of speech transmitted over a channel.

logic 1. The formal principles of reasoning. 2. The process by which incoming digital information is translated into outgoing digital information or signals.

logic, wired A circuit which makes logic decisions based on the way the circuit has been strapped or wired. Changes in program necessitate physical changes to wires, straps or components.

logic analyser Test instrument used in the monitoring of states and state sequences, e.g., in microprocessor-based systems.

logic element Device which performs a logical function; also known as a logical element or gate.

logic gates The basic decision-making circuits in digital equipment, which usually have two or more binary inputs, and one binary output. Simple functions can be implemented by single gates, but several gates of different types, together with different forms of memory unit, are often combined together to form complex decision-making networks.

logic levels (of a gate) The voltages that must exist for the circuit to operate correctly. Typically, logic 0 could be 0.2v and up to 0.4v; logic 1 could be 2.4v and up to 3.3v.

logic one The digit 1 as used in a binary notation, used as the mathematical equivalent of 'true' in Boolean algebra.

logical compatibility Compatibility at the functional level such that two devices can be interconnected and will work in harmony without requiring functional modifications.

logical instruction An instruction in a computer program which is specifically logical and not purely arithmetic.

log-periodic antenna See antenna, log-periodic.

lone signal unit (LSU) A signal unit in common channel signaling system CCITT No. 6 carrying a one unit message.

long distance A toll call, or toll telephone service.

long-distance local An off-premises extension on a PBX.

long distance terminal Subscriber's line (for a major user of long distance services) which terminates directly on a long distance switchboard.

long distance transit exchange An exchange used as a switching point for long distance traffic within a national network; it is not a local, local transit or international exchange.

Long Distance Xerography (LDX) The Xerox high speed facsimile service which uses a wideband channel between terminals.

long feeder route analysis program (LFRAP) An AT&T software routine used in planning long feeder routes with subscribers fed off at different points using carrier techniques.

long-haul communications Communications which permit users to convey information on a worldwide basis. Long-haul communications are characterized by higher levels of users, more stringent performance requirements, longer distances between users,

higher traffic volume and density, and fixed or recoverable assets.

long lines All forms of physical conductors used for communication purposes, such as open wire systems, underground and overhead cables, and submarine cables, but not local connections.

Long Lines (LL) Department The AT&T department which was responsible for the operation of interstate toll services. See Bell System entry.

long persistence Quality of a cathode ray tube which has phosphorescent compounds on its screen (in addition to flourescent compounds) so that the image continues to glow after the original electron beam has ceased to create it by producing the usual flourescence. Long persistence is often used in radar screens or where photographic evidence is needed of a display.

long term bit error rate Bit error rate measured over a sufficiently long time period, such as one month.

long wavelength In fiber optics, operation at wavelengths in the range 1,100 nanometers to 1,700 nanometers. (Some work is being done at longer wavelengths of up to 30,000 nanometers).

longitudinal balance See balance, longitudinal.

longitudinal current Current that travels in the same direction on both wires of a pair: the return current either flows in another pair or via a ground return path.

longitudinal offset loss See gap loss.

longitudinal redundancy check (LRC) Method of error correction. An 8-bit check character is added to each block (a byte per block) to make the total number of 1s in each significant bit position in the block an even number. Syn.: block sum check

longwave Radio broadcast signals with wavelengths longer than 600 meters, frequencies of less than 500 kHz.

loop 1. Go and return conductors of an electric circuit; a closed circuit. 2. A single connection from a switching center or an individual message distribution point to the terminals of an end instrument. 3. A closed path under measurement in a resistance test. 4. A type of antenna used extensively in direction finding equipment. 5. In computer systems, repetition of a group of instructions in a computer routine. 6. In telephone systems, a pair of wires from a central office to the subscriber's telephone.

loop, LD Line from a subscriber's instrument or PBX directly to the local toll switchboard.

loop, local Line connecting subscriber's instrument or PBX to the local end office.

loop, open-wire Subscriber's loop which utilizes open wire on a pole route.

loop, polar telegraph Telegraph circuit in which positive and negative batteries of the same voltage are used for the mark and signal conditions.

loop, program supply Circuit from point of origin of a radio broadcast program to the radio transmitter. Such circuits usually are conditioned to give a flat response over a wide bandwidth (10 or 15 kHz).

loop, radio Circuit utilized for program supply.

loop, subscriber's Pair of wires providing a circuit from a subscriber's telephone instrument or PBX to the central office.

loop, Varley An arrangement of a Wheatstone bridge circuit which provides the difference in resistance between the two wires of a loop in one measurement, and hence enables ground faults to be found easily.

loop and leak Testing procedure used when checking make-break ratio of received loop-disconnect dial pulses.

loop back A method of performing transmission tests of access lines from the serving switching center that does not require the assistance of personnel at the served terminal. A connection is established over one access line from the serving switching center through the transmission testing and switching equipment and back to the serving switching center over another access line.

loop carrier analysis program (LCAP) An AT&T software routine used in planning route relief, with carrier systems or with cables.

loop gain 1. Total usable power gain of a carrier terminal or two-wire repeater. The maximum usable gain is determined by and may not exceed the losses in the closed path. 2. The product of the gain values acting on a signal passing around a closed path loop.

loop-mile 1. One mile of two-wire line. 2. A measure of resistance: the resistance in ohms of a mile of two-conductor line.

loop noise The noise contribution of the line loops in a system.

loop options Choice of dc circuit arrangements for a teletypewriter, such as full duplex, half duplex, polar, or neutral.

loop resistance The ohmic resistance of a subscriber's cable pair measured at the exchange. This normally includes the resistance of the off-hook telephone instrument.

loop start Method of calling a central office by applying a DC closed loop across the line.

loop test A method of testing employed to locate a fault in the insulation of a conductor when the conductor can be arranged to form part of a closed circuit or loop.

loop wire Wire which links together several terminals on adjacent components.

looping A programming technique where a portion of a program is repeated over and over until a certain result is obtained.

loran A long-range radionavigation position fixing system using the time difference of reception of pulse type transmissions from two or more fixed stations.

loss Transmission loss, usually expressed in decibels, or power dissipated in a circuit.

loss, attenuation Fall in energy due to a wave or signal passing through space or along a transmission line.

loss, bridging The loss which follows from inserting a bridging path across a transmission line.

loss, copper The heat loss or I^2R loss: heat produced by the passage of current through a resistance.

loss, dielectric hysteresis Energy used in continually reversing the electric field in a dielectric or insulator.

loss, diffraction Loss experienced by radio waves which have been diffracted.

loss, echo return The loss which must be inserted in an echo path in order to reduce an echo to an acceptable level: the greater the delay in hearing the echo the higher the echo return loss needs to be.

loss, eddy current Energy loss caused by eddy currents flowing in cores of transformers or inductances.

loss, effective transmission Overall loss for a complete telephone channel, including impairments caused by noise and restriction of bandwidth as well as attenuation.

loss, expected measured (EML) See expected measured loss.

loss, free space Theoretical loss between two antennae, both assumed to be isotropic, so that the loss will depend on distance and frequency.

loss, hybrid insertion The unwanted loss between two-wire and four-wire paths through a hybrid.

loss, inserted connection (ICL) The loss on a trunk measured at 1000 Hz.

loss, insertion Loss in a circuit at a specified frequency due to the insertion of a particular piece of apparatus.

loss, iron Losses in the iron core of a transformer or inductance, due to hysteresis and sometimes eddy currents.

loss, junction Loss due to the joining together of two circuits with different impedances.

loss, line Total energy loss in a line.

loss, magnetic hysteresis Energy loss from repeatedly reversing the magnetic fields concerned.

loss, net Sum of all losses in a circuit minus the sum of all the gains in the same circuit.

loss, obstruction Loss in a radio path when the radio signal does not pass solely through free space, e.g. a grazing path just over ground level will often result in an obstruction loss of about 10dB.

loss, polarization Loss in a radio path when the receiving antenna is not adjusted to the same angle of polarization as the incoming radio signal.

loss, power Ratio of total power delivered to a line to the power from line into load, expressed in dB.

loss, reflection Loss in passing from transmission line of one impedance to a line of a different impedance.

loss, return Loss due to impedance mismatch; the reflection loss.

loss, shadow Loss in radio signals which are obstructed by such things as hills.

loss, terminal net (TNL) The total of the VNL (via net loss) and the losses of switching pads which are in the circuit when it is in a terminal condition.

loss, toll terminal Transmission loss that occurs from a toll center to the subscriber's instrument.

loss, transformer Ratio of the power that an ideal transformer would provide for a load to the power that a real transformer of identical impedance ratio would provide for a load of equal impedance. Expressed in dB.

loss, trans-hybrid Loss between transmit and receive branches of a four-wire line, at a hybrid.

loss, transition Ratio, in dB, of power actually transferred from one circuit to another to the power which would be transferred if the source and load impedances had been matched together, i.e., had been conjugates.

loss, transmission Ratio, in dB, of power at input to power at output for a transmission line.

loss, tributary Loss between a subscriber's instrument and the toll center, including the subscriber's loop to the end office, the loss through the end office, and the loss in the tributary trunk to the toll center.

loss, trunk The portion of loss attributable to the trunk used for the call.

loss, via net (VNL) The lowest loss at which telephone trunk facilities may be operated, depending on their particular characteristics and practical return losses.

loss, wet weather The transmission loss of an open-wire pair when wet; much of the cause of difference in transmission over open-wire circuits during rain is due to current shunting the normal path, either across wet insulators or across wet leaves touching wires.

loss of frame alignment detector A monitoring unit designed to indicate to the signaling terminal that frame alignment of the PCM system has been lost.

loss system A system in which bids that cannot be immediately served are lost.

lossy Conditions when line loss per unit length is significantly greater than normal.

lossy cable Coaxial cable constructed to have high transmission loss so it can be used as an artificial load or as an attenuator.

lost block A block successfully transferred across the source user/telecommunication system functional interface, but not delivered to any user within the specified maximum end-to-end block transfer time.

lost call A request for a connection which is rejected due to network congestion.

lost traffic That part of the traffic offered to a pool of resources which is not carried, and no additional resource is provided to handle such traffic.

loudness Subjective sensation of intensity of sound power.

loudness, sound The magnitude of the sensation in the brain produced by sound.

loudness contour Graph showing sound pressure, frequency and loudness sensation relationships for a typical person.

loudness level Numerically equal to the median sound pressure level relative to 2×10^{-5} newton per square meter, of a 1000 Hz signal which is subjectively considered to be equally loud.

loudspeaker A transducer which converts electrical energy into acoustic energy.

loudspeaker baffle A plane screen on which a loudspeaker is mounted so that sound waves from the front of the speaker cannot travel directly to the back of the speaker so as to reduce its efficiency.

loudspeaker telephone A hands-free telephone.

louver Slots or holes on the front of a loudspeaker which permit sound to pass but give mechanical projection.

low band TV channels 2 through 6.

low frequency (LF) The band of radio frequencies between 30 and 300 kHz.

low-level keying Use of low levels of voltage and current on keying contacts between the limits of positive or negative 6 volts.

low-level language A programming language that reflects the structure of a computer or that of a given class of computers. A language which consists of instructions which are converted directly into machine code.

low-level signaling Use of low levels of voltage and current on signal lines between the limits of positive or negative 6 volts.

low loss Circuits in which energy losses due to heat production are low.

low loss (optic fiber) Sometimes used to define optical waveguide with attenuation of less than 10dB/km. Standards are, however, continually being improved: yesterday's low loss fibers become tomorrow's high loss scrap!

low pass filter See filter, low pass.

low performance equipment 1. Equipment with insufficiently exacting characteristics to permit their use in trunks or links. 2. Tactical ground and airborne equipment whose size, weight, and complexity must be kept to a minimum and where the primary requirement is to operate in nets with similar minimum performance standards.

low power flux-density In the broadcasting-satellite service, a power flux-density lower than the medium power flux-density which enables the necessary grade of reception quality to be obtained only by using more specialized transmission and reception techniques than those required for medium power flux density reception.

low-power television TV distribution system providing TV (often pay-TV) services over a restricted area.

low speed Data transmission speed of 300 bit/s or less.

low tension Low voltage.

low VHF band The part of the frequency band allocated in the U.S. by the FCC for VHF broadcast television, including television channels 2 through 6, or 54 through 108 MHz.

low voltage It can mean an operating voltage of less than 100v, but if used to describe a public power supply route, it may mean anything up to 750 volts.

low voltage alarm See alarm, low voltage.

low-voltage enclosed air circuit breaker switchgear A dead front switchgear assembly for service up to

600 v alternating current and up to 750 v direct current, containing air circuit breakers, buses, and connections, with an enclosure on ends, back and top. The air circuit breakers are contained in individual compartments and controlled remotely or from the front panels.

lower sideband Sideband containing frequencies below the carrier frequency produced by amplitude modulation.

lowest useful high frequency (LUF) The lowest frequency in the high frequency band at which the received field intensity is sufficient to provide the required signal-to-noise ratio on 90 percent of the undisturbed days of the month.

L-pad A volume control that has essentially the same impedance at all settings. It consists of an L network arranged so that both of its elements can be adjusted simultaneously.

LR group A 48 kHz bandwidth carrier group (a 12 audio channels FDM system) established over a radio path in tandem with a cable section.

lubricant, cable pulling When pulling lead cable into a duct, the cable is painted freely with a heavy grease often bearing powdered soapstone or colloidal silica, to make it easier to pull the cable in. The use of plastic sheathed cables and PVC ducts eliminates the need for such lubricants.

lug A tag or projecting terminal onto which a wire may be connected by wrapping or soldering.

lug, solder Method of terminating a heavy stranded power conductor: the conductor is forced into one end of a short copper tube filled with molten solder, the other end of the tube is flattened and a hole drilled through it so the lug may be bolted to a power terminal.

lug, solderless Terminal which compresses and holds conductors by tightening a screw fitting.

lug, spade A solder lug with an open-ended two-prong spade instead of a round hole which can be slipped under the head of a retaining screw or binding post terminal.

lug, terminal Threaded stud with nuts which are used to clamp wires into firm contact.

lumen The derived SI unit of luminous flux, equal to the flux on a unit surface, all points of which are at a unit distance from a source of one candela.

luminance The photometry quality of luminous flux, as compared with the subjective or sensory response word; brightness.

luminance signal That portion of the color television signal which contains the luminance or brightness information.

lumped At a point, rather than distributed.

lumped constant Resistance, inductance or capacitance connected at a point, not distributed uniformly throughout the length of a route or circuit.

lumped loading Loading coils as normally used, inserted at regular intervals in cable circuits in order to improve transmission characteristics. In Latin-

influenced countries: Pupinized (after the inventor of lumped loading).

lumpy A line in which inductance, capacitance or resistance are not regularly and uniformly distributed along its whole length.

lunk Combined line and trunk circuit used in some crossbar offices for terminating circuits out to PBXs.

M regions The areas of the sun's surface which appear to be responsible for many of the electromagnetic disturbances experienced on Earth.

machine, cable lashing Small wrapping machine pulled along a messenger strand to wrap or lash the messenger to a suspended aerial cable.

machine code The instruction code designed into the hardware of a processor; the direct representation of the computer instruction in memory.

machine language A low level language whose instructions consist only of computer instructions; a program of binary coded instructions stored in a computer's memory.

machine ringing An automatic ringing current supply which continues until the called subscriber goes off hook or the call attempt is abandoned.

machine-sensible In machine language, or capable of being interpreted by a machine.

machinery, cable The apparatus on a cableship for the pick-up and pay-out of cable.

macro Short for macroinstruction. An instruction that generates a larger sequence of instructions.

macrobend loss In an optical waveguide, that portion of the total loss attributable to macrobending. Syn.: curvature loss

macrobending In an optical waveguide, all macroscopic axial deviations of the waveguide from a straight line, as opposed to microbending.

macro-instruction A mnemonic instruction denoting a group of machine-code instructions; an instruction in a source language that is to be replaced by a defined sequence of instructions in the same source language.

magic eye A small cathode ray tube used as an indicator of received signal strength, used to tune in radio receivers accurately. Also, a light-sensitive or infrared sensitive device used in a burglar alarm or other devices, to trigger action when a beam is interrupted.

Magicall   A proprietary repertory dialer.

magnet   Device which produces a magnetic field, and can attract iron and attract or repel other magnets.

magnet, permanent   Bar or device made of high-permeability steel which retains its magnetism.

magnet, release   In an electromechanical step-by-step switch, an electromagnet which, when operated, causes the wipers of the switch to turn and/or drop back to their rest position.

magnet, stepping   An electromagnet in a step-by-step switch which, when operated, causes the wipers of the switch to step up or round to a selected outlet.

magnetic bubble   A solid state memory device which depends on the magnetic polarization of "domains," usually in a garnet-type material. The magnetic bubble is caused to move through the material by an external field, and the direction of polarization of individual domains, read off as each passes a sensing device, represents the two logical states of 0 or 1. See also bubble.

magnetic card   A card with a magnetizable surface layer on which data can be stored by magnetic recording.

magnetic core   A piece of ferromagnetic material which may be polarized by current-carrying conductors placed near it or through it. The two alternative polarizations allow a magnetic core to assume two states, enabling it to be used as a memory-storage medium.

magnetic disc   A memory device employing magnetic material coated on a circular base. Data is stored by changing the direction of magnetization of small localized areas on concentric tracks on the surface of the disc. A write/read head can be moved radially to operate on any of the tracks. The disc rotates like a phonograph record, permitting fast access to large amounts of data in store. See disk

magnetic drum   Large capacity memory storage device which operates on the same principle as a magnetic disc. The magnetic material is on the outer surface of a rigid cylinder, usually mounted vertically, and capable of rapid rotation. The read/write heads are fixed. Drums are usually capable of more rapid access than discs.

magnetic field   The space surrounding a magnet which contains magnetic flux.

magnetic field strength   The strength of a magnetic field at a point following the direction of the line of force at that point.

magnetic flux   The field produced in the area surrounding magnets or electric currents. The SI unit of flux is the weber.

magnetic flux density.   A vector quantity measured by an SI unit called the tesla. Basically, it is the number of magnetic lines of force per unit area, at right angles to the lines.

magnetic hysteresis.   A property of magnets leading to heat production and based on the fact that magne-tization of a ferromagnetic material does not vary on a linear basis with the strength of the magnetic field applied. It may be likened to internal friction which occurs between molecules as they are each in turn made to align themselves with the applied field.

magnetic ink character recognition (MICR)   The use of printed characters containing particles of magnetic material: these can be read automatically by a scanner and converted to a computer-readable form.

magnetic leakage   Magnetic flux which does not follow a useful path.

magnetic medium   Any data-storage medium, and related technology, including disks, diskettes and tapes, in which different patterns of magnetization are used to represent bit values.

magnetic pole   Point which appears from the outside to be the center of magnetic attraction or repulsion at or near one end of a magnet.

magnetic recording   A method of recording signals in a magnetic medium. The most common method is that of the ordinary tape recorder in which plastic tape with a ferrous coating is passed close to an electromagnetic head with pole pieces separated by a small gap immediately touching the tape as it passes the head. As currents vary in the coil controlling the head, the flux between the pole pieces varies. The flux pattern is, in effect, written into the tape at the instant a particular element of tape is opposite the gap in the head.

magnetic storm   Violent local variations in the Earth's magnetic field, usually due to sunspot activity.

magnetic stripe   A strip of magnetic material affixed to a credit card, I.D. badge, etc., on which data is recorded and from which data can be read.

magnetic tuning   The tuning of a microwave cavity-type oscillator by varying the magnetic flux density of a ferrite rod in the resonator.

magnetism   A property of iron and some other materials by which external magnetic fields are maintained, other magnets being thereby attracted or repelled.

magnetism, residual   See residual magnetism.

magnetization   Exposure of a magnetic material to a magnetizing current, field or force.

magnetizing force   Force producing magnetization.

magneto   In telephone practice, a simple hand-cranked AC generator used to produce ringing currents to operate bells.

magneto line   A subscriber line with magneto telephone instruments.

magneto operation   Type of telephone system utilizing the hand cranking of a magneto for making calls. Local batteries at each station produce the microphone current.

magneto-optic   Describing a change in a material's refractive index under the influence of a magnetic field. Magneto-optic materials are used to rotate the plane of polarization.

magneto switchboard A manual switchboard in which all signaling is ringdown. It is normally used only in rural areas. Local batteries are required to energize subscribers' telephones.

magneto telephone A telephone with a local AC generator (usually hand-cranked) for calling out, and a local DC battery for microphone current.

magnetomotive force The force which tends to produce lines of force in a magnetic circuit. It bears the same relation to a magnetic circuit that voltage does to an electrical circuit. The SI unit is the ampere but the term "ampere-turn" is sometimes used.

magnetostriction, negative A decrease in the length of a rod with an accompanying increase in magnetic flux density.

magnetostriction, positive An increase in the length of a rod when an axial magnetic flux is applied to it.

magnetostriction oscillator Device which produces ultrasonic energy. One practical application is in cleaning equipment: changing dimensions at ultrasonic frequencies can produce strong pressure/rarefaction waves which when passed through a cleaning fluid loosen ingrained dirt and grease.

magnetron A high-power, ultra high frequency electron tube oscillator which employs the interaction of a strong electric field between anode and cathode with the field of a strong permanent magnet to cause oscillatory electron flow through multiple internal cavity resonators.

magnetron, pulsed A magnetron which is pulsed rather than run continuously in order to produce very large power levels.

Magsat A U.S. research satellite used for mapping the Earth's magnetic field.

mailbox System for the transmission of text messages between terminals using one or more central file bases in which a sender deposits a message for subsequent retrieval by the receiving terminal. Mailbox systems often provide notification of incoming message and of confirmed delivery.

mailboxing The technique of using a timesharing service for the exchange of messages. A store and retrieve system.

mailgram An overnight service combining the electronic transmission of messages with physical delivery by postal personnel during the next day's regular delivery.

main cable In submarine systems, the portion of a total cable system which is laid in the deeper reaches of the route.

main distribution frame (MDF) A distribution frame on one part of which terminate the permanent outside lines entering the central office building and on another part of which terminate the subscriber line multiple cabling, trunk multiple cabling, etc. It is used for associating any outside line with any desired terminal in such a multiple or with any outside line. Syn.: main frame

main exchange An automatic exchange with other exchanges dependent upon it for their principal traffic outlets.

main frame In telephone terminology, a main distribution frame; in computer terminology, a large computer installation.

main lobe In an antenna radiation pattern, the lobe containing the direction of maximum radiation intensity.

main repeater station A station — always the terminal of a line link — where direct line filtering or demodulation, or both together, may take place. As a consequence, in such a station there are equalizers and it is possible to find points (called "flat points") which are of uniform relative level independent of frequency.

main section The portion of a group, supergroup, etc. link between two adjacent stations having control functions. Often these two stations are in different countries.

main station 1. Telephone instrument directly connected to a central office. 2. Primary extension in a PBX numbering plan. 3. Point at which a leased line terminates on to a transmission device.

main store That area of store which is random access, meaning that it takes equal time to access any word held in the store.

main trunk (in CATV systems) The major link from the headend to feeder lines.

main (worker) and standby operations A method of improving the reliability of a processor installation by operating two processors such that one performs the workload while the other performs checks on the main (worker) and replaces it should it fail.

mains A term used in many countries when referring to the public electric power supply (called "the mains").

mains frequency A frequency of 60 Hz in the U.S., 50 Hz in Europe.

maintainability A characteristic of design and installation expressed as the probability that an item will be retained in, or restored to, a specified condition within a given period of time, when the maintenance is performed in accordance with prescribed procedures.

maintenance Any activity — such as tests, measurements, replacements, adjustments and repairs — intended to eliminate faults or to keep a functional unit in a specified state.

maintenance, preventive See preventive maintenance.

maintenance control circuit (MCC) A voice circuit used over microwave links for maintenance personnel coordination.

maintenance time Term usually restricted to mean the time used for hardware maintenance, both preventive and corrective.

major relay center In telegraphy, a tape relay center providing alternative routes for traffic.

make When pertaining to electrical contacts, it describes contacts which are normally open but which close when a device is put in operation.

make, percent The percentage of make in a repeated make-and-break contact device such as a loop-disconnect dial.

make contacts Contacts which are open when a device is at rest and make when the device is operated.

make-before-break contacts Contacts which make with new contacts before they break with the contacts with which they are associated in the rest position of a relay or key.

make-break ratio The percentage of make in a repeated make-and-break contact device.

malfunction An equipment failure or a fault.

malicious call trace Assistance given, at the discretion of an administration, to ascertain the origin of malicious, nuisance or obscene calls.

management domain (MD) The set of Message Handling System entities managed by a telecommunications administration or by an organization that includes at least one Message Transfer Agent (MTA).

management information system (MIS) Computerized system for economic management control, especially of a manufacturing process.

management signals Signals sent over a signalling system which concern the management or maintenance of the speech circuit network and the signaling network.

Manchester encoding Digital encoding technique in which each bit period is divided into two halves, a negative-to-positive transition in the middle represents binary "1", a positive-to-negative transition a binary "0".

mandrel Hard cylindrical unit about 30cm long, pulled through a duct to prove that the duct is suitable to be used for a telephone cable. The mandrel's diameter depends on the type of ducts being tested.

manganin A metal alloy used for precision resistors because its resistance is little changed by temperature.

manhole An underground cable-jointing chamber with duct routes in and out which serves a cable distribution network.

manifold, dry air Method of supplying dry air under pressure to cable systems. It acts as a reservoir for air supply to several cables.

man-machine interface (MMI) The series of commands and responses used by maintenance personnel and test positions particularly for computer-controlled (SPC) exchanges.

man-machine language A language designed to facilitate direct user control of a computer. It contains inputs (commands), outputs, control actions, and procedures sufficient to ensure the performance of all functions relevant to the operation, maintenance and installation testing of stored program controlled systems.

man-machine system A system in which the functions of a man and a machine are interrelated. Syn: man-machine interface

manometer A test device for measuring gas pressure. In telephony this usually applies to the pressure inside a pressurized cable system. Manometer readings are plotted as an aid to fault localization.

manual Operated by hand.

manual answering In data transmission, a facility whereby a call is established only if the called party signals his readiness to receive the call by means of a manual operation.

manual calling In data transmission, a facility which permits the entry of selection signals from a calling terminal installation at an undefined character rate. The characters may be generated at the data terminal equipment or the data circuit-terminating equipment.

manual exchange A telephone central office where switching functions are manually controlled by an operator.

manual office Telephone office where switching is performed by human operators at manual switchboards.

manual position Point at which switching or assistance functions are carried out by an operator or attendant.

manual start In power systems, a standby-power system which must be started by hand when the public power supply fails.

map To establish a set of values having a defined correspondence with the quantities or values of another set.

map, topographic A scaled map having contour lines which show degrees of land elevation above sea level. It is used in preparing profiles of microwave radio propagation paths.

mapping The logical association of one set of values with the values of another set.

margin 1. The extra gain which could be added before a transmission circuit became unstable and began to sing. 2. The extra loss which could be introduced into a radio communication circuit before the circuit failed.

margin, fade The margin which must be designed into a radio communication circuit to ensure that the required grade of service is provided, despite fading, during normal propagation conditions for a given percentage of the time (typically 99.99%).

margin, singing Extra gain which, if added to a circuit, would result in its singing.

marginal Describing a circuit which is right on the margin of acceptability, particularly for its transmission parameters.

marginal capacity The increment of traffic intensity that can be offered per circuit added to a given circuit group while maintaining the same grade of service.

marginal occupancy The increment in carried traffic intensity per circuit added to a circuit group while the offered traffic intensity is held constant. Syn. marginal utility

marginal overflow The increment in overflow traffic intensity per unit of traffic intensity offered to a circuit group while the number of circuits is held constant.

marginal relay A relay with a small specified and dependable difference between the current needed to operate it and that current which will not operate it.

MARISAT A satellite communications system for maritime use. Both common channel signaling and working channel signaling are employed.

maritime mobile (terrestrial) service Conventional radio service to ships at sea, using HF and VHF radio via coast stations.

maritime mobile satellite service Radio service to ships at sea, via geostationary satellites and shore stations.

mark 1. In binary transmission, one of the two significant conditions of modulation, normally corresponding to state "1". 2. In teletypewriter operation, the closed circuit condition. 3. To place a potential, lead or ground on a terminal so that a hunting circuit may find that marking condition and connect other equipment to the line so marked.

mark, class of services An indication, often in software, which shows the class of service to which a line is entitled, such as residential, business or public coin phone.

mark-hold The steady no-traffic condition in a transmission system such as a teletype line.

mark-sense Process by which pencil marks on data cards are read by machine and the appropriate holes punched automatically in the card so that the card may be used as input to a processor.

marker 1. A switching control unit in common control offices which locates the calling line and selects, marks and tests paths through the office to the called line. 2. A marking post indicating the exact location of buried plant, such as cables, joints and loading coils.

marker, buried cable Small posts, usually made of concrete, indicating the route of a directly buried cable.

marker, crossbar Circuit unit which controls the setting up of calls in a common-control crossbar exchange.

marker, defective pair Plastic identifying tube which is placed over a cable pair in a multipair cable when tests at the factory show that the pair is not up to specification.

marker, tandem Marker which controls the selection of trunk routes at tandem offices.

marking bias The uniform lengthening of all marking signal pulses at the expense of all spacing pulses.

MAROTS A satellite communications system for maritime use with fully automatic working both for ship- and shore-originated calls, using the specified telephone signaling and call set-up procedures. An acronym for Maritime Operational Telecoms Satellite.

M-ary code The generic name applied to all multilevel codes.

M-ary signaling The transmission of digital data in a way that permits the pulses to take on with equal probability any of m rather than two amplitude levels. This multilevel system implies three or more transmitted conditions. Syn. multilevel modulation

MASER (Microwave Amplification by Stimulated Emission of Radiation) A microwave amplifier capable of detecting and amplifying very weak signals. They are low-noise devices which operate by molecular interaction.

maser, cavity Maser material, often paramagnetic material such as ruby, placed inside a cavity resonator.

maser, gas Early type of maser using a tube containing ammonia or other gas. Now largely replaced by masers using solid state material.

maser, traveling-wave Maser material placed in a waveguide into which the operating frequency is pumped. As in the ordinary traveling wave tube, the signal to be impressed on the carrier frequency travels slowly along the tube, providing effective amplification.

mask 1. To hide, obscure, make less noticeable. 2. Device used in the production of thin-film circuits and other components as a means of restricting patterns or deposits.

mass calling Network management control which provides a means of limiting the traffic carried through the system offered to a specific directory number wherever an excessive number of calls are routed to it, e.g., a radio or TV talk-show number.

mast Vertical pole used to support antennae.

mast, service Pole next to a pay telephone booth which supports both the telephone drop wire for the pay phone and the power cable for lighting the booth.

master antenna television (MATV) A single efficient antenna array connected via amplifiers and coaxial feed to sockets in rooms in hotels or large apartment blocks.

master clock A clock which generates accurate timing signals for the control of other clocks or equipment.

master frequency generator In frequency division multiplex, equipment used to provide system end-to-end carrier frequency synchronization and frequency accuracy of tones transmitted over the system.

master ground bar (MGB) A copper ground bar generally located on the lowest floor level of the building.

master intercom station An intercom station, usually manned by an attendant, which deals with incoming calls and can switch these to any of the extensions on the system.

master number hunting A PBX feature by which dialing a preset digit causes the equipment to hunt over lines in a pre-determined sequence.

master office A central office which acts as parent office for smaller offices and through which all calls to the smaller offices are routed.

master oscillator A high quality oscillator which provides a standard frequency signal for other carrier terminals.

master station   In a data network, the station that has been requested by the control station to ensure data transfer to one or more slave stations.

mastergroup   A mastergroup link terminated at each end by terminal equipment. The terminal equipment provides for the setting up of five supergroup links occupying frequency bands separated by 8 kHz in a 1232 kHz band. The basic mastergroup consists of supergroups 4, 5, 6, 7 and 8 within the band of frequencies from 812 kHz to 2044 kHz.

mastergroup link   The whole of the means of transmission using a frequency band of specified width (1232 kHz) connecting two pieces of terminal equipment. The ends of the link are the points on mastergroup distribution frames, or their equivalent, to which the terminal equipment is connected.

mastergroup section   The whole of the means of transmission using a frequency band of specified width (1232 kHz) connecting two consecutive mastergroup distribution frames or equivalent points.

master-slave timing   A system wherein one station or node supplies the timing reference for all other interconnected stations or nodes.

match, impedance   To design or adjust impedances so that there is no reflection loss when signals pass from one circuit unit to another.

matched pulse   A method of synchronization of equipment with ringing signals. A typical application is to permit calls to individual party line instruments, using coded ringing, to be intercepted and, if necessary, diverted.

matched termination   A termination which absorbs all the incident power and so produces no reflected waves or mismatch loss.

matching   Joining together circuits in such a way that there is maximum transfer of power from one to the other.

matching, antenna   Adjusting antenna impedance to that of the transmission feeder line.

material dispersion   In optoelectronics, that dispersion attributable to the wavelength dependence of the refractive index of material used to form the waveguide. See material dispersion parameter.

material dispersion parameter   In optoelectronics, a value, M, signified by the following equation:

$$M(\lambda) = -1/c \, (dN/d\lambda) = \lambda/c \, (d^2n/d\lambda^2)$$

where n is the refractive index, N is the material group index: $N = n - \lambda(dn/d\lambda)$, λ is the wavelength, and c is the velocity of light in vacuum. For many present optical waveguide materials, M is zero at a specific wavelength λ_0, usually found in the 1.2 to 1.5 μm range. The sign convention is such that M is positive for wavelengths shorter than λ_0 and negative for wavelengths longer than λ_0. Pulse broadening caused by material dispersion in an optical fiber is given by M times spectral linewidth ($\Delta\lambda$), except at $\lambda \cong \lambda_0$.

material scattering   In an optical waveguide, that part of the total scattering attributable to the properties of the bulk materials used for waveguide fabrication. Material scattering may be either intrinsic scattering resulting from frozen-in inhomogeneities, or extrinsic scattering resulting from impurities.

matrices   Plural of matrix.

matrix   A simple switching network in which a specified inlet (matrix row) has access to a specified outlet (matrix column) via a crosspoint placed at the intersection of the row and column in question. Syns. connecting matrix, switching matrix

matrix, antenna switch   Matrix of coaxial switches which enables receivers and high power high frequency band radiocommunication transmitters to be switched to different antennae, cut for different frequencies, or made to work in different directions.

matrix, complete   A matrix in which each inlet has access to every outlet. Syns. full matrix, connecting multiple

matrix, incomplete   A matrix in which each inlet has access to only some of the outlets. Syn. partial matrix

matrix, nonblocking   A switching matrix through which a path is always available until all terminals are in use.

matrix, reed relay   A matrix of reed relays providing space-division switching. It is used in many semi-electronic switching systems.

maximal-ratio combiner   A diversity combiner in which the signals from each channel are added together. The gain of each channel is made proportional to the rms signal and inversely proportional to the mean square noise in that channel, with the same proportionality constant for all channels. Syn. ratio-squared combiner

maximum access time   Maximum allowable waiting time between initiation of an access attempt and successful access.

maximum block transfer time   Maximum allowable waiting time between initiation of a block transfer attempt and successful block transfer.

maximum calling area   Geographical calling limits permitted to a particular access line. Calling limits are not assigned; rather they are based on requirements for the particular line. Such restrictions are imposed for purposes of network control.

maximum disengagement time   Maximum allowable waiting time between initiation of a disengagement attempt and successful disengagement.

maximum justification rate   In pulse code modulation multiplexing, the maximum rate at which justifying digits can be inserted or deleted. Syn. maximum stuffing rate

maximum keying frequency   In facsimile systems, the frequency, expressed in hertz, which is numerically equal to the spot speed divided by twice the X-dimension of the scanning spot.

maximum modulating frequency   The highest picture frequency required for a given facsimile transmission system.

maximum observed frequency (MOF)   Highest frequency at which radio signals are observable on an oblique-incidence ionogram.

maximum power transfer The condition under which there is no power loss due to reflection at a source/load junction owing to the fact that the load impedance is a conjugate of the source impedance.

maximum stuffing rate The maximum rate at which bits can be inserted or deleted. Syn. maximum justification rate

maximum usable frequency (MUF) The upper limit of the frequencies that can be used at a specified time for radio transmission between two points and involving propagation by reflection from the regular ionized layers of the ionosphere. A higher frequency would penetrate the ionosphere and be lost in space.

maximum user signaling rate The maximum rate, in bits per second, at which binary information could be transferred in a given direction between users over the telecommunication system facilities dedicated to a particular information transfer transaction, under conditions of continuous transmission and no overhead information.

maxwell An obsolete unit of flux density equal to the number of lines of force, assumed to be one line per square centimeter normal to a field of one gauss.

Maxwell bridge An AC bridge used for measuring inductance.

Maxwell's equations Mathematical equations which express the effect of varying electric and magnetic fields at different points.

Mbit/s Megabits (millions of bits) per second.

MCDU A system used in Europe to indicate a telephone number by thousands (M), hundreds (C), tens (D) and units (U). Sometimes these are preceded by OPQ to indicate an exchange code. The American counterpart of OPQMCDU is N1X NNX XXXX where N is any digit from 2 to 9 and X is any digit from 0 to 9.

mean An arithmetic average in which values are added and divided by the number of such values.

mean busy hour The mean busy hour of a group of circuits, switches, or an exchange is the uninterrupted 60-minute-period for which the total traffic of a sample is the maximum. If it is not known which 60-minute-period constitutes the mean busy hour, a sample measurement taken over 10 days can be used to establish its position.

mean life Average time between going into service and failing in service.

mean life, observed The arithmatic mean of actually observed "times to failure" of samples under specified conditions.

mean power See power, mean.

mean-power-of-the-talker volume distribution The mean-power-talker-volume less a conversion factor used to convert from vu to dBm.

mean time between failures (MTBF) For a particular interval, the total functioning life of a population of an item divided by the total number of failures within the population during the measurement interval. The definition holds for time, cycles, miles, events, or other measure-of-life units.

mean time between outages (MTBO) The mean time between equipment failures that result in loss of system continuity or unacceptable degradation as expressed in the equation:

$$MTBO = MTBF/(1-FFAS)$$

where MTBO is the mean time between outages, MTBF is the nonredundant mean time between failures, and FFAS is the fraction of failures automatically switched.

mean time to repair (MTTR) The total corrective maintenance time divided by the total number of corrective maintenance actions during a given period of time.

mean time to service restoral (MTSR) The mean time needed to restore service following system failures that result in a service outage.

mean-volume talker The median volume of a group of talkers measured at a given location. It is the value in an ordered set of values below and above which there are an equal number of values.

mean waiting time (average delay) The total waiting time of all bids divided by the total number of bids, including those not delayed.

measured rate service Telephone service for which charges are made in accordance with the use made of the line, in contrast to flat rate service in which a single rental payment permits an unlimited number of local calls to be made without further charge.

mechanical splice An optical waveguide splice accomplished by external fixtures or materials rather than by thermal fusion. Index matching material may be applied between the two fiber ends.

mechanized wire centering/cross section (MWC/CS) An AT&T software routine which facilitates the selection of optimum wire center sites for new switching offices.

media access unit (MAU) Circuitry used in LANs to enable data terminal equipment to access the transmission medium.

median A value in a series which has as many readings or values above it as below.

medium Any material substance that can be used for the propagation of signals from one point to another.

medium frequency (MF) Frequencies between 300 and 3000 kHz, corresponding to wavelengths between 100 and 1000 m. In many parts of the world it is the only practicable broadcast radio receiving band.

medium power flux-density In the broadcasting satellite service, a power flux-density which enables signals radiated by broadcasting satellite space stations to be received either by simple receiving installations with a secondary grade of reception quality, or by more sensitive receiving arrangements with a primary grade of reception quality.

medium scale integration (MSI) A medium density integrated circuit containing logic functions more complex than small scale integration but less complex than large scale integration. Most four-bit counters, latches and data multiplexers are considered to be MSI devices.

191

medium speed A data transmission rate between 300 bit/s and 2400 bit/s.

medium term bit error rate Bit error rate that can be encountered for relatively short time periods (measured in minutes) due to temporary malfunctions of transmission equipment.

meet-me conference A conference call facility by which the dialing of a particular number routes the caller to a multi-line bridge enabling him to join the conference.

mega- Prefix meaning one million, or 10^6.

megabit One million binary digits.

megabyte One million bytes.

Megacom AT&T-C service providing WATS-type distance and usage-sensitive outward calling facilities for business customers who use phone lines for at least 1000 users per month.

Megacom 800 AT&T-C service providing inwards WATS-type service for business customers.

megaflops Millions of floating point operations per second.

megahertz One million Hertz.

megastream British Telecom name for digital transmission services at 2 Mbit/s or higher bit rates.

megger A meter calibrated to record the resistance of a ground connection.

megohm One million ohms.

memory Equipment which holds information in store for later use.

memory, external Memory unit which is not part of the main computer.

memory, internal Memory unit which is under the direct control of a processor and which forms part of the centralized complex.

memory, permanent Memory containing information which is likely to be needed for a considerable period of time.

memory, random access Memory system for which the address of the particular section holding the information in question, or into which new information is to be stored, must be specified for each action.

memory, scratch pad Memory into which information is entered for a comparatively short time. It is the computer equivalent of a pencilled note.

memory, semi-permanent Memory used for the stored program of a central office, including routing and translation information.

memory, temporary Memory used to store information to be held only during the processing of telephone call connections.

memory capacity Total number of bits that can be stored in a particular device.

memory map A listing of addresses, or symbolic representations of addresses, which define the boundaries of the memory address space occupied by a program or a series of programs.

memo ticket A special call-charging ticket prepared by an operator for services which are not completely straightforward, such as credit card calls.

menu Videotex frame giving index numbers for detailed information held in the database: the user keys the selected number and is presented with the information he requires on the screen.

meridian plane Any plane that contains the optical axis.

meridional ray A ray which crosses through the optical axis of an optical waveguide.

mesh network A data network with a grid type topology in which some nodes are connected to many others, providing alternative paths across the network.

mesochronous The relationship between two signals whose corresponding significant instants occur at the same average rate. The phase relationship between corresponding significant instants usually varies between specified limits.

message 1. A completed telephone call. 2. A teletypewriter message. 3. Any idea expressed briefly in a plain or secret language and prepared in a form suitable for transmission by any means of communication. 4. In data transmission, a defined entity of information from the subscriber to the exchange pertaining to a call or a control operation for a service sent in one sequence over the signaling medium. 5. (In message handling systems) The unit of information transferred by the Message Transfer System. It consists of an envelope and a content.

message, multiple address A message which is to be transmitted and delivered to more than one addressee.

message, service A message from one telecommunications station to another regarding a service problem.

message alignment indication In data signaling, this indicates whether the signal unit (1) contains a one-unit message; (2) is the initial signal unit of a multi-unit message; (3) is an intermediate signal unit of a multi-unit message; (4) is the final signal unit of a multi-unit message.

message center The portion of a communications center responsible for acceptance and processing of outgoing messages and for receipt and delivery of incoming messages.

message discrimination In common-channel signaling, the process which decides, in the case of each incoming message, whether the signaling point is the destination point or the signaling transfer point and, accordingly, whether the message should be handled according to (signaling) message routing functions.

message distribution In common-channel signalling, the process of determining, upon receipt of a signaling message at its destination point, to which user part the message is to be delivered.

message format Rules for the placement of such portions of a message as its heading, address, text, and end.

message handling address The address to which a message is to be delivered. It consists of an Administration Management Domain name, a country name, and a set of User Attributes.

message handling system (MHS) The set of User Agents (UAs) plus the Message Transfer System (MTS).

message numbering The identification of each message within a system by the assignment of a sequential number.

message oriented text interchange system (MOTIS) Protocol in a data network which allows store and forward message switching between systems.

message rate charge service Service-related feature that, for a fixed monthly charge, allocates a fixed number of message units to a subscriber to permit a limited number of calls to be completed to a specified group of destination codes.

message relay System whereby a caller may dictate a message into recording equipment and require that it be passed to a particular telephone number by the following morning.

message retrieval The process of locating a message which has been entered in a telecommunications system.

message route In common-channel signaling, a link or consecutive links connected in tandem and used to convey a signaling message from an originating point to its destination point.

message routing In signaling, the process of selecting, for each signaling message to be sent, the common-channel signaling link to be used.

message sequence chart Chart showing the order in which messages can be transferred from one process to another in order to achieve synchronism.

message signal unit A common-channel signaling signal unit containing a service information octet and a signaling information field which is retransmitted by the signaling link control if it is received in error.

message suffix In data transmission, the character indicating the end of the message.

message switching Transmission method in which messages are transmitted to an intermediate point, stored temporarily and then transmitted later to a final destination. The destination of the message is usually indicated in an internal address field (or header) in the message itself.

message switching network Public datacommunications network over which subscribers send textual messages, e.g., Telex, TWX.

message telecommunication service (MTS) The long distance telephone service = toll or trunk services.

message toll service See direct distance dialing.

message transfer agent (MTA) The functional component that, together with other MTAs, constitutes the message transfer system. MTAs provide message transfer service elements by (1) interacting with originating user agents (UAs) via the submission dialogue, (2) relaying messages to other MTAs based upon recipient designations and (3) interact-

ing with recipient user agents via the delivery dialogue.

message transfer agent entity (MTAE) An entity located in a message transfer agent (MTA) that is responsible for controlling the message transfer layer (MTL). It controls the operation of one protocol to other peer entities in the MTL.

message transfer layer (MTL) A layer in the application layer that provides message transfer system (MTS) services elements.

message transfer part The functional part of a common-channel signaling system which transfers signaling messages as required by all the users, and which performs the necessary subsidiary functions such as error control and signaling security.

message transfer protocol The protocol which defines the relaying of messages between message transfer agents (MTAs) and other interactions necessary to provide MTL service.

message transfer service The set of optional service elements provided by the message transfer system.

message transfer system (MTS) The collection of message transfer agents (MTAs) which provide the message transfer service elements.

message unit A unit which serves as a method of charging for messages. It is based on time and distance.

message waiting service A service provided by many hotels whereby a lamp directly associated with a room telephone shows whether a message came in for a guest during his absence and is being held for him by reception.

messenger 1. Person who delivers a telegram. 2. Steel strand used to support an aerial cable.

meta-language A symbolic method for defining man-machine language input and output syntax.

metal, nitride, oxide semiconductor (MNOS) A form of "read mostly" electrically alterable device. Ordinary erasable MOS cells must be radiated with ultraviolet light to erase programs; MNOS devices can be erased by electric means only.

metal-clad switchgear An indoor or outdoor metal structure containing switching equipment and other associated equipment such as instrument transformers, buses and connections. The pieces of equipment are insulated and placed in separate grounded metal compartments.

metal film resistor A resistor made of metallic material deposited on film.

metal-oxide semiconductor (MOS) A semiconductor which depends on the conditions at the interface between a semiconductor layer, usually silicon, and an insulator layer, usually silicon dioxide.

metal rectifier Rectifying device using metallic plates in contact with plates of other materials, typically copper on selenium or copper on copper oxide.

metal semiconductor field effect transistor (MESFET) A junction field effect transistor with a Schottky barrier instead of a normal semiconductor junction.

metallic circuit A circuit in which metallic conductors are used and in which the ground or earth forms no part.

metallic circuit currents Currents which flow in opposite directions in the two wires of a pair.

metallic test access A hardware unit providing metallic connections between test access points (e.g., in subscriber line circuits in a digital switching center) and various types of test equipment.

metallic Varley A Varley test using a third wire instead of a ground to localize a contact fault or check an unbalance.

meteor burst links Radio links over long distances (e.g., 2000 km) utilizing the presence of meteor trails (about 100 km above the earth's surface) to reflect radio transmissions back to earth, using frequencies in the 40-50 MHz band. About 10^9 meteors, of different sizes, enter the earth's atmosphere daily so transmission, which takes place in bursts when a suitable meteor trail has been located, is proving fairly reliable.

meteors, meteorites Metallic or stone bodies which enter the Earth's atmosphere and burn themselves up during entry. They produce ionization which affects long distance radiocommunications.

meter 1. The SI unit of length. (Spelled metre in Europe.) 2. An instrument for measuring the value of some quantity.

meter, all trunks busy (ATB) Meter which counts and records the number of times all the trunks in a group are busy.

meter, ampere-hour Meter which integrates current and time to indicate the number of ampere-hours of power to be paid for by the consumer.

meter, bias Milliammeter with center zero used to measure teletypewriter currents and indicate signal bias.

meter, field strength Combination radio receiver and meter calibrated for use with a particular antenna and designed to give a direct reading of the strength of a radio signal at a given point.

meter, gas volume A totalizing gas meter which measures the total volume of dry air (or nitrogen) being pumped into a cable distribution system.

meter, last trunk busy (LTB) Meter which counts and records the number of times the last trunk in a group is busy.

meter, overflow Meter which counts and records the number of calls offered to a section of an office which could not be connected because all trunks were busy.

meter, peg-count Meter which counts the number of calls being handled at a particular stage in an office.

meter, pipe alarm Meter which monitors pressure and volume of gas passing into a pressurized cable system. Alarm contacts are fitted to these meters to signal wide divergence from the normal.

meter, running time Totalizing clock which runs whenever a device is in operation. Such meters are used at standby power plants so that maintenance work can be carried out at appropriate times.

meter, usage Traffic meter which scans different equipment at different frequencies and gives readings or printouts of total usage (in CCS or erlangs).

meter panel, dry air A panel in a central office containing a bank of individual meters showing pressure and air flow for each main pressurized cable, as well as totalizing meters showing the total outflow of dry gas or air into the cable network.

metering, periodic pulse (PPM) A method of metering calls by pulsing the subscriber's meter at intervals which depend on the distance between the two subscribers and on the time of day. Syn: zone registration

metering-over-junction signal In Great Britain, a signal passed back from a group switching center to the originating local exchange to signify a periodic metering pulse.

metering service, message Metering service provided for hotel guests in which meters associated with telephones in guest rooms show the number and cost of all calls which will be included in the guest's bill.

metric system A decimal system of measurement based on the meter, the kilogram and the second. See SI system (Appendix B).

metropolitan area network (MAN) A data network linking together terminals, memories and other resources at many sites within a city area. Each site may have its own local area network (LAN). Links between sites are usually on digital circuits rented from the local telephone company using a bit-rate appropriate to traffic requirements (e.g., 56 kbit/s, 64 kbit/s, 1.5 Mbit/s, 2 Mbit/s, 8 Mbit/s).

Metropolitan Area Transmission Facility Analysis Program (MATFAP) An AT&T software routine which is an aid in long-term planning of interoffice transmission networks in major cities.

mho The old unit of conductance, or reciprocal ohms. Now replaced by the siemen.

micro- A prefix meaning one-millionth part, or 10^{-6}.

Micro Switch A brand name for a type of miniature switch.

microbend loss In an optical waveguide, that portion of the total loss attributable to microbending, which see.

microbending In an optical waveguide, sharp curvatures involving local axial displacements of a few micrometers and spatial wavelengths of a few millimeters. Such bends may result from waveguide coating, cabling, packaging or installation.

microcircuit 1. Any small circuit. 2. A device in which a number of passive and active circuit elements are considered as indivisibly associated on or within a continuous structure to perform the function of a circuit.

microcircuit, film-type An array of circuit elements and metallic interconnections formed by deposition of a thick or thin film on an insulating substrate.

microcircuit, hybrid A microcircuit consisting of a combination of film-type microcircuits and monolithic integrated circuit elements or combinations of either with discrete components and/or circuit elements.

microcode Set of control functions performed by the instruction decoding and execution logic of a computer and which defines the instruction repertoire of that computer.

microcomputer An electronic logic unit capable of executing instructions from internal memory, on data contained in the internal memory. In certain cases the internal memory may be augmented by an external memory.

microelectronics The study, development and production of large scale integrated circuits, semiconductor devices generally, and hybrid circuits.

microinstruction An instruction of a microprogram.

microm A read-only-memory integrated circuit containing microcode.

micromicrofarad One-trillionth (10^{-12}) of a farad, now called a picofarad.

micron The unit previously used for specifying the wavelength of light. It is equal to one-millionth of a meter. It has been replaced by the nanometer (10^{-9}m).

microphone Transducer which converts sound waves into electric signals.

microphone, carbon A carbon granule microphone used in telephone instruments in which pressure on the granules from the diaphragm produces resistance changes.

microphone, crystal Microphone in which the diaphragm's movements deform a piezoelectric crystal. This produces voltages which are directly related to the original audio frequency signal.

microphone, foil A microphone having a plastic diaphragm with a thin metallic deposit. Vibration of the diaphragm produces an electric field which is used to generate the required voice-frequency current.

microphone, moving coil Microphone whose diaphragm moves a small coil through a strong magnetic field to generate the required voice-frequency currents.

microphone, moving conductor A microphone in which sound waves act on a ribbon-shaped diaphragm which then generates voice-frequency currents by moving in a magnetic field.

microphonic Likely to produce noise in a circuit as a direct result of component movement.

microprocessor A single package (normally a single chip) electronic logic unit capable of executing from external memory a series of general purpose instructions contained in the external memory. The unit does not contain integral user memory although internal memory on the chip may be present for internal utilization by the chip in performing its logic function. User memory is customarily provided via an external memory chip although some products have some limited on-board memory that is normally used for purposes other than program storage.

microprogram 1. A sequence of elementary instructions, maintained in special storage, which corresponds to a specific computer operation. Execution is initiated by the introduction of a computer instruction into an instruction register of a computer. 2. A program implemented as microcode.

microvolt One-millionth (10^{-6}) of a volt.

microvolts per meter A measure of the field intensity of a radio signal.

microwatt One-millionth (10^{-6}) of a watt.

microwave A term loosely applied to those radio frequency wavelengths which are sufficiently short to exhibit some of the properties of light. Commonly used for frequencies from about 1 GHz to 30 GHz.

Microwave Communications Inc. (MCI) First and largest of the alternative long-haul carriers in competition with AT&T Communications.

microwave frequency Any of the frequencies suitable for microwave communication. Most common are the 2 GHz, 4 GHz and 6 GHz bands.

mid band The part of the frequency band that lies between television channels 6 and 7, reserved by the FCC for air, maritime and land mobile units, FM radio and aeronautical and maritime navigation. Mid-band frequencies, 108 to 174 MHz, can also be used to provide additional channels on cable television systems.

mid-section Located halfway between two loading coils in a lump-loaded cable system.

mid-span Between poles in a pole route. Also sometimes used to describe fault localization midway between manholes in an underground cable system.

mil 1. One-thousandth part of one inch. 2. The angle subtended by a 1 foot arc at a distance of 1000 feet (17.5 mils equal 1 degree).

mile, cable For submarine cables: 1855.3 meters, or 6087 feet.

mile, nautical 1852 meters, or 6076 feet.

mile, wire One mile of a single conductor.

mile of standard cable (MSC) The original transmission unit used when all telephone transmission was at audio frequencies. It is the loss at 800 Hz of one mile of paired conductor with loop resistance of 88 ohms and capacitance of 0.54 mfd.

mileage Distance used in tariff calculations. It is locally defined and usually refers to airline distance, rather than actual route miles.

Military Affiliated Radio System (MARS) An amateur radio group whose services would be made available for military communications in the event of an emergency.

milli- Prefix meaning one-thousandth (10^{-3}).

milliammeter Meter which measures in thousandths of an ampere

millihenry One-thousandth of a henry.

millivolt One-thousandth of a volt.

milliwatt One-thousandth of a watt.

minicomputer A complete computer including input-output devices, memory stores, processors, etc., but intended to deal only with comparatively simple tasks.

minimum charge duration (MCD) The period of long-distance call-time for which a call charge is always applied, regardless of the actual length of the call.

minimum discernible signal The smallest input which will produce a discernible change in output.

minimum standard antenna A directional antenna having the specified minimum characteristics as regards directive gain and service sector at its operating frequency or frequencies.

minimum usable field strength Minimum value of the transmitter field strength necessary to permit a desired reception quality, under specified receiving conditions, in the presence of natural and man-made noise, but in the absence of interference from other transmitters. Syn. minimum field-strength.

minor relay center In telegraphy, a tape relay center with no alternative routing facilities.

minor switch See switch, minor.

mirror, splicer's Small mirror used by cable splicers to check on the quality of the wipe made around a lead cable splice.

misalignment Fault in a long transmission line resulting if each repeater section does not insert sufficient gain to offset losses. If actual conditions vary from those expected there can be significant overall gains (making crosstalk higher than it should be) or losses (making signal/noise ratio worse than it should be).

misalignment loss See angular misalignment loss, gap loss, lateral offset loss.

miscellaneous common carrier A common carrier offering TV or radio facilities, not a public landline telephone service or a public message telegraph service.

misdelivered block A block received by a user other than the one intended.

mismatch, impedance Difference between impedances at a point. These differences introduce reflection losses and make maximum power transfer impossible.

mission bit stream The totality of subscriber information bits being passed through a system. This excludes framing, stuffing, control and service channel bits.

mitigation, electrolysis Method of reducing the severity of damage to lead cable sheaths by attempting to control the direction of current flow to ground from the sheath so that the metal of the sheath is not eaten away. See cathodic protection, galvanic anode, negative return.

mixer 1. A circuit with two or more input signals and an output that is some desired function of the input. 2. In sound transmission, recording or reproducing, a device that combines two or more input signals, usually adjustable, into one output signal. 3. The stage in a superheterodyne radio receiver at which the incoming signal is modulated with the signal from the local oscillator to produce the intermediate frequency signal. Syn. first detector.

mixing ratio The ratio of the mass of water vapor to the mass of dry air in a given volume of air (frequently expressed in g/kg). The mixing ratio affects radio propagation.

MKS. Old system of internationally standardized physical measurements based on the meter, kilogram and second. Now replaced by the SI system.

mnemonic A memory aid in which an abbreviation or arrangement of symbols has an easily remembered relationship to the subject.

mnemonic address A simple address code with some easily remembered relationship to the actual name of the destination, often using initials or other letters from the name to make up a pronounceable word.

mobile control terminal The fixed base radio station which controls a network of mobile radio stations. The control terminal sends out coded pulses to call the required mobile, arranges for frequency/channel allocation and provides supervisory services during a call.

mobile telephone A public radiotelephone service which, by means of transportable equipment, gives both-way access to (1) the public telephone network via one base station, (2) other mobile telephone stations via one or more base stations.

mobile telephone exchange A switching center controlling calls in a cellular mobile radio system. (The exchange itself is not normally mobile.)

modal dispersion Pulse spreading caused by different optical path lengths in a multi-mode fiber.

modal distance In telephony, the distance between the center of the grid of a telephone handset transmitter cap and the center of the mouth of a person talking, when the handset is in the modal position.

modal distortion See multimode distortion.

modal noise Disturbance in multi-mode fibers fed by laser diodes, occurring when fibers contain elements with mode-dependent attenuation (e.g., imperfect splices).

modal position In telephony, the position of a handset when its receiver is held in close contact with the ear of a person with head dimensions that are modal for a population.

mode 1. In any transmission line, one of the permitted electromagnetic field distributions. The field pattern of a given mode depends on wavelength, refractive index and waveguide geometry. 2. One of the modes of vibration of a piezoelectric crystal. 3. The most frequent value in a frequency distribution.

mode, break-in Method of echo suppression operation. Suppression is removed when the receiving end attempts to interupt the voice from the sending end.

mode, dominant waveguide The simplest mode for propagation of radio waves through a particular waveguide. The longest possible wavelength can be transmitted when using the dominant mode.

mode, suppressor Method of echo suppression operation in which the receiving end is unable to break in and interrupt the sending end. The suppression lasts as long as there is a signal arriving from the sending end.

mode, TE Transverse electric mode. In a homogeneous isotropic medium, it is an electromagnetic wave in which the electric field vector is everywhere perpendicular to the direction of propagation.

mode, TEM Transverse electromagnetic wave. In a homogeneous isotropic medium, it is a wave in which both the electric and magnetic field vectors are everywhere perpendicular to the direction of propagation.

mode, TM Transverse magnetic wave. In a homogeneous isotropic medium, it is an electromagnetic wave in which the magnetic field vector is everywhere perpendicular to the direction of propagation.

mode coupling In an optical waveguide, the exchange of power among modes. Mode coupling reaches equilibrium after propagation over a finite distance that is designated the equilibrium length.

mode distortion See multimode distortion.

mode filter A device used to attenuate certain modes.

mode mixer See mode scrambler.

mode mixing Exchanging of energy between modes in an optic fiber system, due to inhomogeneities of fiber geometry.

mode of communication identification Information used to give an instruction to the switching equipment to select the required network or mode of communication.

mode scrambler A device for inducing mode coupling. Syn. mode mixer

mode volume 1. The number of propagating modes which an optical waveguide will support. For $V \geqq$ 5 it is approximately given by $V^2/2$ and $(V^2/2)(g/(g+2))$ for step index and power-law profile waveguides respectively, where g is the profile parameter. 2. Product of cross-sectional area and solid angle available in an optic waveguide for light propagation.

modem Acronym for modulator-demodulator. 1. A device that modulates and demodulates signals. Modems are used primarily for converting digital signals into quasi-analog signals for transmission and for reconverting the quasi-analog signals into digital signals; although many additional functions may be added to provide for customer service and control features. Syn. signal conversion equipment. 2. In frequency division carrier systems, devices that change the frequency of a signal: voice band on one side, and the carrier frequency with either erect or inverted sidebands on the other side.

modem, fixed frequency A modem operating on a predetermined fixed frequency or frequencies.

modem, frequency agile A modem capable of being shifted from one operating frequency to another, under software control.

modem patch A method of electrically connecting paths of a circuit through employment of back-to-back modems.

modes Discrete optical waves that can propagate in optical waveguides. There can be several hundred modes in multimode fibers, but only one (the fundamental) in mono-mode fibers.

modes I through V Different methods of operating teletypewriters or data terminals, variants of half-duplex or full-duplex, synchronous, or non-synchronous, with or without automatic message numbering or acknowledgement.

modified alternate mark inversion An alternate mark inversion signal which does not strictly conform with alternate mark inversion but includes violations in accordance with a defined set of rules.

modified chemical vapor deposition (MCVD) A process for manufacture of low-loss optical waveguides.

Modified Final Judgment Settlement (associated with 1982 Consent Decree) between AT&T and the U.S. Government involving the separation of Bell Operating Companies from Western Electric, AT&T Long Lines and the Bell Labs.

modified Julian date Julian date less 2,400,000.5 days (January 1, 1900 was Julian date 2,415,020).

modified refractive index The sum of the refractive index of the air at a given height above sea level and the ratio of this height to the radius of the Earth.

modular 1. Having dimensions which are integral multiples of a unit of length called a module. 2. In switching equipment, designed and manufactured in functional modules so that individual modules may later be replaced by newer versions. 3. Telephone terminal equipment with readily separable parts which the subscriber may himself replace, reducing the need for repair service calls.

modulate To vary characteristics of a carrier signal in order to carry information supplied by a modulating signal. See modulation.

modulation 1. The process whereby the amplitude or frequency or phase of a single-frequency wave (called the carrier wave) is varied in step with the instantaneous value of, or samples of, a complex wave (called the modulating wave). 2. The intelligence carried on a modulated carrier wave. 3. A controlled variation with time of any property of a wave for the purpose of transferring information.

modulation, amplitude A type of modulation in which the amplitude of the carrier is varied in accordance with the instantaneous value of the modulating signal.

modulation, angle A type of modulation in which the angle of the sine wave is varied in accordance with the modulating signal. The two forms of angle modulation are phase modulation (PM) and frequency modulation (FM).

modulation, delta A code representing the difference between the amplitude of a sample and the amplitude of the previous sample. It operates well in the presence of noise but requires a wide frequency band. See delta mod.

modulation, differential Type of modulation in which the choice of the significant condition for any signal element is dependent on the choice for the previous signal element.

modulation, double Modulation carried out in two stages. First the sub-carrier is modulated by the information-carrying modulating signal, then the modulated sub-carrier is used to modulate a higher frequency carrier.

modulation, duobinary Modulation of a carrier by a three-level signal, typically $+1, 0, -1$.

modulation, frequency (FM) A type of modulation in which the instantaneous frequency of a sine wave carrier is made to depart from the carrier frequency by an amount proportional to the instantaneous value of the modulating signal.

modulation, frequency shift A type of frequency modulation in which the carrier frequency is shifted between a mark frequency and a space frequency in response to the impressed intelligence signal.

modulation, high-level Modulation carried out at a power level comparable with the final output power of the unit.

modulation, incorrect or defective As applied to telegraph distortion, modulation containing one or more elements, the significant condition of which differs from that corresponding to the kind prescribed by the code. It is similar to incorrect restitution.

modulation, low-level Modulation carried out at a low power level point in the system so that final stages of amplification operate on the modulated signal.

modulation, multiple Double modulation.

modulation, perfect As applied to telegraph distortion, modulation such that all significant intervals are associated with correct significant conditions and conform accurately to their theoretical durations. It is similar to perfect restitution.

modulation, phase (PM) A type of modulation in which the angle relative to the unmodulated carrier is varied in accordance with the instantaneous value of the amplitude of the modulating signal.

modulation, pulse amplitude (PAM) See pulse amplitude modulation.

modulation, pulse code (PCM) See pulse code modulation.

modulation, pulse duration (PDM) See pulse duration modulation.

modulation, pulse frequency See pulse frequency modulation.

modulation, pulse length See pulse duration modulation.

modulation, pulse position See pulse position modulation.

modulation, pulse time See pulse time modulation.

modulation, pulse width See pulse duration modulation.

modulation, single-sideband (SSB) Type of modulation in which the spectrum of the modulating function is translated in frequency by a specified amount either with or without inversion.

modulation, spread spectrum See spread spectrum modulation.

modulation, velocity Modulation of an electron stream by varying the field acting on the stream so as to accelerate or decelerate it, causing the bunching of electrons. See klystron.

modulation capability Maximum percentage modulation that can be carried by a transmitter without introducing unacceptable distortion.

modulation depth The modulation factor expressed as a percentage. Syn.: depth of modulation

modulation factor In amplitude modulation, the ratio of the peak variation actually used to the maximum design variation in a given type of modulation.

modulation index In angle modulation, the ratio of the frequency deviation of the modulated signal to the frequency of a sinusoidal modulating signal.

modulation plan The manner in which audio channels, or groups of channels, are modulated to their final frequencies for transmission by line or radio.

modulation rate The reciprocal of unit signal time figured in seconds and expressed in bauds.

modulator Device which enables the intelligence in an information carrying modulating wave to be carried by a signal at a higher frequency.

modulator, balanced A modulator circuit which transmits an output which does not include the original carrier frequency itself. Only information carrying sidebands are produced.

modulator, FM A low power FM transmitter used in CATV systems.

modulator, ring See ring modulator.

modulator, television Low power TV transmitter designed to feed programs into a CATV system.

modulator, television sound Device used to feed a sound program into a CATV system, utilizing an unused TV channel.

module 1. An assembly replaceable as an entity, not normally capable of being disassembled, often an interchangeable plug-in item. 2. A program unit that is discrete and identifiable with respect to compiling, combining with other modules, and loading.

moisture-repellant Made or treated in such a way that moisture will not penetrate.

moisture-resistant Made or treated in such a way that exposure to high humidity will not cause faulty operation.

molded case circuit breaker A circuit breaker assembled as an integral unit and enclosed in an insulated housing. Molded case circuit breakers are for use on systems rated at 600 V and less.

mole The base SI unit of the amount of substance. It is the amount of substance of a system which contains as many elementary entities as there are atoms in 0.012 kg of carbon-12.

molecular beam epitaxy Method of producing layers of crystal material on a substrate by bombarding it with beams of atoms.

Molnya A series of Russian asynchronous low-altitude telecommunications satellites.

monaural Describing sound reproduction using only one source.

monitor 1. To listen to a voice communication for the purpose of determining its quality and trouble-free operation. 2. To check on the quality of a teletypewriter circuit by means of a monitoring printer. 3. To listen in on a cord circuit at a switchboard to determine whether the circuit is busy. 4. A functional unit that observes and records selected activities within a system for analysis.

monitor, ringing Testing device which continually monitors the ringing supply at a central office and gives an alarm if this fails or if the periodicity of interruptions becomes unacceptable.

monitor, tone Testing device which continually monitors the busy tone supply at a central office and gives an alarm if the tone supply or the interruption circuit fails.

monitor jack A jack which permits access to communications circuits for the purpose of observing the circuit signal conditions without interrupting the service provided.

monitor printer Device which prints out a copy of all data being transmitted over the circuit under observation.

monitoring 1. Listening to a communication service for the purpose of determining its quality or whether or not it is free from trouble or interference. 2. In military communications, listening to, reviewing, and/or recording enemy, one's own, or friendly forces' communications for the purpose of maintaining standards, improving communications, or for reference.

monochromatic An idealized concept meaning a single frequency or wavelength. In practice, radiation is never perfectly monochromatic but, at best, displays a narrow band of wavelengths.

monochromatic radiation Radiation of a narrow band of frequencies.

monocontainer Describing a submarine cable system repeater configuration that consists of a single cylindrical chamber to which cable is attached at each end, either flexibly or inflexibly.

monolithic integrated circuit A microcircuit fabricated as a single component consisting of elements formed in or on a single semi-conducting substrate by diffusion, implantation or deposition.

monomode optical waveguide Single mode optical waveguide.

monophonic Sound over a single channel. Monaural.

monostable Describing a device which is stable in one state only; an input pulse causes it to change its state, but it reverts immediately to its stable state.

moon The natural satellite of the Earth.

Morse telegraph code The dot-dash code used in early line telegraphy and still used in radio telegraphy where machine codes cannot yet be economically brought into use.

motherboard Printed circuit board on to which subordinate boards (daughter boards) may be connected.

motor, electric Machine which converts electrical energy into mechanical energy.

motor, series Electric motor with its armature and field windings in series.

motor, shunt-wound Electric motor with its armature and field windings in parallel.

motor, stepper A type of rotary motor which converts pulses of DC current into rotary steps, one step per pulse.

motor, synchronous An AC motor which operates at a speed controlled by the frequency of the AC supply.

motor-boating A low frequency oscillation which sometimes occurs under fault conditions in electron tube circuits.

motor effect The repulsion force exerted between adjacent conductors carrying currents in opposite directions.

moulding, ground wire Wooden or plastic cover which protects a vertical copper ground wire running up a telephone pole.

mounting A rack or frame used as a support for equipment.

mounting, dial Fitting used to mount telephone dials on the faces or desks of switchboard operating positions.

mounting, handset The part of a telephone instrument on which the handset rests. The cradle switch, which gives on-hook and off-hook signals, is located inside.

mounting, jack Mounting strip used to hold jacks in a switchboard face panel.

mounting, key Mounting plate used to hold rows of keys on switchboard operating positions.

mounting, protector An insulating base with fuses and lightning arrestors, and sometimes heat coils, used at outdoor telephone installations.

mouse Hand-held device which translates movement or click-button instructions into movement of the cursor on a VDT screen, replacing the more laborious ordinary typewriter keyboard instructions.

mouth The open end of a horn antenna.

mouth, artificial Laboratory equipment used in the design and testing of telephone instruments.

moving coil Any device which utilizes a coil of wire in a magnetic field in such a way that the coil is made to move by varying currents, or itself produces varying voltages because of its movement.

mu-law (μ) Method of encoding audio signals used in U.S. standard 1544 kbit/s 24 channel PCM systems.

muldex Pulse code modulation multiplex/demultiplex equipment with analog voice on one side and coded binary signals on the other.

multiaccess telephone service Special arrangement designed for extra reliability which permits major users of a telephone service to be given direct access from their PBXs to several central offices.

multiaddress calling facility A facility which permits a user to nominate more than one addressee for the same data. The network may accomplish this sequentially or simultaneously.

multi-alternate routing A routing plan which provides for more than one possible routing of calls to a destination.

multiblock A group of 8 blocks or 96 signal units on the signaling channel in common-channel signaling system CCITT No 6.

multiblock synchronization signal unit (MBS) A signal unit carrying a signal concerning the multiblock synchronization of common-channel signaling system CCITT No 6.

multichip microcircuit A microcircuit containing two or more monolithic integrated circuit chips bonded to a common substrate.

multiconductor Containing more than one pair of conductors in the same cable sheath.

multicontainer A submarine cable system repeater configuration consisting of several containers coupled flexibly end-to-end. The cable connects at the end containers.

multicoupler, receiving Coupling device using isolating amplifiers and enabling one antenna to feed several radio receivers.

multidrop A communications arrangement where multiple devices share a common transmission channel, though only one may transmit at a time. Syn.: multipoint line

multidrop line A multipoint line, one which has more than one terminal or station connected to it.

multielement dipole antenna An arrangement of a number of dipole antennas.

multielement service plan Pricing plan for local telephone service which divides costs into segments for wiring charges, jack charges, visit to premises charge, line connection charge, etc.

multiemitter transistor (MET) A device which saves chip space by replacing several individual transistors.

multifiber cable An optical cable that contains two or more optical waveguides, each of which provides a separate information channel.

multifiber connector An optical connector designed to mate two multifiber cables, providing simultaneous optical alignment of all individual waveguides.

multiframe Describing a set of consecutive frames in pulse code modulation multiplexing in which the position of each frame can be identified by reference to a multiframe alignment signal. The multiframe alignment signal does not necessarily occur, in whole or in part, in each multiframe.

multifrequency signaling A signaling method using combinations of two-out-of-six (MF 2/6) or two-out-of-eight (MF 2/8) voiceband frequencies to indicate telephone address digits, precedence ranks, and line or trunk busy. "MF 2/6" uses frequencies of 700, 900, 1100, 1300, 1500, and 1700 Hz; "MF 2/8" uses 697, 770, 852, 941, 1209, 1336, 1447, and 1633 Hz.

multifunction facilities Facilities which can provide both intra-LATA and inter-LATA services and so

can be shared by a Bell Operating Company and AT&T Communications.

multihop 1. A long distance radiocommunication service which operates via more than one reflection from the ionosphere. 2. A microwave system with several repeaters between the two terminal stations.

multilayer A type of printed circuit board which has several layers of pattern interconnected by electroplated holes from one plane to another.

multilayer dielectric filter An optical filter consisting of a sequence of thin layers of transparent material with controlled thicknesses and refractive indices.

multiline telephone Telephone instrument with keys or buttons which enable it to be connected to any of a number of lines.

multiline variety package Bell System software package providing Centrex-type services.

multilink A circuit or microwave system made up of two or more links in tandem.

multimeter A meter which can be used to measure potential or current with a wide range of values.

multimode Term used to describe an optical waveguide that permits the propagation of more than one mode.

multimode distortion In an optical waveguide, that distortion resulting from the superposition of modes having differential modal delays. The term "multimode dispersion" is often used as a synonym but is erroneous since the mechanism is not dispersive in nature. Syns. intermodal distortion, modal distortion

multimode fiber Optical waveguide whose core diameter is large compared with the optical wavelength and in which a large number of modes is capable of propagation.

multimode group delay Differential modal delay.

multimode laser A laser which produces simultaneous emission at two or more discrete wavelengths and/or in two or more transverse modes.

multimode optical waveguide An optical waveguide which will allow more than one bound mode to propagate. It may be either a graded index or step index waveguide.

multioffice exchange An exchange area served by several central offices.

multiparty line A subscriber's line which serves more than one subscriber.

multipath The propagation phenomenon that results in radio signals reaching the receiving antenna by two or more paths.

multipath fading Fading of a received radio communications signal because incoming signals are received via more than one path. Since path lengths differ, the different incoming signals do not always assist each other; they sometimes cancel each other out.

multipath propagation Propagation by way of a number of transmission paths.

multiple Describing the wiring up normally in parallel of a number of identical connecting devices (such as jacks) on operating positions or outlets from a bank of automatic switches.

multiple, level A set of multiples connecting the outlets of a switching stage.

multiple, subscriber's A bank of jacks on manual operating positions, each jack leading to a subscriber's line and the whole pattern of jacks repeated on a fully paralleled basis so that all operators have access to appearances of all lines.

multiple, switchboard panel The number of panels of multiple jacks in a complete set of appearances of all the lines concerned, after which the multiple is repeated.

multiple, transposed A form of skipped grading used particularly for subscriber concentration stages in which a subscriber shares some outlets with one group of subscribers and other outlets with other groups of subscribers.

multiple access 1. Describing the connection of a subscriber to two or more switching centers by separate access lines using a single message routing indicator or telephone number. 2. In satellite communications, the capability of a communications satellite to function as a portion of a communications link between more than one pair of satellite terminals simultaneously.

multiple address Describing a message which is to be transmitted to several addresses.

multiple circuit Circuit with several elements connected in parallel.

multiple frequency shift keying (MFSK) A form of frequency shift keying in which multiple frequency codes are used in the transmission of digital signals.

multiple homing 1. The connection of a terminal facility so that it can be served by any one of several switching centers. This service may use a single directory number. 2. The connection of a terminal facility to more than one switching center by separate access lines. Separate directory numbers are applicable to each accessed switching center.

multiple key telephone set A telephone instrument with keys which enable direct connections to be established with several other users, in addition to a normal dial or DTMF push-button set for calls to the public network.

multiple lines at the same address In data transmission, a service permitting a user to receive calls to a single address on more than one access circuit.

multiple marking Special marking associated with a multiple jack on a manual switchboard indicating whether there are special features connected to this line or if special priority service must be given to all callers.

multiple spot scanning In facsimile systems, scanning carried on simultaneously by two or more scanning spots, each one analyzing its fraction of the total scanned area of the subject copy.

multiple subnyquist-sample encoding Process by which a 20 MHz bandwidth high definition television signal is reduced to an 8 MHz bandwidth for transmission.

multiple switchboard See switchboard, multiple.

multiple system operator (MSO) A company which owns more than one CATV system.

multiple unit steerable antenna (MUSA) A directional antenna made up of fixed units. The direction of the main power lobes can be varied by adjusting the phase relationship of the leads to the different units.

multiplex (MUX) 1. Use of a common channel to make two or more channels. This is done either by splitting of the common channel frequency band into narrower bands, each of which is used to constitute a distinct channel (frequency division multiplex), or by allotting this common channel to multiple users in turn, to constitute different intermittent channels (time division multiplex). 2. In telegraphy, simultaneous transmission of two or more messages in the same or in opposite directions, using the same transmission path.

multiplex, frequency division (FDM) A multiplexing system in which different frequency bands are used by different channels, enabling many different channels to be carried by one bearer.

multiplex, time division (TDM) A multiplexing system in which the original analog signals are converted into digital form, and these digital signals for each of many channels transmitted sequentially at different time instants.

multiplex aggregate bit rate The bit rate in a time division multiplexer which is equal to the sum of the input channel data signaling rates available to the user plus the rate of the overhead bits required.

multiplex baseband The frequency band occupied by the aggregate of the signals in the line interconnecting the multiplexing and radio or line equipment.

multiplex baseband receive terminals The point in the baseband circuit nearest the multiplex equipment from which connection is normally made to the radio baseband receive terminals or intermediate facility.

multiplex baseband send terminals The point in the baseband circuit nearest the multiplex equipment from which connection is normally made to the radio baseband send terminals or intermediate facility.

multiplexer, intelligent A processor-controlled multiplexer that allocates bandwidth utilization.

multiplexing System by which several individual information-carrying channels are combined for transmission over one bearer (line, fiber or radio), using frequency division or time division techniques.

multiplexing, phase Method of encoding more than one information-carrying channel so that all the information may be carried by a single tone.

multiplier, frequency Circuit which accepts a signal of one frequency as input, and produces as output a signal of a multiple of that frequency.

multiplier, instrument Precision resistor which enables a meter with a low voltage range to be used to measure high voltages.

multipling Connecting together in parallel.

multipoint circuit A circuit providing simultaneous transmission among three or more separate points.

multipoint distribution service (MDS) A point-multipoint public radio service on microwave frequencies.

multipoint distribution system Distribution of telecommunication facilities using microwave techniques, typically with each main antenna covering a 90° distribution angle and each outstation able to use 64 kbit/s or some other share of the total available bandwidth.

multipoint link A data communication link connecting two or more stations.

multipolar Describing a circuit which has more than one pair of magnetic poles.

multiport Describing two or more stations sharing the same facility by use of frequency or time channels.

multiprocessor A processing method in which program tasks are logically and/or functionally divided among a number of independent central processing units, with the programming tasks being simultaneously executed. Syn. cluster system

multisatellite support system A complex multicomputer system (based in Darmstadt, Germany) providing monitor, control, data storage and retrieval and data analysis facilities for ESA (European Space Agency) satellites.

multischedule private line (MPL) rate AT&T's rate schedule for voice-grade private lines under which the customer leases a line for a certain amount per mile per month. One of three different rate schedules applies, based on whether the line runs between two large cities ("A" rate centers), two small cities ("B" rate centers), or between a large and a small.

multiservice plant Plant intended and put to use so as to be able to support more than one type of service.

multislot connection Time slots associated with two or more digital circuits switched in parallel through a digital exchange for use on the same call to provide a wideband service.

multiswitched Toll calls established by connecting circuits together at two or more switching centers.

multitap Cable TV device which enables a coaxial feeder cable to be connected to several coaxial feeders to customers' premises, while maintaining the correct impedance matching and isolating each customer's line so that faults on one line will not affect others.

multiunit message (MUM) A message that is transmitted using more than one signal unit, in common channel signaling.

multiunit message rate (MUME) Service which permits telephone companies to charge for local calls at different rates.

multivibrator A relaxation oscillator with outputs from each of two amplifying stages fed back, in phase, to the input stages of the other amplifier to produce oscillation.

multivibrator, free running A multivibrator which operates without external triggering or synchronization pulses.

multivibrator, one-shot A multivibrator which gives one output pulse for each input pulse received.

multiwheel gear In a submarine cable-laying ship, a description of a laying engine containing a number of pairs of pneumatic-tired wheels running tangent to each other and all in the same vertical fore-and-aft plane. Each pair of wheels is pressed together in order to grip the cable as it is led between them in a straight line. The wheels are equipped with drive and brake mechanisms that limit the shear forces, and arranged so that each pair may be successively parted to allow the passage of a repeater. Syn. paired-wheel gear

multiwink (MW) A method of signaling used for coin control of payphones: typically five winks each of about 100 ms duration and 100 ms interval between winks.

M-unit A unit in terms of which refractive modulus is expressed.

Murray loop A testing bridge used to localize faults in external distribution plant.

mushroom anchor In submarine cable systems, an anchor the head of which is a portion of a hollow iron sphere with the stock affixed inside at the "pole." Placed in mud or sand, it tends to become imbedded and is used principally as a buoy anchor or a mooring anchor.

music-on-hold Facility provided by many PABX and key systems; music is sent back to the calling subscriber until the required extension answers.

mutilation A transmission defect in which a signal element becomes changed from one significant condition to another.

mutual impedance Between any two pairs of terminals of a network, this is the ratio of the open-circuit potential difference between either pair of terminals to the current applied at the other pair of terminals, all other terminals being open.

mutually synchronized network A condition under which each clock in the network exerts a degree of control on all others.

nail, pole dating Galvanized steel nail, with the head marked to show the last two digits of the year, driven into a pole to indicate the date of erection.

nail, wiring A type of nail incorporating a clip, protective disc or clamping strip and used to secure wires or small cables to walls.

nailed-up circuit A voice circuit semipermanently connected through a digital switch in both directions of transmission.

nailed-up connection A connection established on a semi-permanent basis through a switching center.

NAND gate A solid state device which is the equivalent of an AND gate followed by an invertor or NOT gate.

nano- prefix for one-billionth (10^{-9}).

nano second one-billionth (10^{-9}) of a second.

Napierian logarithm A natural logarithm with base e ($=2.718282$).

narrative traffic Teletypewriter messages originated on paper tape and prepared in accordance with standardized procedures.

narrowband In data transmission, describing circuits able to carry digital transmission systems up to 2400 bits per second.

narrowband modem A modem whose modulated output signal has an essential frequency spectrum limited to that which can be wholly contained within, and faithfully transmitted through, a voice channel with a nominal 4 kHz bandwidth.

narrowband signal Any analog signal, or analog representation of a digital signal, whose essential spectral content is limited to that which can be contained within a voice channel with a nominal 4 kHz bandwidth.

narrowcast TV cable or radio programs aimed at a small segment of the market: specialist programs likely to be watched only by those with direct personal interest.

n-ary digital signals Digital signals in which a signal element may assume n discrete states.

n-ary information element An information element permitting the representation of n distinct states.

National Aeronautics and Space Administration (NASA) U.S. official body which has been responsible for most of the western world's space activities.

National Association of Broadcasters (NAB) U.S. organization responsible for audio tape and disc standards.

National Association of Regulatory Utility Commissioners (NARUC) Association of those responsible for regulating intrastate operations of telephone and other utility companies.

national circuit (sound program) A circuit connecting the International Sound Program Center (ISPC) to the broadcasting authority. A national circuit may also interconnect two ISPCs within the same country.

national circuit-group-congestion signal In data transmission, a signal sent in the backward direction indicating that the call cannot be completed.

National Electrical Code A code giving rules for the installation of electric wiring and equipment in public and private buildings and published by the National Fire Protection Assn. This code has been adopted as law by many states and municipalities. (UK equivalent are wiring regulations published by the Institution of Electrical Engineers).

National Electrical Manufacturer's Assn. (NEMA) An industry association in the U.S. which standardizes specifications for electrical components and power wires and cables.

National Electrical Safety Code A code of safety rules for the installation and maintenance of electric supply and communication lines and published by the National Bureau of Standards. This code has been adopted as law by many states and municipalities.

national extension That part of an international connection which extends from the national side of the international exchange to the subscriber.

national indicator Information within a signaling message which permits a distinction to be made between national and international messages.

national/international call indicator In data transmission, information sent in the forward direction indicating in the national network whether the call is a national or an incoming international call.

national numbering plan A plan whereby each subscriber's line in the network can be identified.

national section The group, supergroup, etc., sections between a station with control or subcontrol functions and a frontier station within the same country. A national section will usually comprise several group or supergroup sections.

national (significant) number The number to be dialed following the trunk prefix to obtain a subscriber in the same country but outside the same numbering area. It may consist of the trunk code followed by the subscriber number, the numbering area code followed by an exchange code and directory number, or the numbering area code followed simply by a directory number.

national switching-equipment-congestion signal In data transmission, a signal sent in the backward direction indicating the failure of the call setup attempt due to congestion encountered at the switching equipment in the national network.

National Telecommunications and Information Administration (NTIA) Unit within U.S. Dept of Commerce which advises the executive branch on telecommunications matters.

National Telephone Cooperative Association (NTCA) Trade association for smaller telephone companies, particularly those serving rural areas.

National Yellow Pages Service (NYPS) A centralized service which permits single-contract advertising in a number of telephone directories throughout the country.

nationwide cost averaging Averaging of costs so that similar prices are charged in rural as well as in metropolitan areas.

nationwide dialing Direct distance dialing.

nautical mile See mile, nautical.

NAVSTAR/GPS A satellite radionavigation system under development by the United States. This global positioning system will ultimately consist of 24 satellites in three circular 12-hour orbits inclined 63° to the equator. This configuration will ensure that at least three satellites will be visible at all times from anywhere on Earth. The system will give accurate position determination in all three dimensions.

N-channel metal oxide semiconductor (NMOS) A semiconductor which has the advantages of speed and power over P-channel metal oxide semiconductors. It operates with a single low voltage.

N-channel MOS See N-channel metal oxide semiconductor.

NC signal An information signal indicating that there is no circuit available to route a call towards its destination. A busy or congestion signal.

nearby telephone number A telephone number at an address nearby to the address of a person who is not a telephone subscriber. Such lists are maintained at information offices for reaching non-subscribers in emergency situations.

near-end crosstalk Crosstalk propagated in a disturbed channel in the direction opposite to the direction of propagation of the current in the disturbing channel.

near field diffraction pattern The diffraction pattern observed close to a source or aperture. Syn.: fresnel diffraction pattern

near-field radiation pattern Distribution of the irradiance over an emitting surface, e.g., over the cross-section of an optical waveguide.

near field region 1. The region of the field of an antenna between the close-in reactive field region and the far field region wherein the angular field distribution is dependent upon distance from the antenna. 2. The region close to a source or aperture. The diffraction pattern in this region typically differs significantly from that observed at infinity and varies with distance from the source.

near-sing Describing an amplified circuit which is close to instability.

necessary bandwidth For a given class of emission, the minimum value of the occupied bandwidth sufficient to insure the transmission of information at the rate and with the quality required for the system employed, under specified conditions.

needle, cable sewing Flat metal needle used when lacing up cable forms.

needles, test point Fine-pointed test probes with insulated shanks used for making contact with conductors through their insulation, for test purposes.

negative 1. Voltage source to which the conventional DC current returns. 2. Battery terminal with an excess of electrons. If an external conducting path is provided, the electrons will move to the positive terminal.

negative acknowledge character (NAK) A transmission control character sent to a transmitting station by a receiving station to indicate receipt of a block con-

taining one or more errors. Upon receipt of the NAK, the transmitting station will resend the block of information.

negative battery The negative terminal of a battery. Most central offices use a 50 V storage battery with the positive side grounded. The negative side is then usually called simply "battery."

negative impedance An impedance characterized by a decrease in voltage drop across a device as current through it is increased, or a decrease in current through the device as voltage across it is increased.

negative justification In digital multiplexing, the controlled deletion of digits from the tributary digital signal so that the digit rates of the individual tributaries correspond to a rate determined by the multiplex equipment. The deleted information is transmitted by means of a separate low-capacity time slot. Syn.: negative pulse stuffing

negative resistance See resistance, negative.

negative return See return, negative.

negative terminal 1. Terminal from which electrons flow. 2. The negative output terminal of a DC power source.

nematic A type of liquid crystal, used in display units, in which the molecules are aligned parallel to one another but with centers of gravity arranged at random.

neoprene A synthetic rubber used as outer insulation on drop wires.

neper Dimensionless unit used to express the ratio between two powers, currents, voltages, etc. It is based on Napierian (natural, or base e) logarithms. One neper is the Napierian log of the square root of the power ratio.

nested program A program which is contained as a part of another larger program.

net A network, particularly a group of radio communication stations, using the same frequencies.

net gain The overall gain of a transmission circuit. It is measured by applying a test signal of some convenient power at the beginning of a circuit, measuring the power delivered at the other end, and taking the ratio of these powers as expressed in dB.

net loss The overall loss of a transmission circuit. It is measured by applying a test signal of some convenient power at the beginning of the circuit, measuring the power delivered at the other end, and taking the ratio of these powers as expressed in dB.

net loss variation The maximum change in net loss occurring in a specified portion of a communication system during a specified period.

net operation The operation of an organization of stations capable of direct communications on a common channel or frequency.

net radio interface (NRI) An interface between single channel radio users and switched communication systems.

net weekly circulation (NWC) The estimated number of television households viewing a particular station at least once per week, Monday-Sunday, 6 am to 2 am, EST. Used in the U.S. to determine whether a station is "significantly viewed" in an area and must be carried by a cable system operating in that area.

network 1. An organization of stations capable of intercommunication but not necessarily on the same channel. 2. Two or more interrelated circuits. 3. A combination of terminals and circuits in which transmission facilities interconnect the user stations directly. 4. A combination of circuits and terminals serviced by a single switching or processing center. 5. An interconnected group of computers or terminals.

network, active Network which includes a source of energy, such as a battery.

network, annulling Circuit which can be added to a network in replacement of one filter in a series of channel filters. The network simulates the missing filter and annuls response changes in filters for adjoining channels.

network, antenna coupling A network which employs a radio frequency circulator: a device which permits two separate radio transmitters to use the same antenna at the same time.

network, balanced Network in which series elements in both legs are symmetrical with respect to ground.

network, balancing Network used at a two-wire/four-wire hybrid as a balance for the impedance of the two-wire line. The closer the balance, the more efficient the hybrid.

network, bilateral Network which passes signals in both directions.

network, bridged-T A three-element T-network with one extra element in parallel with the two series elements.

network, C A three-element network with one element in each leg and one bridged across the input or output.

network, carrier line filter balancing Network simulating a carrier line filter, used when one side of a group has carrier line filters in use and it is desired to maintain group balance in order to be able to continue using a phantom-derived circuit.

network, compromise Network designed as a standard for use in many different locations. It has an impedance/frequency characteristic which is a compromise, giving acceptable performance in all the locations concerned.

network, crossover In a high fidelity system, a pair of filters — one high-pass, the other low-pass — which separate audio frequency band signals into two separate groups, each of which is fed to a special loudspeaker: a tweeter for the higher frequencies, a woofer for the lower frequencies.

network, decoupling Filter placed in a power feed circuit to prevent interaction between different circuits fed from the same power supply.

network, de-emphasis Network with frequency/gain characteristics which are the reverse of those used in a pre-emphasis network in the same transmission

system, thus restoring the overall signal to its original relationship.

network, distribution External plant network connecting subscribers' premises with a telephone exchange/central office.

network, equalizing A network with frequency/attenuation characteristics complementary to those of the line to be equalized. In other words, the sum of line attenuation plus network attenuation is substantially independent of frequency.

network, equivalent A network which is electrically equivalent to another network insofar as external effects are concerned.

network, heterogeneous switching A switching network in which different connections between inlets and outlets may use different numbers of crosspoints.

network, hierarchal A network which includes two or more classes of switching exchanges such that (1) the highest class is totally interconnected by last choice trunk groups, and (2) each exchange of a class which is not the highest is connected to a unique exchange of a higher class, called its home office, by a last choice trunk group.

network, homogeneous switching A switching network in which every connection between an inlet and an outlet uses the same number of crosspoints.

network, hybrid A four-port network which provides a low impedance path between adjacent ports (ie, two-wire line to outgoing direction and incoming direction to two-wire line) but high impedance between opposite ports (ie, incoming through to outgoing, and two-wire line to balance).

network, L Two-port circuit with one impedance element bridged across the input or output and a second impedance element in series with one leg or the other of the circuit.

network, ladder A two-port circuit made up of a repeated sequence of L, T, pi or H networks.

network, lattice A network made up of four components in a square, with input across one diagonal and output across the other.

network, line filter balancing A carrier line filter balancing network.

network, linear A network with characteristics that are independent of the voltage applied.

network, number Network used in a type of calling line identification equipment; lamps enable the calling lines to be identified.

network, passive A network which does not include a source of energy.

network, pi A four-terminal network with three impedance elements: one in series with one leg of the line and the other two across the line, one at the output and the other at the input.

network, precision A specially made-up balance network simulating the impedance of a two-wire line and used as an accurate balance at a hybrid coil termination of a toll circuit.

network, private A network of communications channels whose use is restricted to one customer.

network, ringer Small encapsulated circuit fixed near or inside a telephone instrument and used to convert the bell from a normal ringing to a chime operation, or to a louder or softer ring.

network, shaping Network inserted in a telegraph circuit to improve the shape of the received signals and reduce errors.

network, switched services Network used by common control switching arrangement customers for such purposes as switching calls between PBXs.

network, switching An arrangement of switches which connect inlets to outlets. Several connections can exist simultaneously through one switching network. Syn.: connecting network

network, symmetrical Two-port network with the same impedances looking in at A when B is open as looking in at B when A is open, and the same impedances at each end when the distant ends are closed.

network, T Four-terminal network with three impedance elements: two equal in series in one leg and the third across the network from the junction of the two equal impedances to the other leg.

network, telephone set The transmission circuit of a telephone instrument (apart from the microphone and earpiece receiver), normally mounted on a board or card.

network, transmission The collection of transmission nodes and their interconnecting transmission sections.

network, unilateral Two-port network containing circuit elements which have directional properties such that the network does not pass signals equally well in both directions.

network, weighting See weighting network.

network, Y A three-pronged star network with one terminal common to each of the three branches.

network architecture Reference framework for the definition and development of protocols and products for interworking between data processing systems. In switching, the relationship between subsystems and the way in which the design of a complete network is influenced by the roles of these subsystems.

network busy hour (NBH) The busy hour for an entire network.

network channel termination equipment (NCTE) Equipment terminal used on digital data lines which performs signal shaping and loop-back testing services.

network cluster A final circuit group plus all the high usage circuit groups which have at least one terminus in common with it, and for which the final circuit group is in the last-choice route.

network connectivity The topological description of a network which specifies the interconnection of the transmission nodes in terms of circuit termination locations and quantities.

network control phase In data transmission, that phase of a data call during which network control signals are exchanged between data terminal equipment and the network for the purpose of call establishment, call disconnection, or for control signaling during the data phase.

network harms Features which could damage telephone plant, or personnel, or users e.g., excessive power, high voltage, unbalanced lines, non-standard signaling.

network integration Close cooperation between telephone companies, not just interconnection but technical and economic integration.

network inward dialing Ability to dial from the public switched telephone network directly into an extension on a PABX.

network layer The logical network entity (in the OSI seven-layer data system model) that passes data from the transport layer to and through the network.

network management The management of a switched network to provide greatest possible efficiency with economy; avoiding congestion on public networks by switching to alternate routes on long distance calls. Where there is no alternative routing available around a heavily loaded section, telephone system planners with network management availability try to give busy tones to a carefully calculated proportion of callers before their calls reach a difficult section, thus avoiding congestion and keeping traffic flowing.

network out dialing Ability to dial into the public switched telephone network directly from an extension telephone on a PBX.

network resources 1. Means of supplying a need. 2. In telecommunication networks, the capacity for sending recorded announcements, traffic service positions, network integrated data banks, etc.

network topology The physical and logical relationships between nodes in a network: typically star, bus, tree, ring or hybrid.

neutral 1. Neither positive nor negative. 2. The return path condition in some types of power distribution which, if conditions are correctly balanced, carries no net current.

neutral, floating power The neutral of a power system which is not connected to ground potential.

neutral conductor A conductor in a power distribution system connected to a point in the system which is designed to be at neutral potential. In a balanced system the neutral conductor carries no current.

neutral direct current telegraph system A telegraph system employing current during marking intervals and no current during spacing intervals for transmission of signals over the line ("neutral" here means that the direction of current flow is immaterial). Syns. single-current system, single-current transmission system, single Morse system

neutral ground An intentional ground applied to the neutral conductor or neutral point of a circuit, transformer, machine, apparatus or system.

neutral operation A method of teletype operation in which marking signals are formed by current pulses of one polarity, either positive or negative, and spacing signals are formed by reducing the current to, or near, zero.

neutral relay   A relay in which the movement of the armature does not depend upon the direction of the current in the circuit controlling the armature.

neutral transmission   A form of signaling employing two distinct states, one of which is the absence of current.

neutron   An elementary atomic particle with zero charge and a rest mass nearly equal to that of the proton. Protons and neutrons together make up the nuclei of atoms.

new sinc   Feature of some data sets which permits a rapid resynchronization during transition from one transmitter to another on a multipoint private line or leased circuit data network.

newton   The derived SI unit of force. It is the force which, when applied to a body having a mass of 1 kg, gives it an acceleration of 1 m per second per second.

next, fext   Near-end crosstalk, far-end crosstalk. Usually, crosstalk into one circuit from signals on another circuit, but also used for crosstalk between go and return channels of the same circuit; with the near/far designation depending upon where in the circuit the interference is induced.

nibble   Half a byte: a digital word of only four bits.

Nicopress sleeves   A brand of splicing sleeve used in making compression joints on open-wire routes.

night alarm circuit   An alarm circuit which can be made audible during periods of low traffic in order to alert a switchboard to incoming calls.

night answer   A common arrangement at PABXs whereby incoming calls, which are normally answered by an operator, ring a bell to alert extensions on the system when the board is unattended.

night transfer   A load transfer arrangement in a large manual operating center whereby calls are transferred to a small number of operating positions during periods of light traffic.

Nimbus   A series of meteorological satellites.

nitrogen   A gas widely used under pressure in telephone cables. If a small puncture occurs in a cable sheath, the nitrogen keeps moisture out so that service is not adversely affected.

Nixie tubes   Small gas glow tubes with cathodes, usually designed in the form of the digits 0 to 9, in order to give illuminated readings of digital values.

NNX   A general way of referring to the three digits representing the central office code in a telephone number. N may be any digit 2 thru 9; X may be any digit 0 thru 9. See NXX.

NOAA   United States research satellite relaying data from meteorological balloons and oceanographic buoys. (National Oceanic Atmospheric Administration)

no break power unit   A power unit which utilizes the public power supply plus a standby plant with special equipment to ensure that, if the public supply fails, the standby unit will take over.

nodal operation   Refers to remote location and control of line groups.

nodal period   The period of time elapsing between two consecutive passages of a satellite through the ascending node.

node   1. In network topology, a terminal of any branch of a network, or a terminal common to two or more branches of a network. 2. In a switched communications network, the switching points, including patching and control facilities. 3. In a data network, the location of a data station which interconnects data transmission lines. Syn.: junction point, nodal point, vertex 4. A point in a standing wave at which the amplitude is at a minimum. Syn.: nodal point, null

node, ascending   Point where a satellite crosses the plane of the Earth's equator when moving north.

node, current   In a transmission system with standing waves, points at which the current is at a minimum.

node, descending   Point where a satellite crosses the plane of the Earth's equator when moving south.

node, voltage   In a transmission system with standing waves, points at which voltage is at a minimum.

noise   Any random disturbance or unwanted signal in a communication system which tends to obscure the clarity of a signal in relation to its intended use. The CCITT system of noise measurement is based on absolute values of noise-power level expressed in dBm (i.e. referred to 1 mw) if measured unweighted (flat frequency response) or in dBmp if the CCITT weighting network is used. The suffix p in the latter indicates psophometric weighting. The nuisance value of noise on a telephone circuit is clearly a subjective matter — noises at some frequencies over a telephone annoy humans more than noises at others. The CCITT weighting network has been designed to cope with this. The CCITT psophometer used with its weighting network measures the psophometric voltage at a point in a telephone system. Its meter is scaled in millivolts. This voltage is defined as the voltage of a tone at 800 Hz which, if it replaced the noise voltage, would produce the same degree of subjective interference to a telephone conversation as the noise voltage. As noise power levels in telephone systems are usually very low, 10^{-12}W (the picowatt) is the commonly used unit. A weighted noise power level of 1 picowatt, referred to 1 mW would have an absolute value of

$$\frac{10 \ \log 10^{12}}{\log 10^{3}}$$

or —90 dBmp. At the U.S. reference frequency of 1000 Hz the CCITT weighting is +1 dB so that a reading equivalent to —90 dBmp will be obtained on a CCITT psophometer for an absolute power level of —91 dBM if 1000 Hz is used.

noise, ambient   Acoustic noise which is part of the environment in which a system transducer is located.

noise, atmospheric   A component of sky noise arising from natural phenomena within the atmosphere, such as lightning discharge.

noise, background   System noise in the absence of a signal.

noise, battery Noise which is sometimes fed into telephone circuits on the DC power supply leads, e.g. a low hum from the rectifier or impulse noise from relay contact circuits.

noise, broadband Noise with a spectral intensity significant over a band where the upper frequency is many times greater than the lower. Syn.: wideband noise

noise, carrier Noise resulting from variations in carrier level.

noise, circuit The sum total of all the noise on a telephone circuit.

noise, contact Noise resulting from current passing through a contact of varying resistance value.

noise, cosmic Random noise picked up at microwave frequencies, particularly between 20 and 100 MHz when antennae are directed towards the Milky Way.

noise, diffusion A type of shot noise which arises when charges diffuse through a region where carrier recombination and generation exist.

noise, flicker A low-frequency current noise above the background of the shot noise. It is usually a surface phenomenon, such as contact noise.

noise, front-end Thermal noise in the early radio frequency amplifying stages of a radio receiver, which varies inversely with radio frequency input level.

noise, galactic A component of sky noise. It is a continous background noise whose intensity varies with celestial direction, due to noise sources in our own galaxy.

noise, granular A quantizing noise resulting from the transition from a signal continuously varying in amplitude into one which has discrete sampling steps.

noise, idle Noise in the absence of a signal. Intrinsic noise or background noise.

noise, impulsive A nonoverlapping random succession of transient disturbances resulting from such sources as ignition systems, switching centers, lightning discharges and dial pulses.

noise, induced power Harmonics of the frequency of a public power supply (particularly the third harmonic of 60 Hz) induced into telecommunications transmission lines or telephone circuits.

noise, intermodulation Sum-and-difference frequencies produced by every nonlinear device through which a multichannel carrier signal passes. There are usually many of these spurious frequencies at low power levels with the result that intermodulation noise sometimes sounds very like white noise.

noise, intrinsic Noise normally present in a device or circuit which is not caused by modulation and is not affected by input signal. Idle noise or background noise.

noise, ionospheric Various radio noises from rockets and satellites, including auroral hiss, very low frequency hiss and whistler-mode noise.

noise, magnetospheric Noise levels substantially higher than those attributable to cosmic noise which are picked up by satellite-borne receivers. There are

several types associated with aurora, geomagnetic disturbances, plasma instabilities and other complex factors.

noise, man-made Noise which is heard on radio communications circuits due to electrical machines, badly suppressed automobile ignition circuits, etc.

noise, metallic The weighted noise current in a metallic circuit through the 100 kilohm circuit of a circuit noise meter connected between one or more telephone wires and ground.

noise, modulation Noise in a modulation system which is associated with the signal and varies in strength with the strength of the signal.

noise, narrowband Noise with a spectral intensity significant over a band whose width is small compared with the center frequency.

noise, partition A random noise arising when an electron stream divides among a number of electrodes.

noise, pink Noise with a power spectral intensity which is inversely proportional to frequency over a specified range, therefore dissipating in a resistance equal power in any octave bandwidth in that range.

noise, power line Noise concentrated at power line frequencies (normally 50 Hz or 60 Hz) and harmonics (usually odd harmonics) of these. Syn.: induced power noise

noise, quantizing See quantizing noise.

noise, random See random noise.

noise, recombination A random noise in a semiconductor due to the generation and recombination of charge carriers.

noise, reference See reference noise.

noise, room The noise at the receiving end of a telephone circuit which, fed back from the local microphone into the local receiver as sidetone, makes it difficult for the listener to hear a faint incoming signal.

noise, shot A type of random noise associated with the discrete contribution of moving charges forming a current.

noise, sky The noise, expressed in equivalent noise temperature received when an aerial is pointing vertically upward, giving minimum atmospheric absorption. (At lower angles the absorption is greater but the atmospheric "temperature" is also greater.)

noise, solar Random noise at radio frequencies detected when a receiver's antenna is directed towards the sun.

noise, static Radio noise from lightning or other atmospheric electricity phenomena.

noise, thermal A type of random noise in which the elementary disturbance is the intercollision transit of free electrons within a conductor.

noise, valve Shot noise in a valve due to random emission of electrons at the cathode.

noise, white See white noise.

noise band, system In submarine cable systems, a frequency band outside the regular transmission band which is monitored continuously for the occurrence of excessive system noise.

noise band, width of the effective Of a radio receiver, the width of a rectangular frequency response curve, having a height equal to the maximum height of the receiver frequency response curve and corresponding to the same total noise power.

noise equivalent power (NEP) In optoelectronics, the radiant power, at a given modulation frequency and for a given bandwidth, that produces a signal-to-noise ratio of 1 at the output of a given detector. In this sense, it is the minimum detectable power at the given frequency and for the given bandwidth.

noise factor (NF) Of a radio receiver, the ratio of noise power measured at the output of the receiver to the noise power which would be present at the output if the thermal noise due to the resistive component of the source impedance were the only source of noise in the system. Both noise powers are determined at an absolute temperature of the source equal to $T = 290K$.

noise figure A measure of the noise in dB generated at the input of an amplifier as compared with the noise generated by an impedance-method resistor at a specified temperature.

noise filter Network which attenuates noise frequencies.

noise generator Generator of wideband random noise.

noise immunity 1. The ability of a device to discern valid data in the presence of noise. 2. (of a gate). The maximum noise voltage that can appear at its input terminals without producing a change in the output state.

noise level The volume of noise power, measured in decibels, referred to a base.

noise load ratio (NLR) Ratio, expressed in decibels, of the power of a wideband test signal applied to a baseband to the noise power which results in a single channel.

noise margin Syn.: noise immunity

noise measurement units The following units are used to express weighted and unweighted noise power: (a) dba (Fla) - Fla weighted circuit noise power, in db referred to 3.16 picowatts (-85 dbm), which is zero dba. (b) dbrnc - C-message weighted circuit noise power, in db referred to 1.0 picowatt (-90 dbm), which is zero dbrn. (c) dbrn (flat) - Noise power in db referred to 1.0 picowatt (-90 dbm) with no weighting except exclusion of all frequencies except 30-3000 Hz. (d) dbpwp - Noise power in picowatts, psophometrically weighted. See also noise.

noise-operated gain-adjusting device Device which reduces circuit gain when there is no speech signal present.

noise power The mean power supplied to the antenna transmission line by a radio transmitter when loaded with white noise having a Gaussian amplitude distribution.

noise power ratio (NPR) Ratio, expressed in decibels, of signal power to intermodulation product power

plus residual noise power, measured at baseband level.

noise suppressor Filter or digital signal processing circuity in a receiver or transmitter that automatically reduces or eliminates noise.

noise temperature The temperature, expressed in Kelvins, at which a resistor will develop a particular noise voltage. The noise temperature of a radio receiver is the value by which the temperature of the resistive component of the source impedance should be increased, if it were the only source of noise in the system, to cause the noise power at the output of the receiver to be the same as in the real system.

noise to ground Reading of weighted noise current between a telephone wire or group of wires and ground via a high resistance noise meter

noise transmission impairment (NTI) The loss which must be added to a noise-free circuit to produce the same transmission degradation as a measured amount of noise.

Amount of Noise	NTI
17 dBa	0 dB
20 dBa	1 dB
23 dBa	2 dB
25 dBa	3 dB
26 dBa	4 dB
27 dBa	5 dB

noise weighting A specific amplitude-frequency characteristic which permits a measuring set to give numerical readings which approximate the interfering effects to any listener using a particular class of telephone instrument.

nominal alternating discharge current through a gas discharge protector For currents with a frequency of 15 Hz to 62 Hz, this is the alternating discharge current which the protector is designed to carry for a defined time.

nominal bandwidth The widest band of frequencies, inclusive of guard bands, assigned to a channel.

nominal bit stuffing rate The rate at which stuffing bits are inserted (or deleted) when both the input and output bit rates are at their nominal values.

nominal impulse discharge current through a gas discharge protector The peak value of the impulse current with a defined curve shape with respect to time for which the protector is rated.

nominal justification rate In pulse code modulation multiplexing, the rate at which justifying digits are inserted (or deleted) when both the tributary and the multiplex digit rates are at their nominal values. Syn.: nominal stuffing rate

nominal linewidth In facsimile systems, the average separation between centers of adjacent scanning or recording lines.

nominal usable field strength The agreed value of the usable field strength that can serve as a basis for frequency planning. Depending on the receiving conditions and the quality required, there may be several nominal usable field strength values for the same service.

nominal value A specified or intended value independent of any uncertainty in its realization. In a device that realizes a physical quantity, it is the value of such a quantity specified by the manufacturer.

nomogram A chart showing three or more scales across which a straight edge may be held in order to read off a graphical solution to a three-variable equation.

nomograph A nomogram.

non-associated mode of signaling The mode where common channel signaling messages for a signaling relation involving two non-adjacent signaling points are conveyed, between those signaling points, over two or more signaling links in tandem passing through one or more signaling transfer points.

nonblocking Describing a switching network with enough paths across it so that an originated call can always reach an idle line without encountering a busy tone on the way.

nonbridging Describing wipers or moving contacts which leave one fixed contact before they make with the next.

noncentralized operation A control discipline for multipoint data communication links in which transmission may be between tributary stations or between the control station and a tributary station or stations.

nonconductor A substance which does not transmit a given form of energy such as electricity, heat or sound.

noncritical technical load That part of the technical load not required for synchronous operation.

nondestructive read A special read method whereby the information read from a store by a computer is written back in immediately for future reuse.

nonequilibrium mode distribution That distribution of modes prevailing in a length of waveguide shorter than the equilibrium length.

nonerasable store A type of store which contains information which is not automatically destroyed when it is read into a processor.

non-file-structured device A device — such as paper tape, a printer or a data terminal — in which data cannot be referenced, as it can be in a file.

nonhoming switch A rotary switch which remains at the last position reached when the circuit is released. When the circuit is next activated, it starts hunting again from the point last reached.

noninductive Without significant inductance.

non-isochronous Describing devices which are not synchronized but operate on a start-stop basis, as do teletypewriters.

nonlinear Any device in which output does not have a single constant relationship with input.

nonlinear distortion Distortion caused by a deviation from a linear relationship between specified input and output parameters of a system or component.

nonlinear resistor A resistor whose resistance varies as a function of the instantaneous current flowing through it, or of the instantaneous voltage across its terminals.

nonlinear scattering In optoelectronics, direct conversion of a single photon from one wavelength to one or more other wavelengths.

nonlinear singing A defect which can occur in long submarine cable systems having a large number of repeaters with common amplifiers for both directions of transmission. It is caused by irregularities in the system which result in surplus gain, if the nonlinear singing margin of the repeaters is insufficient.

nonlist A telephone number not available in the published directory.

nonloaded A circuit which is not loaded with inductances at regular intervals.

nonlocking A key which returns to its rest position as soon as it is released.

nonmagnetic Material which has no effect on a magnetic field.

nonnumerical A switch which operates automatically rather than under the direct control of a dial.

nonoperate current Maximum current which can be passed through a relay without causing the relay to operate.

nonoperational load Administrative, support and housing power requirements. Syn.: utility load

nonpole pair On a pole route, any pair of wires which is not separated by the pole itself.

nonprinting character A control or functional character transmitted as part of a message but not reproduced in printed copy.

nonregulated 1. Having speed, voltage or current not maintained at a predetermined standard. 2. A communication service the charges for which are not controlled or regulated by the FCC or by a State Utilities Commission.

non-return-to-zero-change-at-logic-1 (NRZ1) **recording** A method of magnetic recording in which 1's are represented by a change in condition and 0's are represented by no change.

non-return-to-zero (NRZ) code A code form having two states — zero and one — with no neutral or rest condition.

nonsecure mode Unencrypted.

nonsimultaneous transmission Half-duplex transmission, i.e., transmission through a system or channel which cannot transmit in two directions at the same time.

nonsynchronous data transmission channel A data channel in which no separate timing information is transferred between the data terminal equipment and the data circuit-terminating equipment.

nonsynchronous network A network in which the clocks need not be synchronous or mesochronous. Syn.: asynchronous network

nontechnical load That part of the total operational load used for such purposes as general lighting, air conditioning and ventilating equipment during normal operation.

nontraffic sensitive (NTS) plant Facilities whose costs do not change directly with variations in traffic levels.

nonuniform encoding In pulse code modulation, the generation of character signals representing nonuniformly quantized samples.

nonuniform quantizing In pulse code modulation, quantizing in which the intervals are not all equal.

nonvolatile memory A form of computer memory that will store information for an indefinite period of time with no power applied.

nonworking number Telephone number which is connected to office equipment but is not in actual service.

NOR gate A solid state logical device which only gives a "1" output if all inputs are "0". It is equivalent to an OR gate followed by a NOT gate.

Nordic network Packet-switching network established by the telecommunications administrations of the Scandinavian countries. Also, the cellular radio system in the same group of countries.

normal 1. The rest position of a key, dial or relay. 2. A line perpendicular to another line or to a surface.

normal post springs A mechanically operated springset on a Strowger two-motion switch which is operated when the switch steps to a particular level.

normalized frequency The ratio between the actual frequency and its nominal value.

normalized frequency departure The frequency departure divided by the nominal frequency value. Syn.: relative frequency departure

normalized frequency difference The algebraic difference between two normalized frequencies. The two nominal values can be identical or different. Syn.: relative frequency difference

normalized frequency drift The frequency drift divided by the nominal frequency value. Syn.: relative frequency drift

normalized offset The offset divided by the nominal value. Syn.: relative offset

normally closed Contacts which open when a relay or key is operated.

normally open Contacts which close when a relay or key is operated.

normals The springs which rest in contact with moveable springs in a key or relay springset, but whose contacts break as soon as the key or relay moves from its non-operated position.

North American Telecommunications Association (NATA) A U.S. trade association of telephone equipment distributors and manufacturers who are not associated with telephone companies or carriers. NATA is thus the "interconnect" association.

north pole The pole of a magnet which seeks the north magnetic pole of the earth.

NOSFER New Master System for the Determination of Reference Equivalents (translated from the French).

no-such-number signal Previously a continuous high tone rising and falling in pitch, indicating that one has dialed an unused office code or a telephone number in an unused series. Now largely replaced by an automatic voice announcement.

no-test trunk Trunk in a dial office which is only available to operators and which provides access to a required line even if the line is busy.

NOT gate An inverter. See logic gates. A solid-state logical device which inverts its input, e.g., "1" input produces "0" output.

not-obtainable signal In data transmission or in signalling, a signal sent in the backward direction indicating that the call cannot be completed because the called number is not in use or is in a different user class.

not-ready condition A steady-state condition at the data terminal equipment (DTE) interface with the data circuit-terminating equipment (DCE) that denotes that the DCE is not ready to accept a call request signal or that the DTE is not ready to accept an incoming call.

notched noise Noise in which a narrow band of frequencies has been removed.

notice of inquiry FCC notice inviting comments and asking for information.

notice of proposed rulemaking FCC notice inviting comments on proposed new or amended rules.

N-plus-1 One spare, backing up N workers. Typically, applied to multiprocessors where the mean time between failures is so long that it would be wasteful to provide one backup spare unit for every working unit.

nth harmonic Harmonic with a frequency n times the fundamental frequency.

n-type semiconductor Semiconductor in which there is an excess of electrons making electrons the majority carriers.

nucleus The central part of an atom which is responsible for most of its mass and which has a positive charge normally equal to the total negative charge of its electrons.

nuisance call A telephone call which is annoying, obscene, malicious or harassing.

null A zero or minimum position.

number, binary A number expressed to the base two.

number, directory The full number required to designate a telephone subscriber.

number, extheo Numbers on the third of three offices which share the use of the same 10,000 groups of numbers, but with different central office codes.

number, non-discriminating A telephone number with two central office codes. The dialing of either number reaches the same phone.

number, physical Numbers on the first of three offices which share the use of the same 10,000 group of numbers, but with different central office codes.

number, pin The numerical position of a pin on the crossarm of a pole route. If 10-pin arms are used, the pins on the top arm are 1-10, on the second arm they are 11-20, etc.

number, telephone A seven-digit telephone address consisting of a three-digit central office code plus a four-digit station number.

number, theoretical Numbers on the second of the three offices which share the use of the same 10,000 group of numbers, but with different central office codes.

number, unassigned Telephone number not yet assigned to a subscriber.

number assignment, paystation A system whereby a distinctive telephone number indicates to an operator that the telephone is a pay phone.

number 2-5 A seven-digit telephone number consisting of two letters and five numerals. It has been superseded by numbers consisting of seven numerals.

Numbering Plan Area (NPA) Any of the 215 geographical divisions of the United States, Canada, Bermuda, the Caribbean, Northwestern Mexico, Alaska and Hawaii within which no two telephones will have the same seven-digit telephone number. Each numbering plan area could have the same number of telephones ultimately (up to 8 million), and each has been assigned a distinctive three-digit "area code."

numbering scheme An arrangement of numbers each of which uniquely identifies each terminal or group of terminals of a telecommunications service.

numbers shift The key or code which indicates that the following characters are numeric. Syn.: figures shift

numerical aperture (NA) In optoelectronics, NA = n sin θ where θ is, at a specified point, half the vertex angle of the largest cone of meridional rays that can enter or leave an optical element or system, and n is the refractive index of the homogeneous isotropic space that contains the specified point. The specified point is usually an object or image point. In practice the sine of the acceptance angle is referred to as the numerical aperture.

numerical switch A switch which operates under the control of dialed pulses from a telephone instrument.

numericals The digits which identify a subscriber in a central office.

N-unit A unit in terms of which N (refractivity) is expressed.

nut, insulation crushing A nut used on some binding posts. It has an annular ring which cuts into the insulation around a wire and makes good contact without having to strip the insulation from the end of a wire before terminating it on the binding post.

NXX An American Central Office Code (N=2-9, X=0-9). CO codes were previously in NNX form but the available number of codes has now been increased by permitting 0 and 1 as second digits in some circumstances (previously their use was restricted to Numbering Plan Area codes).

NYNEX Regional Holding Company for New York and New England Bell Operating Companies.

Nyquist interval Maximum time interval between regularly spaced instantaneous samples of a wave of bandwidth W for complete determination of the waveform of the signal. Numerically equal to W/2 seconds. Common illustration of this is the sampling of voice signals (up to 4 kHz) at a sampling rate of 8 kHz in pulse code modulation (PCM) systems.

Nyquist rate The reciprocal of the Nyquist interval.

object program A program that is in machine executable form. This means that the program is most likely to be in some sort of a binary pattern, ready to be used by the computer.

objective, design See design objective.

OCC terminal Location from which an other common carrier provides service to its customers.

Occam A concurrent programming language (e.g., for the Inmos transputer). Occam was the name of the 14th century European philosopher whose most remembered saying is, roughly, "simplest explanations are the best."

occupancy 1. Refers to the percentage of time a circuit or facility is in use. One circuit in continual use represents 1 erlang of traffic, or 36 hundred call seconds. 2. Ratio of operator-time spent actually handling calls to the total time spent on duty at operating positions.

occupancy date, beneficial (BOD) Date on which a building under construction is to be ready for equipment installation to commence.

occupied bandwidth The frequency bandwidth such that, below its lower and above its upper frequency limits, the mean powers radiated are each equal to 0.5% of the total mean power radiated.

ocean block. Portion of a seacable system between equalizers, typically consisting of 10 or more repeater sections and two equalizer half-sections.

octal base An eight-pin base, used on some electron tubes.

octal numeral A numeral in the octal (base 8) numbering system, represented by the characters 0, 1, 2, 3, 4, 5, 6, 7, optionally preceded by 0' (0 apostrophe).

octave Frequency interval between two signals such that the higher frequency signal is twice the frequency of the lower.

octet A group of eight binary digits, usually operated upon as an entity.

octet alignment Alignment of bits into sequences of 8 binary digits.

octet timing signal A signal that identifies each octet in a contiguous sequence of serially transmitted octets.

odd-even check See parity check.

odd parity A data error detection method: One extra bit, the parity bit, is added to the code signal for each data character such that the total number of 1s in the data, including the parity bit, is an odd number.

oersted Unit of magnetic field strength in the cgs system.

off-hook 1. In telephone operations, the conditions existing when the receiver or handset is removed from its switch. 2. One of two possible signaling states such as tone or no tone, ground connection or battery connection. 3. The active state (closed loop) of a subscriber or PBX user's loop.

off-hook service The automatic establishment of a connection between specified subscribers as a result of lifting the handset off the hook.

off-hook signal In telephone switching, a signal indicating seizure, request for service, or a busy condition.

office A telephone switching center or central office. Sometimes the building housing a central office.

office, automatic A dial office. A telephone switching center where connections are made automatically in accordance with instructions dialed or keyed in from a telephone instrument.

office, central A 10,000-line switching center.

office, circuit control An office which controls maintenance activities affecting toll circuits.

office, common battery A central office with a main storage battery which supplies signalling voltage and microphone currents to all the lines on the office.

office, community dial (CDO) A small automatic office, particularly one in a rural area.

office, crossbar central An automatic central office which uses electromechanical crossbar switches.

office, dial An automatic telephone switching center.

office, dial central 1. An automatic telephone switching center. 2. A 10,000-line office.

office, end 1. A Class 5 office in the North American direct distance dialing network. 2. A local office serving telephone subscribers whose lines are terminated thereon.

office, local central An end office serving local telephone subscribers.

office, magneto A manual telephone switch where signaling is by magneto ringing and local batteries are required at the subscribers' instruments.

office, manual A telephone office where operators handle the switching of calls.

office, master A parent telephone office which has satellite offices connected to it.

office, rotary central A central office using rotary switches and selectors.

office, satellite A minor telephone office which routes all toll calls through its parent office.

office, step-by-step central An automatic office where switching is done by step-by-step electromechanical selectors.

office, tandem An automatic switching center which switches calls between other offices and normally has no subscribers' loops directly terminated.

office, toll A toll switching center or toll point.

office, tributary An end office with access to the toll network through a toll center.

office, unattended dial A small automatic telephone office which is visited only for maintenance purposes.

office alarm system System which reports trouble in an office or exchange to the maintenance control center.

office balance Process of adjusting the impedances in the transmission paths of connections between inter-toll trunks in order to minimize the echo return to either party from that office.

office busy hour The hour, not necessarily a clock hour, when an office or exchange carries most traffic.

office classification See Class 1 office, Class 2 office, etc.

office code The three-digit code that designates a central office within a Numbering Plan Area.

office impedance Nominal impedance value selected for an office for the interconnection of trunks and loops. A single value is selected to permit universal interconnection with minimum irregularities.

office of record. The office responsible for retaining file copies of orders and record correspondence.

Office of Telecommunications Policy (OTP) Division of the Executive Office of the President's staff that advises the Executive Branch on communications policy, studies policy questions and develops legislative proposals.

official board The company PBX of the local telephone company.

off-line 1. That condition wherein devices or subsystems are not connected into, do not form a part of, and are not subject to the same controls as an operational system. 2. A device which is not permanently connected to the processor or to external circuits.

off-line operation Method of operation in which the encryption and decryption of restricted messages are handled separately from their transmissions, with the crypted messages passed out for transmission and the received messages passed on for decryption.

off-line recovery The process of recovering nonprotected message traffic by use of an off-line processor or central processing unit.

off-line storage Memory not directly accessible or under the control of a central processor, e.g., on a reel of magnetic tape.

off-net calling That process wherein telephone calls which originate in or pass through private switching systems in transmission networks are extended to stations in the commercial telephone system.

off-network access line (ONAL) A circuit which enables the customer to complete calls to and from points on the direct distance dialing network.

off-network access trunk (ONAT) Trunks which permit telephone subscribers in the U.S. to use the facilities of specialized other common carriers without degraded transmission and supervisory services.

off-normal Springsets or contacts which operate when a switch or dial is moved away from its rest position.

off-premises extension A PBX extension located in a building other than the one housing the PBX itself.

offset An intentional difference between the realized value and the nominal value.

offset, frequency Difference between the frequency of a tone at the input of a system and its frequency at the output.

off-the-air Reception of a TV signal that has been broadcast through the air.

ohm The derived SI unit of electric resistance. It is the resistance between two points of a conductor when a constant potential difference of 1 volt, applied between these two points, produces in this conductor a current of 1 ampere, the conductor not being the source of any electromotive force.

ohm-centimeter Unit of resistivity. It is the resistance in ohms between opposite faces of a 1 cm cube.

ohmic loss Power dissipation due to electrical resistance.

ohmmeter Test set which gives resistance readings in ohms.

ohm-pound/mile The resistance of a wire one mile in length and weighing one pound. A unit now seldom used.

Ohm's Law The law which says that the current in a linear constant-current circuit is inversely proportional to the resistance of the circuit, and directly proportional to the electromotive force in the circuit.

ohms-per-volt A measure of the sensitivity of a voltmeter.

oil switch A power switch whose contacts break while they are immersed in a nonconducting oil which rapidly quenches any arcs.

oligarchic network Said of a synchronized network in which the synchronization is in the hands of a selected few clocks, the rest of the clocks being controlled by the few.

omni-directional antenna. An antenna whose pattern is nondirectional in azimuth.

on-call circuit A permanently designated circuit, usually privately leased, which can be activated immediately when this is requested by the customer.

one-per-desk (OPD) Intergrated telephone and personal computer terminal (ICL, UK).

one-plus dialing Procedure used in areas where digit 1 must be dialed before that of the required tele-

phone, particularly for calls to other numbering plan areas.

one state Condition of a binary cell storing a "1."

one-unit message A signal message in common channel signaling system CCITT No. 6 which is transmitted entirely within one signal unit.

one way A qualification applying to traffic which indicates that the call setups always occur in one direction.

one-way communication A mode of communication such that information is always transferred in one pre-assigned direction only.

one-way-only channel A channel capable of operation in only one direction, which is fixed and cannot be reversed. Syn. unidirectional channel.

one-way reversible operation See half-duplex operation.

one-way splitting PBX feature which permits the operator or attendant to talk with the inside extension without being heard by the caller on an external line.

one-way trunk A trunk between switching centers which is used for traffic in one pre-assigned direction.

ongoing interface In submarine cable systems, the junction between the sea cable system and the facilities of the inland telecommunications network on shore.

on-hook 1. In telephone operation, the conditions existing when the receiver or handset is resting on the switch. 2. One of two possible signaling states such as tone or no tone, ground connection or battery connection. 3. The idle state (open loop) of a subscriber or PBX line loop.

on-hook signal In telephone switching, a signal indicating a disconnect, an unanswered call, or an idle condition.

on-line 1. Describing a method of operating whereby a teletypewriter message is transmitted and processed simultaneously. 2. Describing a device that is permanently connected to a computer or processor so that no delay is experienced in accessing the device.

only-route circuit group A circuit group which is the only route for all the parcels of traffic it carries.

on-off Signaling method in which a transmitter is, in effect, switched on and off repeatedly.

opaque A transmission line which cannot pass a digital signal.

open Describing a break in a circuit.

open air transmission Transmission type that uses no physical communications medium other than air, e.g., radio systems. Syn.: free space transmission.

open circuit 1. In electrical engineering, a defined loop or path that closes on itself and that contains an infinite impedance. 2. In communications, a circuit available for use.

Open System Interconnection (OSI) An internationally agreed model for data systems which interwork: "Open" means that the systems use a standard set of protocols, e.g., those developed by the International Standards Organization (ISO) and the CCITT. The OSI model defines a hierarchy of seven layers of communications functions, to enable any OSI-compliant computer or terminal to communicate meaningfully with any other OSI-compliant unit, anywhere in the world.

The OSI Layers are divided up into:

(a) three network dependent layers (1-3) which ensure that data can be transmitted between systems over a chain of physical networks,

(b) a single 'transport' layer (4) which ensures that the physical networks as available provide the services required by the applications, and

(c) three application-oriented layers (5-7) which provide general services and specific protocols for common applications.

The layers are:

1.	Physical	the electrical and mechanical interface
2.	Link	moves data between nodes connected directly together
3.	Network	interfaces to packet network, packetizes and reassembles message, flow control
4.	Transport	moves message from originating source node to destination node
5.	Session	establishes, maintains and terminates the logical link
6.	Presentation	editing, mapping, translation
7.	Application	application programs and dialogue

open wire Describing conductors, usually copper or copper-steel, erected on poles using arms and insulators. Nowadays open wire is largely restricted to rural areas, having been largely replaced by underground cables.

OPEN world Open Protocol Enhanced Networks: Northern Telecom (Canada) system for interlinking data processors.

open-circuit voltage Voltage measured at the terminals when there is no load and, hence, no current is flowing.

open-circuit working In telegraphy, a method of operation whereby no current flows during idle periods.

open-wire carrier equipment Equipment which enables a number of speech channels (typically three or 12) to share an open wire circuit using frequency division multiplex techniques.

operand Any of the qualities arising out of or resulting from the execution of a computer instruction, a constant, a parameter, the address of any of these quantities, or the next instruction to be executed.

operate time Time interval between current flowing in a relay coil and the operation of the relay's contacts.

operating costs Recurrent costs (largely salaries, wages and power bills) associated with the operation of plant to provide service.

operating lifetime Period of time during which the principal parameters of a continuously operated component (e.g., a laser diode in an optic fiber system) remain within a prescribed range.

operating system 1. Software that controls the management and execution of programs. 2. A collection of programs used to enhance the utility of a processor by providing facilities during its operation which are not readily built into the hardware. An example of such a facility: the automatic loading, when necessary, of designated programs.

operating telephone company (OTC) Bell System term for an operating company.

operation The method, act, process or effect of operating a device or system.

operation, broadcast The transmission of information which may be received by many stations simultaneously.

operation, call circuit An order wire method of manual operation of long distance circuits. The operator uses a separate call circuit to advise the distant operator of the number required.

operation, CLR (combined line and recording) A method of manual long distance operation in which the operator answers the subscriber's line, records the number required for charging purposes, and establishes the call.

operation, conference In telegraphy, a method of operation in which several stations exchange information on a simplex basis.

operation, direct dial Method of operation in which an operator plugs into a trunk to the required distant office and dials the distant number herself.

operation, full-duplex In telegraphy, a method of operation whereby information may be sent simultaneously in both directions over a single circuit.

operation, half-duplex In telegraphy, a method of operation whereby information may be sent in both directions over a single circuit, but only in one direction at a time.

operation, multiplex Simultaneous transmission of two or more messages in the same direction over the same transmission line.

operation, neutral In telegraphy, a method of operation in which signals are transmitted by making and breaking a looped circuit.

operation, push-to-talk Method of operation, particularly in radio systems, which requires the operation of a key at the transmitting end. The key activates the transmitter at that end of the circuit.

operation, ring-down Method of operation in which an operator sends out a ringing signal to capture the attention of the distant operator. At the end of the call, a ringing signal is again required to signal the distant operator to pull out the plugs.

operation, simplex Operation in one direction at a time.

operation, single channel System used in smaller telephone networks by which a single dialed code (often "0") calls in the operator when assistance or long distance services are required.

operation, straight forward Manual call handling method by which incoming calls go straight to an incoming operator who then completes the call.

operation, unidirectional Method of operation which permits information flow in one direction only.

operation and maintenance center processor A centralized processor for operation and maintenance purposes which serves one or more switching centers.

operational Ready for use.

operational load The total power requirements for communication facilities.

operational MUF (maximum useable frequency) Highest frequency that permits acceptable radiocommunication operation between given points at a given time, and under specified working conditions. Also called simply MUF.

operational service period A performance measurement period, or succession of performance measurement periods, during which a telecommunications service remains in an operational service state.

operational service state A telecommunication service condition which exists during any performance measurement period over which the calculated values of specified supported performance parameters are equal to, or better than, their associated outage thresholds.

operations support systems Software programs which help operating companies manage their systems and networks.

operator Person who operates a telephone switchboard.

operator, A Operator who answers local calls and connects them to toll circuits.

operator, amateur radio A person holding a license issued by the Federal Communications Commission authorizing him to operate an amateur radio station.

operator, B Operator who answers incoming toll calls and connects them to local subscribers' lines.

operator, checking Operator who is called in automatically to check the caller's number when a subscriber-dialed toll call is being toll ticketed.

operator, CLR (combined line and recording) Toll operator who attempts to make an on-demand connection of a long distance call. If this proves impossible, the call ticket is passed to a terminating toll operator for completion.

operator, DSA (dial service assistance) Operator who provides operator assistance to subscribers served by an automatic central office.

operator, long distance Operator who works on a toll board.

operator, toll switching Operator who connects incoming toll calls to local subscribers' lines.

operator, TX (terminating toll) Outward toll operator who deals with long distance calls requiring personal handling such as delayed calls and "leave word" calls.

operator-assisted calls Calls which require and are given help by telephone company operators, e.g., credit card calls, collect calls.

operator distance dialing (ODD) Dialing by an operator direct to a distant subscriber without the assistance of an intermediate circuit operator.

operator number identification Equipment used to bring a checking operator into the circuit to check the calling line number when a subscriber has direct dialed a long distance call which is to be charged on an itemized bill basis by CAMA equipment.

operator office A central office which provides operator assistance for subscribers linked to several offices in a local area.

operator's telephone number These are standard as follows: (NNX is the appropriate central office code.)

Inward operator	NNX-1211
Information	NNX-1311
Conference	NNX-1411
Fault/Trouble reporting	NNX-1611
Zero operator	"0"

optical axis The axis of symmetry of an optical system.

optical blank A casting made from an optical material molded into the desired form for grinding, polishing, or (in the case of optical waveguides) drawing to the final optical/mechanical specifications. See preform.

optical cable A fiber or fiber bundle in a structure fabricated to meet optical, mechanical and environmental specifications. Syn. optical fiber cable

optical cable assembly A cable that is connector terminated, generally by a manufacturer, and ready for installation.

optical cavity A region bounded by two or more reflecting surfaces whose elements are aligned to provide multiple reflections. Example: the resonator in a laser. Syn. resonant cavity

optical character recognition (OCR) The machine recognition of printed or written characters based on inputs from photoelectric transducers. Cf. magnetic ink character recognition (MICR).

optical conductor Optical waveguide (deprecated).

optical connector See optical waveguide connector.

optical coupler See optical waveguide coupler.

optical data bus A data bus using optical waveguides and optical waveguide components.

optical density A measure of the transmittance of an optical element expressed by $\log_{10}(1/T)$ or $-\log_{10}T$, where T is transmittance. The analogous term $\log_{10}(1/R)$ is called reflection density. The higher the optical density, the lower the transmittance.

optical detector A transducer that generates an electrical signal which is a function of irradiance.

optical fiber Any fiber, made of dielectric material, that guides light. One of its uses is the transmission of signals.

optical fiber cable Optical cable.

optical fiber waveguide Optical waveguide.

optical filter An element which transmits a range of wavelengths while blocking adjacent wavelengths.

optical link Any optical transmission channel designed to interconnect two end terminals or to be connected in series, as part of a circuit, with other channels.

optical path 1. A visual path from transmitting antenna to receiving antenna. 2. An above-the-horizon radio path between transmitting and receiving antennas, assuming normal refraction of the radio wave in the Earth's atmosphere and first Fresnel zone clearance over the intervening terrain. Definition No. 2 is the usual meaning for an optical or "line-of-sight path" for a radio link.

optical path length   In a medium of constant refractive index n, the product of the geometrical distance and the refractive index. If n is a function of position, then

$$\text{optical path length} = \int n ds,$$

where ds is an element of length along the path. This expression simplifies to n if the medium has a constant refractive index.

optical power   Radiant power.

optical receiver   Device for converting optical signals into electrical signals.

optical repeater   In an optical waveguide communication system, an optoelectronic device which receives an input optical signal, converts it into an electrical signal, amplifies this signal (or, in the case of a digital signal, reshapes, retimes or otherwise reconstructs it) and reconverts it into an optical signal for retransmission.

optical spectrum   As most commonly employed, the term refers to the portion of the electromagnetic spectrum within the wavelength region extending from the vacuum ultraviolet at 40 nm to the far infrared at 1 mm. There is some vagueness of meaning because bandwidth limits have not been rigidly defined.

optical switching   The switching of visual or quasivisual (e.g., infra red) signals. Term sometimes used for wavelength division multiplexing.

optical thickness   As applied to thin films, a measurement of the physical thickness times the refractive index.

optical time-domain reflectometer (OTDR)   Instrument used to locate faults in an optical waveguide by sending out a short pulse then timing the arrival of backscattered signals originating at discontinuities in the fiber.

optical transmitter   Device for converting electrical signals into optical signals.

optical waveguide   Any structure capable of guiding optical power. In optical communications, the term generally refers to a fiber designed to transmit optical signals. Syns. lightguide, optical conductor (deprecated), optical fiber waveguide, optical waveguide fiber.

optical waveguide connector   Component giving easy and rapid connection of two optical waveguides; insertion loss is normally higher than that of a well-made splice.

optical waveguide coupler   1. A device designed to distribute optical power among three or more ports. 2. A device designed to couple optical power between a waveguide and a source or detector.

optical waveguide fiber   Optical waveguide.

optical waveguide splice   A permanent joint designed to couple optical power between two waveguides.

optically active material.   A material with the ability to rotate the plane of polarization of light sent through it. Such a material exhibits different refractive indices for left and right circular polarizations.

optimize   To adjust for best output or maximum response.

optimum traffic frequency (FOT)   The highest frequency that is predicted to be available for skywave transmission over a particular path at a particular hour for 90% of the days of the month.

optoelectronic device   A device responsive to electromagnetic radiation (light) in the visible, infrared or ultraviolet spectral regions, emits or modifies noncoherent or coherent electromagnetic radiation in these same regions, or utilizes such radiation for its internal operation.

OR gate   A solid state device which gives a "1" output if any of its inputs are "1".

O/R name   Descriptive name for a user agent in a message handling system (originator/recipient).

ORACLE   Proprietary name for the IBA (British) teletext service. (Optional Reception of Announcements by Coded Line Electronics).

orbit   (1) The path, relative to a specified frame of reference, described by the center of mass of a satellite or other object in space, subjected solely to natural forces, mainly gravitational attraction. (2) By extension, the path described by the center of mass of an object in space subjected to natural forces and occasional low-energy corrective forces exerted by a propulsive device in order to achieve and maintain a desired path.

orbit, equatorial   Orbit parallel to a plane through the earth's equator.

orbit, polar satellite   Orbit in plane running north and south: a satellite in a polar orbit is continually changing its position over the earth.

orbit, subsynchronous satellite   A satellite orbit which is at a lower altitude than the 35,880 km required for a synchronous equatorial orbit.

orbit, synchronous   The path followed by a communication satellite which is at such a distance above the Earth (about 35,880 km) that the satellite keeps pace with the earth's rotation, and thus hovers above a particular point on the earth.

orbital elements   Of a satellite, the parameters by which the shape, dimensions and position of the orbit of a body in space can be defined in relation to a specified frame of reference.

orbital plane   Of a satellite, the plane containing the center of mass of the primary body and the velocity vector of a satellite, the frame of reference being that specified for defining the orbital elements.

order   The information sent by the originating operator to the distant operator indicating what trunk or station line she wishes to be connected to.

order, circuit   Formal instruction to rearrange or provide toll circuits.

order, commercial Formal order to install, remove or rearrange a customer's lines or equipment.

order, delayed Order for service which has been received from a customer but which cannot be immediately completed.

order, keep cost An order for work to be done which directs that a careful record be kept of all costs incurred.

order, patch Interim order for rapid provision of a new circuit on a patched-up basis.

order, service Order to install, remove or rearrange a customer's service.

order, work An order to perform certain work, together with the authority to incur expenditure.

order of an intermodulation component In a radio transmitter for amplitude-modulated emissions, the sum of the two positive integral coefficients determining the frequency of an intermodulation component at the output of an amplitude-modulated radio-transmitter with a given carrier frequency, as a function of the frequencies of two sinusoidal oscillations applied simultaneously at the input to the transmitter.

order of diversity The number of independently fading propagation paths or frequencies, or both, used in diversity reception.

order tone Short bursts of high tone indicating to an operator that an order should be passed on.

orderwire circuit Voice or data circuit used by technical controller and maintenance personnel for coordination and control actions relative to activation, deactivation, change, rerouting, reporting and maintenance of communications systems and services. Order wire circuits usually use frequency bands located outside the regular transmission bands. Syns. engineering channel, service channel

orderwire multiplex A multiplex carrier set specifically designed for carrying order wire traffic as opposed to one designed for carrying mission traffic.

ordinary relay contacts Contacts on a relay which are not special early-operate contacts.

orientation The pointing of a directional radio antenna in a required direction.

origin and extremity of an international sound-program circuit The origin of an international sound-program circuit is considered to be the output of the first amplifier and the extremity is considered to be the output of the last amplifier of the circuit. In the case of a circuit on a carrier system for program transmissions, the origin of the circuit is the input of the modulating equipment and the extremity is the output of the demodulating equipment.

original equipment manufacturer (OEM) The maker of equipment that is marketed by another vendor.

originating office The switching center which serves the subscriber initiating a call.

originating point In a signalling network, a signaling point in which a common channel signaling message is generated.

originating point code A part of the label in a common channel signaling message which uniquely identifies the originating point of the message in a signaling network.

originating toll center The toll center serving the calling subscriber.

originating traffic Traffic generated by sources located within the network under consideration regardless of the final destination.

originating user The user who initiates a particular information transfer transaction. This person may be either the source user or the destination user.

originator In a message handling system: the user, either a human being or a computer process from whom the message handling system accepts a message.

orthogonal multiplex A method of time division multiplexing in which pulses with orthogonal properties are used in order to avoid intersymbol interference.

oscillation Variation with time of the magnitude of a quantity with respect to a specified reference when the magnitude is alternately greater and smaller than the reference.

oscillation, parasitic Oscillation, usually unwanted, which occurs in a self-resonant section of a device.

oscillator A nonrotating device for producing alternating current, the output frequency of which is determined by the characteristics of the device.

oscillator, audio Device for producing audio frequency alternating current.

oscillator, balanced Oscillator designed with the electrical center of output coil at ground potential so that the voltages developed between the two ends and the center are equal but opposite in phase.

oscillator, beat-frequency An oscillator used to generate a signal which is combined with a received radio signal in order to produce an audio frequency beat signal.

oscillator, crystal controlled Oscillator with frequency accurately controlled by a quartz crystal device.

oscillator, feedback Oscillator made from an amplifier in which a portion of the output is fed back, on a positive phase basis, to the input.

oscillator, local An oscillator in a domestic radio receiver whose signal is mixed with the received radio signal to produce the intermediate frequency or IF, as used in superhet reception.

oscillator, magnetostriction A positive feedback amplifier/oscillator, with frequency of oscillation determined by the mechanical resonance of the magnetostrictive rod which provides the feedback.

oscillator, master A carefully constructed and stable oscillator used to control carrier frequency oscillators at a transmission equipment station.

oscillator, RC Amplifier/oscillator with controlled positive feedback, with frequency of oscillation determined by the phase of the feedback current through a variable resistance-capacitance network.

oscillator, relaxation Oscillator with frequency dependent on the charging time of a capacitor.

oscillograph   A device designed for recording instantaneous values of one or more rapidly varying electrical quantities, as a function of time or of another electrical or mechanical quantity. See oscilloscope

oscilloscope, cathode ray (CRO, or CRT, cathode ray tube)   Cathode ray tube used to produce a visible pattern which is a graphic representation of electric signals. An oscilloscope does not of itself have the ability to produce a record of signals so is not strictly speaking an oscillograph but there is some misuse of both words.

OSP (Outside Plant)   An abbreviation used by telephone companies which adopt the practices and specifications of the Rural Electrification Administration.

other common carriers (OCC)   Specialized common carriers, domestic and international record carriers, and domestic satellite carriers engaged in providing private line voice, data, audio or video services, or other services as such carriers may be authorized by the Federal Communications Commission to provide.

other line charge   Charge levied on a message which must be passed via a line belonging to another company, since that company must be paid for providing the service.

outage   A service disruption.

outage duration   Average value of elapsed user information transfer time between the start and the end of an outage period.

outage period   A performance measurement period, or succession of performance measurement periods, during which a telecommunication service remains in an outage state.

outage probability   The probability that the outage state will occur within a specified time period. In the absence of specific known causes of outages, the outage probability is the sum of all the outage durations divided by the time period of measurement.

outage state   A telecommunication service condition wherein a user is completely deprived of service due to any cause within the communication system.

outage threshold   A defined value for a supported performance parameter which establishes the minimum operational service performance level for that parameter.

out-band signaling   A signaling method in which signals are sent over the same transmission channel as the user's communication, but in a different frequency band.

outboard channel   The channel farthest from the carrier frequency in an independent sideband or single sideband transmission.

out-dialing, network   Dialing from a private network (including a PABX extension) into the public switched network.

outer conductor   In submarine cables, copper or aluminum tape surrounding the dielectric.

outer jacket   In submarine cables, a plastic jacket around the outer conductor of armorless cable. It is made of high-density polyethylene, with or without pigment.

outgoing   Describing a trunk used for calls going out to a distant toll center.

outgoing traffic   Traffic leaving the network considered, destined for sinks located outside it, whatever its origin.

outgoing WATS   See outward wide area telephone service.

outlet   Point through which the outgoing traffic flow leaves a switching stage.

outlet, TV/FM   A splitting circuit which allows an FM radio and a TV receiver to be connected to a single CATV outlet.

outline code   A symbolic representation of source code not containing sufficient detail to be compiled as it stands.

out-of-area line   A telephone line connecting a subscriber to a distant central office or exchange rather than to the office which normally serves the area. Syn.: foreign exchange line

out-of-band emission   Emission on a frequency of frequencies or the out-of-band spectrum.

out-of-band power   When applied to an emission, the total power emitted at the frequencies of the out-of-band spectrum.

out-of-band power, permissable   For a given class of emission, the permissable level of mean power emitted at frequencies above and below the limits of the necessary bandwidth.

out-of-band signaling   Signaling which utilizes frequencies within the guard band between channels, or utilizes bits other than information bits in a digital system. This term is also used to indicate the use of a portion of the channel bandwidth provided by the medium such as the carrier channel, but denied to the speech or intelligence path by filters. It results in a reduction of the effective available bandwidth.

out-of-band spectrum   When applied to an emission, the part of the power density spectrum (or the power spectrum when the spectrum consists of discrete components) of an emission which is outside the necessary bandwidth, with the exception of spurious emissions.

out-of-band spectrum of an emission, permissable   For a given class of emission, the permissable level of the power density (or the power of discrete components) at frequencies above and below the limits of the necessary bandwidth.

out-of-frame alignment time   The time during which frame alignment is effectively lost. It includes the time to detect loss of frame alignment and the alignment recovery time.

out-of-order signal   In data transmission, a signal sent in the backward direction indicating that the call cannot be completed because either the called terminal or the called terminal's access line is out of service or is faulty.

out-of-order tone Information tone indicating that a wanted line or service is out of order.

out-of-phase Periodic signals which are of the same frequency but do not pass through the same peaks at the same instant.

out-of-service sequence A predesignated code sequence indicating that some element of the network is malfunctioning.

outpulsing The process of transmitting digital address information over a trunk from one switching center to another.

output 1. In man-machine language, the process that consists of the delivery of data from a data processing system or from any part thereof. 2. In specification description language, an action within a transition which generates a signal which, in turn, acts as an input elsewhere. 3. Power, current or voltage at the output of a device.

output, balanced Output which is symmetrical with respect to ground.

output, rated Power, voltage or current which can be provided as output from a device for an unlimited period without danger of overheating.

output, unbalanced Output one of whose terminals is at ground potential.

output impedance The impedance presented at the output of a circuit component.

output rating 1. The power available at the output terminals of a transmitter when connected to the normal load or its equivalent. 2. Under specified ambient conditions, the power that can be delivered by a device over a long period of time without overheating.

outside broadcast Radio or TV program which originates outside the studio. If the program is presented live, the signals must be sent back to the permanent control equipment by temporary links.

outside plant That portion of intrabase communications systems extending from the main distribution frame outward to the telephone instrument or the terminal connections for other technical components.

out-slot signaling In pulse code modulation, signaling associated with a channel but transmitted in one or more separate digit time slots rather than within the channel time slot.

out-trunk switch Method of connecting trunks in a dial office by concentrating these into groups of unidirectional circuits and increasing availability.

outward wide area telephone service (OUTWATS) A telephony service provided over one or more dedicated access lines to the serving central office which permits customers to make calls to specified service areas on a direct dialing basis for a flat monthly charge. Sometimes abbreviated to WATS, sometimes to outgoing WATS (see also WATS).

overcoupling Coupling greater than critical coupling between two resonant circuits. It produces a wide bandwidth with two peaks in the response curve.

overflow 1. Describing switching equipment which operates when the traffic load exceeds the capacity of the regular equipment. 2. Traffic which is handled on overflow equipment. 3. Traffic which exceeds the capacity of the switching equipment, and is therefore lost. 4. The carry digit in a digital computer. 5. The generation of a number which is too large for the capacity of the register or location available to store it. 6. Intermediate message storage which serves as an extension of in-transit storage to preclude system saturation. Magnetic tape is an example.

overhead bit A bit other than an information bit.

overhead information Digital information transferred across the functional interface separating a user and a telecommunication system (or between functional entities within a telecommunication system) for the purpose of directing or controlling the transfer of user information.

overhead line Almost the same as open wire, but purists might say that if the conductors were insulated — for example, with a heavy sheath — the wires were not open even though they certainly were overhead.

overlap In teletypewriter practice, the selecting of a code group while the printing of a previously selected code group is taking place.

overlap, tape The amount by which the trailing edge laps over the leading edge of a spirally wrapped tape.

overlapping markets A case where a TV station's Predicted Grade B Contour is overlapped by the Predicted Grade B Contour of a television station located in a different market.

overlay 1. In a computer program, a segment that is not permanently maintained in internal storage. 2. The technique of repeatedly using the same areas of internal storage during different stages of the execution of a program. 3. In the execution of a computer program, to load a segment of the program in a storage area hitherto occupied by parts of the program that are not currently needed.

overlay method Method of introducing new technology services, such as integrated digital transmission and switching, by providing a network of switching centers and transmission paths using the new technology alongside the existing analogue space division and frequency division multiplex network; and ensuring that, as growth occurs in any area, the new technology routes and centers are expanded.

overlay programs Less-used programs which are allocated a common area of memory into which they are loaded as required. Fault reporting and tracing programs are examples.

overlay splice In submarine cable systems, a means of preserving the strength and protective capacity of the armor wires at a cable junction. It is accomplished by cutting the coaxials shorter than the armor and, after conclusion of the joint, replacing the wires beyond the joint over the undisturbed wires of each section of cable.

overload 1. A load greater than that for which the power source has been designed. 2. On a transmission system, a higher level than that for which the system was designed. Overload produces distortion.

overmodulation A condition whereby the mean level of the modulating signal is such that the peak value of the signal exceeds the value necessary to produce 100% modulation, resulting in distortion of the output signal.

override To obtain access to a circuit even though it tests busy.

overrun Loss of data because a receiving device is unable to accept data at the rate it is being transmitted.

overshoot 1. The result of an unusual atmospheric condition that causes microwave signals to be received where not intended. 2. In an amplifier, the increased amplitude of a portion of a nonsinusoidal wave due to the particular characteristics of the circuit.

overtime 1. Conversation time on a charged toll call which is greater than the initial period (normally three minutes). 2. Describing occupancy time of WATS circuits in excess of the time contracted for in each period.

overvoltage relay Alarm relay which operates when voltage across it exceeds a preset level.

oxide In the chemical sense, a compound composed of an element plus oxygen. In electronic circuitry, the term is often used to describe silicon dioxide (silica).

Pacific Telesis Regional Holding Company for Bell Operating Companies in the Pacific Coast area.

pack To store data in a compact form in a storage medium in such a way that the original form of the data can be recovered.

package, program A program written to cover the possible requirements of many users. A package is likely to be less efficient than a purposely-designed program tailored to suit the special requirements of one user but will have the advantages of ready availability and lower cost.

packet 1. A collection of data and control characters in a specified format, which are transferred as a whole. 2. A group of binary digits, including data and call control signals, which is switched as a composite whole. The data, all control signals, and any error control information are arranged in a specific format.

packet assembly/disassembly device (PAD) Device which provides interface between data terminals and a packet-switched network.

packet disassembly A user facility which enables packets of data destined for delivery to a nonpacket mode terminal to be delivered in the appropriate form (for example, in character form) at the applicable rate.

packet format A set of rules governing the structure of data control information in a packet. The packet format defines the size and content of the various fields which make up a packet.

packet length selection A user facility whereby data terminal equipment may select a certain maximum user data field length out of a defined set.

packet switching 1. Describing a system whereby messages are broken down into smaller units called packets which are then individually addressed and routed through the network. 2. The process of routing and transferring data by means of addressed packets so that a channel is occupied only during the transmission of the packet. Upon completion of the transmission the channel becomes available for the transfer of other traffic.

packet-mode terminal Data terminal equipment which can control and format packets as well as transmitting and receiving them.

packet-switched data transmission service A service handling the transmission, and sometimes the assembly and disassembly, of data in packet form.

packet-switched network A switched network which provides connection by forwarding standard data packets between user parties. Packets may be delivered either individually (e.g., datagram service) or in original sequence until the connection is released (virtual circuit service).

packing fraction In a fiber bundle, the ratio of the aggregate fiber core area to the total cross-sectional area (usually within the ferrule), including cladding and interstitial areas.

pad A passive network which reduces the power level of a signal and may match impedances. Since its components are purely resistive, it does not introduce different attenuation at different frequencies.

pad, climber One of the leather pads which a lineman inserts under his pole climbers for comfort.

pad, fixed A network giving a specified attenuation.

pad, push button A key set used for sending digital information. It refers especially to the 12-button set commonly used on telephone instruments providing dual tone multi-frequency signals.

pad, receiver Circular pad of tissue sometimes used with an operator's telephone headset to improve comfort.

pad, switching A small attenuating network which is switched into a circuit when a toll call is put through to a local subscriber's line. The switching pad is taken out of the circuit when the same toll call is switched through to another toll circuit.

pad control Method of switching giving control of overall transmission levels by inserting loss in a circuit terminated locally but taking that loss out of the circuit when toll circuits are tandem switched.

pad switching See pad control.

page 1. A measure of computer memory. A memory page can contain up to a given maximum number of bytes. 2. A block of information held on a videotex system's database. It is stored as one or more frames.

page copy A printed copy of machine output in readable form.

page printer Device which types a received message onto paper sheets or rolls.

paging To summon a person, exact whereabouts unknown, to the telephone or otherwise deliver him a message. Some paging systems use a selective radio signal to summon the called person to the nearest telephone. More sophisticated systems deliver voice messages, or text messages displayed on a strip of LCDs, to the person carrying the paging unit.

paint, asphalt An asphalt-based paint used to give protection to buried equipment.

paintbox System by which TV designers can draw or paint directly on to a TV monitor screen, mixing colors as desired.

pair The two wires of a circuit, particularly those providing the subscriber's loop.

pair, alarm A cable pair used to carry alarm circuits, particularly circuits from remote pressure contactors to centrally located alarm equipment in a cable pressurization system.

pair, battery A pair of wires used to carry dc power which is tapped from the central office storage battery.

pair, dead A cable pair in a distribution network which has not been spliced through to the central office main distribution frame.

pair, extra An extra cable pair enclosed in the cable's outer sheath to ensure that, in the event of faults in different drum lengths of the cable, the full number of pairs could still be spliced through the spliced cable system. A 2000 pair cable often has up to 16 spare pairs available for this purpose.

pair, generator A specially insulated or screened cable pair designed to carry ringing current. One use is to carry machine generated ringing current out to a PBX so that the operator does not require access to a local ringing current source.

pair, multiple Pair which is terminated at several different places, all in parallel.

pair, non-pole Wire pair on an overhead pole line which is not divided by pole placement.

pair, pole A separated pair of wires on an open wire route. One member of the pair runs along each side of the pole.

pair, shielded A pair of wires wrapped with an electrostatic shield to minimize induced interference.

pair, spare 1. A distributed and terminated cable pair which is available at both the main distributing frame end and at a distribution point at the other end. 2. Extra pairs made up in the cable as spares, to be used if numbered pairs have been found to be faulty.

pair, twisted Pair of wires with the conductors twisted together to reduce the effect of inductive interference.

pair, universally bad A cable pair which is faulty in many locations and cannot be economically repaired for use as a through circuit.

pair, video A low-loss cable pair used for TV signals.

pair-kilometer Length of cable in kilometers multiplied by number of pairs in the cable.

pair of complementary channels Two channels, one in each direction, which provide a bidirectional communication.

paired cable A cable made up of one or more separately insulated twisted pairs, none of which is arranged with others to form quads.

paired-disparity code (PCM) A code in which some or all of the digits or characters are represented by two assemblies of digits, of opposite disparity, which are used in a sequence so as to minimize the total disparity of a longer sequence of digits. An alternate mark inversion signal is an example of a paired-disparity code. Sometimes called alternative or alternating code.

paired-wheel See multi-wheel gear.

pairing (TV) A partial or complete failure of interlace in which the scanning lines of alternate fields do not fall exactly between one another but tend to fall (in pairs) one on top of the other.

pairs, bunched 1. Cable pairs joined in parallel to reduce resistance. 2. Pairs bound together loosely at a cable splicing location to facilitate planned splicing.

palladium A metal sometimes used for contact points on relays or keys.

pan See cable pan.

panel 1. A plate, usually vertical, on which components, keys or meters are mounted. 2. The unit area on the face of a manual operating position in which subscribers' lines are terminated.

panel, dual pressure Unit which is fed by high pressure gas or air which it sends out at a pressure appropriate to the type of cable system being pressurized: higher pressure for buried cables and a slightly lower one for aerial cables.

panel, jack A panel on a manual operating position on which jacks are mounted to give operators access to the circuits terminated thereon.

panel, patch A jack panel used for temporary circuit rearrangements.

panel dial system Automatic telephone system of a matrix type used widely after World War I. It used electric motors to drive brushes or wipers to the required outlet contacts which were mounted vertically in flat rectangular panels.

panelboard Circuit breakers mounted in a small cabinet.

paper, electrosensitive Paper which changes color when a small direct current is passed through. It is used on recording meters and facsimile and printing equipment.

paper, pressure sensitive Carbonless paper used for duplicating forms and documents. The pressure applied by typing or writing crushes tiny balls of ink which reproduce the impression made by the pen or type.

paper tape A continuous strip of paper used for storing binary information by means of punched holes. It is most commonly used with teletypewriter-related equipment.

paper tape reader Device which translates coded punched holes in tape into electrical signal impulses.

par meter A display meter of the peak-to-average ratio (P/AR) measurement technique developed by Bell Labs as a quick means of identifying degraded telephone channels. The measurement is very sensitive to envelope delay distortion and is also useful for idle channel noise, nonlinear distortion, and amplitude distortion measurements.

parabola A conic section formed by cutting a cone with the plane parallel to one side of the cone. Directional microwave antennae are usually made with a parabolic reflecting surface which is fed with radio frequency power from a source at the focus.

parabolic antenna An antenna consisting of a parabolic reflector and a radiating or receiving element at or near its focus.

parabolic profile A power-law index profile with the profile parameter, g, equal to 2. Syn.: quadratic profile

paraboloid A reflecting surface formed by rotating a plane parabola about one of its axes of symmetry.

parachute In submarine cable systems, a device affixed to repeaters when they are deployed over the stern. The parachute opens to form a drogue in the water, thus slowing the repeater sink rate to match the cable subsidence.

paragutta An insulant resin compounded of gutta percha and rubber, formerly used in the seacable telegraph industry.

parallax An apparent change in the position of an object due to a change in the viewer's viewing position.

parallel 1. Describing simultaneous transmission on different paths. For example, an eight-bit signal can be sent serially, meaning one bit after another on a single path, or parallel, meaning that each bit is sent on a separate path so all eight are sent at the same time. 2. Circuit elements so connected that they provide alternative paths for the flow of current.

parallel connection Circuit elements so connected that the same voltage appears across each. Current is divided among them in inverse proportion to impedance.

parallel data transmission Transmission method where data is transmitted one byte at a time, each of the bits making up the byte being transmitted simultaneously over a separate path.

parallel processing Concurrent or simultaneous execution of two or more processes or programs within the same processor.

parallel resonance The condition in a circuit whereby inductive and capacitive reactance are in parallel and are of equal magnitude but reverse in sign. The voltage across the parallel circuit and the current within the circuit are at a maximum at the resonant frequency.

parallel-to-serial converter 1. A device which converts a group of parallel inputs, all of which are presented simultaneously, into a corresponding time sequence of signal elements. 2. A device that converts a spatial distribution of signal states representing data into a corresponding time sequence of signal states. Each of the parallel input signals requires a separate channel while the serial output requires only a single channel. Syns. serializer, dynamicizer

paralleling, rectifier Operation of more than one rectifier to carry a load on a parallel load-sharing basis.

paramagnetic Describing a material with magnetic permeability slightly greater than unity.

parameter 1. A variable which has a specific value when applied to a particular component, circuit or other item. 2. In man-machine language, a variable which identifies and contains a piece of necessary information to execute a command. 3. In software, a variable that is given a constant value for a specified application and that may denote the application.

parametric converter A device, usually dependent on reactance, which converts an input signal at one frequency into an output signal at a different frequency.

parasitic 1. Describing an element in a directional antenna array which has no direct connection to the radio frequency feeder but which acts as a reflector or director of energy. 2. Describing unwanted oscillations due to local resonance.

parasitic emissions Spurious emissions accidentally generated at frequencies which are independent of both the carrier or characteristic frequency of an emission and frequencies of oscillations resulting from the generation of the carrier or characteristic frequency.

parasitically excited Describing an element in a directional antenna which is energized by radiation from a nearby element rather than directly from the radio frequency feed from the radio equipment.

paraxial ray In optoelectronics, a ray which is close to and nearly parallel with the optical axis. For purposes of computation, the angle between the ray and the optical axis is small enough for $\sin \theta$ or $\tan \theta$ to be replaced by θ expressed in radians.

parity Describing a self-checking method of minimizing transmission errors in received data signals. An extra binary signal is added to each character signal to make the total number of 1s or 0s even or odd for each character.

parity, even   Describing the addition of a parity bit to the coded bits representing each character in order to make the total number of 1s an even number.

parity, odd   Describing the addition of a parity bit to the coded bits representing each character in order to make the total number of 1s an odd number.

parity bit   See bit, parity.

parity check   See check, parity.

partial   Describing a single-frequency component of a complex tone.

partially perforated tape   Paper tape which is not completely punched through: chadless tape.

partition   Restriction of access to some switched facilities to a particular group of subscriber lines. Partitioning structure is sometimes changed at different times of day or on different dates to permit the usage of a system to be amended.

partitioning   Division of a complex entity such as a subsystem into smaller parts which can be dealt with independently and which will provide the behavior of the entity when they are recomposed.

party line   See line, party.

pascal   The SI unit of pressure. One pascal equals one newton per square meter.

Pascal   A high-level computer language named after Blaise Pascal who, it is claimed, built the first digital calculating machine in 1642.

pass band   1. The number of hertz expressing the difference between the limiting frequencies at which the desired fraction (usually half power) of the maximum output is obtained. 2. A band of frequencies which is passed through a filter essentially unchanged. 3. The band of radio frequencies accepted by a radio receiver for amplitude modulation signals and measured at the detector input. It is limited by the two frequencies for which the attenuation exceeds that of the most favoured frequency by some agreed value. In general this value is 6 dB. In high quality radio telephony receivers the value is 2 dB.

passivation   Describing the treating of material by creation of a chemically-bonded surface layer which protects against corrosion.

passive   Not supplying energy.

passive network   See network, passive.

passive repeater   See repeater, passive.

passive satellite   A satellite designed to transmit radio-communication signals by reflection.

passive station   In a multipoint connection, a tributary station waiting to be polled or selected.

password   A group of characters which, on input to a computer from a terminal, give the user access to information and allow the user defined control over the information.

paste, burnishing   Metal polish used to maintain the plugs on cord-type manual switchboards.

paster, cable   Paper tape designed for wrapping around a lead cable sheath and splicing sleeve prior to

plumbing work, in order to impart a neat finish to the joint.

patch   1. To connect circuits temporarily by means of a cable known as a patch cord. 2. Machine language instructions added to a program to alter it or to correct an error.

patch and test facility   An organic element of a station or terminal facility that functions as a supporting activity under the technical supervision of a designated technical control facility. It performs such functions as quality control checks and tests on equipment, links and circuits; troubleshooting; activation, changing and deactivation of circuits; technical coordination; and reporting.

patch bay   An assembly of hardware so arranged that a number of circuits appear on jacks for monitoring, interconnecting and testing purposes.

patch panel   One segment of a patch bay.

patchboard   A patch bay used alongside a toll testboard.

patched   1. Describing a temporary connection made with a patch cord. 2. An improvised modification.

patching   1. Connecting circuits by means of cords with plugs inserted into appropriate jacks. 2. The insertion of a program patch into a computer program, routine, or subroutine. 3. Minor changes made, using soldered wire leads, to circuits on printed circuit boards (PCBs) during development. Circuit patches of this nature are normally incorporated in production versions of the PCB concerned.

path   1. The route that a signal follows through a circuit or network. 2. The route that a radio signal follows from transmitter to receiver. 3. The implementation of a means of transmission. The path includes the channels used for the transmission and the equipment used to connect them.

path antenna gain   The change in transmission loss when lossless, isotropic antennae are used at the same locations as the actual antennae.

path antenna power gain   The increase in the system loss when lossless, isotropic antennae are used at the same locations as the actual antennae.

path clearance   In microwave line-of-sight communication, the perpendicular distance from the radio beam axis to obstructions such as trees, buildings or terrain. For a particular K-factor, the required path clearance is usually expressed as some fraction of the first Fresnel zone radius.

path intermodulation noise   Noise in a transmission path contingent upon modulation. It results from any nonlinear characteristic of the path.

path loss   1. The decrease in power during transmission from one point to another. 2. Of a radio circuit, the transmission loss expected between ideal, loss-free, isotropic, transmitting and receiving antennae at the same locations as the actual transmitting and receiving antennae.

path profile   A graphic representation of a propagation path, showing the surface features of the Earth — such as trees, buildings, and other features — that may cause obstruction or reflection in the vertical plane containing the path.

path survey An assembling of the geographical and environmental data required to design a microwave communication system.

pathfinding Selecting an available path through a switching network.

pathfinding, progressive Selecting an available path through a multistage switching network, stage by stage.

pattern, radiation See radiation pattern.

pawl The part of an electromechanical switch which controls rotation or other movement.

pawl, holding Pawl which holds a switch until it has received instructions to release it.

pawl, stepping Pawl which imparts vertical or rotary movement to the shaft of an electromechanical switch.

pay-out The process of dispensing cable from a cable ship.

pay-out engine The aft engine on a cable-laying ship. It is used for laying lengths of cable over the stern.

pay phone A public or rented telephone from which calls can be paid for at the time they are made by means of coins, tokens or credit cards.

pay TV A system of television in which scrambled signals are distributed and are unscrambled at the homeowner's set with a decoder that responds upon payment of a fee for each program. Pay TV can also refer to a system where subscribers pay an extra fee for access to a special channel which might offer sports programs, first-run movies or professional training.

paystation A pay phone.

paystation, local prepay Pay phone used for local calls which must be paid for before they can be dialed. Toll calls made from these stations are normally operator-controlled and are paid for after use.

paystation, postpay A pay phone which requires payment for telephone calls before they are made but which provides for no contact with the operator regarding coin refund.

paystation, semi-postpay Pay phone permitting dialing, but not conversation with the called party, before coins are deposited. Toll calls are made on a postpay basis.

paystation, semi-public Coin type pay phone installed for public use but paid for by a subscriber who retains all coins collected and pays the telephone company for the calls made.

PBX complex An arrangement consisting of a main PBX and satellite PBXs. The main PBX will normally be the switching point for access to the public switched telephone network.

PBX connector Automatic switch in a central office which hunts over all the lines leading to a PBX and gives busy advice only if all these lines are engaged.

PBX extension Telephone instrument served by a PBX.

PBX line hunting The automatic selection of a free line from a group of subscriber lines on receipt of a call to the subscriber's telephone number.

PBX tie trunk A direct connection between two PBXs.

PBX trunk A trunk used to interconnect a PBX with its servicing switch.

P-channel metal oxide semiconductor (PMOS) A metal-oxide silicon field effect transistor using holes as majority current carrier. The channel has a predominantly positive charge.

PCM multiplex equipment Equipment for deriving a single digital signal at a defined digit rate from two or more analog channels by a combination of pulse code modulation and time division multiplexing, and also for carrying out the inverse function.

peak A maximum instantaneous value, usually of current or voltage.

peak amplitude Maximum amplitude of a periodically repeating quality.

peak busy hour (bouncing busy hour; post selected busy hour) The busy hour each day; it is usually not the same over a number of days.

peak envelope power Of a radio transmitter, the average power supplied to the antenna transmission line by the transmitter during one radio-frequency cycle at the highest crest of the modulation envelope, taken under conditions of normal operation.

peak forward voltage Maximum instantaneous voltage applied to a device in the direction in which it is designed to pass current with minimum resistance.

peak inverse voltage Maximum instantaneous voltage applied to a device in the direction of maximum resistance to current flow. If this exceeds a specified limit, an avalanche effect occurs in some devices and high currents flow.

peak limiting 1. Describing a process in which the absolute instantaneous value of a signal parameter is prevented from exceeding a specified value. 2. In pulse code modulation, it is the effect caused by the application to an encoder of an input signal whose value exceeds the virtual decision values of the encoder.

peak power output The output power averaged over the radio-frequency cycle having the maximum peak value which can occur under any combination of signals transmitted.

peak value Peak amplitude. For a sinusoidal wave, the peak is 2 times the root-mean-square value.

peak wavelength In optoelectronics, the wavelength at which the radiant intensity of a source is a maximum. It is expressed in nanometers.

peaked traffic A traffic that has a peakedness factor greater than one.

peakedness factor The ratio of variance to mean of traffic (i.e., low for smooth traffic). Note that the variance and the mean refer to the number of resources that would be occupied if this traffic was offered to an infinitely large pool of resources.

peaking Adjusting a circuit for sharp response at a selected frequency.

peaks Bursts of high volume signals of short duration.

peak-signal level An expression of the maximum instantaneous signal power or voltage as measured at any point in a transmitted path.

peak-to-average ratio The ratio of the instantaneous amplitude of a signal to its time averaged value. Peak-to-average ratio can be determined for voltage, current, power, or other parameters.

peak-to-peak value The algebraic difference between the extreme values of a varying quantity.

peavey Wooden lever used for moving telephone poles in a stores yard.

pedestal Base on which a roadside cable terminal box is mounted.

peek Programming language word which instructs a processor to read information from a particular memory store.

peg count A count of the seizure, or attempts at seizure, of telephone trunks, circuits, or switching equipment or of calls handled by an operator during a specified time interval.

pen register A moving paper recording device used for checking dial pulse speed and ratio.

penalty The additional loss on a circuit which would give the same subjective effect as the actual noise or distortion present on the circuit.

penetration 1. The higher the frequency of an alternating current, the less depth below the surface skin of a conductor is utilized. Penetration is a measure of this depth. 2. Number of telephone lines per household, or per unit of commercial, industrial or residential accommodation, or per square meter of floor space, or per hectare of land area. 3. The ratio of the number of subscribers to the total number of households passed by a cable TV system. Penetration is the basis of a system's profitability.

penetration factor Number of telephone lines per unit of accommodation or tenancy.

penetration frequency The highest radio frequency vertically incident on a layer of the ionosphere which is reflected downward by that layer.

penny, conduit Small plug placed in the end of a cable conduit during building construction to prevent the pipe being blocked.

pentachlorophenol Wood preservative sometimes used on telephone poles.

Pentaconta ITT's crossbar switching system.

pentode An electron tube with five electrodes. Used widely in radio broadcast receivers at one time, it has now been largely displaced by the transistor.

percent break In telephone dialing, the ratio of the open circuit time to the sum of the open and closed circuit times allotted to a single dial pulse cycle. It is expressed as a percentage.

percentage modulation 1. In angle modulation, the fraction of a specified reference modulation expressed as a percentage. 2. In amplitude modulation, the modulation factor expressed as a percentage.

percentage of call requests met The ratio n/N expressed as a percentage; n is the number of call requests followed by calls, and N is the total number of call requests in a specified time.

percentage sync The ratio, expressed as a percentage, of the amplitude of the synchronizing signal to the peak-to-peak amplitude of the picture signal between blanking and reference white level.

perforator, tape Device, under the control of an incoming signal or of a local keyboard, which punches coded holes in paper tape, usually five elements per character as in a teletypewriter signal. The paper tape may be used to control a tape transmitter to retransmit the message to another destination.

performance measurement period The time period over which values of the performance parameters are measured. The duration of a performance measurement period is determined by required confidence limits and may vary as a function of the observed parameter values. A user's time is divided into a succession of consecutive performance measurement periods to enable measurement of user information transfer reliability.

performance parameter A quantity whose numerical values characterize a particular aspect of telecommunication system performance.

performance standards (CATV) The minimum technical criteria that must be met by CATV systems, consistent with standards set by the FCC or the local ordinance.

periastron The point in the orbit of a satellite which is at a minimum distance from the center of mass of the primary body.

perigee The point in the orbit of an Earth satellite which is at a minimum distance from the center of the Earth. The perigee is the periastron of an Earth satellite.

period The time interval after which the specified characteristics of a periodic waveform or feature recur.

period, initial A base period, usually of three minutes duration, for which a minimum charge is made on a toll call.

period, overtime The period during which a per minute rate charge is made after the end of the initial period during a toll call.

period, ringing Period during which ringing current is sent out to the called subscriber.

period, silent Break between bursts of ringing current on a called subscriber's line.

period of revolution Of a satellite, the time between two consecutive passages through a characteristic point in its orbit.

periodic Describing a function which occurs at regular intervals of time.

periodic pulse metering (PPM) A system of metering whereby the subscriber's meter is operated at timed intervals throughout the call. The interval is determined by the type of call, such as local, junction or trunk.

periodic table An arrangement of the chemical elements from lightest to heaviest according to their atomic weights. This order so arranges them that the various families of elements with similar properties are grouped together.

peripheral equipment Equipment which works in conjunction with a communication system or a computer but is not a part of it. In particular, equipment which acts under the control of a processor such as a line printer or video display unit. It is the peripherals which link computers with the real world.

periscope antenna See antenna, periscope.

Permalloy A nickel-iron alloy with high permeability, used in cores of coils where high inductance values are required.

permanent Fault condition of a telephone subscriber's line which sends the central office equipment a signal that the subscriber has gone off-hook but sends no dial pulses.

permanent subscriber number A telephone number which will be retained by a subscriber in the event that he moves to another location.

permanent timing System for disconnecting a permanent line from the central office equipment after the lapse of a specified time during which no signal is received from the subscriber. Syn. time out

permanent virtual circuit A user facility whereby a permanent association exists between two units of data terminal equipment which is identical to the data transfer phase of a virtual call. No call setup or clearing procedure is necessary.

permanently locked envelope An envelope that is always separated by a number of bits corresponding to an integer number of envelopes.

permeability The relationship between magnetic induction and magnetizing force. A measure of the magnetic flux created in a material by a magnetizing force compared to the flux which would be created in air by an equal magnetizing force.

permeable Porous to certain materials such as ions or fluids.

permeance A measure of the ease of establishing a magnetic field. It is the ratio of magnetic flux to the magnetomotive force. The reciprocal of reluctance.

Permendur A cobalt-iron alloy with high permeability over a range of high-flux densities.

Perminvar A nickel-iron-cobalt alloy with high permeability which is constant over a range of low flux densities.

permissible out-of-band power For a given class of emission, the permissible level of mean power emitted at frequencies above and below the limits of necessary bandwidth. It is determined for each class of emission and specified as a percentage of total mean power radiated, derived from the limiting curve fixed individually for each class of emission.

permissible out-of-band spectrum For a given class of emission, the permissible level of the power density at frequencies above and below the limits of the necessary bandwidth. The permissible power density may be specified in the form of a limiting curve giving the power density, expressed in decibels relative to the specified reference level, for frequencies outside the necessary bandwidth. The abscissa of the initial point of the limiting curve should coincide with the limiting frequencies of the necessary bandwidth.

permission granted to add (PEGAD) Procedure followed when one customer permits another to access and add to its own service.

permissive connection Connection to privately owned facilities which is permitted by the telephone company but is not guaranteed by them to perform satisfactorily in all respects.

permittivity The product of relative capacitivity and the electric constant or capacitivity of free space appropriate to the system of units used.

permutation Any of the possible combinations of quantities or variables in a group.

persistence Time taken for phosphor on a cathode ray or TV tube to decrease to 10% of its peak intensity.

person-to-person call See call, person-to-person.

personal indentification number (PIN) A code number (usually 1, 2 or 3 digits) dialed by a customer to obtain access to a system, in particular to a least-cost routing service provided by a specialized common carrier.

personnel number Identification number used by a telephone operator.

perturbation Irregularity in the motion of a satellite.

petticoat Annular rings on an insulator, as used on open wire routes, which increase the length of the leakage path.

phantom circuit See circuit, phantom.

phantom coil See coil, phantom.

phantom group See group, phantom.

phase 1. Any distinguishable state in a periodic phenomenon. 2. The time displacement between two currents or two voltages or between a current and a voltage measured in electrical degrees, where an electrical degree is $1/360$ part of a complete cycle. 3. The number of separate voltage waves in a commercial alternating current, designated as "single phase," "three phase," etc. Phase is abbreviated as the Greek letter phi (ϕ).

phase alternate line (PAL) A color TV system developed in Germany in which the relative phases of the chrominance signal, sent in quadrature, are reversed in alternate lines in order to minimize phase errors and improve color performance.

phase coherence See coherent.

phase constant The imaginary component of the propagation constant. A measure of the phase shift of a signal in a transmission line.

phase degenerate modes Modes which have the same phase velocity.

phase delay See delay, phase.

phase deviation See deviation, phase.

phase difference See difference, phase.

phase distortion See delay distortion.

phase encoded recording A method of recording on magnetic tape in which a "1" data bit is a flux reversal to the polarity of the interblock gap, and the "0" data bit is a flux reversal to the polarity opposite to that of the interblock gap (when reading in the forward direction).

phase equalizer See delay equalizer.

phase flux reversal In magnetic tape phase encoded recording, a flux reversal written at the nominal midpoint between successive "1" bits or between successive "0" bits to establish proper polarity.

phase frequency distortion That form of distortion which occurs under either or both of the following conditions: (a) if the phase-frequency characteristic is not linear over the frequency range of interest; (b) if the zero-frequency intercept of the phase-frequency characteristic is not zero or an integral multiple of 2π radians.

phase hit In a transmission channel, a momentary disturbance caused by sudden phase changes in the signal. Defined by Bell as a case where the phase of a 1004 Hz test signal shifts more than 20 degrees. Phase hits are a serious cause of error in data transmission systems using phase shift keying (PSK).

phase interference fading The variation in signal amplitude produced by the interaction of two or more components with different relative phases.

phase inverter A device which changes the phase of a signal by 180°.

phase jitter A form of phase perturbation. Defined as the measurement, in degrees out-of-phase, that an analog signal deviates from the referenced phase of the main data-carrying signal.

phase lock loop An electronic servo system which controls an oscillator so that it maintains a constant phase angle relative to a reference signal source.

phase locked Describing signals of the same frequency whose phases are the same at all times.

phase modulation (PM) See modulation, phase.

phase perturbation That phenomenon, from causes known or unknown, which results in a relative shifting, often quite rapid, in the phase of a signal. The shifting in phase may appear to be random, cyclic, or both.

phase sequence The order in which the different phases of a polyphase power supply reach their positive peaks.

phase shift 1. The continuing change in phase as a wave proceeds in space or along a transmission line. 2. The change in phase of a periodic signal with respect to a reference.

phase shift keying (PSK) A method of modulation used for digital transmission wherein the phase of the carrier is discretely varied in relation to a reference phase, or the phase of the previous signal element, in accordance with the data to be transmitted.

phase-space diagram Representation of the light-guiding properties of an optical waveguide.

phase splitter Device that produces from a single input wave two or more output waves differing in phase from one another.

phase velocity The velocity of propagation of a uniform plane wave, given by the wavelength times the frequency divided by the refractive index of the medium in which the wave is propagating.

phased-array An array antenna whose directivity pattern is controlled largely by the relative phases of the excitation coefficients of the radiating elements.

phasing In facsimile transmission, the adjustment of picture position along the scanning line.

phon The subjective unit of loudness level. It is numerically equal to the sound pressure level in decibels relative to 2×10^{-5} newtons per square meter of a free progressive wave of frequency 1000 Hz presented to listeners facing the source, which on a number of trials is judged by the listeners to be equally loud.

phone Abbreviation for telephone.

phone, panel Telephone mounted on a panel which can be fixed flush to a wall.

phoneme Speech sounds which are alike except as they may be modified at the beginning and end by the sound of adjoining letters.

phosphor Chemical which gives off visible light when bombarded by electrons or other subatomic particles.

photo transistor Junction transistor in which the collector current varies with the light focused on the base.

photocathode Electrode in an electron tube which will emit electrons when bombarded by photons of light.

photocell Device which produces an electrical output from a visible light input.

photoconductivity The conductivity increase exhibited by some nonmetallic materials, resulting from the free carriers generated when photon energy is absorbed in electronic transitions. The rate at which free carriers are generated, the mobility of the carriers, and the length of time they persist in conducting states determine the amount of conductivity change.

photocurrent The current that flows through a photosensitive device, such as a photodiode, as the direct result of exposure to radiant energy. Internal gain mechanisms, such as in an avalanche photodiode, may increase the electron flow but are distinct mechanisms.

photodetector Device which detects (and measures) the intensity of light.

photodiode A diode having a current vs. voltage characteristic that is dependent on the level of optical power incident on the device. Photodiodes are used for the detection of optical power and for the conversion of optical power to electrical power.

photoelectric effect The emission of electrons by a material when it is irradiated by or exposed to light. Syn.: photoemissive effect

photoelectromagnetic effect The production of potential difference by virtue of the interaction of a magnetic field with a photoconductive material subjected to incident radiation.

photoemissive effect See photoelectric effect.

photolithography Technique used in the manufacture of integrated circuits. Complex circuit designs are photographed and reduced size masks are made up prior to circuit manufacture.

photon A quantum of electromagnetic energy. The energy of a photon is $h\nu$ where h is Planck's constant and ν is the optical frequency.

photon noise Noise due to the discrete nature of optical radiation: quantum noise.

photonics Pertaining to that field of science and engineering that deals with photons of light and their utilization.

photoresist Describing various materials used during photolithography. They form polymers on exposure to light and these act as barriers during various processing stages. Positive photoresists act in the reverse direction. They start off as polymers but on exposure to light they no longer act as barriers.

photosensitive recording Facsimile recording created by the exposure of a photosensitive surface to a signal-controlled light beam.

phototube Electron tube in which current output varies with the number of photons striking its photocathode.

photovideotex High definition videotex (e.g. Picture Prestel), a videotex frame which incorporates a picture of normal television quality.

photovoltaic effect The production of a voltage across a pn junction resulting from the absorption of photon energy. The potential is caused by the internal drift of hole-electron pairs; hence, the phenomenon leads to direct conversion of a part of the absorbed energy into a usable voltage.

physical Real, material, tangible.

physical address space The set of memory locations wherein information can actually be stored for program execution. Virtual memory addresses can be mapped, relocated or translated to produce a final memory address that is sent to hardware memory units. The final memory address is the physical address.

physical circuit A circuit with two metallic paths. One pair of wires.

physical layer Within the OSI model for a data system, the lowest level of network processing, concerned with electrical, mechanical and handshaking procedures over the interface between a device and the transmission medium.

physical optics The branch of optics in which light propagation is treated as a wave phenomenon rather than a ray phenomenon, as in geometric optics.

physical record 1. The smallest unit of data that an input/output peripheral device can transfer. The size of a physical record is fixed and depends on the type of device being referenced. 2. The largest unit of data that the read/write hardware of an input/output device can transmit or receive in a single input/output operation.

physical-theoretical office Method of division by which directory numbers in one group may be divided into three different central office code groups, called the physical, theoretical and extra theoretical central offices. The purpose is to give tariff flexibility.

piano wire A taut wire. In submarine cable systems, this is paid out during cable laying for precise measurement of distance travelled.

piccolo A multi-frequency shift keying technique widely used for 75 baud 7.5 unit start-stop telegraphy.

pick Measure used in defining braid filaments. It is the distance, expressed as a fraction of an inch, between two adjacent crossover points in a braided conductor.

pick, test A test prod with a fine needle end able to penetrate insulation and make contact with the conductor.

pickup 1. To in-board cable into a cable ship from the sea bed. 2. To use a remote microphone and amplifier to feed a program from a distant location back to a broadcasting center. 3. A phonograph pick-up head.

pickup, stereo Phonograph pickup head used with stereophonic records. A single stylus can reproduce the acoustic patterns of both left and right audio channels.

pick-up facility A facility whereby a subscriber away from his telephone can pick up a call on his line by dialing his own number (and sometimes a special code) from any other telephone, after having been informed of the incoming call.

pico- Prefix indicating 10^{-12} (one-millionth of a millionth).

picofarad 10^{-12} farad. One millionth of a microfarad. Previously called a micromicrofarad.

pictorial element (PE) One of a number of standardized graphical entities used within state pictures to represent switching system concepts.

picture frequencies In facsimile systems, the frequencies which result solely from scanning a subject copy.

picture tube A cathode-ray tube used to produce an image by variation of the intensity of a scanning beam.

Picturephone Bell System name for its "see while you talk" telephone set. It consists of three units: (a) a standard telephone with a dual tone multifrequency dial, (b) a combined camera, loud speaker, and TV screen unit, and (c) a control unit with an integral microphone.

Picturephone trunk Equipment originally used with the Bell System Picturephone. It consisted of a four-wire equalized trunk for the video signal plus a two-wire trunk for signaling and talking.

piezoelectric effect Property of some crystals of producing electric charges on the crystal faces when compressed in certain directions.

pigtail 1. A short, flexible braided or stranded wire used to carry current from a moveable member, such as a generator brush. 2. The splice made by twisting together the bared ends of two conductors laid side by side. 3. A short length of optical fiber used to couple power between an optoelectronic component and the transmission fiber.

pileup, spring   Assembly of springsets and their associated contacts which is operated by the armature of a relay or by the insertion of a jack.

pilot   1. A signal, usually a single frequency, transmitted over a system for supervisory, control, synchronization or reference purposes. 2. A pilot lamp. 3. A pilot channel. 4. A pilot regulator. 5. A pilot tone.

pilot, cable   In submarine cable systems, supervisory frequency inserted near the termination of the seacable and extracted at the other end.

pilot, system   In submarine cable systems, supervisory frequency inserted and extracted near the ongoing interface.

pilot cell   One of the cells of a main central office storage battery the specific gravity (SG) of which is regularly tested and recorded.

pilot frequency   Single tone frequency, used in telephony and radio transmission, which is sent out at a specified level to enable receive side equipment to be kept in close adjustment with the transmitter, and for output signal levels to be kept within permitted limits.

pilot make busy (PMB) circuit   A circuit arrangement by which trunks provided over a carrier system are made busy to the switching equipment in the event of carrier system failure, or during a fade of the radio system.

pilot regulator   A device which monitors the level of a received pilot tone and adjusts amplifiers to keep overall transmission loss constant.

pilot tone   An unmodulated tone of a specified frequency appropriate to the transmission system concerned. The pilot tone is transmitted over the system together with the information channels. Monitoring the level of the received pilot tone permits automatic adjustment of the levels of the information channels.

pin   1. A terminal on a base, such as an electron tube. 2. An insulator pin, which is bolted to cross-arms to carry insulators for open wire routes. 3. A personal identity number.

pin, base   A metal prong protruding from the base of an electron tube. It makes contact with a spring-loaded socket.

pin, conduit dowel   An alignment pin used to join adjacent sections of multiple-way conduit.

pin, steel insulator   Pin used to carry open wire insulator in long-span construction where wooden pins are insufficiently strong.

pin, transposition   Special insulator pin for use at a transposition point.

pin, wooden insulator   An 8-inch-long pin used to mount glass or ceramic insulators on a cross-arm in open wire construction.

pin-fet receiver   Optical receiver with a PIN photodiode and a low-noise amplifier with a high-impedance output whose first stage incorporates a field-effect transistor (FET).

PIN photodiode   A diode with a large intrinsic region sandwiched between p and n doped semiconducting regions. Photons absorbed in this region create electron-hole pairs which are then separated by an electric field, thus generating an electric current in a load circuit.

ping pong   Two way communication whereby digital packages are sent in alternate directions on a two-wire circuit. Syns. burst mode, time compression multiplex (TCM), time division duplexing (TDD)

pink noise   See noise, pink.

pipeline, gas   A feeder pipe supplying dry air at about 10 pounds per square inch.

pit, beach   In submarine cable systems, the excavation just above the water's edge made to accommodate the junction of the sea and land portions of a cable as well as the beach anchor.

pitch   A subjective property of sound determined mainly by its frequency.

pitch, standard acoustical   The standard used to pitch musical instruments. It is centered on a frequency of 440 Hz for the A above middle C.

pixel   Picture element: basic unit making up pictures in TV, plasma display panel or LCD visual displays.

plain old telephone service (POTS)   Basic telephone service, with no frills or special facilities.

plain text   Words which are not in code or crypt.

Planck's constant   The constant of proportionality in the quantum theory in which radiant energy is composed of quanta proportional to the frequency of the radiation. The value of this constant, k, is 6.626×10^{-34} Joule seconds.

plane polarized wave   Electromagnetic wave in which the vibrations are rectilinear and parallel to a plane that is transverse to the direction of propagation of the wave. Syn. linearly polarized wave

plane wave   A wave whose surfaces of constant phase are parallel planes normal to the direction of propagation and infinite in extent.

planning cable fill (PCF)   Forecast demand for telephone lines divided by the number of distributed cable pairs to be provided by a development scheme. 75% PCF is a typical figure.

plant   General term for all equipment used by a telephone company in providing telecommunications services, usually divided into external plant and inside plant.

plant, distribution   The duct, pole, cable and wire network linking central offices with subscribers.

plant, exchange   All inside and outside equipment used to provide subscribers with local, as opposed to toll, exchange services. In Britain exchange plant normally means switching equipment only.

plant, feeder   Primary distribution cables from central offices to points (often flexibility cabinets) where secondary distribution cables extend the network to subscribers' premises.

plant, inside   See inside plant.

plant, outside Out-of-doors telephone equipment such as poles, cables and ducts. Syn. external plant

plant, toll Inside and outside telephone equipment used to provide toll service, as distinguished from local exchange service.

plant unit Identifiable item of telephone plant as used for costing purposes, e.g., one meter of 2000 pair cable, one telephone instrument, one manhole.

Plante plate Type of lead plate used in large lead-acid storage batteries. Such a plate, normally used only as the positive plate, has deep, finely divided vertical grooves cut into its surface to make the surface area large in comparison with its superficial area. In this way, its active dimensions may be increased by a factor of 10.

plasma 1. Ionized gas in an arc-discharge tube that provides a conducting path for the discharge. 2. The ionized gas at extremely high temperature found in the sun. Similar effects are obtained in research laboratories conducting nuclear research.

plasma display Type of flat visual display device, usually depends on ionization of gas for light emission.

plasma display panel (PDP) Light-emitting device used for alphanumeric display panels.

plastic film capacitor A capacitor the dielectric of which is a plastic-based film.

plate 1. The anode of an electron tube. 2. An electrode in a storage battery. 3. One of the surfaces in a capacitor. 4. A mounting plate to which equipment may be fastened.

plate, number The plate under the finger wheel of a rotary dial bearing the numbers 1 through 0.

plate dynamometer In submarine cable ships, a tension-measuring device by which the cable's straight line trajectory from engine to sheave (or chute) is slightly distorted by the cable sliding over a raised plate resting on a load cell, the output of which gives a measure of tension.

plates, face Flat plates surrounding keys, switches, etc. to provide protection and improve appearance.

platform, cableman's A work platform which can be suspended from a portable crane. It is used by cable splicers working on splices in aerial cables.

platform, ladder A platform which can be added to a rung of a ladder to provide a working area.

plating The application, by electrolysis, of a coating of one metal on the surface of another.

plesiochronous The relationship between two signals whose corresponding significant instants occur at nominally the same rate, any variation in rate being constrained within specified limits. Two signals having the same nominal digit rate, but not stemming from the same clock or homochronous clocks, are usually plesiochronous.

pliers, cable sheath A hand tool for opening and closing split openings in cable sheaths, particularly in lead cables.

pliers, combination Small hand tool with pivoted jaws for holding round or flat objects and for cutting wires.

pliers, diagonal Pliers with cutting jaws at an angle to the handles, designed for cutting wires close to terminals.

pliers, duckbill Pliers with wide flat jaws, designed for straightening or bending springsets carrying contacts.

pliers, gas A type of large combination pliers.

pliers, heat coil Pliers with special jaws for holding the cylindrical body of a heat coil.

pliers, lamp-cap Pliers with special jaws for holding lamp caps. The tool is used to extract lamp caps from jacks.

pliers, lineman's Heavy combination pliers also used for side cutting.

pliers, long nose Pliers with long narrow jaws, used when wrapping wires around closely spaced terminal pins.

pliers, side cutting Pliers with wire cutting jaws placed close to the hinge for maximum leverage.

plow, cable laying Device which enables a cable to be plowed into the ground. The plow cuts a slot as it is pulled forward; the cable is fed in and pushed to the bottom of the slot, and the slot closes on it as the machine proceeds forward.

plow, sabre Cable-laying plow with a vibrating cutting edge which saws through ground and needs less power to pull forward than does a standard plow. The vibrating action also more efficiently closes the slot behind the machine as it proceeds forward.

plowing In submarine cables, the process of imbedding the cable into the sea bottom as it is being laid.

plug A terminal on the end of a cord, designed to provide good electrical contact with the socket for which it is mated. In telephony, the most common plug is a unit of three elements called tip, ring and sleeve.

plug, cable Multi-contact rectangular plug with up to 50 contacts. It is used in conjunction with a cable connector and a clamp.

plug, coaxial Plug on the end of a coaxial cable which permits rapid connection to a coaxial jack.

plug, conduit Soft rubber plug used to fill unused conduits and to provide a water-tight seal.

plug, dummy Plug, usually made of insulating material, which makes no electrical connections itself but holds associated mechanical springsets in their operated position.

plug, gas pressure A closure or dam inside a cable sheath which prevents the flow of air or gas past that point.

plug, open A dummy plug inserted in a jack in order to operate the associated mechanical springset contacts.

plug, out-of-service Plug which busies out a circuit thus making it inoperative.

plug, phone A standard telephone plug designed for insertion in a standard telephone jack.

plug, pin A single contact plug used to make temporary connections with a pin jack.

plug, polarized Multi-contact plug so made that it can be inserted into its associated jack in only one particular position.

plug, resistance Telephone plug with a simple resistance connected across the terminals. It can be inserted into a jack to provide a 600 ohm impedance termination while a circuit is under test.

plug, reversing Special telephone plug in which tip and ring connections are reversed. It is used when testing circuits.

plug, short and ground Special telephone plug which directly connects together all three circuits (tip, ring and sleeve) and provides a button control to extend them to ground.

plug, shorting Special telephone plug which directly connects together all three circuits (tip, ring and sleeve).

plug, switchboard Standard telephone plug used on manual operating positions. It usually gives a three-wire connection (tip, ring and sleeve) to a switchboard jack.

plug, test Plug designed to be used with the type of main distribution frame and protective devices installed in a central office. It provides a connection for testing a subscriber's line.

plug, three-conductor Standard telephone switchboard plug.

plug, twin Two telephone plugs mounted in one shell so that both can be inserted at once. This device was used for four-wire switching of long distance circuits.

plug, two conductor Telephone switchboard plug for only two conductors, the tip and the sleeve. It is sometimes used on the simpler types of manual operating positions.

plug and socket Basic method of connecting two electrical devices. Usually the plug is "male" and the socket "female", but some plugs and sockets are made so that the two mating parts are identical and can mesh like two two-pronged forks. These are called hermaphroditic or sexless connectors.

plug-ended Describing a trunk which is terminated on a cord rather than on a jack. The single cord may be plugged directly into a jack to establish a connection.

plug-in Describing any device having plug type terminals enabling it to be activated simply by insertion into a socket.

plug-shelf The rear part of the horizontal shelf in a manual operating position. Normally, the pairs of cords protrude through holes in this section of the shelf.

plug-to-plug compatible Equipment which can be interconnected without needing special adapters or interface conversion.

plugging-up circuit In telegraphy, a temporary termination of a spare telegraph circuit which gives audible or visual indication of change on the circuit.

plugging-up cord In telephony, a cord and plug used to disconnect a permanent loop subscriber's line from the central office switching equipment. When the subscriber goes back on hook, an indicator gives a signal and the plug is withdrawn.

plumbing 1. Splicing and jointing lead sheathed cables. The work is similar to that done by plumbers working with lead solder on water pipe joints. 2. Waveguide assemblies used at microwave frequencies.

plunger Moveable piston used in a resonant cavity for tuning or for the introduction of attenuation.

p-n device Transistor utilizing p-type and n-type semiconductor material.

pocket pager Small radio receiver which can be used to receive coded messages.

pocketing, call A traffic routing abnormality which can occur when circuits to a destination do not switch on the basis of full availability. A call may be directed to a group of circuits all of which are busy, although an alternate group of circuits to the same destination may have idle trunks.

point, access See access points.

point, code Point in an electromechanical marker corresponding to a particular central office code of up to three digits.

point, control Junction between main feeder (primary) cables and branch feeder (secondary) cables. These junctions were at one time largely spliced joints, but jumpering through at a cabinet is now becoming standard.

point, load Point at which loading coils are inserted.

point, pick-up Point at which an outside or remote broadcast program joins the network.

point, singing Point of instability where losses around a loop are just exceeded by gains and the circuit begins to howl or sing.

point, transmission level (TLP) See transmission level point.

point, vacant code A code point in an electromechanical marker which is not assigned. Calls to vacant code points are routed to an operator or to a "number unobtainable" tone.

point, zero transmission level (OTLP) See zero transmission level reference point.

point of interface (POI) Connection between inter-LATA facilities and LATA access.

point of presence Point at which responsibility for handling inter-LATA traffic changes over from the local telephone operating company to the interexchange carrier.

pointer 1. The indicating needle of a meter. 2. An address of a memory location. 3. A data pointer, procedure pointer or label pointer.

pointer chain A series of pointers.

point-to-point A circuit connecting two (and only two) points.

point-to-point link A data communication link connecting only two stations.

point-to-point transmission Transmission between two designated stations.

232

poisoning, cathode Reduction in efficiency of electron emission from a cathode in an electron tube, normally due to impurities released during operation.

Poisson distribution Mathematical formula which indicates the probability of the occurence of an event.

Poisson tables Tables based on mathematical probabilities. They can be used to calculate the numbers of circuits needed in a group in order to provide a specified grade of service for a given level of traffic.

Poisson traffic A traffic that has a Poisson distribution of arrivals.

poke Programming language word which instructs a processor to move information into an addressable memory store.

poking tape Preparing a punched tape for transmission by an automatic teletypewriter.

polar Polarized.

polar direct-current telegraph transmission A form of binary telegraph transmission in which positive and negative direct currents denote the significant conditions. Syn.: double-current transmission

polar operation A system whereby marking signals are formed by current or voltage pulses of one polarity and spacing signals by current or voltage pulses of equal magnitude but opposite polarity.

polar orbit Of a satellite, a satellite orbit the plane of which contains the polar axis of the primary body.

polar relay See relay, polar.

polarential telegraph system A direct current telegraph system employing polar transmission in one direction and a form of differential duplex transmission in the other.

polarity The property of terminals of a battery (positive or negative) or of the poles of a magnet (north or south).

polarization 1. The inhibition of current through an electric cell because of the formation of chemical products near the electrodes. 2. That property of a radiated electromagnetic wave describing the time-varying direction and amplitude of the electric field vector; specifically, the figure traced as a function of time by the extremity of the vector at a fixed location in space, as observed along the direction of propagation.

polarization, circular Of an electromagnetic wave, the rotation of the plane of polarization through 360 degrees as the wave propagates forward. Created by combining equal magnitudes of vertically and horizontally plane polarized waves, with the phase of one exactly 90 degrees ahead of or behind the other. Depending on the sequence, this creates right-hand or left-hand rotation.

polarization, elliptical Of an electromagnetic wave, the rotation of the plane of polarization through 360 degrees as the wave propagates forward. Created by combining vertical and horizontal plane polarized waves whose amplitudes are not equal, or if equal whose vertical and horizontal components are not separated by exactly 90 degrees in phase. See also polarization, circular.

polarization diversity Any method of diversity transmission and reception wherein the same information signal is transmitted and received simultaneously on orthogonally polarized waves with fade-independent propagation characteristics.

polarization fading Fading which occurs as a result of changes in the direction of polarization of the downcoming wave, relative to the orientation of the receiving antenna, due to random fluctuations in the electron density along the path of propagation. Polarization fading lasts for a fraction of a second to a few seconds.

polarized 1. Acting or flowing in a single direction, or sensitive only to a particular direction of current flow. 2. In a primary cell, restriction of current flow by the formation of chemical products near the electrodes.

polarized wave, horizontal Electromagnetic wave polarized so that the electric field lies in a plane parallel to the earth's surface.

polarized wave, left-hand A circularly or elliptically polarized electromagnetic wave in which the rotation of the electric field-intensity vector is counter-clockwise when looking in the direction of wave propagation.

polarized wave, plane Electromagnetic wave in which the vibration is restricted to a single plane. The wave's electric field may be horizontal, vertical, or at any angle in between.

polarized wave, right-hand Circularly or elliptically polarized electromagnetic wave such that the rotation of the electric field-intensity vector is clockwise, looking in the direction of wave propagation.

polarized wave, vertical Electromagnetic wave polarized so that the electric field lies in a plane perpendicular to the earth's surface.

pole 1. A tapering column of wood, steel or concrete used to support overhead wires and aerial cables. 2. One of the terminals of a battery. 3. One of the poles of a magnet. 4. One side of a switch.

pole, butt treated Telephone pole treated with preservative only at the butt end, that is, the end to be buried.

pole, dead-end Pole at which a cable route or an open wire route ends.

pole, gin See gin pole.

pole, joint A pole shared by the telephone company and by one or more other utility companies.

pole, junction A pole which demands special staying because stresses and tensions are unbalanced.

pole, pike Steel-tipped wooden rod about 10 feet long used when setting telephone poles in the ground or straightening existing poles.

pole, pressure treated A pole which has been impregnated with preservative while under pressure in a tank. Impregnation permits poles to be made of lighter timber with low risk of the timber rotting.

pole, sky-grey A pole which has been impregnated with preservative with a color additive which gives a blue-grey color for improved appearance.

pole, terminal A pole on which a cable terminal or distribution point has been mounted, or one on which an open wire route ends and all circuits are connected through into a cable.

pole, treated A pole treated with a preservative. Creosote and pentachlorophenol are the most commonly used preservatives.

pole attachment contract Agreement between companies concerned by which, for example, cable TV systems use existing pole lines maintained by telephone or power utility companies.

pole face The flat piece at the end of a relay core to which the armature is attracted when current flows through the relay coil.

pole iine A line of poles supporting open wires or aerial cables.

pole pair See pair, pole.

pole piece Ends of a magnet or a relay core at which the magnetic field is concentrated.

poling 1. Adjustment of polarity and use of transpositions on open wire routes to reduce overall crosstalk from one circuit to another. 2. Erection of telephone poles along a route.

polisher, plug A specially designed cylindrical polisher which removes tarnish from telephone switchboard plugs.

polite search The act of searching in sequence all high usage and recognized alternate routes in the automatic alternate routing program, and giving a "no circuit" tone if no idle circuit is found available.

polling The process by which a processor invites one of several external units to feed it with information. The initiative for the transfer of information rests with the processor.

polyethylene The thermoplastic synthetic polymer which, in a highly refined state, is used, in high molecular weight form, for cable insulant and in high density form, for cable jacket material. It is a tough, waxy-appearing material and has low dielectric losses even at radio frequencies.

polygrid network A non-hierarchic switching network in which each switching center is interconnected with many other centers over separately trunked paths.

polygrid routing plan A toll switching network with interconnected home grids plus an overlay of long haul trunks.

polymer A long chain molecule, usually a good insulator, made by combining together several smaller molecules.

polyphase Describing electrical equipment which produces or uses two or more phases. Three-phase power distribution is a common standard.

polyphase circuit A power circuit, normally balanced, with two or more phases.

polystyrene A plastic frequently used as an insulator in electric work.

polythene British designation for polyethylene.

polyurethane A rubber-type polymer.

polyurethane gel A combination of liquid polyurethane, powdered fiberglass and a hardener. It is used to construct a gas plug inside a cable. It hardens within one hour.

polyvinyl chloride (PVC) A thermoplastic made of polymers. It is tough, nonflammable and water resistant and is used as an insulant. It has higher dielectric losses than polyethelene.

pony circuit A purely local teletypewriter circuit in a building complex, often a local extension at the end of an external circuit.

porcelain Ceramic material used as an insulator on open wire routes and on aerial feeder lines.

port A place of access to a device or network where energy may be supplied or withdrawn, or where the device or network variables may be measured.

position The part of a manual operating center designed to be controlled by one switchboard operator.

position, inward and through toll Toll position used to answer incoming toll circuits and connect these to outgoing toll circuits or to local subscribers.

position, inward toll Toll position which answers incoming toll circuits and connects these to local subscribers.

position, outward toll Toll position used to answer local customers and establish their outgoing toll calls.

position, pin Position of the insulators carrying a circuit on a pole route. The top arm usually carries pins 1—10, the second arm carries pins 11—20, etc.

position, through toll Toll position which handles only connections on inter-toll trunks.

position, traffic service (TSP) See traffic service position.

position-independent code A code which can execute properly wherever it is loaded in memory without modification or relinking. Generally this code uses addressing modes which form an effective memory address relative to the central processor's program counter.

positive 1. Describing the terminal of a storage battery from which the conventional current leaves the battery. 2. Having the ability to attract electrons. 3. The opposite of negative.

positive justification In digital multiplexing, the provision of a fixed number of dedicated time slots (normally at regular intervals) in the output digital signal. These time slots are used to transmit either information from the tributaries or no information, according to the relative digit rates of the individual tributaries and the output digital signal. Syn. positive pulse stuffing

positive/negative justification A combination of positive and negative justification in which justifying digits in pulse code modulation multiplexing are transmitted (positive) or information bits are deleted (negative) at each justification opportunity. Syn.: positive/negative pulse stuffing

positive terminal The terminal towards which electrons flow, and from which a conventional current flows in the external circuit.

234

positive/zero/negative justification A combination of positive and negative justification in pulse code modulation multiplexing in which noninformation bits are transmitted (positive) or information bits are deleted (negative) only when it is essential that this be done in order to avoid loss or mutilation of information.

positron An elementary particle with the same mass as an electron but a positive charge.

possible crosstalk components Transferred speech currents which do not intrude into the channels of other systems at the point considered but which may do so elsewhere.

post, binding See binding post.

Post, Telegraph and Telephone (PTT) Administrations, usually Government-controlled, which manage and operate postal and telecommunications services in many countries. These services are usually monopolies outside North America but competitive organizations exist in some countries.

post answer delay The period between the lifting of the handset by the called party and the establishment of a speech path.

post-detection combiner See combiner, post-detection.

post-dialing delay See delay, post-dialing.

posted delay The expected delay in establishing a toll call during periods of congestion. The expected delay is normally posted so that the same information will be given to all callers by the operators.

Postfax The public document facsimile transmission service provided by British Telecom at selected main post offices.

pot A potentiometer.

pot, paraffin A lidded kettle used to heat paraffin for drying splices in paper-insulated, lead-sheathed cables.

pot, solder A heavy cast-steel pot used to melt solder for the wiping and jointing of lead cable sleeves and joints.

potential The difference between voltage at a specified point and some other point, usually ground.

potential, Earth In submarine cable systems, the difference in potential between the points on the Earth's surface at the terminals of a cable system.

potential, ground Zero potential.

potential difference Difference in voltage existing between two points in a circuit.

potentiometer A three-terminal adjustable voltage divider.

potentiometer, telegraph hub Resistors which supply battery potential to the hub. Normally, the +130 V battery feeds through a 5300 ohm resistor, while a 4400 ohm resistor connects through a diode to ground.

potentiometer, telegraph loop Variable resistance adjusted to bring all telegraph loops to the same standard value, usually 2880 ohms.

pothead 1. The terminal of a cable, particularly of a power cable feeding into an overhead power route.

2. The terminating joint (usually below a Main Distribution Frame) at which a large main primary distribution cable is jointed on to several smaller cables, often one cable per vertical of the frame on which access blocks are mounted.

POTS (Plain Ordinary (or Old, Telephone Service) Service restricted to push-button dialing, nationwide and international direct dialing, and accurate and regular telephone bills.

potting Sealing of components under a plastic cover to keep out moisture.

powder-powered Describing small handtools used to shoot bolts into masonry in order to secure equipment to walls or floors.

powdered-metal Finely powdered iron or other material in which the particles have been insulated with varnish, compressed into moulds, and heated to make cores for transformers, etc. It reduces eddy current losses, thus improving efficiency.

power The rate of transfer or absorption of energy in a system.

power (alternating current) The product of the effective voltage, the effective current, and the cosine of the phase angle between them.

power (direct current) The product of the voltage and the current, or of the resistance and the square of the current.

power, apparent The product of the RMS voltage and the RMS current. The real power is obtained by multiplying this product by the power factor.

power, auxiliary Power at a telecommunications station which does not have to be maintained on a no-break basis.

power, available The maximum power available from a source by a suitable adjustment of the load.

power, average In a pulsed laser, the energy per pulse (joules) times the pulse repetition rate (Hertz).

power, average speech The total speech energy over a period of time, divided by the length of the period.

power, carrier Of a radio transmitter, the average power supplied to the antenna transmission line by the radio transmitter during one radio frequency cycle under conditions of no modulation. For each class of emission the condition of no modulation should be specified.

power, effective radiated (ERP) The product of the power supplied to an antenna and the antenna gain relative to a half-wave dipole. Some confusion has been caused by the use of ERP also to mean, for the higher frequency bands, the product of the power supplied to the antenna and the antenna gain in a given direction relative to an isotropic antenna. The latter should really be considered a definition of the equivalent isotropically radiated power (EIRP).

power, mean Of a radio transmitter, the power supplied to the antenna transmission line by a transmitter during normal operation averaged over a time sufficiently long compared with the period of the lowest frequency encountered in the modulation. A time of 0.1 seconds during which the mean power is greatest will normally be selected.

power, peak   In a pulsed laser, the maximum power emitted.

power, peak envelope   Of a radio transmission, the average power supplied to the antenna transmission line by a transmitter during one radio frequency cycle at the highest crest of the modulation envelope taken under conditions of normal operation.

power, real   The power in an alternating current circuit which is used in doing work. It is the product of the RMS voltage times the RMS current times the power factor (the cosine of the angle between voltage and current).

power, ringing   Power sent out to a line to ring the bell of a called subscriber's telephone, usually about 70 V at 17 to 20 Hz.

power amplifier   See amplifier, power.

power circuit breaker   1. A circuit breaker for use on ac circuits rated in excess of 1500 V. 2. The primary switch used to apply and remove power from equipment.

power control unit (PCU)   The control equipment (normally mounted together with the rectifiers which provide dc power for the central office) which monitors batteries and loads and controls charging operations and end cell switching.

power density   1. Power in watts per hertz, or the total power in a band of frequencies divided by the bandwidth in hertz. 2. In optoelectronics, a colloquial synonym for irradiance.

power efficiency   Ratio of emitted optical power of a source to the electrical input power.

power factor   The ratio of the true power (in watts) to the apparent power (the product of volts times amps). If waveforms are sinusoidal, the power factor is equal to the cosine of the phase angle between current and potential.

power feed equipment   In submarine cable systems, equipment designed to energize the in-water portion of a seacable system. It normally consists of a constant current source of high reliability and double redundancy.

power flux density, high   Flux density provided at a station receiving broadcast programs direct from a satellite, sufficient to produce a quality of reception of emission which is subjectively comparable to that provided by a terrestrial broadcasting station in its main service area. Medium or low power flux densities provide signals of lower quality.

power frequency   Frequency at which electricity is distributed for domestic and industrial use. In the United States it is normally 60 Hz.

power gain of an antenna   The ratio of the power required at the input of a reference antenna to the power supplied to the input of the given antenna to produce, in a given direction, the same field at the same distance.

power-law index profile   In optoelectronics, a class of graded index profiles characterized by the following equations:

$$n(r) = n_1(1 - 2\Delta(r/a)^g)^{1/2} \quad r \leq a$$

$$n(r) = n_2 = n_1(1 - 2\Delta)^{1/2} \quad r \geq a$$
$$\text{where } \Delta = \frac{n_1^2 - n_2^2}{2n_1^2}$$

where $n(r)$ is the refractive index as a function of the radius, n_1 is the refractive index on axis, n_2 is the refractive index of the cladding, a is the core radius, and g is a parameter that defines the shape of the profile. For this class of profiles, multimode distortion is smallest when g takes a particular value depending on the material used. For most materials, this optimum value falls around 2. When g is very large, the profile becomes a step index profile.

power level   At any point in a transmission system, the ratio of the power at that point to some arbitrary amount of power chosen as a reference. This ratio is usually expressed either in decibels referred to one milliwatt, abbreviated dBm, or in decibels referred to one watt, abbreviated dBW.

power-off condition   A condition in which power is not available within a unit of equipment.

power plant   A station which provides electric power. In a telecommunications system, power usually comes from a public ac main supply and is supported by local standby engines and generators.

power separation filters   In submarine cable systems, networks designed to separate the energizing current from the transmission signals. They are located in repeaters, equalizers, and the power feed equipment.

power station   A complete generation station together with all the plant needed to produce electric power and feed it to the national system.

practices   Set of standard instructions issued by major telephone organizations describing work procedures.

preamble   Beginning section of a telegraphic message which gives such information as the file number of the message, the date-time, and the address.

preamplifier   Amplifier designed to receive a very low level input signal and amplify this to a level which may be carried in local wiring without risk of interference or distortion and accepted as input by a standard amplifier.

precedence   A designation assigned to a message by the originator to indicate to communications personnel the relative order of handling, and to the addressee the order in which the message is to be noted.

precedence level   Relative priority level assigned to a telephone user, particularly in military communications systems, in which high ranking callers may pre-empt circuits already in use by lower priority users. Standard precedence levels in systems using these are:

top priority	Flash Override	FO
	Flash	F
	Immediate	I
	Priority	P
bottom priority	Routine	—

precedence pro-sign   A priority indication at the beginning of a message.

precipitation attenuation The loss in electromagnetic energy by scattering, refraction and absorption during passage through a volume of the atmosphere containing precipitation in the form of rain, snow, hail or sleet.

precipitation-scatter propagation Propagation by scattering from precipitation particules.

precipitation static A type of interference experienced in a receiver during snowstorms, rainstorms and duststorms. It is caused by the impact of charged particles against an antenna.

precise frequency A frequency which is maintained to the known accuracy of an accepted reference frequency standard.

precise time A time mark, the position of which (or for a time interval, the duration of which) is known accurately with reference to an accepted reference time standard.

precision 1. A measure of the ability to distinguish between nearly equal values. 2. The degree of agreement among repeated measurements of the same object or event. 3. A measure of the "spread" or deviation of the results from a mean.

precision balance Balance at a two-wire/four-wire hybrid designed to match the two-wire line exactly and the building-out capacitor adjusted to suit the exact line end section capacitance. Precision work of this type can produce a much greater return loss than is usually provided.

precision depth recorder (PDR) On cable ships, an instrument with a higher order of accuracy, and usually a much larger chart trace, than the ship's regular echosounder. Using the echosounder's pulse transmission and echo reception, it produces a depth profile of desirable accuracy and size.

pre-detection combiner A circuit or device for combining two or more signals prior to demodulation.

pre-detection combining A technique used to obtain an improved signal from multiple radio receivers involved in diversity reception.

predicted grade A contour (TV) The line on a map representing the service area in which a good picture is computed to be available 90 percent of the time at 70 percent of the receiver locations. Signal contours determine what educational channels are carried on a cable system and, in smaller markets, what stations must be carried from other small markets.

predicted grade B contour (TV) The line on a map bounding a television station's service area in which a good picture is computed to be available 90 percent of the time at 50 percent of the receiver locations.

prediction chart, radio Chart based on sunspot predictions, used for selecting frequencies for radio communication links.

pre-emphasis A process designed to increase the magnitude of some frequency components in a system with respect to the magnitude of others in order to reduce adverse effects, such as noise, in subsequent parts of the system.

pre-emphasis improvement The improvement in the signal-to-noise ratio of the high-frequency end of the baseband resulting from passing the modulating signal (at the transmitter) through a preemphasis network, which increases the magnitude of the higher signal frequencies, and then passing the output of the discriminator through a de-emphasis network to restore the original signal power distribution.

pre-emphasis network A network inserted into a system in order to increase the magnitude of one range of frequencies with respect to another.

preempt search The searching procedure followed for automatic alternate routing for high precedence calls. All available trunk groups, both high usage and alternate route, are searched in sequence for an idle circuit. If no idle circuit is found, the most direct circuit carrying the lowest precedence call will be preempted.

pre-emption The seizure (usually automatic) of system facilities which are being used to serve a lower precedence call, in order to serve a higher precedence call.

preemption, ruthless Seizure of a busy circuit by a high priority caller without prior warning to the original users of the circuit.

pre-equalization The introduction of distortion into a circuit at its input to offset the distortion which the signal will suffer in passing through the circuit.

preferential answering Priority given to certain classes of calls, such as incoming long distance calls, even though this means that some types of calls may have to wait longer for attention.

preferential hunting A system employed in electromechanical offices to reduce hunting time and wear on wipers/multiples when switches have to search over several levels of outlet trunks. Outlets are divided into two groups and one of these is connected on the multiple banks in reverse order.

preferential jack A specially marked jack on a manual operating position which indicates to the operators that testing should commence with that line.

prefix 1 A prefix used on subscriber dialed, direct distance dialed station-to-station calls.

prefix 112 A prefix used on subscriber dialed, direct distance dialed calls from central offices not yet converted to prefix 1 calling.

prefix 0 A prefix used on subscriber dialed, direct distance dialed calls which require operator intervention and assistance, such as person-to-person, collect and credit card calls.

prefixing, digit The act of sending out extra routing from a central office, or prefixing digits before sending out the called subscriber's number as received from the calling subscriber.

preform A glass structure from which an optical fiber waveguide may be drawn.

pregroup, carrier A subgroup of four channels individually modulated to fill the band 8—24 kHz, used in some 12-channel carrier systems with more than one stage of modulation, before reaching the final

237

60—108 kHz standard for the whole 12 channel group.

prejumpering Provision in cabinets/flexibility points (in local distribution networks) of jumpers between pairs in primary cables and those in secondary cables, in advance of specific individual requirements.

prelash To lash an aerial cable to its messenger or suspension wire before the messenger is pulled onto its poles.

preliminary pulse Commonly experienced false digit 1 which is often the result of failing to lift a handset cleanly off its rest.

preliminary relay contacts Contacts on a relay which make or break before the early contacts whenever the relay operates.

premises In telecommunications usage the buildings or offices of the user, not divided or separated by a public thoroughfare.

premium Centrex Enhanced Centrex services, e.g., including data transmission, least cost routing, etc.

preparation operating In international telephony, after the request is recorded by an operator in the outgoing international exchange, another operator in the exchange sets up the call. After the requests have been put in order at the exchange, the controlling operator sees that the calling station is connected on the international circuit without loss of time.

prepay coin telephone See paystation, prepay.

preplanned number control (PPLN) A network management control which applies remote dynamic overload controls when an external input indicates that these are needed.

prepostpay coin telephone See paystation, semipostpay.

preroute peg count A network management control for studying traffic levels to various destination codes, giving an indication of when restrictions need to be applied in order to keep traffic flowing freely.

presbycusis Hearing loss due to aging.

preselected alternate master-slave (PAMS) A master-slave network with designated primary and preselected secondary timing links for each node in the network. It is used to provide synchronization between digital centers.

preselecting Describing a nonhoming electromechanical switch which keeps its wipers standing where last used, so that when the switch is next seized, it does not have to spend time hunting.

present worth Funds it would be necessary to invest now in order to ensure that a specified sum, including interest, would be available at a specified future date.

presentation layer Within the OSI model for a data system, that layer of processing that provides services to the application layer.

press-to-talk Describing a type of handset with a button to activate the transmitter. It is especially useful in high noise areas where the sidetone resulting from local noise would otherwise make it difficult to hear an incoming voice signal.

presser, connector Crimping pliers used with wire connectors when jointing pairs in a cable splice.

pressure coefficient of attenuation In submarine cable systems, a change of attenuation of cable per increment of a pressure unit in pounds per square inch or kilograms per square centimeter. The pressure coefficient itself may be a function of pressure and frequency.

pressurization Pumping dry air or an inert gas such as nitrogen into cables, at a pressure of about 5 pounds above atmospheric pressure. This prevents moisture from entering at small sheath faults, thus improving service.

Prestel The public videotex service provided by British Telecom.

presubscription Choice by a customer of his selected inter-LATA carrier. At end offices providing equal access all long distance calls will be switched to a non-Bell carrier if the customer requires this to be done.

pretranslation Examination of digits as they are received by a central office and determining from the first three digits how many more digits the equipment may expect to receive for that particular call.

pretranslator, crossbar Device which carries out pretranslation.

preventive cyclic retransmission method A noncompelled, positive acknowledgement, cyclic retransmission forward error correction system.

preventive maintenance Maintenance — including tests, measurements, adjustments and parts replacement — performed specifically to prevent faults from occurring.

pre-wiring The practice of laying telephone cabling in a building during building construction.

primary First or fundamental, such as the primary winding of a transformer, the primary cables in a distribution network, or a primary cell.

primary block A basic group of pulse code modulation (PCM) channels assembled by time division multiplexing. Syn. digroup. The following conventions could be useful:
Primary block μ — a basic group of PCM channels derived from 1544 kbit/s PCM multiplex equipment.
Primary block A — a basic group of PCM channels derived from 2048 kbit/s PCM multiplex equipment.

primary body The attracting body which primarily determines the motion of a satellite.

primary cable distribution Part of the exchange area local distribution network, comprising the main cables between exchange and cabinets (flexibility points) or remote line units (RLUs). Large RLUs of more than 500 lines may themselves require primary distribution to cabinets.

primary cell See cell, primary.

Primary Center 1. Any of the third rank toll switching points in the distance dialing network. It may home on Sectional Centers, Regional Centers, or both. It is also known as a Class 3 toll office. 2. An exchange to which local exchanges are connected and by means of which long distance calls are established.

primary channel 1. The channel that is designated as a prime transmission channel and is used as the first choice in restoring priority circuits. 2. In a data communications network, a transmission channel having the highest signaling rate capability of all the channels sharing a common interface.

primary coating Plastic coating applied directly to the cladding surface of an optical waveguide during manufacture to preserve the integrity of the surface.

primary distribution system A system of alternating current distribution for supplying the primaries of distribution transformers from the generating station or substation distribution buses.

primary frequency standard A frequency standard whose frequency corresponds to the adopted definition of the second, with its specified accuracy achieved without calibration of the device.

primary grade of reception quality In the broadcasting-satellite service, a quality of reception of emissions from a broadcasting-satellite space station which is subjectively comparable to that provided by a terrestrial broadcasting station in its main service area.

primary ground electrode (PGE) The ground electrode in contact with the earth.

primary group The lowest level of a multiplexing hierarchy, e.g., 12-channel (FDM) Group, 24- or 30-channel PCM (TDM) primary system.

primary inter-LATA carrier (PIC) Long distance carrier designated by a telephone customer to provide him with inter-LATA service without having to dial a special access code.

primary power A reliable source of power normally serving the station's main bus. The source may be a government-owned generating plant or a utility system.

primary power circuit Power supply at a voltage higher than the distribution voltage.

primary relay station A major radio relay station which is responsible for net control in its area.

primary route See route, primary.

primary standard A standard that is used widely, nationally or internationally, as the basis for a physical unit.

primary station In a data communication network, the station responsible for control of a data link. It generates commands and interprets responses, and is responsible for initialization of data and control information interchange, organization and control of data flow, retransmission control, and all recovery functions at the link level.

primary substation Equipment that switches or modifies voltage, frequency, or other characteristics of primary power.

primary switching center A switching center at a level immediately above that of end offices or local exchanges in the national switching hierarchy.

primary time standard A time standard which operates according to the adopted definition of the second without calibration of the device.

primary trunk See trunk, primary.

primary winding Winding of a transformer which receives energy from an outside source and passes it on by induction into the secondary winding.

primitive functions Those language terms and macros which a programmer may use in his source code.

primitive name A name assigned by a Message Handling System (MHS) authority. Primitive names are components of MHS Descriptive Names.

principal outlet Old name for a primary center in the toll switching hierarchy.

print-control character A control character which instructs the receiving terminal on the format to be used in print-out.

printed circuit A copper foil circuit formed on one or both faces of an insulating board to which circuit components are soldered. The copper foil pattern serves to connect components and is produced either by etching or plating. Technology now permits printed circuits to be made up of several planes, with electrical circuits accurately connected from one plane to another at required points. Syn.: printed circuit board (PCB) or printed wiring board (PWB). Note that some manufacturers use "board" to mean the basic board including the printed circuits but with no components mounted. The board including its components is then called an "assembly."

printed record on duration and charge of call A service which presents the calling subscriber with a printed record of charging data for outgoing calls or for the use of special services and facilities, after the subscriber has activated the required device.

printer A teletypewriter or Telex machine.

printer, page Teletypewriter which produces typed copy on sheets of paper rather than on tape.

printer, tape Teletypewriter which produces typed copy on paper tape.

printer, toll ticketing Machine which prints toll tickets under the control of the automatic message accounting subsystem of the central office.

printer, traffic Printing unit operated by telephone traffic recording euqipment.

printroller Device which prints out, usually on tape, a record of all fault conditions detected by the routiner.

print-through The transfer of magnetic signals from one layer of tape to an adjoining layer when the tape has been left on the reel for a long period of time.

prioritization The process of assigning different values to users such that a user with a higher priority value will be offered access before a user with lower value.

priority Provision made in telephone exchanges to give preferential treatment to certain calls in the order of path or circuit selection. Priority is the fourth highest ranking message precedence. See precedence.

priority facility In data transmission, a facility which gives one user preference over other users. Priority may be given to the handling of the call, packet transfers, or other services provided by the network.

priority indicator Character or group of characters which determines its position in any queue which may build up at a receiving station.

priority interrupt An interrupt that is given preference over other interrupts within the system.

privacy 1. The protection afforded to information transmitted in a communications system in order to conceal it. Syn. segregation. 2. Short-term protection afforded those unclassified communications which require safeguarding, within existing laws, from unauthorized persons. 3. The protection afforded by a communication system against unauthorized disclosure of the information in the system.

privacy equipment Equipment which so scrambles the outgoing signal that it is unintelligible to a listener who does not have access to the appropriate unscrambling equipment.

privacy override Facility permitting nominated extensions on a PABX to butt in on already-established calls.

private automatic branch exchange (PABX) See exchange, private automatic branch.

private automatic exchange (PAX) See exchange, private automatic.

private branch exchange (PBX) See exchange, private branch.

private line See line, private.

private line service (PLS) Provision of leased circuits to heavy users.

private line teletypewriter service (PLTTY) Leased line teletypewriter with no direct access to the dial teletypewriter exchange service switched network.

private management domain (PRMD) A message handling system (MHS) management domain managed by a company or by a non-commercial organization.

private manual branch exchange (PMBX) A PBX with manual switching under the control of an operator.

private videotex system A videotex system run by a private organization or group of users, not available to the general public.

private wire A circuit provided for a customer's private use.

privates In an automatic switching center, the third and fourth wires associated with a particular trunk. These are usually designated the C and EC wires (for control and extra control).

privatization Transfer of a utility organization (e.g., a telecommunications administration) from government ownership to public shareholding ownership; transfer of financial responsibility from the public sector to the private sector.

probabilistic forecasting A forecasting method used where it is not appropriate to use past history as the sole guide to future growth. Basically, the factor to be forecast is broken down into as many smaller units as is practicable and subjective probability forecasts are made for each of these, on the principle that forecasting errors overall will, with reasonable foresight and luck, cancel each other out. Syn. risk analysis

probability traffic table Table which indicates the probability that a designated percentage of originated calls will find all trunks busy during the busy hour.

probe 1. A test prod used to test components for the presence of signals. 2. A wire loop inserted in a cavity for coupling with an external circuit.

procedure call A call employing the use of a procedure name which causes the execution of the procedure when encountered.

proceed-to-select An event in the call establishment phase of a data call which confirms the receipt of a call request signal and advises the calling data terminal equipment to proceed with the transmission of the "selection signals."

proceed-to-send signal A signal returned from the incoming exchange following the receipt of a seizure signal, to indicate that circuit conditions have been established for receiving a further signal or signals, usually containing digital information.

process 1. In a data processing system, a course of events occurring according to an intended purpose or effect. 2. In specification description language (SDL), the performance of a logic function that requires a series of information items to proceed, where these items become available at different points in time. In the context of SDL, a process is an object that either is in a state awaiting an input or in a transition.

processing, data See data processing.

processor 1. A device capable of performing systematic execution of operations upon data. 2. For an exchange, a device which controls the functions of a stored program controlled exchange by reference to information stored in memory.

processor, regional In an stored program controlled switching system, a logic unit or microprocessor which is subordinate to the central processor and assists it by carrying out simpler tasks.

prod, pole Steel probe for jabbing into a pole to test the soundness of the wood.

prod, test A fine needle with all except the point covered with insulation. It is used when making test connections.

profile chart Chart used when planning line-of-sight radio links. The elevation of hills and other raised features is marked thereon.

profile dispersion In an optical waveguide, that dispersion attributable to the variation of refractive index profile with wavelength.

profile dispersion parameter (P)

$$P(\lambda) = \frac{n_0 \ \lambda \ d\Delta}{N_0 \ \Delta \ d\lambda}$$

where n_0, N_0 are respectively the refractive and group indices at the core center, and $n_0 \sqrt{1-2\Delta}$ is the phase index at the core edge or cladding. The expression is uniquely specified for power-law index profiles. Sometimes it is defined with the factor (-2) in the numerator.

profile parameter The shape-defining parameter, g, for a power-law index profile.

program 1. Instructions placed in the memory of a stored program controlled switching system. 2. A series of predetermined instructions which control the functions of a processor.

program, emergency action Special routine followed when an electronic switching system fails to operate correctly.

program, generic Program designed to fit all the stored program controlled offices of a specified type, thus controlling all functions which are common to all offices.

program, stored The instructions in the memory of a stored program controlled office and to which it refers while processing a call.

program, wired Program making use of logic devices which are interconnected by hard wiring. Changes in the program necessitate changes in the wiring of the logic devices.

program circuit Telephone-type circuit which has been given a wider bandwidth than is needed for normal telephony in order to produce a high quality circuit for sound broadcasting of either speech or of music.

program compatibility The ability of one computer system to execute programs written for another computer system and to obtain identical results. Compatibility can be obtained through similar instruction repertoires, or through emulators, simulators, translators, or common source language coding.

program counter A register in the central processing unit that holds the address of the current instruction being executed plus one. In other words, it holds the address of the next instruction, unless the current instruction causes a jump.

program development The process of writing, entering, translating and debugging source programs.

program evaluation and review technique (PERT) A management procedure, often computerized, to provide close control of all operations needed to complete a project on time and within budget. The procedure enables actual progress to be compared with progress originally planned. Control charts prepared are called PERT diagrams.

program library A collection of available computer programs and routines.

program network diagram (PND) Specialized form of block diagram in which symbology has been adapted for software structure representation.

program nonduplication (CATV) Under FCC rules, a cable system must black out the programming of a distant television station it carries, when the system would duplicate a local station's programming, on the request of the local station.

program patch A temporary change in a computer program, routine, or subroutine. Syn.: computer patch

programmable array logic (PAL) Programmable equipment used to provide the discrete logic required to link large scale integration circuit packages and to tie a total system together.

programmable logic array (PLA) A general purpose integrated circuit logic circuit containing an array of logic gates which can be programmed to perform various functions.

programmable multiplex (PMUX) A device which gives flexibility of routing of outputs. Once programed for a particular pattern (normally by blowing diode fusible links), it cannot be changed.

programmable read-only memory (PROM) Memory which can be programmed after manufacture by external equipment. The fusible link connecting a memory cell can be disconnected to produce a logic 0 instead of 1.

programmer 1. That part of digital apparatus which controls the timing and sequencing of operations. 2. A person who prepares sequences of instructions for a computer.

programming 1. The act of defining objectives and making schedules for achieving them, segregating activities sharing the same objective into programs, and estimating resource requirements. 2. Entering the assembly of instructions into a computer to enable it to carry out a particular job. Also refers generally to the various stages and operations necessary to produce a software program.

programming language A language that is designed to be understood by a computer. A high-level programming language is converted into the required machine code by a program called a compiler.

prolog Algorithm enabling formal logic to be used as a programming language.

propagation The traveling of waves through or along a medium.

propagation, anomalous Abnormal radio propagation caused by the trapping of radio waves in ionized ducts.

propagation, radio Electromagnetic radiation at radio frequencies.

propagation, velocity of Speed with which electromagnetic waves or electric currents travel through a particular transmission medium. Some of these speeds are:

Light through space	299,792 km/s
Radio through air	228,478 km/s
Wideband systems on coaxial cable	214,000 km/s
Carrier systems on open wire	190,000 km/s
Carrier system on cable	170,000 km/s
Voice signals on nonloaded cable	75,000 km/s
Voice signals on loaded cable	16,000 km/s

propagation constant 1. Radio: the negative of the natural logarithmic partial derivative, with respect to distance in the direction of the wave normal, of the phasor quantity describing the wave. The real part of this complex quantity is the attenuation constant (nepers per unit length) and the imaginary part is the phase constant (radians per unit length). 2. Per unit length of a uniform transmission line: the natural logarithm of the ratio of the phasor current at a point of the line, to the phasor current at a second point, at unit distance from the first point along the line in the direction of transmission, when the line is infinite in length or is terminated in its characteristic impedance.

propagation delay Time lost between the transmission of a signal and the time it is received due only to delays in the transmission medium itself.

propagation loss The system loss which would be expected if the antennae gain and circuit resistances were the same as if the antennae were located in free space.

propagation mode 1. The manner in which radio signals travel from a transmitting antenna to a receiving antenna, such as a ground wave, sky wave, direct wave, ground reflection or scatter. 2. One of the electric and magnetic field configurations in which energy propagates in a waveguide or along a transmission line.

propagation path obstruction A man-made or natural physical feature that lies near enough to a radio path to cause a sensible effect on path loss, exclusive of reflection effects.

propagation time delay The time required for a signal to travel from one point to another.

proper time The concept of time inherent to a specific location. Example: The second is defined in the proper time of the cesium atom.

property, expendable Items which are used up during the maintenance or operation of a system and are therefore not considered to be part of the fixed assets of a company.

prorate To divide into proportionate amounts. (UK: to divide pro rata)

proration 1. The distribution or allocation of parameters, such as noise power, proportionally among a number of tandem connected items, such as units of equipment, links or trunks, in order to balance the performance of communications circuits. 2. In a telephone switching center, the allocation of equipment or components proportionally among a number of functions in order to provide a requisite grade of service.

protect 1. To provide protective devices which will shield equipment from damage. 2. To encrypt a message or voice signal.

protected location A reserved storage location in which data to be stored must undergo a screening procedure to establish suitability.

protected wireline distribution system A wireline distribution system and/or fiber optics distribution system to which adequate electrical, electromagnetic and physical safeguards have been applied to permit its use for the transmission of unencrypted classified and sensitive national security information. Syn.: approved circuit

protection, conduit Protection given to a nest of conduits against accidental damage. It often consists of a layer of concrete above the ducts, sometimes complete encasement in concrete of the duct formation.

protection, overload A circuit breaker or mechanical fuse which opens when the current is greater than

that which the equipment has been designed to carry.

protective ground system A grounding system provided for protection of personnel and equipment and for the reduction of noise. Any ground lead that is not normally used in a current-carrying capacity can be considered to be a protective ground.

protective reservation equipment (PRE) A network management control which restricts the flow of alternate-routed traffic offered to a trunk group (but does not restrict directly routed traffic on the same group).

protector Device to protect telephone facilities from high voltages or currents.

protector, central office Devices on the main distribution frame of a central office to provide protection against high voltage and high currents and against currents which are too low to blow fuses but which would be harmful if continued for long periods. Typically these are heat coils or delayed action fuses.

protector, drawing-in A horn-shaped protective device placed in a conduit at a manhole while a cable is being pulled in. Its purpose is to avoid damaging the cable at any sharp lip.

protector, fused station Device designed to protect telephone equipment and personnel at a subscriber's premises from dangerous voltages or currents coming from the external plant. Carbon blocks or gas-discharge tubes give high voltage surges a low impedance path to ground. Fuses in series with the line protect against possible contact with domestic power supplies.

protector, fuseless station Simple type of protector used to protect telephone equipment and personnel on a subscriber's premises from excessive voltages coming from the external plant. It utilizes grounding protectors which connect the line to ground.

protector, gas-discharge tube Small gas filled tube either a three-electrode tube (ground plus the two wires) or two two-electrode tubes (ground and one wire). The gas ionizes and provides a low impedance path to earth for protection from lightning, etc.

protector, guy wire A wooden, metal or plastic shield fixed to a guy wire to make it more visible to pedestrians.

protector, open wire Lightning arresters fitted on crossarms of a pole route, to drain lightning and static voltages from open wires.

protector, power contact Heavy duty over-voltage protector, installed on open wire routes, especially in areas exposed to danger from power line contacts.

protector, station Small protector block used on subscriber's premises, often at the junction between the external drop wire and an internal cable, to protect from excessive voltage. Station protectors may be fused or fuseless.

protector color code Air-gap protectors are sometimes color coded to show the breakdown voltage needed. The code:

Red	2500 V

Yellow	1100 V
Blue	700 V
White	350 V

protector mounting Small shelter, usually with hinged metal doors, in which protectors are housed. It is often mounted on a telephone pole.

protector unit Small screw-in device which protects against lightning. It usually comprises a pair of carbon blocks spaced for breakdown at a specific voltage.

protocol The rules for communication system operation which must be followed if communication is to be effected; the complete interaction of all possible series of messages across an interface. Protocols may govern portions of a network, types of service, or administrative procedures.

protocol converter Device for translating the protocols of one terminal or system to those of another, enabling equipment with different formats and procedures to intercommunicate.

proton Positively charged elementary particle which is part of the nucleus of an atom.

prototype filter See filter, prototype.

proximity effect Nonuniform current distribution in a conductor, caused by current flow in a nearby conductor.

pseudocode A code which must be translated before it can be acted upon.

pseudo-random A sequence of signals which appears to be completely random but has, in fact, been carefully drawn up and repeats itself after a significant time interval.

pseudo-random noise Noise which satisfies one or more of the standard tests for statistical randomness. Although it seems to lack any definite pattern, there is a sequence of pulses which repeats after a long time interval.

pseudo-random number sequence A sequence of numbers which satisfies one or more of the standard tests for statistical randomness. Although it seems to lack any definite pattern, there is a sequence which repeats after a long time interval.

psophometer An instrument which gives visual indication corresponding to the aural effect of disturbing voltages of various frequencies. It usually incorporates a weighting network, the characteristics of which differ according to the type of circuit under consideration, for example, high-quality music or commercial speech circuits.

psophometric voltage Circuit noise voltage measured in a line with a psophometer which includes a CCITT standard (International Consultative Committee for Telephony & Telegraphy) weighting network.

psophometric weighting A noise weighting recommended by the CCITT for use in a noise measuring set or psophometer.

PTT A Postal and Telecommunications authority (originally Post, Telegraph & Telephone).

p-type semiconductor Semiconductor with a deficiency of electrons. The holes present where there should be electrons represent the equivalent of a positive charge. It is the movement of these holes which gives the majority current.

public access (CATV) To ensure that divergent community opinion is aired on cable television in the U.S., FCC rules require systems in the top 100 markets to set aside one public-access channel along with the education and government channels. The public-access channel is free and available at all times on a first-come, first-served basis for noncommercial use by the general public.

public data network A network established and operated by a telecommunications administration, or a recognized private operating agency, for the purpose of providing data transmission services for the public.

public data transmission service A data transmission service established and operated by a telecommunications administration, or a recognized private operating agency, that uses a public data network.

public message service (PMS) Electronic transmission of messages, given in oral and written form, and their delivery to addressees.

public recorded information service Recorded information of public interest which is available to subscribers who call the appropriate service number. The information is sometimes provided in cooperation with appropriate public or private institutions.

Public Service Commission (PSC) A regulatory authority at state level.

public switched digital capability Service providing alternate voice and data on a two-wire user loop.

public switched telephone network (PSTN) The ordinary dial-up telephone system. A phrase often used when referring to data or other nontelephone services carried over a path initially established using normal telephone signaling and ordinary switched long distance telephone circuits.

public television Television broadcasting supported by contributions from the general public.

Public Utilities Commission (PUC) A regulatory authority at state level.

public videotex system A videotex system provided as a service for the general public.

puff Slang term for one picofarad.

pull Unbalanced stress on a telephone pole at a corner or where some of the lines are terminated.

pull-box Small box with above-ground access which is inserted in a long run single conduit to facilitate pulling a cable through the duct.

pull lead A wire lead which carries current for operating a relay which is then held by a separate holding current.

pull-through Cable-jointing manhole through which a cable is pulled without a cable splice being made at that point.

puller, strand Gripping device designed to be attached to a messenger wire for the purpose of pulling it taut.

pulling, frequency Change in the frequency of an oscillator due to a change in the impedance of the output load.

243

pulsating direct current Current that is changing in value at regular or irregular intervals but which has the same direction at all times.

pulse One of the elements of a repetitive signal characterized by the rise and decay in time of its magnitude. It is usually short in relation to the time span of interest.

pulse, dial See dial pulse.

pulse, enabling A pulse which prepares a circuit for later action.

pulse, inhibit See inhibit pulse.

pulse, read A pulse which instructs a processor to read out information from a memory store.

pulse, synchronizing Timed pulse which keeps two or more circuits operating in synchronism.

pulse, write A pulse which leads to information being stored in memory.

pulse address A selective calling arrangement by means of which each user station is assigned a unique combination of time and frequency slots.

pulse-address multiple access The ability of a communication satellite to receive signals from several Earth terminals simultaneously and to amplify, translate, and relay the signals back to Earth, based on the addressing of each station by an assignment of a unique combination of time and frequency slots.

pulse amplitude See amplitude, pulse.

pulse amplitude modulation (PAM) That form of modulation in which the amplitude of the pulse carrier is varied in accordance with some characteristic of the modulating signal.

pulse broadening An increase in pulse duration.

pulse checking Method of providing ringing at specified frequencies to give selective calling to party line telephone subscribers.

pulse code modulation (PCM) That form of modulation in which the modulating signal is sampled and the sample quantized and coded, so that each element of information consists of different kinds or numbers of pulses and spaces.

pulse correction Circuit element which receives dial pulses which may be distorted and sends them out without distortions.

pulse decay time The time required for the pulse amplitude to go from 90% to 10% of the peak value. Syn.: fall time

pulse duration 1. The time interval between the points on the leading and trailing edges at which the instantaneous value bears a specified relation to the peak pulse amplitude. 2. In radar, measurement of pulse transmission time in microseconds, that is, the time the radar's transmitter is energized during each cycle. Syns. pulse length, pulse width. 3. In optoelectronics, the time between a specified reference point on the first transition of a pulse waveform and a similarly specified point on the last transition. The time between the 10%, 50%, or 1/e power points is commonly used, as is the rms pulse duration. Therefore, the measurement level must be stated in quantitative use of the term.

pulse duration modulation (PDM) That form of modulation in which the duration of a pulse is varied in accordance with some characteristic of the modulating signal. Syn.: pulse-length modulation, pulse width modulation

pulse frequency modulation (PFM) That form of modulation in which the pulse repetition frequency of the carrier is varied in accordance with some characteristic of the modulating signal.

pulse interval Time between the start of one pulse and the start of the next.

pulse length A term often used as a synonym for pulse duration.

pulse link, auxiliary A type of relay set which repeats pulse signals from one signaling section to another.

pulse-link repeater A repeater used in telephone signaling systems for receiving pulses from one "E" and "M" signaling circuit and retransmitting corresponding pulses into another "E" and "M" signaling circuit.

pulse modulation systems These include: pulse amplitude modulation; pulse duration modulation; pulse width modulation; pulse frequency modulation; pulse code modulation; pulse position modulation; and pulse time modulation.

pulse period The time required for a single break operation of a rotary dial, for example, the length of time of a single interruption of the loop.

pulse position modulation (PPM) That form of modulation in which the positions in time of the pulses are varied, in accordance with some characteristic of the modulating signals, without a modification of the pulse width.

pulse ratio Percentage of the pulse interval during which the pulsing contacts are closed.

pulse regeneration See regeneration, pulse.

pulse repetition frequency In radar, the number of pulses that occur each second. This is not to be confused with transmission frequency which is determined by the rate at which cycles are repeated within the transmitted pulse.

pulse return pattern A method of locating faults in an aerial or buried plant whereby pulses sent out from the test set are reflected back by any impedance irregularity or discontinuities. This provides accurate localization and determination of the nature of the fault. Syn.: time domain reflectometry

pulse rise time The time required for the pulse amplitude to go from 10% to 90% of the peak value.

pulse slot That space allotted on a time division basis for the inclusion or noninclusion of the binary information, in the form of the presence or absence of a pulse.

pulse spreading An increase in pulse width in a given length of fiber due to the cumulative effect of material dispersion and modal dispersion.

pulse string See pulse train.

pulse time modulation (PTM) Those forms of modulation in which the time of occurrence of some characteristic of the pulse carrier is varied with respect to some characteristic of the modulating signal.

pulse train A series of pulses having similar characteristics. Syn.: pulse string

pulse width See pulse duration.

pulse width modulation See pulse duration modulation.

pulsed laser A laser that emits light in pulses rather than continuously.

pulser A pulse generator.

pulsing Transmission of address information in digital form by sending pulses of current in the circuit. These can be either pulses of dc current, as with make-and-break signals from a dial, or pulses of voice frequency tone, as for example with a 2600 Hz single frequency signaling system.

pulsing, dial Transmission of address information by breaking a dc path, the number of breaks corresponding to the decimal digit dialed. Dial speed is normally 10 impulses per second. Impulses are usually controlled by a rotary dial which makes and breaks the loop while it is returning to its rest position, at a speed controlled by a mechanical governor.

pulsing, dual-tone multifrequency (DTMF) A method of sending numerical information from a telephone or PBX switchboard by sending specific pairs of voice frequencies — one from a group of four low frequencies and the other from a group of four high frequencies — to indicate the ten digits and other call instructions.

pulsing, frequency shift Method of signaling using two frequencies for the two states, often 1070 Hz and 1270 Hz tones.

pulsing, key See key pulsing.

pulsing, loop Making and breaking a subscriber's loop, thereby interrupting loop current and transmitting address information from the subscriber's station to the central office, as with a rotary dial.

pulsing, multi-frequency A method of transmitting address signals at voice frequency in which the identity of the ten digits, 0 through 9, are each determined by various combinations of two each of six frequencies. The two frequencies representing each digit are transmitted simultaneously over the trunk.

pulsing, panel call indicator A system of coded dc pulses using polarized and marginal signals. It is faster than 10 ips loop dialing but not usually so reliable as dual tone multi-frequency signaling.

pulsing, revertive See revertive pulsing.

pump 1. To add energy to electrons, as in a maser or laser, thus exciting them to a higher energy level. 2. The energy source that drives the amplification in the active medium of a laser.

pump, manhole Pump used to drain water out of manholes and jointing chambers.

punch To perforate (e.g., a paper tape) in order to code information into machine-readable form.

punch, card See card punch.

punch, tape Keyboard unit which punches five-unit coded holes into a paper tape for teletypewriter transmission.

punched paper tape A strip of paper on which characters are represented by combinations of punched holes.

puncture A breakdown of insulation or of a dielectric, such as in a cable sheath or in the insulant around a conductor.

pure chance traffic A Poisson traffic which has a negative exponential distribution of holding time.

purge To clear a condition.

push-back Describing an insulant which can be easily pushed back from around a conductor to expose the conductor so that it may be terminated or jointed.

pushbutton A normally open and normally nonlocking switch which is operated by pressing a button.

pushbutton data transmission The transmission of data by means of a keyphone to another terminal via the public telephone network.

pushbutton dialing Use of buttons on a telephone instrument instead of a rotary dial. Signaling generated is usually DTMF (dual tone multifrequency) but in some administrations where central offices will not respond to DTMF signals, pushbutton dials are arranged to give a 10-impulse-per-second loop/disconnect output.

push-pull amplifier See amplifier, push-pull.

push-to-talk operation In telephone or two-way radio systems, that method of communication over a speech circuit in which transmission occurs from only one station at a time, the talker being required to keep a switch operated while talking. Syn.: press-to-talk operation

push-to-type operation In telegraph or data transmission systems, that method of communication in which the operator at a station must keep a switch operated in order to send messages. It is generally used in radio systems where the same frequency is employed for transmission and reception. Syn.: press-to-type operation

PVC sleeve (PS) A nonmetallic insulated material that surrounds a vertical riser where it goes through floor or wall openings.

pW. Abbreviation for picowatt. A unit of power equal to 10^{-12} W (-90 dBm). It is commonly used for both weighted and unweighted noise measurements. Context must be observed.

pWp. A psophometrically weighted picowatt.

pWp0 Psophometrically weighted power, in picowatts, referred to a zero transmission level point.

Q-factor A measure of the efficiency of a coil as an inductance. It equals the coil reactance divided by coil resistance.

Q-signal A three-letter code used in radiotelegraphy to represent a common service-oriented sentence. An example: QRN: I am troubled by static.

Q-switch A device which prohibits oscillation of a laser until the energy stored in the active medium increases to a desired level. In a pulsed laser, a Q-switch increases peak power by shortening pulse duration. The device provides shorter and more powerful pulses than would be possible by direct electrical or optical switching.

quad A group of four wires composed of two pairs twisted together. Usually the pairs have a fairly long length of twist and the quad a fairly short length of twist.

quad, spiral four A cable, originally used for military carrier routes, with four conductors twisted into a spiral. Spiral four cables are often equipped with plug-and-socket couplers for rapid installation.

quadded cable See cable, quadded.

quad-in-line (QUIL) A method of packaging large scale integrated circuits with two rows of staggered pins on each side, providing for 48 or more pins for one packaged chip.

quadrant In submarine cable systems, a portable mechanical guide consisting of a framework carrying many grooved rollers, which has the shape of a quarter circle. It is used during cable loading to guide the cable through a 90 degree change of direction.

quadraphonic A sound reproduction system with four sound channels feeding four separate loudspeakers.

quadratic profile See parabolic profile.

quadrature Waves or alternating signals separated by one quarter of a complete cycle.

quadrature amplitude modulation Modulation system which greatly increases the amount of information which can be carried within a given bandwidth.

quadrature component Component of voltage or current which is at an angle of 90 degrees to a reference signal and is entirely due to inductive or capacitive reactance.

quadrature phase shift keying (QPSK) Phase shift keying using four phase states. Syns.: quadriphase, quaternary phase shift keying

quadriphase See quadrature phase shift keying.

quadruple diversity The simultaneous combining of, or selection from, four independently fading signals and their detection through the use of space, frequency, angle, time, or polarization characteristics or combinations thereof.

quadruplex circuit In telegraphy, a doubled duplex circuit, that is, one which can carry two transmissions in each direction simultaneously.

quality Absence of objectionable distortion.

quality assurance (QA) A planned and systematic pattern of all steps necessary to ensure that a product conforms to established technical requirements.

quality control (QC) A function whereby management exercises control of the quality of raw material or intermediate products in order to prevent production of defective materiel.

Quality Control Circle (QCC) Group of workers exercising the quality control function in a particular section of a manufacturing process.

quality factor See Q-factor.

quality of service 1. A measure of the service provided to a subscriber. The characteristics of this measure must be declared when specifying a quality of service, and may include such characteristics as transmission quality, faults, congestion, delays, etc. 2. The percentage of call requests which cannot be immediately satisfied during the busy hour because there is no circuit free in the desired direction.

quantity of electric charge The coulomb is the SI unit of electrical charge. It measures the quantity of charge that passes any cross section of a conductor in one second when the current is maintained constant at one ampere.

quantization A process by which the continuous range of values of a signal is divided into nonoverlapping, but not necessarily equal, subranges, and a discrete value of the output uniquely assigned to each subrange. Whenever the signal value falls within a given subrange, the output has the corresponding discrete value.

quantization, non-linear Method of quantization used in PCM systems, with small quantizing steps for low amplitude signals so as to minimize quantizing distortion.

quantization level The discrete value of the output designating a particular subrange of the input.

quantizing distortion The distortion resulting from the quantization process.

quantizing distortion power The power of the distortion component of the output signal resulting from the process of quantizing.

quantizing interval (for pulse code modulation) The interval between two adjacent decision values.

quantizing noise An undesirable random signal caused by the error of approximation in a quantizing process. It may be regarded as noise arising in the pulse code modulation process due to the code-derived facsimile not exactly matching the waveform of the original message.

quantum A discrete package as opposed to a continuously variable process. In the electronic field, the quantum of electromagnetic radiation is the photon, but this unit is so small that, for all large scale purposes, it may be considered that energy is carried by continuously varying electromagnetic fields obeying classical laws.

quantum efficiency A dimensionless measure of the efficiency of conversion or utilization of optical energy, the measure being the average number of charged carriers produced for each incident photon.

quantum limited operation Operation wherein the minimum detectable signal is limited by quantum noise.

quantum noise Any noise attributable to the discrete nature of electromagnetic radiation. Examples are shot noise, photon noise and recombination noise. See shot noise.

quark An elementary particle that is believed to be a constituent of other supposedly elementary particles.

quarter speed An international leased circuit capable of carrying information at 16 ⅔ words per minute.

quarter-wave Having an electrical length measuring one-quarter of a wavelength.

quartz Mineral occurring in nature in hexagonal crystals and having piezoelectric properties much in demand in telecommunications equipment. When excited electrically quartz crystals vibrate and maintain extremely accurate and stable frequencies.

quasar A quasi-stellar radio source. A star-like body near the limit of the presently observable universe which emits radio and visible light radiation.

quasi-analog signal A digital signal which has been converted to a form suitable for transmission over a specified analog channel.

quasi-associated mode A form of common channel signaling in which the route followed by signaling messages is not necessarily geographically the same as the route which will be used by the voice connection which is being established. The message route is determined for each signaling message by information contained in the message's own routing label.

quaternary phase shift keying See quadrature phase shift keying.

quench A circuit which suddenly stops an oscillation.

queue indicator A circuit device or part of a program which indicates the number of calls waiting for access to a particular circuit element or trunk.

queue traffic 1. In a store and forward switching center, the outgoing messages awaiting transmission at the outgoing line position. 2. A series of calls awaiting service.

queuing Holding of calls in order of their arrival and presenting them automatically, in the same order, to an operator or to a subsystem for attention.

queuing delay The delay incurred by a common channel signaling system signal message as a result of the sequential transmission of signal units on the signaling channel.

queuing theory Probability theory as applied to the study of delays.

quick-acting relay A relay which both operates and releases with very little delay.

quick-break fuse Fuse with the fusible link under tension for rapid operation.

quick-break switch A two-stage blade switch sometimes used for interrupting large direct currents. The second blade is operated by a spring which gives it a rapid break, thus minimizing arcing.

quick-connect Terminal block into which insulated wires can be pushed. The insulation is cut as the wires are pushed in, and a reliable low-resistance contact is established.

quick-connect panel Type of central office main distribution frame in which jumper wires are terminated by pushing them into split cylinders which cut through insulants to make good contact with the wires themselves.

quiescent Having no input signal. Inactive.

quiescent current Current which flows in a device in the absence of an applied signal.

quiet battery A central office battery providing microphone currents. The battery is of low internal resistance, and rectified dc supply across it is well filtered.

quinning Making up of jumper wires in twisted-together units of five separately insulated wires.

raceway Covered trough or channel for internal wiring and cabling.

rack, battery Wooden or steel rack used to support storage battery cells in a central office.

rack, cable Light steel runway or ladder on which internal cables are laid in a central office.

rack, plastic tubing Grooved strip used to hold pipes carrying pressurized dry air to cables.

rack, power Mild steel rack on which such items as rectifiers, tone generators and meters are mounted in a small central office.

rack, relay Mild steel rack designed to carry relay sets in a central office.

rack, underground cable Galvanized steel channel bolted to the sides of manholes in which cable hooks can be placed to support cables. (UK: cable bearer)

rack, wall Mild steel rack mounted against a wall in a central office and designed to carry relay sets.

radar Radio detection and ranging equipment that determines the distance, and usually the direction, of objects by transmission and return of electromagnetic energy.

radiac Detection, identification and measurement of nuclear radiation. The term is an acronym for radioactivity detection, identification and computation.

radial lead A wire lead from the side of a component rather than axially from the end.

radiance Radiant power, in a given direction, per unit solid angle per unit of projected area of the source, as viewed from that given direction. Expressed in watts per steradian per square meter.

radiance theorem   See conservation of radiance.

radiant emittance   Radiant power emitted into a full sphere (4 pi steradians) by a unit area of a source. Expressed in watts per square meter. Syn. radiant exitance.

radiant energy   1. Energy which radiates in the form of radio waves, infrared (heat) waves, light waves, x-rays, etc. 2. Energy that is transferred via electromagnetic waves.

radiant exitance   Radiant emittance.

radiant flux   Radiant power (obs.)

radiant incidence   See irradiance.

radiant intensity   Applied to a point source only, the time rate of transfer of radiant energy per unit solid angle. Expressed in watts per steradian.

radiant power   The time rate of flow of radiant energy, expressed in watts. Syns. flux, optical power, power, radiant flux

radiate   1. To spread out from a central source. 2. To emit electromagnetic energy.

radiation   1. In radio communication, the emission of energy in the form of electromagnetic waves. 2. The outward flow of radio frequency energy from a source. 3. Energy flowing in a medium in the form of radio waves.

radiation, spurious   Emission from a radio transmitter which is outside its permitted frequency band and power limitation.

radiation angle   Half the vertex angle of that cone within which can be found a specified fraction of the total radiated power at any distance in the far field. Syn. output angle.

radiation field   See far field region.

radiation pattern   1. The variation of the field intensity of an antenna as a function of direction. 2. The output radiation of an optical waveguide, specified as a function of angle or distance from the waveguide axis. Far field radiation pattern is specified as a function of angle. Near field radiation pattern is specified as a function of distance from the waveguide axis. Radiation pattern is a function of the length of waveguide measured, the manner in which the waveguide is excited, and the wavelength.

radiation scattering   The diversion of thermal, electromagnetic or nuclear radiation from its original path as a result of interactions or collisions with atoms, molecules or large particles in the atmosphere or other media between the source of radiation and a point some distance away. As a result of scattering, radiation (especially gamma rays and neutrons) will be received at such a point from many directions rather than only from the direction of the source.

radiative mode   Unbound mode.

radiator   Any part of an antenna which radiates electromagnetic waves.

radiator, tower   See tower radiator.

radio   1. A method of communicating over a distance by modulating and radiating electromagnetic waves. 2. A radio receiver or transmitter.

radio, microwave   Radio communication using frequencies between approximately 1GHz and 1000GHz.

radio, mobile   Vehicle-mounted or otherwise transportable radio transmitter and receiver.

radio, urban mobile   Telephone communication within any exchange limit which uses a radio path to the central office.

radio and wire integration (RWI)   The combining of wire circuits with radio facilities.

radio and wire integration (RWI) device   An interface device that permits the combining of wire circuits with radio facilities.

radio atmosphere, standard   An atmosphere having the standard refractivity gradient.

radio baseband   The baseband of a radio.

radio baseband receive terminal   The point in the baseband circuit nearest the radio receiver from which connection is normally made to the multiplex baseband receiver terminal or intermediate facility.

radio baseband send terminal   The point in the baseband circuit nearest the radio transmitter from which connection is normally made to the multiplex baseband send terminal or intermediate facility.

radio beam   A radiation pattern from a directional antenna such that the energy of the transmitted electromagnetic wave is confined to a small angle in at least one dimension.

radio broadcasting   Transmission by radio of program material intended for general reception by the public.

radio channel   See channel, radio.

radio circuit   Communications circuit established between two points using a radio bearer.

radio common carrier   Common carrier licensed by the FCC to receive from and transmit to domestic mobile stations.

radio compass   A radio receiver with a directional antenna, installed in a vehicle or aircraft. The antenna is rotated to provide an accurate bearing to the transmitting beacon or radio transmitter.

radio deception   The employment of radio to deceive by sending false dispatches, using deceptive headings, employing enemy call signs, etc.

radio detection   The detection of the presence of an object by radio location without precise determination of its position.

radio direction finder (RDF)   A radio receiver with a steerable antenna which can determine the direction of arrival of a radio signal.

radio direction finding   Radio location in which only the direction of a station is determined by means of its emissions.

radio emissions, classification and designation of   A standard scheme indicating bandwidth and classification of radio emissions, drawn up by the CCIR.

Three numerals plus one letter indicate bandwidth; three (optionally five) letters or symbols give classification, e.g., type of modulation, nature of signal and type of information transmitted.

radio fadeout The situation in which incoming radio signals fade away because a sudden and unusual ionization change in the ionosphere has increased the absorption of radio waves which would normally have been reflected by the ionosphere. Syn. radio blackout.

radio field intensity See field strength.

radio fix 1. The location of a radio transmitter by determining its direction from two or more listening stations. 2. The location of a ship or aircraft by determining the direction of radio signals reaching it from two or more sending stations, the locations of which are known.

radio frequency (RF) Those frequencies of the electromagnetic spectrum normally associated with radio wave propagation. Sometimes defined as transmission at any frequency at which coherent electromagnetic energy radiation is possible, usually above 150 kHz. See spectrum designation of frequency.

radio frequency interference (RFI) 1. Equipment-induced noises, resonant at radio frequencies, which interfere with radio reception. 2. The intrusion of unwanted signals or electromagnetic noise into a submarine cable system cable, for which shielding is required.

radio frequency (RF) protection ratio Value of the radio-frequency wanted-to-interfering signal ratio that, under specified conditions, enables the audio-frequency protection to be obtained at the output of a receiver. These specified conditions include such diverse parameters as spacing of the wanted and interfering carrier, emission characteristics, and receiver input and output levels, as well as such receiver characteristics as selectivity and susceptibility to cross-modulation.

radio-frequency (RF) wanted-to-interfering signal ratio Ratio, expressed in dB, between the values of the radio-frequency voltage of the wanted signal and the interfering signal, measured at the input of the receiver under specified conditions.

radio guard A ship, aircraft or radio station assigned to listening for and recording transmissions, and to handling traffic on a designated frequency for a specified unit or units.

radio horizon The locus of points at which direct rays from the antenna become tangential to the Earth's surface, taking into account the curvature due to refraction.

radio horizon, standard Radio horizon in the case of standard refraction.

radio link Radio system established between two points.

radio location Original (British) name for radar.

radio navigation Radio location intended for the determination of position or direction or for obstruction warning in navigation.

radio paging service A service that allows transmitting a signal—usually a "buzz" or "beep" tone—via ra-dio from any telephone in the public network to a personal, portable receiving device in a defined operating area. More sophisticated systems provide audible or visual display messages.

radio propagation predictions Charts available on a regular basis indicating future probable best bands of frequencies for providing reliable radiocommunication services. They are based on an evaluation of such transmission-affecting factors as sunspot activity, day/night variations in ionospheric levels, seasonal changes, and geographic factors.

radio recognition In military communications, the determination by radio means of the friendly or enemy character, or the individuality, of another.

radio recognition and identification See identification, friend or foe.

radio relay system A point-to-point radio transmission system in which the signals are received, amplified and retransmitted by one or more intermediate radio stations.

radio sky A conception of what the sky would look like if our eyes were sensitive to electromagnetic radiation at radio frequencies rather than to the visible light range.

radio telegraphy The transmission of telegraphic codes by means of radio waves.

radio telephone A radio transmitter and receiver used for two-way voice communications.

radio telephony The transmission of speech by means of modulated radio waves.

radio transmitter See transmitter, radio.

radio transmitting system A device consisting of a radio transmitter connected to its antenna or antennae, or several transmitters connected to a common antenna.

radio window Radio frequencies which are not reflected by the ionosphere and so cannot be used for long-range terrestrial radio communications links. The window ranges from approximately 10 MHz to 40 GHz.

radio-wire integration (RWI) See radio and wire integration.

radiogram Message transmitted by radio telegraphy.

radiometry The science of radiation measurement.

radiosonde Small balloon-carried radio transmitters which send out coded messages indicating atmospheric conditions, for use in compiling weather forecasts.

radiotelescope A steerable antenna, usually a large dish-shaped unit, similar to, but larger than, ordinary communications ground stations and used to detect and study radio emissions from nonterrestrial sources such as quasars.

radioteletype (RTTY) Radio communication with a teletypewriter/teleprinter directly associated with the radio receiver and transmitter.

radius, effective Earth's Because, for radio purposes, the effective radius of the Earth is 4/3rds its actual geometric radius, all calculations regarding the establishment of line of sight radio links use this 4/3rds value to correct for atmospheric refraction.

radix The base of a number system, e.g., 2 for the binary system, 10 for the decimal system.

radome Dome-shaped cover, usually of plastic or fiber glass, which protects a parabolic antenna from extremes of climate.

Raduga A series of Russian geostationary communications satellites.

rail, guard Angle iron frame, often found in central offices, about 6 in. above the floor and projecting along the fronts of equipment racks to protect equipment from accidental physical damage.

rail, piling Wooden or plastic block found at the bottom of each panel on the face equipment of a switchboard and on which jack strips are piled.

rain barrel effect The echo effect on a sound system.

raintight External plant construction feature which prevents rain from entering and damaging equipment.

rake To slant a pole away from the vertical (against the direction of pull), thus enabling a component of the pull to be resisted by the compression resistance of the pole.

Rand Corporation A nongovernmental U.S. research body which has conducted much fundamental and applied research in telecommunications and defense.

random access Pertains to the storage of data in a manner which permits access to any memory location independently, as opposed to serial access memory which requires that the memory locations be accessed in sequence.

random access discrete address (RADA) Radio communication system employing pulse modulation and a broadband carrier. Selective coding permits addressees to receive only those messages addressed to them.

random access memory (RAM) A solid state memory device used for transient memory stores. Information can be entered and retrieved from any storage position.

random errors Errors distributed over the digital signal that can be considered statistically independent of one another.

random noise Noise consisting of a large number of transient disturbances with a statistically random distribution.

random number A number formed by a set of digits in which each successive digit is equally likely to be any of the digits in a specified set.

random wound Method of coil winding in which wire is wound in irregular layers, this means that a given volume is able to accommodate fewer turns.

randomizer A device used to invert the sense of pseudorandomly selected bits of a bit stream in order to avoid long sequences of bits of the same sense. The same selection pattern must be used on the receive terminal in order to restore the original bit stream.

range, dynamic See dynamic range.

range adjustment, teletypewriter Control on a teletypewriter which permits compensation for signal distortion by allowing the machine to use any chosen part of the received pulse signals for signal recognition.

rank, switching center One of the six levels in the switching center heirarchy in the North American Nationwide Toll Switching and direct distance dialing plan. The six levels, according to rank, are:

Class 1	— Regional Center
Class 2	— Sectional Center
Class 3	— Primary Center
Class 4C	— Toll Center
Class 4P	— Toll Point
Class 5	— End Office

raster A predetermined pattern of scanning lines within a display space. Example: The pattern followed by an electron beam scanning the screen of a television camera or receiver.

rate 1. Price charged the customer for a particular service. 2. Price charged for the initial period (normally three minutes) of a manually established toll call. The rate depends on the distance between toll centers, whether the call is station-to-station or person-to-person, and on what day of the week and/or time of day the call is made.

rate, bit See bit rate.

rate, charging Current in amperes used for charging a storage battery.

rate, error See error rate.

rate, modulation See modulation rate.

rate, overtime 1. Charge made for each extra minute after the initial time period (normally three minutes) of an operator-controlled toll call. 2. Charge levied for each extra minute of usage of WATS facilities after reaching the monthly figure covered by the chosen tariff.

rate, time Current in amperes which will discharge a storage battery to a given final voltage during a specified time period.

rate averaging Regulatory practice providing for uniform rates for the same service.

rate base The total investment on which a regulated telephone company in the U.S. is entitled to earn a profit.

rate base regulation Limitation of a carrier's operations to ensure that his receipts permit only a specified return on invested capital (rate base).

rate center A geographic location with specific V and H coordinates (vertical and horizontal on a map, i.e., distances north and east of an origin point) used for mileage determination in traffic rate calculations.

rate of information transfer (RIT) The amount of information that can be communicated from sender to receiver in a given length of time.

rate of transmission See effective speed of transmission.

rate period The specific portion of a 24-hour period or the specific day of the week during which a particular rate is charged for toll calls.

rate step (RS) Numbers used in calculating long distance/toll call charges. These are either assigned to distance bands radiating from the serving office, or arbitrarily on a location-to-location basis.

rate treatment number Code number which indicates the particular rates which apply to a particular type of toll call.

rated output power That power available at a specified output under specified conditions of operation.

rater An automatic device which allocates the applicable rate treatment number to each call after analyzing date, time, called number, and station-to-station or person-to-person status of the call.

rating, storage battery Number of ampere-hours delivered by a storage battery when discharged to a given final voltage (usually 1.85 V per cell) during a specified time period (usually either 8 or 10 hours).

ratio, bandwidth See bandwidth ratio.

ratio, carrier-to-noise Ratio of carrier to noise magnitude (usually RMS voltages of these are compared).

ratio, front-to-back See front-to-back ratio.

ratio, LC See LC ratio.

ratio, M-L Ratio of metallic circuit noise in an exposed telephone route to the longitudinal noise.

ratio, signal-plus-noise-to-noise Ratio between signal and noise, expressed in dB, as actually measured. When signal level is being measured it is inevitably accompanied by noise; it is not possible to measure noise and signal separately.

ratio, signal-to-noise See signal-to-noise ratio.

ratio, standing wave (SWR) See standing wave ratio.

ratio-squared combiner See maximal-ratio combiner.

rat race A hybrid ring, a ring method of interconnecting several waveguides or coaxial lines.

rawin (from radio or radar wind) Determination of wind parameters by observation of a radio sonde or balloon, either by radar or by radio direction finding.

ray A geometric representation of a light path through an optic device.

Rayleigh distribution A mathematical statement of the frequency distribution of random variables, for the case where the variables have the same variance and are not correlated.

Rayleigh fading Phase interference fading due to multipath which is approximated by the Rayleigh distribution.

Rayleigh scattering Scattering by submicroscopic inhomogeneities (fluctuations in material density or composition) in refractive index. The scattered field is inversely proportional to the fourth power of the wavelength.

RC constant The time constant of a resistor-capacitor circuit. It is the time in seconds required for current in an RC circuit to rise to 63% of its final steady value or fall to 37% of its original steady value, obtained by multiplying resistance value in ohms by capacitance value in farads.

RC network A circuit that contains resistors and capacitors, normally connected in series.

RC oscillator Oscillator using a resistance capacitance tuning circuit.

reactance That part of the impedance of a network which is due to inductance or capacitance. Reactance of components varies with the frequency of the signal.

reactance, capacitive Reactance due to capacitance in a circuit.

reactance, inductive Reactance due to inductance in a circuit.

reactive Possessing reactance.

reactive power Power circulating in an ac circuit. It is delivered to the circuit during part of the cycle but is returned during the other half of the cycle. Obtained by multiplying voltage, current, and the sine of the phase angle between them.

reactor In electronics, usually signifies a component with inductive reactance.

reactor, saturable An iron-core inductance with two coils, one carrying dc which varies the saturation of the core, the other carrying ac. The inductance of the ac winding is controlled by the level of dc through the dc winding.

read 1. To obtain information from a storage device. 2. To sense, or obtain the state of the data either in memory, or from an input device.

read wires Wires threaded through ferrite cores. Pulses of current through the read wire will induce voltage pulses on the sense wire if logic 1 has been stored in the core.

reader Device which converts coded information into signals for transmission or interprets stored data and converts it into a different format.

reader, card See card reader.

reader, magnetic tape Machine that reads data which has been recorded on magnetic tape and transmits it to a computer's electrical memory store.

reader, optical character (OCR) Device which reads printed copy (sometimes using special type faces) and produces data code signals representing the words read, for recording on magnetic tape or transmission to the memory store of a word processor or computer.

reader, paper tape Machine which reads paper tape punched with a standard code and converts this to a digital electrical output, with bits sent either on a parallel or serial basis.

read-only memory (ROM) A solid state memory device which has information permanently written into the memory during manufacture.

readout Visual display of the output of a meter or of a memory store.

readout, destructive Reading of a memory cell wherein the information content is destroyed by the act of interrogation by a read pulse.

readout, nondestructive Reading of the content of a memory cell which can be interrogated by a read pulse without affecting the information in storage.

251

read-write cycle Sequence of operations required to read and write (restore) memory data.

read/write (R/W) memory A memory system in which data may be stored and from which data may be extracted.

ready A steady-state condition at the data terminal equipment/data circuit terminating equipment (DTE/DCE) interface which denotes that the DCE is ready to accept a call request signal or that the DTE is ready to accept an incoming call.

ready for data A call control signal transmitted by the data circuit terminating equipment (DCE) to the data terminal equipment (DTE) to indicate that the data connection is available for data transfer between both DTEs.

real time Pertaining to the actual time during which a process takes place or to the performance of a computation during a period short in comparison with the actual time that the related physical process takes, in order that the results of the computation may be used in guiding the physical process itself. An expression used in discussing a type of computer operation in which the computer is interacting with events in the world of people, rather than of circuits. Generally speaking, the interaction must take place in a fast enough time so as to be able to influence or react to the particular "people" event in progress.

real-time system An information system whose performance rate is determined by external factors.

real-time transmission Transmission by means in which there is no significant delay, as opposed to record and retransmit, store-and-forward, or off-line operations.

rearrangement The disconnection of one or more existing connections in a switching network and their re-establishment via new paths with the objective of obtaining a free path for an additional connection.

recall To bring an operator into an already established circuit.

recall, attendant See attendant recall.

recall signal A signal at a switchboard which calls the attention of an operator to the circuit concerned.

receive interruption The interruption of a normal transmission in order to use the facilities for a higher priority message.

receive only (RO) Pertaining to teletypewriter equipment arranged to receive and print signals without the capability to transmit.

receive only typing reperforator (ROTR) A reperforator which also prints incoming messages in characters alongside their punched holes.

receive-after-transmit time delay The time interval from keying off the local transmitter until the local receiver output has increased to 90% of its steady state value in response to a radio frequency signal from a distant transmitter.

received noise power 1. The calculated or measured noise power at the receive end of a channel, link, or system within the bandwidth being used. 2. The absolute power of the noise calculated or measured at a receive point. When expressing this value, the related bandwidth and the noise weighting must also be specified. 3. The value of noise power from all sources measured at the line terminals of a listener telephone set. Either flat weighting or some other specific amplitude frequency or noise weighting characteristic must be associated with the measurement.

received signal level (RSL) The value of a specified bandwidth of signals at the receiver input terminals relative to an established reference.

receiver 1. Telephone receiver, either found in the handset or as a loudspeaker. 2. A radio communication receiver for demodulating radio signals. 3. A TV receiver equipped for visual and audio reception. 4. A terminal that includes a detector and signal processing electronics to convert electrical signals to optical or audio signals, or to both of these.

receiver, paging Small, lightweight FM radio receiver designed for carrying in a coat pocket by persons desiring to be paged when away from their phone. The receiver decodes the paging signal and produces repeated beeps so that the person being paged can respond by calling from a nearby telephone. See: radio paging service.

receiver, radio Equipment which receives incoming radio signals, demodulates them after selection and amplification, and produces an audio output corresponding to the original modulating signal information input.

receiver, superheterodyne radio Radio receiver in which the input signal is converted to an intermediate frequency which is amplified and detected to give an audio frequency output.

receiver, telephone The capsule which is held close to the ear in a telephone instrument. It converts an alternating electric current into sound waves, usually by the action of an electromagnet on a moveable diaphragm in a permanent magnetic field.

receiver, tone A receiver tuned to the various audio frequencies used in dual tone multifrequency pushbutton calling. If the correct pairs of frequencies are received, a coded output indicates the digit keyed.

receiver, touch calling A tone receiver for dual tone multifrequency signals.

receiver, tuned radio frequency (TRF) See tuned radio frequency.

receiver, watch case A separate telephone receiver attached to a standard telephone which enables a second person to hear incoming signals.

receiver attack-time delay The time interval from the time of application of a step input RF signal (of a level equal to the receiver sensitivity) to the receiver input until the receiver output amplitude reaches 90% of its steady-state value.

receiver isolation The attenuation between any two receivers connected to a CATV system.

receiver lockout system See lockout.

receiver release-time delay The time interval from removal of RF energy at the receiver input until the receiver output is squelched.

receiver sensitivity Optical power required by a receiver for low-error signal transmission. Usually quoted as optical power in dBm at which a bit error rate of 10^{-9} is attained.

receptacle An electrical socket designed to receive a special mating plug.

receptacle, coin A cash container in a pay telephone. Most types self-seal as soon as they are removed from the telephone instrument casing.

reception The act of receiving, listening to, or watching information carrying signals.

reception congestion A network congestion condition occurring at a data switching center.

recipient In a Message Handling System (MHS), the user, either a human being or a computer process, who receives a message from the MHS.

recipient User Agent The User Agent in a message handling system to which a message is delivered or that is specified for delivery.

recloser, automatic circuit A type of circuit breaker which automatically restores itself to normal a few seconds after operation (in the expectation that an unusual surge caused the breaker to trip). If the fault condition remains, the recloser normally locks open.

recognition In military communications, the determination by any means of the friendly or enemy character or of the individuality of another, or of objects such as aircraft, ships, or tanks, or of phenomena such as communications-electronics patterns.

reconstructed sample An analog sample generated at the output of a pulse code modulation decoder when a specified character signal is applied at its input. The amplitude of this sample is proportional to the quantized value of the corresponding encoded sample.

record An assemblage of a number of data elements that are all in some way related and are handled as a unit.

record, circuit layout Record kept in a maintenance center listing the transmission equipment used to provide the facilities concerned and giving levels, pair numbers, etc. to facilitate testing.

record, line card Record card kept in the repair service center or test desk area, listing subscriber's name, address and telephone number plus such information as routing and loop resistance of the line, and providing space for recording faults reported and action taken.

record communications Any type of communications which provides as output a semi-permanent or permanent message, usually in typed or printed form.

record-completing trunk Outgoing trunk from an end office to a toll office over which call charging information is passed automatically.

record medium In facsimile transmission, the physical medium on which the recorder forms an image of the subject copy. The record medium and the record sheet may be identical.

record sheet In facsimile transmission, the medium which is used to produce a visible image of the subject copy in record form.

record traffic 1. Traffic that is recorded, in permanent or semipermanent form, by the originator, the addressee, or both. 2. Traffic that is permanently or semipermanently recorded in response to administrative procedures or public law.

recorded spot In facsimile transmission, the image left by the recording spot on the record sheet.

recorder A meter which records varying readings, often on moving paper tape or on magnetic tape.

recorder, AMA Punched tape controller used in some Automatic Message Accounting units to record call data on toll calls.

recorder, answering time A totalizing meter which can be connected to a group of trunks to record answering times. Time intervals can be preset, so that the recorder can give totals for the numbers of calls not answered within each of these time intervals.

recorder, depth On cable ships, that part of an echo-sounder which produces a continuous plot of the depth versus time.

recorder, dial tone delay Portable test set which can be connected to a central office to measure the number of calls initiated by the set which receive a dial tone within the permitted time, normally a three second maximum. Recorders can give readings for the total number of calls broken down according to delay time.

recorder, integrated telephone A telephone instrument incorporating a casette type recorder to enable conversations to be recorded.

recorder, magnetic tape Device which can record sound on plastic tape coated with a magnetizable material.

recorder, traffic A device for measuring the amount of telephone traffic carried by a group or several groups of switches or trunks, and for periodically printing a record of that traffic. The electromechanical type consists of rotary stepping switches which every 100 seconds scan the "C lead" of each group of switches or trunks and register the number of busies encountered, then periodically print out the busy count. Since the scanning period is 100 seconds, the registers read directly in CCS. Microprocessor controlled devices are also available.

recorder, video tape (VTR) Tape recorder capable of recording a color television signal for later playback.

recorder-announcer, intercept Voice tape recorder used in conjunction with an intercept service and serving as an announcing machine to advise callers of the reason why their calls cannot be connected.

recorder-connector Unit which permits a subscriber to connect a tape recorder across his telephone line to record conversations. Recording is indicated by a warning beep tone every 15 seconds.

recording In facsimile systems, the process of converting the electrical signal to an image on the record medium.

recording spot In facsimile transmission, the source image formed on the record medium by the recorder.

recording trunk See trunk recording.

records, assignment Record cards or log books recording the allocation of cable pairs and giving details of interconnections.

records, plant assignment Record cards listing pairs allocated to each subscriber's loop together with all interconnections and terminations.

records arrival date (RAD) Bell System term used in Universal System Service orders. The date on which the central office, station installment force, or other work group associated with implementation functions is to have received all design and assignment information. It is a minimum of one working day prior to the designed, verified and assigned date.

records issue date (RID) Bell System term used in Universal System Service orders. It is the date on which the CPB is to send all design and assignment information (WORD/CLRD, station diagrams, etc.) to the central office and station installation forces. RID is a positive report date. The record medium and the record sheet may be identical.

recovery procedure In data communications, a process whereby a data station attempts to resolve conflicting or erroneous conditions arising during the transfer of data.

recovery time When used regarding echo suppressors, the time between the cessation of the signal causing operation of the suppressor and the cessation of suppressor operation.

rectification Conversion of alternating current into current flowing in only one direction.

rectifier Device for converting alternating current into direct current. A rectifier normally includes filters so that its output is, within specified limits, smooth and free of AC frequency noise.

rectifier, bridge A full-wave rectifier with four rectifying units arranged in a bridge so that when AC is connected across one diagonal, DC is available across the other diagonal.

rectifier, controlled Rectifier utilizing silicon controlled rectifier elements.

rectifier, copper oxide Disc-pile rectifier using alternate discs of copper and cuprous oxide, which are tightly bolted together.

rectifier, diode Rectifier utilizing a diode, either of electron tube or semiconductor type, as the rectifying element.

rectifier, dry disc Rectifier with discs of different materials tightly bolted together. Copper/copper oxide and selenium/iron rectifiers are typical examples.

rectifier, full-wave Rectifier which converts both the negative and positive half cycles of ac into direct current.

rectifier, half-wave Rectifier which uses either the positive or the negative half cycle of ac to give a dc output.

rectifier, mercury arc Rectifier which uses an electron tube filled with mercury vapor as a rectifying device, the electrons moving from cathode to anode.

rectifier, selenium Dry disc rectifier using rectification at the junction between selenium and iron.

rectifier, silicon-controlled (SCR) Rectifier sometimes used in central office battery chargers. It employs a p-n-p silicon transistor as rectifying element.

rectifier, silicon power Rectifier using a silicon diode as rectifying device.

rectifier, thyristor Rectifier using a thyristor as rectifier and as voltage controlling device.

rectifier enclosure unit (REU) A complete packaged power supply unit including two rectifiers and their associated meters, alarms and fuses housed in one cabinet.

rectify To convert alternating current into direct current.

recursive Pertaining to a process in which each step makes use of the results of earlier steps.

recycle The act of reattempting to establish a path across a switching center.

recycling circuit breaker Circuit breaker which will attempt to reset itself several times before locking open.

red area Area in which classified communications do not need to be encrypted.

red/black concept The concept that electrical and electronic circuits, components, systems, etc., which handle classified plain language information in electrical signal form (red) be separate from those which handle encrypted or unclassified information (black). Under this concept, red and black terminology is used to clarify specific criteria relating to — and to differentiate between — such circuits, components, systems, etc., and the areas in which they are contained.

red circuit Circuit which can carry classified or sensitive communications in clear language form.

red designation See red/black concept.

redirect-to-new-address signal A signal sent in the backward direction indicating that the called data service customer has requested redirection of calls to another address.

redirected-call indicator Information sent in the forward direction indicating that a data service call is a redirected call.

redirected-to-new-address signal A signal sent in the backward direction indicating that the call has been redirected to an address other than the destination selected by the calling data service customer.

redirection address Information sent in the backward direction consisting of a number of address signals indicating the complete address to which the data call is to be, or has been, redirected.

redirection of calls A facility which permits a called user of a data service to request the network to transfer a call to another nominated address. This may pertain to all calls following the request, or individual calls only.

reduced carrier transmission See suppressed carrier transmission.

reducer, acoustic shock A device, usually a varistor, in parallel with an operator's earpiece, which shunts high voltages and so limits acoustic shock.

reducer, noise Voice-operated device which reduces gain in the absence of an incoming voice signal and so makes a circuit appear less noisy.

reduction, data Changing of raw data into a more directly useable and useful form.

redundancy Signals additional to those actually needed to carry the specified information. They are the components of a message which may be ignored or eliminated with no loss of essential information.

redundancy checking An error detection technique involving the transmission of additional data related to the base data in such a way that the receiving terminal, by comparing the two sets of data, can determine to a specified degree of probability whether an error has occurred in transmission.

redundant 1. More than is actually needed for intelligibility. 2. Describing items of equipment which are provided in duplicate or triplicate so that a required grade of service may be reliably achieved.

redundant code See code, redundant.

redundant n-ary signal A digital signal whose elements can assume n discrete states and which has an average information transmission capacity of less than $\log_2 n$. This can also be expressed in terms of the number of binary digits which can be transmitted by an element of a particular line code.

reed capsule A sealed glass tube, sometimes filled with nitrogen and sometimes partially evacuated, containing one or more pairs of flat ferromagnetic reeds which are fixed at the ends of the tube and overlapping at the center. The overlap region is where the reeds make contact. Tips are often plated with a precious metal. The glass capsule is normally mounted inside a coil assembly. When current flows in the coil, the reeds come together to complete a magnetic path.

reel, cable Steel or wooden reel on which telephone cable is wound for storage and shipping.

reel, pay out Reel with vertical axis on which a coil of open wire can be mounted and from which the wire can be pulled during construction.

reel, shipping Device upon which short lengths of submarine cable are spooled for transportation or to facilitate placement.

reel, take-up Power-driven reel on a telephone construction truck, used to reel in salvaged wire or messenger strand.

reference circuit A hypothetical circuit of specified length and configuration with a defined transmission characteristic, used primarily as a reference for the performance of other circuits and as a guide for planning and engineering of circuits and networks.

reference circuit, voltage A circuit with a steady DC voltage used as a reference voltage to control the operation of other circuits.

reference clock A clock, usually of high stability and accuracy, used to govern a network of mutually synchronized clocks of lower stability.

reference edge The edge of a data carrier used to establish specifications or measurements in or on the data carrier. Syn.: guide edge

reference equivalent (RE) A measure of the loudness of telephone speech. The higher the reference equivalent, the less loud is the telephone speech. The electro-acoustic properties of subscribers' instruments and local lines (including the transmission bridge in the local telephone exchange) are described in terms of reference equivalents. The RE of an assembly of a telephone instrument and a local line is a quantity obtained by balancing the loudness of received speech signals and is expressed in decibels relative to an internationally-agreed-upon reference system. Separate figures relate to the sending and receiving directions of transmission.

reference frequency 1. A standard fixed frequency from which operational frequencies may be derived or with which they may be compared. 2. A frequency that has a fixed and specified position in the frequency spectrum with respect to the assigned frequency or another reference frequency.

reference level Power in a circuit at a reference point. For audio circuits, this is normally 1 mW (zero dBm) of 1000Hz tone.

reference noise The magnitude of circuit noise that will produce a circuit noise-meter reading equal to that produced by 10^{-12} W (-90 dBm) of electrical power at 1000 Hz for noise meters calibrated in dBrn(144-line) or dBrnC. For noise meters calibrated in dBa (F1A), the reference noise is adjusted to -85 dBm.

reference transmission level point See transmission level.

reference voltage A voltage used for control or comparison.

reflectance Ratio of total luminous flux reflected by a given surface to the incident flux. Syn. reflection factor.

reflected binary Synonym for gray code.

reflected impedance See impedance, reflected.

reflection 1. Abrupt change in direction of a light beam at an interface between two dissimilar media so that the light beam returns to the medium from which it originated. 2. Similar changes of direction for other wave phenomena, such as sound or radio waves. Reflection of electromagnetic waves can occur at an impedance mismatch in a transmission facility. It is not necessary for different media to be involved at the reflection point.

reflection, diffuse Reflection from a rough surface producing reflected rays in many different directions.

reflection, radio wave Sometimes used to describe the bending of radio waves providing over-the-horizon or troposcatter transmission links, although "refraction" is a better description.

reflection, specular Reflection from a smooth surface, not necessarily plane, obeying normal optical laws of reflection.

reflection coefficient 1. In radio propagation, the ratio between the amplitude of the reflected wave and the amplitude of the incident wave. For large smooth surfaces, the reflection coefficient may be near unity. 2. At any specified place in a transmission line between a source of power and an absorber of power, the vector ratio of the electric field associated with the reflected wave to that associated with the incident wave. The reflection coefficient (RC) is given by the formulae:

$$RC = (Z_2 - Z_1)/(Z_2 + Z_1)$$
$$= (SWR - 1)/(SWR + 1)$$

where Z_1 is the impedance toward the source, Z_2 is the impedance toward the load, and SWR is the standing wave ratio.

reflection loss 1. The ratio in dB between the incident and the reflected wave at any discontinuity or impedance mismatch. 2. Total loss from reflections at the junction between two optical components. 3. The reflection loss for a given frequency at the junction of a source of power and a load is given by the formula:

$$\text{Reflection loss} = 20 \log_{10} |(Z_1 + Z_2)/(4Z_1Z_2)^{1/2}|$$

where the reflection loss is in dB, the vertical bars designate absolute magnitude, Z_1 and Z_2 are the impedances of the source of power and the load. The ratio, expressed in dB, is the same as that of the scalar values of the volt-amperes delivered to the load, to the volt-amperes that would be delivered to a load of the same impedance as the source. The reflection loss is equal to the number of decibels which correspond to the scalar value of the reciprocal of the reflection factor.

reflections (TV) Signals which arrive at the receiver at different times from the primary signals. These produce "echoes" or "ghosts" on the TV screen.

reflectivity The reflectance of the surface of a material so thick that the reflectance does not change with increasing thickness. The intrinsic reflectance of the surface, irrespective of other parameters such as the reflectance of the rear surface. (No longer in common usage.)

reflectometer Device which measures energy traveling in each direction in a waveguide and used in determining standing wave ratio.

reflector In radio, one or more conductors or conducting surfaces for reflecting radiant energy. Also the metal elements placed behind the active element of an antenna in order to make it directive.

reflector, parabolic Symmetrical metal or mesh antenna reflector with parabolic cross section used to provide a directional antenna. Power originating at the focus of a parabola is reflected as a parallel beam.

reflector, periscope Passive reflector made up of two plane reflectors facing each other so that a beam

from a distant transmitter is reflected by one, then the other, in order to turn the beam a few degrees from its original direction. Periscope reflectors are sometimes used on difficult-to-reach mountain top sites which interrupt a direct line of sight microwave path.

reflector electrode Electrode in a reflex klyston which reverses the flow of the electron stream.

refraction 1. The bending of a sound, radio or light wave as it passes obliquely from a medium of one density to a medium of another density which varies its speed. 2. The bending of a beam of light in transmission through an interface between two dissimilar media or in a medium whose refractive index is a continuous function of position (graded index medium).

refraction, standard The refraction which would occur in a standard radio atmosphere.

refraction, sub- Refraction for which the refractivity gradient is greater than standard.

refraction, super- Refraction for which the refractivity gradient is less than standard.

refractive index The ratio of the velocity of light in vacuum to the phase velocity in a medium. A function of wavelength denoted by n. May also be defined as the square root of relative permittivity. Syn. index of refraction.

refractive index (radio) Ratio of the speed of radio waves in a vacuum to the speed in the medium under consideration.

refractive index difference Difference between the maximum refractive index occurring in the core of an optical waveguide and the refractive index of the cladding.

refractive index profile See index profile.

refractive modulus In radio, one million times the amount by which the modified refractive index exceeds unity.

refractivity, standard vertical gradient A standard (-40N/km) used for comparison in phase and refraction studies. It corresponds approximately to the median value of the gradient in the first kilometer of altitude in temperate climates.

reframing time In data transmission, the time that elapses between the moment a valid frame alignment signal becomes available at the receive terminal and the moment frame alignment is established. Also called frame alignment recovery time.

refrangible Capable of being refracted, as are radio and light waves.

refresh rate The rate per unit of time at which a displayed image is renewed in order to appear stable: 60 times per second in U.S. and 50 times per second in Europe.

refund key Key at a manual telephone operating center by means of which coins held on deposit in a pay telephone are returned to the caller.

regenerate To restore pulses to their original shape or to restore information in memory to its original form.

regeneration 1. The gain that results from coupling the output of an amplifier to its input. The increase of signal power in a circuit by the use of positive feedback. 2. The process of recognizing and reconstructing a digital signal so that the amplitude, waveform and timing are constrained within stated limits. 3. The action of a regenerative repeater in which digital signals are amplified, reshaped, retimed and retransmitted. 4. In a storage device whose information storing state may deteriorate, the process of restoring the device to its latest undeteriorated state.

regeneration, memory Restoration of a memory device to its original state.

regeneration, pulse Action of a device which receives a train of pulses (which may or may not be distorted by an earlier transmission path) and retransmits the train with all the pulses correctly shaped and with correct amplitude.

regenerative repeater A repeater in which the pulse signals are amplified, reshaped, retimed and retransmitted. 2. A repeater that is designed for digital transmission. 3. A device which regenerates digital signals.

regenerator A device which regenerates signals.

regenerator, pulse Device which provides pulse regeneration.

regenerator section A regenerator together with its preceding transmission path in a digital network.

Regional Bell Companies The seven holding or operating companies into which the Bell System's local telephone companies were assigned.

Regional Center Any of the highest rank (Class 1) switching centers in the Nationwide Distance Dialing Plan.

register 1. A device, accessible to a number of input circuits, which accepts and stores information relating to a called number or service. 2. The apparatus, in an automatic system, which receives the dialed impulses and controls the subsequent switching operations. 3. A message register or subscriber's meter. This can be an electromechanical counter, or an accumulator controlled by software. 4. In a computer, an electronic circuit within the microprocessing unit that is capable of storing one or more bytes of information.

register, incoming Register in a common control switching office which receives digital address information on an incoming trunk, and passes this on to markers for action.

register, message A meter or counting device which steps up one unit every time it receives a DC pulse.

register, originating Register in a common control switching office which receives digital address information from subscribers' lines and passes this on to a marker for action.

register, pen A device employing moving paper and a stationary pen used to monitor dial pulses.

register, shift Register used for multiplication. It shifts bits of information one place to the left resulting in multiplication by the radix.

register circuit Circuit which stores information until it is needed by another circuit.

register control Control of switching stages by registers which receive and interpret the digits from line.

register function Any of the functions including receiving, storing, analyzing, translating and transmitting address and other information in order to control the setting up of a call.

register key Key controlled by the operator at a switchboard and used to place a unit call in the register of a calling line.

register-sender Equipment which performs the functions of both a register and a sender.

register signaling End-to-end or link-by-link, multifrequency, in-band pulse signaling used for the transmission of address information.

register signals Signals passed between register-controlled exchanges during the establishment of a call. Registers accept and store information relating to a called number or service.

register-translator A device in which the functions of a register and a translator are combined.

registered roamer A mobile radio telephone station which has been registered for service with a cellular radio authority outside its home office service area.

registration A process by which a vendor's telephone equipment may be tested and certified "harmless" to the public switched telephone network.

registration progress The FCC authority for the interconnection of data terminal equipment to the telephone network, under the interconnect docket.

regrade To re-arrange trunks between switching stages in order to improve efficiency of operation of a switching center.

regression analysis A forecasting method used where the factor to be predicted can be expressed as a function of one or more variables. These variables may themselves have to be forecasted.

regression testing Repetition of some or all of the tests which took place on an earlier version of an item after a defect has been corrected or a modification has been made to a program.

regular signaling link The signaling link which normally carries some particular parcel of signaling traffic.

regulated Controlled for uniformity.

regulated line section In a carrier transmission system, a line section in which the line regulating pilot or pilots are transmitted from end to end without any associated intermediate amplitude regulation.

regulation 1. Voltage regulation. 2. The regulation of levels over a transmission system. 3. The control of sag of an open wire route. 4. The process by which it is ensured that public utilities operate in accordance with legal authorized rules.

regulation, baseband Continuous adjustment of the level of baseband output from a radio system despite significant variations in actual receiver output.

regulation, voltage The ability of a supply system to maintain a specified output voltage despite significant changes in load.

regulator, exchange voltage A voltage-regulated, low-voltage rectifier used in some central office power systems when connected in series with the main storage battery. Load changes result in changes in storage battery voltage, but these are offset by changes in the voltage produced by the regulator, bringing the net voltage available to the load to a standard value whatever the drain current.

regulator, feedback controlled Regulator which compares the level of a pilot signal to a reference level and makes appropriate adjustments.

regulator, gas pressure Regulator which acts as interface between gas in a high pressure tank and the gas at only a few pounds above atmospheric pressure which is introduced into a cable pressurization system.

regulator, induction voltage An exchange voltage regulator.

regulator, pilot-wire An early type of pilot regulation in which the variation of resistance with temperature of a pilot wire in a cable was employed to control gain of repeaters for circuits in the same cable, on the assumption that the resistances of conductors carrying signals would vary in step with the resistance of the pilot wire.

regulator, telephone A regulator sometimes used on telephone instruments so close to the central office/exchange that an unusually large line voltage is present. Many telecommunications administrations insist on telephone instruments being equipped with such regulators which insert extra loss in both send and receive paths. In this way, the output will not overload the input of transmission equipment, and the user will not find incoming calls unacceptably loud. Some administrations are, however, now changing over to constant-current subscriber's line circuits, in these circumstances telephone regulators are not required to be operative.

regulator, voltage A device which maintains its output voltage at a constant level. Such devices often depend on saturated core operation for control.

re-initialization The process whereby upon detection of a software fault, the decision is made to restart the whole of the affected functional area. This involves the regeneration of read-write data areas from the backing store and restarting all the processes in that functional area. The extent of the re-initialization procedure will depend upon the severity of the software malfunction and will normally be in stages. See rollback.

rejection, image See image rejection.

relation, telephone A relation which exists between two terminal countries when there is an exchange of telephone traffic between them (and, normally, a settlement of accounts).

relative humidity with respect to water (or ice). Percentage ratio of the vapor pressure of water vapor in moist air to the saturation vapor pressure with respect to water (or ice) at the same temperature and pressure. This ratio affects radio propagation.

relative transmission level The ratio of the signal power in a transmission system to the signal power at some point chosen as a reference. The ratio is usually determined by applying a standard test tone at zero transmission level point (or adjusted test tone power at any other point) and measuring the gain or loss to the location of interest.

relay 1. A device, usually electromagnetic, by which current flowing in one circuit causes contacts to operate which control the flow of currents in other circuits. 2. An intermediate station on a multihop radio system. 3. To retransmit a message through an intermediate point: a repeater or regenerator.

relay, alarm Relay which operates a circuit by giving an alarm.

relay, automatic tape Message relay in which message headers in a tape record of a received message automatically ensure that the tape is sent out by an appropriate transmitter.

relay, biased High speed polarized relay with a biasing winding and an operating winding, sometimes used in dc telegraph operations.

relay, bridge cut-off Relay in a subscriber's line circuit in an electromechanical office which ensures that when the line is in use for incoming calls it is disconnected from that point in the office which deals with outgoing calls.

relay, codel Relay, normally of reed type, used as a digital storage device in some electromechanical translators. It operates and releases by pulses through set and reset windings.

relay, coin Relay in some types of coin telephone which holds deposited coins in a temporary store until they are either dropped into the coin box or refunded to the caller.

relay, combined line and cut-off Relay in an electromechanical office which operates when a subscriber's line goes off-hook and simultaneously cuts off access to the line relay and prepares the circuit for dialing.

relay, cut-off A relay in an electromechanical central office which disconnects line relay from line during incoming or outgoing calls.

relay, dash-pot An early method of providing delayed action in which the relay does not operate immediately when current flows, but slowly moves a piston in an oil-filled cylinder. At the end of the piston's movement, which can take several minutes, external contacts are made, e.g. to give delayed alarms.

relay, differential Relay with two windings in which operation is dependent on current flow in both.

relay, hermetically sealed Relay in which the contacts are sealed in an airtight glass or metal enclosure.

relay, latching Relay which latches into its operated position and is held there without the need for a holding current. Release is accomplished by energizing a release winding.

relay, line Relay in a subscriber's line circuit in an electromechanical office which operates as soon as the instrument on the line goes off-hook.

relay, locking Relay which closes contacts in a second, holding winding which holds the relay operated until other contacts break the holding winding circuit.

relay, mercury Relay which tilts a small tube of mercury to bridge a gap between two contacts inside the tube.

relay, mercury-wetted Relay with contacts in sealed enclosures constantly wetted by mercury in order to provide low-resistance contact.

relay, multi-contact Relay in which a large number of springsets are all operated simultaneously.

relay, neutral Relay which operates with current in either direction, the operation depending only on the magnitude of the current.

relay, nonadjustable Relay in a sealed case, sometimes with mercury-wetted contacts.

relay, overload Relay which operates only when current in a circuit exceeds a specified value.

relay, pad control Relay at a toll switching center which takes a 2 dB switching pad out of circuit, or inserts it in circuit, as required by the transmission plan.

relay, polar Relay which operates only when current flows in a particular direction.

relay, polarized Relay of the type used in dc telegraph operation. It has a permanent magnet, a line winding and a biasing winding.

relay, rare gas A gas discharge tube filled with argon or another gas which ionizes readily to permit discharge, used to drain high voltages from overhead routes. It usually has one electrode per line and one to ground.

relay, reed Relay in which ferromagnetic reeds are sealed in small glass tubes and surrounded by operating coils. Used in semi-electronic switching systems. See reed capsule.

relay, resonant-reed Relay with vibrating reed tuned to a particular frequency. Sometimes used in selective ringing systems for mobile radiotelephones in vehicles.

relay, sensitive Relay, often with a galvanometer type operation, capable of operating on currents less than 10 mA.

relay, slow Relay so designed, through use of copper slugs or sleeves, to operate and/or release slowly.

relay, solid state A device in which current or voltage in one circuit controls the switching on or off of another circuit, but which involves no mechanical movement, armatures, moving contacts or reeds.

relay, supervisory Relay which is, in effect, operated by the line current supplied to an off-hook subscriber. When the line goes back on-hook, the relay indicates that the call has ended.

relay, tape Relay station in a telegraph system in which punched tape received on one channel is transferred physically to a tape-controlled transmitter which transmits the message toward the addressee.

relay, thermal A heat-operated relay which normally depends on the bending of a bimetallic strip.

relay, transfer See transfer relay.

relay, two step Relay in which one set of contact springs operates first, and the second set later.

relay, undercurrent Relay which operates when the current being monitored falls below a specified value.

relay, undervoltage Relay which operates when voltage being monitored falls below a specified value.

relay, voltage alarm Relay constructed on a voltmeter, with alarm contacts which close whenever the voltage falls below a specified minimum or rises above a specified maximum.

relay, voltmeter A voltage alarm relay.

relay, wire-spring Standard type of electromechanical relay used in many American and Japanese switching systems.

relay armature The movable part of an electromechanical relay, usually coupled to springsets on which contacts are mounted.

relay center A point at which messages are regenerated and retransmitted or passed from one circuit to another.

relay configuration An operating configuration in which a circuit is established between two stations via an intermediate relay station. Two links are utilized simultaneously and the channel connections at the relay station are effected completely within the station.

relay rack ground (RRG) A conductor which is a combination of return and framework ground. Generally it consists of a busbar mounted on the rack.

relay station, primary A main relay station which has net control responsibility.

relaying In a message handling system, the interaction by which one Message Transfer Agent transfers to another the content of a message plus the relaying envelope.

relaying envelope In a message handling system, the envelope which contains the information related to one operation of the Message Transfer System plus the service elements requested by the originating User Agent.

release 1. Describing any of several mechanisms which operate to return automatic switching equipment to its idle condition when the subscriber hangs up his telephone handset. 2. The event which ends a busy state.

release, timed Release of a circuit after a delay interval.

release alarm An alarm signal in an electromechanical office which indicates that one or more selectors have failed to release after calls have been completed and both parties have gone back on-hook.

release current Value to which relay current must fall in order for a previously operated relay to release.

release-guard signal A signal sent in the backward direction in response to the clear-forward signal when the circuit concerned is brought into the idle condition.

release time 1. Time interval between a calling instrument going back on-hook and switches being restored to normal. 2. Time interval between de-energization of a relay coil and the end of contact closure.

reliability The ability of a system or subsystem to perform within the prescribed parameters of quality of service. Often expressed in a more restrictive way to indicate the probability that the system or subsystem will perform its intended function for a specified interval under stated conditions.

reliability, circuit Percentage of time the facilities of the circuit were available for use during a particular period.

relocatable address In computer usage, an address that is adjusted when the computer program containing it is relocated.

relocate To move a computer program, or part thereof, and to adjust the address references involved so that the program can be executed after relocation.

reluctance Ratio of magnetomotive force to the magnetic flux.

remanence Magnetic flux which remains in a magnetic circuit after the magnetomotive force has been removed.

remendur An alloy with very high remanence developed by Bell Laboratories.

remote access Communication with a data processing facility through a data link.

remote call forwarding A service offered by some telephone companies by which all calls to a given number (normally one serving no line or instrument) are automatically transferred to a different number.

remote concentrator unit A unit which concentrates originating traffic from a number of local lines on to a smaller number of circuits to its parent local exchange, and similarly deconcentrates terminating traffic. It behaves as an outstationed part of its parent exchange, being controlled by the parent software.

remote control Describing a system for control of a device from a distance.

remote exchange concentrator A switching stage wherein a number of subscriber lines carrying relatively low traffic can be through-connected to a fewer number of circuits carrying high traffic, and vice versa. The switching stage is located remote from the exchange that controls it and to which its higher traffic volume circuits are connected. Such switching stages normally have no capability to directly interconnect subscriber lines terminating in that concentrator.

remote job entry In computer operations, that mode of operation that allows computer input of a job from a remote site and receipt of the output at a remote site via a communications link.

remote line unit The "front end" of a telephone central office/exchange on which subscribers' lines are terminated and in which the first stage of concentration switching takes place. These front ends can be located either in the same building as the main switching apparatus or remote from the central office. Some of the remote units have the ability to connect subscribers without reference to the parent exchange. See remotely controlled exchange and remote concentrator unit.

remote order wire An order wire extended to a point more convenient for personnel to perform required monitoring functions.

remote station A station or terminal which is physically remote from a main station or computer but can gain access through communication channels.

remote station answering Facility on some PABXs by which calls to one extension may be answered by different extensions.

remote switching stage A switching stage associated with, and controlled by, an exchange in a different location.

remotely controlled exchange An exchange whose switching functions are wholly or partially controlled by a unit in another location. It provides local switching, with limited routing facilities, but is dependent upon a parent exchange for some of its functions.

reorder signal A tone (usually repeated at 120 pulses per minute) indicating that the subscriber should hang up and redial later because the attempt to complete the call has run into a busy route.

repair cable In submarine cable systems, cable of lesser attenuation than the main cable. It is used in deep water repairs to permit lengthening a repeater section without upsetting the repeater gain/cable loss relationship.

repair clerk Person at repair desk who receives reports of trouble from customers and passes the information on to test personnel for localization and clearance.

repair person Person who checks subscribers' drops and instruments and clears faults, working in conjunction with local test desk personnel.

repair repeater In submarine cable systems, one of a group of spares manufactured concurrently with the system repeaters and whose circuitry permits its substitution for any repeater in the system.

repeat coil A one-to-one ratio audio frequency transformer for insertion in telephone lines. It permits the establishment of simplex and phantom circuits.

repeated call attempt (reattempt) Any of the subsequent call attempts related to a first call attempt.

repeater 1. A device which serves as an interface between two circuits, receiving signals from one circuit and transmitting them to the other. 2. A device which amplifies an input signal or — in the case of pulses — amplifies, reshapes, retimes or performs a combination of any of these functions on an input signal for retransmission. It may be either a one-way or two-way type. 3. In submarine cable systems, an electronic device, placed at regular intervals along the cable, whose purpose it is to amplify system transmission signals.

repeater, baseband Microwave radio repeater station which brings the received signal down to baseband frequency level between reception and retransmission.

repeater, battery and ground A dial pulse repeater using both battery and ground, for dc dialing over substantial distances.

repeater, carrier Amplifier used for carrier circuits, cable or open wire. Two-wire FDM carrier, common on open wire routes, uses different frequencies of transmission in the two directions. Separation is accomplished by filters.

repeater, digit-sending A dial pulse repeater which automatically sends digital information over the circuit.

repeater, duplex A dial pulse repeater using duplex signaling.

repeater, duplex telegraph Repeater in a telegraph circuit which uses different paths for the two directions of transmission.

repeater, E & M A line signal and dial pulse signaling repeater using E & M codes. E signals coming in from one direction control M signals being sent out in the other direction.

repeater, extra-pulse A dial pulse repeater which sends one signal pulse forward as soon as it has been accessed.

repeater, flexible submarine One-way, broadband repeater constructed in a long flexible case of a diameter only slightly greater than the cable itself. It can be laid by a cable ship without the need for slowing down.

repeater, four wire Telephone repeater for amplifying signals in both directions. Two separate amplifiers are needed. There is no connection between the two directions of transmission at the repeater.

repeater, frogging See repeater, low-high.

repeater, half-duplex telegraph Two-way telegraph repeater which can only operate in one direction at a time.

repeater, heterodyne See heterodyne repeater.

repeater, IF heterodyne A heterodyne repeater in which the incoming signal is shifted down to an intermediate frequency for amplification, then shifted up again to the required radio frequency for retransmission.

repeater, intermediate Repeater installed at regular intervals along a long transmission system in order to maintain proper signal levels.

repeater, knuckle-joint submarine Two-way broadband repeater with flexible pressure-tight joints at both ends of a short repeater housing. It is designed to go over the large diameter sheave on a cable ship so that it can be laid without slowing down the ship.

repeater, low-high A carrier repeater which receives low frequency signals, amplifies them and translates them to a high frequency band for retransmission. The reverse process occurs in the opposite direction.

repeater, microwave Repeater station in a multi-hop microwave system. Repeaters are typically spaced about 30 miles apart.

repeater, negative impedance Repeater which operates by inserting negative resistance into a circuit, thus reducing impedance and giving the effect of transmission gain. Since it can be used in two-wire circuits which retain their DC continuity, it permits loops for loop-disconnect dialing.

repeater, passive A microwave reflector, designed as either a flat metal screen which acts as a mirror to the radio beam or a pair of parabolic antennae connected back-to-back by a short length of waveguide.

repeater, pulse Relay set which receives dial pulses from one source and repeats them to another.

repeater, pulse correcting Relay set which receives dial pulses, which may or may not be distorted, and retransmits them with corrected amplitudes, lengths and shapes, without modifying the speed of pulsing or spaces between pulses.

repeater, pulse-link An E & M signaling repeater.

repeater, radio An intermediate repeater station on a multi-hop VHF, UHF or microwave radio system.

repeater, regenerative Repeater which receives digital signals, which may or may not be distorted, and retransmits corrected versions of the original signal.

repeater, regenerative telegraph Device which receives incoming start-stop teletypewriter signals, in which the code pulses may or may not be distorted, and retransmits a corrected version of each character signal as it is received.

repeater, rigid submarine Two-way broadband repeater contained in a rigid case for application to submarine cables. These repeaters cannot go over cable ship sheaves in the normal manner and are handled by special cable engines.

repeater, series-shunt negative impedance A negative impedance repeater which provides improved stability and greater gain by combining features of both series-connected and shunt-connected devices.

repeater, series-type negative impedance Negative impedance repeater connected in series with the circuit and permitting dc to flow through unchanged.

repeater, shunt-type negative impedance Negative impedance repeater connected across the circuit to be amplified.

repeater, simplex telegraph A telegraph relay for one-direction-at-a-time working, designed to repeat telegraph signals from one single wire circuit to another.

repeater, telegraph A relay device which receives weak telegraph signals and recreates them for retransmission. Duplex, half-duplex and simplex repeaters are available.

repeater, telephone An amplifier which receives an attenuated voiceband signal and retransmits it at a higher power level.

repeater, terminal Repeater at the end of a transmission system. Carrier terminal will include multiplexing to bring the channel down to audio frequency. Terminal on a voice frequency system involving four-wire repeaters will include hybrid coils and balances to produce two-wire circuits. Signalling units will normally be needed for all channels.

repeater, two-wire Voice frequency band repeater which separates the two directions of transmission by hybrid coils facing in each direction. It provides separate amplifiers for the two directions.

repeater section The cable between two repeaters plus one of the repeaters.

repeater station A station in which equipment for transmission flexibility, amplification and regeneration is housed.

repeater supervision The electrical monitoring of repeater performance in a submarine cable system, from the terminal station.

repeating coil See coil, repeating.

repeller The negative reflector electrode in a reflex klyston which reverses the flow of the electron beam so that it passes twice through the resonator, thus providing feedback.

reperforator (Abbreviation for receiving perforator). A device which is capable of receiving incoming signals and punching a tape in accordance with these so that the tape may be fed to a transmitter and the message re-transmitted.

repertory dialer See dialer, repertory.

repetition rate Rate at which regularly recurring pulses are repeated.

replicate An identical unit (provided for security reasons and/or to handle extra traffic).

replication Provision of more than one of a function to provide increased security or to enable additional traffic to be carried. Replication now usually means action in software; duplication, triplication, etc., usually means more than one unit of hardware.

reply A transmitted message which is a direct response to an original message.

reproducibility A measure of the ease with which an object, event, or image can be reproduced, or a known result can be obtained.

reproduction speed In facsimile systems, the area of copy recorded per unit time.

reprogrammable ROM (REPROM) A read-only memory that can be readily reprogrammed in the field without requiring factory-level test equipment.

repulsion Mechanical force which tends to separate like magnetic poles, like electric charges, or conductors carrying currents in opposite directions.

request data transfer A call control signal sent by the data terminal equipment to the data circuit-terminating equipment to request the establishment of data connection in leased circuit service.

request for information (RFI) General notification of an intended purchase of equipment, sent to potential suppliers to determine interest and solicit product information.

request for price quotation (RPQ) Solicitation for pricing for a specific component, software product, service or system.

request for proposal (RFP) Follow-up to an RFI, sent to interested vendors to solicit a priced configuration proposal that meets a user's requirements.

request-repeat system See automatic request-repeat system.

request to send (RTS) Part of modem handshaking during the establishment of a connection.

Requisition Received Date (RRD) Bell System term used in Universal System Service orders. It is the date that the requisition for material and equipment is scheduled to be received by the supplier. The RRD is assigned on projects and close supply coordination cases when RRD and assignment date do not coincide.

rering A recall signal sent by an operator at the calling end to get the called operator back onto the circuit.

reroute An alternative transmission path.

rerouting Using an alternate transmission path because of congestion or circuit failure.

rerun To run a part or the whole of a program through a computer again as a check.

resale carrier A company which hires circuits or services from a major carrier and resells them to individual users.

reserve, battery busy-hour The number of busy hours during which the voltage of the storage battery will fall to 1 V higher than the minimum voltage necessary to operate the office, if all external sources of power are disconnected.

reserve signaling link A signaling link which can be used to carry all, or part, of the signaling traffic of a regular signaling link when the latter has failed or has been withdrawn from service.

reset To restore to the original state. With binary devices, this means placing cells in their initial, or zero, state.

resident programs Call programs which are present in memory at all times.

residual charge Small charge remaining after a single discharge of a capacitor.

residual current See current, residual.

residual error rate The ratio of the number of bits (or unit elements, characters or blocks) incorrectly received but undetected or uncorrected by the error-control equipment, to the total number of bits (or unit elements, characters or blocks) sent. Syn. undetected error rate.

residual flux Magnetic flux which remains after the magnetomotive force has been removed.

residual gap The gap between pole face and relay armature when a relay is fully operated.

residual magnetism Magnetism or flux which remains in a core after current ceases to flow in the coil producing the magnetomotive force.

residual modulation See carrier noise level.

residual pin Nonadjustable button on a relay armature which maintains a minimum residual gap when a relay is fully operated.

residual screw Nonmagnetic adjustable screw on a relay armature which enables the residual gap to be adjusted.

residual stud A residual pin or button.

residual voltage 1. Vector sum of the voltages in all the phase wires of an unbalanced polyphase power system. 2. In a gas-discharge protector, the instantaneous voltage appearing across the terminals of a protector during the passage of a discharge current.

resin, epoxy A resin made from two separate chemical compounds, usually both liquid, which when mixed together form a strong, hard compound capable of forming water-tight joints in some types of plastic-sheathed cables.

resistance Property of a conductor which determines the current which will flow through it when a particular voltage is applied. The SI unit of resistance is the ohm.

resistance, antenna Quantity obtained by dividing the power in watts supplied to the antenna by the square of the rms antenna current.

resistance, contact Resistance measured across relay contacts. Mercury-wetted contacts will normally have zero contact resistance. Good dry contacts have contact resistances measured in milliohms.

resistance, effective 1. At high and radio frequencies, the effective ohmic resistance of a conductor; this is substantially greater than that at dc because of the skin effect. 2. At all alternating frequencies, the total resistance, including ordinary dc resistance, resistance due to skin effect, resistance due to eddy currents, resistance due to hysteresis, and resistance due to dielectric losses.

resistance, insulation Resistance offered to the flow of dc through insulation.

resistance, leakage Resistance of a path, normally to ground, over which a leakage current flows.

resistance, negative Property of some specialized circuits by means of which current increases as voltage is decreased.

resistance, pneumatic Resistance to the flow of gas through a pipe or cable. Pressurized cables with leaks exhibit pressure gradients analogous to voltage gradients caused by electrical resistance.

resistance, protective A limiting resistance connected in series with a power supply unit. It puts a ceiling value on current which would flow under fault conditions.

resistance, radiation Antenna resistance.

resistance, subscriber's loop Total resistance from the central office to, and including, the subscriber's instruments. Loop resistance in North America normally is taken to be the resistance of the outside plant up to the subscriber's premises at 68°F, plus 100 ohms for heat coils and internal wiring on the subscriber's premises and an allowance for temperature variation, plus 200 ohms for the subscriber's instrument.

resistance battery In testing circuits, the storage battery negative voltage as measured from a specified resistor.

resistance box Box containing precision resistors, wound noninductively, with heavy low-resistance brass connectors. Precision resistors are intended for testing purposes only, using balanced bridges.

resistance coupling See coupling, resistance.

resistance design of loops One method of designing distribution cable networks, based solely on loop resistance limits for signaling, in which loading coils are inserted on all loops over 18,000 ft to improve transmission.

resistance drop Fall in potential (in volts) between two points. It is the product of the current (in amps) and the resistance (in ohms). Also called IR voltage drop.

resistance-grounded Grounded for safety through a resistance, which then limits the value of the current flowing through the circuit in the event of a fault.

resistance pad Assembly of resistors made up in balanced and unbalanced attenuation networks giving various losses, terminated on an electron tube type base so that different pads may be plugged into a circuit with an appropriate socket.

resistivity Resistance per unit volume or per unit area.

resistivity, temperature coefficient of Factor based on the amount resistance in a conductor increases as temperature increases. The factor calculations to be made of resistance of a route at a given temperature varies from one conductive material to another; for example, in silver and copper resistance increases by a factor 1.0038 per 1° C increase in temperature, aluminum 1.0034.

resistor A device the primary function of which is to introduce resistance into an electrical circuit.

resistor, bias Resistor in the cathode circuit of an electron tube. A voltage drop in this resistor due to the current flowing through the tube provides bias voltage for the grid.

resistor, carbon Small cylindrical resistor made of carbon.

resistor, current-limiting Resistor inserted in a circuit for security or safety reasons which limits the value of the current which would flow under fault conditions.

resistor, dropping Resistor placed in series with a power supply and load in order to reduce the voltage across the load.

resistor, noninductive Resistor wound in such a way as to minimize inductance.

resistor, wire Resistor made of wire wound on a core.

resistor color code Colored markings on a resistor which indicate the value and tolerance.

resistor-transistor logic A logic family which was sometimes used to provide NOR and NAND gates before integrated circuits became readily available. Input to a pair of transistors is via resistors.

resolution Of a measuring instrument, the accuracy with which it can be read.

resolution (facsimile) The number of times each millimeter of copy is scanned, e.g., 3.78 lines per mm for Group I and II, 3.85 and 7.7 lines per mm for Group III equipment.

resolution, horizontal (TV) The amount of resolvable detail in the horizontal direction in a picture. It is usually expressed as the number of distinct vertical

263

lines, alternately black and white, which can be seen in three-quarters of the width of the picture. This information usually is derived by observation of the vertical wedge of a test pattern. A picture which is sharp and clear and shows small details has good, or high, resolution. If the picture is soft and blurred and small details are indistinct it has poor, or low, resolution. Horizontal resolution depends upon the high-frequency amplitude and phase response of the pickup equipment, the transmission medium and the picture monitor, as well as the size of the scanning spots.

resolution, vertical The amount of resolvable detail in the vertical direction in a picture. It is usually expressed as the number of distinct horizontal lines, alternately black and white, which can be seen in a test pattern. Vertical resolution is primarily fixed by the number of horizontal scanning lines per frame. Beyond this, vertical resolution depends on the size and shape of the scanning spots of the pickup equipment and picture monitor and does not depend upon the high-frequency response or bandwidth of the transmission medium or picture monitor.

resonance Tuned condition conducive to oscillation, when reactance due to capacitance in a circuit is equal in value to reactance due to inductance.

resonance, parallel The condition in a circuit with capacitance and inductance in parallel, when the frequency is such that the current entering the circuit from outside is in phase with the voltage across the parallel circuit.

resonance, series The condition in a circuit with capacitance and inductance in series, when the frequency is such that the current through the circuit is in phase with the voltage across the circuit.

resonant Related to or exhibiting resonance.

resonant frequency Frequency at which the inductive reactance and capacitive reactance of a series circuit are equal.

resonator A resonant cavity.

resource ceiling The maximum value that a particular resource may have as a function of time.

resources Staff, computer resources and any other facilities or items that may be needed in order to carry out a project satisfactorily.

respond opportunity In data transmission, the link-level logical control condition during which a given secondary station may transmit a response.

response 1. A reply to a query. 2. In data transmission, the content of the control field of a response frame advising the primary station concerning the processing by the secondary station of one or more command frames.

response, audio 1. Fidelity with which equipment reproduces an audio signal. 2. Equipment which gives an audible response, such as a synthesized voice.

response, frequency Gain or loss of a system over its specified frequency band.

response, spurious Response of tuned equipment to an undesired signal at a frequency to which the equipment is not tuned.

response, transient Response of an amplifier or other circuit to high frequency signals, such as those represented by a square wave test signal.

response frame In data transmission, a frame that may be transmitted by a secondary station. (Videotex) A frame through which the user can send a message or other response to the information provider who provides the frame being displayed.

response message A data service message sent in the backward direction containing an indication of the called terminal line condition (or of a network condition) and information relating to user and network facilities. In some cases, the message contains an address or identity.

response time The time which elapses between the generation of an inquiry and the receipt of a reply. It includes transmission time, processing time, time for searching records and files to obtain relevant data, and transmission time back to the inquirer. In a data system, it is the elapsed time between the end of transmission of an inquiry message and the beginning of the receipt of a response message, measured at the inquiry originating station.

responsivity In optoelectronics, The ratio of an optical detector's electrical output to its optical input, the precise definition depending on the type of detector. The ratio is generally expressed in amps per watt or volts per watt of incident radiant power. "Sensitivity" is often incorrectly used as a synonym.

restart To re-establish the process of executing a routine after a program or data error or machine malfunction.

restoration, circuit Re-establishment of services, on a priority basis, following the breakdown of a system.

restore 1. To repair and bring back into service. 2. To return to the non-operated position.

restorer, direct current Device by means of which a dc component is added to an ac signal after its reception.

restoring spring Spring which moves the armature of a relay back from its operated position when the relay is de-energized. This is often accomplished by the springsets carrying the relay's contacts.

restriction The condition under which a PABX is strapped or programmed so that certain extensions are not permitted to initiate certain classes of calls.

restriction, full Removal of the ability of all PABX extensions to make external calls other than operator-generated calls.

restriction, partial Removal of the ability of certain PABX extensions to make external calls, other than via the operator.

restriction, toll Removal of the ability of some or all PABX extensions to make toll or long distance calls other than via the operator.

restrictive station control PBX facility which enables extensions to be put on to different service levels, e.g., a phone in an unoccupied hotel room may not be permitted to initiate long distance calls.

restrictor Device in an outgoing dial trunk from a PABX which counts the digits dialed and forbids calls to certain designated codes. Such calls are diverted to an announcing machine or to the operator.

restrictor, toll code Device in an outgoing dial trunk from a PABX which permits local calls to enter the public switched telephone network but forbids toll calls.

retentivity Measure of the amount of magnetic flux which remains in a material after it has been saturated with flux and the responsible magneto-motive force has then been removed.

retiming Adjustment of the intervals between corresponding significant instants of a digital signal, using a timing signal as the reference.

retractile cord A coiled cord so wound that it will extend to its full length when under tension but will return to a tightly wound helix when tension is released, as with a telephone handset cord.

retransmission buffer Storage capacity in the signaling link control for signal units transmitted but not yet positively acknowledged.

retransmission on error Facility, incorporated in many data transmission systems, which provides automatic correction of errors introduced in transmission.

retransmissive star In optical fiber transmission, a passive component that permits a light signal on an input fiber to be retransmitted on several output fibers.

retrial A reattempt, after a failed attempt, to establish a telephone call.

retrieval The process of transferring all messages in the retransmission buffer of a signaling link which have not yet been positively acknowledged, to the transmission buffer of an alternative signaling path.

retry In bisync protocol, the process of resending the current block of data a prescribed number of times or until it is accepted.

return 1. A return path for current, sometimes through ground. 2. To refund coins deposited in a coin telephone.

return, common A return path common to two or more circuits.

return, negative A wire between two external plant points for equalizing their potential in order to minimize damage to cable sheaths by electrolysis.

return key Switchboard key by means of which an operator releases coins held in deposit in a coin telephone box and refunds them to the caller.

return loss The ratio, expressed in decibels, of the amplitude of the reflected wave to the amplitude of the incident wave at the junction of a transmission line and a terminating impedance. More broadly, the return loss is a measure of the dissimilarity between two impedances.

return-to-zero (RZ) code A code form having two information states called "zero" and "one", and having a third state to which each signal returns during each period.

reusable program (routine). A program (routine) that can be loaded once and executed repeatedly subject to the requirements that any instructions which are modified during its execution are returned to their states and its external program parameters are preserved unchanged.

reuse The sharing of reusing of the same radio channels in two or more cells of a cellular radio system. Sufficient cell separation is required to obtain adequate signal-to-noise ratio.

reverberation Persistence of sound due to repeated echos, as in a large hall, after the sound source has stopped.

reverberation time Time required for sound energy to drop by 60 dB after the sound source has been cut off.

reversal A change in magnetic polarity, in the direction of current flow, or of binary state.

reversal, polar Change in polarity of a voltage, or in the direction of current flow.

reverse battery supervision A method of indicating to an originating office that the called telephone has gone off-hook. Line voltage in the backwards direction is reversed.

reverse channel A simultaneous data path in the reverse direction. It has a lower bandwidth than the main, forward data path and is usually used for positive or negative acknowledgements of previously received data blocks.

reverse current Small current which flows through a diode when voltage across it is such that normal forward current does not flow.

reverse interrupt (RVI) In bisync protocol, a control character sequence sent by a receiving station to request premature termination of a transmission in progress.

reverse preemption Preemption of a circuit by a terminal which is normally the receiving terminal.

reverse voltage 1. Voltage in the reverse direction from that normally applied. 2. Voltage polarity which produces smaller current through a device.

reverting call Call from one telephone served by a party line to another telephone on the same party line.

reverting call switch Switch in a central office which is accessed when a reverting call is to be established.

revertive pulsing Pulsing back to the originating office from the incoming office. The originating office signals the train of pulses to stop when the desired number has been reached.

rewrite To write in again to a memory all the information which was in it before it was read out.

rf bandwidth The difference between the highest and the lowest emission frequencies in the region of the carrier or principal carrier frequency.

R-F pattern A fine herringbone pattern in a TV picture. May also cause a slight horizontal displacement of scanning lines resulting in a rough or ragged vertical edge of the picture. Caused by high-frequency interference.

rf power margin Extra transmitter power that may be specified by a designer because of uncertainties in the empirical components of the prediction method, the terrain characteristics, atmospheric variability, or equipment performance parameters. Syn. design margin.

rheostat A variable resistor.

rhombic antenna A horizontal directional high frequency band (HF) antenna needing four masts and a diamond-shaped array of radiating wires, fed at one apex and terminated at the opposite apex with an impedance, usually a resistor. It is much used for long-distance HF radiocommunication. When unterminated, it is bidirectional.

rhumb line Line intersecting meridians of longitude at the same angle.

ribbon, bonding Tinned copper strapping used for bonding the sheaths of lead-covered cables.

ridged Drop-wire or other one-pair insulated cable which has a ridge along one side of the outer insulation to enable a particular conductor to be identified.

rigging, manhole Steel framework erected above a manhole entrance, used to feed cable from cable reels into ducts in manholes.

right, pole Permission granted by a pole owner for a fitting to be attached to the pole, often for a small annual rental.

right hand rule When the thumb and the first two fingers of the right hand are all held at right angles to one another with the thumb representing the direction of motion and the first finger that of the magnetic flux, the second finger will point in the direction of the induced conventional current.

right-of-way Permission to have access to specified land and to use a specific portion of it.

right-of-way companies (ROW) Utility companies or agencies which maintain their own communication systems.

ring 1. To activate a called subscriber's telephone bell. 2. The second line wire in a telephone office (tip, ring, sleeve are the three contacts on a standard three-way switchboard plug and jack)

ring, cable Spring wire hook at one time widely used to secure aerial cable to a messenger wire. Now largely superseded by self-supporting aerial cables and by lashing wires.

ring, cable identification Colored plastic ring placed over cable pairs in a splice to aid in identifying individual pairs. Syn.: collett

ring, distributing Insulated smooth ring through which jumper wires are fed on a main distribution frame. Also called a jumper ring.

ring, manhole adjusting Cast iron ring used when the height of a manhole cover has to be adjusted to accommodate road level changes.

ring around Incorrect routing action whereby a call is routed back through an office through which it has previously passed.

ring circuit Ring-shaped cable network with half of the circuits going in each direction so that a cut at any

one point will still leave all locations with limited access to the central office.

ring connection Connection of components in series with the first connected also to last so as to form a closed ring.

ring cut Cut in an aerial cable sheath caused by abrasion of the supporting ring.

ring ground (RG) A bare, solid tinned-copper conductor forming a ring around the central office. The RG may be located within the wall or a minimum of two feet from the wall on the outside.

ring main 1. A cable, usually a power cable, fed from both ends so that connection may be made at any point on the ring. 2. The same technique can also be employed in subscribers' distribution networks using PCM multiplexing.

ring modulator A modulator with two pairs of diode elements so arranged that current flows in one direction, or its reverse, under the control of the modulating current. Ring modulators are much used in FDM multiplexing equipment.

ring network Data network with circular topology in which each node is connected to its neighbor to form an unbroken ring. A ring network in which one of the nodes exercises central control is often called a loop.

ring wire The second line wire (tip, ring, sleeve) of a telephone line, inside the office.

ringback Signal used by the operator at the receiving end of an established connection to recall operator at originating end.

ringdown 1. In telephone switching, that method of signaling an operator in which telephone ringing current is sent over the line to operate a lamp and the drop of a self-locking relay. 2. The type of signaling employed in manual operation, as differentiated from dial signaling. Ringdown signaling utilizes a continuous or pulsing AC signal transmitted over the line from a switchboard to the user. It can also apply when no switchboard is involved.

ringdown circuit A circuit on which the signaling is manually applied.

ringdown signaling The application of a signal to a line for the purpose of operating a line signal lamp or supervisory signal lamp at a switchboard, or for ringing a user's instrument.

ringer 1. Bell at a subscriber's premises which warns of an incoming call. 2. The device at the central office which provides the current to ring subscribers' bells.

ringer, biased Polarized bell designed so that it can only be rung by pulsating current of a particular polarity. Sometimes used on party lines.

ringer, high-impedance Biased bell used on party lines between one conductor and ground. It has high impedance at voice frequencies in order not to shunt an unacceptable amount of power to ground during conversation.

ringer, polarized Normal telephone bell designed to operate at about 20 Hz. It has a small permanent magnet to provide bias.

266

ringer, straight-line A party line bell unit which is polarized but not tuned.

ringer, telephone Standard polarized telephone bell.

ringer, tone Transducer producing an audible sound on receipt of the ringing current. Used on some telephones instead of a bell.

ringer, tuned Party line bell tuned to respond to ringing current at a particular frequency only. Frequencies from 17 Hz to 67 Hz are available.

ringer, unbiased Ringer without bias designed to operate satisfactorily on AC current but not on varying DC ringing current.

ringing 1. AC ringing current, usually 17 Hz to 20 Hz, sent out from a central office. 2. (In TV) An oscillatory transient occurring in the output of a system as a result of a sudden change in input. Results in close-spaced multiple reflections.

ringing, code Selective calling of subscribers by coded ringing pulses.

ringing, DC Activation of a ring key at a switchboard, which operates a normal DC relay circuit, the contacts of which in turn send AC ringing out to the line.

ringing, decimonic Party line selective calling using ringing frequencies of 20, 30, 40, 50 and 60 Hz.

ringing, divided Connection of bells on a party line so that half the bells are from the A line to ground and half from the B line to ground.

ringing, harmonic Party line ringing using tuned ringers. Frequencies commonly used are 16 ⅔, 25, 33 ⅓, 50 and 66 ⅔ Hz.

ringing, immediate A feature of some switching systems which provides an immediate ring to the called line, without waiting to pick up the normal ringing cycle.

ringing, manual Switchboard ringing which is sent out only while the ring key is operated.

ringing, selective A system which permits ringing of a single subscriber on a party line.

ringing, semi-selective Party line ringing for four-party lines with two having bells from A line to ground, the other two running from B line to ground. Coded ringing pulses are used to distinguish between the two who receive the same ringing signals.

ringing, synchronomic frequency Selective calling ringing using 16, 30, 42, 54 or 66 Hz.

ringing, thousand/twenty Ringing using a 1000 Hz oscillator interupted 20 times per second.

ringing, 20-cycle Ringing using a nominal 20 Hz signal to ring bells or operate calling indicators.

ringing current Current — usually about 70v, AC, and about 20 Hz — sent out on a subscriber's line in order to ring his bell.

ringing generator (RG) A programmable generator which can output a variety of ringing waveforms on receipt of a suitable drive signal.

ringing key Key thrown by an operator to transmit ringing current to a line.

ringing period The one second period during which (in the U.S.) ringing current is applied to a line.

ringing signal Tone signal sent back to the calling subscriber to indicate that ringing current is being sent out to ring the bell of the called subscriber.

ringing tone An intermittent low tone, one second on and three seconds off, which indicates that the ringing function is being applied at the called end.

ripple AC voltage superimposed on DC storage battery voltage, usually due to imperfect filters or rectifier faults.

ripple counter Type of nonsynchronous counter made up with flip-flop devices.

ripple voltage The alternating component of the unidirectional voltage from a rectifier or generator used as a source of DC power.

rise time Time needed for a pulse to rise from 10% to 90% of its peak value.

riser Duct, conduit or cable which runs vertically in a building from floor to floor.

risk analysis The process of assigning probabilities, e.g. of calculating the overall likelihood that a particular delivery will fail to materialize or will materialize late or deficient.

road side The side of a pole route next to a street or highway.

roamer A mobile radio telephone station being used outside the cellular areas served by its home system.

roaming subscriber A mobile radio telephone station user requiring service in a cellular system outside his own home office area.

rod To push or pull a device through a duct which cleans it of stones or rubble and prepares it to receive a cable.

rod, anchor Galvanized steel rod designed to be attached to a buried anchor. The upper end has an eye to which a guy may be attached to give stability to the pole.

rod, duct Hardwood rods in sections which can be joined for pushing into duct lines to clear them and to prepare the duct line for the pulling in of a cable. The lead rod can be equipped with a brush.

rod, ground Steel or copper rod driven into the ground to provide a low resistance ground connection.

rod, lightning Metal strip mounted above the highest point of a building and designed to discharge static electric charges to ground without lightning damaging the building.

rod, sectional ground Copper-weld ground rod made in sections which can be joined to reach considerable depth in order to give low resistance ground contact with moist earth.

rod, triple eye An anchor rod with three grooves in its eye, permitting the attachment of three separate guys.

rod, twin-eye An anchor rod with two grooves in its eye, permitting the attachment of two separate guys.

rodding Pushing sectional rods, taking with them a draw wire, through a duct line. The draw wire is left in the duct for subsequent use in pulling in cable. Sometimes the draw wire is pulled in to the duct line as rods are removed.

rod-in-tube technique A method of fabricating an optical waveguide by placing a rod in a tube and drawing the rod and the tube apart to form a fiber.

rods, flexible duct Flexible steel sectional rods, each about 1 m long, which can be joined together and pushed into a duct. When fitted with a brush, they can be used to push rubble out of the duct. Short cylindrically shaped mandrels are then pulled through to ensure that a duct has not collapsed and that there is adequate room for a cable of the required diameter.

rods, ranging Light pointed rods, about 3 m long, painted in red and white bands and used for the laying out of routes for new pole lines.

roger 1. Phonetic alphabet word signifying the letter R. 2. A voice communication meaning that the message has been received.

rollback The procedure by which a central processing unit recovers automatically from a fault which has led to a software corruption. The complexity of the procedure, and the resultant temporary effect on the service of the system, depend on the nature of the fault. The procedure will usually involve the process of re-initialization.

roll-off A gradual attenuation of gain-frequency response at either or both ends of the transmission pass band.

roof, pole Slanting cut on top of a wooden pole.

roof filter A low pass filter (with a roof representing the top frequency of the pass band.)

room noise level See ambient noise level.

root mean square (rms) Effective value of an alternating wave. For AC, this is numerically equal to the DC value of the current with the same heating effect. The rms value is 0.707 times the peak value of a sinusoidal variation.

root-mean-square (rms) pulse duration A measure of the duration of a pulse waveform.

rope-lay Describing a stranded cable in which the individual wires are stranded into groups before the groups are stranded together to form the whole.

rosin joint A dry joint in which the wire is held in place by dry flux and there is poor, or no, electrical contact.

rotary dial Common telephone calling device which makes and breaks the loop provided by the instrument, thus transmitting the digits of the called telephone subscriber's number to the central office/exchange.

rotary hunt group A term used in a step-by-step office to indicate that when a call is placed to a directory number, the equipment will hunt consecutively over other lines on the subscriber's PBX in order to find an idle circuit.

rotary step One of the 11 rotary positions at one horizontal level which can be taken by wipers of a two-motion selector. One of the steps of a rotary stepping switch.

rotary stepping switch A uniselector or single motion rotary switch which goes around only (ie not also up to horizontal levels as in a two-motion selector). Some rotary switches step under dial control, some hunt rapidly and automatically for idle trunks.

rotary switching system A common-control switching system previously used extensively. All dial pulses are received by a register which controls the group selector and final selector switches. The rotary finder and rotary selector switches are power driven from a continuously rotating shaft. Instead of clutches it uses a flexible driven gear which can be driven into and out of mesh by a control magnet. Ringing, busy signal, and closing of the talking path are accomplished by a motor-driven sequence switch.

rotator, antenna Motor which rotates a directional antenna to a required bearing.

rotator, Faraday Ferrite block used in a waveguide as a switch which lets radio energy through in one direction but blocks the other direction of transmission.

rotor The rotating part of an electric generator or motor.

round trip In satellite communications, the distance from an originating station through a satellite to a receiving station and return via the satellite to the originating station.

round-trip delay time In satellite systems, the time required for a signal to complete a round trip (which see).

route 1. The particular circuit group, or interconnected circuit groups, between two reference points used to establish a path for a call. 2. The route followed for the transmission of a telegram or the establishment of a connection. 3. The means of transmission (wire, cable, radio) used for the transmission of a telegram or the establishment of a connection. 4. In submarine cable systems, the actual location of the cable. Also, loosely, the names of the cable landing points or the system termini.

route, alternate A second route, for use when the first-choice route is busy or out of order.

route, primary The first choice and usual routing for traffic to a particular destination.

route, toll 1. Pole route or duct route which carries toll circuits on open wires or in cables. 2. Circuit designation between toll centers.

route control Controlling the routing of a toll call by limiting the number of links in tandem and giving preference to direct circuits and forward routing.

route control digit Digit transmitted, before the telephone number itself, from one toll center to the next to prevent toll calls from being routed around in loops and to limit lateral routings.

route diversity See dispersion.

routes In international telephony, the routes followed by international telephone traffic are designated by agreement between administrations. A distinction is made between primary routes, secondary routes and emergency routes. Primary routes are the circuits normally used in a given relation. Secondary routes are the circuits to be used when the primary routes are congested, when the transmission on the primary routes is not sufficiently good, or at a time outside the normal hours of service on the primary routes. Emergency routes are the circuits to be used in case of complete interruption or major breakdown of the primary and secondary routes.

Routine The lowest (fifth) precedence level (R) used on systems such as Autovon which make use of preemption for high priority calls.

routine A group of program instructions which accomplishes a particular task that needs to be done frequently. A group of instructions can be used over and over again, simply by referring to it from another part of the program.

routiner, automatic equipment Apparatus which automatically tests switches and connections through a dial central office to different test numbers. Any test calls which encounter difficulty hold the faulty switch for investigation by maintenance personnel.

routiner, automatic line insulation Apparatus which automatically tests subscribers' lines for low insulation, from a dial central office. Any lines found with ground faults are marked by the apparatus for investigation by maintenance personnel.

routiner, automatic trunk Apparatus which automatically tests interoffice trunks, first for access then for transmission, by checking the received level of a test tone connected automatically at the distant end on command from the routiner.

routing The assignment of a communications path through a network, or the process of selecting such path.

routing, alternative Provision of a second or third choice route for calls if the primary choice route is not available.

routing, avoidance Routing which avoids the use of circuits which are known to be possible sources of difficulty.

routing, diverse Routing of some interoffice circuits on different physical paths from others in order to minimize the risk of one plant fault isolating a central office.

routing, traffic The selection of a route or routes for a given traffic stream. This term is applicable to the selection of routes by switching systems or operators, or to the planning of routes.

routing code The digits needed to route a call to its office of destination.

routing indicator Coded indication preceding a teletypewritten message showing transmission routing.

routing label The part of the message label that is used for message routing in the signaling network. It includes the destination point code, the originating point code, and the signaling link selection field.

row, connecting All those crosspoints directly accessible from one inlet. At any instant, only one connection can be established via a connecting row. Also called switching row.

rubber Natural or synthetic material once important as an insulant, especially in power cables. Now rarely encountered in telecommunications equipment.

ruby laser An early type of laser, based on the use of a ruby crystal. It produces a narrow beam of coherent red light.

rudder, active On cableships, a rudder containing a motor-driven propeller with its shaft in the plane of the rudder blade. It is used as a maneuvering aid.

rule, pressure testing A circular slide rule used in cable pressurization calculations.

run, cable Route followed by cables, either directly buried or in ducts.

run, conduit Route followed by cable conduits. Also called a duct run.

running open In telegraphy, the condition under which a teletypewriter on an open line appears to be running because the type hammer is continually striking the type-box, but there is no movement across the page. The open line is decoded as the signal for "blank."

Rural Electrification Administration (REA) Agency within the U.S. Dept of Agriculture which helps to finance rural area telephone services.

rural line A multiparty line in a rural area. It often uses code ringing and sometimes still uses magneto calling.

Rural Telephone Coalition (RTC) Trade association of minor telephone companies and associated organizations.

ruthless preemption The act of seizing a busy toll circuit without warning to the parties using it, disconnecting them, and connecting high priority users to the circuit.

sabin A unit of sound absorption which is the equivalent of one square foot of surface of perfectly absorptive material.

safety factor The overload allowance included in equipment design. It is the amount of stress, above normal working conditions, which an item can bear without breaking down. Quantitatively, it is breakdown stress divided by working stress.

safety ground Connection made to ground, as a protective measure, usually from the frame or chassis of a piece of equipment.

sag Vertical distance from a conductor to the straight line joining its two points of support, as measured at the midpoint of a span.

salvage Property that is in such worn, damaged or deteriorated condition that it is not usable without major repairs or alterations.

Salyut A series of Russian space stations.

sample 1. The value of a particular characteristic of a signal at a chosen instant. 2. To select samples only of an electrical signal which, for one reason or another, cannot be transmitted in toto. The sample is itself often used as the basis for deciding on a discrete value for the parameter with the result that, instead of a continuously variable (analog) signal, a series of discrete values is obtained. In coding systems such as pulse code modulation the discrete value is itself represented by a binary word, usually 8 bits long, so that the original analog signal is now represented by a series of binary digits: the digits representing the number of the amplitude level which is close to the actual instantaneous value of the original signal.

sampling The process of taking samples, usually at equal time intervals. In pulse code modulation, the act of selecting samples of an analog wave at recurring intervals such that the original wave can later be reconstructed with reasonable fidelity from the samples.

sampling frequency The rate at which signals in an individual channel are sampled for subsequent modulation, coding, quantization, or any combination of these functions. The sampling frequency is usually specified as the number of samples per unit time. Syn.: sampling rate

sampling rate See sampling frequency.

sampling time The reciprocal of the sampling frequency.

SATCOM RCA's domestic satellite system, U.S.

satellite 1. A body which revolves around another body of greater mass and which has a motion primarily determined by the force of attraction of the more massive body. 2. A man-made object designed to orbit the Earth, the moon or another celestial body. 3. A dependent central office or exchange.

satellite, active communication A satellite with a self-contained energy source, designed to transmit radio communication signals back to earth.

satellite, Early Bird The 1965 geo-stationary satellite used for transatlantic telecommunications.

satellite, Echo A type of aluminum-coated plastic balloon satellite used by the U.S. in the 1960s to reflect radio waves for long-distance communication.

satellite, Explorer I The first U.S. satellite, launched in January 1958.

satellite, geo-stationary See geostationary satellite.

satellite, IDCSP Equatorially-positioned near-synchronous satellite operated by the U.S. Defense Communication System to provide limited wideband digital communications.

satellite, INTELSAT A range of communications satellites used for international and domestic telecommunications services.
INTELSAT IV satellites carry 4000 voice circuits plus 2 TV channels.
INTELSAT IV A satellites carry 6000 voice circuits plus 2 TV channels.
INTELSAT V satellites carry 12,000 voice circuits plus 2 TV channels.
INTELSAT VI satellites will carry 33,000 voice circuits plus 4 TV channels.

satellite, Molniya A series of Soviet communications satellites, the first of which was launched on April 23, 1965. All have been in 12-hour non-synchronous orbits.

satellite, Nimbus One of the series of weather satellites.

satellite, passive communication A purely reflective satellite without its own power supply.

satellite, stationary A synchronous satellite with an equatorial, circular and direct orbit. A stationary satellite remains fixed in relation to the surface of the primary body.

satellite, stationkeeping A satellite, the position of the center of mass of which is designed to follow a specified law, either in relation to the positions of other satellites belonging to the same system or in relation to a point on Earth. The Earth point can be fixed or can move in a specified manner.

satellite, sub-synchronous (super-synchronous) A satellite for which the mean sidereal period of revolution about the primary body is a submultiple (an integral multiple) of the sidereal period of rotation of the primary body about its own axis.

satellite, synchronized A satellite controlled so as to have a nodal period equal to that of another satellite, or to the period of a given phenomenon, and to pass a characteristic point in its orbit at specified instants.

satellite, synchronous A satellite for which the mean sidereal period of revolution is equal to the sidereal period of rotation of the primary body about its own axis.

satellite, Telstar One of AT&T's active communications satellites.

satellite, weather Man-made satellite designed to send information back to Earth about cloud conditions.

satellite acquisition The process of lining up a ground station with a telecommunications satellite so that the antenna locks on and tracks the satellite.

satellite availability Probability that a satellite will be on station and available for a particular task.

Satellite Business Systems (SBS) IBM-owned domestic long distance carrier, U.S.

satellite communications The use of orbiting satellites to relay transmissions from one earth station to another or to several other earth stations.

satellite conjunction The passing of a man-made satellite directly in front of the sun or moon. This has the effect of introducing much interference into the received signals.

satellite downlink Microwave radio link from a satellite to a ground station on earth.

satellite earth station A complete ground station designed to work to a satellite system and to interconnect this system with a terrestrial system.

satellite earth terminal That portion of a satellite link which receives, processes and transmits communications between the Earth and a satellite.

satellite eclipse The blocking of the path to a satellite by another body so that the solar cells receive no power and the satellite must run its systems on its own batteries.

satellite handover The transfer of radio links from one satellite to another. This is of great importance for nonsynchronous satellites which do not hover over one point on Earth but instead move around the Earth so that before one satellite sets another has risen. The ground stations concerned all change from one satellite to another.

satellite interconnect facility The interface equipment at a ground station which enables terrestrial services to interwork with satellite systems.

satellite master antenna TV Distribution of TV signals to users via a small dish receiver feeding by cables to subscribers in the service area concerned.

satellite multiple access The ability of a satellite to work simultaneously with more than one ground station.

satellite mutual visibility Zone within which a satellite is visible from two or more ground stations.

satellite office (exchange) A local end office (or exchange) on a low level of the telephone network hierarchy, which is associated with another local office and which has no route switching functions except those towards the associated higher level local office. A satellite exchange normally has the capability to connect local subscribers' lines terminating in it, without switching through to the higher level exchange.

satellite operation System with a subordinate PBX operating at a remote location, connected to the main PBX by tie lines.

satellite orbit, equatorial An orbit in a equatorial plane.

satellite relay An active or passive satellite repeater that relays signals between two Earth terminals.

satellite space sector That part of a satellite communications system which is physically part of the satellite itself. The ground sector, located on Earth, is the other part of the system.

satellite system, DCSS U.S. Department of Defense satellite system.

satellite uplink Microwave link from a ground station to a satellite.

saturation Condition in which any further change of input no longer results in appreciable change of output.

save Postponement of recognition of a signal when a process is in a state in which recognition of that signal does not occur.

SAW chip Surface acoustic wave device which can act as a frequency selective element and can operate as a delay line, a resonator or a filter.

saw-toothed wave Wave shape with a slow linear rise and an abrupt fall. It is used to produce line scan for TV.

SB A submarine cable system type designation. SB designates the design of the first transatlantic telephone cable system. SD, SF, and SG represent later generations of development.

scan 1. To sample periodically, as with traffic recorders. 2. To sweep an electron beam, an optical device, or a radar signal in a repeated pattern.

scanner 1. A device that performs a scanning operation. 2. In facsimile systems, that part of the transmitter which systematically translates the densities of the subject copy into a signal.

scanner, traffic Device which, at specified intervals, tests circuits or equipment and is able as a result to provide such information as the traffic carried, traffic overflowed and lost, register occupancy, and call holding times.

scanning Periodic examination of the activity taking place in incoming and outgoing lines to determine whether or not further processing is required. 2. In television, facsimile and picture transmission, the process of analyzing successively the colors and densities of the subject copy according to the elements of a predetermined pattern.

scanning line (TV) A single continuous narrow strip of the picture area containing highlights, shadows, and halftones determined by the process of scanning.

scanning line length In facsimile systems, the total length of a scanning line. It is equal to the spot speed divided by the scanning line frequency. This is generally greater than the length of the available line.

scanning rate In facsimile systems, the rate of linear displacement of the scanning spot.

scanning spot In facsimile systems, the area on the subject copy viewed instantaneously by the pickup system of the scanner.

scanpoint Small ferrite core, threaded by sense and interrogate wires. It is used in some electronic switching systems as an off-hook detection device.

SCARAB Unmanned miniature submarine used for locating submarine cables, to effect repairs.

scatter 1. The process whereby the direction, frequency or polarization of waves is changed when the waves encounter one or more discontinuities in the medium which have lengths on the order of a wavelength. 2. The term is frequently used to imply a disordered change in the incident energy.

scatter, forward See forward scatter.

scatter, ionospheric See ionospheric scatter.

scatter, tropospheric See tropospheric scatter.

scatter propagation Broadband radio communication over distances much greater than line-of-sight, accomplished by beaming powerful radio signals to layers above the earth which reflect or refract the signals forward to their destined stations.

scattering 1. In radio, the diffusion of radio waves when they encounter air masses or ionized layers in the troposphere or ionosphere. The new waves so produced have random direction and polarization. 2. In optoelectronics, the deflection of light from the path it would follow if the refractive index were uniform or gradually graded. Scattering is caused primarily by microscopic or submicroscopic fluctuations in the refractive index. Scattering is the principal cause of the attenuation of an optical waveguide. 3. In acoustics, the irregular diffuse dispersion of sound in many directions, caused by reflection, refraction or diffraction.

scattering region In the ionosphere, the elliptical patch in the ionized E layer by day and F layer by night which is illuminated by the beam from a transmitting antenna and toward which a receiving antenna is directed. In the troposphere, the generally similar elliptical patch, about 10 km above ground, in which air turbulence causes radio wave refraction.

schedule, circuit Schedule of the transmission characteristics of a circuit in order to indicate its suitability for use for any particular data or wideband service.

schematic A circuit diagram, usually in a simplified version indicating functions rather than complete details of components and wiring.

scheme A design proposed for implementation, especially for civil works (including construction of ducts and manholes), main (primary) cable works or distribution (secondary) cable works.

Schering bridge An ac bridge used for measuring capacitance.

Schmitt trigger A bistable circuit which, in effect, sends out 0 until the input level has reached a specified value. It then continues to send 1 as long as the input exceeds that level. It is, therefore, analog-in, binary-out in the simplest possible manner.

Schottky barrier The reduction in the minimum energy needed to liberate an electron from a semiconductor when the surface is in direct contact with a metal surface. This effect is the basis of action of many solid-state devices.

Schottky circuit Circuit which takes advantage of the Schottky effect which occurs in junctions between metals and semiconductors and results in an increase in the operating speed of the device.

Schottky effect See Schottky barrier.

Schottky transistor A bipolar transistor with a Schottky diode internally connected between base and collector, to provide increased switching speed.

Schottky TTL A family of logic gates which uses Schottky transistors to provide very rapid switching. Typical propagation delays are 3 ns (7 ns for low-power versions) compared with 10-30 ns for non-Schottky ttl gates.

Science Research Council (SRC) Official body in Great Britain, which is involved with satellite design and with the funding of space projects.

scintillation In radio propagation, a random fluctuation of the received field about its mean value. The deviations are usually relatively small. The effects of this phenomenon become more significant as frequency increases.

scope A cathode ray oscilloscope.

scrambler A device that provides privacy to telephone communications by distorting the signals so as to be incomprehensible; a similar device at the receiving end "unscrambles" the signals.

scratch pad A read/write random access memory space used for the temporary storage of data; the working area of a memory unit. Scratch pad memories are often high-speed integrated circuits which are addressed as internal registers.

screen 1. A metal mesh or sheet used to isolate a device from external fields. 2. A thin metallic wrapping, such as aluminum foil, used to isolate cable pairs from interference.

screen grid Grid in an electron tube which improves performance of the tube by shielding the control grid from the plate.

screening A telephony function that determines the eligibility to complete a call as dialed, based on class-of-service information associated with the line.

screening, tandem Examination in a dial office of the calling subscriber's number to check that his class of service entitles him to make the toll call which he has dialed.

screening, tandem code Barring of access to certain trunks for incoming calls on specific trunk groups.

screw, cable suspension A 4-inch long wood screw with a stud about 1-inch long extending from its head. The screw can be screwed into a pole. The protruding stud is then used to support a suspension clamp holding an aerial cable and its messenger strand.

screw, Phillips Screw with an indented cross, rather than a slot, in its head, much used for communications equipment.

screw eye, insulated A screw eye with its porcelain lined eye slotted, so that drop wire may be inserted without the need for threading.

sea-earth cable Cable connecting the earth terminal of the power feed equipment with the sea-earth electrode of a submarine cable system.

sea-earth electrode Electrode, or set of electrodes, connected to the end of a sea-earth cable. It is designed to free the sea-cable system from station earth potential disturbances, and is generally essential in direct current cable telegraphy using submarine cables.

sea plow Cable plow used to place underwater cable.

sealing, duct The closing up of a spare duct where it enters a manhole or cable vault. Various types of duct sealing are available.

search pattern The order of choice in which groups of toll circuits are tested in an alternate routing system.

seat, pole Steel seat which can be attached to a pole for use when cable specialists have to work for long periods on a pole-top cable terminal or distribution point.

SECAM The French-designed color television system used in France, Russia, and in some other countries. It is an acronym from the words sequential couleur a memoire.

second The second is the SI unit of time. It is the duration of 9,192,631,770 periods of the radiation corresponding to the transition between the two hyperfine levels of the ground state of the cesium-133 atom.

second order beat An unwanted carrier created by two separate carriers beating against each other. These beating carriers may have the same or different frequencies.

second window Characteristic of an optical fiber having a relatively high transmittance surrounded by regions of low transmittance in the wavelength range 1,200 to 1,350 nanometers.

secondary The output winding of a transformer.

secondary cell See cell, secondary.

secondary center A switching center at the second level in the long distance call routing hierarchy. A Class 2 office.

secondary channel A data transmission channel having a lower signaling rate capability than the primary channel in a system in which two channels share a common interface.

secondary distribution network See distribution network.

secondary electron Electron driven from a material as a result of bombardment by other electrons or heavier atomic particles.

secondary frequency standard A frequency standard which is calibrated with respect to a primary frequency standard. The term describes the position of the standard in a hierarchy, not the quality of its performance.

secondary grade of reception quality In the broadcasting-satellite service, a quality of reception of emissions from a broadcasting-satellite space station which is subjectively inferior to the primary grade of reception quality but is still acceptable.

secondary route The circuits to be used when the primary routes are congested. In manual and semiautomatic operations, secondary routes may also be used when the transmission on the primary route is not of sufficiently good quality, or if traffic is to be handled outside the normal hours of service on the primary routes.

secondary station In a data communication network, the station responsible for performing unbalanced link-level operations, as instructed by the primary station. A secondary station interprets received commands and generates responses.

secondary switching center A switching center at a level above the primary centers in the national switching hierarchy.

secondary time standard A time standard which requires calibration.

secondary trunk An intra-office trunk leading to a secondary level switch.

secondary voltage 1. Voltage used for power distribution to premises, in America usually 110 V at 60 Hz. 2. Operating voltage on the load side of a transformer.

secondary winding Output winding of a transformer. It receives power by induction from the current flowing in the primary winding.

secretarial unit Type of telephone instrument with keys and switches to enable a secretary to filter calls to one or more executives.

section, building-out A group of small capacitors which may be built-up to the best total value for a particular installation, in order to bring the capacitance of an end section of cable up to a standard value.

section, cable regulator Length of repeatered cable, including several repeaters, over which the overall transmission loss is adjusted by automatic checking against the level of a pilot signal.

section, duct Length of duct route between adjacent jointing chambers or manholes. Also called a manhole section.

section, loading Length of cable between adjacent loading coils in a loaded cable system.

section, regenerator See regenerator section.

section, regulator Section of cable controlled by a pilot regulator.

section, repeater See repeater section.

section, switchboard Old-type manual operating suites were normally made in sections of up to three operating positions.

Sectional Center The next to the highest rank (Class 2) toll switching center which homes on a Regional Center (Class 1 office).

secure voice Encrypted voice signals.

secure voice cord board (SECORD) A desk-mounted patch panel which provides the capability for controlling 16 wideband (50 kb/s) or narrowband (2400 b/s) subscriber lines and five narrowband trunks to AUTOVON or other Defense Communications System narrowband facilities.

security, communications (COMSEC) Procedure for the processing of telecommunications traffic signals in such a way as to make eavesdropping either impossible or prohibitively expensive.

security and fault recovery Security is the means provided within a system for allowing it to carry on working when parts of it have failed. Fault recovery is the process of reconfiguring the system or omitting parts of it which the system has discovered are faulty.

security arrangements In a common channel signaling system, the measures provided to ensure continuity of service of the signaling system in the event of the failure of one or both of the data channels.

security techniques General term covering procedures such as redundancy in design, replication, the encryption of messages, the use of passwords and personal identity numbers (PINs) and the scrambling of voice signals.

seek time   The length of time required to find a record of data, generally used in reference to disk files.

segment   1. To divide a program into parts so that each may be stored separately but with the necessary instructions to link the different parts together when needed. 2. Basic charging unit for packet-switched data: a packet of up to a maximum of 64 octets of data, i.e., 512 bits. See kilosegment. 3. That portion of a message that can be contained in a buffer.

segmented encoding law   An encoding law in which an approximation to a smooth law is obtained by a number of linear segments. Syn.: piece-wise linear encoding

seize   To access a circuit and make it busy in order to prevent others from seizing it.

seizure   A successful bid.

seizure signal   In telephone systems, a signal used by the calling end of a trunk or line to indicate a request for service. This is usually a sustained off-hook transmitted from calling end towards the called end of a circuit.

selectance   Measure of the falling off in the response of a resonant device with departure from resonance.

selection   1. The process by which a computer contacts a peripheral or station to send it a message. 2. The process of indicating the number of the terminal being called. In telephony this is synonymous with dialing but the term "selection" is now preferred by some administrations as the called number may be indicated by other means, e.g., pushing buttons on a key pad.

selection, conjugate   Method of establishing a connection through a switching network, whereby the link(s) which can connect to the calling inlet are seized in a single logical operation, which can also include the choice of the outlet.

selection position   See decision instant.

selection signals   The sequence of characters which includes all the information necessary to establish a data call. Selection signals consist of two elements: the facility request and the address. In some cases one of the two elements may be omitted.

selection stage   An aggregate of switches enabling an inlet to access one of a number of outlets and designed to operate as a single unit from a traffic handling standpoint.

selective accounting   The process whereby the account to which charges for a particular call are to be debited is identified when the call is made.

selective calling   See calling, selective.

selective combiner   A circuit or device for selecting one of two or more diversity signals in such a way that only the signal having the most desirable characteristics is used.

selective editing   The ability of some facsimile equipment to send only part of a complete page, e.g., a paragraph or line.

selective fading   Fading at different rates of signals of different frequencies within the same modulation envelope.

selective ringing   See selective calling.

selective signaling   A selective calling method by means of which only one bell is rung on a party line. Frequency timed ringing can select and ring the bell at one out of eight stations across a line.

selective trunk reservation   A network management feature which dynamically observes the occupancy of an outgoing trunk group and applies level filters (i.e., constraints) to offered traffic at high levels of group occupancy.

selectivity   Ability of equipment to select and operate upon a signal of a particular frequency despite the presence of other signals at frequencies close to that to which the equipment is tuned.

selector   An electromechanical switching device in a central office. The most common types are rotary, two motion and crossbar.

selector, absence-of-ground searching   A selector which hunts for idle trunks in a step-by-step office. It has wipers which rotate until they reach an outlet without a ground condition; the switch stops at this point and puts the line through to the next switching stage.

selector, battery searching   A selector which hunts for idle trunks by searching for the presence of full or resistance battery voltage. Only when this is identified is the line put through.

selector, digit absorbing   A step-by-step Strowger selector with the ability to mark levels so that after the selector has received and analyzed the first digit, it can, if necessary, drop to the start position again to receive the remaining digits.

selector, Gill   Selector switch used on some railway control waystation phone circuits. All switches step as the first digit is sent out from control, but only the ones which are programmed for this digit stay on the line to receive the second digit. The third digit narrows the selection so that only one bell is actually rung.

selector, group   In a step-by-step office, a selector which switches traffic to a level representing a group of trunks corresponding to the early digits of the telephone number. A group switch.

selector, MDA   (multiple digit absorbing) A step-by-step Strowger selector with the ability to mark levels and act on the digits received in accordance with the programs strapped, eg: cut into bank, absorb and start again, absorb repeatedly, etc.

selector, special third   Selector giving access to service lines after only three digits have been received. Examples: 113 for Information, 114 for Repair Service.

selector, step-by-step   An electromechanical switch in a dial central office which operates directly under the control of the subscriber's dial pulses, stepping to the level dialed then hunting for an idle trunk on that level.

selector, Strowger   A step-by-step selector named after the American inventor of automatic telephony.

selector, X-Y   A step-by-step switch in which wipers and banks are in the same plane. The wiper first goes along the required set of contacts, then advances into the bank to hunt for an idle trunk.

selenium rectifier See rectifier, selenium.

self-bias Provision of bias in an electron tube by the voltage drop in the cathode circuit.

self-checking code Syn.: error detecting code

self-excited Provision by a generator of current for its own field coils.

self-focusing fiber A graded index optical waveguide.

self-impedance The ratio of the applied voltage to the resultant current at a pair of terminals of a network, all other terminals being open.

self-inductance Property of a coil or other circuit element which determines the electromotive force induced in the circuit by a given rate of change of current in the circuit.

self-induction Generation of an opposing electromotive force in a circuit by changing the current flowing in the circuit.

self-interrupted Describing a self-driven step-by-step switch which steps around to hunt for an idle trunk without requiring additional incoming digital signals.

selsyn A self-synchronous rotary switch used to convert mechanical position into electrical signals and vice versa. It is used for remote setting of directional devices.

semi-automatic system A system in which the calling subscriber's order is given to an operator who completes the call through automatic switches.

semiconductor A material whose resistivity is between that of conductors and insulators, and can sometimes be changed by light, an electric field, or a magnetic field. Current flow can be either by movement of negative electrons or by transfer of positive holes. Semiconductors are used in a variety of solid-state devices including transistors, integrated circuits and light-emitting diodes. Silicon and germanium are at present the most commonly used semiconductor materials.

semiconductor, extrinsic Semiconductor whose properties are dependent on the impurities in the material.

semiconductor, intrinsic Semiconductor with no significant amount of impurities. A pure crystal. Also called I-type.

semiconductor, n-type A semiconductor material which has a small amount of impurity, such as antimony or arsenic, added to increase the supply of free electrons. Such a material conducts electricity through movement of electrons.

semiconductor, p-type A semiconductor material which has been doped so that it has a net deficiency of free electrons. It therefore conducts electricity through the transfer of positive holes.

semiconductor device A device, such as a transistor, which utilizes the properties of a semiconductor.

semiconductor integrated circuit A combination of several interconnected circuits mounted on a semiconductor substrate.

semiconductor laser An injection laser. A laser in which lasing action occurs at the junction of n-type and p-type semiconductor materials. Semiconductor lasers are about the size of a grain of sand and are therefore size-compatible with optical waveguides.

semiconductor memory Memory with storage elements formed by integrated semiconductor devices. Semiconductor read/write memories are usually low cost but data volatile. Semiconductor read only memories are nonvolatile.

semi-duplex Describing a communication circuit where one end is duplex and the other end (usually a minor or mobile station) is simplex.

semi-electronic switch A switch whose matrix is composed of some form of relay and whose common control equipment is electronic.

semi-permanent connection A connection established part of the time for one user. At other times the connection may be released and made available for use in handling traffic of the switched network.

semi-permanently switched circuit A circuit carried through the switch block of a digital switching office such that a fixed connection is maintained in relation to the circuit. Private circuits are one example of its use.

semi-postpay coin telephone A coin telephone combining post payment local automatic services with prepayment toll service. Numbers can be dialed without inserting coins but conversation is not possible until coins have been deposited. Operator-checked deposition of coins is necessary before toll calls are connected.

send/receive 1. Switch which affects changeover of direction of operation of a device. 2. Sometimes teleprinters (at a computer's man-machine interface) which have authority to input commands as well as to receive responses.

sender Equipment in a signaling terminal which outpulses routing digits and the called subscriber's number.

sender, incoming Equipment which receives incoming 10 i p s or multifrequency pulses and obtains all necessary code translations before outpulsing digits in correct form direct to trunks or to outgoing senders.

sender signal Light signal on a manually operated switchboard which indicates when a sender has been associated with the trunk so that digits may be outpulsed.

sending-end crossfire In teletypewriter systems, the interfering current in a channel from one or more adjacent teletypewriter channels transmitting from the end at which the crossfire is measured.

senior multiple Type of switchboard position used when multiple repeats every fifth panel. There are two jack panels per operating position.

sensation, auditory Perception of sound by the human ear.

sense To detect the presense or absence of a tone in a signal or on a tape, or of a punched hole in a tape or card.

275

sense wire One of the wires threaded through memory cores to detect the state of the store.

sensitivity 1. In a radio receiver or similar device, the minimum input signal required to produce a specified output signal having a specified signal-to-noise ratio. 2. Of a voltmeter or other meter, the current required to produce a full-scale deflection. 3. In optoelectricity, an imprecise synonym for responsivity.

sensitivity, klystron deflection The linearity of a klystron. The variation with frequency of the ratio of frequency deviation to repeller voltage.

sensitivity, maximum usable Of a linear radio receiver, the larger of the minimum input signal levels (expressed as the electromotive force of the carrier) which must be applied in series with the specified source impedance (dummy antenna) to the input of the receiver to produce at the output: (a) the signal level or (b) the signal-to-noise ratio necessary for normal operation when the normal degree of modulation is applied to the carrier.

sensitivity, optical receiver The optical power required by a receiver for low-error signal transmission. In the case of digital signal transmission the mean optical power is usually quoted at which a bit error rate of 10^{-9} is attained.

sensitivity, radio receiver Minimum input signal, in microvolts, at which the receiver will have a specified output signal with an acceptable signal-to-noise ratio.

sensor Detection device which is sensitive to changes in level or state.

separate channel signaling Signaling arrangement on a carrier system whereby the signaling leads for all channels are combined and multiplexed on one of the voice channels.

separately excited Describing a generator whose field current is fed from a different source.

separation filter A double-section filter used to separate one frequency band from another, usually having a high-pass section together with a low-pass section.

separations The division of telephone company costs and revenues between interstate and intrastate jurisdictions for regulatory purposes.

separator A character in man-machine language used to delimit syntax elements.

separator, battery Sheets of nonconducting porous material between positive and negative plates in a storage battery. They prevent short circuits caused by plates bending and touching.

sequence The order in which bits, bytes, records or files are arranged.

sequence control register A register holding the address of the next instruction.

sequencing The process of dividing a message into smaller frames, blocks or packets for transmission.

sequential Events which occur in a specific time order or code order.

sequential answering Answering incoming calls at a switchboard in the order of arrival, as used in various automatic call distribution systems.

sequential pulsing Transfer of digital information by interrupting a steady tone or a direct current. The number of interruptions indicates the digit.

sequential transmission Information transfer method in which the bits making up a character are sent one after another, in sequence. Syn.: serial transmission

sequentially controlled automatic transmitted start (SCATS) A multipoint teletypewriter arrangement providing automatic message transmission between all terminals on the system without station contention.

serial 1. Said of pulses which are sent separately, one after the other. The opposite of "parallel." 2. An arrangement whereby one element of data is linked to the next so that progress must proceed from the first element through the next, then the next, etc., without skipping.

serial access 1. Pertaining to the sequential or consecutive transmission of data to or from storage. 2. That process wherein data are obtained from, or entered into, a storage device in such a way that the process depends on the location of those data and on a reference to data previously accessed.

serial data transmission A transmission method where data characters or bytes are transmitted one bit at a time over a single path.

serial-to-parallel converter 1. A device that accepts a single time sequence of signal elements and distributes them among multiple parallel outputs. 2. A device that accepts a single time sequence of signal states representing data and translates these states into a spatial distribution of digits, all of which are presented simultaneously. Also called a deserializer or staticizer.

serial transmission See transmission, serial.

serializer 1. A device which converts a group of digits, all of which are presented simultaneously, into a corresponding sequence of signal elements. 2. A parallel-to-serial converter.

series, number In telephony, a group of ten thousand or less numbers, from 0000 to 9999, which may be associated with up to three central office codes.

series circuit See circuit, series.

series completion A service-related telephony feature that permits calls to a busy line to be routed to another specified directory number.

series connection See connection, series.

series excitation Situation in which the field current is in series with the armature winding in a rotating machine.

series feed 1. Power supply arrangement such that the same current flows through all the units concerned. 2. Describing a radio transmitter or receiver connected to one end of an antenna.

series multiple Method of wiring jack strips on small switchboards so that insertion of a plug in one jack breaks the connection to other multiple appearances.

series-parallel connection Connection of cells or other components in such a way that several series-connected groups are themselves connected in parallel.

series wound Field winding and armature windings connected in series.

service In telecommunications, the sum of all acts and facilities necessary to exchange information.

service, extended area See extended area service.

service, fixed Radio communication service between two fixed stations.

service, flat-rate Telephone service with a monthly fee which permits unlimited local calls.

service, foreign exchange Connection of a telephone subscriber to a central office other than the one which normally serves his area.

service, full period A facility provided on a 24 hours per day basis.

service, grade of See grade of service.

service, measured rate See measured rate service.

service, mobile Telephone service made available to vehicle-mounted or boat-mounted radiotelephone terminals.

service, private line Service employing a point-to-point, nonswitched line for the use of one customer.

service, short period Special type of private line service established for use during short periods on specified days, rather than on a permanent basis.

service, standard frequency Transmissions at accurate radio frequencies, modulated by accurate audio frequencies and made from various radio transmitting stations under the control of the National Bureau of Standards.

service, telephone Installation and maintenance of telephones to provide communications needs.

service, teletypewriter exchange (TWX) See teletypewriter exchange service.

service access code Code of "NOO" type (N is any digit 2 thru 9) which replaces the usual NPA code in accessing special services provided by different carriers.

service arc The arc of a geo-stationary satellite orbit within which the space station could provide the required service to all of its associated Earth stations in the service area.

service area The geographic area served by a telecommunications provider.

service bit An overhead bit, such as one used for a numbering sequence, which is not a check bit.

service bureau An office where a user can lease processing time on a processor and its peripheral equipment.

service channel See channel, service.

service code A numerical code designating a supplementary telephone service.

service digits Digits which are added at regular time intervals to a digital signal in order to enable the equipment associated with that digital signal to function correctly. Also called housekeeping digits.

service indicator Information within a common channel signaling message identifying the user to whom a message belongs.

service information Eight bits, contained in a message signal unit, comprising the service indicator and the sub-service field, in the CCITT common channel signaling system.

service observation Statistical measurement of the quality of service provided for telephone subscribers. Some service observation work is done by senior operators who time and record such features as time taken for an operator to answer. Some is done by computers.

service probability The probability of obtaining a specified grade of service, or better, during a given period of time.

service sector Of a directional antenna, the horizontal sector containing the main beam of the antenna radiation and including the direction required for service.

service signals Signals that enable data systems equipment to function correctly. Syn.. housekeeping signals

service specific plant Plant designed and installed to support only one type of service.

servicing Inspection, readjusting and repairing of equipment to keep it working satisfactorily.

serving A wrapping of wire or jute around a cable to stop armor wire strands from unraveling or give a measure of extra protection.

serving area 1. Region surrounding a radio broadcasting station where the signal strength is at or above a stated minimum. 2. The geographic area handled by a local telephone central office or exchange.

servo system Electromechanical system for relaying positional or angular information.

session A connection between two stations that allows them to communicate.

session layer In the OSI model, the network processing layer responsible for binding and unbinding logical links between end users and maintaining an orderly dialog between them.

set, antisidetone See antisidetone telephone set.

set, common battery Telephone set powered from the central office in such a way that no local battery is needed to provided microphone current.

set, composite Arrangement of capacitors and chokes which uses each of the two individual legs of a two-wire telephone circuit to provide dc telegraph circuits.

set, current-flow test Simple meter assembly which enables relays to be adjusted to operate at designated current levels.

set, data See data set.

set, explosion-proof telephone A mine telephone so designed that no sparks can be generated which might cause fires.

set, fire-reporting telephone Telephone, usually with no provision for incoming calls, connected directly to a fire-control center and mounted conspicuously to facilitate rapid notification of fires.

set, four-wire A hybrid type unit interconnecting a four-wire circuit and a two-wire circuit together with their signaling facilities.

set, hand telephone Telephone designed for holding in the hand as contrasted with a desk or loud speaking telephone.

set, head telephone Microphone and receiver set equipped with a cord and plug as worn by telephone operators.

set, key telephone See key telephone set.

set, lineman's test A portable telephone and basic test set used by linemen for outside work.

set, local battery A telephone which relies on a local power supply to provide microphone current.

set, loudspeaker A loudspeaking telephone used primarily to enable more than one person at one end to participate in telephone discussions.

set, magneto telephone See magneto telephone.

set, noise measuring Test set used for noise measurement. It usually incorporates weighting networks of appropriate types.

set, operator's telephone An operator's headset, with lightweight receiver and transmitter designed for continuous wearing.

set, sidetone telephone A telephone instrument designed with no balancing network so that all the speech power being sent out to line also goes through the local earpiece receiver.

set, sound-powered telephone Telephone instrument with a microphone which generates voice-frequency ac without requiring an externally fed microphone current.

set, telephone A complete telephone instrument including the bell.

set, telephone answering Tape-recorder-based unit which can give callers a brief message when the subscriber himself is unavailable to answer. Some types can also record incoming messages.

set, Touch-Tone telephone Bell system name for a dual-tone multi-frequency push button dialing telephone.

set, wall telephone Telephone instrument designed to be mounted on a vertical surface. Some models can be countersunk, leaving only push-buttons and the answering key standing clear of the surface, with the loudspeaker and the microphone both behind grills.

set, weather-proof telephone Telephone instrument in a weatherproof casing, designed for outside use.

settlements Procedure by which revenues are distributed between local telephone companies and long-haul carriers.

seven layer system The Open Systems Interconnection (OSI) protocols permitting general interconnection of computers and data terminals regardless of manufacturer.

sf signal unit See single frequency (SF) signaling system.

shaft A rotating axis.

shape To change a waveform by filtering or restricting the bandwidth.

shaping, pulse Passing a square wave pulse through a filter network to remove the high frequency signals, which produce the sharp rise and fall of the pulse, without restricting the fundamental frequency of the pulse itself. This enables the signal to be transmitted more economically, using a relatively narrow bandwidth.

shaping network See network, shaping.

shared access line Group of lines from a public central office to a switching center of an other common carrier (OCC). This is normally a hunting group of circuits using a single access code or directory number and is used to obtain dial-in access to an OCC network.

shared disc A magnetic disc that may be used for information storage by two or more systems at the same time.

shared service Use on a shared basis by more than one telephone customer of the same pair of wires leading to the central office.

shared tenant services Centralized telecommunications services provided for all the occupants of a building or complex.

shaving Process in submarine coaxial cable manufacture which involves sizing the extruded dielectric precisely within tolerances as specified, and the required degree of concentricity with respect to the center conductor, before application of the outer conductor.

shears, cable Strong shears used to cut salvaged cables into short lengths for ease of handling.

sheath The outer covering of a cable. Its main function is to provide protection for the insulated conductors which make up the cable.

sheath-kilometers Total lengths of cables involved (i.e., not dependent on the pair-capacity of the cables).

sheave, manhole Large diameter pulley used when feeding cables into ducts via manholes.

sheaves Grooved wheels or pulleys. In telecommunications: 1. The wheels at the bow of all cable ships, and at the stern of some, over which the cable passes in laying or recovery operations. 2. Variously sized grooved wheels on other devices designed to carry cable.

sheet, cut Written instructions defining work to be done at a cable joint, flexibility point or main distribution frame.

shelf, distribution frame The horizontal side of a main distribution frame in a central office.

shelf, plug The horizontal section of a switchboard position, with rows of holes supporting pairs of plug-ended cords.

shelf, switch A shelf, found particularly in an electromechanical central office, which carries multiway jacks into which selectors or other switches may be plugged.

shelf life Time interval during which components may be safely stored without losing quality.

278

shield Metal covering protecting sensitive equipment from external magnetic or electric fields, or surrounding equipment which might otherwise produce interference in other units.

shield, cable Metallic layer around a cable core designed to minimize interference from external sources.

shield, manhole Cylindrical extension placed above a manhole entrance designed to minimize entry of rain or snow while cable testing or splicing is being performed in the manhole.

shielded pair Pair of conductors in a cable which are wrapped with metallic foil or braid designed to insulate the pair from interference and provide high quality, noise-free transmission.

shielding The provision of covering materials such as iron or copper tapes or lead extrusions on the outside of a coaxial submarine cable, under the armor, in order to reduce electromagnetic interference.

shift The movement of the bit pattern of a word, either to the left or to the right.

shift, case Key and signal which change a teletypewriter from letters to figures.

shift, cyclic A shift whereby the bits that leave one end of the word re-enter at the other so that the pattern moves in a cycle.

shift, logical A shift whereby the bits that leave the end of the word are lost.

shift character Control character which determines whether alphabetic or numeric form will be used in printout, or whether lower or upper case characters will be produced.

shift-down modem Modem used with digital facsimile equipment which permits the bit rate to be reduced (e.g., from 9.6 kbit/s to 2.4 kbit/s) in order to suit the transmission characteristics of a particular line.

shift register A storage device in which a serially ordered set of data may be moved, as a unit, a discrete number of storage locations. Shift registers may be configured so that the stored data may be moved in more than one direction or entered from multiple inputs.

shifter, strand A winch used to pull up an aerial cable which is supported on a strand wire.

shiner End of a wire from which the insulation has been stripped just prior to the wire being terminated.

ship station identity A nine-digit number giving a ship's identification as transmitted in the maritime mobile satellite service. The first three digits, the national identification digits or NIDs, define the nationality of the ship.

ship station number Number which identifies a ship for access from a public telephone network and forms part of the international number to be dialed by a public network subscriber.

shock, acoustic Physical pain or discomfort caused by extremely loud or persistent noise. Operators' headsets are usually protected by limiting devices so that amplitudes of incoming signals are restricted.

shock excitation Impulse excitation. The application of a steep-wavefront voltage to a circuit, inducing oscillation.

shock mounting Suspension cradle used for insulation of delicate instruments from physical or mechanical shock.

shoe, reversing Device which can be inserted on a main distribution frame in a central office in order to reverse the connection of wires in a pair leading from the equipment to a distribution cable.

shoe, test Device, usually spring loaded, to give contact with the lugs on a main distribution frame protection strip or cable termination strip.

shore station Earth station on land working via a satellite to ships at sea using the maritime mobile satellite service.

shore-end cable Sea cable with heavy armor for mechanical protection in shallow water, and containing shielding to reduce electromagnetic interference.

short A short circuit. A direct low resistance connection between conductors not normally in contact with each other.

short circuit See short.

short-haul Type of equipment and tariff used for circuits of length normally less than 25 airline-miles.

short-haul carrier See carrier, short-haul.

short-period circuit See circuit, short-period.

short wave A layman's term for high frequency band radio. Radio in the 10 meter to 100 meter band.

shorted A circuit prevented from operating normally because two or more conductors have provided a low resistance path, often in parallel with the device concerned, which stops operation.

shot noise Noise caused by current fluctuations due to the discrete nature of charge carriers and random and/or unpredictable emission of charged particles from an emitter.

shunt 1. A parallel and alternative path for current. 2. A resistor of known value connected in parallel with a meter to enable the meter to be used to measure current values higher than those which would otherwise provide a full scale deflection for the meter.

shunt, instrument Accurate resistor of low resistance connected in parallel across a millivoltmeter to enable the meter to be used to measure current in amperes.

shunt, magnetic Moveable iron bar which can be placed in such a way as to give an alternative path for a magnetic field, thus reducing the flux through its usual path.

shunt feed Applying plate voltage to an electron tube through a high inductance coil but taking the output from between the plate and the coil.

shunt-wound Describing a motor or generator with armature and field coils connected in parallel.

shuttle The act of routing a toll call back on the same group of circuits as it came in on. It is an incorrect routing.

shuttle, space Reusable U.S. spacecraft which first orbited the Earth in 1981. Used to lift telecommunications satellites from earth and launch them into geostationary orbits.

SI units See Appendix B.

SID (scheduled issue date) Term used in the Bell System's Universal System Service order. It is the date that the primary order document is to enter the order distribution system.

side, field See field side.

side circuit A one-pair metallic circuit, used to provide one leg of a phantom circuit by center-tapping a 1:1 ratio repeating coil in the metallic pair.

side stable relay A polar relay that remains in the last signaled contact position.

sideband Frequency bands on both sides of the carrier frequency within which fall the frequencies of the wave produced by the process of amplitude modulation.

sideband, double Amplitude modulation with the transmission of both the upper and lower sidebands.

sideband, independent Amplitude modulation of a radio communication frequency using different modulating frequencies for the production of upper and lower sidebands.

sideband, lower The sideband represented by the difference between carrier frequency and the instantaneous value of the modulating frequency.

sideband, single (SSB) Amplitude modulation in which only one of the information carrying sidebands is transmitted.

sideband, upper The sideband represented by the sum of the carrier frequency and the instantaneous value of the modulating frequency.

sideband, vestigial See vestigial sideband.

sideband transmission That method of transmission in which frequencies produced by amplitude modulation occur above and below the carrier frequency. The frequencies above the carrier are called "upper sideband" and those below are called "lower sideband."

sidereal period of revolution Of a satellite, the time elapsing between two consecutive intersections of the projection of the satellite on a reference plane which passes through the center of mass of the primary body with a line in that plane extending from the center of mass to infinity. Both the normal to the reference plane and the direction of the line are fixed in relation to the stars.

sidereal period of rotation Period of rotation around its own axis of an object in space, such as a natural satellite or a spacecraft, in a frame of reference fixed in relation to the stars.

sidetone The tone produced when one hears in one's own telephone receiver an output produced from the input signal to one's own microphone.

sidetone path Special bypass circuit in a four-wire telephone instrument which feeds into the local receiver circuit part of the signal produced by the local microphone.

siemens The SI unit of conductance which replaces and is identical in value to the "mho." It is the reciprocal of resistance in ohms.

sigma Group of telephone wires, usually most or all of the wires of a line, which is treated as a unit in the computation of noise or in arranging connections to ground for the measurement of noise or current balance ratio.

signal 1. The information that is transferred over a communications system by electrical or optical means. 2. An alerting signal. 3. An acoustic or visual device which attracts attention by lighting up or emitting sound. 4. In a Specification & Description Language, a flow of data conveying information to a process. 5. A time-dependent phenomenon carrying information. 6. As applied to electronics, any transmitted electrical impulse. 7. A type of message, the text of which consists of one or more letters, words, characters, signal flags, visual displays or special sounds with prearranged meanings and which is conveyed or transmitted by visual, acoustical, or electrical means.

signal, alerting A bell, buzzer or lamp which draws attention to a switchboard, a calling line, or an incoming telephone call.

signal, analog See analog signal.

signal, audible ringing Tone signal of 440 Hz + 480 Hz at a level of -19 dBm per frequency, which indicates that the called line has been reached and is being rung. Different frequencies (and periodicities) are used by some telephone administrations and telephone companies. Also called ringing tone.

signal, bipolar A three-state signal (+, 0, -) used for the transmission of binary code signals and utilizing positive and negative signals alternately for the same binary state.

signal, busy See busy signal.

signal, coin collect A dc signal, usually with a high positive voltage relative to ground on both wires of the loop, which is sent out from an operator's position to a coinbox to transfer coins, previously held in temporary store, into the cash collection box.

signal, coin return A dc signal, usually with a high negative voltage relative to ground on both wires of the loop, sent out from an operator's position to a coinbox to refund to the caller the coins previously held in temporary store.

signal, common audible A feature employing a single tone ringing device on a multi-line telephone instrument so that the same tone is used to indicate incoming calls on all lines.

signal, delay pulsing Signal returned to the calling end to indicate that equipment has not been allocated to the line concerned to receive the expected digits, and the calling end must wait for a "proceed to send" signal.

signal, disconnect An on-hook signal longer than 300 ms which instructs the called end to release the switches previously used for a call.

signal, error Signal used to correct a previously received incorrect signal.

280

signal, facsimile framing Signal used for adjusting the position of a picture in facsimile transmission.

signal, go A "proceed to send" on-hook signal sent back to the calling end, following the "delay pulsing" signal, as soon as the necessary signal receiving equipment has been associated with the circuit.

signal, high level digital A digital signal of the type generally used by teletypewriters, e.g. plus or minus 60 V.

signal, incoming call lamp A signal employing a lamp flashing once per second, used on some key phone systems to indicate an incoming call.

signal, intelligence See intelligence signal.

signal, line Signal indicating an incoming call. It employs a steady glow of the calling lamp associated with an incoming circuit jack.

signal, low level digital A digital signal of low current value used where high level signals could be expected to produce unwanted crosstalk between circuits. Typical voltages used are plus and minus 6 volts.

signal, monochrome Signal which controls luminance of a television picture.

signal, no circuit (NC) Low tone interrupted at 120 impulses per minute indicating that toll circuits towards the called destination are all engaged.

signal, permanent Timed signal given after a subscriber's line has gone off-hook for a specified period, usually about half a minute, without any digital information being received by the central office.

signal, pseudo-random A signal which appears to be a random number indication but which does in fact repeat itself after a substantial time interval. It is used in cryptography as a key to be added to a voice or data signal.

signal, quasi-analog A digital information signal suitable for transmission over an analog transmission path.

signal, recall See recall signal.

signal, ringing See ringing signal.

signal, ringing start Signal which starts up a ringing signal, e.g. as soon as called subscriber's line has been tested and found free, and a path established through the switch for the conversation.

signal, speech-simulated A made-up voice frequency signal sometimes used to test voice-immunity of inband signaling systems.

signal, spurious An unwanted signal.

signal, stop Signal sent back to the calling end during the transmission of digits representing the called subscriber's number, to indicate that the called end equipment is not ready to receive this information.

signal, supervisory A signal to indicate the state of a circuit. It is used in particular to indicate that the parties to a call have gone back on-hook and the connection may be released.

signal, television See television signal.

signal center A combination of signal communication facilities usually consisting of a communications center, telephone switching central, and appropriate means of signal communications.

signal constellation Two-dimensional signal-space diagram indicating modulation levels of the sine and cosine carriers in Quadrature Amplitude Modulation (QAM).

signal contrast In facsimile systems, the ratio, expressed in decibels, between white signal and black signal.

signal conversion equipment See modem.

signal converter A device in which the input and output signals are formed according to the same code, but not according to the same type of electrical modulation.

signal distance 1. The number of digit positions in which the corresponding digits of two binary words of the same length are different. 2. The number of digit positions in which the corresponding digits of two words of the same length in any radix are different. For example, the signal distance between 21415926 and 11475916 is 3. Syn.: hamming distance

signal element A portion of a telegraph or data signal that makes up the smallest unit of a signaling code. Elements are distinguished by their nature, magnitude, duration and relative position, or by some combination of these features.

signal frequency shift In facsimile systems, the numerical difference between the frequencies corresponding to white signal and black signal at any point in the system.

signal-guard circuit A guard circuit designed to provide immunity from voice-operation of inband voice frequency signaling systems.

signal level The rms voltage measured during the r-f signal peak. It is, for TV, usually expressed in microvolts referred to an impedance of 75 ohms, or in dBmV, the value in decibels with respect to a reference level of 0 dBmV, which is one millivolt across 75 ohms.

signal-plus-noise to noise ratio ((S+N)/N) The ratio of the amplitude of the desired signal plus the noise to the amplitude of the noise at a given point. It is usually expressed in decibels.

signal regeneration The restoration, to the extent practical, of a signal to an original predetermined configuration or position in time or space.

signal sample The value of a particular characteristic of a signal at a chosen instant.

signal sampling The process of taking samples of a particular characteristic of a signal, usually at equal time intervals.

signal security A combination of communications security and electronic security.

signal-to-noise ratio The ratio of the amplitude of the desired signal to the amplitude of noise signals at a given point in time. It is usually expressed in decibels.

signal transfer point (STP) In a common channel signaling system each STP examines each received signaling message to determine its destination and routes the message to the appropriate outgoing link.

signal transition The change from one signaling condition to another, for example, the change from "mark" to "space" or vice versa.

signal unit alignment A condition existing when flags are received at intervals which correspond to integral numbers of octets and which fall within certain upper and lower limits.

signal unit error rate monitoring A procedure whereby the error rate of an active signaling link is measured on the basis of a count of correctly checking and erroneous signal units.

signal units A group of bits forming a separately transferable entity used to convey information on a signaling link.

signaling 1. The use of signals for communication. 2. A method of conveying signals over a circuit. 3. The exchange of electrical information other than by speech. It is specifically concerned with the establishment and control of connections in a communication network.

signaling, battery-and-ground A dc signaling system employed on some interoffice trunks which uses battery and ground at both ends of the circuit and may thus be utilized over greater distances than is possible with ordinary loop-disconnect signaling.

signaling, carrier See carrier signaling.

signaling, CCITT No. 4, 5, 6, 7 See CCITT Signaling Systems.

signaling, closed circuit A dc signaling system using different levels of current for information signals but no complete breaks in flow of current.

signaling, common-battery See common battery signaling.

signaling, compelled See compelled signaling.

signaling, composite See composite signaling.

signaling, confirmation See confirmation signaling.

signaling, dual tone multifrequency (DTMF) See dual tone multifrequency signaling.

signaling, duplex A dc signaling system using polarized relays and the same metallic pair as is used for speech.

signaling, E & M Signaling system widely used between central offices, particularly where the speech signal is carried on a transmission system using some form of multiplexing. Two leads carry the signaling information. The M lead gives the outgoing information and the E lead receives incoming information. Usual conditions for the M lead are ground for on-hook, battery for off-hook; for the E lead they are open for on-hook, ground for off-hook.

signaling, E-type Use of a single-frequency inband tone at 2600 Hz for signaling. Control of the tone pulses can be by either E & M or loop-disconnect.

signaling, in-band See in-band signaling.

signaling, loop DC signaling using the loop circuit of a metallic pair of wires. Making and breaking the loop is a standard method worldwide for passing signals from subscriber to central office. Reversal

of battery polarity on the pair of wires is a common method for sending supervisory signals. Changing the value of the resistance in the loop to give signal information is a loop signalling method used in some special devices.

signaling, magneto Ringdown signaling using a hand-cranked magneto at the subscriber's instrument.

signaling, multifrequency See multifrequency signaling.

signaling, 1000 Hz Ringdown signaling using bursts of 1000 Hz tone, modulated at 20 Hz.

signaling, open-circuit DC loop signaling with no current flowing during idle time.

signaling, out-of-band Signaling outside of the commercial voice band of 300-3400 Hz. The most commonly used frequency is 3825 Hz, when transmission is by standard frequency division multiplexing so that signals such as periodic pulse metering charging pulses can be sent during calls without interfering with conversations.

signaling, PCM The special circuit arrangements that are made for the transmission of signals over pulse code modulation (PCM) systems, using an in-band time slot in American 24 channel PCM and a separate out-band time slot in European 30 channel PCM.

signaling, positive-negative See battery-and-ground signaling.

signaling, reverse battery Common loop signaling system utilizing making and breaking of the loop for digital address information, and the reversing of the polarity of the battery when the called subscriber goes off-hook.

signaling, ringdown See ringdown signaling.

signaling, separate channel Use of voice frequency telegraphy (VFT) techniques to carry signaling for voice circuits, in place of multiplexed voice channels.

signaling, single frequency Signaling, using a 2600 Hz tone. The normal practice is for the tone to be on when the circuit is idle, to be pulsed to give digital address information, and to be off during calls.

signaling, T-carrier Signaling used with 24-channel pulse code modulated systems, utilizing pulses to indicate signaling states.

signaling, time division Signaling used in pulse code modulated transmission.

signaling, tone See tone signaling.

signaling, 20Hz Ringdown signaling, especially over metallic loops.

signaling, two-state A signaling system transmitting two signals only, corresponding to off-hook and on-hook signals.

signaling, voice frequency The passing of information relating to the establishment of a telephone call, using tones in the voice frequency band. From central office to subscriber various information tone signals are used (busy tone, ringing tone etc.). From subscriber to central office dual tone multifrequency (DTMF) pulses are used to indicate the wanted number. Between central offices there are many different types of voice frequency signaling system, e.g., CCITT No. 5, R1, R2.

signaling channel   (for Common Channel Signaling systems) A data channel in combination with the associated signaling terminal equipment at each end.

signaling data link   A combination of two data channels operating together in a single signaling system. The channels operate in opposite directions and at the same data rate.

signaling destination point   A signaling point to which a message is destined.

signaling information   The information content of a signaling message related to a call control, management action, etc. The message alignment and service indications are not part of the signaling information.

signaling information field   The bits of a message signal unit which carry information specific to a particular user transaction and always contain a label.

signaling link   A method of transmission consisting of a signaling data link and its transfer control functions.

signaling link blocking   The unavailability of a signaling link caused by a "processor outage" condition at one end of the link.

signaling link code   A field of the label in the signaling network management messages, which indicates the particular signaling link to which the message refers among those interconnecting the two involved signaling points.

signaling link error monitoring   Monitoring consisting of two functions: initial alignment error rate monitoring and signal unit error rate monitoring.

signaling link failure   The unavailability of a signaling link caused by a failure in signaling terminal equipment or in the signaling data link.

signaling link group   A set of signaling links directly connecting two signaling points and having the same bit rate, propagation delay, and other physical characteristics.

signaling link management functions   Functions that control the integrity of locally connected signaling links.

signaling link restoration   The completion of the initial alignment procedure on a signaling link following the removal of the previous causes of failure.

signaling link selection field   A field of the routing label which is typically used by the message routing function to perform load sharing among different signaling links or link sets.

signaling link set   A set of signaling links directly connecting two signaling points.

signaling link unblocking   The removal of the previous causes of signaling link blocking. If no other causes of unavailability exist, the signaling link becomes available.

signaling message   An assembly of signaling information, including the associated message alignment and service indications, that is transferred as an entity via the message transfer part.

signaling message handling functions   Functions that, at the actual transfer of a message, direct it to the proper signaling link or user part.

signaling message route   The signaling link, or consecutive links connected in tandem, used to convey a signaling message from an originating point to its destination point.

signaling network   A network used for signaling and consisting of signaling points and connecting signaling links. The term applies particularly to the collection of digital exchanges and signal transfer points together with their interconnecting signalling paths for a common channel signalling system.

signaling network functions   The functions performed by the message transfer part and which are common to, and independent of, the operation of the individual signaling links. They include the signaling message handling functions and the signaling network management functions.

signaling network management   Automatic and manual procedures which can be overriden by ad hoc manual intervention, to vary the signal routing in accordance with prevailing conditions.

signaling network management functions   Functions which, on the basis of predetermined information about the status of the signaling network, control the current message routing and the configuration of signaling network facilities.

signaling originating point   A signaling point at which a message is generated.

signaling point   A node in a signaling network which originates or receives signal messages, or transfers them from one signaling link to another.

signaling point code   A binary code uniquely identifying a signaling point in a signaling network. It is used, according to its position in the label, either as a destination point code or as an originating point code.

signaling protocol   Method of coordinating signaling between two central offices or telephone switches.

signaling relation   A relation between two signaling points involving the possibility of information interchange between corresponding user part functions.

signaling route   A predetermined path described by a succession of signaling points that may be traversed by signaling messages directed by a signaling point towards a specific destination point.

signaling route management functions   Functions that transfer information about changes in the availability of signaling routes in the signaling network.

signaling route set   A combination of all the permitted signaling routes that may be used to pass signaling messages from a signaling point to a specific destination.

signaling route set test procedure   A procedure which is used to test the availability of a given signaling route previously declared unavailable.

signaling routing   Procedures for directing the allocation of signaling paths.

signaling system The procedures involved in the interpretation and use of a repertoire of signals, together with the hardware and/or software needed for their generation, transmission, and reception.

signaling time slot A time slot starting at a particular phase in each frame and allocated to the transmission of supervisory and control data.

signaling traffic management functions Functions that control and, when required, modify routing information used by the message routing function.

signaling transfer point A signaling point which transfers signaling messages from one signaling link to another.

significant condition of modulation A condition assumed by the appropriate device corresponding to the quantized value(s) of the characteristic(s) chosen to form the modulation.

significant instants During modulation or restitution of a signal, the instants at which successive significant conditions of the signal begin. The significant conditions are those recognized by an appropriate device. Each of the significant instants is determined at the moment the device assumes a state usable for performing a specific function, such as recording, processing, or gating.

significant interval Time interval between two consecutive significant instants.

silence, radio 1. Period of silence maintained every hour by maritime radio stations to facilitate reception of weak emergency signals. 2. As used in a military sense, enforced cessation of radio transmission for security reasons.

silent periods Time gaps between bursts of ringing current from a central office to called subscribers.

silica Silicon dioxide, which occurs naturally as rock crystal or quartz.

silica gel Chemical which absorbs moisture from the air. It is often packed in small bags inserted in crates of telecommunications equipment to minimize corrosion due to high humidity during shipping or storage.

silicon Chemical element used as the basis of most semiconductors. It is also used as an additive to iron in making silicon-steel cores. It is a dark-grey, hard-crystalline solid, the second-most abundant element in the earth's crust and is widely used in the manufacture of optic waveguides, called glass fibers.

silicon-controlled rectifier A pnpn device which operates as a gate and is used in solid state relays.

silicon controlled switch A pnpn semiconductor used as a control gate.

silicon diode Silicon semiconductor used as a diode rectifier.

silicon-on-sapphire (SOS) A sapphire substrate used as the base on which a thin layer of silicon is deposited. Complementary metal oxide semiconductor (CMOS) circuits are then fabricated in this silicon layer. The insulation and low capacitance provided by the sapphire, after the silicon layer has been partially etched away, enable these CMOS-SOS

chips to operate at a faster speed than conventional CMOS chips.

silicon-on-something-else (SOSE) At one time regarded as the 1980s replacement for silicon-on-sapphire.

silicon tetrachloride The major constituent in most optic waveguides or fibers.

silk-and-cotton cable Cable enamelled and insulated with layers of silk and cotton which was once widely used for terminating other cables on main distribution frames.

simple computing service A service providing simple arithmetic computations for telephone subscribers, using the pushbuttons of their telephones to input figures with answers sent back by synthesized human voice.

simple scanning In facsimile transmission, scanning using only one spot at a time.

simplex One-way data transmission, with no capability for changing direction.

simplex circuit A circuit using ground return and permitting communication in either direction, but in only one direction at a time. The circuit may be a single wire with ground return, or a signaling path over a dry talking circuit which uses the two sides of the circuit in parallel, derived by connecting to the midpoints of repeating coils which are across the circuit. Also called phantom circuit.

simplex operation A type of operation which permits the transmission of signals in either direction alternately.

simplex signaling (SX) Signaling using two conductors for a single channel. A center tapped coil or its equivalent is used at both ends for this purpose.

simplexed circuit A two-wire circuit from which a simplex circuit is derived. The two-wire circuit and the simplex circuit may be used simultaneously.

simplified modular frame assignment system (SMFAS/ESS) An AT&T software routine used to reduce main distribution frame (MDF) congestion by optimizing re-use of jumper wires on MDFs serving main electronic exchanges.

simulation A mathematical model which employs physical and mathematical quantities to portray a real-life situation.

simulator Device or program that simulates the operation of another device.

simultaneous Pertains to the occurrence of two or more events at the same instant of time.

SINAD A method of obtaining reference output for such factors as sensitivity, selectivity and spurious response of a radio receiver. The name is derived from the ratio Signal + Noise + Distortion/Noise + Distortion.

sine wave Wave with its amplitude varying in proportion with the sine of an angle. This goes from 0 at 0° to 1 at 90°, 0 again at 180°, -1 at 270°, and 0 again at 360°.

singing An undesired self-sustaining audio oscillation in a circuit, usually caused by excessive gain or unbalance of a hybrid termination. It occurs when the sum of the repeater gains exceeds the sum of the circuit losses.

singing, repeater Unwanted oscillation of a repeater circuit caused by either excessive gain or by a poor balance at the two-wire/four-wire hybrid.

singing arc An arc which gives out a musical tone. It is caused by oscillatory changes in heating effects.

singing margin The difference in level between the singing point and the operating gain of a system or component.

singing point The threshold point at which additional gain in the system will cause self-oscillations.

single armor One layer of steel wires placed around a cable for protection.

single channel per carrier (SCPC) A system employed where traffic routes are not very heavy and circuits are provided by satellite, especially to small dish standard "B" stations.

single frequency (SF) A method of signaling over long-haul telephone lines in which the presence of a 2600 Hz continuous tone indicates to the far end that the line is idle. Removal of the tone indicates to the far end that a call is being placed over that line.

single harmonic distortion See distortion, single harmonic.

single hop 1. A radio system with no intermediate repeater stations. 2. A radio link which is reflected down from the ionosphere once between transmitter and receiver.

single-address message A message to be delivered to only one destination.

single-channel office Telephone central office which uses the one digit "0" to reach both assistance operators and toll operators.

single-current telegraphy Telegraphy using a unidirectional current for the transmission of signals.

single-digit dialing Procedure which enables the dialing of a single digit to be sufficient to access a special service.

single-ended control See single-ended synchronization.

single-ended synchronization A synchronization control method used between two locations in which phase error signals used to control the clock at one location are derived from a comparison of the phase of the incoming signals and the phase of the internal clock of the same location. Syn.: single-ended control

single-frequency interference That interference caused by a single frequency source. Example: interference in a data transmission line induced by a 60 Hz source.

single-frequency signaling In telephone communications, a method of conveying dialing or supervisory signals, or both, with one or more specified single frequencies.

single-frequency signaling system In telephone communications, a system that uses single-frequency signaling.

single-mode optical waveguide An optical waveguide in which only one bound mode can propagate at the wavelength of interest. In step index guides, this occurs when the normalized frequency, V, is less than 2.405. For power-law profiles, single mode operation occurs for normalized frequency, V, less than approximately $2.405\sqrt{(g+2)/2}$, where g is the profile parameter. The diameter of the core of single mode waveguides is between two and ten microns. Single-mode operation is attractive because all modes except the lowest and simplest mode are excluded. This reduces time distortion of signals and dispersion and extends distances between repeaters. Syn.: monomode optical waveguide

single-phase A circuit in which there is only one sinusoidal voltage variation.

single-point connection (SPC) A connection that blocks current from flowing through any part of the isolated ground circuit.

single-pole double-throw (SPDT) Switch which can connect one input to either one of two outputs.

single-pole single-throw (SPST) Switch which either makes or breaks a single contact.

single-rate area An area designated by the telephone company as a fixed-rate designation and assigned a single-rate step.

single-sideband equipment reference level The power of one of two equal tones which, when used together to modulate a transmitter, cause it to develop its full rated peak power output.

single-sideband modulation See modulation, single-sideband.

single-sideband noise power ratio The ratio of the power in the notch bandwidth with the notch in to the power in the notch bandwidth with the notch out. Both are measured at the output and the notch is applied to an input sufficient to maintain the total system mean noise power output constant.

single-sideband suppressed carrier (SSBSC) That method of single-sideband transmission wherein the carrier is suppressed.

single-sideband transmission (SST) Sideband transmission in which only one sideband is transmitted. The other sideband is suppressed and the main carrier wave itself may be either transmitted or suppressed.

single-tone interference An undesired discrete frequency appearing in a signal channel.

single-wire line A communications circuit using a single wire plus an earth return path.

sink 1. In communications, that part of a system in which messages are considered to be received. 2. A memory or recording device in which information can be stored for future use.

sink, heat See heat sink.

sinusoidal Varying in proportion to the sine of an angle.

SITA World-wide data network used by airlines for ticketing, seat reservation, flight operations, aircraft movements, etc. Abbreviation for Societe Internationale de Telecommunications Aeronautiques.

skew In facsimile systems, the angular deviation of the received frame from rectangularity due to asynchronism between scanner and recorder. It is expressed numerically as the tangent of the angle of deviation.

skew ray A ray which does not intersect the optical axis of a system, in contrast to a meridional ray.

skiatron A type of cathode ray tube with a normally white base color. The screen is coated with a halide compound which darkens on bombardment by an electron stream.

skin effect Tendency for higher frequency currents to travel near the outside surface of a conducting wire rather than distributed throughout the entire cross section of the wire.

skinner Section of a cable form where individual wires are fanned out to their connection tags.

skinner connection, extended A connection made by wire in a formed-out cable being terminated on more than one connection tag.

skinner length Length of wire in a fanned out cable form from the butt of the cable to the end of the wire itself.

skip control A network management control which reroutes a percentage of direct-routed and alternate-routed traffic offered to an outgoing group by skipping over the specified group to the next in the routing chain.

skip distance The minimum distance between the transmitting station and the point of return to the earth of a transmitted wave reflected from the ionosphere.

skip fading A type of fading which may be observed at receiving locations near the skip distance at about sunrise and sunset when, because of the instability of the ionization density of the ionosphere, maximum usable frequency for a transmission path may oscillate around the actual frequency. The amplitude of the received signal may fall abruptly when the skip distance crosses over the receiving point and may suddenly increase with the decrease in the skip distance, thus causing the fading.

skip zone A ring-shaped region within the transmission range wherein signals from a transmitter are not received. It is the area between the farthest points reached by the ground wave and the nearest points at which the reflected sky waves come back to earth. Syn.: zone of silence

skipping The interconnection of identically numbered choices of nonadjacent grading groups.

skirt Flared-out base of a glass or ceramic insulator used on open wire routes in order to increase leakage path resistance.

sky wave See wave, sky.

Skylab U.S. space station which returned to earth in July 1979.

Skynet Digital transmission service by AT&T using groundstations and a 1.5 Mbit/s bearer.

slack 1. Any excess length of wire or aerial cable between two supports which allows excessive sag. 2.

An excess length of cable, compared with the length of its supporting messenger strand, which permits the cable to snake. 3. The difference between the length of submarine cable paid out and the geographic distance along the bottom contour.

slack puller Clamp used to take tension off an aerial cable by pulling the messenger wire in until it is slack between clamp and pole.

slant distance Actual distance between two points which are not at the same elevation.

slant range The line-of-sight distance from a transmitting antenna to a specified object, such as a target or receiving antenna.

slave intercom Intercom set able to receive calls from and initiate calls to a master station on the same system.

slave station In a data communication network, the station which is selected and controlled by a master station. The slave station can usually only call, or be called by, a master station.

sleeve A tubular covering. Lead sleeves are used in cable splices. Plastic sleeves are used to slip over individual jointed conductors in a cable splice (these are sometimes called colletts, pair). Bronze sleeves are used to make compression joints in conductors. Copper sleeves, which form the outer contacts in telephone switchboard plugs, usually carry control and supervisory circuits.

sleeve, filled plastic splice Small plastic tube filled with silicone grease and designed to be slipped over jointed conductors as a cable splice.

sleeve, heat shrinkable Sleeve of polyethylene placed around a plastic cable splice. Application of heat by a hot air blower shrinks the sleeve and makes a tight fit around the cable on each side of the splice. A thin layer of a mastic compound usually acts as a seal and a glue.

sleeve, lead Length of lead piping used as an outer sleeve around a lead cable splice. The ends are plumbed to the cable sheath with molten solder.

sleeve, relay Copper sleeve placed around a relay coil where it acts as a short-circuited winding and delays changes.

sleeve, splicing Bronze or copper sleeve used with a special crimping tool to splice open wire lines or pairs in multi-pair cables.

sleeve, split A lead sleeve split down one side. It can be opened and placed on an existing cable splice when it is necessary to re-open and then reclose the splice.

sleeve control 1. The use in a switching office of a third wire to control operation of a device external to the switching system but directly associated with the line. Syn.: sleeve lead control. 2. Some early designs of manual operating positions were called sleeve control boards because supervisory signals were controlled by relays in the sleeve or third-wire circuit.

sleeve wire The third wire in a circuit, used for control functions in most manual and automatic switching centers.

sleeving, lead Long lengths of lead piping which may be cut to the length required to make up individual sleeves.

slicer A limiting circuit which suppresses both positive and negative peaks of a signal.

slicing level Of a binary signal, the decision level between "0" and "1," for example, between mark and space for a particular bit.

slide, chassis Runners placed so that rack mounted equipment may be pulled out of its mounting.

slide rule A ruler for hand calculation based on logarithmic scales and on the fact that adding logs is equivalent to multiplying antilogs. In recent years slide rules have been largely replaced by digital calculators.

sling, rope Type of safety belt worn by linemen when working in trees to clear a path for a pole route.

sling, wire rope Length of wire rope, looped at both ends, used in storage yards when moving heavy poles.

slip 1. Difference between the synchronous speed of an induction motor and its actual speed. It varies with the load. 2. Loss or gain of digit position or positions in a digital signal.

slip, controlled The controlled irretrievable loss or gain of a set of consecutive digit positions in a digital signal to enable the signal to accord with a different rate.

slip, uncontrolled The uncontrolled loss of a digit position or a set of consecutive digit positions resulting from an aberration of timing processes.

slip rates North American objectives for slip rates in the national digital networks are less than one slip per five hours on an end-to-end connection or less than one slip in 10 hours for each of the end offices involved in the connection. Under temporary trouble conditions where a synchronization link is broken and the exchange clock is free-running, the slip must be less than one in 40 hours if the receiving switch is a toll switch or less than 255 in 24 hours if the receiving switch is an end office or local exchange.

slip rings Concentric copper rings used to make connection with rotor windings in rotating electric machines.

slipping The interconnection of differently numbered choices of adjacent or nonadjacent grading groups.

slitter, figure-8 Special pliers used for cutting the messenger wire section away from the cable section without damage to either in a figure-8 type aerial cable.

slitter, parallel wire Hand tool for separating the two insulated conductors of a one-pair plastic-sheathed drop wire.

slitter, switchboard cable Hand tool with a blade for insertion under the outer sheath of a switchboard cable. It cuts the sheath lengthwise as it is pulled out.

slope The rate of change, with respect to frequency, of attenuation of a transmission line over the frequency spectrum.

slope compensation The action of a slope-compensated gain control, whereby slope of amplifier equalization is simultaneously changed with the gain so as to provide the correct cable equalization for different lengths of cable; normally specified by range and tolerance.

slope equalizer A device or circuit used to achieve a specified slope in a transmission line.

slope resistance As pertaining to electrode ac resistance, the ratio of the voltage increment applied to an electrode to the current increment of the same electrode, all other electrode voltages remaining unchanged.

slot 1. A narrow band of frequencies. 2. A time slot.

slot antenna A radiating element formed by a slot in a conducting surface or in the wall of a waveguide.

slow-operate relay Relay with a copper slug at the armature end of its core, making the relay slow both to operate and to release.

slow-release relay Relay with a copper slug at the heel end of its core, making the relay slow to release but having little effect on the operating lag.

slow scan TV A process through which TV pictures of good quality are transmitted over ordinary telephone circuits in a few seconds. Pictures do not move, but can be updated every few seconds, depending on the condition of the various sections of the circuit between the two terminals.

slug Copper sleeve placed over a relay coil, or at one end of the core, in order to modify the operate and release timing of the relay.

slug tuning Adjustment of resonant frequency by changing the inductance of a coil by moving a ferrite slug into or out of the coil.

small area network Cable network linking together microprocessor-controlled devices in a domestic environment.

small scale integration (SSI) Logic gates made up as single units which perform a single basic function.

smart card Plastic card resembling a credit card but carrying embedded components and circuitry providing programmable storage capacity; used for financial transactions, shopping, etc.

smart terminal Terminal that has local processing facilities as well as communications capabilities. Syn.: Intelligent terminal

smear Picture condition in which objects appear to be extended horizontally beyond their normal boundaries in a blurred or "smeared" manner.

smectic A type of liquid crystal with molecules arranged in layers about 2 nm thick which is used in display units.

smooth When used of a transmission line, one with smoothly distributed characteristics such as impedance, capacitance, etc.

smooth earth Idealized surfaces, such as water surfaces or very level terrain, having radio horizons that are not formed by high terrain but are determined solely as a function of antenna height above ground and the effective earth radius.

smooth traffic Traffic that has a peakedness factor less than one.

smoothing circuit A filter-type circuit designed to reduce the amount of ripple in a circuit, usually a dc power supply. The smoothing circuit has the prime function of removing public power-frequency ripple.

sneak current See current, sneak.

sneak current fuse A heat coil which operates like a delay-action fuse by breaking the circuit if a current, although not high enough to blow a normal quick-acting fuse, is high enough over time that equipment might be endangered.

Snell's Law A law which applies to light traveling obliquely from one medium into another, whereby part of the light is reflected and part is refracted as it passes into the second medium. By Snell's Law $n_1 \sin \theta_1 = n_2 \sin \theta_2$ where n_1 and n_2 are the two refractive indices and θ_1 and θ_2 are the angles of incidence and of refraction respectively. If n_2 is lower than n_1, it follows that, above a critical angle, no refraction escapes and all the light is internally reflected.

sniffer, cable Microphone on a probe feeding a small portable amplifier and frequency converter. It is used to detect leaks in pressurized cables by picking up the sound of escaping gas or air.

sniffer, radio-frequency Broadband radio frequency receiver used to detect radio frequency leaks due to faults or imperfections in the radio shielding

snow (TV) Heavy random noise.

snow pack telemetry radio (SNOTEL) Radio system operated by the U.S. Dept. of Agriculture to monitor winter snow storms. Uses meteor trail reflections (meteor burst transmission).

soak 1. To pass dc through a relay coil until the core is magnetically saturated. 2. To run a program under closely supervised operating conditions to uncover any problems which might develop.

Society of Motion Picture and Television Engineers (SMPTE) The group responsible for setting U.S. technical standards for film and videotape.

socket, lamp Device made up into strips and designed to hold tubular lamps used in switchboard panels.

socket, tube Base in which an electron tube is mounted.

soft Of an electron tube, not reduced to a low vacuum.

soft data Data which is semipermanent in the system, such as routing data, and which is under the control of the administrative program.

software The totality of programs, procedures, rules and routines used to extend the capabilities of computers, and including such things as compilers, assemblers, and narrators. These functions are normally defined by a pattern of words stored as electrical states in hardware. The function is implemented by the interpretation of these words by the processor. Software enables the function of a device to be changed without changing physical wiring or equipment hardware.

software, time critical A software item which contributes to the overall runtime or throughput of the machine to a significant extent. If its runtime is too long, the overall system is unlikely to meet its performance requirements.

software defined network service AT&T facility by which major customers can use part of the AT&T Communications network for their own dedicated internal voice and data systems.

software engineering The production of the "linguistic machinery" required to harness and utilize the capability of computers.

software maintenance Continual improvements and changes needed to keep programs working at top efficiency.

software master library A computer file in which software items and software parts list are deposited, kept secure and can be retrieved and built into other systems as required.

software message generator A program used for process level testing which simulates a task interface.

software process An independently schedulable item of code running on a target machine.

solar cell See cell, solar.

solder A lead or tin alloy which melts readily and is used in a wide variety of wire, terminal and component connecting applications.

solder, aluminum tinning Solder used with aluminum sheath cables, usually composed of 90% tin and 10% zinc.

solder, arsenical wiping Solder used for lead cable plumbing, usually composed of 63% lead, 37% tin and small amount of arsenic. Syn. wiping solder

solder, eutectic Solder, in various metal combinations, which have in common a low melting point: 63%/37% tin/lead solder melts at about 180°C; 96%/4% tin/silver solder melts at about 220°C.

solder, hard Solder with a high melting point used for joints which must withstand heat or stress. One type, made of 55% copper and 45% zinc, melts at about 870°C.

solder, rosin core Solder composed of hollow wire, with rosin flux inside. It is wound on spools.

solder, sealing A solder sometimes used on top of an ordinary lead plumped splice. It flows into and seals small holes which occur during wiping. Its usual composition is 40% bismuth, 20% tin and 40% lead.

solder, seam Solder used for fixing copper wires to tinned tags or for joining the sides of a split lead sleeve. It is composed of 50% tin and 50% lead.

solder, silver Hard solder designed to be used where stress must be resisted. It is composed of 57% silver, 33% copper and 10% zinc.

solder, soft General purpose solder, with a 60%/40% lead/tin composition.

solder, stearine flux A tin/lead solder combination in hollow wire form with flux in the center. It is used for soldering lead sheaths.

soldering Process of joining metals by fusing them by means of a molten metal with a relatively low melting point.

solenoid An induction coil which acts as a relay coil. It can also pull a moveable iron core into the coil when current is passed through the coil.

solid 1. A single wire conductor, as contrasted with a stranded, braided or rope-type wire. 2. A connection made direct to ground with no impedance inserted.

solid-state Pertaining to the use of semiconductors rather than electromechanical relays and electron tubes.

solid-state circuit An integrated circuit with all elements formed as a single block of semiconductor material.

solid-state device A device which depends on the movement of charged particles rather than on mechanical movement.

solid-state laser A laser whose active medium is glass or crystal.

solid-state memory A memory unit formed in a semiconductor chip.

solution, pressure testing A soap solution containing alcohol and glycerine. When applied to the outside of a pressurized cable, even the smallest pinprick in the cable sheath will be indicated by the presence of bubbles.

SONAR A sonic and ultrasonic underwater ranging, sounding and communications system.

sonic Pertaining to sound.

sonic delay line See acoustic delay line.

soot technique A method of fabricating an optical waveguide preform by forming small glass particles (soot) and depositing them on the surface of a cylinder.

sort To arrange and group items according to a system of classification.

SOS The international radiotelegraph distress signal. In Morse code it consists of three dots, three dashes, three dots.

sound An oscillation in pressure, stress, particle displacement and particle velocity in a medium with internal forces.

sound analyzer A microphone, amplifier and wave analyzer used to measure the amplitude and frequency of the components of a complex sound.

sound-powered telephone A telephone in which the operating power is derived from the speech input only.

sound-program circuit section Part of an international sound-program circuit between two stations, at which the program is transmitted at audio frequencies. Carrier sound-program equipment is normally used to provide a sound-program circuit section in the international network.

sounder, ionospheric Device sometimes used when planning high frequency radio links. A transmitter sends pulses at each of the various radio frequencies under consideration while a receiver at the distant location changes tuning in synchronism with the transmitter and produces a visual display of the results of all the transmissions. The process permits an analysis of the suitability of the frequencies.

sounder, telegraph Receiver which was once widely used in railway and other hand-speed telegraph systems. It gave distinctive click signals so that an operator could recognise his own call sign and take down the message without wearing earphones.

source That part of a system from which messages are considered to originate.

source code Text written in an appropriate language which can by automatic means be translated into a program which will run on a target machine. The running program is known as an object code.

source efficiency In optoelectrics, the ratio of emitted optical power of a source to the input electrical power.

source user The user providing the information to be transferred to a destination user during a particular information transfer. Syn.: information source; message source

south pole Pole of a magnet which seeks the south magnetic pole of the Earth.

Southwestern Bell Corp. One of the seven Regional Holding Companies resulting from the divestiture of AT&T.

Soyuz A series of Russian spacecraft.

space 1. In binary modulation, the significant condition of modulation that is not specified as the "mark," e.g., when any of the following conditions exist: (a) Current or tone is "off." (b) The loop is open. (c) There is a negative voltage, line to ground. (d) FSK system is on the upper frequency. (e) There is no hole in the paper tape. 2. The continuous three-dimensional expanse outside the Earth's atmosphere. 3. The expanse between electrodes in a vacuum tube.

space, climbing Vertical space on a pole or tower which is kept clear of fixtures so that a lineman will have access to do maintenance or repair work at all levels.

space, deep See deep space.

space, near The zone in which the Earth's gravitational field controls telecommunications satellites and other orbiting vehicles.

space charge A charge density greater than zero, found, for example, at the junction region in a semiconductor junction transistor or in the region around the glowing cathode of a thermionic valve.

space diversity A method of transmission or reception, or both, employed to minimize the effects of fading by the simultaneous use of two or more antennas spaced a number of wavelengths apart.

space division The separation in the space domain of a plurality of transmission channels between two points.

space-division switching 1. A method whereby a switch utilizing a physically separated set of matrix contacts is used to determine single transmission path routing. 2. The switching of inlets to outlets using space division techniques.

space probe A spacecraft designed for making observations or measurements in space.

space-time-space (STS) Where the number of ports on an exchange is too great to give complete switching flexibility by a time-slot change only it is usual to sandwich one or more space switches between two time switches but some manufacturers prefer to use the reverse technique, sandwiching the time switch between two space switches. Some digital exchanges use reeds in their subscribers' concentration stages; it is not usual to call such electromechanical analog devices "space switches." The name should be reserved for high speed solid state switching.

spacecraft A man-made vehicle designed to go beyond the major part of the Earth's atmosphere.

spacer, cable A separator used to hold apart two cables entering a mechanical splice case, or to maintain separation between aerial cable and messenger wire at splices.

spacer, jack A wooden or plastic strip used to fill a space in a switchboard panel reserved for later use by a strip of jacks.

spacer, wire Plastic spacer used midspan on open wire routes to reduce the risk of wires touching in extremely long spans.

spacing (CATV) Length of cable between amplifiers expressed as dB loss at the highest TV channel provided for in a system, equal to amplifier gain in main trunks.

spacing, load coil Distance between adjoining loading coils in a cable system. There are many different spacings designed for use with different kinds of cable and also to give different transmission characteristics. Spacings utilized are between 640 and 9000 feet.

spacing, repeater Distance between adjoining repeater stations in a transmission system.

spacing bias The uniform lengthening of all spacing signal pulses at the expense of all marking signal pulses.

S-pad Attenuation pad sometimes inserted in intertoll connections.

span 1. Length of suspended wire or cable between adjoining poles. 2. The length of cable in ducts between two manholes. 3. Distance between line extenders or distribution amplifiers; also, distance between taps.

spare Available but not at present in use.

spare pair Pair in a cable in a distribution network which is terminated at both ends and ready for connection when needed with no need for cable splicing.

spare wire Spare pair placed in a cable for use when a regular pair develops a fault.

spares, running Holding of spare parts needed to keep equipment working satisfactorily. Most manufacturers produce formulae to determine recommended backup equipment based on expected mean time between failures and mean time to repair.

spark The flash of light seen when there is a sudden discharge of electricity between two points.

spark gap Gap between two electrodes designed to produce a spark under given conditions.

spark-over of a gas discharge protector A condition which occurs on electrical breakdown of the respective discharge gap.

spark suppression The connection of high resistance resistors or small capacitors across contacts which break currents in inductive circuits. These act as spark suppressors.

spark test Test of insulation around a wire by grounding the conductor and drawing the insulated wire between highly charged electrodes.

sparking Spark discharges between the fixed brushes and the rotating commutator of a motor or generator, usually indicating that maintenance is needed.

spatial coherence See coherent.

spatially aligned bundle See aligned bundle.

spatially coherent radiation See coherent.

speaker 1. An order wire circuit between repeaters or terminal stations on a transmission system. 2. A loudspeaker.

speakerphone A loudspeaking (hands-free) telephone.

spec A specification.

special billing telephone number Telephone number outside the normal central office numbering series, allocated to certain customers for billing purposes.

special-effects generator A device permitting combinations of images on a television screen supplied by one or more video inputs.

special grade Describing a transmission circuit which has been specially prepared to enable it to carry special signals or services which could not satisfactorily be carried on ordinary local, toll or long distance circuits.

special grade access line An automatic voice network access line specially conditioned to give it characteristics suitable for handling special services.

special grade service In the automatic voice network, a communications service which utilizes specially conditioned interswitch trunks and access lines to provide the required capability for secure voice, data, and facsimile transmission.

special instruction call Toll call requiring special handling, such as a call to be charged to a credit card.

special service trunk Trunk connecting a caller to services such as weather or information.

specialized common carrier (SCC) A common carrier, not a telephone company, offering new and different services to meet public needs not adequately met by existing "traditional" carriers.

specialized mobile radio system Two-way mobile radio telephone systems used mostly for dispatch services.

specific detectivity See D*.

specific gravity Density. The ratio of the weight of a volume of a liquid or solid to the weight of the same volume of water, usually at 4°C.

specification 1. A document intended primarily for use in procurement, which clearly describes the essential technical requirements for items, materials or services, including the procedures by which it will be determined that the requirements have been met. 2. A description of the general parameters required of the system and of its behavior.

specification and description language (SDL) The CCITT language used in the presentation of the functional specification and functional description of the internal logic processes in a stored programmed control switching system.

speckle noise See modal noise.

spectral bandwidth (between half power points) Wavelength interval in which a radiated spectral quantity is not less than half its maximum value.

spectral irradiance Irradiance per unit wavelength interval at a given wavelength, expressed in watts per square meter per micrometer.

spectral line A narrow range of emitted or absorbed wavelengths.

spectral linewidth A measure of the purity of a spectral line occurring in a line spectrum. It can be specified as the full width at half maximum, specifically: the difference between the wavelengths at which the spectral emittance or absorption drops to one half of its maximum value. This method may be difficult to apply when the line has a complex shape. It can also be specified as the rms width. The relative spectral linewidth is also frequently used.

spectral radiance Radiance per unit wavelength interval at a given wavelength, expressed in watts per steradian per square centimeter per micrometer.

spectral responsivity Responsivity per unit wavelength interval at a given wavelength.

spectral window A wavelength region at which relatively minimal attenuation of an optical signal is experienced. Syn.: transmission window

spectrum A range of frequencies of radiant phenomena. See Appendix B.

spectrum, audio Range of sound frequencies which can be detected by the normal human ear. It ranges from about 20 Hz to 16 to 20 kHz.

spectrum, microwave Range of radio frequencies not strictly defined but usually taken as between 200 MHz and 100 GHz.

spectrum, radio Electromagentic frequencies used for radiocommunication ranging from extremely low frequency (ELF) at about 200 Hz up to tremendously high frequency (THF) at 3000 GHz.

spectrum, voice The range of fundamental frequency for human voices are from 78 Hz for the lowest bass notes, to 1397 Hz for the highest soprano notes. Harmonics which give quality and timbre to a voice are several times these fundamental frequencies.

spectrum designation of frequency A method of referring to a range of communication frequencies. In American practice, the designation is a two or three letter acronym for the name. In International Telecommunications Union practice, the designation is numeric. These ranges are:

Frequency Range (Lower Limit Exclusive, Upper Limit Inclusive)	American Designation	ITU Designation
Below 300 Hz	ELF (Extremely Low Frequency)	—
300 — 3000 Hz	ILF (Infra Low Frequency)	—
3 — 30 kHz	VLF (Very Low Frequency)	4
30 — 300 kHz	LF (Low Frequency)	5
300 — 3000 kHz	MF (Medium Frequency)	6
3 — 30 MHz	HF (High Frequency)	7
30 — 300 MHz	VHF (Very High Frequency)	8
300 — 3000 MHz	UHF (Ultra High Frequency)	9
3 — 30 GHz	SHF (Super High Frequency)	10
30 — 300 GHz	EHF (Extremely High Frequency)	11
300 — 3000 GHz	THF (Tremendously High Frequency)	12

spectrum signature The pattern of radio signal frequencies, amplitudes and phases which characterizes the output of a particular device, and tends to distinguish it from other devices.

specular reflection See reflection, specular.

speech, coded Speech signals which have been quantized and levels put into digital format using binary notation.

speech-digit signaling Signaling in a pulse code modulation multiplexing system in which digit time slots primarily used for the transmission of encoded speech are periodically used for signaling.

speech-only transmission Describing a circuit which is adequate for the transmission of speech signals but of insufficient bandwidth for the transmission of music signals.

speech operated noise adjusting device (SONAD) A device used in radio receivers to reduce equipment gain during the absence of received signals.

speech-plus Describing various types of equipment which enable one or more duplex teletypewriter circuits to be provided over one voice circuit, usually by restricting the frequency band available for the voice signal. One type filters the band from 1500 Hz to 2000 Hz from the voice circuit and uses this for bothway voice frequency telegraphy. The degradation of the voice circuit by the removal of this 500-Hz-wide band cannot be detected by most listeners.

speech-plus-duplex operation That method of operation in which speech and telegraphy (duplex or simplex) are transmitted simultaneously over the same circuit, using filters to prevent mutual interference.

speech-plus signaling An arrangement of equipment that permits the use of part of a speech band for transmission of signaling.

speech power See volume unit.

speech predictive encoding (SPEC) A digital form of TASI.

speech scrambler A device in which speech signals are converted into unintelligible form before transmission and are restored to intelligible form at reception. Used for security reasons.

speed, data transmission Rate of transmission of digital information.

speed, telegraph signaling Rate of transfer of information over a teletypewriter channel. Given the standard assumption of 6 character intervals per word, 75 words per minute equals 55.9 bauds, and 100 words per minute equals 74.5 bauds.

speed calling See abbreviated dialing.

speed dialing Dialing at a rate faster than 10 impulses per second.

speed matching The ability of several types of switched data services to interconnect terminals with different operating speeds.

speed number The assignation of a calling number with a reduced number of digits so that frequently-called lines can more easily be dialed.

speed of answer The length of time an average call is delayed before being answered by an operator.

speed of service (SOS) 1. The time elapsed from release of a message by the originator to the receipt by the addressee. 2. The elapsed time from the entry of a message into a communications system until receipt at the terminating communications facility.

spike A high amplitude, short duration pulse superimposed on an otherwise regular waveform. See impulse hit.

spill Digits sent forward from one telephone office to the next.

spill-forward feature A feature whereby an intermediate office can assume routing control of a call from the originating office. This increases the chances of completion by offering the call to more trunk groups than are available in the originating office.

spill office Intermediate toll office which receives the full telephone address of the called line from another office, and treats the call as if it had been originated locally.

spilling, no-skip Sending forward on an outgoing trunk all the digits received from an incoming trunk or a subscriber.

spilling, skip-3 Sending forward on an outgoing trunk all the digits received except the first three. This results in sending 7 members of a 10-digit code or 4 of a 7-digit code.

spilling, skip-6 Sending forward on an outgoing trunk only the last four digits after eliminating both the area code and the central office code.

spilling, variable Deleting a variable number of the received digits, depending on the route selected, before sending digits forward on an outgoing trunk.

spin, electron The rotation of an electron about its own axis. This is independent of the orbital movement of the electron around the atomic nucleus.

spinner A cable lasher which lashes aerial cable to the messenger as it is pulled along messenger wire.

spin-stabilized Said of a communications satellite whose body rotates to provide stability and whose antennae are mounted on a rotating platform which constantly directs the antennae towards the Earth.

spiral-four A quad type cable with opposite conductors used as pairs.

spiral-wrap Tape or binder wrapped helically over a cable core.

splash ring An immediate burst of ringing provided as soon as the circuit is put through to the called line, instead of a wait to pick up the normal interrupted ringing current.

splash tone See tone, splash.

splice To join together .permanently, to provide an electric or optic path from one wire or waveguide to another.

splice, beach The junction between the sea portion and the land portion of a submarine cable system.

splice, cable See cable splice.

splice, compression Splice incorporating a bronze or copper tube which is compressed around both conductors by means of a special crimping tool.

splice, conductor An electrically and mechanically sound joint between conductors.

splice, duct Long, thin cable splice made in such a way that the cable can be pulled back into a conduit. The procedure is normally adopted only when a cable fault has occurred in a conduit, midway between two manholes.

splice, fiber See optical waveguide splice.

splice, final In a submarine cable system, the junction between the seaward end of the previously-laid shore-end cable and the bitter end of the main cable.

splice, final test In a balanced audio cable, the splice made at the mid-point of a loading section which endeavours to offset capacitance unbalance irregularities on one side against compensating irregularities on the other side.

splice, first In a submarine cable system, the cable junction between the seaward end of a previously-installed shore-end cable and the first end of the cable in the cable-laying ship. It is the splice that commences the cable-laying operation.

splice, overlay A splice by means of which the strength of the armor wires is maintained at the junction between two sections of armored cable in a submarine cable.

splice, preformed Spirally wound steel wires which can be wound individually around a steel strand in order to splice the strand.

splice, random Method of cable joining whereby pairs are spliced together, within each group, with no regard for their theoretical pair numbers.

splice, rotation Method of splicing cables in which pairs are joined starting at a marker pair and progressing around the cable core in sequence.

splice, semi-final test A splice in quad type cable at the one-quarter and three-quarter points of a loading section (the final splice is at the half-way point) in order to balance capacitance effects or prepare for these to be balanced at the final splice.

splice, tag Cable splice with every pair identified and labelled with its pair number. (UK: a numbered joint)

splice, Y Cable splice in which one cable is joined to two smaller cables.

splice loss See insertion loss.

splicing chamber A miniature cable vault to which access is gained by lifting floorboards in front of the main distribution frame. In older installations splices were often made here to tip cables which could be safely fanned out and terminated on the frame. Increased use of plastic insulated and sheathed cables has eliminated the need for such joints.

split homing   The connection of a terminal facility to more than one switching center by separate access lines, each having separate directory numbers.

split-phase   Describing a device which derives a second phase from a single phase power supply by passing it through a capacitive or inductive reactor.

splitter   A radio frequency hybrid which connects one 75 ohm input with up to eight 75 ohm outputs for CATV distribution.

splitter, band   A network which accepts a broadcast radio signal and divides it into two bands, at different frequencies, for CATV distribution.

splitting   The action of an operator in dividing a cord circuit or link circuit into two parts, so that she may talk to only one of the two parties to the connection.

S-pole   Pole at the end of open wire transposition sections.

spontaneous emission   Radiation emitted when the internal energy of a quantum mechanical system drops from an excited level to a lower level without regard to the simultaneous presence of similar radiation.

spoon   A long-handled shovel used when digging holes for poles.

sporadic   Occurring at random and infrequent intervals.

sporadic E propagation   Radio wave propagation by means of returns from irregular ionization that appears at heights of about 90 km to 120 km. Syn.: sporadic E

sports blackout   Federal law requires cable systems and television stations in the U.S. to delete coverage of local sports events to protect gate receipts.

spot beam   In satellite communications, a narrow and focused downlink transmission that allows the satellite to use different frequencies, or reuse the same frequencies, in other downlinks. A spot beam covers a much smaller geographic area, or footprint, than was possible with older satellite downlink transmissions.

spot speed   In facsimile systems, the speed of the scanning or recording spot along the available line.

spread spectrum (SS) modulation   A radiocommunication system in which the average energy of the transmitted signal, under the control of a random or pseudo random bit stream, is spread over a bandwidth which is much wider than the information bandwidth. It is wider by a factor of at least two for double-sideband amplitude modulation, typically four or more for narrowband FM, and typically 100 for a linear SS system. SS systems have the potential of sharing the frequency spectrum with conventional narrowband systems because of the potentially low power transmitted in the narrowband receiver's passband, or by allocating different bit streams to different subscribers transmitting simultaneously significantly greater use can be made of available bandwidth. In addition, high levels of interference are rejected by SS receiving systems.

spring, contact   Flat strip of metal, usually a phosphor bronze alloy, used in springsets, relays or keys. Contacts, often made of a semi-precious metal, are mounted at one end.

spring adjusting tool   Hand tool used for bending springs to provide specified contact pressure.

spring pile up   The complete assembly of contacts operated by a single relay armature or by a single key. Syn.: spring pile

springs, impulse   The pair of springs in a rotary dial whose contacts open and close 10 times per second to provide the makes and breaks of the subscriber's loop which provide telephone address information.

springs, shunt   Contacts which close as soon as a rotary dial is moved from its rest position, shorting out the receiver and thus preventing annoying dialing clicks in the receiver during dialing. Syn.: off-normal contacts

spud   A long handled shovel used to loosen soil deep in a hole excavated for a pole or anchor.

spurious emission   See emission, spurious.

spurious frequency conversion products   Spurious emissions, not including harmonic emissions, at the frequencies, or whole multiples thereof, of any oscillations generated to produce the carrier or characteristic frequency of an emission.

spurious intermodulation products   1. Intermodulation products at frequencies resulting from: intermodulation between (a) the oscillations at the carrier, characteristic, or harmonic frequencies of an emission; or the oscillations resulting from the generation of the carrier or characteristic frequency, and (b) oscillations of the same nature, of one or several other emissions, originating from the same transmitting system or from other transmitters or transmitting systems. 2. Intermodulation products at frequencies resulting from intermodulation between any oscillation generated to produce the carrier or characteristic frequency of an emission.

spurious radiation   See radiation, spurious.

spurious response   See response, spurious.

spurious response rejection ratio   Of a radio receiver, the ratio of the input level at the interfering frequency required to produce a specified output power from the receiver to the level of the wanted signal to produce the same output signal.

spurs   The sharp points on the climbers used by linemen.

sputter   Rapid fading experienced on radio signals which traverse high latitudes. It is caused by rapid ionization changes at these latitudes.

sputtering, cathode   Formation of a thin film, usually by depositing metal atoms on an insulating base in an integrated circuit. This is done by connecting a very high negative voltage to a metal electrode in an argon atmosphere, causing it to initiate a glow discharge which takes metal atoms out from the electrode and deposits them on the insulator base.

square wave testing   Using a square wave containing many odd harmonics of the fundamental frequency as an input signal to a device. Visual examination of the output signal on a cathode ray tube will indicate the amount of distortion introduced.

squelch The reduction of radio receiver gain when no modulated input signal is being received.

squelch circuit A radio circuit which reduces background noise in the absence of desired input signals.

stability Ability to remain stable in frequency, power level, etc.

stability, frequency See frequency stability.

stability test When applied to transmission circuits, a device which automatically applies worst case conditions (such as open circuit and short circuit) and standard terminated conditions in order to test the stability of repeated circuits.

stabilizer, voltage See voltage stabilizer.

stack 1. A pileup of plates in a dry-disc rectifier. 2. A last in, first out store made up of registers or main memory locations. A temporary store. 3. A ferrite core store: a stack of matrices or planes.

stacked Describing carrier channels placed one above the other in the frequency spectrum.

stacked array An array of VHF or microwave antennae placed one above the other and connected in correct phase relationship so that the combined effect is that of increasing antenna gain.

stage A single section of a multi-stage device.

stage, concentration The stage in a central office where traffic is concentrated before switching. A concentration ratio of 10 to 1 is common, but variations are wide.

stage, connecting In a homogeneous switching network using a number of crosspoints in series, if the crosspoints included in a connection are numbered starting with the closest to the network inlets, then all connecting matrices containing crosspoints that are given the same number in such connections are said to belong to a connecting stage.

stage, originating The "front end" of a central office, including the subscribers' line unit and the concentration stage.

stage, switching The main switching matrix of a central office. It will normally have equal numbers of inlets and outlets.

stage, terminating The outgoing section in a central office running from the outlet of the switching stage through an expansion stage to the line circuit of the called line.

stagger In facsimile systems, periodic error in the position of the recorded spot along the recorded line.

stagger tuning See tuning, staggered.

staggered twist Method of cable construction in which adjacent pairs are made up with a different length twist, in order to reduce crosstalk.

staggering, carrier frequency Use of one carrier frequency for one direction of transmission for all systems on a route, and another carrier frequency for all the systems in the other direction.

stalpeth Cable with a corrugated aluminum sheath covered with corrugated steel with a soldered seam, plus a polyethylene oversheath.

stand, desk A candlestick-shaped telephone with a microphone mounted at the top and a separate receiver resting on a hook.

stand-alone remote line unit A remote satellite or switching unit not dependent on its parent end office or exchange for local switching. It is provided with limited processing capability so that limited service continues even if the link to the parent exchange is interrupted.

standard, primary A standard of voltage, current, frequency, etc., precisely defined by the National Bureau of Standards, against which portable secondary standards are calibrated.

standard cubic meters per day (Sm³/d) Unit used in calculating air consumption of continuous flow cable pressurization systems. Cables in good condition are likely to need less than 0.1 Sm³/day per sheath-kilometer. During fault conditions consumption could easily rise to 3 Sm³/day or more.

standard frequency See frequency, standard.

standard frequency emission An emission which disseminates one or more standard frequencies at regular intervals with a specified average daily frequency accuracy. The International Radio Consultive Committee recommends a normalized departure of less than 1.10^{-10}.

standard frequency and/or time signal station A station whose primary purpose is to provide a standard frequency and/or time signal emission.

standard frequency satellite service A radiocommunication service using space stations on Earth satellites for the same purpose as those of the standard frequency service.

standard metropolitan statistical area Cellular radio areas in the U.S. as allocated by the FCC.

standard network interface (SNI) The point inside a customer's premises at which an access line terminates.

standard radio atmosphere An atmosphere having the standard refractivity gradient.

standard refractivity vertical gradient A standard used for comparison in phase and refraction studies, namely -40 N/km. It corresponds approximately to the median value of the gradient in the first kilometer of altitude in temperate regions.

standard telegraph level (STL) The power per individual telegraph channel required to yield the standard composite data level. For example, for a composite data level of -13 dBm at 0 dBm transmission level point, the STL would be -25.0 dBm for a 16 channel voice frequency carrier telegraph terminal from the relation:

$$STL = -13 -10 \log_{10}n = -(13 + 10 \log_{10}n)$$

where n is the number of telegraph channels and the STL is in dBm.

standard test signal A single-frequency signal with a standardized level, generally used for testing the peak power transmission capability and for measuring the total harmonic distortion of circuits or parts of a circuit.

standard test tone A single-frequency signal with a standardized level, generally used for level alignment of single links and of links in tandem.

standard time and frequency signal 1. A time-controlled radio signal broadcast at scheduled intervals on a number of different frequencies by government-operated radio stations. 2. A carrier frequency and time signal emitted in allocated bands in conformity with International Radio Consultive Committee Recommendation 460. Standard time and frequency signals are broadcast by the U.S. Naval Observatory and the National Bureau of Standards.

standard-time-signal emission An emission which disseminates a sequence of time signals at regular intervals with a specified accuracy. The International Radio Consultive Committee recommends that standard time signals be emitted with an accuracy not deviating from coordinated universal time by more than 1ms.

standard work seconds (SWS) Unit of work-time used in some operating positions. Calculations are based on the type of call handled and the time spent on other activities.

standard work time Average work-time per call over all operators in a location.

standby communication facility A facility kept in good working condition for use, when necessary, as a substitute for a normally used facility.

standing wave Pattern caused by two waves of the same frequency moving in opposite directions along a transmission line. Where voltages add together, the pattern will form a voltage antinode; where they subtract, it will form a voltage node. The nodes and antinodes are stationary in a standing wave.

standing wave ratio Ratio of the amplitude of a standing wave at an antinode to its amplitude at a node.

star connection Three network elements with one end of each connected to a common node.

star coupler In optoelectronics, a passive coupler whose purpose is to distribute optical power from one port to all other ports or to a set of all other ports.

star network Data network with a radial topology in which a central control node is the point to which all other nodes join.

star-LAN A local area network design characterized by 1 Mbit/s baseband data transmission over two-pair, twisted-pair wiring.

start bit In asynchronous transmission, the first element in each character that prepares the receiving device to recognize the incoming information elements.

start dial The generation of a signal to the far end to request that signaling be sent.

start dialing signal Lamp signal that indicates to an operator that she may dial out on a particular line.

start lead Wire or lead on which a signal is given to start an activity, for example, to instruct a linefinder to begin to hunt for a marked line.

start-of-heading (SOH) character A transmission control character used as the first character of a message heading.

start of message (SOM) Control character or group of characters transmitted by a polled station which indicates to other stations that what follows are addresses of stations to receive the following message.

start-of-text (STX) character A transmission control character that precedes a text and may be used to terminate the message heading.

start-record signal In facsimile systems, a signal used for starting the process of converting the electrical signal to an image on the record sheet.

start signal In start-stop transmission, one or more bits that precede each character transmitted.

start-stop Asynchronous mode of operation for teletypewriters in which the receiver stops on completion of each character and starts again on receipt of the next character, each character signal being complete with its own start and stop signals, i.e., each character or byte is transmitted as a self-contained piece of information needing no additional synchronizing or timing information to be transmitted.

stat- Prefix used in the now-obsolete cgs system for naming electrostatic units. Examples: statvolt, statohm.

state 1. Either of the two conditions of a bistable device; the "1" state or the "0" state. 2. In specification and description language, the condition in which the action of a process is suspended and awaiting an input.

statement In a high-level language, a single unit of program command.

static Describing a non-moving electric charge, such as the charge on a capacitor plate.

static, precipitation Radio noise caused by electrically charged rain drops or dust falling onto a receiving antenna.

static charge An electric charge on the surface of an object, particularly a dielectric.

station One of the input or output points in a communications system.

station, extension Telephone instrument connected to the same main line as a main station. It is reached by dialing the number of the main station.

station, left-in Telephone instrument left installed on a vacated premises but disconnected from service to the central office. When the premises are reinhabited, service can be reactivated from the central office.

station, main Telephone instrument connected directly to a central office and having its own unique telephone number.

station, mobile base Any of the fixed radio communication stations which transmit to and receive from mobile phones in vehicles.

station, off-premises Telephone instrument served from a PBX but not on the same premises as the PBX.

station, public telephone Telephone instrument available for general public use on payment of specified charges to a coinbox or to an attendant.

station, pushbutton   Telephone instrument equipped with a set of push buttons for sending digital information to a central office.

station, radio relay   A station with radio and power equipment which receives signals from one or more directions and retransmits them, usually on different frequencies.

station, receiver   A station equipped with receiving antennae and radio and power equipment which receives radio signals from fixed, mobile, maritime or aeronautical stations. Receiver stations are usually also the centers from which operators control associated remote transmitters.

station, repeater   See repeater station.

station, telephone   An installed telephone instrument together with its internal wiring and test and protection devices.

station, teletex   An installed teletex machine with its associated equipment.

station, teletypewriter   An installed teletypewriter. It may include a keyboard and a transmitter/receiver with a tape perforator and a tape-controlled transmitter, or it can be a simple receive-only unit.

station, telex   A teletypewriter station using Western Union's Telex equipment or its equivalent.

station, terminal   In a submarine cable system, the physical plant comprising shelter, utilities, and the system's terminal equipment.

station, transmitter   A station housing radio transmitters and the associated power plant. Most HF radio transmitters are controlled remotely, often from the receiving equipment complex.

station battery   A separate battery power source within a facility which provides the necessary dc input power associated with the facility. Such a capability is often centrally located. The batteries may power radio and telephone equipment as well as provide controls for equipment and emergency lighting.

station cable   The section of cable between the cable vault and the terminal equipment in a submarine cable system.

station call   See station-to-station call.

station clock   A clock that controls all station equipment requiring time control.

station engineering control office (SECO)   Bell System term used in Universal System Service orders. On an international switching center order the SECO is the office which has station design coordination responsibility for provision of equipment for one of the circuit locations and overall station compatibility.

station keeping   Describing the ability of a geostationary communications satellite to keep itself at or near a particular position relative to the Earth.

station load   The total power requirements of the integrated station facilities.

station set   An individual telephone.

station signaling rack   The equipment rack in a central office providing ringing current and the various tones needed for service advice.

station-to-station call   A telephone call to a particular number, not to a specified person or department at that number.

statistical multiplexing   Multiplexing in which channels are established on a statistical basis. For example, connections can be made according to anticipated need. Channels are allocated time slots on a main transmission facility dynamically according to actual demand rather than on a predefined basis.

stator   The stationary part of a rotating electric machine.

status channel   A channel indicating whether a group of bits is for data or control use.

status field   The bits of a link status signal unit in a common channel signaling system which indicate one of the major signaling link states.

status reports   Automatically generated reports from a switching center reporting service conditions and conditions of terminals.

statute mile   A unit of distance equal to 1.609 km or 5280 ft.

steady-state condition   1. In a communication circuit, a condition in which some specified characteristic of a condition — such as value, rate, periodicity or amplitude — exhibits only negligible change over an arbitrarily long period of time. 2. In an electrical circuit, a condition occurring after all initial transient or fluctuating conditions have damped out, in which currents, voltages or fields remain essentially constant or oscillate uniformly without changes in characteristics such as amplitude, frequency, or wave shape. 3. Equilibrium mode distribution.

stearin   A waxy compound used as flux by cable splicers working on lead sheathed cable joints.

steep wavefront   A rapid rise in voltage, indicating the presence of high frequency odd harmonics of a fundamental wave frequency.

steerable antenna   See antenna, steerable.

step   1. In automatic telephony, one unit of movement of an electromechanical switch, usually corresponding to one impulse from a rotary dial. 2. One of a series of galvanized steps screwed into a telephone pole to provide easy access to a cable terminal or distribution point.

step, detachable pole   A movable pole step which clips onto a plate which has been permanently fixed to a telephone pole. It provides easy pole climbing but only by authorized personnel in possession of the necessary step.

step-by-step exchange   An electromechanical exchange comprising switching stages working one after the other independently of the state of the following stages.

step-by-step selection   Selection, one after the other, of the links from switching stage to switching stage independently of the state of following stages.

step down   To decrease the range or value of an electrical parameter, for example, to decrease voltage from 110 V to 6 V.

step index optical waveguide An optical waveguide having a step index profile.

step index profile An index profile characterized by a uniform refractive index within the core and a sharp decrease in refractive index at the core-cladding interface. It corresponds to a power-law profile with profile parameter, g, approaching infinity.

step up To increase the range or value of an electrical parameter.

stereo Pertaining to a system for the reproduction of stereophonic sound.

stern chute A guide which is the last mechanical element a cable or repeater passes before entering the water during laying operations. Its radius of curvature should exceed the bending radius of the cable.

stern sheave The wheel at the stern of a cable laying ship over which cable or repeaters are deployed into the sea.

stile strip Vertical strip on a switchboard face which separates panels and gives circuits designations.

stimulated emission Radiation emitted when the internal energy of a quantum mechanical system drops from an excited level to a lower level as induced by the simultaneous presence of radiant energy at the same frequency. An example is the radiation from an injection laser diode above lasing threshold.

stop, finger Cresent-shaped stop on a rotary telephone dial which limits travel of the finger wheel.

stop band The frequency band between the two transmission bands of a bidirectional system.

stop bit In asynchronous transmission, the last transmitted element in each character, which permits the receiver to come to an idle condition before accepting another character.

stop-dial Off-hook signal received by an operator or by automatic equipment from the distant end of a toll circuit as an indication that no more digits should be transmitted.

stop-go signaling Signaling procedure between electromechanical central offices by which transmission of digits is stopped until switches are confirmed as available to receive and act upon them.

stop-record signal In facsimile systems, a signal used for stopping the process of converting the electrical signal to an image on the record sheet.

stop-send signal A signal returned from the incoming exchange following receipt of signals, usually containing numerical information, to indicate that the information has been received and that the repeated transmission may stop.

stop signal In start-stop transmission, one or more bits that terminate each character transmitted.

stopper A device by means of which a submarine cable can be made fast without cutting or bending. In one such system, a piece of grapnel rope is attached to the cable using the overlay splicing technique, and the bight of the rope is terminated in a thimble which can easily be connected to more rope or to a chain or anchor. Other devices include self-tightening grips and preformed wire stoppers.

storage Describing equipment which stores digital information for later use.

storage, dynamic Storage containing information that is moving in time and is thus not always instantly available.

storage, erasable See erasable storage.

storage, non-erasable See non-erasable store.

storage, non-volatile Information which can be retained in storage despite the absence of power and which becomes available again as soon as power has been restored. Magnetic tape is an example.

storage, parallel Storage in which all bits, characters or words are equally available, access time not being dependent upon the order in which they were stored.

storage, serial Storage in which access time varies according to the order in which information was stored. Storage can be serial by word, by character or by bit.

storage, volatile Storage employing media such that if the applied power is cut off the stored information is lost.

storage battery See battery, storage.

storage capacity The quantity of information that can be retained in a memory system, usually measured in kilobits or megabits.

storage register A device into which information can be inserted and retained, and from which it can be retrieved.

storage tank Large cylindrical tank in a cable factory or depot for the storage of sea cable.

storage tube An electronic tube which can be used to store information.

store 1. A storage memory unit in which information may be held until needed. 2. A material storage center holding cables, instruments, spares, etc. needed for maintenance and system development.

store, backing A store of much larger size than an immediate access store, but one requiring longer access time.

store, call A temporary working store in which information is retained until it can, if necessary, be transferred to a semipermanent store.

store, immediate access See immediate access store.

store, magnetic disc A store in which the data is stored in the magnetic material on the flat surfaces of discs.

store, magnetic drum Store in which data is stored in the magnetic material on the curved surface of a cylindrical drum.

store, random access A store designed to give roughly equal access times to all items stored.

store, working See immediate access store.

store-and-forward Method of transmitting data which does not require an immediate reply. The data is held in store until it is convenient for it to be sent on its destination.

store-and-forward switching center A message switching center in which a message is accepted immediately from the sender, held in a physical storage, and forwarded to the receiver in accordance with the priority placed upon the message by the sender.

stored program control (SPC) 1. Control of an automatic switching arrangement in which the call processing is determined by a program stored in an alterable memory. 2. Control of a function by sequentially interpreting information stored in a memory whose structure is independent of the function to be performed.

storm, ionospheric Ionospheric disturbance, usually due to unusual sunspot activity, which affects radiocommunication services, especially those in high terrestrial latitudes.

storm, magnetic See magnetic storm.

straggler message Teletypewriter message delivered to the wrong address because the preceding message did not bear a clear end-of-message indication, causing the switch to retain the connection to the earlier addressee.

straightforward call completion Operating procedure by means of which number-required information travels over the same circuit as is used to set up the call.

straightforward trunk Trunk circuit between two manual switchboards in which the insertion of a calling plug at one end automatically gives a calling indication at the distant end.

straight-line capacitance Capacitance employing a variable capacitor with plates so shaped that capacitance varies directly with the angle of rotation.

straight-line frequency Capacitance employing a variable capacitor with plates so shaped that the frequency to which the circuit is tuned varies directly with the angle of rotation.

strained layer superlattice Device made up from alternating layers of different semiconductor materials, with layers under electric strain: technique used in some advanced lasers.

strand Wires twisted together to form a strong and flexible cable.

strand, guy Wire rope, often of galvanized steel, used for making poles secure.

strand, messenger Cable strand used to support an aerial cable.

strand, suspension A messenger strand.

stranding, concentric Forming bare wires into a stranded cable with the wires wound in concentric helically wrapped layers. Used when a flexible conductor is desired but high tensile strength is not required.

strandlink A compression sleeve for joining steel strand cables.

strandvise A sleeve and steel wedge device for terminating a guy strand or messenger strand.

strap A connection link between two tags, terminals or items of equipment.

strap, cable A cleat used for supporting a cable against a flat surface.

strap, safety Leather or nylon harness strap worn by riggers and linemen working on antenna towers or telephone poles.

stratosphere The atmospheric layer above the troposphere, its outer limits being about 90 km above the Earth's surface.

stray capacitance Unintended capacitance between wires and components, which has a significant effect at high radio frequencies.

stray current Current through a path other than the intended one.

stray magnetic field Magnetic flux from nearby components, such as inductors, which can induce noise into a circuit.

streaking Picture condition in which objects appear to be extended horizontally beyond their normal boundaries. This will be more apparent at vertical edges of objects when there is a transition from black to white or white to black. The change in luminance is carried beyond the transition, and may be either negative or positive. For example, if the tonal degradation is an opposite shade to the original figure (white following black), the streaking is called negative; however, if the shade is the same as the original figure (white following white), the streaking is called positive. Long streaking may extend to the right edge of the picture, and in extreme cases of low-frequency distortion, can extend over a whole line interval.

stress, voltage Electric stress on a dielectric in an electric field.

striking voltage The voltage needed in order to start the flow of current across a gas discharge tube.

stringing Erecting overhead wires or aerial cables along a pole route.

strip, busy-designation Translucent strip placed over a circuit-busy lamp on a manual switchboard.

strip, designation Strip on a manual switchboard into which circuit designation labels are placed.

strip, fanning Plastic or wooden strip with regularly spaced holes which permit wires to be led out neatly from a cable form for termination on tag blocks.

strip, stile Vertical strip separating panels on a manual switchboard.

strip, terminal Block carrying terminals or tags for the connection of cables or jumper wires.

stripe, hash mark Colored identification stripe applied helically to conductor insulation in a multipair cable.

stripper Section at the end of a cable from which the outer sheath has been removed.

stripper, insulation Hand tool used to strip insulation from a conductor.

stripping Removal of the outer sheath from a cable prior to splicing or termination.

stroke speed In facsimile systems, the number of times per minute that a fixed line perpendicular to the direction of scanning is crossed in one direction by a scanning or recording spot. Syn.: scanning or recording line frequency

strowger See switch, strowger.

stub, cable Short length of cable containing pairs leading out from a main cable. The pairs are not terminated but are available for future use as needed.

stub, guy Guy used when local conditions make it impossible for a normal guy to be attached directly to a pole: horizontal guy goes from the pole in question to a shorter pole which is itself guyed in the normal way.

stub, matching Short length of transmission line used to match a feeder line to an antenna.

stub, quarter wave Length of radio frequency transmission line one-quarter wavelength long and short-circuited at the distant end. It has a high impedance at the operating frequency but effectively shunts even harmonics.

stub out To divert some cable pairs from a main cable and join them to a stub cable so that they will be available for future use without the need to open up the main cable splice.

stud, powder activated Hardened steel bolt designed to be shot into a masonry wall, leaving a threaded section protruding. It is used for mounting equipment on wall racks.

stuffing (pulse stuffing, stuffable digit time slot, stuffing service digits, stuffing digit, stuffing ratio, etc). See under justifiable, justifying, justification, bit stuffing; de-stuffing.

stuffing character A character used on isochronous transmission links to take account of differences in clock frequencies.

stunt box A device for controlling the nonprinting functions of a teleprinter. Control characters can be sent to it over the communications channel.

stylus Phonograph pickup.

subassembly Functional unit of equipment.

subband (cable TV) The frequency band from 6 MHz to 54 MHz, which may be used for two-way data transmission.

subcarrier A carrier which is applied as modulation on another carrier, or on an intermediate subcarrier.

subframe A sequence of non-contiguous sets of digits assembled within a pulse code modulation multiplexing frame, each set being repeated at n times the frame repetition rate where n is an integer > 1.

subgroup, trunk Section of a group of selectors with access to circuits leading to the same destination.

subharmonic A frequency equal to the fundamental frequency divided by a whole number.

submission The interaction by which an originating user agent in a message handling system transfers to a message transfer agent the content of a message plus the submission envelope.

submission and delivery entity The entity in the Message Transfer Layer (MTL) of a message handling system that is responsible for controlling submission of a message and delivery interactions with a Message Transfer Agency Entity (MTAE).

sub-office Name sometimes given to remote line concentrators, part of switching equipment physically separated from the main central office itself.

subrefraction Refraction for which the refractivity gradient is greater than standard.

subroutine A sequence set of statements that may be used in one or more programs and at one or more points in a program.

subscriber An individual station set user.

subscriber calling rate The call intensity of a subscriber line. Note that it should not be used to mean traffic rate, and that it should be made clear whether the rate refers to the originating calling rate, or to the terminating calling rate or to the sum of both.

subscriber distribution frame (SDF) A cable-pair distribution frame on a subscriber's/customer's premises.

subscriber line usage (SLU) Registers monitor peg and usage counts on any line or group of lines: the total time plant in use is used in calculating non-traffic-sensitive (NTS) costs, used in the allocation of revenues between carriers.

subscriber number The number which must be dialed to reach a telephone subscriber in the same local network.

subscriber plant factor (SPF) Factor used in apportioning costs of non-traffic sensitive (NTS) plant and message-related plant between intrastate and interstate operations.

subscriber traffic rate The traffic intensity of a subscriber line. May refer to the originating traffic rate, to the terminating traffic rate or to the sum of both.

subscriber trunk dialing (STD) The dialing by the subscriber of long distance calls within the country.

subscriber's alpha-numerical display The visual display at a telephone subscriber's terminal of information sent to or received from the public telephone network.

subscriber's line The telephone line connecting the end office to the subscriber's station.

subscriber's line free, charge A backward register signal indicating that the called subscriber's line is free and that the call is to be charged on answer.

subscriber's line free, no charge A backward register signal indicating that the called subscriber's line is free and that the call is not to be charged on answer. This signal is used only for calls to special destinations.

subscriber's line interface circuit (SLIC) A unit interfacing a subscriber's line with a digital telephone switching center.

subscriber's line use system (SLUS) An AT&T software routine used as a database for collection and analysis of customer usage practices.

subscriber's loop Pair of wires between the end office and the subscriber's telephone.

subscriber's meter showing call charge. Meter on the subscriber's premises showing call charge units debited.

299

subscriber's national telex number. Number used to reach a telex subscriber.

subscriber's telephone set A telephone instrument.

subscription TV TV broadcast in a scrambled form so that only those users with special equipment are able to watch the programs being transmitted.

subsequent address message (SAM) An address message, either one-unit or multi-unit, sent following the initial address message, in a Common Channel Signaling System.

subsequent signal unit (SSU) A signal unit of a multi-unit message other than the initial signal unit, in a Common Channel Signaling System.

subset 1. Subscriber's set. 2. A modem or data set. 3. A modulation-demodulation device designed to make the output of data processing equipment compatible with communication transmission facilities.

subsidiary carrier authorization (SCA) modulator Device which translates the audio frequencies of a sound program to a higher frequency band and uses this higher band to modulate an FM radio transmitter which is already being modulated by its normal audio program. The transmitter sends out both programs simultaneously, but special equipment is needed to demodulate the SCA program.

substrate The support of an integrated circuit, either a semiconductor or an insulator.

subsystem A functional unit of a system.

successful block delivery The transfer of a nonduplicate user information block between the source user and the intended destination user.

successful block transfer The transfer of a correct, nonduplicate user information block between the source user and the intended destination user.

successful call A call that has reached the wanted number and allows the conversation to proceed.

successful disengagement The termination of user information transfer between a source user and a destination user in response to a disengagement request.

sudden ionospheric disturbance (SID) Abnormally high ionization densities in the D region caused by an occasional sudden outburst of ultraviolet light on the sun (solar flare). This results in a sudden increase in radio wave absorption which is most severe in the upper MF and lower HF frequencies.

sulfuric acid A compound used in dilute form as an electrolyte in lead-acid storage cells.

sump, manhole A fitting at the lowest point of a manhole floor used to pump a manhole dry before working in it.

sunspot Variations in the surface condition on the sun leading, in turn, to variations in the ionosphere and affecting terrestrial radio communication services.

sunspot cycle A cycle of about 11 years between periods of maximum sunspot activity.

sunspot number Number of sunspots visible on the face of the sun at a given time.

super high frequency (SHF) Frequencies from 3 GHz to 30 GHz.

superband (TV) The frequency band from 216 to 600 MHz, used for fixed and mobile radios and additional television channels on a cable system.

superconductivity Reduction to zero of resistance to the flow of electric current which occurs in some metals when they are cooled to temperatures close to absolute zero (-273°C).

supergroup A supergroup link (which see) connected at each end to terminal equipment. This terminal equipment provides for the setting up of five group links occupying adjacent frequency bands in a 240 kHz band, or for one or more data transmission or facsimile channels. The basic supergroup occupies the band consisting of 312 to 552 kHz. In a submarine cable system, the five groups of a supergroup provide 60 voice channels at 4 kHz spacing or 80 at 3 kHz spacing.

supergroup distribution frame (SGDF) In frequency division multiplex, the distribution frame that provides terminating and interconnecting facilities for group modulator output, group demodulator input, supergroup modulator input, and supergroup demodulator output circuits of the basic supergroup spectrum of 312 kHz to 552 kHz.

supergroup link The whole of the means of transmission using a frequency band of specified width (240 kHz) connecting two terminal equipments, for example, group translating equipments, wideband sending and receiving equipments (modems, etc). The ends of the link are the points on supergroup distribution frames, or their equivalent, to which the terminal equipment is connected. It can include one or more supergroup sections.

15-supergroup assembly A 15-supergroup assembly link terminated at each end by terminal equipment. This terminal equipment provides for the setting up of 15-supergroup links or sections separated by free spaces of 8 kHz and occupying a band whose total width is 3716 kHz. The basic 15-supergroup assembly is made up of supergroups 2 to 16 occupying the frequency band consisting of 312-4028 kHz (or 8620 to 12336 kHz).

15-supergroup assembly link The whole of the means of transmission using a frequency band of specified width (3716 kHz) connecting two terminal equipments (supergroup modems permitting the setting up of a 15-supergroup assembly). The ends of the link are the points on 15-supergroup assembly distribution frames, or their equivalents, to which the terminal equipment is connected. As the frequency band occupied by a 15-supergroup assembly (8620 to 12336 kHz) lies within the band occupied by a basic supermaster group (8516 to 12388 kHz), a basic supermastergroup link can transmit one supermastergroup or an assembly of 15-supergroups.

15-supergroup assembly section The whole of the means of transmission using a frequency band of specified width (3716 kHz) connecting two consecutive 15-supergroup assembly distribution frames, or equivalent points, and connected at one or both ends to through 15-supergroup assembly connection equipment.

superheterodyne receiver See receiver, superheterodyne radio.

superimposed ringing Sending out normal interrupted ac ringing current to a called subscriber's line together with a continuous dc voltage from the central office storage battery. It is used on some party line selective-calling systems.

superluminescent diode A solid state light-producer with narrower angle of emission than a light emitting diode (LED) but wider emission angle than a laser.

supermastergroup A supermastergroup link (which see) connected at each end to terminal equipment. This terminal equipment provides for the setting up of three mastergroup links or sections separated by two free spaces of 88 kHz and occupying a band whose total width is 3872 kHz. The basic supermastergroup is composed of mastergroups, 7, 8 and 9 occupying the frequency band 8516 to 12388 kHz.

supermastergroup link The whole of the means of transmission using a frequency band of specified width (3872 kHz) connecting terminal equipment at both ends, for example, mastergroup translating equipments, wideband sending and receiving equipment (modems, etc.). The ends of the link are the points on supermastergroup distribution frames, or their equivalent, to which the terminal equipment is connected. The link can include one or more supermastergroup sections.

supermastergroup section The whole of the means of transmission using a frequency band of specified width (3872 kHz) connecting two consecutive supermastergroup distribution frames, or equivalent points.

superradiance Amplification of spontaneously emitted radiation in a gain medium, characterized by moderate line narrowing and moderate directionality. This process is generally distinguished from lasing action by the absence of positive feedback and, hence, the absence of well defined modes of oscillation.

superrefraction Refraction for which the refractivity gradient is less than standard.

supersonic Traveling at a speed greater than the speed of sound. Not to be confused with "ultrasonic" which describes sound at frequencies above the human audio range.

supervision The function of monitoring and controlling the status of a call.

supervision, answer The receipt of a signal at the originating office indicating that the called subscriber has answered. His off-hook signal produces a battery reversal which is sent back to the calling end.

supervision, disconnect Receipt of a signal at all of the offices involved indicating that the parties to the call have gone back on-hook at the end of the call.

supervisor A senior operator or one who controls the work of operators or other employees.

supervisory equipment 1. Equipment which controls a connection after it has been established. 2. Equipment located at the terminal station of a submarine cable system for the purpose of monitoring submarine repeater performance.

supervisory lamp Lamp in a cord circuit which shows the operator whether or not either party has gone back on-hook.

supervisory relay See relay, supervisory.

supervisory signals Signals used to indicate and control the various operating states of the circuits or circuit combinations involved in a particular connection.

supplement 1. A separate publication, related to a basic publication and prepared for the purpose of disseminating additional information. 2. When referring to Bell's Universal System Service orders, a modification of an order to correct, change or cancel the original order.

supplementary ground field (SGF) General description of a manmade ground field, usually employing an arrangement of a ring ground.

supplementary information Any information, which must be sent by the subscriber to the exchange for the performance of a control operation.

supplementary services Services beyond the basic services which are offered to telephone subscribers.

supply Procurement procedures such as purchasing, storage, issue, stock control and stores accounting.

supply, power Power company ac power or some other form of prime power supply fed into an office or station.

supply voltage Voltage at which the local power company supplies electric power for consumer use.

support, handset Plastic clip which can be affixed to a handset so that the handset can be supported on the shoulder while talking, leaving the hands free.

support, tangent Steel spiral wire which can be wound around a drop wire to give extra reinforcement where the wire passes a pole.

suppressed carrier Carrier restricted to a power level more than 32 dB below the peak envelope power, and preferably 40 dB or more below the peak envelope power.

suppressed carrier transmission Transmission, particularly radiocommunication, in which only one or both sidebands, not the carrier frequency component itself, are transmitted.

suppressed zero Said of a meter with the zero position outside of the marked scale.

suppression, noise Reduction of gain in a radio receiver in the absence of an incoming signal.

suppression hangover time Of an echo suppressor, the time interval between the instant when defined test signals applied to the send and/or receive input ports are altered in a defined manner, and the instant when the suppression loss is removed from the send path.

suppression loss Of an echo suppressor, the specified minimum loss which is introduced into the send path of the echo suppressor to reduce the effect of echo currents.

suppression operate time Of an echo suppressor, the time interval between the instant when defined test signals applied to the send and/or receive input ports are altered in a defined manner, and the instant when the suppression loss is introduced into the send path of the echo suppressor.

suppressor, blockless echo An echo canceller.

suppressor, full echo Echo suppressor installed at one end only of a long circuit. It can insert loss in either direction of transmission and so can suppress the return of echos to talkers at both ends.

suppressor, noise A noise limiter. In radio receivers, a squelch circuit reducing receiver gain in the absence of an incoming radio frequency signal.

suppressor, radio frequency A circuit element which absorbs radio frequency energy (such as that generated by automobile ignition circuits) so that interfering radiation is minimized.

suppressor, split echo A half-echo suppressor which is able to insert loss only in the transmit leg of a circuit. An equivalent suppressor must also be fitted at the distant end.

suppressor grid The fifth grid of a pentode electron tube which provides screening between plate and screen grid.

surface acoustic wave Advanced technology used in compact microwave devices including frequency-selective elements in networks, filters, delay lines and resonators.

surface leakage Leakage current from line to ground over the face of an insulator supporting an open wire route.

surface-mounted assembly Method of mounting components/devices onto the surface of printed circuit boards (i.e., instead of mounting them by inserting component leads through holes punched in boards).

surface refractivity The refractive index, calculated from observations of pressure, temperature and humidity at the Earth's surface.

surface wave See wave, surface.

surge A rapid rise in current or voltage, usually followed by a fall back to the normal value.

surround tapes The copper tapes which used to be applied helically to a central copper wire in a submarine cable to form a composite center conductor consisting of the central wire and three tapes. An obsolete configuration.

survey, route Detailed survey for a new buried cable route or a new pole line (indicating where poles are to be placed).

survivability Ability of a communications network to continue to provide service after major damage to any part of the system.

susceptance The reciprocal of reactance, and the imaginary component of admittance. It is expressed in siemens. See also admittance.

susceptibility 1. The degree to which a device, equipment, or a weapons system is vulnerable to effective attack due to one or more inherent weaknesses. 2. In electronic warfare, the degree to which electronic equipment is affected by electromagnetic energy radiated by enemy equipment, such as jamming transmitters.

susceptibility, electric The ratio of dielectric polarization to electric field strength, or the relative ease of polarization of a dielectric.

susceptibility, magnetic The ratio of magnetization to magnetic field strength, or the response of a material to a magnetic field.

susceptiveness In telephone systems, the tendency of circuits to pick up noise and low frequency induction from power systems. This tendency depends on telephone circuit balance, transpositions, wiring spacing, and isolation from ground.

swamp anchor See anchor, swamp.

sweep To vary the frequency of a signal over a whole band as a means of checking the response of equipment under test.

sweep circuit Saw-tooth waveform generator used to produce a scanning pattern for television tubes.

sweep frequency Type of oscillator whose frequency is varied over a band.

sweep voltage Voltage generated within a cathode ray tube which, when applied to the deflecting control plates or coils, causes the electron beam to be deflected.

SWIFT Packet-switching network used by banks to effect financial transactions with security. Abbreviation for Society for Worldwide Interbank Financial Telecommunications.

swing Variation between maximum and minimum readings of voltage, current or frequency.

swinging short An intermittent short circuit, such as that between two wires swinging together in a wind.

switch A mechanical or solid state device which opens or closes circuits, changes operating parameters, or selects paths or circuits, either on a space or time division basis.

switch, analog See analog switch.

switch, band Switch used in a radio receiver to change from one frequency band to another.

switch, circuit Switch which interconnects circuits, lines or trunks.

switch, coaxial A switching matrix used at radio stations to connect radio transmitters to different antennae.

switch, crossbar See crossbar switch.

switch, digital Solid state switch which switches digital signals by means of a time division matrix.

switch, end-cell Switch which brings additional cells into operation, when necessary, to supplement a central office storage battery.

switch, homing See homing type switch.

switch, instrument Switch on a power or control panel which puts a voltmeter across a load or introduces an ammeter into a circuit.

switch, knife Switch with hinged open brass blades which can be connected to stationary spring mounted contacts.

switch, mercury Small glass tube containing liquid mercury. When it is tilted, the mercury provides a low resistance path between two contacts inside the tube.

switch, message Switching center which receives messages on incoming circuits and retransmits them on outgoing circuits on a store-and-forward basis. It is used for data messages. It cannot normally be used when interactive simultaneous both way communications must be established.

switch, minor A uniselector or rotary-only stepping switch used for minor line-finding or other functions in a dial office. It has 10 outlets per switch. The wipers return to home position when not in use.

switch, reverting call Switch in a dial office used to send ringing current back out to a line to enable repairmen to adjust ringers.

switch, rotary A uniselector or rotary-only switch in a dial office, usually driven by self-energized steps. It has 25 or 50 sets of contacts around a semicircular arc and is used for line finding and for traffic concentration.

switch, series-parallel Double-pole double-throw switch used to change a connection from series to parallel.

switch, step-by-step Electromechanical switch with wipers which move over banks of contacts either on a step-at-a-time basis or automatically. Each switch deals with one of the digits dialed until the call gets through to the last or final switch which steps both vertically and horizontally under the direct control of the last two dialed pulses.

switch, stepping Electromechanical switch which steps its wipers across a bank of contacts.

switch, Strowger Electromechanical step-by-step switching system using both uniselectors and two-motion selectors. Each section of the office deals with one of the digits dialed. After a concentration stage, the call is passed on to connector switches dealing with a particular first digit, then to connector switches for a particular second digit, and so on until the call reaches the final connector switch which selects the called line by acting on the last two (or sometimes three) digits dialed.

switch, 3-digit An electromechanical Strowger type switch with a 200-outlet multiple. It is first switched to one of the banks of 100, then to the required level, then to the required rotary position.

switch, time Clock-controlled switch which activates circuits at specified times.

switch, wideband See wideband switch.

switch, X-Y Two-motion switch with banks in a horizontal plane such that the first digit moves the wipers horizontally to the desired column and the next movement pushes them forward to enter the bank and make contact.

switch busy hour The busy hour for a single switch. toll calls.

switchboard A manual operating position. Cord type boards have pairs of cords and keys on a horizontal desk and strips of jacks in panels on a vertical face.

Cordless boards are often microprocessor-controlled and enable similar operator tasks to be performed.

switchboard, A Switchboard at which subscribers' lines are answered.

switchboard, B Switchboard at which incoming toll calls are answered.

switchboard, central battery signaling Early type of switchboard using a central battery for signaling but requiring a local battery at each subscriber's premises to provide microphone current.

switchboard, common battery A manual switchboard which serves non-dial subscribers, using a central storage battery to provide microphone current.

switchboard, cordless See cordless switchboard.

switchboard, dial service auxilary (DSA) Switchboard associated with a dial office and used to provide operator assistance to subscribers.

switchboard, dial system B (DSB) Switchboard associated with a dial office and used for completing incoming calls via other operators.

switchboard, emergency reporting Small PBX serving police and fire call boxes.

switchboard, magneto See magneto switchboard.

switchboard, multiple Large manual switchboard with multiple appearances for every line so that all operators can obtain access to all lines.

switchboard, non-multiple Small manual switchboard with only one appearance of each circuit.

switchboard, power Cabinet or board for controlling the main ac power, including standby engines.

switchboard, telephone answering Special type of switchboard used at telephone answering services. Incoming calls are answered and messages recorded for later delivery.

switchboard, teletypewriter Switchboard used for control of teletypewriter circuits and providing TWX service.

switchboard, toll Switchboard controlling

switched circuit A circuit that may be temporarily established at the request of one or more of the connected stations.

switched digital capability The ability of a network to carry analog or digital signals even though some of the switching centers concerned may use space division rather than time division switching.

switched line Communications link for which the physical path, established by dialing, may vary with each use, e.g., a dial-up telephone circuit.

switched network A network providing switched communications service, i.e., the network is shared among many users, any of whom may establish communication between desired points when required.

switched network backup An option in some links and devices where a dial-up switched path is used as an alternative path if the primary path is unavailable.

switched service (CATV) A cable communications service in which each subscriber has a terminal and may communicate with any other subscriber.

switched transit country A transit country through which traffic is routed by switching in an international transit exchange.

switcher (TV) A control which permits the selection of one image from any of several cameras to be fed into the television display or recording system.

switcher/fader (TV) A device permitting gradual, overlapping transition from the image of one camera to another. Sometimes incorporated as part of a special effects generator.

switchhook The switch on a telephone which is controlled by the lifting of the handset, this goes "off-hook" to call the local central office.

switching Interconnecting circuits in order to establish a temporary connection between two or more stations.

switching, circuit Method of handling data traffic by interconnecting circuits to provide direct connection between calling and called terminals.

switching, machine Automatic switching.

switching, message See switch, message.

switching arrangement, common control (CCSA) See common control switching arrangement.

switching center See center, switching.

switching delay The time required to set up a call in a switching office or exchange.

switching-in not permitted The barring of network access to an engaged subscriber's line.

switching matrix An array of crosspoints in a space division exchange which operates as a switch from a traffic point of view.

switching network See network, switching.

switching node A point in a network where inlets and outlets can be interconnected.

switching-selector-repeater Step-by-step switch in a satellite office which picks up a junction circuit to the parent office as soon as the local calling line goes off-hook. Switches at both the parent office and the local satellite office step up under the control of dial impulses. If the call is for a line served from the local office, the junction is dropped and the call is switched through at the satellite. Syn: discriminating selector repeater, DSR

switching stage One or more connecting stages in a switching system which serve a particular switching function. This function is to be defined in each case. It may refer, for example, to one of a number of such stages within an exchange (with further qualification), or to an exchange within a national network. Also called selection stage.

switching system Any electronic or digital system which processes any input call to any output port.

switchover The transfer of a function to an alternative component in the event of a failure. Switchover may be manual or automatic.

switchroom Section of a central office housing automatic switching equipment.

switchtrain The series of switches used to move a telephone call through an office, particularly a step-by-step office.

symbol 1. Conventional representation of a concept. 2. A design used on diagrams to represent plant elements or circuit components. 3. In specification and description language, a representation of a state, input, task, output decision, or save.

symbol rate The reciprocal of the unit interval in seconds. It is expressed in bauds.

symmetrical binary code A pulse code derived from a binary code in which the sign of the quantized value, positive or negative, is represented by one digit, and in which the remaining digits constitute a binary number representing the magnitude. In a particular symmetrical binary code, the order of the digits and the use made of the symbols "0" and "1" in the various digit positions must be specified.

symmetrical channel A network parameter used to indicate that the send and receive directions of transmission have the same data signaling rate.

symmetrical pairs A balanced transmission line in a multipair cable having equal conductor resistances per unit length, equal impedances from each conductor to earth, and equal impedances to other lines.

sync Synchronization.

sync compression (TV) The reduction in the amplitude of the sync signal, with respect to the picture signal, occurring between two points of a circuit.

sync generator A device used to supply a common or master sync signal to a system of several cameras. This ensures that their scanning pulses will be in phase. Scanning pulses out of phase produce distortion or rolling, sometimes called "sync loss."

sync level (TV) The level of the tips of the synchronizing pulses.

synchro A synchronous device, such as a selsyn.

synchromonic frequencies Frequencies used for selective calling on party lines in the series 16, 30, 42, 54, 66 Hz.

synchronization The process of adjusting the corresponding significant instants of two signals to obtain the desired phase relationship between these instants. The words which describe different timing conditions use Greek prefixes with the base word, "chronous" (=time):

iso	=	equal
syn	=	together
homo	=	the same
meso	=	middle
plesio	=	near
hetero	=	different

synchronization bit A binary digit which is used for character synchronization.

synchronization network The collection of digital switching exchanges and their interconnecting synchronization links which carry information to ensure that all clock rates conform to the network standard.

synchronization signal unit (SYU) A signal unit in Common Channel Signaling System CCITT No. 6 containing a bit pattern and information designed to facilitate rapid synchronization, and which is sent on the signaling channel when synchronizing or when no signal messages are available for transmission.

synchronize To cause two systems to operate at the same speed.

synchronized network See synchronous network.

synchronizing In facsimile systems, the maintenance of predetermined speed relations between the scanning spot and the recording spot within each scanning line.

synchronizing level A processor timing level during which repetitive programs are run as initiated by a real time clock which interrupts the processor at a regular rate.

synchronizing pilot In frequency division multiplex, a reference frequency used for maintaining the synchronization of the oscillators of a carrier system, or for comparing the frequencies or phases of the currents generated by those oscillators.

synchronizing pulse Pulse causing two or more circuits to operate in synchronization.

synchronizing signal In facsimile systems, the signal which maintains predetermined speed relations between the scanning spot and the recording spot within each facsimile scanning line.

synchronous Signals whose corresponding significant instants have a desired phase relationship.

synchronous crypto-operation A method of on-line crypto-operation in which various parts of the terminal crypto equipment have timing systems to keep them in step, synchronism of the system being independent of the traffic passing through the channel concerned.

synchronous data channel See isochronous data channel.

synchronous data link control (SLDC) A bit-oriented IBM version of HDLC protocol, as used in IBM's Systems Network Architecture.

synchronous data network A data network which uses a method of synchronization between data circuit-terminating equipment and the data switching exchange, as well as between data switching exchanges. The data signaling rates are controlled by timing equipment within the network.

synchronous modem A modem which is able to transmit timing information in addition to data; a modem which must be synchronized with the associated data terminal equipment by timing signals. Sometimes called an "isochronous modem."

synchronous network A network in which clocks are controlled so as to run at identical rates, or at the same mean rate with limited relative phase displacement. Ideally the clocks are synchronous, but they may be mesochronous in practice. Syn.: synchronized network

synchronous operation Method of transmission of data or telegraph signals in which both terminals are kept in step by timing pulses, whether or not live traffic is being carried by the circuit.

synchronous orbit An orbit in which a satellite has an orbital angular velocity synchronized with the rotational angular velocity of the Earth, and thus remains directly above a fixed point on the Earth's surface.

synchronous satellite A satellite in a synchronous orbit.

synchronous system A system in which the transmitter and receiver are operating in a fixed time relationship.

synchronous terminal A terminal requiring the transmission of timing information from the data source for proper reception of data.

synchronous time division multiplex A multiplex in which timing is obtained from a clock which, in turn, controls both the multiplexer and the channel source.

synchronous transmission A transmission process such that between any two significant instants in the overall bit stream, there is always an integral number of unit intervals. Transmitting and receiving equipment are maintained in step by timing signals, which eliminates the need for start and stop bits (as used in asynchronous transmission) and significantly increases data throughput rates.

syncompex (synchronized compressor and expander) A form of linked compressor and expander which provides an improved signal to noise ratio on radiocommunication links by using digital instead of analog techniques to modulate the control channel. This gives the system great tolerance to end-to-end frequency error.

syntax The relationships among characters or groups of characters, independent of their meanings or the manner of their interpretation and use.

syntax diagram In man-machine language, a method of depicting the syntax of the input and output language by pictorial representation.

system 1. A collection of personnel, machines and methods organized to accomplish a set of specific functions. 2. In man-machine language, a stored program control switching system and its man-machine communication facility.

system, all relay Switching system using electromechanical relays for all functions.

system, automatic switching Switching system using analog or digital automatic devices to establish and supervise calls or to forward messages.

system, automatic transmission measuring (ATMS) System which makes transmission and noise measurements automatically on toll and long distance circuits, and provides a printout of results.

system, balanced polyphase A polyphase power system in which currents and voltages are symmetrical.

system, balanced three-wire A three-wire power system in which no current flows in the neutral conductor.

system, carrier See carrier system.

system, crossbar See crossbar system.

system, dynamic gas pressure Cable pressurization system which continuously feeds dry air to the cables of a network.

system, equivalent four-wire carrier Carrier system working on a two-wire path but using different frequency bands for the two directions of transmission, thus providing, in effect, a four-wire transmission path.

system, gas pressure Cable pressurization system which fills cables with dry air or nitrogen under pressure greater than normal atmospheric pressure. This improves customer service by facilitating rapid location of cable faults.

system, grounded An electrical system in which a neutral point is intentionally grounded.

system, group alerting Emergency alerting to a selected group of lines. Those which are busy at the time of alerting are given a special warning tone and are picked up as soon as they become idle. All transmission is one way: from the control station to the alerted stations.

system, key telephone See key system.

system, loading System of arranging loading coils at regular intervals along a cable route in order to improve transmission characteristics.

system, microwave relay Microwave radio system providing multichannel communications and incorporating the use of repeaters.

system, panel dial Electromechanical automatic switching system using continuously driven electric motors and vertical driving shafts. It is now obsolete.

system, public address A system of microphones, amplifiers and loudspeakers for addressing large audiences.

system, radio relay Microwave or VHF radio multihop multi-channel radio system.

system, static gas pressure Cable pressurization system in which pressure is built up to a steady overall level. All leaks must be repaired so that there is no day-to-day change in pressure. No air need be pumped into the system unless a cable fault occurs.

system, step-by-step Electromechanical automatic switching system in which calls are passed from stage to stage through the central office, the switches at each stage responding to one or more of the digits dialed by the calling subscriber.

system, synchronous System in which sending and receiving equipment, or digital central offices at both ends of long distance circuits, are kept operating at the same frequency on a synchronized basis.

system, transposition A system for interchanging the position of wires in pairs and of pairs in groups along an open wire route in order to minimize crosstalk and induction problems.

system blocking See access denial.

system blocking signal A control message generated within a telecommunication system to indicate temporary unavailability of the resources required to complete a requested access.

system busy 1. A busy state automatically imposed on equipment in response to a fault condition. 2. The status of trunk circuits that have failed the tests performed by automatic trunk testers and have been taken out of service pending testing by maintenance personnel.

system control signal unit (SCU) A signal unit carrying a signal concerning the operation of common channel signaling system CCITT No. 6, e.g. changeover, load-transfer.

system level The level of signal in a CATV system at the output of each amplifier. Must be carefully chosen and maintained for least distortion and noise.

system loading In a frequency division multiplex transmission system, the absolute power level, referred to zero transmission level point, of the composite signal transmitted in one direction.

system loss Of a radio circuit consisting of a transmitting antenna, a receiving antenna, and the intervening propagation medium, system loss is defined as the ratio Pt/Pa where Pt is the radio frequency power input to the terminals of the transmitting antenna and Pa is the resultant radio-frequency signal power available at the terminals of the receiving antenna. Both Pt and Pa are expressed in watts. System loss is expressed in decibels.

system noise Random energy generated by thermal and shot effects in the system, specified in terms of its rms level as measured in the U.S. in a four-MHz bandwidth centered within a six-MHz cable television channel.

system noise band A frequency band outside the regular transmission band of a submarine cable system which is monitored continuously for detection of excessive system noise.

system pilot In submarine cable systems, a supervisory pilot frequency which is inserted near the ongoing interface of one terminal station and extracted near the ongoing interface of the other terminal stations.

system programmer Programmer who plans, generates, maintains, extends and controls the use of an operating system for the purpose of improving the overall productivity of an installation.

system software A group of programs and subroutines all related to processing for a particular application area.

system standards 1. The minimum required electrical performance characteristics of communication circuits, which are based on measured performance of developed circuits under the various operating conditions for which the circuits were designed. 2. The specific characteristics which, although not dictated by electrical performance requirements, are necessary in order to permit interoperation. For example: the values for center frequencies for telegraph channels or test tones.

systems analysis Analysis of an activity to determine precisely what must be accomplished and how it is to be done.

systems network architecture (SNA) Layered protocol for data communications, developed by IBM.

systolic array VLSI architecture utilizing a regular array of identical processing elements with common control and timing so that each element performs the same function simultaneously but on different data items.

T flip-flop   A flip-flop unit with only one input wire, the T wire. When T is pulsed the device changes state.

T network   Network with two series branches and one shunt branch. The shunt joins the network at the point between the two series units.

T 1   The basic 24-channel 1.544 Mb/s pulse code modulation system as used in the United States.

table, trunk adjustment   Traffic table indicating the number of trunks needed in a group to carry traffic with a specified grade of service, based on the number of existing trunks and meter readings for "all trunks busy" (ATB), "last trunk busy" (LTB) and "overflows registered" (OF).

table, truth   See Boolean algebra.

table driven   The logical process by which a user-entered variable is matched against an array of predefined values; freqently used in network routing.

tabulator   Automatic toll ticketing equipment device which sorts the output before passing it on as punched tape.

tacker   Stapling device for fixing small cables to walls.

tactical automatic digital switching (TADS)   A transportable store-and-forward message switching system designed for rapid deployment in support of tactical forces.

tactical communications   A method of conveying information among tactical forces by means of electronic equipment.

tactical communications system   A military communications system of fixed-size, self-contained assemblages designed to meet the requirements of ever changing situations among tactical forces.

tactical digital information link (TADIL)   A class of data links with rigidly controlled protocol.

tactical load   That part of the operational load consisting of weapons, detection, command control systems, and related functions.

tag   1. Number or other label identifying a cable, cable pair, or pole. 2. Flattened brass pin on terminal block, used for termination of wires either by soldering or wrapping.

tag, cable   Label identifying cables in manholes.

tag, duct splice   Special label marking either end of a duct section containing a cable splice.

tag, pair identification   Plastic label giving circuit or line designation which can be snapped over wires at binding posts.

tag, pole   Lead or aluminum label giving the serial number, and sometimes the date of installation, of a telephone pole.

tag, terminal   Label on a stub cable giving details of pairs therein.

tag board   1. A terminal board on which cable pairs can be terminated. 2. Strip of plastic, fiber or leather with numbered holes into which cable pairs are placed as they are identified during splicing operations.

tagging   Numbering cable pairs by feeding them through holes in a tag board.

tailing   In facsimile systems, the excessive prolongation of the decay of the signal. Syn.: hangover

takeoff angle   See departure angle.

talk-back circuit   Device used with 4-wire telephones to feed a proportion of the outgoing signal back into the local receiver circuit in order to provide sidetone.

talk-off   Disconnection of circuits caused by voice signals of large amplitude simulating a momentary off-hook condition.

tandem   The connection of the output terminals of one network, circuit, or link directly to the input terminals of another.

tandem center   1. An installation in a communication system in which switching equipment connects trunks to trunks, but does not connect customer loops. 2. An exchange used primarily as a switching point for traffic between other exchanges.

tandem cross-section program   (TCSP) An American Telephone and Telegraph software routine used for economic evaluation of alternative tandem exchange configuration.

tandem data circuit   A data channel passing through more than two data circuit-terminating equipment (DCE) devices in series.

tandem exchange   A switching center on which inter-office or inter-exchange circuits are terminated.

tandem office   A major switching center linking together end offices particularly in a densely-settled area where it is uneconomic for direct interconnection to be provided between all end offices.

tandem switching system   A network which permits calls to be routed through two or more relay points or switching centers in tandem.

tandem trunk   Trunk to or from a tandem office or tandem exchange. Also, a trunk between end offices which can be used to pass traffic to other end offices.

tank, cable   Cylindrical cable stowage spaces aboard a cable ship. Also found in submarine cable factories and depots.

tank, gas   1. Tank of nitrogen or dry air under pressure used in a cable pressurization system. 2. Tank of liquid petroleum gas used as an energy source.

tank circuit Circuit with inductance and capacitance which stores energy at frequencies near its resonant frequency.

tap 1. A branch or intermediate circuit. 2. To monitor conversations on a line.

tap, bridged Part of a cable pair which is bypassed because the equipment is connected to the pair at a point other than its end. Such a tap represents an impairment to transmission, the capacitance between wires providing a shunt for high frequency signals.

tap, CATV A device used on CATV cables for impedance matching or connection of subscribers' drops.

tap, compression Clamp used to connect a power lead to a main power conductor without breaking and terminating the conductor.

tap, distribution wire Terminal box used for joining drop wires to rural distribution wires without breaking and terminating the latter.

tap, gutter A compression tap insulated by tape and left in a wiring gutter.

tape 1. Magnetic or punched paper tape used for information storage. 2. Narrow strip of material used for wrapping, insulating, or shielding.

tape, aluminum Tape of soft aluminum wrapped around cable splice to provide continuity of electrostatic shielding.

tape, AMA Paper tape used in some automatic message accounting machines. It is 3 inches wide and accomodates 28 holes punched crosswise for giving call information.

tape, chad Paper tape with punched holes representing characters.

tape, chadless Paper tape with character holes which, although only partially punched out, can still be sensed by a tape reader.

tape, cigaret wrap Longitudinally applied cable insulating tape.

tape, closure sealing Sealing-compound tape designed to provide a moisture-proof seal around the outer sheath of a cable at the point at which it enters a splice case.

tape, DR A double layer of rubber tape used as a temporary outer wrapping for large, incomplete cable splices.

tape, electrical An insulating tape used to provide good temporary insulation around telephone cable splices or permanent insulation for domestic wiring splices.

tape, fish Narrow, springy steel tape designed for pushing through short lengths of ducting in order to pull cables in from the far end.

tape, friction Cloth adhesive tape for use with electrical insulation tape.

tape, laminated Insulating tape consisting of two or more layers of different materials.

tape, lashing Nylon tape used for lashing aerially suspended cables to messenger wires.

tape, lead serving Lead tape for wrapping around the outside of lead-sheathed cables to increase diameter at clamps.

tape, magnetic Plastic tape with a thin coating of magnetizable particles, used in tape recording.

tape, marker Identifying tape placed around the core under the outer sheath of a cable, giving the cable's manufacturer and specification.

tape, measuring Tape of fiber or metal marked with a linear scale. Usually supplied in a carrying case into which the tape may be rewound when not in use.

tape, paper Tape made of paper and used to record and store information by means of punched holes, partially punched holes and/or imprinting.

tape, plastic electrical Insulating adhesive tape used around spliced conductor joints.

tape, punched Paper tape for storing data by means of punched holes.

tape, rubber Tape made of rubber with two-sided adhesive finish used to insulate power conductor splices.

tape, sealing Tape used to make a gas-tight seal where a cable enters a mechanical splice-case.

tape armor Steel tape helically applied to submarine and land cables to protect them from damage from the back-filling of land cable trenches in direct-burial installations.

tape reader A device which reads messages which have been coded onto punched or magnetic tape.

tape recorder Device that feeds tape past a control head at a constant speed so that signals can be recorded by selective magnetization of tape particles.

tape relay A method of retransmitting teletypewriter traffic from one channel to another, in which incoming messages are recorded on perforated tape which is then either fed automatically into an outgoing channel or manually transferred to a position with an automatic transmitter on an outgoing channel.

tape relay, automatic A store-and-forward message switch which receives incoming messages on tape and automatically retransmits them to address of destination.

tape relay, semiautomatic A store-and-forward message switch which permits an operator to punch in the necessary routing instructions for retransmission of an incoming taped message.

tape relay, torn tape A store-and-forward message switch which requires that an incoming taped message be torn from the receiver and transferred to a separate transmitter for forwarding.

tape speed Speed at which magnetic tape is moved past the recording or playback head.

tape wrap Tape wrapped along or around a wire or cable for insulation or protection.

taper, potentiometer Distribution of resistance around a rotation-type potentiometer to give finer control of change at high end, at low end or uniformly over the whole angle of turn.

tapered fiber waveguide A waveguide tapered along its length. Syn. tapered transmission line.

tapes, intercalated Tapes applied helically in an overlapping pattern around a cable core.

tapped resistor   A resistor with intermediate terminals tapped off so that it can be used as a potential divider.

tapping element   A T-coupler for dividing power between two outgoing optical waveguides, or for combining two inputs into a single fiber.

tar flooding   The application of hot, fluid tar to armored cable at or near the point of armoring.

target language   The language into which a statement is to be translated. Syn. object language

target processor   The principal processor in the system concerned; the term may be used for the processor on which firmware runs and is used, as opposed to the host processor.

target program   A program in a target language which has been translated from a source language. Syn. object program

tariff   1. The schedule of rates and regulations governing the provision of telecommunications services. 2. A document filed with a regulatory body by a common carrier which: defines service offered, establishes rate customer will pay, and states general obligations of the common carrier and customer.

tarnish   Surface corrosion of a material due to exposure.

tarpaulin   Heavy waterproof canvas often used as a canopy to provide shelter.

task   1. A unit of work for a central processing unit. 2. In specification and description language, any action within a transition which is neither a decision nor an output.

taut wire   A small-gauge, high-tensile steel wire which is overboarded with an anchor and paid out with controlled tension over the stern of the cable ship during a submarine cable laying operation; the length of the wire is continuously measured and thus provides the exact distance between the ship and the fixed geographical starting point. This information is used to determine slack, and as a corroborative aid to navigation; called also piano wire.

T-carrier   The general designation of any Bell digital transmission system.

T1: DS-1 formatted digital signal at 1.544 Mbit/s;

T1C: DS-1C formatted digital signal at 3.152 Mbit/s;

T2: DS-2 formatted digital signal at 6.312 Mbit/s.

TE mode   Abbreviation for transverse electric mode.

tearing   A term used to describe a picture condition in which groups of horizontal lines are displaced in an irregular manner. Caused by lack of horizontal synchronization.

tech rep   A manufacturer's technical representative.

technical control facility (TCF)   Physical plant containing the necessary equipment to enable telecommunications systems control personnel to exercise the essential operational control over communications paths and facilities, make quality analyses of communications and communications channels, monitor operations and maintenance functions, recognize and correct deteriorating conditions, restore disrupted communications, provide requested on-call circuitry, and otherwise direct such actions as may be required to ensure the fast, reliable, and secure exchange of defense information.

technical control hubbing repeater   See data conferencing repeater.

technical load   The portion of the operational load required for communication, tactical operations and ancillary equipment. It includes any lighting, air conditioning, or ventilation required for continuity of communication.

Technical Reference   A publication that gives additional descriptive and technical details to supplement a tariff.

tee coupler   In optoelectronics, a passive coupler which connects three ports.

tee junction   A three-way junction, particularly of waveguides.

telco   Generic abbreviation for telephone company or telecommunications administration.

tele-alarm service   A supplementary service which transmits, via the public telephone network to a predetermined terminal, information indicating that abnormal conditions exist at a given location.

telecast   A televised broadcast.

telecommunication   Any process that enables a correspondent to relay written or printed matter, fixed or moving pictures, words, music, or visible or audible signals or signals controlling the functioning of mechanisms, etc. by means of an electromagnetic system.

telecommunication circuit   See circuit, telecommunication.

telecommunication security   The restriction of the information contained in telecommunication signals to those authorized and equipped to extract it. See communications security.

telecommunication service   A specified set of user-information transfer capabilities provided to a group of users by a telecommunication system.

Telecommunication Service Request   (TSR). Data card for transmitting requests for telecommunications services.

telecommunication system   A system which performs the basic functions of acceptance, transmission and delivery of telecommunicated messages.

telecommunication system operator   The organization responsible for providing a telecommunication service.

telecommunications   The technology concerned with communicating at a distance.

Telecommunications Access Method (TCAM)   An IBM software package used for the management of communications networks.

telecommunications center   A facility responsible for receipt, processing, transmission, and distribution of incoming and outgoing messages among a number of terminals.

telecommunications traffic (teletraffic)   A flow of attempts, calls and messages; the science (and art) of designing networks so that the maximum traffic may flow with the minimum of expenditure on equipment.

telecoms Abbreviation for telecommunications.

telecon 1. A teletypewriter conference. 2. A telephone conversation.

teleconference A conference between persons linked by a telecommunications system.

telecopie French term meaning document facsimile telegraphy.

telefax Facsimile service between subscribers' stations via the public switched telephone network or the international Datel network.

telegram Message sent by telegraphy.

telegraph Communication system for the transmission of written messages by manual or machine code.

telegraph, bridge-duplex Two-way telegraph transmission system using a balanced bridge network for separation between sent and received signals.

telegraph, differential duplex Two-way telegraph transmission system using a single wire between stations.

telegraph, inverse-neutral Telegraph transmission system with current flow during spacing intervals but not during mark signals.

telegraph, neutral Telegraph in which messages are transmitted by opening and breaking a looped circuit. The closed circuit transmits a mark signal; the open circuit indicates a space.

telegraph, polar Telegraph transmission system in which the current flow is reversed.

telegraph key Hand-operated key for making and breaking a telegraph circuit, usually by using the Morse code.

telegraph key, automatic Hand-operated telegraph key with spring loaded contact which vibrates to give Morse code dot signals at an appropriate speed.

telegraphy The technology associated with sending messages electrically and providing a written record of the message transferred.

telegraphy, carrier Use of a modulated carrier frequency for the transmission of telegraph signals.

telegraphy, printing Telegraphy in which the signal received is converted into a machine printed message.

Tele-Lecture Educational system providing amplified voice communication between a lecturer and one or more classrooms which may be in remote locations.

telematics (French telematique). Term for the socio-economic significance of the convergence of computers and telecommunications. Syn.: information technology

telemeter To use telemetry.

telemetry The science of transmitting the results of measurements to a distant station where they are interpreted and/or recorded.

telephone Device which converts the human voice into electric signals for transmission by line, radio, or fiber to a distant point where the signals are reconverted to sound waves.

telephone, antisidetone Telephone with a balancing network which ensures that only a small proportion of the originated signal is fed through the receiver at the originating end.

telephone, coin See coin telephone.

telephone, common battery Telephone operated by battery power from the central office

telephone, deaf-aid Telephone with special receiver amplifier. New telephones for the deaf now include alphabetic keyboards and a visual display strip to enable the typing of short messages.

telephone, desk Any telephone designed to be operated while resting on a desk or other horizontal surface.

telephone, field A telephone designed for portable use under severe conditions and working to different types of central offices

telephone, fire reporting Special telephones providing direct access to a fire control or reporting switchboard.

telephone, hand test Special telephone used by maintenance personnel at central office locations. It can be plugged into equipment or clipped to cable pairs for use at cable splices.

telephone, hard-of-hearing Telephone in which an amplifier in the receive circuit is powered from line current. See also telephone, deaf-aid.

telephone, intrinsically safe A telephone in a safety case which protects the surrounding atmosphere from all sources of sparks, thus preventing ignition of atmospheric gases. Also called mine telephone.

telephone, key Telephone with keys which provide immediate access to other telephones on the same key system.

telephone, local battery Telephone which requires a local battery in order to provide microphone current.

telephone, loudspeaker A hands-free telephone with microphone and loudspeaker built in to desk units.

telephone, magneto Telephone which incorporates a magneto, either hand cranked or local battery driven, for calling the local switchboard.

telephone, message waiting Telephone with a small warning light, often used in hotels to signal guests that a message is waiting for them at the reception desk.

telephone, mine See telephone, intrinsically safe.

telephone, multiline Telephone with push-buttons or keys to enable it to be used on two or more separate central office lines or PBX extensions.

telephone, outdoor Telephone protected from exposure to weather by a heavy outer case.

telephone, panel Telephone (often with built-in loudspeaker and microphone) designed for flush-fitting on a wall and without exposed cables or cords.

telephone, railroad Party line magneto telephone used on railroads.

telephone, wall Telephone designed for wall mounting.

telephone answering service (TAS) A company which arranges to answer telephone calls for subscribers who are temporarily away from their offices.

telephone circuit The electrical path whereby a direct connection is made between two manual or automatic exchanges. An international circuit directly connects two international exchanges in different countries. A trunk circuit connects two domestic exchanges.

telephone density Number of telephone instruments per 100 inhabitants. (Sometimes, however, density is quoted as the number of working telephone lines per 100 inhabitants).

telephone engineering center (TEC) The stores, workshops, garages, and technical engineering offices of a telephone organization, especially where these are separate from the administrative and switching offices.

telephone influence factor (TIF) A measure of the harmonic content of a power signal. The TIF of a voltage or current wave in an electric supply circuit is the ratio of the square root of the sum of the squares of the weighted RMS values of all the sine wave components (including fundamental frequency and all harmonics) to the unweighted RMS value of the entire wave.

telephone number The number assigned to a telephone connected to a central exchange.

telephone shop Shop at which telephone bills may be paid, service queries answered and various types of subscribers' apparatus demonstrated and purchased.

telephone sidetone See sidetone. The transmission and reproduction of sounds through a local path from the transmitting transducer to the receiving transducer of the same telephone so that the talker hears his own voice in the receiver.

telephone signal In a common channel signaling system, any signal which pertains to a particular telephone call or to a particular speech circuit.

telephone system A telecommunication system set up for the transmission of speech or other sounds.

telephone tag Repeated attempts by two persons to establish a telephone conversation; each time the call is set up one of the parties has just become unavailable.

telephony The engineering science of converting voices and other sounds into electrical signals which can be transmitted by wire, fiber or radio and reconverted to audible sound upon receipt.

telephoto Transmission of facsimiles of pictures by radio or wire.

teleprinter See teletypewriter.

teleprinter exchange service A service providing interconnected teletypewriters so that customers dialing or keying calls from station to station can communicate by using the teletypewriter equipment rather than telephones. Note that TWX operates at a different speed than telex but because both are start-stop (not synchronous) operations, interworking is possible.

teleprocessing 1. An information transmission system in which telecommunications, automatic data processing, and man-machine interface equipment function as an integrated whole. 2. A form of information handling in which a data processing system utilizes communications facilities. Teleprocessing is distinguished from distributed data processing (DPP) because DPP does not depend on remote communications.

telesoftware Software transmitted by line or broadcast facilities to specially adapted television receivers employing an additional microprocessor. Upon receipt, interactive routines can be undertaken using the keypad without further external transmission.

Teletel French version of videotex. Formerly known as Titan.

teletex A sophisticated update of the international telex service introduced in many countries in the mid-1980s. Teletex reproduces upper and lower case characters so the output resembles a facsimile copy of a well-typed letter. Transmission speed is many times faster than telex. Teletex transmission is memory-to-memory between intelligent devices similar to word processors. Teletex systems provide gateways for interconnection with the older and slower telex systems.

teletext A system which permits a limited number of pages of text to be transmitted by television broadcasting stations together with their program emissions, the special signals being transmitted via two of the unused lines in an ordinary video signal. A special decoding unit in domestic receivers permits selection and display on the screen. All the information is transmitted on a cyclic repetition basis by the broadcasting station without need of a telephone line.

teletraffic Telecommunications traffic.

Teletype Teletypewriter made by the Teletype Corp.

teletypesetter Teletypewriter used to set up messages in type by remote control.

teletypewriter (TTY) A telegraph instrument having a signal-actuated mechanism for automatically printing received messages. It may also have a keyboard similar to that of a typewriter for sending messages.

teletypewriter control unit (TCU) A device that serves as the control and coordination unit between teletypewriter devices and a message switching center.

teletypewriter exchange service (TWX) A service whereby a subscriber's leased teletypewriter is connected to a TWX switchboard, and from there can be connected to the teletypewriter of any other subscriber to the service either in this country or abroad.

teletypewriter signal distortion The shifting of the transition of the signal pulses from their proper positions relative to the beginning of the start pulse. The magnitude of the distortion is expressed in percent of a perfect unit pulse length. Syn.: start-stop TTY distortion

televise To transmit by television.

television Electrical transmission and reception of transient visual images together with directly associated audible signals.

television, community antenna (CATV) A system which supplies television signals to homes by means of a distributed cable network. It enables customers to have access to more channels and better quality pictures than would be possible by means of the television antenna alone.

television, master antenna Television which utilizes special antennae to feed signals by amplified cable circuits to a group of individual television sets.

television camera Device used to convert optical images into electrical signals. The optical image is made to fall on a plate of sensitized material, just as in a normal photographic camera but, instead of chemical processes which make the picture permanent, a series of electrical scans takes place, in effect reading off the picture from its plate and establishing a modulated signal. The signal is fed through amplifiers to TV transmitters. In color TV there are usually three separate camera tubes, each dealing with a particular color spectrum range.

television households A household having one or more television sets. Estimates for each U.S. county are based on an updating of ARB (American Research Bureau) TV penetration estimates. The number of households that watch TV in any locality during prime hours determines the top 100 markets for FCC cable rules.

television market A city or complex of closely associated cities served by commercial TV broadcast signals from one or more TV stations located within the area. The FCC uses television markets for designating what kind of cable services an operator should provide in terms of signal carriage and non-broadcast channel use.

television receiver Device that receives a broadcast television signal and converts it into a picture accompanied by its associated sound.

television signal Radio signal which includes both the video components and the audio channel.

television standards In the United States and many other non-European countries, television pictures use a 525 lines per frame system. European countries and countries using European-based TV systems employ a system with 625 lines per frame.

telex A teletypewriter exchange service available worldwide through various telecom administrations and companies.

telex destination code A set of digits used for routing purposes and characterizing the subscribers or stations of a country or a network.

telex network identification code Letter or two-letter combination identifying the subscribers or stations of a country or network.

Telidon A sophisticated videotex service provided by the Canadian Department of Communications.

Telpak A Bell System bulk private line service that gives large users substantial savings over standard private line rates. Telpak is no longer available to new customers.

Telset Finnish version of videotex.

Telstar The name of the first active communication satellites, launched in 1962 and 1963 and used until 1965. They were low-altitude satellites with perigees of about 600 miles and apogees of up to 7000 miles.

temperature coefficient of attenuation Measurement of change of attenuation per degree centigrade temperature change. The temperature coefficient itself may be a function of temperature and frequency.

temperature compensation Equipment design such that the effect of a temperature change in one component of the system is offset by the effect of the same temperature change in other components of the system, thereby keeping the overall transmission response substantially independent of temperature.

temperature inversion An increase in temperature with height.

temperature profile Graphic representation of the sea bottom temperature along the route of a submarine cable. For the shallow water portion, the seasonal fluctuations of the temperatures are also shown.

tempest An unclassified short name signifying investigations of compromising emanations.

temporal coherence See coherent.

ten high-day busy hour The hour, not necessarily a clock hour, which produces the highest average load for the ten highest business-day loads in a busy season.

ten-party line A common line arranged to serve ten main stations. Ringing selectivity is dependent on central office multiparty ringing arrangements.

tensile strength The pulling stress needed to break a material.

tension 1. Potential difference. 2. Mechanical stress.

tent, splicer's aerial Canvas tent with a light metal frame designed to be erected on an aerial platform to protect a splicer working on an aerial cable.

tent, splicer's ground Canvas tent with a metal frame designed to be erected over a manhole or cable-jointing chamber opening to give extra protection to maintenance workers.

tera- SI prefix signifying one trillion (10^{12}).

terahertz (THz) A unit denoting one trillion (10^{12}) hertz.

terminal 1. A point at which a circuit element may be directly connected to one or more other elements. 2. Input/output device connected to a processor or computer in order to communicate with it and control processing. An intelligent terminal has some local computing power and an associated data store.

terminal, aerial cable A device mounted on or near a pole carrying an aerial cable and on which the pairs of the aerial cable have been terminated.

terminal, buried cable A mounted cable terminal for the pairs of a buried cable.

terminal, cable A set of binding posts connected via a stub cable to the pairs of a distribution cable, the whole mounted on an insulating base and enclosed in a weatherproof housing.

terminal, cross-connecting A flexibility point at which pairs from two or more distribution cables are terminated and may be cross-connected as required to meet service demands.

terminal, data The equipment connected to the end of a transmission line to provide a terminal for the transmission or reception of data.

terminal, dispatch Direct link from a subscriber's telephone to a switchboard controlling a mobile radio telephone system in order to provide the subscriber concerned with superior service to subscribers on the mobile system.

terminal, gas-tight Cable terminal with cable stubs fitted with pressure plugs to permit use of pressurized cables.

terminal, in-and-out Cable terminal or flexibility point at which some or all cable pairs are cut and terminated in both directions. This avoids losses due to bridged taps.

terminal, LD Appearance of a subscriber's line as a direct termination on a long distance or toll switchboard in order to provide superior service.

terminal, looping Method of bringing plastic sheathed cables through a terminal by removing the outer sheath and permitting through circuits to loop through the terminal without being cut or terminated. Only pairs actually needed at the terminal are broken out and terminated there.

terminal, master carrier Carrier terminal whose operating frequencies are controlled by an extremely stable oscillator.

terminal, mobile radio control Control unit at the base radio station for a link or network. It causes the radio and line paths to interface and provides conversion from four-wire to two-wire working.

terminal, protected cable Cable terminal in which all the cable pairs are protected by carbon or gas-discharge protective devices.

terminal, protected cross-connecting A cross-connected cable terminal with a special fuse inserted into each cross-connection path.

terminal, quick connect Special moulded terminal block with spring grip fasteners for jumper wires. By this means, connections can be made without stripping the insulation from the end of a jumper wire by merely inserting a special tool.

terminal, ready access Plastic splice closure for standard terminal blocks.

terminal, slave carrier Carrier terminal with frequency-dependent equipment operating under the control of a stable pilot frequency received from a master station.

terminal, solderless Binding post or other terminal in which conductors are secured by screw compression.

terminal, splice case Aerial cable terminal mounted on a splice case closure.

terminal, strand-mounted Aerial cable terminal fastened to the cable messenger strand rather than to a pole.

terminal, underground cable Waterproof cable terminal designed for installation in a manhole.

terminal, unprotected cable Cable terminal with binding posts but no protective devices.

terminal, watchcase Small one-pair cable terminal designed to be attached to the messenger strand of a suspended multipair distribution wire. It joins one of the pairs to a one-pair drop wire leading to a subscriber's premises.

terminal, wire Terminal strip designed for use with multipair distribution wire.

terminal, wire wrap A tag with sharp corners designed for a wire wrapping tool to be used to make connection.

terminal balance The process of adjusting the impedances in the transmission paths of connections of intertoll trunks to toll-connecting trunks. The purpose is to minimize the echo returned on the intertoll trunk.

terminal charge management A Bell system enhancement to its Dimension® PABXs which provides a customer-operated station reconfiguration system to control calling privileges for the various PABX extensions.

terminal country A terminal country is both the country of origin and the country of destination in a given transmission.

terminal-engaged signal A signal sent in the backward direction indicating that a data call cannot be completed because the called terminal's exchange access line is engaged in another call.

terminal equipment 1. Device which terminates a telecommunications channel and enables the channel to be used by man or machine for the transmission of information. 2. Equipment at a subscriber's or user's terminal including such items as telephones, key systems, and PBXs.

terminal grade circuit A comparatively high-loss circuit between toll center and end office the use of which is restricted to terminal traffic calls.

terminal impedance The complex impedance seen at the unloaded output terminals of a line or transmission equipment which is otherwise in normal operating condition.

terminal interface processor (TIP) A node in a packet-switching network.

terminal international exchange An international exchange which is not connected directly to intercontinental transit circuits, but gains access to the intercontinental transit network through one or more intercontinental transit exchanges.

terminal pair Pair numbering designation on a cable terminal. Numbering normally goes from pair 1 up to the capacity of the terminal, rather than duplicating the numbers of the main cable pairs to which these terminated pairs are ultimately joined.

terminal repeater Repeater at the end of an amplified trunk.

terminal room In a manual central office, the room containing the main distribution frame and relay sets associated with the operating positions and trunks and lines terminated thereon.

terminal station   In a submarine cable system, the physical plant located near the landing point of a sea-cable system and containing the terminal equipment.

terminal-per-line   Party line system with one central office terminal to serve all the parties on the line. Selective ringing is used to ring a particular party.

terminal-per-station   Party line system with one central office terminal for each party on the line.

terminals   Binding posts, tags, or lugs to which an external circuit may be connected.

terminate   1. To connect a line to a terminal or to equipment. 2. To connect an appropriate bridge (normally of the characteristic impedance) across a circuit in order to make it stable.

terminated   1. The situation in which a cable pair is connected to binding posts or to other equipment. 2. The situation in which a circuit is connected to a load of its characteristic impedance or iterative impedance.

terminated line   A telephone circuit with a resistance at the far end equal to the characteristic impedance of the line so no reflections or standing waves are present when a signal is entered at the near end.

terminating exchange   The local end office on which the called subscriber's line is terminated.

terminating link   A toll connecting trunk joining a toll office to a local end office.

terminating priority   Feature which advances a call to an assistance operator if it encounters an all trunks busy or called line busy condition.

terminating set   Network of components designed to give a balanced impedance for connection to a hybrid coil used as a two-wire or four-wire termination.

terminating toll center   Toll center through which the called subscriber's line is reached.

terminating traffic   Traffic destined for the end office concerned, whatever its origin.

termination   1. Connection of a line to equipment. 2. Impedance connected across the end of a circuit to provide stability during testing.

termination, balance test   Device connected to the far end of a line to test for balance.

termination, balanced   Load presenting the same impedance to ground for each output terminal of a device with two such terminals.

termination, idle line   Network connected across the terminals of an amplified circuit to provide balance when the circuit is not in use. This prevents the idle circuit from singing and thereby inducing interference with other circuits.

termination, matched   Terminating a line with its iterative impedance so that it appears electrically to be of infinite length.

termination, midcoil   Use at the terminal of a loading coil of half the inductance of other loading coils

when a loaded circuit is terminated at a loading point.

termination, mid-section   Terminating a loaded cable at a point at which the distance to the first loading coil is one-half of a normal loading section.

termination, mid-series   Termination of a network consisting of "T" sections at the mid-point of a series element.

termination, mid-shunt   Termination of a network made up of "pi" sections such that the termination is a shunt element of double the impedance of the normal shunt element.

terminator (CATV)   A resistive load for an open coaxial line to eliminate reflections. Usually capacitively coupled to avoid shorts in cable-powered systems.

terminus   In a submarine cable system, the name of the place at which the terminal station is located.

tertiary center   A switching center at the third level in the long distance call routing hierarchy.

tertiary circuit   A third circuit into which crosstalk is induced from the disturbing circuit and which induces this same crosstalk into the disturbed circuit.

tertiary winding   A third winding of a coil.

tesla   The SI unit of magnetic flux density equal to one weber per square meter.

test   Sequence of operations designed to establish the mode of operation of a circuit or device, the value of a component, or the behaviour of a circuit.

test, acceptance   1. Test of newly purchased equipment to ensure that the equipment is fully compliant with contractual specifications. 2. A set of tests agreed with the customer for the formal handing over of a system in the field.

test, busy   Manual or automatic test to determine whether or not a line or circuit is busy.

test, capacitance unbalance   A test made in both directions at a test splice joint of a quad-type cable. The two sections will then be spliced together so that the unbalances cancel each other out.

test, coin   Test of circuit condition on a coinbox telephone line to ascertain whether a coin is being held in the box.

test, friendly busy   A normal busy test. The subscriber who is busy receives no indication that his line is being tested for another caller.

test, hostile busy   Busy test followed by pre-empting of the busy circuit for use by a high priority caller.

test, system   (during laying of a submarine cable system) Upon completion of the first splice, the first ocean block of the system is energized and equalizer characteristics are computed from continously conducted transmission tests. The process is then repeated for successive ocean blocks.

test, tank   Cable insulation test in which a length of cable is submerged in a tank of water, and a high voltage is applied between the conductors and the grounded water.

test, tip-busy   The traditional way of testing for a busy line at a manual office. The tip connection of one of the operator's cords is touched to the sleeve of the subscriber's jack on the operator's panel. If the line is busy, the operator hears a loud click.

test and validation Physical measurements taken to verify conclusions obtained from mathematical modeling and analysis, or taken for the purpose of developing mathematical models.

test antenna An antenna of known performance characteristics used in determining transmission characteristics of equipment and associated propagation paths.

test board Manual position equipped for testing outward to the external plant and inward into the switching equipment.

test center A facility for detecting and diagnosing faults and problems with communications services.

test clip Spring-closing, crocodile-jaw type clip which can be used for temporary test connections.

test desk, local End office manual test position equipped with specialized test sets to enable testing of subscriber's lines and control of repair service staff activities.

test desk man Fault control person working on a test desk from which he supervises clearance of faults.

test point Point which provides electrical access to signals and used for the purpose of fault isolation.

test set Any testing device, particularly one designed to enable testing of complex functions.

test set, call-through A routining test set which originates calls through a central office to test the establishment of switched paths and ringing, tone generation and call supervision operations. These routiners usually stop action and hold paths in question whenever a fault is encountered, so that trouble may be cleared.

test set, dial speed and pulse ratio Small test set associated with a local test panel or test desk in a central office. It gives a direct reading of the speed (in impulses per second) and pulse ratio (in percentage make/ break) of rotary dial pulses originated from remote locations.

test tone A tone sent at a predetermined level and frequency through a transmission system to facilitate alignment of the gains and losses of devices in the transmission circuit.

test turret A console or small test desk equipped with basic instruments for local line testing and designed for use at a small central office.

testboard, toll Manual test position at which toll circuits are terminated and semipermanently connected to the central office. In this manner, toll circuits may be tested and patches inserted for any necessary rerouting.

testing, remote Testing subscribers' lines terminated at a location other than the testing location.

tetrode A four-electrode electron tube. Usually a screened-grid tube.

text That part of a message which contains the information to be conveyed.

theoretical exchange Concept used when a group of subscribers is served by a central office outside

their immediate area, and is charged a tariff different from the usual rates for that office.

thermal noise The noise generated by thermal agitation of electrons in a conductor. Syn. Johnson noise

thermal noise limited operation Operation wherein the minimum detected signal is limited by the thermal noise of the detector and by amplifier noise.

thermionic Pertaining to the emission of ions or electrons by a hot material.

thermionic valve British name for an electron tube.

thermistor A thermally-sensitive resistor whose resistance changes with temperature.

thermocouple A device in which two dissimilar metals are joined at one end. When the joined ends are heated to a higher temperature than the other ends, a dc voltage is generated.

thermoelectric cooling Method of cooling in which an electric current is passed through two dissimilar metals joined at two points. Heat is liberated at one junction and absorbed at the other.

thermoelectric junction Junction between dissimilar metals which are capable of generating electricity when heated.

thermoelectric power Power generated by thermoelectricity. Many isolated radio repeater stations are now powered by thermoelectric generators using propane, butane, or natural gas as fuel. The gas burns to heat one end of a thermopile while the other end is kept cool.

thermoelectricity Electricity generated by the application of heat, as with a thermocouple or thermopile.

thermomagnetic Pertains to the effect of temperature on the magnetization of a material, or the heating effect of a magnetic change.

thermometer Device for indicating temperature. In telecommunications centers, the temperature of the electrolyte in storage batteries is recorded regularly using a thermometer standing in a pilot cell.

thermophone A source of sound power of calculable magnitude resulting from the heating of a conductor by an audio-frequency current.

thermopile An assembly of thermoelectric junctions or thermocouples connected in series to produce a usable output.

thermoplastic Resin which can be repeatedly softened when heated.

thermosetting Resin which hardens when cured by heating and cannot be resoftened.

thermostat Device to control temperature. Basic types are activated by the heat-induced bending of a bimetallic strip.

thimble A metal ring or loop placed at the end of a length of rope to prevent wear.

thimble-eye Metal loop fitted to a fiber rope or a steel-strand messenger to equip it for joining to a bolt or rod.

thin-film circuit (or technique) A process in which circuit components are manufactured by means of a thin film (usually only a few micrometers thick) deposited on a ceramic or glass base.

thin film transistor (TFT) A field-effect transistor made on an insulating substrate by a thin-film deposition technique.

thin film waveguide An optically transparent dielectric film that forms a core that guides light when bounded by lower-index material.

third harmonic Frequency three times that of the fundamental frequency of a wave.

third-order beat An unwanted carrier created by three separate carriers beating against each other. These beating carriers may have the same or different frequencies.

third party database (videotex) A database contained in a computer outside a videotex system, where information can be accessed through a videotex gateway.

third wire In an analog or space division telephone central office, two wires are needed to carry speech signals. A third wire, switched through the office at the same time as the two speech wires, can be used for metering, control, and supervision. Syn. sleeve circuit.

three-party service A supplementary telephone service by means of which a subscriber engaged in an established call can simultaneously set up another call to a third party.

three phase Alternating current supply with three sinusoidal voltages differing in phase by 120 degrees.

three pole Any switch which makes, breaks, or changes three circuits at the same time.

three-row keyboard A common type of teletypewriter keyboard with alphabetic characters in the "letter shift" position and numerical characters, punctuation marks and symbols in "figure shift."

three-way calling Telephony feature that permits a talking subscriber to add a third party to the call without operator assistance.

threshold 1. The minimum signal value that can be detected by the system under consideration. 2. A value used to denote predetermined levels pertaining to volume of message storage utilized in a message switching center. 3. The value of the parameter used to activate a device.

threshold, FM improvement The level at which signal peaks entering an FM receiver equal the peaks of internally generated thermal noise power.

threshold, noise In a radio receiver, the radio frequency input level at which signal power equals internally generated thermal noise power.

threshold current The driving current corresponding to lasing threshold.

threshold extension In an FM receiver, the process of lowering the noise threshold by decreasing receiver bandwidth.

threshold of audibility The minimum sound pressure for a particular frequency which can be heard by a particular individual.

threshold of feeling The minimum sound pressure for a particular frequency which produces discomfort or pain.

threshold value (of a gate) The input voltage at which a change in output state is just triggered.

through balance The process of adjusting the impedances in the transmission paths of connections between intertoll trunks to minimize the echo returned to either speaker.

through circuit Circuit suitable for use as part of a connection requiring more than one toll circuit in tandem.

through connection delay The interval from the instant at which the information required for setting up a through connection in an exchange is available for processing to the instant that the switching network through connection is available for carrying traffic.

through dialing PBX feature which enables an operator to connect an extension to an external line on which the extension may dial its own outgoing calls.

through-group A group of frequency division multiplex voice channels, normally 12, which are handled as a signal with a bandwidth of 48 kHz instead of multiplexing down to voice frequency.

through-group connection point Point at which several group sections comprising a group link are connected in tandem by means of through-group filters.

through-group equipment In carrier telephony, equipment which accepts the signal from the group receiver and attenuates it to the proper level for input to a group transmitter. This is accomplished without frequency translation.

through-mastergroup connection point Point at which a mastergroup link, made up of several mastergroup sections, is connected in tandem by means of through-mastergroup filters.

through supergroup A group of 60 voice channels which goes through a repeater as a unit, without frequency translation.

through-15 supergroup assembly connection point Point at which a 15-supergroup assembly link, made up of several 15-supergroup assembly sections, is interconnected in tandem by means of through-15 supergroup assembly filters.

through-supergroup connection point Point at which a supergroup link, made up of several supergroup sections, is connected in tandem by means of through-supergroup filters.

through-supergroup equipment In carrier telephony, equipment which accepts the multiplexed signal from a supergroup receiver, amplifies it, and provides the proper signal level for input to a supergroup transmitter. This is accomplished without frequency translation.

through-supermastergroup connection point The point at which a supermastergroup link is made up of several supermastergroup sections, and connected in tandem by means of through-supermastergroup filters.

through supervision Supervision of a toll call by the originating operator right through to the called line.

throughput The number of bits, characters, or blocks which can pass through a data communications system, or portion thereof, when the system is working at saturation. The throughput, which is expressed in data units per period of time, will vary greatly from its theoretical maximum. For telephone systems throughput is measured in terms of the number of telephone call attempts satisfactorily processed per second.

thruster, bow The motor-driven propeller in a 'thwartship tunnel near the bow of a cable ship.

thump Low frequency noise sometimes heard coming from telegraph circuits operated simultaneously on the same pair of wires.

thunderstorm days The average number of days per year on which there is likely to be a thunderstorm, based on past history.

thyratron A gas-filled electron tube in which plate current flows when grid voltage reaches a predetermined level. At that point the grid has no further control over the current which continues to flow until it is interrupted or reversed.

thyristor A semiconductor used as a gate circuit. The solid state equivalent of a thyratron.

thyrite Material with nonlinear resistance which normally decreases with an increase in applied voltage.

tick tone Clicking noise heard on some PABX trunks indicating that digits dialed will be repeated to the central office.

ticker A receive-only teletypewriter commonly used to report stock exchange transactions.

ticket, cable repair Form used by telephone companies to control repair works on cables.

ticket, mark-sense A special 80-column data card used by toll operators to record details of chargeable calls. A sensing machine then automatically reads the card and punches holes to enable a computer to calculate the charge and enter it in subscriber billing records.

ticket, toll Paper ticket for recording details of toll calls. A calculagraph is often used to stamp the total elapsed time on the back of the ticket.

ticket, trouble Form used to record service faults as they are reported.

ticket filing position One or more positions in a switchroom where call tickets are presorted before being sent to the billing office.

ticket operator An operator in charge of tickets for delayed calls who deals with queries and passes the tickets to operating positions for appropriate action.

ticket position Position manned by the ticket operator.

ticketer Device which prints toll tickets automatically using input from the calling telephone and the charge computer.

ticketing, automatic See automatic toll ticketing.

tickler A feedback or regenerative coil in a radio circuit.

tie 1. A connection or strap between components. 2. Soft wire used to tie open wire lines to insulators on overhead routes.

tie down To terminate a jumper wire on a main or intermediate distribution frame.

tie-line A leased or private dedicated telephone circuit provided by common carriers that links two points together without using the switched network.

tie point Terminal on a base or mounting plate which can be used for interconnection of several conductors.

tie trunk A telephone line directly connecting two private branch exchanges.

tie wire Soft wire cut in lengths about 20 in. long, of same material as line wire, and used to tie line wire to its insulator on an open wire route.

tightly coupled The inter-relationship between two processors that share real storage, are controlled by the same control program, and communicate directly with each other.

tile Hard-baked clay pipe used as conduit for cables.

tile, multiple Hard-baked clay conduit made up in multiple-duct configurations.

tile, sewer Vitrified clay pipe with spigot/socket ends sometimes used as cable conduit.

tile, split Clay conduit split longitudinally into sections for erecting around an existing cable in situ in order to repair a damaged duct.

tilt The angle the axis of an antenna makes with the horizontal.

tilt control Ability to change the gain/frequency characteristic of a broad-band amplifier.

timbre Quality of sound, the presence in different proportions or the absence of harmonics.

time, answering Time elapsing between a signal appearance at a switchboard and the verbal response of an operator.

time, attack Time between the receipt of a signal and the initiation of appropriate action by the equipment in question.

time, build-up Of a telegraph signal, the time during which the telegraph current passes from one-tenth to nine-tenths (or vice versa) of the value reached in the steady state. For asymmetric signals the build-up time at the beginning and end of a signal can be different.

time, build-up, relative Of a telegraph signal, the ratio of the build-up time to the half-amplitude pulse duration.

time, busy Total time a circuit is occupied with a call.

time, compressor attack Time elapsing between a significant increase in input signal level to a compressor and the output signal reaching a specified proportion of its steady-state value.

time, compressor recovery Time elapsing between a significant fall in input signal level to a compressor and the output signal reaching a specified proportion of its steady-state value.

time, connection See connection time.

time, daylight saving A time system used during summer months which is one hour later than standard time. Syn.: summer time

time, decay Time for a signal to decay to a fraction, usually $1/e$ or $\frac{1}{2}.7183$ of its original value.

time, deductible   Time which, at the discretion of the operator, is deducted from the total elapsed toll call time by the calculagraph in order to offset time during which a circuit could not be used.

time, elapsed   1. Time during which a toll circuit is usefully occupied. 2. The time figure provided by a calculagraph.

time, hangover   Time required for equipment to return to normal after an inhibiting situation has been ended.

time, holding   See holding time.

time, isolation   The part of maintenance time during which a least-replaceable unit is isolated.

time, localization   The · part of maintenance time needed to locate the faulty section of the system.

time, memory access   The interval from the time information is requested from a memory unit to the time it begins to be received.

time, net elapsed   Total chargeable time for a toll call (elapsed time less deductible time).

time, pulse decay   Time needed for the instantaneous amplitude of a wave to fall from 90% to 10% of its peak.

time, pulse rise   Time needed for the instantaneous amplitude of a wave to rise from 10% to 90% of its peak.

time, real   Not subject to storage.

time, recovery   Time a condition remains in effect after the control signal which triggered it has stopped.

time, repair   Part of maintenance time during which a least-replaceable unit is replaced or readjusted.

time, reverberation   Time for average sound energy in a room to decay to a level 60 dB below its initial value.

time, rise   Time for a steep wave front or pulse to rise from 10% to 90% of its peak value.

time, standard   Local civil time. Time zones are approximately 15 degrees wide, starting with the Greenwich mean time zone centered on Greenwich, England.

time, transit   For an electron tube, the time taken for an electron to pass from cathode to anode.

time, warm up   Time between connection of power to a system and the time when the system is ready for operation.

time, Zulu   Greenwich mean time.

time announcer   Clock driven machine, usually with magnetic tape drive, which provides callers with the correct time.

time assignment speech interpolation (TASI)   A technique used on certain long FDM links to improve utilization of voice channels by switching an additional user onto a channel temporarily idled because the original user has stopped speaking. When the first user resumes speaking, he will in turn be switched to any channel that happens to be idle. The use of this technique enables a large group of circuits to carry almost twice as many simultaneous conversations as there are two-way circuits.

time availability   See circuit reliability.

time base generator   Circuit in a cathode ray tube device which produces the saw-tooth waveform needed for most cathode ray tube applications.

time block   An arbitrary grouping of several consecutive hours of a day, usually for a particular season, during which propagation data are assumed to be statistically homogeneous.

time code   A time format used for the transmission and identification of time signals.

time coherence   See coherent.

time comparison   The determination of time scale difference.

time compression multiplex   See burst mode, ping-pong.

time congestion   The probability that a system is congested over any particular time period.

time consistent busy hour (mean busy hour)   The 60 consecutive minutes commencing at the same time each day, for which the average traffic volume of the observed exchange or circuit group is greatest over the days of observation.

time constant   Time required to complete 63.2% of the total rise or decay of a current or voltage.

time delay   Time taken for a signal to travel between two points.

time-delay distortion   See delay distortion.

time division   The separation in the time domain of a number of transmission channels between two points.

time-division duplexing   See ping-pong, burst mode.

time-division highway   In switching, a link within a switching stage shared by a number of channels or circuits in the time domain.

time-division multiple access (TDMA)   In satellite communications, the use of time interlacing to provide multiple, and apparently simultaneous, transmissions to a single transponder with a minimum of interference.

time-division multiplex (TDM)   A method of multiplexing in which a common transmission path is shared by a number of channels on a cyclical basis by enabling each channel to use the path exclusively for a short time slot. In this way a circuit capable of a relatively high information transfer rate (in bits) is subdivided into time slots to provide a number of lower speed channels.

time-division switching   A switching system whereby the information content of each incoming time slot may be delayed in time and switched to any of a number of outgoing time slots. The switching of inlets to outlets using time division multiplexing techniques.

time domain reflectometry   (TDR)   A pulse-sending test practice which enables external plant faults to be located.

time gate   Logic circuit which has an output during specified time intervals only.

time guard band   A time interval left vacant on a channel to provide a margin of safety against intersymbol interference in the time domain between sequential operations.

time interval The time duration between two instants read on the same time scale.

time jitter Short-term variation in the duration of a specified interval.

time lag Time between the activation of an input change device and the subsequent change in output response.

time marker A reference signal, often repeated periodically, permitting the assignment of numerical values to events on a time scale.

time-of-day A time announcement service reaching by dialing a special number.

time-out 1. The automatic release of equipment or circuits on calls which do not proceed to completion. 2. A delayed decision in which a circuit waits for a predetermined interval. 3. A network parameter related to an enforced event designed to occur at the conclusion of a predetermined elapsed time. 4. A specified period of time allowed to elapse in a system before a specified event takes place, unless another specified event occurs first.

time quantized control On digital networks, a synchronization control system in which the error signal is derived or utilized only at a number of discrete instants which may or may not be equally spaced in time.

time scale Arbitrary time measuring system graduated in multiples and submultiples of a second.

time scale difference The difference between the readings of two time scales at the same instant.

time scale reading The value read on a time scale at a given instant.

time scale unit The basic time interval measured by a time scale.

time scales in synchronism Two time scales are in synchronism when they assign the same time to an event.

time series analysis A method of forecasting used when relevant data is available and can be plotted on a graph. If the curve of the graph is reasonably smooth, a mathematical formula can be designed to fit the curve and forecast future figures. Also called curve fitting.

time sharing 1. A method whereby a facility is shared by several users at the same time. The processor actually services the different customers in sequence, but its high speed of operation makes it appear that all are being dealt with simultaneously. 2. A mode of operation of a data processing system that provides for the interleaving in time of two or more processes in one processor. Sometimes called time slicing.

time signal Accurate time-of-day signals broadcast at regular intervals by various radio stations.

time signal-satellite service A radiocommunication service using space stations on earth satellites for sending time-of-day signals.

time slicing A method in which two or more processes are assigned interleaving time on the same processor. Sometimes called time sharing.

time slot Any cyclic time interval which can be recognized and uniquely defined.

time slot In switching, an interval in the time domain capable of providing a channel.

time slot interchange The transfer of information from one time slot to another between incoming and outgoing time division highways.

time slot sequence integrity The assurance that the digital information contained in the various time slots of a multislot connection arrives at the output or terminal in the same sequence as it was introduced.

time-space-time (TST) The most common form of switching matrix for small digital telephone exchanges. A time switch followed by a solid state space switch, then another time switch.

time-space-space-space time (TS³T) A three-space element used inside time switches where one space switch is insufficient for the size of the exchange. In this way, many thousands of ports may be interconnected in a switching matrix capable of handling very high traffic levels.

time standard A device used for the determination of a time scale or time unit.

time step An intentional discontinuity introduced in a time scale reading. Time step is positive if the reading is increased and negative if it is decreased.

time switch Device incorporating a clock which arranges to switch equipment on or off at predetermined times.

time tick Time mark output of a clock system.

timed release Release of a circuit after a specified interval of delay.

timed-release disconnect Telephony feature wherein a line is disconnected if the calling party fails to go back on-hook within a specified time after the called party has gone on-hook.

timer, AMA master A master clock which provides automatic message accounting equipment with the correct time of day for recording against chargeable calls.

timer, initial period reminder Electric timer used on manual operating positions to remind operators to challenge a call originating from a coin telephone just before the paid-for time has run out.

timing allowance Deductible time allowance to offset time lost by interruption or breakdown during a toll call.

timing extraction See timing recovery.

timing recovery The derivation of a timing signal from a received signal. Also called timing extraction.

timing signal 1. The output of a clock. 2. A signal used to synchronize interconnected equipment. 3. A cyclic signal used to control the timing of operations.

timing tracking accuracy A measure of the ability of a timing synchronization system to minimize the frequency difference between a master clock and any slave clock.

tinned Covered with a thin layer of metallic tin to inhibit corrosion and facilitate soldering.

tinned wire Copper wire coated or plated with tin to facilitate soldering.

tinsel conductor Flexible conductor made with thin ribbon wires of copper over a nylon cord core for use in telephone cords.

tip 1. The rounded end of a standard telephone switchboard plug. 2. One of the two speech wires in a central office, the other being called a ring wire. 3. Packet switching term used to designate equipment which accepts data from outlying terminals and reformats it into correct network language for retransmission.

tip, silk and cotton A tip cable with conductors insulated by wrappings of cotton and silk.

tip, spade Metal terminal for a wire. It is spade-shaped with an open end for insertion under a binding post nut.

tip cable Short length of cable which can be exposed to the atmosphere without harmful effect. It is spliced to a dry-paper insulated cable so that pairs can be terminated on the main distribution frame.

tip wire One of the speech wires of a pair in a central office.

Tiros A series of weather satellites.

toggle To change state of a flip-flop device.

token bus A local network access mechanism and topology in which all stations actively attached to the bus listen for a broadcast token. Stations wishing to transmit must receive the token before doing so. Bus access is controlled by preassigned priority algorithms.

token passing Method of controlling the use of a communications channel, especially ring networks. A token packet is circulated from node to node when there is no live traffic. Possession of the token gives a node access to the network for transmission of data.

token ring A local network access mechanism and topology in which a token is passed from station to station in sequential order. Stations wishing to transmit must wait for the token to arrive before transmitting data.

tolerance Permissable variation from a standard.

toll 1. Service charge for long distance telephone calls. 2. Plant used primarily for toll call services.

toll, foreign area Toll calls for offices with an area code different than that of the home area.

toll, home area Toll calls originating and terminating within the same area code.

toll, inward Toll calls coming in to an office over intertoll circuits addressed to lines which are on that office or on tributary offices.

toll, outward Toll calls directed outward on intertoll trunks from the toll center serving the locality where the call originated.

toll, through Toll calls coming in over intertoll trunks which are switched out to other intertoll trunks.

toll board A switchboard handling toll traffic.

toll call Call to a point outside the local service area.

toll center (TC) A class 4C toll office where operators give assistance in completing incoming calls in addition to providing other traffic-related operating functions.

toll center office Central office whose main function is the completion and supervision of toll calls.

toll centering and metropolitan sectoring (TCMS) An American Telephone and Telegraph software routine used in long range studies to determine least-cost solutions in developing local city telephone networks.

toll circuit See circuit, toll.

toll connecting trunk A terminating link. A trunk between a toll center and an end office.

toll cord On cord-type switchboard positions; the cord on which a toll circuit is connected.

toll dial assistance operator (TDA operator) Operator who assists in completion of dialed toll calls.

toll dial assistance position (TDA position) Position to which incoming, through, and outward toll calls are routed when operator assistance is required.

toll diversion Adjustment made in local telephone company switching equipment so that attempts by the user to place toll calls will be diverted to the PBX operator.

toll office A toll center. An office at which toll calls are switched.

toll point (TP) A Class 4P toll office where operators handle only outward toll calls, or where there are no operators.

toll restriction Arrangement by which some telephone lines are denied access to long distance circuits.

toll station Telephone instrument connected directly with a toll board.

toll switching operator Operator at a 'B' board who answers incoming toll trunks and connects calls through to subscriber's lines.

toll switching position A 'B' position at which incoming intertoll trunks are terminated, providing connection through to local subscribers lines.

toll switching trunk A trunk connecting one or more end offices to a toll center as the first stage of concentration for intertoll traffic.

toll tandem A type of trunk circuit used to interface with an intertoll or toll tandem trunk.

toll terminal A long distance terminal. A direct line from a toll switchboard to a subscriber's premises.

toll trunk A high-grade circuit between a toll center and a local central office, used for extending incoming toll calls to subscribers.

tone An audible signal of identifiable frequency or periodicity giving information to customers about the status of their call requests.

tone, all trunks busy (ATB tone) Low tone, interrupted twice per second, which indicates that local switching paths or equipments are busy.

tone, busy Low tone, interrupted once per second, indicating that the called line is busy. In other countries different frequencies and periodicities are used.

tone, class of service Short bursts of tone, for the information of the operator only, to indicate the class of service of the calling subscriber's line.

tone, coin collect Low tone which indicates to the originating operator that coins have been collected.

tone, coin return High tone which indicates to the originating operator that coins have been returned.

tone, dial Tone which indicates to the caller that equipment is ready to receive dialed information. A dial tone is usually 350 Hz plus 440 Hz at -13 dBm.

tone, high Tone of 480 Hz, at -17 dBm.

tone, high frequency test Ultrasonic test tone used in conjunction with a frequency-converting detector by splicers working on cables without disturbing service.

tone, hold and trace Tone at 2100 Hz interrupted twice a second. Sometimes used inside a central office while tracing calls.

tone, line busy Busy tone. U.S. standard is a low tone interrupted once per second, indicating that the called line is busy.

tone, low Tone of 480 Hz plus 620 Hz at -24 dBm per frequency. Used for busy, reorder, and no-circuit tones.

tone, no circuit Tone indicating that there are no free outgoing trunks. Usually a low tone interrupted twice per second.

tone, no such number A 'woo woo' tone. Usually 500 Hz and 600/120 Hz interrupted once per second. UK equivalent = number unobtainable tone.

tone, order Short bursts of high tone, for the information of the operator only, to indicate that the number of the called line should be passed on.

tone, pay station identification Short burst of tone, for the information of the operator only, to indicate that the calling line is a pay station or coin box line.

tone, precise dial Mixture of 350 Hz and 440 Hz tones at -13 dBm.

tone, pre-empt Low tone which indicates to users that the toll circuit they are using is being pre-empted for a high priority call.

tone, pure A sinusoidal wave tone with no harmonics.

tone, recorder warning Warning beeps at 1400 Hz fed at 15-second intervals to indicate that a recording device has been connected to the circuit.

tone, reorder Low tone interrupted twice per second to indicate that dialing should be re-attempted because either the office or area code dialed on the first try was unassigned or there were no circuits available leading in the required direction.

tone, ringback Audible ringing tone, normally 440 Hz plus 480 Hz at -19 dBm per frequency.

tone, second dial Dial tone received from an office other than the end office to which the calling line is connected.

tone, splash Short burst of tone, for the information of operators only, indicating class of service or readiness to receive an order. Sometimes called zip tone.

tone, standard test A tone at 1000 Hz used for transmission testing of audio circuits at a standard level of 1 mW across a 600 ohm impedance.

tone, test 1. Tone at 1000 Hz. Used for transmission tests. 2. Low buzzing tones used by cable splicers to make cable identification checks.

tone, trunk busy A low tone interrupted twice per second to signify that all trunks are busy. Sometimes this signal uses a 600 Hz plus 120 Hz tone.

tone diversity A method of voice frequency telegraph transmission wherein two channels of a 16-channel transmitter carry the same information.

tone generator Rotary machine, oscillator, or solid state digital tone synthesizer which produces the information and advice tones needed for telephone service.

tone jacks Set of jacks to which all locally used tones are wired for purposes of demonstration.

tone receiver A detector on an incoming line or trunk which decides if voice-frequency tone signaling is present, and decodes any such signals received.

tone ringer Small tone producer inside a telephone which is sometimes used instead of a bell.

tone signaling Use of voice frequency tones to carry signaling information on an in-band basis.

tones, coin denomination Various tones produced at a telephone coin box station by a mechanism that feeds coins of different denominations by different routes to produce tones that tell the operator what coins have been inserted.

tones, supervisory Tones, such as dial tones and busy tones, which advise customers of circuit conditions.

tool, coinbox sealing Pliers used to compress lead security seals on cash containers collected from coin telephones.

tool, connector crimping Crimping pliers used to compress the connectors used for joining conductors for cable splicing.

tool, impact Power tool which operates drills by a series of impacts rather than by a steady rotation.

tool, powder-powered Pistol-like tool for firing bolts into masonry or steel.

tool, relay blocking Small tool with plastic wedges of different sizes which can be inserted under a relay armature to block its operation.

tool, sheath constriction Hand tool used by cable splicers. It has a semicircular shaped rod which can be beaten lightly into a cable sheath to produce a small constricting groove around the cable.

tool, spring adjusting Hand tool used to bend and adjust the tension of spring sets on relays and keys.

tool, wire raising A tool consisting of a U-shaped metal head at the end of a light pole, used to lift drop wires over obstructions.

tool, wire wrapping Hand tool used to wrap connecting wires around the rectangular solderless terminals used in place of soldering tags on terminal blocks.

top 100 market   Ranking of largest TV broadcast areas in the U.S. by size of market, i.e., the number of viewers and TV households. Used in FCC rulemaking and in the selling of airtime to advertisers.

top down forecast   Forecast of demand for telephone service based generally on broad statistical projections of population, income, housing units and other economically relevant variables.

top hat   See WAL 2, di-phase line code.

topology   The study of the properties of shapes and figures. Used in telecommunications to describe the ways in which different facilities and services interact, and how the introduction of new services can most readily be achieved. For networks: the geometric form of the nodes and connecting links, e.g., ring, star, bus.

torn-tape relay   A tape relay system in which the perforated tape is manually transferred by an operator to the appropriate outgoing transmitter position.

toroid   A coil wound around a doughnut-shaped magnetic core.

torque   Moment of force acting on a body and tending to produce rotation about an axis.

total access communications system (TACS)   An advanced cellular radio system providing mobile telephone service.

totem pole   A type of output stage with two transistors, often used in TTL logic gates. Schematic diagrams of these units look vaguely like American Indian totem poles.

touch calling   See calling, touch.

Touchtone   Bell System method of providing push-button dialing using dual-tone multifrequency signaling.

tower, antenna   A self-supporting steel structure able to carry VHF or microwave antennae at different heights above ground.

tower radiator   A steel tower or mast which acts as the radiating element for a radio transmitter.

trace   Glow left on screen of cathode ray tube by moving beam of electrons.

trace, return   Path taken by the electron scanning spot as it returns to its starting point.

tracer   Special mark on insulation to identify a particular pair or conductor in a multiwire cable.

tracer pair   A pair of wires with distinctive marking designed to provide a starting point for the counting of pairs in each layer of a cable. Syn.: marker pair

tracer quad   A quad with a distinctive marking found in each layer of a quad-type cable designed to provide a starting point for counting the quads in that layer. Syn.: marker quad

tracer stripe   Color coded stripe on the insulant of cables. The main stripe is the base stripe; the narrow stripes are called tracer stripes.

track   That portion of a moving-type storage medium (e.g., film, drum, tape, disc) which is accessible to a given reading station.

tracking   1. The locking on to a satellite by a ground station. 2. Locking of tuned stages in a radio receiver so that all stages are changed appropriately as the receiver's tuning is changed.

tracking, automatic   Locking on of a ground station to a satellite so that movements of the satellite are immediately followed by adjustment of the ground station antenna.

trade-off   Process of weighing conflicting requirements and reaching a compromise decision, e.g., in the design of a component or a subsystem.

traffic   1. Messages sent and received. 2. The engineering of circuit quantities and switching and routing patterns to maximize efficiency.

traffic, effective   The traffic intensity relative to the call durations.

traffic, incoming   Traffic entering a network from outside sources, regardless of its final destination.

traffic, internal   Traffic originating and terminating within the network in question.

traffic, lost   Traffic which is directed to a pool of resources unable to handle it.

traffic, originating   Traffic generated by sources located within the network in question.

traffic, outgoing   Traffic leaving the network under consideration, regardless of its origin.

traffic, overflow   Traffic which must be directed to additional resources because the original resources are unable to handle it.

traffic, own exchange   Traffic originating and terminating within the same exchange.

traffic, peaked   Traffic with a peakedness factor greater than one.

traffic, Poisson   Traffic that has a Poisson distribution of arrivals.

traffic, pure chance   Poisson traffic which has a negative exponential distribution of holding times.

traffic, smooth   Traffic that has a peakedness factor of less than one.

traffic, telecommunications   A flow of calls and messages.

traffic, telephone   Total traffic on a telephone route, usually expressed in CCS or erlangs in the busy hour.

traffic, terminating   Traffic destined for sinks located within the network under consideration, regardless of its origin.

traffic, transit   Traffic passing through the network under consideration, generated by sources outside it and destined for outside sinks.

traffic capacity   The maximum traffic per unit time that can be carried by a specified telecommunication system, subsystem, or device under specified conditions.

traffic carried   1. Traffic that actually occupies a set of circuits or switches. A measure of this in erlangs is the average number of simultaneously occupied circuits or switches. 2. That part of the traffic offered to a pool of resources which is served by the pool.

traffic carried, amount of Total traffic carried by a set of circuits or switches in a defined time period, given by the sum of the holding times of all calls carried in that period.

traffic-carrying device Functional unit used directly or indirectly during the establishment and maintenance of a connection.

traffic circuit A circuit for the transmission of information between two exchanges.

traffic distribution imbalance Condition in a switching center when traffic flow of one incoming unit is unevenly distributed among all the outgoing units.

traffic engineering See engineering, traffic.

traffic flow The amount of traffic divided by the duration of the observation. Traffic flow calculated in this way is expressed in erlangs.

traffic flow control In signaling, procedures designed to limit signaling traffic at its source in the cases where the signaling network is not capable of transferring all incoming traffic owing to network failure or overload.

traffic flow security The protection provided by cryptoequipment, which conceals messages on a communications circuit by causing the circuit to appear busy at all times.

traffic intensity The volume of traffic divided by the duration of the observation. It is equal to the average number of simultaneously busy resources. A traffic intensity of one traffic unit (one erlang) signifies continuous occupancy of a facility during the time period under consideration, regardless of whether or not information is transmitted. Also called traffic load.

traffic intensity, equivalent random The theoretical pure chance traffic intensity that, when offered to a number of theoretical circuits (equivalent random circuits), produces an overflow traffic with a mean and variance equal to that of a given offered traffic. The equivalent random concept permits traffic theories that do not explicitly recognize peakedness to be used in peakedness engineering. See equivalent random circuits group.

traffic load The total traffic carried by a trunk or trunk group during a specified time interval.

traffic load imbalance Condition in an exchange or switching center when traffic load is unevenly distributed among similar units.

traffic matrix An array of crosspoints in the form of a matrix in which a given inlet row has access to a given outlet column via the crosspoint at the intersection of the row and column.

traffic offered 1. (to a set of circuits or switches). It is necessary to distinguish between traffic offered and traffic carried. The traffic carried is only equal to the traffic offered if all calls are immediately handled (by the group of circuits or group of switches being measured) without any call being lost or delayed on account of congestion. The flow of traffic offered, and of traffic carried, is expressed in erlangs. The amount of traffic offered and of traffic carried is expressed in erlang-hours. 2. The traffic that would be served by a pool of resources sufficiently large to serve that traffic without limitation. Its usage is as a calculating quantity similar to traffic intensity.

traffic office 1. Group of operator-attended positions maintained for service-related functions. 2. An office at which the public may pay phone bills and arrange for different services to be installed or provided.

traffic overflow reroute control A network management feature which reroutes traffic to alternative regional centers during high load periods.

traffic peakedness factor The variance to mean ratio of a traffic load.

traffic relation The traffic originating at a source and intended for a destination. Syn. traffic stream, traffic item, parcel of traffic, point to point traffic

traffic route A set of circuits, all of which carry a given class of traffic between two exchanges.

traffic routing The selection of a route or routes for a given traffic stream. This term is applicable to the selection of routes by switching systems or by operators, or to the planning of routes and switching topology.

traffic service position (TSP) A computer-controlled operating position.

traffic unit The erlang. Traffic represented by one circuit continuously occupied.

traffic usage recorder (TUR) A device for measuring the amount of traffic carried by one or more sets of switches or trunks.

traffic volume The sum of the holding times of the traffic carried by a pool of resources over a given period of time.

trailer, cable reel A two-wheeled trailer designed to carry a reel of telephone cable which can be unwound without removal from the trailer. In UK: cable drum trailer.

trailer, pole Two-wheeled trailer designed to carry telephone poles.

train Sequence of signals or similar units of equipment through which calls are routed.

transaction A message destined for an application program. A computer-processed task that accomplishes a particular action or result.

transaction service (videotex) A two-way service that allows a videotex user to send a specific response that activates particular actions by the system.

transceiver A combination of transmitting and receiving equipment in one housing, usually for portable or mobile use. It employs common circuit components for both transmitting and receiving and employs simplex operation. The word is usually used in describing radio equipment but is sometimes also used in data networks to describe a device for transmitting and receiving data signals, usually baseband signals.

transceiver, mobile Transmitter-receiver mounted in an automobile. It can send and receive telephone calls, via a base station, to and from the public switched telephone network.

transcoder Device which enables differently coded transmission systems to be interconnected, e.g., two 30-voice channel systems (each a standard 2 Mbit/s PCM system) interconnected with a single 2 Mbit/s system providing 60 voice-grade circuits, each 32 kbit/s, using adaptive differential PCM.

transconductance Mutual conductance of an electron tube expressed as the change in plate current divided by the change in control grid voltage.

transcriber Equipment which converts recorded data into language used by a computer.

transcutaneous electronic nerve stimulators (TENS) A pulse generator which produces signals that relieve pain when transmitted through the human nervous system.

transducer Device by means of which energy can flow from one transmission system or medium to another transmission system or medium.

transducer, active Transducer whose output waves are dependent on a power source other than that supplied by the actuating waves, although the actuating waves control said source.

transducer, gas pressure Device which can be connected to a pressurized cable to enable the gas pressures to be read off at the central office.

transducer, magnetostriction Underwater loudspeaker device in which electrical energy is converted into sound energy for radiating through the water.

transducer, passive Transducer whose output waves derive their power from the input waves.

transducer, pressure A gas pressure transducer.

transender A type of register sender in a common control dial office.

transfer To change working circuits from one pole route or cable to another.

transfer, automatic power Automatic transfer of a load from one power source to a standby source owing to the failure of the main source.

transfer, call Transfer of a call to another extension on the same PABX.

transfer, information Transmission of a message or signal from one terminal to another.

transfer, maximum power Condition for maximum power transfer exists when impedances of source and load are conjugate.

transfer, parallel information Simultaneous transfer of all the bits comprising one character over a number of parallel paths.

transfer, power failure PABX feature which automatically connects predetermined extensions straight through on to central office lines when local power supply fails.

transfer, serial information Sequential transfer of bits of information.

transfer allowed A procedure included in common channel signaling route management which is used to inform a signaling point that a signaling route has become available.

transfer channel For a common channel signaling system, a voice-frequency channel or a digital channel.

transfer characteristics The intrinsic parameters of a system, subsystem, or unit of equipment which, when applied to the input of the system, subsystem, or unit of equipment, will fully describe its output.

transfer function That operator which, when applied to an input, will describe the output.

transfer link For a common channel signaling system, a combination of two transfer channels operating together in a single signaling system.

transfer of a call in progress See three party service.

transfer of calls to another number A service whereby calls to a subscriber's number can be automatically transferred to another number during a specified period of time.

transfer of technology (TOT) Instructional process by which technical knowledge is systematically transferred from a country or region of advanced technology to a lesser developed country or region.

transfer pairs Pairs in a local distribution network which enable a subscriber or group of subscribers to be transferred from one central office/exchange to another.

transfer prohibited A procedure included in common channel signaling route management which is used to inform signaling points of the unavailability of a signaling route.

transfer rate See data transfer rate.

transfer relay Relay which monitors performance of a device, and in the event of failure transfers loads to a standby device.

transfer service, subscriber Service available to many business subscribers whereby, by turning a key on his telephone when he leaves his office, the subscriber can have incoming calls transferred to his home or answering service.

transform 1. To change information from one code to another without changing its meaning. 2. To pass through a transformer, usually with an accompanying impedance change.

transformer Device with two or more windings wrapped around a single core or linked by a common magnetic circuit.

transformer, audio Transformer designed to be used for signals at audio frequencies.

transformer, bell Small transformer with primary winding fed by public power supply (typically 115 V 60 Hz) and secondary output 10 V ac to feed a small ac bell such as a door bell.

transformer, constant current Transformer which can supply a steady current to a varying load.

transformer, constant voltage Transformer which can supply a steady output voltage despite wide variations in the input voltage.

transformer, coupling Transformer used to couple two circuits via its mutual inductance.

transformer, delta-matched An RF impedance-matching device which uses a delta-matched transmission line.

transformer, impedance-matching Transformer which transforms impedance of one circuit to that of another for maximum power transfer.

transformer, intermediate-frequency Transformer tuned to IF and used as inter-stage coupling in an IF amplifier.

transformer, isolation Transformer used to separate sections of a circuit; for example, to separate an unbalanced section from a balanced section.

transformer, line Transformer in a transmission system providing impedance-matching or isolation, or deriving additional facilities.

transformer, matching Transformer with turns ratio which matches impedances one to another.

transformer, output Transformer which matches power output of an amplifier to impedance of a load.

transformer, potential Transformer used to step down voltage of a high voltage power source to a lower voltage which can be measured on a low-range voltmeter.

transformer, pulse Transformer which can pass a wide band of frequencies, including high harmonics, which enables pulses to be passed without introducing distortion.

transformer, push-pull Transformer with center-tapped and balanced windings for use in push-pull circuits.

transformer, quarter-wave Section of waveguide or co-axial cable used to match a transmission line to an antenna.

transformer, radio-frequency Transformer for use at radio frequencies. Owing to high iron losses at RF, such transformers are likely to have air cores, but special ferrites are sometimes used.

transformer, step-down Transformer producing a lower output, or secondary voltage, than its input, or primary voltage.

transformer, step-up Transformer producing a higher output, or secondary voltage, than its input, or primary voltage.

transformer, tuned Transformer tuned for resonance at a particular frequency, producing high values of secondary voltage; for example, tuned IF transformers in a radio receiver.

transformer, variable Power transformer with variable output voltage.

transformer ratio Ratio of the number of turns in the secondary winding of a transformer to the number of turns in the primary winding. Also called turns ratio.

trans-horizon propagation Propagation over paths extending beyond the normal radio horizon. It may include a variety of mechanisms such as diffraction, forward scatter, specular and diffuse reflection, and ducting.

transient 1. A rapid fluctuation in voltage or current, usually due to changes in circuit switching or loads. 2. A sudden variable which occurs during transition from one steady state condition to another.

transient response Time response of a system under test to a stated input stimulus.

transistor A semiconductor device used in amplifiers, oscillators, and control circuits in which current flow is modulated by voltage or current applied to electrodes. Most transistors are based on the use of silicon.

transistor, field effect Transistor which makes use of the principle that the flow of current in a solid is controlled by an external field. It has a high impedance gate electrode which controls the flow of current through a channel of semi-conductor material.

transistor, junction Transistor with a base electrode and two or more junction electrodes. The emitter and collector are the junctions between p-type and n-type material.

transistor, n-p-n Junction transistor with a thin layer of p-type material (the base) between two sections of n-type material.

transistor, point contact Transistor with a base electrode and two or more point contact electrodes. Emitter and collector electrodes are two fine wires with pointed tips almost touching a semiconductor wafer.

transistor, p-n-p Junction transistor with a thin layer of n-type material (the base) between two sections of p-type material.

transistor-transistor logic (TTL) A family of logic circuits with input via a multi-emitter transistor and output usually via a pair of transistors in push-pull. The TTL logic family is at present the most widely used circuit type.

transistorized Equipped with solid-state devices in lieu of electron tubes.

transit country A country through which traffic is routed between two terminal countries.

transit exchange An exchange used primarily as a switching point for traffic between other exchanges. Also called tandem or tandem switch.

transit route A route between two exchanges which is switched at an intermediate transit exchange.

transit switching center (TSC) An exchange which switches traffic between trunk exchanges.

transit time effect A condition occurring at microwave frequencies where the time it takes an electron to travel from cathode to anode in an ordinary electron tube can be greater than the time it takes the control grid to swing from its maximum to its minimum voltage. For this reason, ordinary electron tubes cannot be used for microwave frequencies; physical structure places an upper frequency limit for each electron tube.

transit traffic Traffic passing through the network considered, generated by sources outside it and destined for sinks outside it.

transition In specification and description language, a sequence of actions which occurs when a process changes from one state to another in response to an input.

transition, signal Changeover point between mark and space signals.

transition zone The zone between the end of the near field region of an antenna and the beginning of the far field region.

translate 1. To change office codes into routing digits. 2. To change frequencies of a band of signals. 3. To convert information from one language or code to another.

translation 1. Conversion of the area and office codes of a telephone address into routing instructions or routing digits. 2. Changing of the frequency spectrum by the building up of channels into groups, groups into supergroups, etc. 3. Retransmission of received digits after deletion, insertion, or change of digits.

translation, digit Conversion of a telephone address into routing instructions for switches.

translation, frequency Translation of a block of frequencies from one band to another without altering bandwidth; for example, translating a voice channel at audio frequency bandwidth 0-4 kHz to a carrier channel with bandwidth 60-64 kHz.

translation, three-digit In a common controlled switching center, analyzing the first three digits dialed to determine the routing. These can represent a central office code in the same area or an area code.

translation, six-digit In a common controlled switching center, analyzing the first six digits dialed on calls to another numbering plan area in order to determine what routing pattern to follow.

translator 1. A network or system so connected that input signals expressed in a certain code cause output signals to appear which represent the same information in a different code. 2. A type of relay system which picks up signals from distant or blocked-out television stations, converts the signals to another channel to avoid interference and retransmits them by radio into areas the original signals could not have reached.

translator AMA Device used in some crossbar offices to convert calling line identification into subscribers' directory numbers for recording in automatic message accounting records.

translator, foreign area Device which provides information on different charge formulas to be applied to calls to different destination codes.

translator, VHF/VHF For CATV: device which receives a VHF television signal, down-converts it to an IF (usually 45 MHz) then, after amplification, converts it back up to a different VHF channel frequency and feeds it into the CATV network.

translator, UHF/VHF For CATV: device which receives a UHF television signal, down-converts it to an IF for amplification, then up-converts it to a spare VHF band for feeding to the local CATV network.

Transmic French digital leased lines network.

transmission 1. The transfer of electrical power from one location to another by means of conductors. 2. The dispatching of a signal, message, or other form of information by means of wire, optical fiber, or radio waves. 3. Act of conveying an object or information by physical means between two points.

transmission, asynchronous See asynchronous transmission.

transmission, beam Directional effect radio antenna designed to concentrate radio power into a small solid angle.

transmission, double-current Polar telegraph transmission with positive and negative current flow.

transmission, parallel Simultaneous transmission of a number of signals; for example, two tones at a time with MF signaling.

transmission, radio Transmission of electromagnetic radiation at radio frequencies.

transmission, serial Transmission of sequential signals.

transmission, stereophonic Method of producing stereophonic radio service by transmitting a main carrier which represents the sum of left and right channels, and a sub-carrier which represents the difference between left and right channels.

transmission, synchronous Digital transmission procedure with synchronization between terminals such that there is always an integral number of unit intervals between any two significant signal transitions.

transmission bandwidth See fiber bandwidth.

transmission bridge See bridge, transmission.

transmission buffer Storage in the signaling link control for signal units not yet transmitted.

transmission channel All of the transmission facilities between the input of an initiating node and the output of a terminating node.

transmission code violation Digits which are not in the transmission code and which, when used in small quantities, can give more information without significantly affecting the spectrum of the signal.

transmission coefficient A number indicating the probable performance of a portion of a transmission circuit. The value of the transmission coefficient is inversely related to the quality of the link or circuit.

transmission control character See data communication control character.

transmission delay Through a digital exchange, the sum of the times necessary for an octet to pass in both directions on a connection due to buffering, frame alignment and time-slot interchange functions for digital-to-digital connections and to A/D conversions for analog-to-analog connections.

transmission distributor, CATV Directional coupler and bridging amplifier used for feeding TV signals to several CATV cables.

transmission flexibility Multiplexing, demultiplexing, encoding, decoding, semipermanent interconnection, or any combination of these. The term excludes amplification and regeneration unless they are combined with one or more of the five functions named above. The term is sometimes used in a sense which excludes audio interconnection and first-order multiplexing and demultiplexing.

transmission level (TL) The power (in dBm) that should be measured at a given point in a transmission system, when a standard test signal (0 dBm, 1000 Hz) is transmitted at some point chosen as a reference point. The transmission level of a point is a function of system design and is a measure of the design (or nominal) gain at 1000 Hz of the system between the chosen reference point (known as the zero transmission level point or 0 TLP) and the test point in question.

transmission level point (TLP) A point in a transmission system evaluated by the ratio (in decibels) of the power of the test signal at that point to the power of the test signal at a reference point. A 0 TLP is an arbitrarily established point in a communications circuit to which all relative levels at other points in the circuit are referred.

transmission line A power line, coaxial cable, paired cable, optic fiber, waveguide, or open wire pair conveying telecommunications signals from one station to another.

transmission loss The reduction in power between any two points in a telecommunications system. Changes in power level are normally expressed in decibels by calculating ten times the logarithm (base 10) of the ratio of the two powers.

transmission maintenance points On an international line, elements within the general maintenance organization located at the terminals of that part of a leased or special circuit.

transmission mode One of the field patterns in a waveguide in a plane transverse to the direction of propagation.

transmission node A point in the transmission network where a transmission section terminates and which provides for the interconnection of transmission sections, multiplexing, demultiplexing, or analog/digital conversion.

transmission of a verbal message A supplementary telephone service whereby, at the request of a caller, a short message is transmitted by an operator at a specified time to a specified person or persons, should such persons call the operator.

transmission routing Assigning a path between two terminal transmission nodes. Transmission routing may be provided over two or more transmission sections connected end to end.

transmission section A path between two transmission nodes which is carried on one particular transmission medium.

transmission security See communications security.

transmission system The set of equipment which provides multichannel telecommunications facilities capable of carrying record traffic, voice signals, and other data.

transmit To send, especially via a radio transmitter.

transmit-after-receive time delay The interval from the time of removal of rf energy at the local receiver input point to the time the local transmitter is automatically keyed on and the transmitted rf signal amplitude has increased to 90% of its steady-state value.

transmit flow control In data communication systems, a means of adjusting the rate at which data may be transmitted from one terminal so that it is equal to the rate at which it can be received by another terminal.

transmittance A measure of the ability of a material to permit light signals to pass through. The optic equivalent of conductance for electricity.

transmitter 1. In telephony, a microphone. 2. In electromagnetic radio transmission, equipment which feeds radio signals to an antenna, for radiation. 3. In telegraphy, a transmitter which reads punched holes on tape and sends out teletypewriter signals. 4. In a fiber optic system the device which converts a modulated electrical signal into an optical signal for transmission through the fiber. A transmitter typically consists of a light source (laser or LED) and the driving electronics.

transmitter, coin signal Small microphone inside some types of coin telephone box which transmits various tones produced by coins striking bells or gongs. This enables an operator to determine what denominations of coins have been inserted in the box.

transmitter, land Fixed radio transmitter which sends signals to mobile stations on land or to ships.

transmitter, mobile Radio transmitter designed for installation in an automobile, boat, or aircraft.

transmitter, noise-canceling Telephone microphone replacement containing openings at the sides of the diaphragm so noise reaching the phone from the sides tends to be canceled out. Sound waves from the user's voice come straight toward the diaphragm and so are transmitted normally.

transmitter, radio Apparatus producing radio frequency energy for the purpose of radio communication.

transmitter, tape Mechanical device which accepts punched paper tape and serially transmits the coded signals on the tape.

transmitter, telephone A transducer which uses voice sound pressure on a diaphragm to compress carbon granules between electrodes. The resulting resistance variation modulates a battery current flowing between the electrodes, thus translating the acoustic message into an analog electrical signal. Capacitance units or electromagnetic units are sometimes used in place of carbon granules.

transmitter attack-time delay The time interval from the keying on of a transmitter until the transmitted rf signal amplitude has increased to 90% of its steady-state value. This delay excludes any necessary time for automatic antenna tuning.

transmitter-distributor Device which reads punched paper tape carrying a teletypewriter message and transmits appropriate electrical signals to one or more receiving teletypewriters.

transmitter module The optical transmitters in an optical waveguide system.

transmitter power output rating A specified power output capability of a radio transmitter.

transmitter release-time delay The time interval from the keying off of a transmitter until the transmitted rf signal amplitude has decreased to 10% of its key-on steady-state value.

transmitting system, radio Device comprising a radio transmitter connected to its antenna or antennae. Also several transmitters connected to a common antenna.

transmultiplexer 1. In pulse code modulation multiplexing, equipment which transforms signals derived from frequency-division-multiplex equipment (such as group or supergroup equipment) to time-division-multiplexed signals having the same structure as those derived from pulse code modulation (PCM) multiplex equipment (such as primary or secondary PCM multiplex signals). The term also describes the reverse transformation. 2. Device which converts digital TDM signals to analog FDM and vice versa, e.g., two 12-channel FDM groups (each 48 kHz bandwidth) to one 24-channel PCM primary system at 1.544 Mbit/s, or two 30-channel PCM primary systems at 2.048 Mbit/s to one 60-channel FDM supergroup (240 kHz bandwidth).

transpac Packet switched data network used in France.

transparency 1. In communication systems, that property which allows transmission of signals without changing their electrical characteristics or coding beyond the specified limits of the system design. 2. An image fixed on a clear base by means of a photographic, chemical, or other process, especially for viewing by transmitted light. 3. That quality of a data communication system or device that uses a bit-oriented link protocol which does not depend on the bit sequence structure used by the data source.

transparency, information Ability of a network or system to transmit information to a remote end and provide an understandable output, i.e., one for which all necessary code conversion or decoding has been handled by the system.

transparency, real time Time transparency with delay kept to a minimum, typically less than a few hundred milliseconds, thereby permitting interactive conversations, as with a telephone call.

transparency, semi-real time Time transparency with delays typical of enquiry response services such as teletype conversations. Delays are typically seconds or tens of seconds.

transparency, space Ability of a network or system to communicate with called parties wherever they may happen to be.

transparency, time Ability of a network or system to permit messages to be put in to the system whether or not the required called party is available at the actual time of origin.

transparency, time shift Time transparency with a deliberate shift of delivery time to a time convenient to the called party. Traffic from one time zone to another, e.g., from the U.S. to Europe may often be of this type.

transparent interface An interface that permits a system, subsystem, or other equipment to connect with another and operate without modification of system characteristics or operational procedures on either side of the interface.

transparent mode The operation of a facility such that the user has complete and free use of the available bandwidth.

transparent network A network having the property of transparency.

transponder In telecommunication satellites, equipment which receives signals from the Earth, amplifies them, changes their frequency band, and retransmits them back to Earth.

transport, tape The mechanical section of a tape recorder which holds the reels of magnetic tape and draws the tape at steady speed past the two heads.

transport layer In the OSI model, the network processing entity responsible (in conjunction with network, data link and physical layers) for the end-to-end control of transmitted data.

transportable Equipment which can be divided up or broken down into units which are readily moved and which, upon arival at the new site, can be rapidly reassembled into a working terminal.

transpose At specified points to interchange pin positions occupied by the line wires of an open wire route in order to reduce crosstalk.

transposition 1. In data transmission, a transmission defect in which, during one character period, one or more signal elements are changed from one significant condition to the other, and an equal number of elements are changed in the reverse manner. 2. In outside plant construction, an interchange of positions of the several conductors of a circuit between successive lengths. This interchange is normally used to reduce inductive interference on communication circuits.

transposition, junction Transposition of wires at a junction pole between two transposition sections.

transposition, phantom circuit A transposition involving a phantom circuit. It is not sufficient merely to periodically interchange the wires of each pair. It is necessary that all four wires of the phantom group be changed in position at carefully spaced intervals, sometimes changing both side circuits, sometimes only one side circuit, sometimes the phantom as well.

transposition, point Interchanging of wire positions at a pole instead of spreading the change over two spans of the route as is the usual procedure.

transposition, rolled A rotating type of open wire transposition in which the relative positions of the four wires of a group, or two wires of a pair, are continuously changed.

transposition, side circuit A phantom group transposition in which only the wires at a side circuit are transposed.

transposition interval The distance between the points at which conductors are interchanged.

transposition section A length of an open wire route which is considered a single unit for transposition design purposes. In such a case, the system designers endeavour to ensure that all possible coupling effects cancel each other out over the complete section. The section must be long enough to make transposition of long routes practicable and to ensure that circuits joining the main route at intermediate points between stations are balanced.

transpositions Interchanging of pin positions of open wire conductors to reduce crosstalk.

transpositions, coordinated A jointly planned project in which overhead power route and parallel telecommunications open wire route are transposed to obtain best possible overall results.

transputer An INMOS design which combines on a single chip CMOS memory, the logic gates of a high performance computer and its own in-built input/output structure.

transverse parity check A type of parity error checking performed in a group of bits in a transverse direction for each frame.

transverter, automatic message accounting (AMA) Device in a common control office which obtains details of outgoing call, converts the format, and forwards information direct to the automatic message accounting tape.

transverter, automatic number identification (ANI) Device which alerts the automatic bill preparation equipment to the equipment number of the calling line, translates this into the directory number, and passes the information on for automatic bill preparation.

trap A fault detection mechanism designed to isolate various abnormal operating conditions.

trap, annoyance call Central office equipment which enables an annoyance call to be held and traced to its origin.

trap, tuned Series resonant circuit bridged across a circuit which, for all practical purposes, short circuits it at the resonant frequency.

trap-interrupt An interrupt generated when there is a hardware or software error.

trapped mode In ducting, a mode of propagation within a radio duct. At sufficiently high frequencies several such modes may exist, as in a waveguide.

tray, cable Steel mesh trough erected above equipment racks to support cable runs in a central office.

tray, storage battery Lead or plastic tray placed under cells in a storage battery as a precaution against leakage.

treatment, acoustic Fixing of sound-absorbing materials to ceilings and walls in order to reduce reverberation.

treatment, service Precedence, circuit conditioning, and other action taken by automatic switching center equipment as a result of analysis of the class of service of a particular subscriber's line.

tree A network configuration in which there are branches but no closed loops or meshes.

trellis code modulation A version of Quadrature Amplitude Modulation (QAM) which enables relatively high bit rate signals to be used on ordinary voice-grade analog circuits, e.g., 9.6 kbit/s over dial-up

PSTN circuits and up to 16.8 kbit/s over 4-wire leased voice grade circuits.

tremendously high frequency (THF) Frequencies from 300 to 3000 GHz.

trencher Excavating machine, operating either on wheels or a track, which excavates a trench for burying cables or ducts by means of a steel digging wheel or chain.

triac A gated switching device which will conduct in either direction.

triangulation A method of conducting tests and fault localization on a circuit from one end only, with assistance at the other end only in looping or disconnecting circuit ends. By measuring the loop resistance of three different loops made up from three wires arranged in three different sets of two, it is possible to calculate the individual resistance of each leg.

tribo-electric Charging with static electricity by movement or rubbing of one surface on another.

tributary circuit Circuit connecting an individual terminal to a switching center, or a minor data terminal to a data network.

tributary office A Class 5 or end office.

tributary station In a data network, a station other than the control station.

trigger To initiate action which cannot be stopped by another action on the trigger circuit.

trigger circuit Circuit in which a small input change can cause an abrupt change in circuit characteristics or operation.

trimmer Small mechanically-adjustable component connected in parallel or series with a major component so that the net value of the two can be finely adjusted for tuning purposes.

trimming, tree Cutting trees or bushes to provide clearance for a pole route.

triode A three-electrode electron tube.

trip free A power circuit breaker which will trip and break a faulty circuit even if the operating handle is held closed.

triple 1. Three insulated wires twisted together. 2. Method of routing used in the AUTOVON system. Trunk groups are in sets of three, leading to different destinations, all of which may be used as transit switching points on the way to the final destination.

triple beat A third-order beat whose three beating carriers all have different frequencies, but are spaced at equal frequency separations.

triple harmonics See harmonics third.

tripper Device that holds a circuit breaker in the active position until the breaker is operated. The tripping device may then be operated manually to restore the breaker to its operating position.

trombone (loop) connection The use for a single call of two circuits in tandem between a remote switching stage and its controlling entity; for example, to the main office and back again.

tropopause The upper boundary of the troposphere above which the temperature either increases slightly or remains constant.

troposcatter   See tropospheric scatter.

troposphere   The layer of the earth's atmosphere, between the earth's surface and the stratosphere, in which about 80% of the total mass of atmospheric air is concentrated and in which temperature normally decreases with altitude.

tropospheric radio duct   A quasi-horizontal layer in the troposphere within which radio energy of a sufficiently high frequency is substantially confined and propagated with abnormally low attenuation.

tropospheric scatter   1. The propagation of radio waves by scattering as a result of irregularities or discontinuities in the physical properties of the troposphere. 2. A method of transhorizon communications utilizing frequencies from approximately 350 MHz to 8400 MHz. Syn. troposcatter

tropospheric scatter propagation   Propagation involving scattering from many inhomogeneities and discontinuities in the refractive index of the atmosphere.

tropospheric wave   A radio wave propagated by reflection from a point of abrupt change in the dielectric constant or dielectric gradient in the troposphere.

trouble   Failure or fault affecting the service provided by a system.

trouble operator   Operator who busies out faulty lines or circuits or cooperates with test desk and fault clearance personnel in clearing trouble.

trouble-shoot   To investigate a fault, localize it, and, if possible, correct it.

trough   1. A guide structure for the transit of cable and repeaters along the deck of a cableship. 2. In a cable factory, a facility for cooling the extruded core to ambient temperature.

trunk   1. A single- or multichannel communications medium between terminal facilities where channels can be tested, rerouted, dropped out, or switched to another route. 2. A circuit between two ranks of switching equipment in the same office, or between different switching centers or different central offices.

trunk, combined line and recording (CLR)   Trunk seized when calling an operator to place a toll call. The combined line and recording circuit is used for booking the call, recording details, and establishing the call itself while the caller waits on the line.

trunk, common   A trunk accessible from all groups of a grading.

trunk, end office toll   Toll trunk joining two Class 5 end offices.

trunk, extendable information   Facility on some PABXs by which an operator can connect an extension not normally permitted to initiate city calls to a trunk direct to the central office so that calls can be dialed from the extension.

trunk, foreign exchange   Trunk connecting a PABX to a central office other than the one which normally serves the area concerned.

trunk, incoming   Trunk used for calls coming into an office for connection to local subscribers terminated at the office.

trunk, individual   A trunk which serves only one group of a grading.

trunk, intermarker group   A trunk used to interconnect two common control offices in the same building.

trunk, inter-office   Trunk connecting two telephone offices.

trunk, interposition   See interposition trunk.

trunk, interswitch   A trunk between switching centers.

trunk, intertandem   A trunk between two tandem switching centers or tandem offices.

trunk, intertoll   A trunk between two toll offices.

trunk, intraoffice   1. A trunk between ranks or groups of switches within a central office. 2. A trunk between two central office units in the same switching center complex.

trunk, outgoing   A trunk used for telephone calls to offices outside the home switching center.

trunk, partial common   A trunk accessible from more than one, but not all, groups of a grading.

trunk, PBX   A trunk between a PBX and its home central office.

trunk, primary   A trunk from the outlet of the concentrator stage in an electromechanical office to the first switching stage.

trunk, recording   Trunk to a toll operator used only for booking and recording toll calls. Since calls are not set up on recording trunks, there is no stringent transmission limit set for such trunks.

trunk, recording-completing   Trunk to the toll operator used for recording and completion of toll calls.

trunk, reverting call   Trunk used to establish a talking circuit for a reverting call from one subscriber on a party line to another party on the same line.

trunk, secondary   Trunk to a secondary switch, such as a secondary line switch.

trunk, secondary intertoll   Trunk between an automatic toll switch and its associated manual assistance switchboard.

trunk, tandem   Trunk to or from a tandem office.

trunk, tie   Trunk between two PBXs.

trunk, toll   Circuit used for intertoll calls.

trunk, toll connecting   Trunk from an end office to a toll office higher in the hierarchy.

trunk, toll switching   Trunk from toll office to end office used to establish incoming toll calls.

trunk, tributary toll   Trunk between a toll center and a tributary office.

trunk, trouble intercepting   A trunk formed by patching the office side of an out-of-order line to a special switchboard trunk so that calls to the line may be intercepted and appropriate information provided to callers.

trunk, trouble observation and test   A trunk formed by patching the line side of an out-of-order line to a test desk or test panel so that the test desk staff is alerted to any attempt to initiate calls.

trunk answer any station   PBX feature which enables any extension to answer incoming calls (e.g., when the PBX attendant is temporarily absent or during nights).

trunk busy tone   Tone used to inform operators that a trunk is busy. It operates at 600/120 Hz interrupted twice per second.

trunk circuit   A pair of complementary channels with associated equipment terminating in two exchanges.

trunk code   A digit or digit combination (not including the trunk prefix) which characterizes a particular numbering area within a country. The trunk code must be dialed before the called subscriber's number is dialed when the calling and called subscribers are in different numbering areas.

trunk distribution frame (TDF)   A distribution frame dedicated to trunk (long distance) circuits. In some administrations a TDF is used for crossconnection at high (PCM) bit rates.

trunk encryption device (TED)   A bulk encryption device used to provide secure communication over a wide band digital transmission link. It is usually located between the output of a trunk group multiplexer and a wideband radio or cable facility.

trunk exchange   A central office exchange the principal function of which is to control the switching of trunk traffic.

trunk-free   A condition sent forward or backward on the interexchange data channel when a data circuit has been released by the sending exchange (i.e. when the data circuit is considered to be idle) or while waiting for release by the other exchange. This condition appears as a clearing signal at the customer interface.

trunk group   Group of trunks terminated at the same two points.

trunk group multiplexer (TGM)   A time-division multiplexer for combining individual digital trunk groups into a higher rate bit stream for transmission over wideband digital communication links.

trunk-hunting   Describing a connector or selector switch which hunts over a group of trunks until it finds and seizes an idle circuit.

trunk line (CATV)   The major distribution cable of a CATV network. It divides into feeder lines which are tapped for service to subscribers.

trunk network   The network whose nodes are trunk exchanges interconnected by trunk circuits.

trunk prefix   A digit or combination of digits which must be dialed in order to reach a subscriber with a different area code. It provides access to the automatic outgoing trunk equipment.

trunk-seized   A condition sent forward in the interexchange data channel when a data circuit is seized. This condition appears as the connection-in-progress state at the called customer interface.

trunked radio systems   Radio systems using common frequencies.

trunking   1. The interconnection among the various ranks of equipment. 2. The branch of telephony or telegraphy concerned with providing equipment to carry traffic with a specified grade of service.

trunks, final   Trunks designed to carry overflow traffic from high-usage trunks. Trunks used for final choice routing.

trunks, high-usage   Trunks which provide first-choice routing between points. Circuit quantities are calculated on the basis of heavy loading and efficient use, with overflow traffic spilling over to a final trunk or final choice group.

trunkside connection   Method used by inter-LATA carriers to obtain access to local telephone networks using trunk lines.

T-span   A telephone cable or route through which a T-carrier (PCM system) runs.

T-tap   A passive line interface for extracting data from a circuit or optical signals from a fiber.

tube, cable   (or cable tunnel) Deep tube system used in some cities for telephone cables. These tunnels are sometimes 30 meters below ground level and provide a measure of security.

tube, cathode ray   (CRT) An electron beam tube used for display of changing electrical phenomena, generally similar to a television picture tube.

tube, cold-cathode   Electron tube with cathode which emits electrons without the need of a heating filament

tube, electron   Evacuated or gas-filled tube enclosed in a glass or metal case in which the electrodes are maintained at different voltages, giving rise to a flow of electrons from the cathode to the anode.

tube, gas   Gas-filled electron tube in which the gas plays an essential role in the tube's operation.

tube, hard   Highly evacuated electron tube.

tube, mercury-vapor   Tube filled with mercury vapor at low pressure. Used as a rectifying device.

tube, metal   Electron tube enclosed in a metal case.

tube, plastic entrance   Short length of plastic pipe used to feed cables or drop wires into a building.

tube, soft   Electron tube containing residual gas.

tube, traveling wave   (TWT) Wideband microwave amplifier in which a stream of electrons interacts with a guided electromagnetic wave moving substantially in synchronism with the electron stream, resulting in a net transfer of energy from electron stream to wave.

tube, vacuum   A hard tube. An electron tube in which the residual gases, which adversely affect performance, have been removed.

tube, velocity-modulated   Electron tube in which the velocity of the electron stream is continually changing. A klystron.

tube, voltage regulator   Gas tube with voltage drop substantially constant over a given range of currents.

tubing, heat shrinkable   Tubing which shrinks to about half its original diameter when heat (such as from a hot air blower) is applied. Sometimes used to provide oversheaths for cable splices.

tubing, lead   Lead pipe of various diameters used for various external-plant purposes; for example, air pressurization systems.

tubing, plastic Polyethylene pipe used for various purposes; for example, dry air runs into pressurized cable systems.

tubing, poly-cor Outer plastic sheath designed to cover several plastic pipes used to transmit dry air or gas between a cable vault and cable pressurization monitoring equipment.

tubing, temporary closure Temporary plastic cable splice sleeve which can be closed by a zip-type fastener.

tune To adjust frequency. In particular, to adjust for resonance or for maximum response to a particular incoming signal.

tuned circuit See circuit, tuned.

tuned radio frequency (TRF) The condition describing a radio receiver with amplification and selectivity obtained by stages operating at the radio frequency itself (not, as with a superheterodyne receiver, by stages operating at an intermediate frequency or IF).

tuned relay Relay tuned to operate by mechanical resonance at a particular frequency.

tuned ringer Bell designed to be used with a party line telephone which uses selective calling. Available systems use harmonic, decimonic or synchromonic frequencies between 16 Hz and 66 Hz.

tuner, radio The radio frequency and intermediate frequency parts of a radio receiver which produce a low level audio output signal.

tuner, waveguide Device which permits adjustment of the impedance of a waveguide.

tungar tube A rectifier tube used for battery charging and which utilizes a tungsten cathode and argon filling.

tuning Adjusting a circuit for resonance or for best possible performance.

tuning, broad Tuning which is not sharply resonant.

tuning, electron Adjusting the frequency of a device by changing the potentials of electrodes.

tuning, ganged Simultaneous tuning of several circuits by a single control. Often involves varying capacitors mounted on a common shaft.

tuning, permeability Tuning to resonance by moving a core (usually ferrite) into or out of a coil in order to change its inductance.

tuning, sharp Tuning which is sharply resonant over a narrow frequency bandwidth.

tuning, staggered Method of obtaining a broader effective bandwidth by tuning each individual IF stage to a slightly different frequency.

tuning core Ferrite or powdered iron core which can be moved relative to a coil in order to change its inductance.

tuning indicator Voltmeter or a voltage-sensitive gas tube device which enables a radio receiver to be tuned to a particular transmission.

tuning screw Impedance-adjusting rod which tunes by penetrating a cavity or waveguide.

tunnel effect The effect which enables electrons or other particles to find their way through potential barriers which they were once believed unable to penetrate. Although it defies classical theory, the tunnel effect model was invented because it has practical application.

tunneling mode See leaky modes.

turn over In a telephone line, to reverse the connection of the tip and ring conductors.

turnaround time 1. Time needed to reverse the direction of transmission on a half-duplex circuit. 2. Time lost in the processing and return of data over a channel.

turns ratio Number of turns in the secondary winding of a transformer divided by the number of turns in the primary winding.

turret, attendant's Same as attendant's console.

turret, test A console providing basic test features an suitable for use on the test desk of a small central office.

TV penetration The percentage of homes having one or more television sets at the time of an ARB (American Research Bureau) survey. The ARB surveys local markets from October through July. The number of surveys in a year depends on the size of the market.

tweeter Loudspeaker designed to handle high audio frequencies efficiently (3 kHz to 20 kHz).

twinaxial cable A shielded coaxial cable with two central conducting leads.

twin-channel carrier Carrier system used on minor routes. Upper and lower sidebands of the same carrier frequency are used on an independent sideband basis.

twin-line Feeder cable with two parallel, insulated conductors. It is widely used to carry TV signals to a receiver.

twine, lacing A waxed linen cord used to lace together cables on cable ladders inside central office or equipment rooms.

twist, waveguide Waveguide section in which the cross-section rotates about the longitudinal axis.

twisted pair A pair of insulated wires which are twisted together but are not covered with an outer sheath.

twister A magnetic memory device using thin magnetic tape and small solenoids.

twister, sleeve Hand tool used for sleeve jointing line wires in an open wire route. The tool grips the sleeve and twists it at both ends to make a secure joint.

two-out-of-five code A positional notation in which each decimal digit is represented by five bits: two of one kind (1s) plus three of another (0s). CCITT Signaling System No. 5 uses a code of this type.

two-party service A party line, with two users having their telephones connected to the same pair of wires. Selective ringing is normally used on these lines.

two-phase An alternating current circuit with two sinusoidal voltages which are 90 degrees apart.

two-pilot regulation In frequency division multiplex systems, the use of two pilot frequencies within a transmitted band in order that the change in attenuation due to twist can be detected and compensated for by a regulator.

two-source frequency keying See frequency exchange signaling.

two-tone keying 1. In telegraphy, a system employing a transmission path composed of two channels traveling in the same direction, one for transmitting the space of binary modulation, the other for transmitting the mark of the same modulation. 2. That form of keying in which the modulating wave causes the carrier to be modulated with a single tone for the marking condition and with a different single tone for the spacing condition.

two-tone telegraph See two-tone keying.

two-way Trunk which can be used for calls originated from either end.

two-way alternate operation See half-duplex operation.

two-way capacity A CATV system which can conduct signals to the headend as well as away from it. Two-way or bi-directional systems now carry data, and may in due course carry full audio and video television signals in either direction.

two-way simultaneous operation See duplex operation.

two-way splitting PBX feature by which the attendant may break in to a call and converse with either party without the other overhearing.

two-wire circuit See circuit, two-wire.

two-wire line A circuit provided by two metallic paths, both directions of which use the same two wires.

two-wire switching Switching using the same path, frequency band, or time interval for transmission in both directions.

TX operator An operator who attempts to complete delayed calls.

tying-in Using a tie wire to bind a line wire to an insulator on an open wire route.

type approval Procedure by which equipment is authorized as suitable in particular for connection to the public switched telephone network.

typing perforator A reperforator which types received characters onto paper tape alongside the punched holes which represent these characters.

UL approved Tested and approved by the Underwriters' Laboratories, Inc.

ultra high frequency See frequency, ultra high.

ultrasonic Acoustic signals at frequencies higher than can be heard by a human ear: above 20 kHz.

ultraviolet radiation Electromagnetic radiation in a frequency range between visible light and high frequency x-rays. Radiation from the region of the electromagnetic spectrum between the short wavelength extreme of the visible spectrum (about 0.4 μm) and 0.04 μm.

unaffected level Of a compandor, the absolute level (at a point of zero relative level on the line between the compressor and the expander of a signal at 800 Hz) which remains unchanged whether the circuit is operated with the compressor or not.

unattended Equipment designed to operate without a human attendant.

unattended operation System which permits a station to receive and transmit messages without the presence of an attendant or operator.

unavailability A measure of the degree to which a system, subsystem, or piece of equipment is not operable and not in a committable state at the start of a mission, when the mission is called for at a random point in time.

unbalanced circuit A two-wire circuit with legs which differ from one another in resistance, capacity to earth or to other conductors, leakage, or inductance.

unbalanced line A transmission line in which the magnitudes of the voltages on the two conductors are not equal with respect to ground; for example, a coaxial line.

unbalanced modulator A modulator in which the modulation factor is different for the alternate half-cycles of the carrier. Syn.: asymmetrical modulator.

unbalanced output Output with one leg at ground potential.

unbalanced wire circuit A circuit whose two sides are inherently electrically unlike.

unbound mode In fiber optics, any mode which is not a bound mode; a leaky or radiation mode of the waveguide. Syn.: radiative mode

uncertainty An expression of the magnitude of a possible deviation of a measured value from the true value. Frequently it is possible to distinguish two components: the systematic uncertainty and the random uncertainty. The random uncertainty is expressed by the standard deviation or by a multiple of the standard deviation. The systematic uncertainty is generally estimated on the basis of the parameter characteristics.

unclassified Information which need not be restricted in circulation for security reasons.

undamped wave A signal with constant amplitude.

underbunching Traveling wave tube condition wherein the tube is not operating at its optimum bunching rate.

undercurrent relay Relay which operates when the current in a circuit under observation drops below a specified level.

underlying carrier A common carrier whose facilities are rented or leased to other common carriers.

undervoltage alarm Alarm given when the storage battery voltage falls below a specified level.

undervoltage protection Automatic disconnection by circuit breakers of loads from power sources when the incoming voltage is too low for safety.

undervoltage relay Relay which gives an alarm when voltage in the circuit under observation falls below a specified level.

Underwriters Laboratories, Inc A laboratory established by the National Board of Fire Underwriters which tests equipment, materials, and systems which may affect insurance risks, with special reference to fire dangers and other hazards to life.

undetected error rate See residual error rate.

ungrounded Not connected to ground.

UNICCAP (universal cable circuit analysis program) A Bell System software program which provides quick and precise calculations of transmission line characteristics.

unicoupler Device used to couple a balanced circuit to an unbalanced circuit. A balun.

unidirectional The transmission of information in one direction only.

unidirectional channel See one-way-only channel.

unidirectional operation Operation in one direction only.

unidirectional repeater Repeaters designed for transmission in one direction only. Twin facilities are required for two-way service.

unifilar Suspension of an instrument's moving part by a single wire or strip.

uniform encoding 1. The generation of character signals representing uniformly quantized samples. 2. An analog-to-digital conversion process in which all of the quantization subrange values are equal. Syn.: uniform quantizing

uniform line Transmission line with uniform electrical properties along its entire length.

uniform quantizing See uniform encoding.

Uniform Service Order Code(USOC) Bell System term used on Universal System service orders. An alpha-numeric code used to identify a tariff item on a universal service order.

uniform system of accounts (USOA) Standardized system of accounting prescribed by the FCC to be used by common carriers.

uniform system of accounts revision (USOAR) USOA is to be revised to meet all new regulatory requirements. Telecommunications Industry Advisory Group (TIAG) has been established to recommend revisions.

uniform waveguide A waveguide with constant dimensions and electrical characteristics along its entire length.

unigauge system Method sometimes used for the design of distribution cable networks providing subscribers' loops. The system puts all customers up to 30,000 feet away from the central office on the same gauge of cable (26 AWG), with range extenders on loops over 15,000 ft. and inductive loading, after 15,000 ft., on all loops over 24,000 ft.

unigrounded neutral A three-wire power system with its source wye connected with the neutral of the transformer, but grounded in the one originating location.

unijunction transistor A bipolar transistor with two bases and one emitter, sometimes used in relaxation oscillator circuits.

unilateral control A synchronization control system between two exchanges in which the clocks at both locations are controlled from one of the two exchanges.

unintelligible crosstalk components Transferred speech currents which can introduce unintelligible crosstalk into certain channels.

Uni-Pair Plastic-insulated cable in which the insulated pairs are twisted and bound together.

unipolar 1. Neutral transmission of teletypewriter signals, where current indicates a mark and no current means a space signal. 2. Transistor formed from a single type of semiconductor material, N or P, as employed in field effect transistors.

unit A specified quantity in terms of which other quantities can be measured.

unit, answer back Device used with a teletypewriter which automatically transmits a predetermined code so that a calling machine will have a record, on its message copy, of the fact that the outgoing message has reached the required destination.

unit, balancing Adjustable capacitor used in toll offices for balancing the capacitance of local office wiring.

unit, battery test load A device made up of heavy-duty resistors used in testing the capacity of cells in a storage battery.

unit, cable Groups of pairs stranded together in a cable.

unit, central office A unit usually composed of 10,000 lines.

unit, central processing (CPU) A computer which controls the whole, or some part, of the switching activity in a central office. CPUs can be operated on a main-plus-standby basis, a load sharing basis, a multiprocessor basis, or a cluster basis.

unit, cgs The standard units in the original metric centimeter-gram-second system, now replaced by the SI system.

unit, crosstalk Unit of crosstalk coupling. The current in the disturbed circuit of one-millionth of that in the disturbing circuit, i.e., a 120 dB current difference, or a 60 dB power difference.

unit, DX signaling Signaling unit which accepts separate E & M signals and converts these to signals on the cable pair carrying both-way voice.

unit, key service Centralized apparatus containing relays and power supply equipment which enable key telephones to be installed.

unit, L A coaxial cable facility providing a broad based transmission path.

unit, labor Length of time the average trained man will take to carry out a particular task.

unit, line building-out Unit used on a repeated circuit to adjust the impedance of the circuit in order to improve matching.

unit, message Method of charging for telephone calls. Locally, a message unit usually represents a call lasting three minutes. On calls to offices at greater distances, a complete message unit will span a shorter time.

unit, power A component or module which provides a power supply at an appropriate voltage for a particular item or items of equipment.

unit, protector Unit on customer's premises or at the ends of cable routes to protect against damage by lightning or contact voltages.

unit, R A radio facility providing a broad band transmission path.

unit, receiver The receiving capsule in a telephone handset.

unit, ring and talk (RT) Small power unit with ac public supply input at 60 Hz. It gives output power at 24 V dc for microphone currents and 20 Hz ac to ring telephone bells.

unit, secretarial answer Key cabinet which bridges up to 20 lines, enabling a secretary to pick up any of them.

unit, signaling Device which produces signaling codes or currents, or which operates upon receipt of appropriate signaling currents or codes.

unit, T Small power unit. Operating on ac public mains input, it provides a 24 volt dc output for microphone current.

unit, telegraph coupling Interface units between telegraph circuits of different types, such as duplex, simplex, or half duplex.

unit, traffic Old name for the erlang. The telephone traffic represented by one circuit continuously occupied. Traffic unit figures are the same as percentage occupancy figures divided by 100.

unit, traffic work Unit used in calculating the work loads of operators who have to deal with several different types of call. One standard traffic work unit is figured as taking 15.65 seconds to complete.

unit, transmitter The microphone capsule in a telephone handset.

unit, trouble Number of cases of trouble, on a percentage basis, which can be anticipated from a particular class of plant during a year.

unit call One hundred call seconds (CCS). Used as a traffic unit.

unit element error rate for isochronous modulation The ratio of the number of incorrectly received elements to the number of emitted elements.

unit interval In a system using isochronous modulation, that interval of time such that the theoretical duration of the significant intervals of a telegraph modulation are all whole multiples of this interval.

unit of property An accounting term defined in the Uniform System of Accounts: a pole, a manhole, etc.

United States Telephone Association (USTA) Originally the trade association of the Independent Telephone Companies (USITA) but after divestiture of AT&T the former Bell operating companies became eligible for membership and "Independent" was dropped from the name.

United States Telephone Suppliers Association (USTSA) An affiliate of the United States Independent Telephone Association.

United Telecom A major U.S. corporation controlling local telephone companies and manufacturers.

unitized power equipment Power equipment as used in telephone central offices. This is often broken down into three units: the rectifiers and control circuits, supervisory equipment, and signaling equipment.

units, international system Most countries in the world have now agreed to use an internationally standardized system of units for expressing the values of physical quantities, known as the SI Units (from the French, Systeme International d'Unites). See Appendix B.

units, SI See Appendix B.

unity coupling Coupling between primary and secondary windings of a perfect transformer.

unity power factor A power factor of 1, which means that the load is, in effect, a pure resistance with ac voltage and current completely in phase.

universal access number A single number which, when dialed from anywhere in the country, will reach a customer with several installations in different parts of the country. Calls from subscribers on exchanges in predetermined areas will be routed to installations chosen (within certain restrictions) for the area in question by the customer having the facility.

universal information services AT&T term for fully integrated network providing voice and non-voice services.

universal motor Motor which will operate on dc or on single-phase ac.

universal service concept U.S. government policy of making telephone services available to all, at a reasonable cost.

universal service fund   Funds levied on inter-LATA carriers and made available to support telephone companies in high-cost areas.

universal service order (USO)   Bell System term for a document initiated by the negotiator, authorizing the implementation, modification or discontinuance of a tariffed customer service. Sometimes a USO will involve work being done by an independent telephone company.

universal synchronous/asynchronous receiver/transmitter (USART)   Integrated circuitry provided in many data communications devices, which converts data in parallel form from a processor into serial form for transmission.

universal time   See Greenwich mean time.

unloaded   Circuit from which loading coils have been removed or which has not had these coils inserted.

unmodulated   A carrier signal which is not modulated by an information-carrying signal.

unnumbered command   In data transmission, a command that does not contain sequence numbers in the control field.

unnumbered response   In data transmission, a response that does not contain sequence numbers in the control field.

unperturbed orbit   The orbit of a satellite under ideal conditions, in which the satellite is subjected only to the attraction of the primary body, effectively concentrated at its center of mass.

unreasonable message   A common channel signaling system message with an inappropriate signal content, an incorrect signal direction, or an inappropriate place in the signal sequence.

unstable   A circuit in which net gains are so nearly equal to net losses that a small change in loss can make the circuit sing.

unsuccessful call   A call attempt that does not result in a connection.

unvoiced sounds   Sounds made by the human voice without using the vocal cords; for example, light "s" sounds or explosive "p" sounds as opposed to voiced "b" or "z" sounds. These sounds do not have a fundamental frequency.

unwanted emission   Spurious or out-of-band emissions.

unweighted   Unadjusted.

upconverter   Modulation circuit giving a radio frequency as its output.

up-counter   Counting device which counts from zero upwards.

update   To make current, as a program that has been modified to comply with current information or some other change in circumstances.

up-down counter   Counting device capable of counting up from zero or down from a predetermined level towards zero.

upgrade   To improve service by offering better facilities.

uplink   In satellites, that portion of a communications link used for transmission of signals from an earth terminal to a satellite or airborne platform. It is the converse of downlink.

upper sideband   Higher of the two bands of frequencies produced by amplitude modulation. The sum of the carrier and the modulating signal frequencies.

upset duplex   In telegraphy, duplex telegraphy sending polar signals in one direction and neutral signals (by opening and closing one line) in the other direction.

upstream   In a broadband network, a signal from a transmitting station to a headend.

up-time   Uninterrupted period of time that network or computer resources are accessible and available to a user.

upward compatible   A computer's capability to execute programs written for another computer without major alteration.

U S West   One of the seven Regional Holding Companies resulting from the divestiture of AT&T. The regional company for Mountain Bell, Northwestern Bell and Pacific Northwest Bell operating companies.

usable field-strength   Minimum value of the field strength necessary to permit a desired quality of reception under specified receiving conditions in the presence of noise and interference, either in an existing situation or as determined by agreements or frequency plans. For fluctuating interference or noise, the percentage of time during which the required quality must be ensured should be specified.

usage   Percentage of time that a circuit is actually in use. Usage is sometimes expressed in CCS for a group of circuits.

usage, air   Volume of air (or sometimes nitrogen) fed into a pressurized cable system.

usage count   Count by a scanning device of the number of times a circuit is busy during a specified time interval. If the count is made every 100 seconds, the reading will be in CCS.

usage sensitive pricing   Method of charging for local calls based on their duration. Syn.: local measured service

use, joint   Joint utilization of poles by power, telephone, CATV, and street lighting authorities. Joint trenching and joint use of space under sidewalks is also now becoming accepted practice in some areas.

useful bandwidth   The bandwidth available for information-carrying signals, bearing in mind the fact that guard bands are needed between channels and groups.

useful life   This is the period during which a constant failure rate can be expected: in early days the failure rate is often high, because some components must be expected to fail early, and again when the design life of the components is being approached it is expected that failures will increase again, as items wear out. The period in the middle is usually the period of least failures and is the useful life of the item.

user   1. A person or an automatic device making or receiving a call. 2. A person, organization, or other group that employs the services of a telecommunication system for transfer of information to others. 3. Of a signaling system, a telecommunication service which uses a signaling network to transfer information. 4. In a message handling system, the person or computer application or process who makes use of the system.

user-action frame   See response frame (videotex).

User Agent (UA)   In a message handling system, the set of computer processes that is used to create, inspect and manage the storage of messages. The UA is typically an editor, a file system, a word processor. During message preparation the originator communicates with his UA via an input/output device (for example a keyboard, display, printer, facsimile machine, or telephone). Messages received from the Message Transfer System (MTS) are passed to the user via the UA. To send and receive messages the UA interacts with the MTS via the submission and delivery protocol.

User Agent Entity (UAE)   In a message handling system, an entity in the User Agent Layer of the Application Layer that controls the protocol associated with cooperating User Agent Layer Services. The UAE exchanges control information with the Message Transfer Agent Entity (MTAE) or with the Submission and Delivery Entity (SDE). This information is needed to create the appropriate envelope and thus provide the desired message transfer service elements.

user-class indicator   Information sent in the forward direction in a data system indicating the user class of the calling customer.

user class of service   A category of data transmission service provided in a public data network in which the data signaling rate, the terminal operating mode, and the code structure are standardized.

user information   Information transferred across the functional interface between a source user and a telecommunication system for ultimate delivery to a destination user. In data telecommunication systems, user information includes user overhead information.

user information bit   A bit transferred from a source user to a telecommunication system for ultimate delivery to a destination user. User information bits do not include those overhead bits which originate, or have their primary functional effect, within the telecommunication system.

user information block   A block that contains at least one user information bit.

user part   A functional part of a common channel signaling system which transfers signaling messages via the message transfer part.

user service or facility   A service or facility available to a user on demand and provided as part of a public data network transmission service. Some facilities are available on a per call basis and others are assigned for an agreed period of time at the request of the user. On certain assigned facilities per call options may also be available.

utility power load   That part of the power load of a telephone central office building (such as air conditioning) which can be interrupted without an adverse effect on real time communications.

utility routine   Standard routine, usually part of a larger software package, which performs a service and/or program maintenance function such as file maintenance, file storage, file retrieval, media conversions, production of memory, and file printouts.

utilization factor   Ratio of maximum demand for power to the rated capacity of the system.

V & H coordinates   Vertical (i.e., north) and horizontal (i.e., east) coordinates of a city or central office, used to calculate the airline distance between centers and so the rate to be charged for toll calls.

V   number   In optoelectronics, a synonym for normalized frequency.

vacancy   An empty site in a crystal which would normally be occupied by an atomic nucleus. A vacancy is not a hole. A hole is an empty site normally occupied by an electron, or one which represents an empty energy level in the valence band.

vacant number   Telephone number which cannot be used until more equipment is provided in the central office.

vacant number intercept   Feature which routes calls addressed to unallocated numbers straight to a recorded announcement or to an information operator.

vacuum   Space from which enough air has been pumped that the operation of a nominally invacuo device is not adversely affected.

vacuum evaporation   A manufacturing technique in which material to be deposited in a thin layer on another material is heated in a vacuum in the presence of the base material, which remains cool. Atoms evaporate from the heated solid and condense on the base material in a thin film of readily controllable thickness.

vacuum fluorescent display   (VFD) A light-emitting triode utilizing fluorescent phosphors which can be used in alphanumeric display panels.

vacuum relay   Relay with contacts enclosed in an evacuated space, usually to give reliable long-term operation.

vacuum switch A switch with contacts contained in an evacuated container so that spark formation is discouraged. Sometimes called a vacuum relay.

vacuum tube An electron tube such as a diode, triode, tetrode, or pentode.

valence band The energy band for electrons which are bound to individual atoms. In metals, such electrons can easily be jumped to the conduction band of the atom in order to facilitate electric conduction.

valence electron An electron found in the valence band where it is free to move from atom to atom. The energies of the valence electrons bind the atoms or molecules in a crystal together.

validity check A test designed to ensure that quality of transmission is maintained.

value, analog A continuously variable parameter.

value-added carrier Common carrier which uses basic services leased from other carriers and adds special features to these, (usually computer-oriented) before renting these out to users.

value-added network (VAN) A data communications system in which the services provided for users greatly enhance the usefulness of the basic facilities utilized.

valve The original British word for an electron tube. The name is based on the properties of the earliest glass bottled devices which rectified currents rather as a valve is used to turn on the flow of water.

valve, gas admission An inlet valve in an air pressure circuit which is used for recharging reservoir tanks in a cable pressurization system.

valve, gas pressure relief A safety valve in a pressurized cable which operates at a preset value if the gas pressure in the cable becomes unnecessarily high.

valve, pressure testing A tire-type valve which can be screwed into a lead-sheathed pressurized cable.

valve voltmeter A once-much-used high impedance electron tube amplifier followed by an output stage, sometimes with rectification, and a measuring instrument. The digital multimeter is today's near equivalent.

Van Allen belts Belts of charged particles surrounding the Earth and found in two concentrations: the first about 3000 km high, the second about 15000 km high.

vapor phase axial deposition (VAD) technique A method of fabricating an optical waveguide preform by forming small glass particles and depositing the particles on the end of a rod. See also chemical vapor deposition technique, double crucible technique, ion exchange technique, rod-in-tube technique, and soot technique.

varactor A variable reactor. A semiconductor that behaves like a capacitor.

varactor tuning Tuning, especially in TV receivers, in which the variable capacitance is provided by a varactor.

variable Symbol or mnemonic whose value changes from the execution of one program to another or during the execution of a single program.

variable capacitor See capacitor, variable.

variable frequency oscillator (VFO) An oscillator whose frequency can be set to any required value in a given range of frequencies.

variable impedances Capacitors, inductors, or resistors which are adjustable in value.

variable length numbering scheme A numbering scheme in which the length of subscribers' numbers varies within a given numbering area.

variable-mu Device in which the amplification factor (mu) can be varied.

variable quantizing level (VQL) A speech-encoding technique that quantizes and encodes an analog conversation for transmission nominally at 32 kbit/s but other data rates may be used.

variable rate adaptive multiplexing (VRAM) A digital refinement of the TASI concept. When the number of active speakers exceeds the channel capacity during short-duration overflow peaks, performance degradation is shared among all users by reducing the sampling rate.

variable-reluctance A transducer in which the input (usually a mechanical movement) varies the magnetic reluctance of a device.

variable spilling The ability of a switch, or of circuit elements in a telephone system, to absorb specified digits rather than to send them on to the next switching stage.

variation, net loss See net loss variation.

variation, seasonal Change in circuit parameters following seasonal change in environmental conditions.

variation monitors Devices for sensing deviations in voltage, current, or frequency and capable of providing an alarm and/or initiating transfer to other power sources when programmed limits of voltage, frequency, current, or time are exceeded.

varicap A diode used as a variable capacitor. A varactor.

varindor An inductor with variable inductance which changes with current value.

variocoupler Radio frequency device used in very early radio equipment design. Its large coils could be physically moved along their axes to increase coupling between stages.

variolosser A word, rarely encountered, for a variable attenuation unit whose actual attenuation depends on the rectification of an incoming signal. It may be used to increase or to decrease attenuation as signal strength increases.

variometer Variable radio frequency inductor with variation obtained by rotating one coil inside another, the two coils being connected in series.

varistor A variable resistor. A device which does not have a linear resistance characteristic.

Varley Measurement of resistance imbalance between two sides of a two-wire metallic circuit.

Varley, three wire Measurement of the resistance unbalance when a third wire is used instead of a ground connection.

Varley loop A testing bridge used to localize an earth or contact fault in a cable.

vault, cable Room under the main distribution frame in a large central office building. Outside plant cables enter the building here, usually in nests of ducts, and are routed through to the main distribution frame. Pressurization equipment and leads are often connected to external cables in the cable vault.

vector Quantity which may be represented by a line having length and direction.

vector diagram Diagram using vectors to indicate relationships between voltages and currents in a circuit.

vector sum A sum of two vectors which, when they are at right angles to each other, equals the length of the hypotenuse of the right triangle so formed. In the general case, the vector sum of the two vectors equals the diagonal of the parallelogram formed on the two vectors.

velocity, escape See escape velocity.

velocity, group Velocity of a modulated signal through a circuit. Velocity of propagation of an envelope, provided this moves without significant change of shape.

velocity, phase Velocity of an equiphase surface along the wave normal.

velocity modulation See modulation, velocity.

velocity of light The velocity of light in a vacuum is 299,792 km/s (186,280 mi/s). (For rough calculations, the figure of 300,000 km/s is used.)

velocity of propagation Velocity of signal transmission. In free space electromagnetic waves travel with the speed of light. In cables speed is substantially lower.

velocity of sound The velocity of sound varies with different materials and under different conditions, e.g.:

Air at 0°C 331m/s
Air at 20°C 344m/s
Water at 15°C 1437m/s
Steel at 20°C 4987m/s
Glass at 20°C 5486m/s

Verband Deutsche Elektrotechniker, (VDE) The German standardization body responsible for electrical safety procedures and for certification.

verification 1. Of a telephone number, verification by an operator who dials into a busy line to confirm that the number quoted by a calling subscriber is correct 2. Of a punched card, verification that the data punched into the card is the same as the data first presented to the operator. If the punched holes are confirmed, the card goes forward.

verification office A central office in which new equipment or features are tested prior to general release.

verification trunk Trunk to which an operator has access and which will switch through to a called line even if the line is busy. Also called test access selectors/lines or no-test trunks.

verified off-hook In telephone systems, a service provided by a unit which is inserted on each end of a transmission circuit for verifying supervisory signals on the circuit.

verifier A machine which checks the accuracy with which data has been punched into cards or tape.

vernier Device which enables precision reading of a measuring set or gauge or setting a dial with fine adjustment.

vertical, crossbar The vertical bar associated with one of the operating electromagnets of a crossbar switch.

vertical, distribution frame Vertical steel bar in a main distribution frame on which the terminations of cable pairs and protective devices are mounted.

vertical block line A development of the magnetic bubble which greatly increases storage capacity, giving the potential of storing 1.6 Gbit of data on a single chip one centimeter square. The chip, its driver and ancillary circuits will occupy a single printed circuit board about 10 cm x 10 cm.

vertical-horizontal (V-H) coordinate system A grid of north-south and east-west lines used in calculating rate steps (telephone call charges).

vertical redundancy check A parity check performed on each character of a block of received information.

vertical riser (VR) A heavy-gauge conductor that effectively extends earth potential to the floor ground window (FGW).

vertical services Optional services available at additional charge, e.g., extra telephones, special instruments, extra lines.

vertical step One of the vertical steps of a two-motion step-by-step or Strowger switch.

very high frequency (VHF) A radio frequency in the band 30 MHz to 300 MHz. Also television channels 2 through 13.

very large scale integration (VLSI) The use of chips each providing a very large number (typically several thousand) of active elements or logic gates. $VLSI^2$ is VLSI with more than 1,000,000 logic gates, sometimes called "very large scale integration indeed."

very low frequency (VLF) A radio frequency in the band 3 kHz to 300 kHz.

vestigial sideband A form of transmission in which one sideband is much attenuated, the other transmitted without attenuation. In this way, there is some trade-off between transmitter power and filter complexity. Used for the composite picture and synchronizing signal in television broadcasting.

via net loss (VNL) Lowest loss at which a trunk facility may be operated. VNL in decibels is obtained by multiplying the VNL factor (for the type of circuit) by the circuit length in miles and adding 0.4 dB.

via net loss factor (VNL factor) A constant which varies for each type of transmission facility. Expressed in decibels per mile, dependent on the speed of propagation of signals over the specified facility.

vibrating rectifier An ac-driven device, resembling a mechanical buzzer, which reverses connections in step with the alternation of the input ac in order to give a dc output.

vibrating ringer A dc-driven device, resembling a mechanical buzzer, which gives an ac output which may be used to ring telephone bells.

vibration Motion like that of a pendulum.

vibration testing Testing whereby subsystems are mounted on a test base which vibrates rapidly, thereby showing up any faults due to badly soldered joints or other poor mechanical design features.

video The picture signal of a television transmission.

video amplifier Amplifier designed to operate over the band of frequencies used for television signals.

video band The frequency band utilized to transmit a composite video signal.

video display A computer output device which presents data to the user in the form of a television picture, either printed characters or an image.

video display unit (VDU) A cathode ray tube often associated with a keyboard to make a computer terminal.

video frequency Frequencies covered by the output from a television camera. These normally contain a very wide band from about 10 Hz to 2 MHz no more.

video pair Balanced pairs which are carefully manufactured and screened and used to carry TV signals.

video teleconferencing Real time and usually two-way transmission of video images between two or more locations. Transmitted images may be of full TV standard or freeze-frame, where the picture is repainted every few seconds. Bandwidth requirements can go as low as 56 kbit/s for freeze-frame transmission.

videoconferencing System providing audio and video links between locations.

videodisc Hard disc used for the recording of TV programs.

videophone A viewphone. A telephone incorporating a small TV camera and screen which enables the user to see as well as hear the subscriber at the other end of the circuit.

videotape Magnetic tape used for recording TV programs. There are at present two standards in use for VTR equipment which are not compatible.

videotape recorder (VTR) A device which allows the recording and playing back via magnetic tape of sound and video programs.

videotex A new information service, operational in England since 1979 under the name Prestel, which uses a slightly modified domestic television receiver in conjunction with a normal public telephone line in order to provide an interactive computerized data retrieval service for homes and offices. Systems providing generally similar facilities have also been designed in France and Canada; these are engineered in slightly different ways. Services of these types were originally called viewdata services, they are now operational in many areas, in many countries.

videotext Alternative word for videotex.

view In satellite communications, the ability of a satellite station to observe a satellite, the latter being sufficiently above the horizon and clear of other obstructions so that it is within the free line of sight from the satellite station.

viewdata See videotex.

viewphone A device which provides real-time picture transmission in direct combination with telephone calls.

Viewtron An American videotex service operating in Florida.

virtual call A user facility in which a datacall set-up procedure and a call clearing procedure will determine a period of communication between two units of data terminal equipment (DTE) in which user's data will be transferred in the network in the packet mode of operation. All the user's data is delivered from the network in the same order in which it is received by the network. This facility requires end-to-end transfer control of packets within the network. Data may be delivered to the network before the call set-up has been completed but it will not be delivered to the destination address if the call set-up attempt is unsuccessful. Multi-access DTEs may have several virtual calls in operation at the same time.

virtual carrier The location in the frequency spectrum that carrier energy would occupy if it were present.

virtual circuit A communication arrangement in which data from one point may be passed to another over various real circuit configurations during a single period of communication. Syn. logical circuit

virtual height The apparent height of a reflecting layer in the ionosphere. The reflection or refraction of radio waves in the ionosphere is not the simple reflection of a wave by a plane surface, but is a more complex process.

virtual storage Auxiliary storage mapped into real addresses so that a computer user views it as an addressable main store.

virtual switching point (VSP) In an international exchange, one of the points in the international circuit that divides the international chain from the national system. The VSPs of an international circuit are fixed by convention at points where the nominal relative levels at the reference frequency are -3.5dBr (sending) and -4.0dBr (receiving).

visible arc The part of the arc of the geostationary satellite orbit over which the space station is visible above the local horizon from each associated earth station in the service area.

visible spectrum See light.

Vista Bell Canada version of videotex.

visual display unit (VDU) See video display unit.

vitreous silica Glass consisting of almost pure silicon dioxide. Syn.: fused silica

vocoder (voice-operated coder) A device which does not encode the human voice directly, as in pulse code modulation, but rather attempts to measure certain parameters of the voice and transmit these via a slow digital bit stream. The receiving end then uses these measurements to synthesize key features of the original voice, using an electronic circuit to imitate the human vocal cord and throat.

voice, artificial A small loudspeaker used in laboratories as a standard when calibrating telephone instruments.

voice, secure Telephone communications that have been encrypted to render the messages confidential.

voice circuit A circuit for the interchange of human speech. Normally, the standard band provided is 300 Hz to 3400 Hz, but narrower bands also provide commercially acceptable circuits in some circumstances.

voice coil The electromagnetic coil which drives a loudspeaker and produces sound.

voice connecting arrangement (VCA) A device to protect the public telephone network from interference caused by customer-provided equipment. Telephone companies lease these devices to customers. The need for most two-wire VCAs has been eliminated, but VCAs are still required on four-wire circuits.

voice digitization Conversion of analog voice into digital symbols for storage or transmission.

voice frequency See frequency, voice.

voice frequency channel A transmission path suitable for carrying analog signals and quasianalog signals under certain conditions.

voice frequency telegraph (VFTG) A system whereby one or more dc telegraph channels are multiplexed into a composite nominal 4 kHz voice frequency channel for further processing through a metallic or radio network. Syn. voice frequency carrier telegraph.

voice grade Suitable for transmitting a voice signal (300 Hz to 3400 Hz).

voice mail A voice messaging system.

voice mailbox The code number determining the store in which incoming voice messages have been placed for later retrieval.

voice message exchange (VMX) A switching unit designed for use in voice messaging systems.

voice messaging System by which spoken messages are recorded for playing back when the addressee becomes available.

voice operated device, anti-singing (VODAS) A device similar in concept to the echo suppressor. The presence of voice signals in one direction deactivates simultaneous use of the channel in the other direction.

voice operated gain adjusting device (VOGAD) A volume compressor which maintains optimum modulation in a transmission circuit but restores original levels at the far end.

voice operated loss control and echo/singing suppression circuit (VOLCAS) A device in the echo suppressor family which controls singing by varying the loss inserted in the nonactive direction of transmission.

voice switched Describing a device which responds to voice signals. Usually a channel stays on receive un-

til a local voice signal is heard; the transmit side then switches on and the receive side is muted.

voiced sounds Sounds produced by the vocal cords and including all vowels plus those consonants which involve the vocal cords rather than only the lips or tongue.

volatile A term applied to information held in a memory store which depends on power being continuously available.

volatile display The non-permanent image which appears on the screen of a visual display unit.

volatile memory A read/write memory whose content is irretrievably lost when operating power is removed.

volt The derived SI unit of electrical potential difference. It is the difference in potential between two points of a conducting wire carrying a constant current of one ampere when the power dissipated between these two points is equal to one watt.

voltage Potential difference between two points.

voltage, average For a sine wave, the average value is 0.637 times peak voltage or 0.901 times the effective voltage.

voltage, breakdown Voltage at which insulation breaks down and there is a current discharge.

voltage, bucking Voltage in opposition to a reference voltage.

voltage, effective The root-mean-square (rms) voltage. For a sine wave, the rms value is 0.7071 times peak value.

voltage, forward Voltage which produces a net current flow of the same polarity.

voltage, high A voltage higher than that used for power distribution. The lower limit is usually taken as either 5000v (Bell) or 8700v (National Electrical Safety Code).

voltage, inverse Voltage which is the reverse of that which is indicated by the actual current flow.

voltage, peak Maximum voltage occurring during a cycle. For a sine wave, peak value is 1.4142 times the rms value.

voltage, plate The anode voltage. The voltage between cathode and anode in a electron tube.

voltage, residual The vector sum of the voltages of the various phases. In a well balanced polyphase power distribution system, residual voltage will be close to zero.

voltage, reverse An inverse voltage.

voltage, rms The effective voltage expressed as root-mean-square voltage.

voltage, striking Voltage required to make current begin to flow through a discharge tube.

voltage amplification Ratio of voltage across the load to voltage at the input to an amplifier, expressed in decibels.

voltage amplifier See amplifier, voltage.

voltage/discharge current curve (for a gas-discharge protector) For alternating currents of frequencies from 15 Hz to 62 Hz, it is a curve that indicates the relationship between the instantaneous values of voltage and current during the passage of discharge current.

voltage drop Decrease in potential as current flows across a resistance.

voltage fed An antenna fed with power at a point of maximum voltage.

voltage gradient Voltage per unit distance across an insulator, or voltages across capacitors connected in cascade.

voltage level Ratio of the voltage at a point to the voltage at an arbitrary reference point, called the point of zero level: in audio systems this zero level point is usually a point at which the power level is 1 mW across a 600 ohm termination.

voltage reference circuit A stable voltage reference source.

voltage stabilizer Device which produces a constant or substantially constant output voltage despite variations in input voltage or output load current.

voltage standing wave ratio (VSWR) The ratio of maximum to minimum voltage in the standing wave pattern that appears along a transmission line. It is used as a measure of impedance mismatch between the transmission line and its load.

voltage to ground Voltage between any given portion of a piece of equipment and the ground potential.

voltage transformer An instrument which enables high voltages to be measured by way of a meter which gives full-scale deflection at comparatively low voltage levels.

voltages, telegraph hub Voltages used for telegraphy which appear in send and receive hubs (normally +60v mark, -30v space, -60v double space).

voltaic cell A primary cell which produces electricity by chemical changes. An ordinary dry battery is sometimes called a voltaic cell.

voltammeter A measuring instrument with scales and terminals enabling it to be used to measure potential (volts) or current (amperes).

volt-ampere (VA) Apparent power in an ac circuit (volts times amperes).

voltmeter An instrument calibrated to read potential differences in volts.

voltmeter, frequency selective A voltmeter combined with a tunable narrow band filter which enables the meter to measure voltages of signals at particular frequencies.

voltmeter, vacuum tube (VTVM) An electron tube amplifier and rectifier with high input impedance which is used to measure voltage.

volt-ohm-milliammeter General purpose multirange test meter with terminals and switches for range selection.

volume 1. Loudness of sound. 2. A certain portion of data, together with its data carrier, that can be handled conveniently as a unit. 3. A data carrier that can be mounted and dismounted as a unit, e.g., a reel of magnetic tape. 4. That portion of a single

unit of storage that is accessible to a single read/write mechanism.

volume compressor Device which reduces the range of amplitude variations in a transmission system. A compressor at the sending end and an expander at the receiving end enable signal-noise ratio to be improved significantly in radiocommunication systems. The pair of units mounted together is called a compandor.

volume control Voltage divider usually found at the input to an audio frequency amplifier and used to adjust the loudness of an audio signal

volume indicator Meter calibrated in volume units (VUs) and having specified impedance and dynamic characteristics.

volume unit (VU) The unit of measurement of electrical speech power as read from a VU meter in a prescribed manner. The VU meter is a volume indicator designed in accordance with American National Standard C16.5-1942. It has a scale and specified dynamic and other characteristics which enable it to obtain correlated readings of speech power. Zero VU equals zero dBm (1 mW) in measurements of sine wave test tone power.

Von Neumann architecture Design principle adopted by most current computers involving sequential completion of tasks and centralized memories for both program and data.

vox A voice-operated relay circuit that permits the equivalent of push-to-talk operation of a transmitter by the operator.

wafer Thin slice of semiconductor material on which integrated circuits are built.

wait on "busy" A service enabling a subscriber making a call to a busy number to be connected when the number is free.

waiting-in-progress signal A signal sent in the backward direction in a data service indicating that the called customer, having the connect-when-free facility, is busy and that the call has been placed in a queue.

walkie-talkie Hand-held radio transmitter/receiver.

walkthrough A critical examination of a design or product undertaken to ensure that it is of adequate quality.

wall distribution frame Distribution frame for a small telephone office which is designed to be installed against a wall.

wand, tuning Plastic rod with ferrite at one end, and brass at the other which is used to check the tuning of radio receivers.

wander Long-term variation of the significant instant of a digital signal from its ideal position in time (jitter is short-term variation of this).

warble-tone Audio tone which is varied in frequency at a slow enough rate (about 15 Hz) to be described as a warble.

warm restart An initialization phase for computer-controlled central offices. Transient calls are usually dropped while calls in the talking state continue.

wash item Informal term for an item which is common to each of several possible ways of carrying out a task. A study purely for comparison purposes does not therefore need to include costs of wash items.

washer, round A general-purpose round galvanized steel washer used in pole line construction.

washer, sealing Washers used in cable splicing, particularly when mechanical splice cases are utilized and cables of several different diameters must be sealed therein. They are made of a variety of materials and come in a variety of sizes.

washer, square A square galvanized steel washer used in pole line construction where a large bearing surface against wood is required.

washer, square curved A square washer curved so that it will fit flush against a telephone pole, used in pole line construction.

Washington Legislative Council for Telecommunications Washington D.C. group representing Bell and Independent telephone companies, labor unions and other bodies interested in issues of concern to the telecommunications industry.

water, battery Demineralized or distilled water suitable for use in a lead-acid storage battery.

water pipe ground (WPG) A metallic underground water pipe system of at least 10 feet of metal pipe in direct contact with the earth.

WATS, inward A Wide Area Telephone Service arrangement whereby a customer, by paying a charge, can authorize a telephone company to route calls in to the customer from calling lines in prescribed areas without the callers being required to pay for their calls.

WATS, outward A Wide Area Telephone Service arrangement whereby payment by the customer of a monthly fee enables the customer to make an unlimited number of long distance calls to stations in prescribed areas up to a preset time limit.

watt The derived SI unit of power. It is equivalent to one joule per second, or one volt-ampere.

watt-hour Work done by one watt over a one hour period.

watt-hour constant For a kilowatt hour totalizing meter, the number of watt-hours contained in one complete revolution of the counting rotor.

wattless power Reactive power.

wattmeter Meter indicating in watts the rate of consumption of electrical energy.

wave A disturbance which is a function of time or space or both and is propagated in a medium or through space.

wave, backward Wave with phase velocity in the reverse direction to the direction of the electron flow.

wave, carrier A single frequency wave which may be used to carry the information of a modulating wave.

wave, damped A wave whose amplitude is reduced with every cycle.

wave, direct A radio wave which travels directly from transmitting antenna to receiving antenna with no reflections or refractions.

wave, elecromagnetic A wave propagated through space (and through many material substances) consisting of varying electric and magnetic fields. Some of the properties of these waves depend on frequency. Heat, light and radio waves are different frequency band manifestations of the same electro-magnetic wave mechanism.

wave, forward Wave with group velocity in the same direction as the direction of electron flow.

wave, ground See ground wave.

wave, ground reflected A radio wave which is reflected from the surface of the earth.

wave, guided See guided wave.

wave, horizontally polarized Electromagnetic wave with its electric field component parallel to the surface of the Earth.

wave, intelligence A wave carrying a message.

wave, interrupted continuous See interrupted continuous wave.

wave, ionosphere A radio wave reflected back to the Earth's surface by the ionosphere.

wave, modulated A wave, one or more of whose characteristics have been varied in accordance with a modulating wave.

wave, modulating The information-carrying wave which modulates a carrier in order that the information may be transmitted by line, fiber or radio more efficiently than would be possible at the original signal frequency.

wave, periodic A wave which repeats the same pattern at regular intervals.

wave, plane polarized Wave whose electric intensity (or electric field vector) lies at all times in a plane that contains the direction of propagation.

wave, radio An electromagnetic wave with electric and magnetic field components. It is radiated by an antenna, travels at the speed of light and may be picked up by another antenna.

wave, sine See sine wave.

wave, sky Radio wave which travels from the earth's surface toward the sky. At some frequencies and in some circumstances the wave can be reflected or refracted down to earth again.

wave, sound   A wave carried in an elastic medium by audio frequency vibrations.

wave, space   Component of a radio wave that travels through the atmosphere just above the earth's surface.

wave, square   A periodic wave with the characteristic of suddenly changing from negative to positive, maintaining a steady period, and then suddenly reversing from positive to negative. A square wave contains odd harmonics of the fundamental frequency.

wave, surface   That part of ground wave radiation which travels above the surface of the earth.

wave, transverse electric   (TE) Mode of propagation in a waveguide in which the electric field vector is in all places perpendicular to the direction of propagation.

wave, transverse electromagnetic   (TEM) Mode of propagation in coaxial cables and in open feeders in which the electric field vector and the magnetic field vector are both perpendicular to the direction of propagation.

wave, transverse magnetic   (TM) Mode of propagation in a waveguide in which the magnetic field vector is in all places perpendicular to the direction of propagation.

wave, traveling plane   A plane wave each of whose frequency components has an exponential variation of amplitude and a linear variation of phase in the direction of propagation.

wave, tropospheric   See tropospheric wave.

wave, undamped   A wave with constant amplitude.

wave, vertically polarized   An electromagnetic wave whose electric field component is in all places perpendicular to the earth's surface.

wave number   The reciprocal of wavelength. The number of wave lengths per unit distance in the direction of propagation of a wave.

wave trap   A tuned circuit which attenuates greatly an undesired frequency.

waveform   The characteristic shape of a periodic wave, determined by the frequencies present and their amplitudes and relative phases.

waveform analyzer   A voltmeter which, together with a frequency-selective device, permits the determination of the frequency and amplitude of all the sine-wave components of a complex wave.

wavefront   A continuous surface that is a locus of points having the same phase at a given instant. A surface at right angles to rays which proceed from the wave source. The surface passes through those parts of the wave which are in the same phase and travel in the same direction. For parallel rays, the wave front is a plane; for rays which radiate from a point, the wave front is spherical.

waveguide   A transmission line consisting of a hollow metallic conductor—generally rectangular, elliptical, or circular—within which electromagnetic waves may be propagated.

waveguide, circular   A waveguide with a circular cross section.

waveguide, dielectric   A waveguide which is primarily dependent on a dielectric in its construction.

waveguide bend   A section of a rectangular waveguide in which the longitudinal axis is bent. An E-plane bend is one in which the narrow side is bent. An H-plane bend is one in which the wide side of the waveguide is bent.

waveguide dispersion   For each mode in an optical waveguide, that portion of total dispersion attributable to the dependence of the phase and group velocities on the geometric properties of the waveguide in particular, for circular waveguides, on the ratio (a/λ), where a is core radius and λ is wavelength.

waveguide scattering   Scattering (other than material scattering) that is attributable to waveguide design and fabrication.

waveguide slug tuner   A dielectric slug used for fine tuning a waveguide. Tuning is accomplished by varying the penetration of the slug into the waveguide.

waveguide stub tuner   A piston arrangement whereby dimensions of a waveguide stub may be adjusted for fine tuning.

wavelength   For a sinusoidal wave, the distance between points of corresponding phase of two consecutive cycles.

wavelength, cutoff   Wavelength which corresponds to a waveguide's cutoff frequency.

wavelength, effective   Wavelength in a broadband radiation such that energy in the band at longer wavelengths is the same as the energy in the band at shorter wavelengths.

wavelength, radio   Wavelength in meters of a radio signal is approximately 300,000 divided by the frequency of the signal in kHz, or 300 divided by frequency in MHz : wavelength is numerically equal to velocity of propagation divided by frequency.

wavelength division multiplexing   (WDM) If signals of different optical wavelengths are used in a fiber optic communications system these different wavelengths can be multiplexed together or demultiplexed, by comparatively cheap passive elements (needing no power supplies) such as optical gratings.

wavemeter   Device for measuring wavelength of a radio signal.

waystation   A telephone on a multiparty line, e.g. dispatch phones on a minor railway system.

weakly guiding optical waveguide   A fiber waveguide for which the difference in refractive index between the core and cladding is small, usually less than 1 per cent.

weatherproof   In the case of a telephone instrument, designed so that it can be used out-of-doors under any of a number of specified climatic conditions.

weber   The derived SI unit of magnetic flux. It is the flux which, linking a circuit of one turn, produces in it an electromotive force of one volt as it is reduced to zero at a uniform rate in one second.

wedges, multiple Pairs of smooth wooden wedges, 16 inches by 3 inches, used to separate switchboard multiple cables in order to get behind them to repair a jack strip or lamp strip.

weight, handline Small lead weight to which a handline can be attached and used for throwing over minor obstacles.

weighting Evaluation of the relative interfering effects of the various frequencies in the voice frequency spectrum as compared with the reference frequency (1000 Hz).

weighting, C-message Weighting network used when noise testing lines terminated by a standard 500-type telephone instrument.

weighting, FIA line Type of noise-measurement weighting network designed to be used with 302-type telephone instrument terminations.

weighting, flat A network with flat characteristic over a specific bandwidth, used when testing noise on circuits terminated on equipment other than telephone instruments.

weighting, psophometric Noise-measurement weighting network designed by the CCITT, for general use.

weighting networks Network with special frequency--attenuation characteristics enabling noise measurements to be made on different types of telephone instruments.

West Ford A 1963 passive satellite project in which many copper needles 18 mm long were placed in orbit around the earth with the intention of improving the usefulness of the 8 GHz band for terrestrial communications.

west terminal The terminal on the west or south end of a circuit.

Westar Western Union geostationary satellite designed to provide voice and TV channels for interstate use within the United States.

wet circuit Circuit which functionally carries ac, but on which a small direct current has been superimposed to improve contact performance and reduce contact noise.

wetted contacts 1. Contacts through which a small superimposed direct current is passed in order to reduce to a minimum the resistance offered by the contacts to the ac signal current passing through them. 2. Contacts which are coated with mercury to improve contact performance.

wetting, contact Procedure for obtaining low-resistance closure of base-metal contacts, surface oxidation of which could otherwise result in contact resistance being of significant (and unsteady) value and introducing noise into the circuit. The most common method is passing low direct current through the contacts.

wetting, mercury contact Coating of relay contacts with mercury to give a low resistance contact.

wetting agent Chemical which reduces surface tension, thus encouraging liquid to spread on a surface.

wheel, jockey Small wheel used in cable ship equipment which rides in the groove of a larger wheel and helps keep cable securely running in the groove.

wheel, measuring Wheel, designed like a bicycle wheel, which is fitted with a handle and a meter for recording distance when rolled along a path. Used for measurement of cable tracks, duct routes, etc.

whip antenna See antenna, whip.

whispering gallery propagation A mode of propagation of radio waves attributed to channelling in the ionosphere, as in a waveguide.

white facsimile transmission 1. In an amplitude-modulated facsimile system, that form of transmission in which the maximum transmitted power corresponds to the minimum density of the subject copy. 2. In a frequency-modulated system, that form of transmission in which the lowest transmitted frequency corresponds to the minimum density of the subject copy.

white noise Noise whose frequency spectrum is continuous and uniform over a wide frequency range.

white signal In facsimile transmission, the signal resulting from the scanning of a minimum-density area of the subject copy.

white space skip Feature of some facsimile machines which speed up transmission by skipping blank spaces.

whiting A chalk and water slurry applied to submarine cable that has been previously tar-flooded, to discourage sticking of adjacent turns or adjacent flakes in the ship's cable tanks during laying.

who are you (WRU) Teletypewriter character which, when transmitted, is a request for the receiving station to identify itself.

wicking The situation in which solder and flux flow under the insulation covering a wire while the end of the wire is being soldered.

Wide Area Data Service (WADS) A service using ordinary telephone lines for data transmission: establishing the call on a dial-up basis then going over to data signals.

wide area network (WAN) A communications network serving geographically separate áreas. A linking together of metropolitan area networks (MANs) which enables data terminals in one city to access data resources in another city or country. Inter-city links are usually digital circuits (e.g., at 56 kbit/s, 64 kbit/s, 1.5 Mbit/s, 2 Mbit/s) leased from common carriers or telecommunications administrations.

Wide Area Telephone Service (WATS) Telephone company service allowing reduced costs for certain telephone call arrangements; may be In-WATS, or 800-number service where calls can be placed from anywhere in the continental U.S. to the called party at no cost to the calling party, or Out-WATS, a service whereby, for a flat-rate charge, dependent on the total duration of all such calls, a subscriber may make an unlimited number of calls within a prescribed area from a particular telephone terminal without the registration of individual call charges.

wideband Passing or processing a wide range of frequencies. The meaning varies with the context. In an audio system, wideband can mean a band of up to 20 kHz wide, but in a television system the band can be many megahertz wide.

wideband modem 1. A modem whose modulated output signal can have an essential frequency spectrum that is broader than that which can be wholly contained within, and faithfully transmitted through, a voice channel with a nominal 4 kHz bandwidth. 2. A modem whose bandwidth capability is greater than that of a narrowband modem.

wideband signal Any analog signal, or analog representation of a digital signal, whose essential spectral content is broader than that which can be contained within a voice channel of nominal 4 kHz bandwidth.

wideband switch Switch capable of handling a wide bandwidth. One of the most commonly encountered wideband switches is associated with switching radio transmitters to different antennae, usually by switching RF outputs from one coaxial feeder line to another.

wideband system A system with a multichannel bandwidth of 4 kHz or more.

Wien bridge An ac bridge used to measure capacitance or inductance.

willful intercept Diversion of messages to another station when the addressee station is malfunctioning.

winch A power-driven drum associated with an outside plant vehicle used to pull cables through ducts or to assist in pole erection.

Winchester Type of hard disk drive magnetic memory store. Originally developed by IBM as a disk unit with 30 Mbytes of storage and 30 ms average access time, so it was called the 3030. To fans of Western films a 3030 was a Winchester rifle so the IBM disk unit was unofficially nicknamed a Winchester. The name stuck, however, and is now used as a general name for hard disk units.

winding Coils of wire, often on ferrous cores, found in a transformer or relay and used to increase inductance.

winding, non-inductive Winding with specified resistance but negligible inductance, usually made by winding two wires at the same time and using one for each direction of current flow so that the inductive effects of the two cancel out.

winding, primary Transformer winding which receives an input signal from a source, thereby creating magnetic flux in a core. This flux induces current in a secondary winding.

winding, secondary Transformer winding which feeds an output. It gets its input by electromagnetic induction from the currents flowing in the primary winding.

window A flow-control mechanism in data communications. The size of the window is equal to the number of packets that may be sent before a positive acknowledgement is required.

window, radio See radio window.

windshield wiper effect Onset of overload in multichannel CATV systems caused by cross-modulation, where the horizontal sync pulses of one or more TV channels are superimposed on the desired channel carrier.

wink In telephone switching systems, a single supervisory pulse.

wink-off A feature permitting the release of switches in last-party-release equipment if the called party does not go back on-hook after a predetermined time interval.

wink operation A timed off-hook signal, normally of 140ms, which indicates the availability of an incoming register for receiving digital information from the calling office.

wink pulsing In telephone switching systems, recurring pulses of a type in which the off-pulse is very short with respect to the on-pulse. An example: On key telephone instruments, the hold position (condition) of a line is often indicated by wink pulsing the associated lamp at 120 impulses per minute, 94 per cent break, 470 ms on, 30 ms off.

wink signal Brief interruptions of current to an indicator lamp resulting in flashes that show, for example, that a line is being held.

wink start Short duration off-hook signal. See wink operation.

wiped joint See joint, wiped.

wiper A moving contact of the type used in switches in step-by-step electromechanical telephone offices.

wiper, bridging See bridging wiper.

wiping contact See contact, wiping.

wire A single metallic conductor, usually solid-drawn and circular in cross section.

wire, aluminum clad A wire with a steel core and an aluminum outer coating used for open wire routes.

wire, annealed copper Copper wire softened by heating after being drawn and used where soft flexibility is needed.

wire, armor Galvanized mild steel wires for application to submarine cables: in one or two layers for installations which are relatively shallow; galvanized high tensile steel wires are used in a single layer for relatively deep water. Armored cables are also used in some terrestrial applications, where a normal unprotected sheath gives insufficient protection against damage to a directly-buried cable.

wire, armored underground A one-pair copper conductor, polyethylene insulated, steel armored light cable used for direct burial of drops to customers' premises.

wire, bare Uninsulated wire.

wire, bell Very light gauge plastic insulated wire (usually 18 AWG) used for bell and buzzer circuits.

wire, bonding Annealed tinned copper wire used to bond metallic cable sheaths in order to maintain sheath continuity during splicing operations.

wire, bridle One-pair or three-wire polyethylene insulated cable used to interconnect open wire circuits or to connect open wire circuits to a cable terminal.

wire, C rural Self-supporting one-pair wire with a copper-steel line core and insulated with polyethylene.

wire, copper line Heavy gauge, hard drawn, bare copper wire, now rarely used for new installations, but which was for many years the backbone of most countries' long distance line communications systems.

wire, copper steel Steel line wire with an outer coating of copper, used where the extra tensile strength of steel is needed. The copper skin permits carrier working with acceptable performance at carrier frequencies.

wire, Copperweld-copper A brand of stranded line wire for use in special long-span operations. Some of the strands are hard drawn copper (for conductivity); some are copper welded on to a steel core (for tensile strength).

wire, D station Polyethylene-insulated, PVC sheathed two-pair cables used for internal wiring on subscribers' premises.

wire, distribution A composite cable used for distribution to subscribers' premises. It consists of a steel messenger wire surrounded by polyethylene-insulated copper pairs and has no overall outer sheath.

wire, drop Insulated one-pair wire used for drops to subscribers' premises.

wire, enameled Wire insulated with enamel, often used in relay coils where insulation of minimum thickness is required.

wire, field Field cable with high tensile strength used by military signal units.

wire, flame-proof Wire insulated with material which has been chemically treated so that it will not support combustion.

wire, flat station Multiconductor wire shaped into flat ribbons with adhesive backing which can be laid down without stapling.

wire, fuse Fine wire of specified current-carrying capacity used to repair factory-made fuses.

wire, galvanized iron Steel wire which has been heavily galvanized to prevent rusting.

wire, hard See hard wire.

wire, hard-drawn copper Copper wire used for open wire routes. Called hard drawn because the final stages of gauge reduction are carried out without annealing.

wire, high strength steel Galavanized steel wire with very high tensile strength. It may safely be used for spans over 100 m long.

wire, jumper Plastic-insulated tinned copper conductors used for jumpers on main and intermediate distribution frames. Most common jumpers are two-wire, but one-wire, three-wire and four-wire varieties are also available.

wire, lashing Soft wire used for lashing an aerially-suspended cable to a messenger wire.

wire, line Copper, copper-steel, copper-aluminum or steel wire strung between poles and used for telephone circuits.

wire, open Uninsulated wire strung between poles and used for telephone circuits.

wire, order A voice or teletypewriter circuit connecting points manned by service maintenance personnel.

wire, pulling-in Galvanized steel wire left in a duct for easy location when a cable has to be pulled in.

wire, reinforcing A hard-drawn wire which is wrapped helically around an open wire conductor before it is tied in to its supporting insulator at a pole.

wire, resistance Wire with high resistivity used in the construction of wire-wound resistors.

wire, rural Self-supporting, one-pair copper steel conductor with polyethylene insulation.

wire, shield A bare and grounded wire sometimes erected along a pole route or buried alongside a cable to shield the working conductors from inductive noise interference.

wire, sleeve The third wire in a central office connection: tip and ring for the two line wires, sleeve for supervisory and testing circuits.

wire, soft-drawn copper Annealed soft copper wire which can be readily bent.

wire, solid Wire made up of a solid conductor rather than being stranded or braided.

wire, station Small capacity cable used for internal cabling in customers' premises.

wire, support Steel messenger wire, polyethylene covered, used as a support for aerially suspended cable.

wire, switchboard Color-coded insulated wire used in switchboard cables.

wire, teletypewriter order A multiappearance teletypewriter circuit used by telephone companies to ensure that all concerned toll testboards are advised of circuit changes.

wire, tie Wire used to bind a line wire to its supporting insulator on an open-wire route.

wire, tinned Wire coated with a thin layer of tin to facilitate soldering and to make the wire corrosion-resistant.

wire, tinsel Flexible conductor made of several strands of very soft copper with a nylon, or other plastic, supporting core.

wire, tree One-pair cable with a particularly abrasion-resistant outer sheath. Used for telephone routes through areas with many trees.

wire, unequipped Wire which is terminated in a central office or transmission station but not yet connected to the planned equipment.

wire, urban A distribution cable with up to 16 pairs of insulated conductors around a central steel messenger wire. The cable is designed without an outer sheath so that conductors may be fed off at any point along the route.

wire mile The resistance of one conductor which is one mile long.

wire stripper Hand tool which enables removal of insulation from a wire without damaging the conductor itself.

wired city The provision to premises by cable or optic fiber of television, telephone and other communications, data, educational material, instructional television, information retrieval and electronic shopping services. Broadcast services must, of necessity, be limited by scarce frequency spectrum space. Wired services have theoretically unlimited channel capacity.

wired logic control Control of an automatic switching arrangement in which the call processing is determined by a program embodied in a pattern of fixed physical interconnections among a group of devices.

wireless Original British word for radio.

wireline Telecommunications services using copper-wire technology (not radio links), e.g., an ordinary residential telephone provided by a telephone company (a "wireline common carrier" or WCC) as opposed to a mobile service provided by a "radio common carrier" or RCC.

wirephoto A photograph transmitted by telephony or telegraphy.

wiretapping Connecting to a telephone circuit in order to monitor conversations.

wire-wrapping Termination of wires on tags by firmly wrapping the wire around a sharp-cornered tag which bites through the insulator to the conductor.

wiring closet Termination point for customer premises wiring.

wiring, embedded All telephone wiring made in customers' premises before Dec. 31, 1983. New wiring is wiring made after that date.

wiring, intersystem All the wiring on the telco's side of the demarcation point (in an office building, etc).

wiring, intrasystem All the wiring on the customer's side of the local telephone company's demarcation point.

wobbulator A test oscillator which is continually changing its frequency between specified limits.

woofer A loudspeaker designed to respond efficiently to lower sound frequencies.

woo-woo tone No such number or number unobtainable tone.

word (1) In data communications, a character string, binary element string, or bit string that is considered as an entity. (2) In telegraph communications, a six-character interval used when computing traffic capacity in words per minute.

word, memory The unit by which bits are read into or out of a memory store.

word, telegraph A word of six character intervals.

word length The number of bits or characters in a word.

work station Input/output equipment at which an operator works. A station at which a user can send data to or receive data from a computer for the purpose of performing a job.

working, closed-circuit In telegraphy, a method of working in which current flows in the circuit when transmitters are idle.

working, open circuit In telegraphy, a method of working in which no current flows in the circuit when transmitters are idle.

working range The permitted range of values of an analog signal over which transmitting or other processing equipment can operate.

working voltage Rated voltage which may safely be applied continuously.

World Administrative Radio Conference (WARC) A conference held periodically by the CCIR/ITU to decide on the international use of the radio spectrum.

worst hour of the year That hour of the year during which the median noise over any radio path is at a maximum. This hour is considered to coincide with the hour during which the greatest transmission loss occurs.

wow An undesired variation in pitch.

wrap To make a connection between a wire and a tag by tightly wrapping the wire around the tag with a special tool.

wrap, longitudinal Tape applied to a cable core in line with the axis of the core.

wrap, spiral Binding tape applied in a helical pattern over a cable core.

write To deposit data into some form of computer memory.

write-after-read Restoring previously read data into a core memory following completion of the read cycle.

write pulse Pulse signal which causes information to be stored in a memory unit.

X-25 CCITT Recommendation covering packet switching.

X-400 CCITT Recommendation covering message handling systems.

x-axis Horizontal axis, as on a graph.

x-band Microwave band from 5.2 GHz to 10.9 GHz.

x-cut Method of cutting a quartz plate for an oscillator, with the x-axis of the crystal perpendicular to the faces of the plate.

x-dimension of recorded spot In facsimile systems, the effective recorded spot dimension measured in the direction of the recorded line. By effective dimension is meant the largest center-to-center spacing between recorded spots which gives minimum peak-to-peak variation of density of the recorded line.

x-dimension of scanning spot In facsimile systems, the effective scanning spot dimension measured in the direction of the scanning line on the subject copy. The actual numerical value depends upon the type of system used.

xerography The form of photocopying or electrophotography in which an electrostatic image is formed on a light-sensitive surface. The charged image areas attract and hold a fine black resinous powder which is transferred to a sheet of paper and fused by heat to make it permanent.

x-guide A surface wave transmission line that consists of a dielectric structure which has an x-shaped cross section.

XOR gate An exclusive OR gate. See logic gates.

x-plate One of the deflection plates used in an electrostatically controlled cathode ray tube or oscilloscope.

x-ray Electromagnetic radiation of from about 100 nm to 0.1 nm, capable of penetrating non-metallic materials.

x-ray procedure A universal requirement in the submarine cable industry that every cable joint be x-rayed with three exposures, each rotated 120°. The results are examined for voids or contaminants in the insulant before the outer conductor is applied.

x-ray television The use of x-ray and closed circuit TV techniques to produce enlarged reproductions of welds and joints for checking purposes.

tween recorded lines which gives minimum peak-to-peak variation of density across the recorded lines.

y-dimension of scanning spot In facsimile systems, the effective scanning spot dimension measured perpendicular to the scanning line on the subject copy. The actual numerical value will depend upon the type of system used. See also scanning.

Yellow Pages Classified section of local telephone directories.

yes or no test A test made to indicate whether a quantity or magnitude would fall above or below a specified limit or boundary defined to distinguish between pass and fail conditions.

yield strength The magnitude of mechanical stress at which a material will begin to deform. From that point on, extension is no longer proportional to stress and rupture is possible.

YIG device An oscillator, filter or microwave amplifier which uses a yttrium-iron garnet crystal in a variable magnetic field to obtain wideband efficiency.

YIG filter A filter with a yttrium-iron garnet crystal positioned in a magnetic field. Tuning control is obtained by varying the dc current through the solenoid producing the magnetic field.

y-network Star network with three branches, each of which goes to one common node.

yoke Material which interconnects magnetic cores.

yttrium-aluminum garnet (YAG) A crystalline material used in some lasers.

yttrium-iron garnet (YIG) A crystalline material used in microwave devices.

Y The symbol for admittance.

Yagi A directional antenna widely used for TV reception. It consists of a dipole with parasitic directors in front and reflectors behind.

YAG-laser A laser using yttrium aluminum garnet.

y-axis The vertical axis, as on a graph.

y-cut Method of cutting a quartz plate for an oscillator, with the y-axis of the crystal perpendicular to the faces of the plate.

y-dimension of recorded spot In facsimile systems, the effective recorded spot dimension measured perpendicular to the recorded line. By effective dimension is meant the largest center-to-center distance be-

Z Symbol for impedance.

z-axis Axis mutually perpendicular to the x and y axes.

zener breakdown Nondestructive breakdown in a semiconductor which occurs when voltage across the barrier region is sufficient to produce a sudden increase in carriers in the region, so that it suddenly becomes conducting.

zener diode A semiconductor which acts as a straightforward rectifier until the applied voltage reaches a specified level. At this point, called the zener voltage or avalanche voltage, the device becomes conducting.

zener voltage Avalanche voltage. The point at which the voltage drop across a device becomes independent of current.

zero Although zero might be thought to mean just that, many computers use $+0$ and -0, to mean different things, e.g. negative zero is in some circumstances indicated by a pulse in every pulse position in a binary word whereas positive zero is a complete absence of pulses.

zero adjuster A screw on some meters which permits tension to be adjusted so that the pointer can be correctly set on zero.

zero beat The condition under which two frequencies are adjusted so that the beat note between them falls in frequency and finally disappears. At this adjustment the two frequencies are the same.

zero bias When used to refer to an electron tube, no potential difference between cathode and control grid.

zero code suppression The insertion of a "1" bit to prevent the transmission of eight or more consecutive "zero" bits, especially in PCM systems and other data transmission systems.

zero level In telecommunications systems, 1 mW of power, normally across a load of 600 ohms. With some broadcast systems utilizing VU meters, zero level is taken as being 6 mW of power.

zero state State of a binary memory cell when 0 is stored therein.

zero suppression Deletion of non-significant zeroes in a numeral signal which has been transmitted, e.g., 00000157 received, but only 157 printed.

zero transmission level reference point (0TLP) Point to which all relative levels at other points in the system are referred. The level at the 0TLP is therefore 0dBr.

Zilog U.S. microprocessor manufacturer.

zinc carbon cell 1. A primary cell or dry battery. 2. A Leclanche cell with positive carbon electrode, negative zinc electrode, a sal ammoniac paste electrolyte and a depolarizer.

zone, far Region distant from a radio transmitting antenna in which the induction field is of no importance and the radiation field intensity varies inversely with the square of the distance from the transmitting antenna.

zone, instrument A zone in which the use of a particular type of telephone instrument is specified. Under such a system, less efficient instruments are used near the central office and more efficient instruments are used at greater distances.

zone, near The area near a radio transmitting antenna in which the induction field predominates, and field intensity does not reduce with the square of the distance.

zone, time One of the zones, each about 15° wide, into which the earth has been divided. Adjacent zones normally have a one-hour time difference.

zone of mutual visibility Area on the earth's surface from which a particular communications satellite can be seen by both of a pair of ground stations.

zone of silence The skip distance in which radio signals from a particular antenna cannot be received either directly or by reflection.

zone registration Method of tallying toll charges by message units. One, two or more units are charged for calls between particular zones. During the call, the message register tallies the proper number of message units for each basic time unit, usually every three or five minutes.

zones, world numbering Geographic areas which have allocated routing codes used in the dialing of international telephone calls.

Zone 1	North America and the Caribbean
Zone 2	Africa
Zone 3)	
)	Europe
Zone 4)	
Zone 5	Central and South America
Zone 6	Australasia and Southeast Asia
Zone 7	Russia
Zone 8	China, Japan and the Far East
Zone 9	India, South Asia and the Middle East

Individual countries have been allotted country codes within zones, e.g., UK = 44, France = 33, Malaysia = 60. In Zones 1 and 7 Numbering Plan Area codes take the place of country codes.

zoning of instruments Policy used by some telephone administrations of using less efficient instruments for lines near the central office and more efficient instruments for lines at a greater distance from the central office.

Zulu time Greenwich mean time. Sometimes called G time.

APPENDIX A
ACRONYMS AND ABBREVIATIONS

A

A   Ampere, Area

AA   Automatic Answer

AAC   Automatic Amplitude Control

AAFE   Advanced Applications Flight Experiments

AAM   Air-Air Missile

AAR   American Association of Railroads, Automatic Alternative Routing

AAS   Automatic Announcement Subsystem

AAU   Automatic Answering Unit

A & CP   Access and Control Point

A & E   Architect and Engineer

ABC   Automatic Brightness Control, Answer-Back Code, Auto Bill Calling (TSPS)

ABCI   Advanced Business Communications Inc.

ABD   Average Business Day

ABH   Average Busy Hour

ABS   Average Busy Season

ABSBH   Average Busy Season Busy Hour

ABT   About

ABU   Asia-Pacific Broadcasting Union

AC, ac   Alternating Current, ACcess, Advice of Charge, Answer Complete, Awaiting Connection, Alternate Call listing

ACA   Automatic Conference Arranger, address complete

ACABQ   Advisory Committee on Administrative and Budgetary Questions

ACAST   Advisory Committee on the Application of Science and Technology to Development

AC & R   American Cable and Radio System

ACC   Automatic Chrominance Control, Administrative Committee on Coordination, Automatic Carrier Control

ACCS   Automatic Calling Card Service

ACCUNET   AT&T switched data service network

Acct   Account

ACD   Automatic Call Distribution, Automatic Call Distributor, Alarm Control & Display

ACD-ESS   Automatic Call Distributor—Electronic Switching System

ACE   Automatic Calling Equipment

ACF   Advanced Communications Function, Access Cost Factor

ACG   Adjacent Charging Group

ACH   Attempts per Circuit per Hour

ACHI   Application Channel Interface

ACITS   Advisory Committee on Information Technology Standardization (Europe)

ACK   Acknowledgement

ACM   Association for Computing Machinery, Alarm Control Module

ACMRR   Advisory Committee on Marine Resources Research

ACNAS   Advanced Cableship Navigation Aid System

ACO   Accounting Control Office

ACOC   Area Communications Operations Center

ACOLI   Advance Circuit Order and Layout Information

ACS   Advanced Communication System, Attitude Control System, Alarm & Control System, Administrative Control System

ACSR   Aluminum Conductors, Steel Reinforced

ACT   Applied Computer Techniques

ACTS   Automated Coin Toll Service, Advanced Communications Technology Satellite

ACU   Automatic Calling Unit, Acknowledgement Unit, Alarm Control Unit

ACUTA   Association of College and University Telecommunications Administrators

AD   Assignment Date, Attendant, Awaiting Disconnection

A/D   Analog/Digital

ADA   After Date of Award of contract, Program language

ADAPSO   Association of Data Processing Service Organizations

ADC   Analog/Digital Converter, Air Defense Command, Address complete, Charge

ADCCP   Advanced Data Communications Control Procedure

ADCSP   Advanced Defense Communications Satellite Project

ADCU   Alarm Display and Control Unit, Association of Data Communications Users

Add   Addendum

ADDER   Automatic Digital Data Error Recorder

Addl   AdditionaL

ADE   Automatic Design Engineering, Audible Doppler Enhancer

ADF   Automatic Direction Finder, Automatic Document Feeder

ADI   Address Incomplete

ADIS   Automatic Data Interchange System

ADIZ   Air Defense Identification Zone

adj   Adjust, Adjusted, Adjustment

ADM   Adaptive Delta Modulation

ADMD   Administration Management Domain

ADMS   Automatic Digital Message Switch

ADMSC   Automatic Digital Message Switching Center

ADN   Address complete, No-charge

ADOIT   Automatically Directed Outgoing Intertoll Trunk

ADONIS   Submarine cable France-Greece

ADP   Automatic Data Processing

ADPC   Automatic Data Processing Center

ADPCM   Differential Pulse Code Modulation with adaptive quantization

ADPE   Automatic Data Processing Equipment

ADPS   Automatic Data Processing System

ADQ code Almost Differential Quasi-ternary code

ADS Automatic Data Set, Administration of Designed Services

ADSI Administrative Design Service Information

ADT American District Telegraph, Automatic Data Transmission

ADTS Automatic Data Test System

ADU Accumulation and Distribution Unit (AUTODIN), Automatic Dialing Unit

ADV Advertising disposition

ad val Ad Valorem, according to value

ADVM Adaptive (variable slope) Delta modulation Voice Modem

ADW Aerial Distribution Wire, Air Defense Warning

ADX Address complete, coin-box

AEC Army Electronics Command

AEGUS Submarine cable Greece-Crete

AEM Applications Explorer Mission

AEN Articulation reference equivalent

AES Auger Electron Spectroscopy

AESC Automatic Electronic Switching Center

AEW Airborne Early Warning

af Audio Frequency

AF Air Force

afc Automatic Frequency Control

AFC Address complete, subscriber free, charge

AFCEA Armed Forces Communications & Electronics Assn.

AFCS Air Force Communications Service

AF DATACOM Air Force Automatic Data Communications Network

AFIPS American Federation of Information Processing Societies

AFN Address complete, subscriber Free, No-charge; All Figure Number(ing)

AFNOR Association Francaise de Normalization

AFR Awaiting Forward Release

AFSATCOM Air Force Satellite Communications system

AFSC Air Force Systems Command

AFT Analog Facility Terminal

AFX Address complete, subscriber Free, coin-box

AG Again, try Again

A/G Air to Ground

AGAMP Automatic Gain adjusting Amplifier

agc Automatic Gain Control, Automatic Gain Correction

AGE Aerospace Ground Equipment

AGS Automated Graphic System

AHT Average Holding Time

AI Address Incomplete, Artificial Intelligence

AIAA American Institute of Aeronautics and Astronautics

AIB Automatic Intercept Bureau

AIC Automatic Intercept Center, Awaiting Incoming Continuity

AIEE American Institute of Electrical Engineers

AIM ATU Interface Module, Awaiting Incoming Message

AIOD Automatic Identification on Out-Dialed calls

AIRCOMNET Air Force Command and Administrative Network

AIRPAP Air Pressurization Analysis Program

AIS Alarm Inhibit Signal, Automatic Intercept System, Alarm Indication Signal, Advanced Information System

AIU Alarm Interface Unit

AJ Analog Junction

AJM Analog Junction Module

AKRO Acknowledge Receipt Of

AL Alarm, Additional Listing, Analog Loop-back, Analog Link

AlAs Aluminum Arsenide

ALBO Automatic Line Build-Out

ALC Automatic Load Control

ALD Automated Logic Diagram

ALDI Associated Long Distance Interstate message

ALF Absorption Limiting Frequency

ALGOL Algorithmic Language

ALICE Alaska Integrated Communications System

ALIT Automatic Line Insulation Test

ALM Alarm

alnico Aluminum-Nickel-Cobalt

ALOHA a random access control technique used with maritime satellite communications systems

ALPAL Submarine cable Algeria-Palma, Spain

ALPC Adaptive Linear Predictive Coding

ALR Active Line Rotation

ALRU Automatic Line Record Update

alt Alternate

ALT radar Altimeter, Automatic Line Testing

ALTS Automatic Line Test Set, Analog Line Termination Subsystem

ALU Arithmetic and Logic Unit

am Ante Meridiem, before noon

AM Amplitude Modulated, Analog Module, Auxiliary Marker

A/M Automatic/Manual

AMA Automatic Message Accounting

AMACS Automatic Message Accounting Collecting System

AMARC Automatic Message Accounting Recording Center

AMARS Automatic Message Accounting Recording System

AMB Auto-Manual Bridge Control

AMC Account Manager Code, Auto-Manual Center

AME Amplitude Modulation Equivalent

AMI Alternate Mark Inversion signal

AMIS Air Movement Identification Service

AMITE Submarine cable France-Morocco

AML Actual Measured Loss

AMM Associated Maintenance Module

amp Amperes

ampl AmpLifier

AMPS Advanced Mobile Phone System/Service

AMRAAM Advanced Medium Range Air-to-Air Missile

AMSAT Amateur Satellite, American Satellite Corp.

AMSU Auto-Manual Switching Unit

AMU Alarm Monitor Unit

ANA Assigned Night Answer, Automatic Number Announcer

ANALIT Analysis of Automatic Line Insulation Tests

ANBFM Adaptive Narrow-Band FM modem

ANC All Number Calling, Answer, Charge

AND Automatic Network Dialing

ANF Automatic Number identification Failure

ANI Automatic Number Identification, Advanced Network Integration

ANL Automatic Noise Limiter

ANN Answer, No-charge

ANNIBAL Submarine cable France-Tunisia

ANR Awaiting Number Received

ANS Answer

ANSI American National Standards Institute

ant Antenna

ANTINEA Submarine cable Senegal-Morocco

ANTIOPE French teletext system

ANZCAN Submarine cable Australia-New Zealand-Canada

AO Answer Only

AOC Awaiting Outgoing Continuity

AOI And-Or-Invert (logic gate)

AONALS type "A" Off-Network Access Lines

AOTT Automatic Outgoing Trunk Test

AP Access Point, Anomalous Propagation, Advance Payment, Administrative Processor

APAD high loss operation

APC Adaptive Predictive Coding, Automatic Phase Control, Average Power Control, AMARC Protocol Converter

APCM Adaptive Pulse Code Modulation

APD Amplitude Probability Distribution, Avalanche Photodiode

APK Amplitude-Phase-Keyed system

APL Average Picture Level, A Programming Language

APNG Submarine cable Australia-Papua New Guinea

APOLLO Submarine cable Greece-Cyprus, European Space Agency (ESA) communications satellite

APP Application date, Auxiliary Power Plant

Apple Type of personal computer

approx Approximate, Approximately

Apricot Type of personal computer

AP(S) Application Process or Program (Structure)

APS Application Process Subsystem, Attached Processor System, Attended Pay Station

APT Automatic Picture Transmission, Apartment, Automatic Progression Testing

APTU African Postal and Telecommunication Union

aq Aqua (water)

AQ Autoquote

aq dest Aqua Destilla (distilled water)

AQL Acceptable Quality Level

A/R Alternate Route

AR Awaiting Reply

ARB All Routes Busy

ARC Alternate Route Cancel

ARCHEDDA Architectures for Heterogeneous European Distributed Databases

AREAN reference system for the determination of the articulation reference equivalents

AREG Apparatus Repair-strategy Evaluation Guidelines

ARIANE French-controlled rocket launching system

ARINC Aeronautical Radio, Inc

ARL Acceptable Reliability Level

arm Armature

ARO After Receipt of Order

A route Alternate Route

ARPA Advanced Research Project Agency (Department of Defense)

ARPANET Advanced Research Project Agency Network

ARQ Automatic Request for repetition, Automatic Request repeat system

arr Arrestor

ARR Automatic Rerouting

ARRE Alarm Receiving and Reporting Equipment

ARRL American Radio Relay League

ARSB Automated Repair Service Bureau

ART Artificial, Automatic Reporting Telephone, Alarm Reporting Telephone, Additional Reference carrier Transmission

ARTCC Air Route Traffic Control Center

ARTEMIS Submarine cable France-Greece

ARTS American Radio Telephone System

ARU Audio Response Unit

AS Articulation Score, Start of Answer

ASA American Standards Association, Army Security Agency

ASAT Anti-Satellite capability

ASBHCA Average Season Busy Hour Call Attempts

ASBHCC Average Season Busy Hour Call Completions

ASC Automatic Sensitivity Control, Automatic digital network Switching Center, American Satellite Corp.

ASCII American Standard Code for Information Interchange

ASCR Asymmetric Silicon Controlled Rectifier (thyristor)

ASEANIS Submarine cable, Association of South East Asian Nations: Indonesia-Singapore

ASEANPS ASEAN submarine cable: Philippines-Singapore

ASK Amplitude Shift Keying

ASM Air-Surface Missile

Asph AspHalt

ASPJ Airborne Self-Protection Jammer

ASR Automatic Send/Receive, Airport Surveillance Radar

ASSD CO Associated Company

assoc Associates, Associated

asst Assistant

assy Assembly

AST Anti-Sidetone

ASTP Apollo-Soyuz Test Project

ASTRA Application of Space Techniques Relating to Aviation

ASU Acknowledgement Signal Unit

ASW Anti-Submarine Warfare

AT Aerial Tape armor

AT&T American Telephone and Telegraph. See entry under Bell System

ATA Automatic Trouble Analysis

ATB All Trunks Busy, Address Translation Buffer

ATBA Automatic Test Break & Access

ATCRBS Air Traffic Control Radar Beacon System

ATD Acceptance and Takeover Date

ATDM Asynchronous Time-Division Multiplexing

ATE Automatic Test Equipment

ATECO Automatic Telegram transmission with Computers

ATIC Time assignment with sample interpolation

ATLANTIS Submarine cable Portugal-Brazil

ATM Automatic Teller Machine

ATME Automatic Transmission Measuring Equipment

ATMS Automatic Transmission Measuring System

ATP Acceptance Test Procedure

ATR Answering Time Recorder

ATS Application Technology Satellite, Automatic Trunk Synchronizer, Alarm Termination Subsystem

ATSU Assn. of Time-Sharing Users

ATT Attended public Telephone, Automatic Toll Ticketing, Attachment

ATTC Automatic Transmission Test and Control

ATTCOM AT&T Communications

ATTI AT&T International

ATTIS AT&T Information Systems

ATTIX AT&T Interexchange Carrier

ATTT AT&T Technologies

ATUR Automatic Telephone Using Radio (cellular system)

ATURS Automatic Traffic Usage Recording System

ATX Automatic Telex exchange

ATU Application Terminal Unit

au Astronomical Unit

aud Audible, Audible range

AUDINET American Electric Power Unified Dial Network

AUSSAT Australian communications satellite system

auto Automatic

AUTODIN Automatic Digital Network

AUTOSEVOCOM Automatic Secure Voice Communications network

AUTOVON Automatic Voice Network

aux Auxiliary

av Average, Avenue

AVC Automatic Volume Control

AVCS Advanced Vidicon Camera System

AVD Alternate Voice Data

aver Average

AVHRR Advanced Very High Resolution Radiometer

avoir Avoirdupois

AWACS Airborne Warning and Control System

AWG American Wire Gauge

AWGN Additive White Gaussian Noise

AWPI American Wood Preservers Institute

AWT Actual Work Time

AXE Digital switching system designed by Ericsson, Sweden

AY Anyone

B

b Bel

B Magnetic induction, Buffer

bal Balancing

balun Balanced-to-Unbalanced line transformer

BANCS Bell Administrative Network Communications Systems

BAPI Submarine cable, Barcelona, Spain to Pisa, Italy

BARGEN Submarine cable, Barcelona, Spain to Genoa, Italy

BARO Submarine cable, Barcelona, Spain to Rome, Italy

BASIC Beginners All-purpose Symbolic Instruction Code

bat Battery

BATE Baseband Adaptive Transversal Equalizer

BATS Bit Access Test System

Baudot Teleprinter code

BB Bunch Block, Bus-Bar layout drawing, Broadband, Backboard, Bit/Byte conversion

BBC British Broadcasting Corp.

BBD   Bucket Brigade Device

BBN   Bolt Beranek and Newman (of ARPANET and TELENET)

BC   Broadcasting, Billing Cease date

BCC   Block Check Character

BCCD   Bulk-Channel Charge-coupled Device

BCD   Binary Coded Decimal

BCD/B   Binary Coded Decimal/Binary

BCD/Q   Binary Coded Decimal/Quaternary

BCF   Billion Conductor Feet

BCH   Bose-Chadhuri-Hocquenhem error correction code, Bids per Circuit per Hour

BCI   Bit Count Integrity

BCM   Binary Coded Matrix, buried coarctate mesa-structure (laser)

BCO   Battery Cut-Off, Bill in Care Of

BCR   Billing-Collecting-Remitting

BCS   Block Control Signal, Business Communication System, Business Customer Services

BCST   Broadcast

BCU   Buffer Control Unit

Bd   Baud

BDIR   Bus Direction

BDLC   Burroughs Data Link Control

BDN   Bell Data Network, Bank Draft Number

BDPSK   Binary Differential Phase-Shift Keying

BDR   Bell Doesn't Ring

BDTS   Bulk Data Transfer Subsystem

BDV   Breakdown Voltage

BDW   Buried Distribution Wire

BE   Band Elimination

BEF   Band Elimination Filter

BELLBOY   Paging service

BER   Bit Error Rate, Submarine cable U.S.-Bermuda__

BERPM   Basic Exchange Rate Planning Model

BERT   Bit Error Rate Test set

BET   Between

BF   Branching Filter

BFO   Beat Frequency Oscillator

BFR   Bridged Frequency Ringing

BG   Billing Group

BH   Brinell Hardness, Busy Hour

BHC   Busy Hour Calls

BHCA   Busy Hour Call Attempts

BHL   Busy Hour Load

BHP   Brake Horsepower

BI   Billing Instructions, Backward Indicator

BIB   Backward Indicator Bit

BICS   Building Industry Consulting Service

BIH   International time bureau

BILBO   Built-in logic block observer

BILGE   Binary Load Generation

BIN   Business Information Network

BIPM   International bureau of weights and measures

B-IR   Bell-Independent Relations

BIS   Business Information System, Brought Into Service

BIS-COBOL   Variant of COBOL program language for business application

BISCOM   Business Information Systems—Communications System

BISCUS   Business Information System/Customer Service

BISTSS   Business Information System/Trunks and Special Services

BI-SYNC   Binary Synchronous protocol

BIT   built-in test (techniques)

bit   Binary digit

BITE   Backward Interworking Telephony Event

BITEL   German videotex telephone (Bildschirm Telefon)

bit/s   Bits per Second

BKI   Break-in

bkr   circuit Breaker

bl   Blue

BLA   Blocking Acknowledgement

BLAST   Blocked Asynchronous Transmission

BLER   block error rate

BLERT   Block Error Rate Test

BLIC   Bimos Line Interface Circuit

blk   Black

BLK   Block

BLO   Blocking

BLU   Basic Link Unit

BMEWS   Ballistic Missile Early Warning System

BNC   Bayonet Neill Concelman

BNR   Bell-Northern Research

BNS   Bill Number Screening (TSPS)

BOAM   Bell Owned And Maintained

BOC   Build Out Capacitor

BOCs   Bell Operating Companies

BOD   Beneficial Occupancy Date

BOI   Blind Operator Interface

BOL   Build Out Lattice

BOM   Bill Of Material

B-ONALS   type "B" Off-Network Access Lines

BORSCHT, BORSHT   Fundamental functions needed per line—Battery feed, Overvoltage protection, Ringing signal sending, Supervisory, (Codec), Hybrid, Test

BOS   Bell Operating System, Business Office Supervisor

BOS11   Basic Operating System PDP-11

BOT   Beginning Of Tape

BOTTS   Busy tone trunks

BP   Band Pass, Binding Post, Block Parity, Breakpoint

BPC   Binding Post Chamber, bilateral private circuit

BPF   Band Pass Filter

bpi   Bits Per Inch

BPO   British Post Office

bps   Bits Per Second

BPSK   Binary Phase Shift Keying

BPSS   Basic Packet Switched System

br   Brown, Bridge, Bridging

BRA   Base Rate Area

BRACAN   Submarine cable Brazil-Canary Islands

BRB   Base Rate Boundary

BRF   Bell Rings Faintly

brg   Bridge, Bridging

BRKT   Bracket

BRL   Balance Return Loss

BRS   Block Received Signal, B-mode Receiving Station

BRUS   Submarine cable Brazil-U.S.

BS   Back Space, Backing Store, Backward Signaling, Building Steel

b/s   Bits per Second

BSC   Binary Synchronous Communications, Base Site Control

BSCL   Bell System Common Language

BSE   Broadcasting Satellite for Experimental purposes (Japan)

BSG   British Standard Gauge

BSGL   Branch Systems General Licence (UK)

BSI   British Standards Institute, Bit Sequence Independence

BSN   Backward Sequence Number

BSP   Bell System Practice, British Standard Pipe thread

BSRFS   Bell System Reference Frequency Standard

BSRS   Bell System Repair Specification

BSTJ   Bell System Technical Journal

BSU   Basic Sounding Unit, Business Service Unit

BT   Buried Tape armor, British Telecom, Busy Tone

BTAM   Basic Telecommunications Access Method (IBM)

BTE   Business Terminal Equipment

BTL   Bell Telephone Laboratories

BTN   Billing Telephone Number

Btry   BatterY

BTS   British Telecommunications Systems, Ltd

BTU   British Thermal Unit, Basic Transmission Unit

bur   Buried

BUREAUFAX   Public international facsimile service

BUSAK   Bus Acknowledgement

BUSRQ   Bus Request

BUV   Backscatter Ultraviolet Spectrometer

BV   Busy Verification

BVW   Backward Volume Wave

BW   Bandwidth, Buried Wire

B/W   Bothway

BWD   Backward

BWG   Birmingham Wire Gauge

bwn   Brown

BWR   Bandwidth Ratio

by   Busy

C

C   Centigrade, Celsius, Control, Capacitance, Confidential, Creosote, velocity of light

C^2 Command and Control

C^3 Command Control and Communications, Computer and Communications in Compact size (i.e., hand-portable intelligent work station)

C^3/CM   Command Control and Communications/Counter Measures

C^3 LASER   Cleaved Coupled Cavity Laser

C^5 Countering (hostile) Counter-C^3

C & C   Computers and Communications (Syn.: information transfer, telematics)

C & W   Cable & Wireless plc

ca   Circa, about

Ca   Cable

CABEX   A computer-based message switching system used in a number of countries

CACS   Centralized Alarm and Control System

CACSP   Coated Aluminium Coated Steel, Polyethylene

CAD   Computer-Aided Design

CADAM   Computer-Aided (or Augmented) Design and Manufacture

CADCAM   Computed-Aided Design and Computer-Aided Manufacture

CADMAT   Computer-Aided Design, Manufacture and Testing

CADW   Civil Air Defense Warning system

CAE   Computer-Aided Engineering

CAI   Computer Assisted Instruction

Cal   Calibrate, Calorie

calc   Calculate

CALC   Customer Access Line Charge

CALMS   Credit and Load Management System

CALRS   Centralized Automatic Loop Reporting System

CAM   Computer-Aided Manufacture; Content Addressable Memory, Submarine cable Portugal-Madeira

CAMA   Centralized Automatic Message Accounting

CAMA-C   Centralized Automatic Message Accounting system—Computerized

CAMA-ONI   Centralized Automatic Message Accounting-Operator Number Identification

CANAL   Command Analysis

CANBER   Submarine cable Canada-Bermuda

canc   Cancellation, Cancel

CANTAT   Submarine cable Canada-U.K.

CANTRAN   Cancel Transmission

CAO Circuit Allocation Order, Communication Authorization Order, Completed As Ordered

Cap Capacitor

CAP Circuit Access Point

CAPM CPU Access Port Monitor

Ca Pr Cable Pair

CAPRI Computerized Area Pricing

CAPS Call Attempts Per Second

CAPTAIN Character And Pattern Telephone Access Information Network (Japan)

CAROT Centralized Automatic Reporting On Trunks

CARP Call Accounting Reconciliation Process

Carr Carrier

CARS Community Antenna Relay Service, Continuous Alarm Reporting Service

CAS Commission for Atmospheric Sciences (World Meteorological Organization), Collision Avoidance System, Cable Activity System, Call Accounting Subsystem

CAT Cumulative Abbreviated Trouble file, Computer-Aided Testing

CATLAS Centralized Automatic Trouble-Locating and Analysis System

CATT Centralized Automatic Toll Ticketing

CATV Community Antenna Television

CAU Crypto Ancillary Unit

CAW Common Aerial Working

CAX Community Automatic exchange

CB Common Battery, Coin Box, Connecting Block, Clear Back, Citizen's Band

CBC Can't Be Called, Canadian Broadcasting Corp.

CBD Call Box Discrimination, Configuration Block Diagram

CBDS Circuit Board Design System

CBDT Can't Break Dial Tone

CBE Centralized Branch Exchange

CBEMA Computer Business Equipment Manufacturers Assn.

CBH Can't Be Heard

CBIC Complementary Bipolar Integrated Circuit

CBMS Computer-Based Message System, Computer-Based Management System

CBR Cavity-Backed Radiator

CBS Common Battery Signaling, Columbia Broadcasting System

CBT Computer-Based Terminal

CBX Computerized private Branch exchange, Computer-Based exchange

CC Coin Collect, Collect Call, Computer Center, Common Control, Central Control, Coin Completing, Code Controller, Control Console

CCAQ Consultative Committee on Administrative Questions

CCB Coin Collecting Box, Common Carrier Bureau, Circuit Concentration Bay, Common Carrier Bureau

CCC Comparative Capital Cost

CCCI(C³I) Command, Control, Communications and Intelligence

CCD Charge Coupled Device, Contract Completion Date

CCDN Corporate Consolidated Data Network (IBM)

CCES Common Control Echo Suppressor

CCF Communications Control Field, Configuration Control Function

CCG Computer Communications Group (Telecom Canada)

CCH Connections per Circuit per Hour

CCIA Computer and Communications Industry Association

CCIF International telephone consultative committee

CCIR International radio consultative committee

CCIS Common Channel Inter-office Signaling

CCIT International telegraph consultative committee

CCITT International telegraph & telephone consultative committee (this absorbed the CCIF and CCIT)

CCL Communications Control Language

CCM Communications Control Module

CCMC Commonwealth Cable Management Committee

CCN Contract Change Notice

CC-NDT Can't Call-No Dial Tone

CCP Cross Connection Point, Computer Central Processing, Call Control Process(ing), Contract Configuration Process

CCR Commitment, concurrency, Recovery services

CCS Hundred call seconds, Continuous Color Sequence, Common Channel Signaling

CCS² Command, Control and Subordinate Systems

CCSA Common Control Switching Arrangement

CCSD Command Communication Service Designator

cct Circuit

CCT Circuit, Coupler Cut Through, Complete Calls To

CCTA Central Computer & Telecommunications Agency (U.K.)

CCTG Configuration Control Task Group

CCTS Coordinating committee on satellite communications

CCTV Closed Circuit Television

CCU Common Control Unit, Communications Control Unit

CCUAP Computerized Cable Upkeep Administration Program

CCW Counter Clockwise, Channel Command Word

CD Circuit Description

C&D Control and Display

CDA Coin Detection and Announcement, Command and Data Acquisition

CDC Characteristic Distortion Compensation, Construction Design Criteria, Call Directing Code

CDCCP Control Data Communications Control Procedure

CDF Combined Distribution Frame, Communications Data Field

CDLRD Confirming Design Layout Report Date

CDM Companded Delta Modulation, Code Division Multiplexing

CDMA Code-Division Multiple Access

CDO Community Dial Office

CDP Communications Data Processor

CDPR Customer Dial Pulse Receiver

CDR Call Detail Recording

CDRR Call Detail Recording and Reporting

CDT Control Data Terminal

CDU Central Display Unit

CDW Civil Defense Warning

C-E Communications-Electronics

CEE International commission on rules for the approval of electrical equipment

CEEFAX UK Teletext service (= see facts)

CEIRD Confirming Engineering Information Report Date

CELTIC Concentrateur Exploitant Les Temps d'Inactivite des Circuits, (French TASI)

CEM Cement Conduit

CEMA Canadian Electrical Manufacturers Association

CEMF Counter Electromotive Force

CEN European committee for standardization

CENELEC European electrotechnical standards coordinating committee

Centrex Centralized PBX services for business customers

CEO Comprehensive electronic office

CEPER Combined Engineering Plant Exchange Record

CEPT European conference of postal and telecommunications administrations

CERG Concrete-Encased Ring Ground

CERMET Ceramic Metallized

CESA Canadian Engineering Standards Association

CF Can't Find, Copy Furnished, Central File, Count Forward

CFA Carrier Frequency Alarm

CFC Coin & Fee Check(ing)

CFE Contractor-Furnished Equipment

CFF Current Fault File

cfh Cubic Feet per Hour

CFL Call Failed or Failure

cfm Cubic Feet per Minute

CFM Companding and Frequency Modulation

CFS Calls For Service signal

CG Common ground

CGA Carrier Group Alarm

CGC Circuit Group Congestion

CGPM General conference on weights and measures

CGRP Circuit Group

cgs Centimeter-Gram-Second

CGSA Cellular Geographic Service Area

CGSET Circuit Group Set

CH Can't Hear, Coastal Harbor

CHAN Channel

CHAPSE CHILL and ADA Programing Support Environment

CHDB Compatible High Density Bipolar code

chez at the home of (French) (In international telegram delivery information details)

CHG Charge

CHILL Programming language developed by CCITT to program stored program controlled exchanges, (CCITT High Level Language)

CHIPS Clearing House Interbank Payment System

CHN Change

Ch Op Chief Operator, supervisor

CHT Call Hold & Trace, Call Holding Time

CI Cast Iron, Community of Interest, Control Interface

CIB Centralized Intercept Bureau

CIC Carrier Identification Code

CICS Customer Information Control System

CID Charge Injection Device

CIE Commission Internationale de l' Eclairage

CIF Captive Installation Function

CIGALE French transmission network serving a packet switched system

CII Call Identity Index

CIM Computer-Integrated Manufacturing

CIMS Computer Integrated Manufacturing System

CIO Confirming Informal Order

CIR Circle, Circular

CIRM International marine radio association

cir mil Circular Mil

CIS Channel & Isolation Supervision

CISPR International special committee on radio interference

CIT Compagnie Industrielle des Telecommunications

CITEL Inter-American telecommunications conference

CIWS Concentrator Isolation Working Subsystem

CKO Checking Operator (in toll ticketing system)

Ckt Circuit

C/kt Carrier to noise power density (independent of bandwidth)

CKT-ID Circuit Identification

CL Center Line, Compatibility List

CLAMS Consumer's Lobby Against Monopolies

CLEO Common Language Equipment Order

CLF Clear Forward

CLI Calling Line Identification, Calling Line Identity

CLIP Cellular Logic Image Processor

Clk   Clerk

CLOAX   Corrugated-Laminated co-axial cable

CLR   Clearance, Combined Line and Recording, Central Logic Rack

CLRC   Circuit Layout Record Card

cm   Centimeter

CM   Command Module, Class Marks, Configuration Management, Continuity Message, Control Memory

C-M   Control-Monitor

CMA   Communications Architecture, Communications Managers Association

CMC   Cable Maintenance Center

CMDS   Centralized Message Data System

cmf   Cymomotive Force

C/MFI   Conversion, Memory and Fault Indication

CMG   Control-Moment Gyroscope

CMI   Code Mark Inversion; joint international committee for tests relating to the protection of telecommunications lines and underground ducts

CML   Current Mode Logic

CMM   Computer Main Memory, Concentration Module Main

CMO   Common Mode Operation

CMOS   Complementary Metal Oxide Semiconductor

CMP   Camp-on

CMR   Communications Moon Relay, Centralized Mail Remittance, CBX Management Reporter (Rolm)

CMRR   Common Mode Rejection Ratio

CMRS   Cellular Mobile Radio Telecommunications Service

CMS   Circuit Maintenance System, Change Management System

cm/sec   Centimeters per Second

CMSG   "C" message weighting

CMTS   Centralized Maintenance Test System

CMTT   Joint study group for television and sound transmissions

CMX   Concentration Module extension

C/N   Carrier-to-noise ratio

CN   Check Not OK, Coin trunk

CNAC   CCIS Network Administration Center

CNC   Computerized Numerical Control

CNCC   Customer Network Control Center

CNI   Changed Number Interception

CNL   Constant Net Loss

CNOTCH   "C" message weighting with notch filter

CNR   Carrier-to-thermal-Noise Ratio

CnS   CCITT n(= 6, 7) Signaling

c/o   Care Of

CO   Central Office, Coinbox line, Crystal Oscillator, Check OK.

COAM   Customer Owned And Maintained equipment

coax   Coaxial cable

COBOL   Common Business-Oriented Language

COC   Compiler Object Code

COCOT   Coin-Operated, Customer-Owned Telephone

CODAN   Carrier-Operated Device, Anti-Noise

CODASYL   Conference On Data Systems Languages

CODEC   Coder plus Decoder

COE   Central Office Equipment

COEES   Central Office Equipment Engineering System

COER   Central Office Equipment Reports

COF   Cause Of Failure

COG   Central Office Ground

COIN   Coin phone Operational and Information Network system

COLR   Circuit Order Layout Record

COLUMBUS   Submarine cable Spain-Venezuela

COM   Common, Communications, Commercial, Commonwealth, Computer Output Microfilm

COMASIII   Computerized Maintenance and Administration Support III

COMFOR   Commercial Wire Center Forecast Program

COMMS   Central Office Maintenance Management System

COMMS-PM   Central Office Maintenance Management System — Preventive Maintenance

comp   Comparator

COMP   Compiler

COMPAC   Commonwealth Pacific Cable, Submarine cable Canada-Fiji-New Zealand-Australia

compandor   Compressor plus expander

COMPTEL   Competitive Telecommunications Assn.

COMSAT   Communications Satellite Corp.

COMSEC   Communications Security

COMTASS   Compact towed sonar system

conc   Concentrated, Concentrator

cond   Conductor, Conditioned

CONECS   Connectorized Exchange Cable Splicing

CONFG   Configuration Process

CONG   Congestion

cont   Control

CONUS   Continental United States, Contiguous United States (excluding Hawaii and Alaska)

conv   Converter

coord   Coordinate, Coordinating

COP   Code Of Practice

COPAN   Command Post Alerting Network

COPES   Consumer-Optimized Product Engineering System

copr   Copyright, Computerized Outside Plant Records

COPUOS   Committee On the Peaceful Uses of Outer Space

COR   Contracting Officer's Representative

CORAL   A high-level programming language.

CORDS   Coordination of Record and Data base System

CORODIM Correlation Of the Recognition Of Degradation with Intelligibility Measurements

CORSA Cosmic Radiation Satellite (Japan)

cos Cosine

COS Class Of Service, Customer's Other Service, Change of Subscribers

COSAM Co-Site Analytical Model

cosh Hyperbolic cosine

COSMIC Common Systems Main InterConnecting

COSMOS Computer System for Main frame Operations

COSP Central Office Signaling Panel

cot Cotangent, Central Office Terminal

COT Continuity

COTC Canadian Overseas Telecommunication Corp.

COTM Customer Owned and Telephone company Maintained

CP Chemically Pure, Corrosion Protection, Central Processor, Calendar Process, Common Process, Connection Pending

CPA Co-Polar Attenuation, Computer Architecture

CPAS Construction Program Administration System

CPC Calling Party's Category

CPCH Calling Party Cannot Hear

CPD Central Pulse Distributor, Cumulative Probability Distribution

CPE Customer Premises Equipment, Central Processing Element, Customer Provided Equipment

CPFF Cost Plus Fixed Fee

CPFR Calling Party Forced Release (on DPO)

CPFT Customer-Premises Facility Terminal

CPH Characters Per Hour

CPI Call Progress Indicator, Characters Per Inch, Cable Pair Identification, Computer-to-PBX Interface

CPIF Cost Plus Incentive Fee

CPL Capability Password Level

CPM Critical Path Method, Cards Per Minute, Call Protocol Message

CPMS Cable Pressure Monitoring System

CPODA Contention Priority-Oriented Demand Assignment (protocol)

CPOL Communications Procedure-Oriented Language

CPP Conductive Plastic Potentiometer

CPPI Consultative Panel on Public Information (United Nations)

CPR Continuous Progress indicator

CPRS Centralized Personnel Record System

cps Cycles Per Second, Characters Per Second

CPS Call Processing Subsystem

CPSK Coherent Phase-Shift Keying

CPU Central Processing Unit, Communications Processor Unit, Communications Processor Utility

CQ Call to all stations

CR Carriage Return, Customer's Report, Clear Record

CRB Customer Records and Billing system

CRBO Centralized Records Business Office

CRC Communications Research Center (Canada), Cyclic Redundancy Checking, Communications Relay Center

CRCC Cyclic Redundancy Check Character

CREG Concentrated Range Extension with Gain

CRISP Cascadable real-time integrated signal processor

crit Critical

CRO Central Radio Office, Central Records Office, Complete with Related Order, Cathode Ray Oscilloscope

CRPL Central Radio Propagation Laboratory

CRQ Call Request

CRS Centralized Results System

CRSA Centralized Repair Service Attendants

CRT Cathode Ray Tube, Continuous Ring Tone

CRV Contact Resistance Variation

CS Currency Sign, Calls per Second, Cast Steel, Call Store, Class of Service

CSA Communications Service Authorization, Commercial Service Authorization, Called Subscriber Answer

CSACCS Customer Service Administration Control Center System

csc Cosecant

CSC Common Signaling Channel, Circuit Switching Center, Central Switching Center

CSDC Circuit Switched Digital Capability

CSE Communications satellite for experimental purposes (Japan)

CSH Called Subscriber Held

CSM Commission for Synoptic Meteorology, Correed Switching Matrix, Call Supervision Module

CSMA-CD/CA Carrier Sense Multiple Access local area network with collision detection/collision avoidance

CSMA-CR Carrier Sense Multiple Access with contention resolution

CSN Common Services Network

CSO Centralized Service Observation

CSOC Consolidated Space Operations Center (Colorado Springs)

CSP Control Switching Point, Communicating sequential processes

CSR Common Services Rack

CSRMP Communications Sales Results Measurement Plan

CSS Customer Switching System, Control Signaling Subsystem, Common Services Subsystem, Computer Sub-System

CST Carrier power Supply, Transistorized

CSTD Committee on Science and Technology for Development

CSU Circuit Switching Unit, Check Signal Unit, Channel Service Unit, Common Services Unit, Cache Store Unit, Customer Service Unit

CSW Channel Status Word

C/T Carrier-to-noise Temperature ratio

CT Transit switching center, Center Tap, Complete Translation, Continuity Transceiver

CTAK Cipher Text Auto Key

CTBM Chief Testboard Man

CTD Charge Transfer Device, Continuity Tone Detector

CTE Channel Translating Equipment

CTI Centralized Ticket Investigation

CTM Complete Treatment Module, Continuity Transceiver Module

CTMS Carrier Transmission Maintenance System

CTO Cut-off, Corporate Telecommunications Operation

CTRAP Customer Trouble Report Analysis Plan

ctrl Control

CTS Communications Technology Satellite (Canada), Cable Turning Section, Cable Terminal Section, Clear To Send

CTSI Central Terminal Signaling Interface

CTT Cable Trouble Ticket

CTU Central Terminal Unit, Cartridge Tape Unit

CTX Centrex system number

CTXCO Centrex Central Office

CTXCU Centrex CUstomer

cu Cubic

Cu Copper

CU Control Unit, Common User, Common Update

cu ft Cubic Feet

CUC Computer Users Committee

CUDAT Common-User Data

CUG Closed User Group

cur Current

CURTS Common User Radio Transmission Sounding system

CUS Common User System, Customer code

CVD Chemical Vapor Deposition

CVGB Cable Vault Ground Bar

CVSDM Continuously Variable Slope Delta Modulation

CW Continuous Wave, Clockwise

CWA Communication Workers of America

CWD Creosoted Wood Duct

CWGC Copper Wire Counterpoise Ground

CWI Call Waiting Indication

CWO Custom Work Order

CWS CopperWeld Steel

cwt Hundredweight

CX Coinbox set, Coin collecting box, pay station, composite signaling

CXR Carrier

CY Calendar Year

CYBERNET Control Data Corporation network

CYCLADES French packet switched network

CZCS Coastal Zone Color Scanner

D

D Delay, Designation for an intermediate dialing center on a toll ticket, Digital

DA Doesn't Answer, Digit Absorbing, Double Armor, Demand Assignment, Data Available, Directory Assistance

D/A Digital to Analog

DAA Data Access Arrangement

DAC Digital to Analog Converter

DACE Data Administration Center Equipment

DACS Digital Access & Crossconnect System

DAIS Defense Automatic Integrated Switching system

DAIV Data Area Initializer & Verifier

DAK Deny All Knowledge

DAL Data Access Line

DAMA Demand Assignment Multiple Access

DAMSU Digital Automanual Switching Unit

DAP Data Access Protocol, Distributed Array Processor, Deformation of vertical Aligned Phases, Digital Access Point

DAPO Digital Advance Production Order

DARPA Defense Advanced Research Project Agency

DAS Directory Assistance System, Data Auxiliary Set, Data Analysis Software

DASD Direct Access Storage Device

DASH Direct Access Storage Handler

DASS Demand Assignment Signaling and Switching

DATALINK Canadian digital data service

DATAPAC Canadian public packet switched network

DATAROUTE Canadian point-to-point data facility

Dataset modulator/demodulator

DATEC Data technical support group

DATEL RCA global communications data transmission service over telephone circuits

DAV Data Above Voice

DAVC Delayed Automatic Volume Control

dB Decibel

DB Dry Bulb temperature, Data-Base

dBa Adjusted-weighted noise power in dB referred to -85dBm

dBa0 Noise power in dBa referred to or measured at 0 transmission level point

dBa(F1A) Noise power measured by a set with F1A-receiver weighting

dBa(H1A) Noise power measured by a set with H1A receiver weighting

dBm dB referred to 1 milliwatt

dBm0 Noise power in dBm referred to or measured at 0 transmission level point

dBm0p Noise power in dBm0 measured by a set with psophometric weighting

dBm(psoph) Noise power in dBm measured by a set with psophometric weighting

dBr Power difference in dB between any point and a reference point

dBrap Decibels above reference acoustic pressure

dBrn Decibels above reference noise

dBrn(144-line) Noise power, in dBrn, measured by a set with 144-line weighting

dBrnC Noise power, in dBrn, measured by a set with C-message weighting

dBrnC0 Noise power, in dBrnC, referred to or measured at 0 transmission level point

dBrn(f₁-f₂) Flat noise power in dBrn

dBW Decibels referred to 1 watt

dBx 90dB-measured coupling loss (crosstalk)

DBC Data Bridging Capability

DBL Double connection, Detailed Billing number required

DBMS Database Management System

DBO Drop Build-Out capacitor

DBS Database Service, Direct Broadcasting by Satellite, Data Bridging Service

dc Direct Current

DC Direct Current, Directional Coupler, Documentation Control(ler)

dcwv Direct Current Working Volts

DCA Defense Communications Agency, Document Content Architecture

DCC Double Cotton Covered, Destination Code Cancel

DCCC Double Current Cable Code

DCD Data Carrier Detector

DC/DC Direct Current/Direct Current

DCDR Data Collection and Data Relay

DCE Data Circuit-terminating Equipment

DCG Dependent Charge Group, Designs Co-ordination Group

DCL Delayed Call Limit(ed)

DCLU Digital Carrier Line Unit

DCM Diagnostic Control Module

DCP Data Collection Platform, Data Collecting Platform, Distribution Common Point

DC-PBH-LD Double-channel planar-buried-heterostructure laser-diode

DCPSK Differential Coherent Phase Shift Keying

DCS Defense Communications System, Data Collection System, Distributed Computing System, Data Collection Subsystem, Digital Crossconnect System (Digital Switch Corp., U.S.)

DCSS Defense Communications Satellite System

DCTL Direct Coupled Transistor Logic

DCTU Directly Corrected Test Unit

DCU Drum Control Unit, Disc Control Unit

DD Due Date, Direct Dialing, Direct Dialed, Discriminating Digit, Data Descriptor

DDA Design Data Administration

DDB Double Declining Balance, Digital Data Bus (or D²B)

DDCMP Digital Data Communications Message Protocol

DDD Direct Distance Dialing, Direct Digital Dialing

DDF Digital Distribution Frame

DDI Direct Dialing In

DDIE Direct Digital Interface Equipment

DDL Digital Data Link, Data Definition Language

DDO Direct Dialing Overseas

DDP Digital Data Processor, Distributed Data Processing

DDR Dialed Digit Receiver

DDS Dataphone Digital Service, Digital Data System, Data Dictionary System, Digital Data Service

DDT Delayed Dialing Tone

DDX Digital Data Exchange, Switched Digital Data service (Japan)

DEC Digital Equipment Corp., Decadic

DECCO Defense Communications Agency Commercial Communications Office

DECEO Defense Communication Engineering Office

DECNET Digital Equipment Corporation Network

DECPSK Differentially Encoded Coherent Phase Shift Keying

DED Distant End Disconnect

de-emph De-emphasis

def Defective

DEFT Dynamic Error-Free Transmission system

deg Degree

Del Delayed

DEL Direct Exchange Line

DeM Delta Modulation

dem Demodulator

demod Demodulator

demux Demultiplexer

DEO Digital End Office

DEPIC Dual-Expanded Plastic-Insulated Conductor

DES Data Encryption Standard, Digital Echo Suppressor

DETAB Decision Table (programming)

DEW Distant Early Warning

DF Douglas Fir (utility pole), Distribution Frame, Direction Finding

DFC Disk File Controller

DFI Digital Facility Interface

DFSG Direct Formed Supergroup

DFSK Double Frequency Shift-Keying

DFT Digital Facility Terminal

DG Differential Gain

DGM Data-Grade Media

dia Diameter

DIA Document Interchange Architecture

diag Diagram

diam Diameter

DIANE Direct Information Access Network for Europe

DIBIT Group of two bits

DIC Digital Concentrator

DID Direct Inward Dialing, Display Interface Device

DIG Design Implementation Guide

DIGIPULSE Keypad sending out loop-disconnect pulses

DILEP Digital Line Engineering Program

DIN Deutsche Industrie Normenausschus

DINA Distributed Information-Processing Network Architecture

DIO Direct Input/Output

DIOSS Distributed Office Support System

DIP Dual In-line Package

DIQd Disc-Insulated Quad

dir Directory

dis Display

disc Disconnect

DISC Digital International Switching Center

DISCON Defense Integrated Secure Communications Network (Australia)

DISPLAY Digital Service Planning Analysis

dist, distr Distribution, Distributor

Dist District

DISU Digital International Switching Unit

Div, divn Division

DIVA Data Inquiry-Voice Answer, Digital Input-Voice Answerback

DJ Digital Junction

DL Document List, Data Link

DLC Data Link Control, Digital Loop Carrier

DLCF Data Link Control Field

DLD Dark Line Defects (LEDs)

DLE Data Link Escape, Direct Line Equipment

DLI Dual Link Interface, Data Link Interface

DLL Dial Long Line units

DLRD Design Layout Report Date

DLS Digital Line System

DLSO Dial Line Service Observing

DLT Digital Line Termination

DLTS Deep-Level Transient Spectroscopy

DLTU Digital Line and Trunk Unit

Dly Delay

DM Delta Modulation, Digital Module

DMA Direct Memory Access

DMAC Direct Memory Access Controller

DMB Disconnect and Make Busy

DME Distance Measuring Equipment

DMEP Data network Modified Emulator Program

DMERT Duplex Multiple-Environment Real Time (operating system)

DMIS Directory Management Information System

DML Data Manipulation Language

DMOS Diffusion Metal-Oxide Semiconductor

DMR Demultiplexing/Mixing/Remultiplexing (device)

DMS Digital Multiplex System

DMW Digital Milliwatt

DN Directory Number

DNA Digital Network Architecture

DNC Direct Numerical Control

DNHR Dynamic Non-Hierarchical Routing

DNI Data Network Interface

DNIC Data Network Identification Code

DNR Dynamic Noise Reduction

DO Design Objective

DOC Dynamic Overload Control

DOCS Document Organization and Control System

DOD Department of Defense, Direct Outward Dialing

DOI Dept. of Industry (U.K.)

DOJ Dept. of Justice (U.S.)

DOMSAT Domestic Satellite system

DON Delayed Order Notice

DOS Disc Operating System

DOV Data Over Voice

DOVAP Doppler Velocity And Position

DP Double Pole, Data Processing, Differential Phase, Distribution Point, Drip Proof, Dial Pulse, Disconnection Pending, Dataport

DPA Dial Pulse Access, Different Premises Address

DPC Data Processing Center, Destination Point Code

DPCM Differential Pulse Code Modulation

DPDT Double Pole, Double Throw

DPE Data Processing Equipment

DPI Different Premises Information

DPLM Domestic Public Land Mobile

DPLMRS Domestic Public Land Mobile Radio Service

DPMA Data Processing Management Assn.

DPNSS Digital Private Network Signaling System (UK)

DPO Dial Pulse Originating

DP(S) Data Packet (Subsystem)

DPS Data Processing System, Different Premises Subscriber

DPSK Differential Phase Shift Keying

DPST Double Pole, Single Throw

DPT Dial Pulse Terminating, Different Premises Telephone number

DQ Directory enquiry service

DQA Design Quality Assurance

DR Data Rate

DRAM Digital Recorded Announcement Module

DRAW Direct Read After Write

DRCS Distress Radio Call System, Dynamically Redefinable character sets (viewdata)

DRE Directional Reservation Equipment

DRGS Direct Readout Ground Station

DRO Destructive Read Operation

DRP Directional Radiated Power, Data Reception Process

DRS Data Rate Selector

DS telephone Disconnected, Direct Sequence

DS-0 Digital Signal level 0 : a 64 kbit/s signal

DS-1 Digital Signal level 1 : a 1.544 Mbit/s signal (T1 carrier)

DS-1C Digital Signal level 1C : a 3.152 Mbit/s signal

DS-2 Digital Signal level 2 : a 6.312 Mbit/s signal (T2 carrier)

DS-3 Digital Signal level 3 : a 44.76 Mbit/s signal

DSA Dial Service Assistance, Digital Serving Area, Dial Service Analysis

DSA board Dial System "A" board, Dial Service Auxiliary switchboard

DSAN Debug Syntax Analysis

DSAP Directory Scope Analysis Program

DSAU DSI Signal Access Unit

DSB Double Sideband, Dial System "B" switchboard

DSBAM Double Sideband Amplitude Modulation

DSBEC Double Sideband, Emitted Carrier

DSBRC Double Sideband, Reduced Carrier

DSBSC Double Sideband, Suppressed Carrier

DSBTC Double Sideband, Transmitted Carrier

dsc Double Silk Covered

DSC Direct Satellite Communications

DSCH Dual Service Channel

DSCS Defense Satellite Communications system

DSD Direct Service Dialing

DSDS Dataphone Switched Digital Service

DSE Data Switching Exchange, Distributed System Environment

DSI Digital Speech Interpolation

DSIR Department of Scientific & Industrial Research (UK)

DSL Deep Scattering Layer

DSLC Data Subscriber Loop Carrier

DSM Direct Signal Monitoring

DSN Deep Space Network

DSR Data Set Ready

DSRV Deep Submergence Rescue Vehicle

DSS Digital Switching System, Direct Station Selection

DSSCS Defense Special Secure Communications System

DSTE Digital Subscriber Terminal Equipment

DSU Disk Storage Unit, Data Service Unit, Digital Service Unit

DSVT Digital Secure Voice Telephone

DSX Digital cross connect frames

DT Digroup Terminal, Dial Tone

DTA Detailed Traffic Analysis

DTAU Digital Test Access Unit

DTBP Dedicated Total Buried Plant

DTC Detection Threshold Computer, Data Test Center

DTDMA Distributed Time Division Multiple Access

DTE Data Terminal Equipment

DTF Dial Tone First, Date To Follow

DTI Distortion Transmission Impairment

DTL Diode Transistor Logic

DTLS Digital Television Lightwave System

DTMF Dual Tone Multi-Frequency

DTO Decentralized Toll Office

DTR Down-Time Ratio

DTS Digital Tandem Switch, Digital Termination Systems, Diffusion Total System (for manufacturing LSI wafers)

DTU Data Terminating Unit, Data Transfer Unit

DTW Dynamic Time Warping

DTWX Dial Teletypewriter exchange service

DU Dimensioning Unit

D/U Delay Unit

DUC Dual-access Utility Circuit

DUCE Denied Usage Channel Evaluator

DUE Detection of Unauthorized Equipment

DUMP Write out contents of register or store

Dup Duplicate

DUP Data User Part

DUS Data User Stations

DUT Device Under Test,

DUV Data Under Voice

DVA Designed, Verified and Assigned date

DVBST Direct View Bi-stable Storage Tube

DVOM Digital Volt-Ohm-Meter

DVM Digital Voltmeter

DVX Digital Voice Exchange

DW Don't Want, Drop and block Wire

DWB Designers' Workbench

dwg Drawing

DWG Drilled Well Ground

DWV Data With Voice

dx Distant, Distance, Distance reception, Duplex

DYP Directory Yellow Pages

E

E voltage, East, Erlang(s)

E10 Family of digital switching offices developed by CIT Alcatel, France

Ea Each

EACSO East Africa Common Services Organization

EADAS Engineering and Administration Data Acquisition System

EAGE   Electrical Aerospace Ground Equipment

E & M   receive and transmit leads of a signaling system

EAM   Electrical Accounting Machines

EAP   Emergency Action Program, Expenditure Analysis Plan

EAROM   Electrically Alterable Read-Only Memory

EAS   Extended Area Service

EASCON   Electronic and Aerospace Systems Conference

EATMS   Electro-Acoustic Transmission Measuring System

EAX   Electronic Automatic telephone exchange

EBCA   External Branch Condition Address

EBCDIC   Extended Binary-Coded Decimal Interchange Code

EBCI   External Branch Condition Input

EBD   Effective Billing Date

EBER   Equivalent Binary Error Rate

EBES   Electron-Beam Exposure System

EB&F   Equipment Blockages and Failures

EBU   European Broadcasting Union

EC   Electrical Conductor (grade of purity of a metal), Extra Control (wire), Eastern Cedar (utility pole), Echo Controller

ECA   Economic Commission for Africa, Exchange Carrier Association

ECAFE   Economic Commission for Asia and the Far East

ECB   Electrically Controlled Birefringence (LCDs)

ECC   Electronic Common Control, Expanded Community Calling

ECCM   Electronic Counter-Counter Measures

ECCS   Economic Hundred Call Seconds

ECD   Error Control Device, Equipment Configuration Data

ECDO   Electronic Community Dial Office

ECE   Economic Commission for Europe, Echo Control Equipment

ECF   Echo Control Factor

ECL   Emitter-Coupled Logic

ECLA   Economic Commission for Latin America

ECM   Electronic Countermeasures

ECMA   European Computer Manufacturers Association

ECN   Emergency Communication Network (highway)

ECO   Electron Coupled Oscillator, Electronic Central Office, Engineering Control Office

ECOM   Electronic Computer Originated Mail

ECOS   Extended Communications Operating System (Harris Corp.)

ECP   Engineering Change Proposal, Equipment Conversion Package, Effective Cable Pairs

ECPGB   Entrance Cable Protector Ground Bar

ECS   Experimental Communications Satellite (Japan), Echo Control Subsystem, European Communications Satellite, Energy Communication Services

ECSA   Exchange Carriers Standards Association

EDA   Embedded Direct Analysis

EDC   Error Detecting Code

EDD   Envelope Delay Distortion

EDP   Electronic Data Processing

EDPE   Electronic Data Processing Equipment

EDPM   Electronic Data Processing Machine

EDS   Electronic Data Switching system (German data network), Electronic Data Systems Corp.

EEC   European Economic Community

EEPROM   Electrically-Erasable Programmable ROM (also EAROM)

EEROM   Electrically Eraseable Read-Only Memory

EET   Equipment Engaged Tone

EFAR   Economic Feeder Administration and Relief

EFB   Error Free Block

EF&I   Engineer, Furnish and Install

EFL   Emitter-Follower Logic

EFRAP   Exchange Feeder Route Analysis Program

EFS   Error Free Seconds

EFT   Electronic Funds Transfer

EFTA   European Free Trade Association

EFTS   Electronic Funds Transfer System

eg   Exempli Gratia, for example

EG   Equipment Ground

EHF   Extremely High Frequency band (30-300 GHz)

EHS   Extra High Strength (steel wire)

EHT   Extra High Tension

EIA   Electronic Industries Association

EIES   Electronic Information Exchange System

EIH   Error Interrupt Handler

EIN   European Information Network

EIP   Equipment Installation Procedure

EIR   Engineering Information Report

EIRD   Engineering Information Report Date

EIRP   Equivalent Isotropically Radiated Power

EIS   Executive Information System; Expanded Inband Signaling

EIU   Equipment Inventory Update

EJ   Electronic Journalism (electronic news gathering)

EJF   Estimated Junction Frequency

EL   Engineering Letter, Exchange Line, Executive Level

ELC   Exchange Line Capacity

ELDO   European Launcher Development Organization

elec   Electric, Electrical

Elec Eng   Electrical Engineer

ELF   Extremely Low Frequency band (300 Hz or lower)

ELINT   Electronic Intelligence

ELMAP   Exchange Line Multiplexing Analysis Program

ELSEC   Electronic Security

ELSI Extra Large Scale Integration

EL-SSC Electronic Switching System Control

ELT Emergency Locator Transmitter

EMC Electromagnetic Compatibility, Engineered Military Circuit

EMC FOM Electromagnetic Compatibility Figure Of Merit

EMCON Electromagnetic emission Control

EME Earth-Moon-Earth

Emer Emergency

emf Electromotive Force

EMI Electromagnetic Interference

EMIRTEL Emirates Telecommunications Corp. (United Arab Emirates)

EML Expected Measured Loss

EMMS Electronic Mail and Message System

EMP ElectroMagnetic Pulse

EMRP Effective Monopole-Radiated Power

EMS Equilibrium Mode Simulator, Electronic Message Service

EMSEC Emanations Security

EMSS Electronic Message Service System

EMT Electrical Metallic Tubing

EMU Electromagnetic Unit

ENADS Enhanced Network Administration System

encl Enclosure, Enclosed

ENFIA Exchange Network Facilities for Interstate Access

ENG Electronic News Gathering

Ent Entrance

ENU Essential/Non-essential/Update

EO End Office

EOA End Of Address

EOD End Of Dialing

EOF End Of File

EOM End Of Message

EOR End Of Run

EOS Earth Observation Satellite, Early Operational Signal

EOT End Of Tape, End Of Transmission

EOW Engineering Orderwire

EP Equipment Practice

EPBX Electronic Private Branch Exchange

EPC Earth Potential Compensation

EPD Exchange Parameter Definitions

EPI Elevation-Position Indicator

EPIRB Emergency Position Indicating Radio Beacon

EPLANS Engineering, Planning and Analysis Systems

EPROM Erasable Programmable Read-Only Memory

EPSCS Enhanced Private Switched Communications Service

EPSS Experimental Packet Switched Service

EQ Equipment, Enquiries

EQUAL Equalizer

equiv Equivalent

ER Easy to Reach, Engineering Route

ERB Earth Radiation Budget

ERD Emergency Recovery Display

ERL Echo Return Loss

EROW Executive Right Of Way

erp Effective Radiated Power

ERS Earth Resources Satellite, Earth Resources Survey, Emergency Reporting System, Electronic Register-Sender, European Remote Sensing (meteorological satellite)

ERSOS Earth Resource Survey Operational System

ERT Estimated Repair Time

ERTS Earth Resources Technology Satellite, Error Rate Test Set

Es Sporadic E-layer

ES Echo Suppressor

ESA European Space Agency

ESAC Electronic Systems Assistance Center

ESC Echo Suppressor Control, Exchange Servicing Center, Engineering Service Circuit

ESCAP Economic and Social Commission for Asia and the Pacific

ESCES Experimental Space Communication Earth Station

ESD Electrostatic Discharge

ESG Exchange Software Generator

ESI Equivalent Step Index

ESL Essential Service Line

ESM/ECM Electronic Support Measures/Electronic Counter Measures

ESMR Electrically Scanning Microwave Radiometer

ESO Echo Suppressor, Originating end

ESOC European Space Operations Center

ESPRIT European Strategic Program for Research in Information Technology

ESR Equipment Supervisory Rack, Extended self-contained ring

ESS Electronic Switching System, Echo Suppression Subsystem

ESSA Environmental Survey Satellite

ESSCIRC European Solid State Circuits Conference

Est Estimate, Estimated

EST Echo Suppressor, Terminating end

ESTEC European Space Research and Technology Center

ESTS Echo Suppressor Testing System

esu Electrostatic Unit

ESU Empty Signal Unit

ET Engaged Tone, Exchange Terminal

ETA Estimated Time of Arrival

et al Et Aliae, and others

ETB End of Transmission Block

etc Et Cetera, and so on

ETD Estimated Time of Departure

ETL Effective Testing Loss

ETN Electronic Tandem Network

ETR Estimated Time of Restoral, Estimated Time to Restore

ETS Engineering Test Satellite (Japan), Electronic Translator System, Electronic Tandem Switching

et seq Et Sequentia, and the following

ETV Educational Television

ETX End of Text or message

EUCATEL European Committee of Associations of Telecommunications Industries

EUREKA European Research Coordinating Agency

EUROSTAR Anglo-french communications satellite

EUTELSAT European Telecommunications Satellite Organization

EVA Extreme Value Engineering

EVX Electronic Voice Exchange

EW Electronic Warfare

EWL Exchange Work List

EWS Electronic Work Station

EWSD Family of digital switching offices developed by Siemens, West Germany

EXC Execute

Exch Exchange, central office

EXTHEO Extra-Theoretical

extn Extension, External,

F

f Frequency

F Farad, Final, Flat, Flash (precedence), Fahrenheit, Failures

FA Fuse Alarm

FAA Federal Aviation Administration (US)

fab Fabricate

fac Facilities

FACD Foreign Area Customer Dialing

FACS Facility Assignment Control System

FADS Force Administration Data Systems, Filtered Attitude Determination System

FAM Final Address Message

FAMOS Floating-gate Avalanche Metal Oxide Semi-conductor

FAP Facility Analysis Plan, Fault Analysis Process

FAS File Access Subsystem, Frame Alignment Signal

FASE Fundamentally Analyzable Simplified English (a program language)

FASS Ford Aerospace Satellite Services Corp.

FAST Fairchild Advanced Schottky TTL

FAT Foreign Area Toll, Foreign Area Translation

FAX Facsimile

FBD Full Business Day, Functional Block Diagram

FC Find Called party, Find Calling party, Fuse Chamber, Forecast Center station, Feature Control

FCC Federal Communications Commission

FCG False Cross or Ground

FCM Fault Control Module

FCP Flat Concurrent Prolog

FD Finished Dialing, Fiber Duct, Frequency Distance, Functional Description

F/D Focal length/Diameter (of parabolic antenna)

FDB Functional Description Block, Fahrenheit Dry Bulb

FDM Frequency Division Multiplex

FDMA Frequency Division Multiple Access

FDM/FM Frequency Division Multiplex/Frequency Modulation

FDN Foreign Directory Name

FD/PSK Frequency-Differential/Phase Shift Keyed system

FDR Frequency Dependent Rejection

FDX Full Duplex

FE Format Effector, Functional Entity

FEC Forward Error Correction

FED STD Federal Standards

FEMF Foreign Electromotive Force

FET Field Effect Transistor

FEX Foreign Exchange

FEXT Far End Cross Talk

FF Form Feed, Field Function

FFT Fast Fourier Transform

FG Framework Ground

FGA Feature Group A (ENFIA)

FGB Feature Group B (ENFIA)

FGC Feature Group C (ENFIA)

FGCS Fifth Generation Computer System

FGGE First GARP Global Experiment

FGMDSS Future Global Maritime Distress & Safety System

FGW Floor Ground Window

FH Frequency Hopping

FHD Fixed Head Disk

FIB Forward Indicator Bit

FIEJ International federation of newspaper publishers

FIFO First In, First Out

fig Figure

FIGS Figures Shift (teleprinter)

fil Filament

FIPS Federal Information Processing Standards

FIR Flight Information Region

FIS Flight Information Service, Functional Interface Specification

FIT Failure In Time, Functional Integration Technology

FITE Forward Interworking Telephony Event

FIU Facilities Interface Unit

FL Foreign Listing

FLINK Flash/wink signal

FLIR Forward Looking Infra-Red

FLORICO Submarine cable Florida-Puerto Rico

FLS Free Line Signal

FLTSATCOM Fleet Satellite Communications System, U.S. Navy

fluor Fluorescent

fm, FM Frequency Modulation, Fault Monitor

FMC Fixed Message Cycle

FMEA Failure Mode and Effect Analysis

FMFB Frequency Modulation Feedback

FMR Frequency Modulation Receiver

FMS Facsimile Mail System, Flexible Manufacturing Systems

FMT Frequency Modulation Transmitter

FMTP File Management Transaction Processor (Bank of America)

FNPA Foreign Numbering Plan Area

FO Flash Override (precedence), Fiber Optics

FOC Fiber Optics Communications

FOM Factor Of Merit, Figure Of Merit

FOOS Force Out Of Service

For Ex Foreign Exchange

FORTRAN Formula Translation (computer language)

FOS Fiberoptic and Optoelectronics Scheme

FOT Optimum traffic frequency, working frequency, Forward Transfer

FOTS Fiber Optics Transmission System

FOV Field Of View

FOX The quick brown fox jumps over the lazy dog (test message using all letters of the alphabet)

FPC Functional Progression Chart

FPIS Forward Propagation Ionospheric Scatter

FPLA Field Programmable Logic Array

FPLF Field Programmable Logic Family

fpm Feet Per Minute

FPMH Failures Per Million Hours

FPR Flat Plate Radiometer

FPTS Forward Propagation Tropospheric Scatter

FPY Failures Per Year

FR Force Release, Frame Reset

FRC Final Routing Center

freq Frequency

freq-mult Frequency Multiplier

FRM Fault Reporting Module

FRP Fault Report Point

FRS Forward Ready Signal

FRT Front

FRXD Fully automatic reperforator transmitter distributor

FS Frequency Shift, message to Follow Sender, Final Splice (in cable), Functional Specification, File Separator, Fast Store

F/S Fetch & Send

FSB Functional Specification Block

FSF Fading Safety Factor

FSK Frequency Shift Keying

FSN Federal Stock Number, Forward Sequence Number

FSP Frequency Shift Pulsing, Fault Servicing Process

FSU Final Signal Unit, Field Support Unit

ft Foot, Feet

FT Forward Transfer, Functional Test

FTA Field To Advise

FTAM File Transfer, Access and Management

ft lb Foot-pound(s)

FTS Federal Telecommunications System

FVW Forward Volume Wave

FWB Fahrenheit Wet Bulb

fwd Forward, Four Wheel Drive

FWS Filter Wedge Spectrometer

FX Foreign exchange

FXC FerroxCube

FX-CCSA Foreign Exchange-Common Control Switching Arrangement

FY Fiscal Year

FYI For Your Information

G

g, gm Gram

G Electrical conductance, Giga- , Gravity

GaAs Gallium Arsenide

gal Gallon

GAPP Geometric-Arithmetic Parallel Processor

GARP Global Atmospheric Research Program

GAS Special autonomous study group

GATT General Agreement on Tariffs and Trade

GBH Group Busy Hour

GC Group Connector, Government Communications

GCE General Certificate of Education

GCF Generation Control Function

GCL Generic Control Language

GDF Group Distribution Frame

GDP Gross Domestic Product

GDPS Global Data Processing System (World Meteorological Organization)

GDSU Global Digital Service Unit

GDX Gated Diode Crosspoint

GE Generic Element, General Electric Corp. (USA)

GEC General Electric Company, plc (Great Britain)

gen Generator

GF Ground Field-earth interface

GFE Government Furnished Equipment

GGG Gadolinium Gallium Garnet

GH/LCD Guest-Host/Liquid Crystal Display

GHz Gigahertz

GI Galvanized Iron, Government Issue, Generic Identifier

GIGO   Garbage In, Garbage Out

GINO   Graphics Input-Output package (CAD)

GL   Ground Line

GME   Generic Macro Expander

GMS   Geostationary Meteorological Satellite, (Japan)

GMT   Greenwich Mean Time

GN   Green

gnd   Ground, earth

GNP   Gross National Product

GNT   Great Northern Telegraph Co. (Denmark)

GOAM   Government Owned And Maintained

GOC   Greatest Overall Coefficient

GOES   Geosynchronous Operational Environmental Satellite

GOS   Global Observing System (World Meteorological Organization), Grade Of Service

GPD   General Purpose Discipline (first IBM data link control)

gpDm   Geopotential Decameter

GPF   General Planning Forecast

gph   Gallons Per Hour

GPL   Group Processing Logic

gpm   Gallons Per Minute

GPO   General Post Office

GPS   Generic Processing System

GPSS   General Purpose Simulation System

grp   Group

GRP   Group Reference Pilot

grt   Gross Registered Tons

GS   Galvanized Steel, Group Separator, Group Selector

GSA   General Services Administration (US)

GSC   Group Switching Center (UK)

GSD   Generic Structure Diagram

GSS   Galvanized Steel Strand

G/T   Gain/Temperature

GT   Gopher Tape armor

GTA   Grading Terminal Assembly, Guam Telephone Authority

GTE   General Telephone & Electronics Corp., Group Translating Equipment

GTEP   General Telephone & Electronics Practice

GTO   Gate Turn-Off switch, Government Telecommunications Organization

GTS   Global Telecommunications System (World Meteorological Organization)

GTU   Group Terminal Unit

GU   Generic Unit

GUARDSMAN   Guidelines And Rules for Data Systems Management

GVHRR   Geosynchronous Very High Resolution Radiometer

GWEN   Ground Wave Energy Network

H

h, H   Henry (measure of inductance)

HAD   Half Amplitude Duration

HAL   High Activity Locations

HAN/LCD   Hybrid Assigned Nematic/Liquid Crystal Display

HASP   Houston Automatic Spooling Priority

HAT   Home Area Toll

HAW 1, 2, 3, 4   Submarine cables continental U.S.-Hawaii

HBT   Heterojunction Bipolar Transistor

HC   House Cable

HCMTS   High Capacity Mobile Telecommunications System

HCP   Hard Copy Printer

HCSDS   High Capacity Satellite Digital Service

HCTDS   High Capacity Terrestrial Digital Service

HDB   High Density Bipolar code, High Density Binary

HDB3   High Density Binary 3 level signal

HDBH   High Day Busy Hour

HDF   Horizontal Distributing Frame

HDLC   High level Data Link Control

HDRSS   High Data Rate Storage System

HDS   Head Set

HDX   Half Duplex

HE   Housekeeping Element

HECI   Human-interface Equipment Catalog Item

HEMT   High Electron Mobility Transistor

HEOS   Highly Eccentric Orbit Satellite

HERF   Hazards of Electromagnetic Radiation to Fuel

HERO   Hazards of Electromagnetic Radiation to Ordnance

HERP   Hazards of Electromagnetic Radiation to Personnel

hex   Hexagon, Hexagonal

HF   High Frequency band

HFC   High Frequency Correction

HFDF   High Frequency Distribution Frame

HH   Hanging Handset

HIA   Human Interface Architecture

HIC   Hybrid Integrated Circuit

HIDF   Horizontal side of an Intermediate Distribution Frame

HiD/LoD   High Density/Low Density tariff

HIGHVISION   Japanese high-definition television system

HIMAIL   Hitachi Integrated Message and Information Library

HIN   Hybrid Integrated Network

HIRS   High resolution Infra-Red Sounder

HL   Hot Line, Half Life (radioactive)

HLL   High Level Language

HLSC   High-Level Service Circuit

HLT   Heterodyne Look-Thru

HMDF   Horizontal side of Main Distribution Frame

HMG   Hardware Message Generator

HN   Host to Network

HNPA   Home Numbering Plan Area

Ho   Hotel

HOBIS   Hotel Billing Information System

HOP   House Operating Tape

HP   Horse Power, High Pass, Hewlett-Packard Corp.

HPBW   Half-Power Beamwidth

HPI   Height Position Indicator

Hq   Headquarters, head office

hr   Hour, Hard to Reach

HRC   Hypothetical Reference Circuit

HRFAX   High Resolution Facsimile

HRIR   High Resolution Infra-Red scanning system

HRT   High Rate Telemetry

HRX   Hypothetical Reference Model (ISDN)

HSCP   High Speed Card Punch

HSCT   High Speed Compound Terminal

HSPTR   High Speed Paper Tape Reader

HSR   Hardware Status Register

HSSDS   High Speed Switched Digital Service

HT   Horizontal Tabulation, Holding Time, High Tension, Height

HTL   Hotel call, time and charges mandatory

HU   High Usage, Hangup

HUD   Head-Up Display

HV   High Voltage

HVAC   Heating, Ventilation and Air Conditioning

HW   Handset, Wall model

H/W   Hardware

HWY   Highway

Hyb   Hybrid

Hz   Hertz

I

i   Instantaneous value of current

I   Current, Immediate (precedence)

IAAB   Inter-American Association of Broadcasting

IAC   International Accounting Center

IACB   Inter-Agency Consultative Board

IAD   Inventory Available Date

IAEA   International Atomic Energy Agency

IAF   International Astronautical Federation, International Aeronautical Federation

IAGA   International Assn. of Geomagnetism and Aeronomy

IAGC   Instantaneous Automatic Gain Control

IAI   Initial Address Information

IALA   International Assn. of Lighthouse Authorities

IAM   Initial Address Message

IAMAP   International Assn. of Meteorology and Atmospheric Physics

IAN   Integrated Analog Network

IARU   International Amateur Radio Union

IAT   International Atomic Time

IATA   International Air Transport Assn.

IATE   International Accounting and Traffic analysis Equipment

IAU   International Astronomical Union

IAW   In Accordance With

IB   Input Buffer, Information Bureau

IBA   Independent Broadcasting Authority (UK)

IBEW   International Brotherhood of Electrical Workers

Ibid   Ibidem, in the same place

IBM   International Business Machine Corp.

IBM TSS   IBM's Timesharing System.

IBRD   International Bank for Reconstruction and Development (The World Bank)

IBTO   International Broadcasting and Television Organization

IC   Integrated Circuit, Interexchange Carrier

I/C   Incoming

I²C   Inter-integrated Circuits

ICA   International Communications Association, International Common Access, Information Content Architecture

ICAN   Individual Circuit Analysis

ICAO   International Civil Aviation Organization

ICB   Incoming Call Barred, Individual Case Basis

ICC   International Chamber of Commerce, International Conference on Communications, International Control Center

ICDR   Inward Call Detail Recording

ICECAN   Submarine cable, Iceland-Canada

ICI   International Commission on Illumination

ICJ   Incoming Junction

ICL   Inserted Connection Loss

ICOT   Institute of New Generation Computer Technology (Japan)

ICP   Incoming (message) Process

ICPO   International Criminal Police Organization (Interpol)

ICRP   International Commission on Radiological Protection

ICS   International Chamber of Shipping, Intercompany Settlement

ICSAB   International Civil Service Advisory Board

ICSC   Interim Communications Satellite Committee (now INTELSAT)

ICST   Institute for Computer Science and Technology

ICSU   International Council of Scientific Unions

ICU   International Communication Unit, Internal Communication Unit, Interface Connecting Unit

ICUP   Individual Circuit Usage and Peg count

ICW   Interrupted Carrier Wave, Interrupted Continuous Wave, In Connection With

Id   Idem, the same

I & D   Integrate and Dump detection

ID Inside Diameter

IDA International Development Assn., Integrated Digital Access

IDB Inter-American Development Bank

IDC Image Dissector Camera, In Due Course

IDCC International Data Communications Center

IDCMA Independent Data Communications Manufacturers Assn.

IDCP International Data Collecting Platform

IDCS Image Dissector Camera System

IDD International Direct Dialing

IDDD International Direct Distance Dialing

IDDF Intermediate Digital Distribution Frame

IDDS International Digital Data Service

IDF Intermediate Distribution Frame

IDN Integrated Digital Network

IDOC Internal Dynamic Overload Control

IDP Integrated Data Processing, Interdigit Pause

IDSCP Initial Defense Satellite Communications Project

IDT Interdigital Transducer

IE Interlocal calling

IEC International Electrotechnical Commission, Interexchange Carrier

IED Interactive Electronic Display

IEE Institution of Electrical Engineers (UK)

IEEE Institution of Electrical and Electronic Engineers

IERE Institute of Electronics and Radio Engineers (UK)

IES Incoming Echo Suppressor

IEV International Electrotechnical Vocabulary

IF Intermediate Frequency

I/F Interface

IFAC International Federation of Automatic Control

IFAM Initial - Final Address Message

IFAX International Facsimile Service

IFD International Federation for Documentation

IFF Intensity Fluctuation Factor, Identification Friend or Foe

IFIP International Federation for Information Processing

IFIPS International Federation of Information Processing Societies

IFL International Frequency List

IFMR Instantaneous Frequency Measurement Receiver

IFOV Instantaneous Field Of View

IFR Instrument Flight Rules

IFRB International Frequency Registration Board

IFS Investment Feasibility Studies, Ionospheric Forward Scatter, Interactive Flow Simulator

IFTC International Film and Television Council

IGES Initial Graphics Exchange Specification (CAM)

IGFET Insulated Gate Field-Effect Transistor

IGOSS Integrated Global Ocean Station System

IGU International Gas Union

IGY International Geophysical Year

IIA Information Interchange Architecture

IIP Implementation and Installation Plan

IIW International Institute of Welding

IKBS Intelligent Knowledge-Based Systems

I²L Integrated Injection Logic

IL Insertion Loss

ILD Injection Laser Diode

ILF Infra Low Frequency

ILS Instrument Landing System

ILTMS International Leased Telegraph Message Switching service

IM Interface Module

IMC International Maintenance Center, Interface Module Cabinet

IMCO Inter-Governmental Maritime Consultative Organization

IMIS Integrated Management Information System

IML Incoming Matching Loss, Intermediate Language

IMP Interface Message Processor

IMPATT Impact ionization Avalanche Transit Time

impreg Impregnated

IMS Information Management System (IBM)

IMT Intermediate Tape, Inter-Machine Trunk

IMTS Improved Mobile Telephone System

in Inch

INAG Ionospheric Network Advisory Group

INC InComing, InComing trunk

INCC International Network Controlling Center

ind Induction, Indicator, Indicated

IND International Number Dialing(-ed)

inf Information

INFONET Information Network (British data system)

INFOPAC Pacific Bell videotex system

INFOTEX Information via Telex (U.S. data system)

ING Integrated Ground

INMARSAT International Maritime Satellite system

INMC International Network Management Center

ins Insulate, Insulation

INS Information Network System (Japan)

INSTN Instruction

int Internal, International, Intermediate, Inter

INTELSAT International Telecommunications Satellite consortium

INTIM Interrupt & Timing (processor handling)

INTR Interrupt

INTUG International Telecommunications User Group

inv Inverse, Invoice, Inverter

INWATS Inward Wide Area Telephone Service

I/O   Input/Output

IOB   Inter-Organization Board for Information Systems and Related Activities, Input/Output Buffer

IOC   Intergovernmental Oceanographic Commission, Input/Output Controller, Initial Operating Capability, Inter-Office Communication, Integrated Optical Circuit

IOCOM   Submarine cable Malaysia-India

IOD   Input/Output Device, Identified Outward Dialing

IOP   In/Out Process

IOU(S)   Input/Output Utility (Subsystem)

IP   Inter-digital Pause, Input Processor

I/P   InPut, Irregular Input Process

IPA   Intermediate Power Amplifier

iph   Inches Per Hour

IPL   Initial Program Load

IPM   Impulses Per Minute, Inches Per Minute, Interpersonal Message

IPN   Instant Private Network

IPP   International Phototelegraph Position, Inter-Processor Process, Interface Package Process

IPR   In Pulse to Register

ips   Impulses Per Second, Inches Per Second

IPS   In Pulse to Sender, Ionospheric Prediction Service

IPSS   Inter-Processor Signaling System

IPTC   International Press Telecommunications Council

IPVC   Irradiated Polyvinyl Chloride

IPY   International Polar Year

IQSY   International Quiet Sun Year

IR   Infra-Red

IRAC   Interdepartmental Radio Advisory Committee

IRC   International Record Carrier, Intermediate Routing Center

IRE   Institute of Radio Engineers

IRIS   Interactive Recorded Information Service (UK)

IRG   Inter-Record Gap

IRLS   Interrogation, Recording and Location System

IROR   Internal Rate Of Return

IRP   International Routing Plan

IRS   Intermediate Reference System, Information Receiving Station, International Repeater Station

IRT   Interrupted Ring Tone

IRU   Indefeasible Right of User

IS   Information Separator, International Standard, Installation Start, In Service

ISAS   Institute of Space and Aeronautical Science (Japan)

ISB   Independent Sideband

ISC   International Switching Center (telephone), Intercompany Services Coordination, International Service Carrier

ISCC   International Service Coordination Center

ISC/USO   Inter-company Service Coordination/Universal Service Order

ISD   International Subscriber Dialing, Information System Development

ISDD   Integrated Systems Development Department

ISDN   Integrated Services Digital Network

ISDS   Integrated Switched Data Service

ISDT   Integrated Services Digital Terminal

ISDX   Integrated Services Digital Exchange

ISF   International Shipping Federation

ISG   Isolated Ground

ISIS   International Satellite for Ionospheric Studies

ISM   Industrial, Scientific and Medical applications

ISMC   International Switching Maintenance Center

ISMX   Integrated Subrate data Multiplexer

ISO   International organization for standardization

ISO-CMOS   Isolated fully recessed Complementary Metal Oxide Semiconductor

ISPABX   Integrated Services Private Automatic Branch Exchange

ISPC   International Sound-Program Center

ISR   Information Storage and Retrieval

ISS   Information Sending Station, Ionosphere Sounding Satellite (Japan), International Switching Symposium

ISSLS   International Symposium on Subscribers' Loops and Services

ISSN   Integrated Special Services Network

IST   Integrated Switching and Transmission, Integrated Services Telephone

ISTC   International Switching and Testing Center

ISTN   Integrated Switching and Transmission Network

ISU   Initial Signal Unit, Independent Signal Unit

ISUP   Integrated Services User Part

IT   Intelligent Terminal, Inter-Toll (trunk), Incomplete Translation, Information Technology

ITA   International Telegraph Alphabet, Independent Television Authority (UK)

ITB   Incoming Trunk Busy

ITC   International Television Center, International Teletraffic Congress, Technological institute for electronics and telecommunications (Colombia), Intermediate Toll Center, Intercept, Intercepted, Independent Telephone Company

ITCS   Integrated Thermionic Circuits

ITDM   Intelligent Time-Division Multiplexer

ITDN   Integrated Telephone & Data Network

ITE   International Telephone Exchange

ITFS   Instructional Television Fixed Service

ITI   Intermittent Trouble Indication

ITMC   International Transmission Maintenance Center

ITOS   Improved Tiros satellite

ITPA   Independent Telephone Pioneers Association

ITPC   International Television Program Center

ITPR   Infra-red Temperature Profile Radiometer

ITR Integrated Telephone Recorder

ITS Insertion Test Signal, Institute of Telecommunication Sciences (US), Inter Time Switch, Invitation To Send, Integrated Test System

ITSC International Telephone Services Center, International Telecommunications Services Complex

ITSU Information Technology Standards Unit

ITT International Telephone and Telegraph Corp.

ITU International Telecommunication Union

ITUSA Information Technology Users Association (UK)

ITV Instructional Television, Industrial Television

IUCAF Inter-Union Commission on Allocation of Frequencies for Radio Astronomy and Space Science

IUGG International Union of Geodesy and Geophysics

IUPAP International Union of Pure and Applied Physics

IUR International Union of Railways

IUWDS International Ursigram and World Days Service

IVDT Integrated Voice-Data Terminal

IW Inside Wire

IWCA Inside Wiring Cable

IWCS Integrated Wideband Communications System

IWU Isolation Working Unit

IXT Interaction CrossTalk

J

JAMS Job Activities Management System

JAMSAT Japanese Satellite for amateur radio use

JAN Joint Army-Navy (specification) (US)

JANAP Joint Army-Navy-Air Force Publications (US)

JARL Japan Amateur Radio League

JAS-1 Japanese amateur radio Satellite

JCENS Joint Communications-Electronics Nomenclature System

JCL Job Control Language

JCP(S) Junction Call Processing (Subsystem)

JCS Joint Chiefs of Staff (US)

JCSAN Joint Chiefs' of Staff Alerting Network (US)

jct Junction

JES Job Entry Subsystem (IBM)

JF Junction Frequency, Junctor Frame

JFET Junction Field-Effect Transistor

JFS Jumbo group Frequency Supply

JGF Junctor Grouping Frame

JISCOS Family of digital switches made by Jeumont-Schneider, France

JIU Joint Inspection Unit

JMOS Job Management Operations System

JMX Jumbo group Multiplex

JO Job Order

JOSS Joint Overseas Switchboard (military)

JP Jute Protection

JPCD Just Perceptible Color Difference

JSD Justification Service Digit

JSEP Joint Services Electronics Program

JSF Junctor Switch Frame

JSS Joint Surveillance System

Jt Joint

JTAC Joint Technical Advisory Committee

JTAM Job Transfer and Management

JTEC Japan Telecommunications Engineering & Consultancy

JTIDS Joint Tactical Information Distribution System

JU Joint User

JUGFET Junction field-effect transistor

JULIE Joint Utility Locating Information for Excavators

K

K one thousand, Kelvin, 2^{10} ($=1024$)

KAK Key-Auto-Key

KAU Keystation Adapter Unit

KBD Keyboard

Kbit/s Kilobits per Second

KCO Keep Cost Order

KCR Key Call Receiver

kcs Kilocycles per Second (kilohertz)

KDD Kokusai Denshin Denwa Co. Ltd (Japanese international carrier)

KDP Key Development Plan

KDS Keyboard Display Station

KE Key Equipment

kHz Kilohertz

kip Kilopound(s)

KIPS Thousand Instructions Per Second

KISS Keep it simple, Sam

km Kilometer

kmc Kilomegacycles (now Gigahertz)

kohm Kilohm

KP Key indicating start of Pulsing in MF signaling, Key Pulsing keys

KPA Key Pulse Adapters

KPF Key Pulse on Front cord

KSR Keyboard Send-Receive

KSU Key Service Unit

KTA Key Telephone Adapter (Rolm)

KTR Keyboard Typing Reperforator

KTS Key Telephone System

KTU Key Telephone Unit

kv Kilovolt(s)

kva Kilovolt-Amperes

kvar Kilovar, reactive kilovolt-ampere

kw Kilowatt
kwh Kilowatt-Hour
kybd Keyboard

L

L Inductance (in henrys), Lamp lead, Local
LA Light Armor, Listed Address
LADT Local Area Data Transport
LAMA Local Automatic Message Accounting
LAMC Language And Mode Converter
LAN Local Area Network
LAP Link Access Procedure, Line Access Point
LAPB Link Access Procedure (Balanced)
LARAM Line-Addressable Random Access Memory
LAS Low-Altitude observation Satellite
LASER Light Amplification by Stimulated Emission of Radiation
LASS Local Area Signaling Services
LAST Large Aperture Scanning Telescope
lat Latitude
LATAs Local Access and Transport Areas
LATIRN Low Altitude Navigation Targeting Infra Red, for night flying
LATIS Loop Activity Tracking Information System
lb Pound
LB Local Battery
LBN Line Balancing Network
LBO Line Build-Out unit
LBRV Low Bit-Rate Voice
LBS Load Balance System
LBT CBS Local-Battery Talking, Common-Battery Signaling
LC Loading Coil
LCAP Loop Carrier Analysis Program
LCC Loading Coil Case, Lead Covered Cable
LCD Liquid Crystal Display
LCM Line Control Module
LCP Local Control Point
LCR Least Cost Routing
LCSU Local Concentrator Switching Unit
LCU Link Control Unit
LD Long Distance, Loaded, Loop-Disconnect, Linker Directive, Laser Diode
LDD Logic Design Data
LDM Limited-Distance Modem
LDN Listed Directory Number
LDPE Low Density Polyethylene
LDR Line Driver-Receiver
LDS Local Digital Switch, Local Distribution System
LDSU Local Digital Service Unit
LDTP Long-Distance Thrift Pak
LDX Long-Distance Xerography
LE Loop Extender, Line Equipment, Local Exchange

LEC Light Energy Converter, Liquid Encapsulated Czochralski technique (lasers)
LECO Local Engineering Control Office
LED Light-Emitting Diode
LEF Left-in telephone
LEM Logical End of Media, Lunar Exploration Module
LEQ Line Equipped, Line of Equipment
LETB Local Exchange Test Bed
LETS Law Enforcement Teletypewriter Service
LEV Loader/Editor/Verifier
LF Low Frequency, Line Feed, Line Filter
LFD Line Fault Detector
LFRAP Long Feeder Route Analysis Program
LFSR Linear Feedback Shift Register
LI Left In place
LID Local Issue Data
LIDAR Light Detection And Ranging
LIDF Line Intermediate Distribution Frame
LIEF Launch Information Exchange Facility
LIFO Last In, First Out
LIM Line Interface Module
LIMS Limb Infra-red Measurements in the Stratosphere
LINCOMPEX Linked ComPressor and Expander
lin ft Linear Feet
LIOP Life In One Position
LIPS Logical Inferences per second
LIS Line Information Store
LISA Local Integrated Software Architecture
LISP List Programming (high level language)
LIST Listening
LIT Local Intelligent Terminal
LIU Line Interface Unit
LJSU Local Junction Switching Unit
LK Looking for party
LK ROUTE Looking for ROUTE
LL Land-Line, Line Leg (telegraph), Long Lines
LLA Low Level Access
LLAR Local Line Automatic Routining
LLE Long Line Equipment, Large Local Exchange
LLF Line Link Frame
LLLTV Low Level Light Television (or L³TV)
LLN Line Link Network (electronic switching system)
LLP Line Link Pulsing
LLSU Low Level Signaling Unit
LM Lunar Module, Leg Multiple (telegraph), Load Monitor
LME L. M. Ericsson (Swedish telecommunications company)
LMF Language Media Format
LMMS Local Message Metering Service
LMOS Loop Maintenance Operations System

LMR Land Mobile Radio

LMS Local Measured Service, Level Measuring Set, Land Mobile Satellite service

LMX L-type Multiplex

ln Natural logarithm

LNA Launch Numerical Aperture, Low Noise Amplifier, Local Numbering Area

LND Local Number Dialed(-ing)

LNR Low Noise Receiver

LNS Linked Numbering Scheme

LO Lock-Out, Local Oscillator, Line Occupancy

LOC Location, Local, Located, Linked Object Code

LOCAP Low Capacitance cable

loc cit Loco Citato, in the place cited

LOF Lowest Observable Frequency

log Logarithm to base 10

LOMAR Local Manual Attempt Recording

LONAL Local Off-Net Access Line

LOPS Lines Of Positions (Omega)

LORAN Long Range Navigational system

LOS Line Of Sight, Line Out of Service

LP Liquified Petroleum, Low Pass filter, Linearly Polarized Logperiodic antenna, Loop, Looping, Load Point, LodgePole pine (utility pole), Longitudinal Parity

LPC Linear Predictive Coding, Linear Predictive Coefficients

LPE Liquid Phase Epitaxy

LPF Low Pass Filter

LPI Longitudinally applied Paper Insulation

LPM Linearly Polarized Mode

LPTV Low Power Television

L/R Local/Remote

LRC Longitudinal Redundancy Check

LRCC Longitudinal Redundancy Check Character

LRF Long Range Facility

LRFAX Low Resolution Facsimile

LRFS Long Range Forecasting System

LRIR Limb Radiance Inversion Radiometer

LRU Least Replaceable Unit

LS Line Switch, Loading Splice, Language System

LSA Limited Space-charge Accumulation

L-sat European Space Agency communications satellite

LSB Lower Sideband, Least Significant Bit

LSCC Local Servicing Control Center

LSCP Low-Speed Card Punch

LSCU Local Servicing Control Unit

LSD Line Signal Detector, Line-Sharing Device

LSF Line Switch Frame (ESS)

LSI Large Scale Integration

LSM Line Selection Module

LSPK Loudspeaker

LSPS Local Service Planning System

LSPTR Low-Speed Paper Tape Reader

LSRP Local Switching Replacement Planning

LSS Local Synchronization Subsystem, Laboratory Support System

LSU Lone Signal Unit, Leading Signal Unit, Local Switching Unit, Local Synchronization Utility

LSV Line Status Verifier

L&T Line and Terminal

LT Link Terminal, Letter Telegram

LTAB Line Test Access Bus

LTB Last Trunk Busy

LTC Local Telephone Circuit, Line Traffic Coordinator, Line Terminating Circuit, Local Test Cabinet

LTD Local Test Desk

LTE Line Termination Equipment

LTM Live Traffic Model

LTP Line and Trunk Peripheral cabinet

LTRS Letters Shift

LU Line Unit

lub Lubricate

LUF Lowest Useful high Frequency

LUS Large Ultimate Size (> 6000 lines)

LVR Low Voltage Relay

LW Leave Word

LWA Light Wire Armored

M

m Meter, noon

M Million, Mutual inductance

ma Milliampere

MAB Metallic Access Bus, Metropolitan Area Business (line)

MAC Measurement & Analysis Center, Multiplexed analog component

mach Machine

MACINTOSH Type of personal computer

MAD Mixed Analog and Digital

MADW Military Air Defense Warning

mag Magnet, Magnetic, Magneto

MAGE Mechanical Aerospace Ground Equipment

MAMI Modified Alternate Mark Inversion

man Manual

MAN Metropolitan Area Network

MAP Minimum Acceptable Performance, Microelectronics Application Project (U.K.)

MAPS Measurement of Air Pollution from Satellites

MARECS Maritime Communication Satellite

MARISAT Maritime Satellite system

MAROTS Maritime Orbital Test Satellite

MARPAL Submarine cable Marseilles, France-Palo, Italy

MARS Military Affiliated Radio System

MARTEL Submarine cable Marseilles, France-Tel Aviv, Israel

MASA Main Store Arrays

MASC Main Store Controller

MASER Microwave Amplification by Stimulated Emission of Radiation

MAT Metropolitan Area Trunks

MATFAP Metropolitan Area Transmission Facility Analysis Program

math Mathematics

MATS Military Air Transport Service

MATV Master Antenna Television

MAU Media Access Unit

max Maximum

MAXIT Maximum Interference Threshold

MB Maintenance Busy

MBA Multiple Beam Antenna

MBDS Modular Building Distribution System

MBE Molecular Beam Epitaxy

MBS Multi-Block Synchronization signal unit

Mb/s Megabits per second (but M bit/s preferred internationally)

Mbit/s Megabits per second

Mc Megacycle

MC Module Control

MCC Maintenance Control Circuit, Master Control Center, Miscellaneous Common Carrier

MCCGL Message Conveying Computers General Licence (UK)

MCD Multiple Concrete Duct, Minimum Charge Duration

MCDU Thousands, hundreds, tens, units (subscriber's telephone number)

MCI Microwave Communications, Inc., Malicious Call Identification

MCL Mercury Communications Ltd.

MCM Thousand circular mils, Maintenance Control Module

MCP Main Call Process

MCPS Mini Core Processing Subsystem, Maintenance Control and Statistics Process

MCS Maintenance Control Subsystem, Maneuver Control System, Microinstruction Control Store

MCTRAP Mechanized Customer Trouble Report Analysis Plan

MCU Multicoupler Unit (antenna), Microprogram Control Unit

MCVD Modified Chemical Vapor Deposition

MCVFT Multichannel Voice Frequency Telegraphy

MCW Modulated Continuous Wave, Memory Card Writer

MD Multiple Dissemination, Management Domain

MDA Multiple Digit Absorbing, Mechanized Directory Assistance

MDF Main Distribution Frame

MDNS Managed Data Network Services

MDP Message Discrimination Process

MDS Multiple Dataset System, Minimum Discernible Signal, Microprocessor Development System, Multipoint Distribution Service

MDT Mean Down Time

MDUS Medium Data Utilization Station (Australia)

MDW Multiple Drop Wire

MDX Modular Digital Exchange

ME Message Element

meas Measured

MEBS Marketing, Engineering, and Business Services

mech Mechanical

med Medium

MED Molecular Electronic Device

MEDARABTEL Regional telecommunications network in Mediterranean and Middle East

meg Megger, Megohm

MEM Memory

MEP Management-Engineering Plan

MERT Multiple Environment Real Time

MESFET Metal, Semiconductor, Field Effect Transistor

MET Multi-Emitter Transistor

METEOSAT European Metereological observation satellite

mf Microfarad

MF Medium Frequency, Multifrequency

MFC Multifrequency signaling, Compelled

mfd Microfarad

mfd Manufactured

MFJ Modified Final Judgment

MFM MF Module

mfr Manufacturer, Multi-Frequency Receiver

MFSK Multiple Frequency Shift Keying, Multilevel Frequency-Shift Keying

MFT Mainframe Termination, Metallic Facility Terminal

mg Messenger, Master-Group

MGB Master Ground Bar

MGN Multi-Grounded Neutral

MGT Master-Group Translator

mh Millihenry(s)

MH Manhole

MHD Moving Head Disk

MHF Message Handling Facilities

MHP Message-Handling Processor

MHS Matra Harris Semiconductors, Message Handling System

MHz Megahertz

mi Mile

mic Microphone

MIC Microwave Integrated Circuit

MICR Magnetic Ink Character Recognition

MICS Maintenance Inventory Control System

MICU Message Interface and Clock Unit

MIFR Master International Frequency Register

mike Microphone

Mil Mileage, Military

MILS Missile Impact Locating System

MIM Message Input Module

MIMD Multiple Instruction stream/Multiple Data stream

min Minimum, Minute

MINET Medical Information Network (GTE Telenet)

MINIT Minimum Interference Threshold

MIOS Modular Input-Output System

MIPS Millions of Instructions Per Second

MIR Model Incident Report

MIS Management Integrated System, Management Information System

misc Miscellaneous

MISD Multiple-Instruction stream/Single Data stream

MISFET Metal, Insulator, Semiconductor, Field Effect Transistor

MISP Microelectronics Industry Support Program (U.K.)

MISR Multiple input signature register

MITER Modular Installation of Telecommunications Equipment Racks

MITI Japanese Ministry of International Trade & Industry

MIU Multistation Interface Unit

MJD Modified Julian Date

MKR Marker

MKS Meter, Kilogram Second system

ML Mean-Life, Message Length

MLAP Metallic Line Access Port

MLE Medium Local Exchange

MLLE Medium Large Local Exchange

MLPP MultiLevel Precedence and Preemption

MLR Mechanized Line Record

MLS Microwave Landing System

MLSE Maximum Likelihood Sequence Estimation

MLSO Mode-Locked Surface-Acousticwave Oscillator

MLT Mechanized Loop Test, Mechanized Line Testing

MLTS Microlevel Test Set

mm Millimeter

MMC Monolithic multicomponents ceramic

MMCS Mass Memory Control Subsystem

mmf Magnetomotive Force, Micromicrofarads

MMIC Monolithic Microwave Integrated Circuits

MMI(S) Man-Machine Interface (Subsystem)

MML Man-Machine Language

MMP Maintenance Message Process, Module Message Processor

MMPP Mechanized Market Programming Procedures

MMU Mass Memory Unit, Multi Message Unit, Metered Message Unit

MMW Millimeterwave Microwave

MN Manual, Main Network

MNC Multiplicative Noise Compensator

MNCS Multipoint Network Control System

MNOS Metal-Nitride-Oxide Semiconductor

MNRU Modulated Noise Reference Unit

MNSC Main Network Switching Center

MOCVD Metallorganic Chemical Vapor Deposition

mod Modulus, Modified, Modification

modem Modulator plus Demodulator

MOF Maximum Observed Frequency

MOJ Metering Over Junction, Material On Job date

MOL Manned Orbital Laboratory

MOM Message Output Module

mon Monitor, Monitoring

MOP Multiple Online Programming

MOPT Mean One way Propagation Time

MOS Metal Oxide Semiconductor

MOSFET Metal Oxide Semiconductor Field Effect Transistor

mot Motor

MOTIS Message Oriented Text Interchange System

MOU Memorandum Of Understanding

MP Multi-Purpose (trunk), Maintenance Process, Management Process, Module Processor

MPC Marker Pulse Conversion, Miniature Protector Connector

MPCC MultiProtocol Communications Controller

MPCS MultiParty Connection Subsystem

MPF Million Pair Feet

mph Miles Per Hour

MPL Multischedule Private Line

MPSK Multiple Phase Shift Keying

MPWD Machine-Prepared Wiring Data

MR Message Rate, Memory address Register

MRD Manual Ringdown

MRF Message Refusal

MRIR Medium Resolution Infra-red Radiometer, market research information system

MRP Message Routing Process

MS Mobile Service, Measured Service pricing, Mechanized Scheduling

MSA Message System Agent

MSB Most Significant Bit

MSBVW Magnostatic Backward Volume Wave

MSC Miles of Standard Cable, Message Switching Center, Message Switching Computer, Message Sequence Chart, Multistrip Coupler, Mobile Switching Center

MSCS Management Scheduling and Control System

ms, msec Millisecond

MSED Minimum Signal Element Duration

MSF Multiscan Function

MSFN   Manned Space Flight Network

MSFVW   Magnetostatic Forward Volume Wave

msg   Message

MSGS   Message Switch

MSI   Medium Scale Integration

MSK   Minimum phase Shift Keying, Minimum Shift Keying

msl   Mean Sea Level

MSO   Multiple System Operator

MSORS   Mechanized Sales Office Record System

MSP   Maintenance Support Plan, Modular Switching Peripheral

MSS   Multispectral Scanner, Management Statistics Subsystem

MSSS   Multisatellite Support System

MSSW   Magnetostatic Surface Wave

MSU   Message Switching Unit, Multiblock Synchronization signal Unit, Multiple Signal Unit, Management Signal Unit, Maintenance Signal Unit, Microwave Sounding Unit, Main Switching Unit, Maintenance & Status Unit, Main Store Update, Metallic Service Unit

MSW   Microwave spectrometer, Magnetostatic Waves

MSYNC   Master Synchronization

MT   Modified Tape armor, Measured Time, Message Transfer

MTA   Message Transfer Agent

MTAE   Message Transfer Agent Entity

MTB   Metallic Test Buses

MTBD   Mean Time Between Degradations

MTBF   Mean Time Between Failures

MTBO   Mean Time Between Outages

MTC   Main Trunk Circuit (World Meteorological Organization), Magnetic Tape Cartridge, Midwestern Telecommunication Conference

MTCE   Maintenance

mtd   Mounted

MTD   Multiple Tile Duct

MTF   Modulation Transfer Function

mtg   Mounting

MTG   Main Traffic Group

MTI   Moving target indicator

MTL   Message Transfer Layer

MTP   Message Transmission Part, Message Transfer Part

MTR   Magnetic Tape Recorder

MTS   Mobile Telephone Service, Main Trunk System, Message Telecommunications Service, Message Transmission Subsystem, Message Transfer System

MTSR   Mean Time to Service Restoral

MTS/WATS   Mobile Telephone Service-Wide Area Telephone Service

MTT   Magnetic Tape Terminal

MTTE   Magnetic Tape Terminal Equipment

MTTF   Mean Time To Failure

MTTR   Mean Time To Repair

MTU   Multi-Terminal Unit

MTWX   Mechanized Teletypewriter exchange

MTX   Mobile Telephone Exchange

M/U   Monitor Unit

MUF   Maximum Useable Frequency

MULDEM   Multiplex/Demultiplex

MULDEX   digital Multiplexer plus Demultiplexer

mult   Multiple

MUM   Multi Unit Message

MUPS   Mechanized Unit Property System

MUR   Radio relay message unit

MUSA   Multiple-Unit Steerable Antenna

MUSE   Monitor of Ultra-violet Solar Energy, Multiple subnyquist-sample encoding (HDTV)

MUT   Mean Up Time

MUX   Multiplex

MUXER   Multiplexer

mv   Millivolts

MV   Move, Moved

MVP   Multiline Variety Package

MVS   Multiple Virtual Storage

mw   Milliwatts

Mw   Megawatts

M/W   Microwave

MWARA   Major World Air Route Area

MWC/CS   Mechanized Wire Centering/Cross Section

MWI   Message Waiting Indication

MWS   Microwave Scatterometer

mx   Matrix

N

N   North, any number from 2 to 9

NA   No Access, Not Applicable, Night Alarm, Numerical Aperture

NAAWS   NORAD Automatic Attack Warning System

NAB   National Association of Broadcasters (US)

NABTS   North American Broadcast Teletext Specification

NACOM   National Communications System (Civil Defense) (US)

NACS   Northern Area Communications System

NAIC   National Astronomy & Ionosphere Center

NAK   Negative Acknowledgement

NAM   Network Access Method

NAMF   North American Multifrequency (signaling)

NAN   Network Access Node

NAND   Not And (Boolean algebra)

NAP   Network Access Pricing

NAPLPS   North American Presentation Level Protocol Syntax (videotex)

NARC   Non-Automatic Relay Center

NARUC National Association of Regulatory Utility Commissioners (US)

NAS Nominal Aggregate Signal

NASA National Aeronautics & Space Administration (US)

NASCOM NASA Communications Network

NASDA National Space Development Agency (Japan)

NATA North American Telecommunications Assn.

NAU Network Addressable Unit

naut Nautical

NAVSAT Satellite Navigation Corp.

NAVSTAR/GPS Navigational satellite global positioning system

NAWAS National Warning System

NB Narrow Band, Narrow Beam

NBC National Broadcasting Corporation (US)

NBFM Narrow Band Frequency Modulation

NBH Network Busy Hour

NBO Network Building Out

NBOC Network Building Out Capacitor

NBOR Network Building Out Resistor

NBS National Bureau of Standards, (US) New British Standard (wire gauge)

NBS/ICST National Bureau of Standards/Institute for Computer Sciences and Technology

NBSV Narrow Band Secure Voice

NBTR Narrow Band Tape Recorder

NBV Net Book Value

NC Normally Closed, No Circuit, Network Connect, Network Congestion, Numerical Control

NCAP Non-linear Circuit Analysis Program

NCC Network Control Center

NCCC Nebraska Consolidated Communications Corp.

NCD Negotiated Critical Dates

NCF National Communications Forum

NCI Office of New Concepts and Initiatives (US Air Force)

NCL Node Compatibility List

NCO Network Control Office

NCP Network Control Program, Network Control Point

NCR No Circuit available, circuit request left

NCR-DNA NCR Corp. Distributed Network Architecture

NCS No Checking Signal, National Communications System (USDOD)

NCT Night Closing Trunks, Network Control and Timing

NCTA National Cable Television Association

NCTE Network Channel Termination Equipment

NDB Non-Directional radiobeacon

NDES Normal Digital Echo Suppressor

NDRO Non-Destructive Readout

NDT Net Data Throughput

NE North-East

NEC National Electrical Code (US), Nippon Electric Co. (Japan)

NECA National Exchange Carrier Assn.

NECOS Network Coordinating Station

neg Negative

NEMA National Electrical Manufacturer's Association

NEMS Microwave spectrometer

NEP Noise Equivalent Power

NESC National Electrical Safety Code

NESS National Environmental Satellite Service

NETEC Network Technical Support Group

NEXT Near End Cross Talk

NF Not Found (telephone listing), Noise Factor

NFET N-channel Junction Field-Effect Transistor

Nfy Notify

Nfyd Notified

NGO Non-Governmental Organization

NHK Nippon Hoso Kyokai (Japan Broadcasting Corp.)

N/I Noise to Interference ratio

NIB Negative Impedance Booster

NIC Negative Impedance Converter, Nearly Instantaneous Companding

NICAM Near-Instantaneous Companded Audio Multiplex

NICSMA NATO Integrated Communications Systems Management Agency

NID Network Inward Dialing

NIMBUS Meteorological satellites

NIOD Network Inward and Outward Dialing

NIP Non-Impact Printer

NIU Network Interface Unit

NJS Noise Jammer Simulator

NL New Line, Non Listed (British)

NLR Noise Load Ratio

NLST Non Listed Name

NLT Not Less Than, Not Lower Than

nm Nautical Mile

NMC National Meteorological Center (WMO), Network Management Center

NMMW Near Millimeter Wave system

NMOS N-channel Metal Oxide Semiconductor

NMS Noise Measuring Set, Network Management Signal

NMT Not More Than, Nordic Mobile Telephone System

NND National Network Dialing

NNR National Number Routed

NNX General code for central offices where N = numbers 2-9, X = numbers 0-9

No Number

NO Normally Open

NOAA National Oceanic Atmospheric Administration

NOD Network Outward Dialing

NODAN Noise-Operated Device for Anti-Noise

NOGAD Noise-Operated Gain Adjusting Device

NOI Notice of Inquiry

nom Nominal

nomen Nomenclature, name

non seq Non Sequitur, it does not follow

NORAD North American Air Defense Command

NORGEN Network Operations Report Generator

NOSFER Nouveau Systeme Fondamental Pour la Determination des Equivalents de Reference (new master system for the determination of reference equivalents)

NOTAL Not sent to All addresses

NOTAM Notice to Airmen

NOTIS Network Operations Trouble Information System

NP Northern Pine (utility pole), No Print

NPA Numbering Plan Area

NPD Network Protection Device

NPL National Physical Laboratory (UK)

NPR Noise Power Ratio

NPRM Notice of Proposed Rulemaking

NQd Non-Quaded

Nr Number

NRC National Research Council (Canada), Non-Recurring Connection, Non-Recurring Change charge

NRI Net Radio Interface, Non-Recurring Installation charge

NRL Naval Research Laboratory

NRRI National Regulatory Research Institute

NRV Net Recovery Value

NRZ Non-Return to Zero

NRZL Non-Return to Zero Level

NRZ1 Non-Return to Zero change at logic 1

NSC Network Switching Center

nsec Nanosecond

NSEP National Security & Emergency Preparedness

NSF National Science Foundation

NSL Net Switching Loss

NSP Network Services Protocol (DEC)

NSPC National Sound-Program Center

NSS Network Synchronization Subsystem

NSSMS NATO Seasparrow Surface Missile System

NSTN Non-Standard Telephone Number

NT Network Termination

NTC National Television Center

NTCA National Telephone Cooperative Association

NTE Network Terminal Equipment

NTI Noise Transmission Impairment

NTIA National Telecommunications & Information Administration

NTP Network Terminating Point

NTPF Number of Terminals Per Failure

NTS Non-Traffic Sensitive

NTSC National Television Standards Committee

NTT, NTTPC Nippon Telegraph & Telephone Public Corp. (Japan)

NTTP Network Test & Termination Point

NTU Network Terminating Unit

NU Number Unobtainable

NUC Nailed-Up Connection

NUDETS National detonation detection and reporting system

NUI Network User Identification

nuit Night delivery

NUL Null

NUT Number Unobtainable Tone

NW North-West

NWC Net Weekly Circulation

NWM Network Management

NYPS National Yellow Pages Service (US)

NYPSC New York Public Service Commission

NZPO New Zealand Post Office

O

O Operational immediate (precedence)

O & M Operation and Maintenance

OA Operational Amplifier, Office Automation

OACSU Off-Air Call Set-Up (mobile telephone systems)

OAM Oscillator Activity Monitor

OAO Orbiting Astronomical Observatory

OAS Organization of American States

OAU Organization of African Unity

OB Outside Broadcast, Output Buffer

OBH Office Busy Hour

OBN Out-of-Band Noise, Office Balancing Network

OCAD Optical Character and Detect

OCB Outgoing Calls Barred

OCC Other Common Carriers (non-Bell)

OCD Office of Civil Defense

OCHC Operator Call Handling Center

OCL Overall Connection Loss

OCMS Optional Calling Measured Service

OCP Overload Control Process

OCR Optical Character Reader, Optical Character Recognition

OCS Overload Control Subsystem

OCTOPUS Control Data Corp. data network

OD Outside Diameter, Out-of-order, Oceanographic Datastation, Overload Detection

ODA Office Document Architecture

ODAS Ocean Data Acquisition Systems (UNESCO)

ODCP One-Digit Code Point

ODD Operator Distance Dialing

ODR Operator Data Register

OE or O/E Own Exchange

OEICS Opto-Electronic Integrated Circuits

OEM Original Equipment Manufacturer

OERS Organization of Senegal Riparian states

OES Outgoing Echo Suppressor

OF Overflow, "One of the Firm"

OFc Office

off Official

Off prem Off Premises

OFHC Oxygen-Free, High-conductivity Copper

oflw Overflow

OFNPS Outstate Facility Network Planning System

OFS Operational Fixed microwave Service

OFTEL Office of Director General of Telecommunications (U.K.)

o/g Outgoing

OGDD Outgoing/Delay Dial

OGID Outgoing/Immediate Dial

OGJ Outgoing Junction

OGO Orbiting Geophysical Observatory

OGP Outgoing (message) Process

OGR Outgoing Repeater

OGT Outgoing Trunk, Outgoing Toll

OGWS Outgoing/Wink Start

OH Overhead

OHS Off-Hook Service

OIA Office Information Architecture

OIRT International radio & television organization

OITT Outpulse Identifier Trunk Test frame

OJ Originating Junctor

OJT On the Job Training

OK Approved, all correct

OKITAI Submarine cable Okinawa-Japan-Taiwan

OL Other Line

OLCR On-Line Character Recognition

OLP On Line Processor

OLR Off-Line Recovery

OLSS On-Line Support Software

OLTE Optical Line Terminating Equipment

OLUD On Line Update

OLUHO Submarine cable Okinawa-Luzon, Philippines-Hong Kong

OLUM On-Line Update control Module

OLYMPUS European Space Agency Communications Satellite

OMB Office of Management and Budget (U.S.)

OMFS Office Master Frequency Supply

OML Outgoing Matching Loss

OMR Optical Mark and Read

ONAL Off-Network Access Line

ONAT Off-Network Access Trunk

ONC Ordinary National Certificate (UK)

ONGA Overseas Number Group Analysis

ONI, ONID Operator Number Identification

OOB Out Of Band

OOO Out Of Order

OOR Operator Override

OOS Operational Operating System, Out Of Service

OP Output, Operation, Output Processor, Operational Process, Office Processor

O/P Output

OPC Optional Calling plans

OPD Operational Programming Department, one-per-desk

OPERATORS Optimization Program for Economical Remote trunk Arrangement and TSPS Operator arrangements

opm Operations Per Minute

OPP Opposite, Opposed, Office of Plans & Policy (FCC)

OPQ Indication of a central office or exchange code (followed by MCDU for the subscriber's number)

opr Operate, Operated, Operator

OPS Off-Premise Station, Operator's Subsystem

OPX Off-Premise extension

OR Orange, Operational Research

O/R Originator/Recipient

Oracle British teletext system

ORAVAC Automatic switchboard routiner

ORD Ordinary (subscriber)

OREP Optical Repeater Equipment

Orig OriGinal, Originated

ORTS Optional Residential Telephone Service

OS Operating System, Operations System, Office System

OSA Office System Architecture

OSC Operator Services Complex, Oscillator, Complete Operational Software

OSCAR Orbiting Satellite Carrying Amateur Radio, Order Status Control And Reporting

osc-mult Oscillator-Multiplier

OSDS Operating System for Distributed Switching

OSEOS Operational Synchronous Earth Observatory Satellite

OSI Open Systems Interconnection

OSO Orbiting Solar Observatory, Origination Screening Office

OSP Outside Plant

OSPS Operator Services Position System

OSS Office of Space Science, Operation Support System

OSSU Operator Services Switching Unit

OST Originating Station Treatment (TSPS), Operating System Trap

OSTEST Operating System Test

OSTL Operating System Table Loader

OSTNPS Operator Services Traffic Network Planning System

OSV Ocean Station Vessel (ICAO or WMO)

OSWS Operating System Workstation

OT Overtime

OTC Originating Toll Center, Overseas Telecommunications Commission (Australia), Operating Telephone Company

OTDR Optical Time Domain Reflectometer

OTE Hellenic telecommunications organization (Greece)

OTH Over The Horizon (scatter)

OTLP Zero dBm Transmission Level Point

OTP Office of Telecommunications Policy (U.S.)

OTP-EPROM One-TimeProgrammable-Electrically Programmable Read-Only Memory

OTS Orbital Test Satellite, Own Time Switch (connection or call)

OTSS Off-The-Shelf System

OTTS Outgoing Trunk Testing System

OUTWATS Outgoing Wide-Area Telephone Service

OW Order Wire, Open Wire

OWF Optimum Working Frequency

P

P Person-to-person, Priority, Pole, Pentachlorophenol

PA Public Address, Power Amplifier, Program Application instructions, Permanently Associated, Process Allocator

PABX Private Automatic Branch exchange

PACUIT Computer Transmission Corp. packet switching system

PAD Packet Assembler/Disassembler

PAL Phase Alternate Line, Programmable Array Logic

PALAPA Communications satellite (Indonesia)

PAM Pulse Amplitude Modulation

PAMA Pulse Address Multiple Access

PAMS Preselected Alternate Master-Slave

PAN switchboard Panel

PANACEA Package for Analysis of Networks of Asynchronous Computers with Extended Asymptotics

PANAFTEL Regional telecommunications network, Africa

PANS Potentially Attractive New Services (often coupled with POTS)

PAR Peak to Average Ratio

paramp Parametric Amplifier

PATO Partial Acceptance and Takeover date

PATROL Program for Administrative Traffic Reports On-Line (now COER)

PATU Pan African Telecommunications Union

PAU Power and Alarm Unit

PAX Private Automatic exchange

PB Pushbutton

PBA Printed Board Assembly

PBC Program Booking Center, Peripheral Bus Computer

PBR Pole Broken

PBX Private Branch exchange

pc(s) Piece(s)

PC Peg Count, Port Control, Printed Circuit, confirmation of delivery date and time of telegram, Process Controller, Peripheral Controller, Private Circuit, Processing Center, Peripheral Control

PCA Protective Clothing Arrangement, Port Communications Area, Polar Cap Absorption, Protective Connecting Arrangement, Printed Circuit Assembly

PCB Printed Circuit Board, Port Check Bit, Public Coin Box, Process Control Block

PCC Processor Control Console

PCCM Private Circuit Control Module

PCD Port Control Diagnostic, Planned Completion Date

PCDDS Private Circuit Digital Data Service

PCF Peripheral Control Facility, Planning Cable Fill

pch Punch, Punched, Punching

PCHK Parity Check

PCI Panel Call Indicator, Pattern Correspondence Index

PCM Pulse Code Modulation, Port Command area, Process Control Module, Plug-Compatible Module/machine, Process Control Module

PCO Plant Control Office

PCP Port Call Processing, Packet Control Process

PCPS Private Carrier Paging System

PCR Pass Card Reader, Preventative Cyclic Retransmission

PCS Port Command Store, Port Control Store, Port Control System, Plastic Coated Silica (optic fiber)

pct Percentage, Pulse Count

PCU Power Control Unit, Paging Control Unit

PD Plane Disagreement, Peripheral Device, Power Distribution

PDB Process Descriptor Base

PDD Post Dialing Delay, PWB Design Data

pdf Probability Density Function

PDF Post-Detection Filter

PDM Pulse Duration Modulation, Pipework Design Management System (CAD)

PDP Programmable Digital Processor (Digital Equipment Co.), Plasma Display Panel

PDR Precision Depth Recorder

PDS Penultimate Digit Storage

Pe Polyethylene insulation

PE Pre-emption, Phase Encoding, Pictorial Element, Processing element

PEARL Performance Evaluation of Amplifiers from Remote Location

PEC Public Extension Circuit

PECC Product Engineering Control Center

PEGAD Permission Granted to Add

PENBAL Submarine cable Peninsular Spain-Balearic Islands

PENCAN Submarine cable Peninsular-Spain-Canary Islands

PEP Peak Envelope Power

perm Permanent

PERT Project Evaluation and Review Technique, Program Evaluation and Reporting Technique

PEV Peak Envelope Voltage

pf Picofarad, Power Factor

PF Power Frame

PFB Provisional Frequency Board

PFD Power Flux Density

PFEP Programmable Front-End Processor

PFM Pulse Frequency Modulation

PFS Primary Frequency Supply

PG Port Group, Program Generic, Permanent Glow

PGC Port Group Control

PGE Primary Ground Electrode

PGH Port Group Highway

PGHTS Port Group Highway Timeslot

PGI Port Group Interface

pH Relative acidity

PH Phantom circuit

PHILSIN Submarine cable-Philippines-Singapore (also ASEAN PS)

PHOTAC Photo Typesetting And Composing

PIB Processor Interface Buffer

PIC Polyethylene Insulated Cable, Polyethylene Insulated Conductor, Position Independent Code, Primary Inter-LATA Carrier

PICB Peripheral Interface Control Bus

PICS Plug-In Control System, Plug-in Inventory Control System

PICS/DCPR Plug-in Inventory Control System/Detailed Continuing Property Record

PID Port Identification

PIDB Peripheral Interface Data Bus

PIM Pulse Interval Modulation

PIN Personal Identification Number, P-type + I-type + N-type, Positive-Intrinsic-Negative (photodiode)

PIO Private I/O

PIPO Parallel-In Parallel-Out

PISA Public Interest Satellite Association

PISO Parallel-In Serial-Out

PITCOM Parliamentary Information Technology Committee (UK)

PIU Path Information Unit, Plug-In Unit

piv Peak Inverse Voltage

PL Program Logic, Pulse, Place, Private Line

PL/1 Programming Language One (IBM)

PLA Plain Language Address, Programmable Logic Array

PLAS Private Line Assured Service

PLL Phase Locked Loop

PLM Pulse Length Modulation

PLR Pulse Link Relay, Pulse Link Repeater

PLS Private Line Service

PLTTY Private Line Teletypewriter service

pm Post Meridiem, afternoon

PM Pulse Modulation, Phase Modulation

PMB Pilot Make Busy circuit

PMBX Private Manual Branch exchange

PMC Private Meter Check

PMD Point of Maximum Definition

PMM Pool Maintenance Module

PMOS P-channel Metal Oxide Semiconductor, Positive-channel Metal Oxide Semiconductor

PMR Pressure Modulated Radiometer, Private Mobile Radio

PMS Public Message Service (Western Union)

PMUX Programmable Multiplex

PMX Packet Multiplexer, Private Manual exchange

PN Pseudo Noise, Part Number

PNA Packet Network Adapter

PND Program Network Diagram

P-NID Precedence Network In-Dialing

pnl PaneL

PNM Pulse Number Modulation

PNX Private Network Exchange

PO Post Office, Postpay coin telephone, Part Of

POB Peripheral Order Buffer

POI Point of Interface

POL Problem Oriented Language, Polarized

POM Pool Operational Module

POP Point of Presence

POPS Program for Operator Scheduling

POPUS Post Office Processing Utility Subsystem

pos Positive, Position

pot Potential, Potentiometer

POTS Plain Old Telephone Service (see PANS)

POTUS President Of The United States

pp Push-Pull, Person to Person, Peak to Peak, Plane Polarized, person and operator

PPC Peak Power Control

PPCS Person to Person: Collect and Special instruction

PPI Plan Position Indicator

ppm Parts Per Million

PPM Periodic Pulse Metering, Pulse Position Modulation

pps Pulses Per Second

PPT Process Page Table

pr Pair

PR Premature Release

PRBS Pseudo-Random Binary-pulse Sequence

PRC Primary Routing Center

PRE Prepayment coin telephone, Protective Reservation Equipment

prep   Preparation, Prepare

PRESTEL   British videotex service

PRF   Pulse Repetition Frequency, Processor Request Flag

PRFD   Pulse Recurrence Frequency Discrimination

PRFU   Processor Ready for Use

pri, prim   Primary

PRM   Pulse Rate Modulation

PRN   Pseudo Random Noise

prob   Probability, Probable

prod   Product, Production

PROLOG   Programming in Logic (high level language)

PROM   Programmable Read-Only Memory

pro tem   Pro Tempore, temporarily

PRRM   Pulse Repetition Rate Modulation

PRTM   Printing Response Time Monitor

PRW   Paired Wire

PS   Program Store, Port Store, Port Strobe, Postscript, Permanent Signal, Process Subsystem, PVC Sleeve

p/s   Pulses per Second

PSA   Port Storage Area, Path Selection Algorithm

PSAP   Public Safety Answering Point

PSC   Public Service Commission, Plant Service Center

PSCM   Process Steering & Control Module

Psd   Passed

PSD   Permanent Signal Detection circuit

PSDC   Public Switched Digital Capability

PSDDS   Pilot or Public Switched Digital Data Service

PSDS   Packet Switched Data Service

PSE   Packet Switching Exchange, Programming Support Environment

PSF   Permanent Signal Finder, Provisional System Feature

psi   Pounds per Square Inch

PSI   Paid Service Indication, Planned Start Installation, Personal Sequential Interface

PSK   Phase Shift Keying

PSL   Power and Signal List, Process Simulation Language

PSN   Public Switched Network

PSS   Program Support System

PSSP   Phone Center Staffing and Sizing Program

PSTC   Public Switched Telephone Circuits

PSTN   Public Switched Telephone Network

PSU   Port Storage Utility, Path Set-Up, Primary Switching Unit, Power Supply Unit

PT   Paper Tape, Port number, Pay Tone

PTC   Pacific Telecommunications Conference

PTD   Plant Test Date

PTF   Programmable Transversal Filters

PTG   Precise Tone Generator

PTI   Party Identity

PTM   Pulse Time Modulation, Portable Traffic Monitor

PTN   Plant Test Number

PTO   Public Telecommunications Operator

PTR   Paper Tape Reader, Printer, Poor Transmission

PTS   Proceed To Send, Proceed To Select, Public Telephone System, Public Telephone Service

PTT   Post Telephone & Telegraph Administration, Party Test

Pty   Party

PU   Processor Utility

PUC   Public Utilities Commission, Peripheral Unit Controller

PUM   Processor Utility Monitor

PUS   Processor Utility Subsystem

PVC   Polyvinyl Chloride, Permanent Virtual Circuit

pW   Picowatt

PWAC   Present Worth of Annual Charges

PWB   Printed Wiring Board, Programmer's Workbench

PWBA   Printed Wiring Board Assembly

PWE   Present Worth Expenditures

PWFG   Primary Waveform Generator

PWM   Pulse Width Modulation

pWp   Picowatt: Psophometrically weighted

pwr   Power

PX   Private exchange

PXML   Private Exchange Master List

Q

Q   Quantity of electricity (in coulombs), quality of a resonant circuit or a capacitor

QA   Quick Acting, Quality Assurance

QAM   Quadrature Amplitude Modulation

QC   Quality Control

QCC   Quality Control Circle

QCP   Quick Connect Panel

qd   Quad

QDPSK   Quaternary Differential Phase Shift Keying

QED   Quod Erat Demonstrandum, that which was demonstrated

QEF   Quod Erat Faciendum, that which was constructed

Q factor   Quality factor

QFM   Quantized Frequency Modulation

QL   Queue Length

QOS   Quality of Service

QPM   Quantized Pulse Modulation

QPRS   Quadrature Partial-Response System

QPSK   Quadrature Phase Shift Keying

QSAM   Quadrature Sideband Amplitude Modulation

QSS   Quasi-stellar radio source: a quasar

QT   Queuing Time

QTAM   Queued Telecommunications Access Method (IBM)

qty   Quantity

quad   Quadruple

qual   Quality, Qualitative

QUAM   Quantized Amplitude Modulation

quan   Quantity

QUARK   Quantizer, Analyzer and Record Keeper

QUIL   Quad In Line

qv   Quod Vide, which see

R

R   Resistance, Ring lead, Routine (precedence), Ring, Remote

R1   Signaling system CCITT R1

R2   Signaling system CCITT R2

R/A   Recorded Announcement

RA   Ready-Access, Repeat Attempt

RAC   Rolm Analysis Center

RACE   Research and Development in Advanced Communications Technology for Europe

rad   Radians, Radial, Radius, Radio

RAD   Random Access Device, Records Arrival Date

RADA   Random Access Discrete Address

radar   Radio Detection And Ranging

RAM   Random Access Memory

RAMS   Random Access Measurement System, Random Access Memory Store

RAMSH   Reliability, Availablity, Maintainability, Safety and Human factors

RAO   Regional Accounting Office

RAPUD   Revenue Analysis from Parametric Usage Descriptions

RARC   Regional Administrative Radio Conference (of the ITU)

RAS   Records & Analysis Subsystem, Route Accounting Subsystem

RATT   Radio Teletypewriter

RAWIN   Radar/Wind, Radio/Wind

RB   Rollback

RBOC   Regional Bell Operating Company

RBRG   Concrete-encased Reinforcing Bar Ring Ground

RBT   Ringback Tone, Remote Batch Terminal

RBV   Return Beam Vidicon

RC   Resistance-Capacitance coupling (or network), Regional Center, Remote Control, Ringing Code, Rate Center, Recording Completing (trunk), Reference Clock, Reply Check

RCAC   Radio Corporation of America Communications, Remote Computer Access Communications service

RCAN   Recorded Announcement

RCC   Rescue Coordination Center, Radio Common Carrier

RCD   Route Control Digit, Receiver-Carrier Detector

RCF   Remote Call Forwarding

RCI   Routing Control Indicator

RCO   Receiver Cuts Out

RCP   Restoration Control Point, Reseau a Commutation par Paquet (French packet switched system)

RCT   Remote Control Terminal

RC/VP   Recent Change/Verify Position

rcvr   Receiver

RD   Ringdown

R & D   Research & Development

RDC   Remote Data Concentrator

RDD   Requisition Due Date

RDF   Radio Direction Finding

RDI   Route Digit Indicator

RDJTF   Rapid Deployment Joint Task Force

RDS   Radio Digital System

re   Regarding

RE   Reference Equivalent

REA   Rural Electrification Administration

reac   Reactive, Reactor

REBUD   Rehabilitation Budgeting program

rec   Receiver, Receive

rect   Rectangle, Rectifier

ref   Referred, Referenced

REG   Register, Range Extender with Gain

REGEN   Regenerator

reinf   Reinforce, Reinforcement

rem   Remove

RENAN   144 kbit/s digital subscriber's service, France

REP   Replication

reperf   Re-perforator

REPL   Replace

REPROM   Reprogrammable Prom

res   Residential, Residence

resis   Resistance

resp   Respectively

RETMA   Radio, Electronic and Television Manufacturers Association

REU   Rectifier Enclosure Unit

rf   Radio Frequency

RFA   Recurrent Fault Analysis

RFB   Reason For Backlog

RFC   Radio Frequency Choke coil

RFI   Radio Frequency Interference, Ready For Installation, Request for Information

RFO   Reason For Outage

RFP   Request for proposal

RFS   Ready For Service, Regional Frequency Supplies

RG   Release Guard, Ringing Generator, Ring Ground

RGB   Red-Green-Blue

RH   Receive Hub (telegraph), Relative Humidity

RHC Regional (Bell) Holding Company

RHCP Right-Hand Circular Polarization

rheo Rheostat

RI Routing Indicator, Repeat Indication

RIAA Recording Industry Assn. of America

RID Records Issue Date

RIM Receiver Intermodulation

RIP Retired In Place, Routing Information Process

RIS Retransmission Identity Signal, Recorded Information Service

RISC Reduced Instruction Set Computers

RISM Radio Interface Switch Module

RIT Rate of Information Transfer

RJE Remote Job Entry

RL Receive Leg, Return Loss, Reflection Loss

RLCU Reference Link Control Unit

RLG Release Guard

rls Release

RLS Remote Line Switch

RLSD Received Line Signal Detector

RLST Release Timer

RLT Remote Line Test

RLU Remote Line Unit

RM Rollback Module

R/M Read/Mostly

R&M Reliability & Maintenance

RMAS Remote Memory Administration System

RMATS-1 Remote Maintenance, Administration and Traffic System-1

RMC Regional Meteorological Center (WMO)

RMCS Remote Maintenance Control System

RMI Route Monitoring Information

rms Root Mean Square

RO Receive Only, Routine Order

ROB Remote Order Buffer

ROH Receiver Off-Hook

ROM Read-Only Memory

ROP Read/receive only printer

ROTL Remote Office Test Lines

ROTR Receive-Only Typing Re-perforation

ROTS Rotary Out Trunk Switch

ROTT Reorder Tone Trunks

ROW Right-of-Way companies

RP Reply Paid (telegram), Restoration Priority, Rollback Process

RPA Re-entrant Process Allocator

RPC Registered Protective Circuit

RPG Report Program Generator

rpm Revolutions Per Minute

RPOA Recognized Private Operating Agency

RPQ Request for Price Quotation

RPS Relative Performance Score, Revolutions Per Second

RPV Remote Piloted Vehicles

RR Relay Rack, Railroad, Release Record, Re-Route

RRD Requisition Received Date, Route/Route Destination

RRE Receive Reference Equivalent

RRG Relay Rack Ground

RRI Re-Route Inhibit

RRL Radio Research Laboratories (Japan)

RRO Responsible Reporting Office

RRRS Route Relief Requirements System

RRS Retransmission Request Signal

RRT Ring-Ring Trip

RRX Railroad Crossing

RS Random Splice, Record Separator, Recommended Standard (EIA), Route Switch(ing), Rate Step

R/S Relay Set

RSA Repair Service Attendant

RSB Repair Service Bureau

RSEU Remote Scanner and Encoder Unit

RSI Register Sender Inward

RSL Received Signal Level

RSM Remote Switching Module

RSO Register Sender Outward

RSP Restoration Priority, Replication Synchronization Process, Robot Support System (U.K.)

RSS Route Switching Subsystem, Remote Switching System

RST Reset

RSU Remote Switching Unit

RT Ring Trip, Route Treatment, Rate, Radiotelephone, Register Traffic, Response Time, Ringing Tone

R/T or RT Register Translator

RTA Remote Test Access, Remote Trunk Arrangement

RTC Real-Time Clock, Rural Telephone Coalition

RTE Route

RTG Radioisotope Thermoelectric Generator

RTH Regional Telecommunications Hub

RTL Resistor-Transistor Logic

RTLP Reference Transmission Level Point

RTNR Ringtone No Reply

RTOS Real Time Operating System

RTS Request To Send, Remote Test System, Return To Service

RTT Radio Teletypewriter

RTTY Radio Teletypewriter

RTU Remote test, Right To Use

RU Request/response Unit, are you?

RURAX Rural Automatic exchange

R/W Rights of Way, Read/Write

RWI Radio and Wire Integration

Rx Receive

RX Through toll operator, Remote exchange

RZ Return to Zero

S

s second

S South, Signal, Salt (pole preservative), Switchboard

SA Service Assistant, Slow Acting, Single Armor, Storage Allocator

SABRE Store Access Bus Recording Equipment

SAC Special Area Code, Serving Area Concept

SAD Space Antennae Diversity

SAFE Store And Forward Element

SAFFI Special Assembly For Fast Installations

SAGE Stratospheric Aerosol and Gas Experiment

SALT Subscribers' Apparatus Line Tester

SAM Stratospheric Aerosol Measurement, Subsequent Address Message, Service Attitude Measurement, Surface-Air Missile

SAMS Stratospheric And Mesospheric Sounder

SAMSARS Satellite-based Maritime Search And Rescue System

SAMT State-of-the-Art Medium Terminal (U.S. Army radio equipment)

SAN Small Area Network

SAR Synthetic Aperture Radar, Successive Approximation Register, Street Address Record, Service Analysis Request, Service Analysis Report, Store Address Register

SARTS Switched Access Remote Test System

SAS Switched Access System, Small Astronomical Satellite, Sequenced Answer Signal

SAT Stepped Atomic Time

SAT 1 Submarine cable Portugal-South Africa

SATT Strowger Automatic Toll Ticketing

SAW Surface Acoustic Wave

SAWO Surface Acoustic Wave Oscillator

SAX Small Automatic exchange

SB Supply Bulletin, Simultaneous Broadcast, Signaling Battery, Special Billing

SBH Strip-Buried Heterostructure

SBS Satellite Business Systems

SBUV/TOMS Solar Backscatter UltraViolet/Total ozone Mapper System

SBX Small Business Exchange

SC Sending Complete, Service Code, Source Code

SCA Subsidiary Carrier Authorization

SCALD Structured Computer-Aided Logic Design

SCAMP State-of-the-art Computer Assisted Machine-tool Project

SCAMS Scanning Microwave Spectrometer

SCAN Switched Circuit Automatic Network

SCAR Scientific Committee on Antarctic Research

SCATHA Spacecraft Charging At High Altitude

SCATS Sequentially Controlled Automatic Transmitter Start

SCC Single Cotton-Covered, Specialized Common Carrier, Satellite Communications Controller, Switching Control Center, Signal(ing) Conversion Circuit, Space Consultative Committee (UK)

SCCF Satellite Communication Control Facility

SCCS Switching Control Center System

SCE Signal Conversion Electronics

SCFD Standard Cubic Feet per Day (gas usage)

SCFM Subcarrier Frequency Modulation

SCH Seizures per Circuit per Hour

sched Schedules, Scheduled

schem Schematic

SCIM Speech Communications Index Meter

SCLog Security Log

SCM Single-Channel Modem, Subscribers' Concentration Module, Supervision Control Module

SCN Self-Compensating Network

SCNA Sudden Cosmic Noise Absorption

SCOM Site Cutover Manager

SCOR Scientific Committee on Ocean Research

SCOSTEP Special Committee On Solar-Terrestrial Physics

SCOTICE Submarine cable Scotland-Iceland

SCOTS Surveillance and Control Of Transmission Systems

SCP Stromberg-Carlson Practices

SCPC Single Channel Per Carrier

SCP(S) Subscribers' Call Processing (Subsystem)

SCR Silicon Controlled Rectifier, System Change Request, Selective Chopper Radiometer, Signal Conversion Relay

SCRT Subscribers' Circuit Routine Tester

SCS Satellite Control Satellite

SCT Subscriber Carrier Terminal

SCU System Control signal Unit, Subscribers' Concentrator Unit, Servicing Control Unit

SD Schematic Drawing, Send Digits

S/D Signal/Distortion ratio (for analog/digital converters)

SDC Switched Digital Capability

SDD Speech Direction Detector

SDE Submission and Delivery Entity

SDF Supergroup Distribution Frame, Satellite Distribution Frame, Software Development Facilities

SDI Strategic Defense Initiative

SDL Specification & Description Language

SDLC Synchronous Data Link Control

SDM Space-Division Multiplex

SDN Subscriber's Directory Number, Synchronized Digital Network, Software Defined Network

SDNS Software Defined Network Service

SDP Signal Dispatch Point

SDS Software Development System

SDT Serial Data Transfer

SDTR Serial Data Transmitter/Receiver

SE Southeast, Service Expansion

SEA Sudden Enhancement of Atmospherics

SEACOM Submarine cable Singapore-Malaysia-Hong Kong-Papua New Guinea-Australia

sec Secant, Secondary, Second(s)

SEC Secondary Emission Control, Secondary, Switching Equipment Congestion

SECAM Sequential Couleur A Memoire (French color TV system)

sech Hyperbolic secant

SECO Station Engineering Control Office

SECORD Secure voice Cord board

SECT Sectional Center (Class 2)

SED Status Entry Devices

SEDM Status Entry Device Multiplexer

SEE Systems Equipment Engineer

SEL Selector, Space Environment Laboratory

SELCAL Selective Calling

SEM Space Environmental Monitor, Station Engineering Manual

SEMS Solar Environment Monitor Subsystem

SEOS Sychronous Earth Observatory Satellite

SES Service Evaluation System, Special Exchange Service

SETA Southeastern Telecommunications Assn.

SETAB Working standard with subscriber's equipment

SETED Working standard with an electrodynamic microphone and receiver

SETI Search for Extra-Terrestrial Intelligence

SF Single Frequency, Straightforward

SFERT Master telephone transmission reference system

SFS Software Facilities & Standards

SFU Status Fill-in Unit

SG SuperGroup, submarine cable transmission system, Signaling Ground, Study Group (e.g., of CCITT)

SGC Supergroup Connector

SGD Signaling Ground

SGDF Supergroup Distribution Frame

SGF Supplementary Ground Field

sgl Single

sgp Supergroup

SGX Selector Group matrix

SH Send Hub (telegraphy), Switch Handler

SHEFA Submarine cable Shetland Islands to Faero Islands

SHF Super High Frequency

sht Short

SHWY Super Highway

SI Special Instruction, Speech Interpolation, Shift-In, international system of physical units, Service Indicator, Supplementary Information

SIAF Service Indicator Associated Field

SIC Service Indicator Code

SID Sudden Ionospheric Disturbance, Scheduled Issue Date, SWIFT Interface Device

SIF Signaling Information Field

sig Signal, Signaling

SIGINT Signal Intelligence

SIL Speech Interference Level

SIM Service Instructions Message, Sequential Interface Machine

SIMD Single-Instruction stream/Multiple Data stream

sin Sine

SINAD Signal Noise And Distortion

SINCGARS Single Channel Ground/Air Radio System (frequency hopping)

sinh Hyperbolic sine

SIMP Satellite Information Message Protocol

SIO Serial Input/Output

SIPO Serial In, Parallel Out

SIRS Satellite Infra-Red Spectrometer

SIS Signaling Interworking Subsystem

SISD Single-Instruction-stream/Single Data-stream

SIT Static Induction Transistor

SITA Societe International de Telecommunications Aeronautique (cooperative providing airline communications)

SITC Satellite International Television Center

SIU Slide-In Unit

SL Slate, Subscriber's Loop, Send Leg (telegraphy)

slc Straight Line Capacitance

SLC Subscriber Loop Carrier, Submarine Laser Communications

SLD Straight Line Depreciation, Superluminescent Diode

SLE Small Local Exchange

SLIC Subscriber's Line Interface Circuit

slf Straight Line Frequency

SLM Subscriber Loop Multiplex

SLOTH Suppressing Line Operands and Translating to Hexadecimal

SLS Signaling Link Selection, Strained Layer Superlattice

SLT Switchman's Local Test

SLU Subscriber Line Usage

SLUS Subscriber's Line Use System

slv Sleeve

slw Straight Line Wavelength

SM Service Monitoring, Signaling Module, Surface Mounting

SMA Surface Mounted Assembly

SMAS Switched Maintenance Access System

SMASH Step-by-step Monitor And Selector Hold

SMATV Satellite Master Antenna Television

SMC Switch Maintenance Center

SMD Surface-Mounted Device

SM³/d Standard Cubic Meters per day (cable pressurization)

SMDR Station Message Detail Recording

SMETDS Standard Message Trunk Design System

SMF Software Maintenance Function

SMFA Simplified Modular Frame Assignment system

SMG Software Message Generator

SML Software Master Library

SMLE Small-Medium Local Exchange

SMMC System Maintenance Monitor Console

SMMR Scanning Multichannel Microwave Radiometer

SMPS Switch-Mode Power Supplies

SMPTE Society of Motion Picture and Television Engineers

SMRS Specialized Mobile Radio System

SMRT Single Message Rate Timing

SMS Synchronous Meteorological Satellite, Speech Mail System

SMSAs Standard Metropolitan Statistical Areas

SMU Super-Module Unit

SN Stock Number

S/N Signal to Noise ratio

SNA Systems Network Architecture (IBM)

SNAP Standard Network Access Protocol, Single Number Access Plan, Steerable Null Antenna Processor

SNAPS Standard Notes And Parts Selection

SNC Synchronous Network Clock

SNI Standard Network Interface

SNL Selected Nodes List

(S+N)/N Ratio of Signal plus Noise to Noise

SNOTEL Snow Pack Telemetry Radio

SNR Signal/Noise Ratio

SO Slow to Operate, Service Order, Shift Out, Send Only

SOA Start Of Address

SOF Satisfactory Operation Factor

SOFRECOM Societe Francaise d'Etudes et de Realisations d'Equipements de Telecommunications (French communications engineering company)

SOH Start Of Header, heading

SOI Silicon-On-Insulator

sol Soluble, Solenoid

soly Solubility

SOM Start Of Message indicator

SONAD Speech Operated Noise Adjusting Device

SOP Standard Operating Procedure

SOS Service Order System, Silicon-On-Sapphire, Speed Of Service

SOSE Silicon-On-Something-Else

SOST Special Operator Service Traffic

SOT Subscriber Originating Trunk, State Of Termination

SP Single Pole, Semi-Public, Singing Point, Signal Processor, Sewer Pipe, Southern Pine (utility pole), Supervisory Process

SPA Semi-Permanently Associated

SPADE Single channel per carrier, Pulse code modulation multiple Access Demand assignment Equipment

SPAG Standard Promotion Application Group (Europe)

SPC Stored Program Controlled, Single Point Connection

SPCS Synchronous Composite Packet Switching

SPDT Single-Pole, Double-Throw

SPDTDB Single-Pole, Double-Throw, Double-Break

SPDTNCDB Single-Pole, Double-Throw, Normally-Closed, Double-Break

SPDTNO Single-Pole, Double-Throw, Normally-Open

SPDTNODB Single-Pole, Double-Throw, Normally-Open, Double-Break

Spec Specification

SPEC Speech Preedictive Coding

SPESS Stored Program Electronic Switching System

SPF System Performance Factor, Subscriber Plant Factor

sp gr Specific Gravity

sph Spherical

SPI Station Program Identification

SPL Splice, Software Parts List

SPM Solar Proton Monitor, Subscriber's Private Meter

SPOOL Simultaneous Peripheral Operation On Line

SPPAY Semipost-Pay pay-station

SPPS Semipost-Pay Pay-Station

SPR Send Priority and Route digit, Spare

SPRINT Special Police Radio Inquiry Network, GTE-owned switched long distance service

SPS System Performance Score, Software Products Scheme (U.K.)

SPST Single-Pole, Single-Throw

SPSTNC Single-Pole, Single-Throw, Normally-Closed

SPSTNO Single-Pole, Single-Throw, Normally-Open

SPT System Page Table

SPUC/DL Serial Peripheral Unit Controller/Data Link

sq Square

SQD Signal Quality Detector

SR Slow Release, Supervisor, Scanning Radiometer

S/R Send/Receive

SRAEN Reference system for the determination of articulation reference equivalents

SRAM Static random access memory

SRATS Solar Radiation And Thermospheric Structure satellite (Japan)

SRC Science Research Council (UK)

SRDM Subrate Data Multiplexer

SRE Site Resident Engineer, Send Reference Equivalent, search radar, surveillance radar, Signaling Range Extender

SRF Surface Roughness Factor

SRL Stability Return Loss, Structural Return Loss, Singing Return Loss

SRS Synchronous Relay Satellite

SS Station to Station, Stainless Steel, Solid State, Spread Spectrum, Semifinal Splice, Special Service, Signaling System, Space Switch, Subscriber Switching

SSAC AC Signaling System

SSADM Structured Systems Analysis & Design Methodology

SSAS Station Signaling & Announcement Subsystem

SSB Single Sideband, Subscriber Busy

SSBAM Single Sideband Amplitude Modulation

SSBSC Single Sideband Suppressed Carrier

SSC Special Service Center, Single Silk-Covered, Sector Switching Center

SSCC Common Channel Signaling System

SSDC DC Signaling System

sse Single Silk covering over Enamel insulation

SSES Small Smart Electronic Switch

SSF Single Sided Frame

SSFC Sequential Single Frequency Code system

SSFS Special Services Forecasting System

SSI Start Signal Indicator, Small Scale Integration

SSL Software Slave Library

SSM Special Safeguarding Measures, Spread Spectrum Modulation, Surface-Surface Missile

SSMA Spread Spectrum Multiple Access

SSMB Special Services Management Bureau

SSMF MF Signaling System

SSN Switched Services Network

SSOG Satellite System Operational Guide

SSP Special Services Protection, Sub-Satellite Point

SSPS Satellite Solar Power Station

SSR Switching Selector Repeater, Senior Site Representative, Secondary Surveillance Radar

SSRA Spread Spectrum Random Access system

SSS Subscribers' Switching Subsystem

SSSP Station to Station Sent Paid (direct distance dialing)

SST Single Sideband Transmission, Supersonic Transport, Subscriber Transferred

SSTV Slow Scan Television

SSU Subsequent Signal Unit, Stratospheric Sounding Unit, Single Signaling Unit, Subscriber Switching Unit

SSW or SS Space Switch

SSWO Special Service Work Order

St Street, Statute

ST end of pulsing signal (Stop), Sidetone, Start, key pulsing Start key, Subscriber's Terminal

sta Station

STAD Start Address

STARTS Software tools for application to large real time systems

STATLIB Statistical Library

STC Station Technical Control, Serving Test Center, Standard Telephones and Cables, plc., Switching and Testing Center, Satellite Television Corp., Society of Telecommunications Consultants

std Standard

STD Subscriber Trunk Dialing (direct distance dialing)

STDN Spaceflight Tracking and Data Network

STE Span Terminating Equipment

STF Some Tests Failed

STL Standard Telegraph Level

STO System Test Objectives

STOL Short Take-Off and Landing

STP Signal Transfer Point, Space Technology Program

STPL Sidetone Path Loss

STPST Stop-Start

STR Sidetone Reduction, Synchronous Transmit/Receive (IBM)

STRATSAT Strategic Satellite system (U.S. Air Force)

STS Space-Time-Space, Shared tenants service

STSK Scandinavian committee for satellite communications

STU Subscribers' Trunk Unit, Signal Transmission Unit

STV Subscription Television

STX Start of Text of message

SU Signaling Unit

sub Subscriber, Substitute character

SUBRATE Digital rates within 64 kbit/s voice band, i.e., 2.4, 4.8, 9.6 and 56 kbit/s

substa Subscriber's Station, Substation

SUM Set-Up (control) Module

sup Supply

Supv, supy Supervisory, Supervisor

SUS Small Ultimate Size (<6000L), Single User System

svc Service

SVC Switched Virtual Circuit, Switched Virtual Call

SVD Simultaneous Voice and Data

SVI Service Interception

SVLog Servicing Log

sw Short Wave

SW Southwest, Switch

S/W Software

swbd Switchboard

SWD Self-Wiring Data

SWDR Single way dynamic range (for Rayleigh backscatter)

SWE Spherical Wave Expansion

SWF Short Wave Fadeout

SWFG Secondary Waveform Generator

SWG Standard Wire Gauge

SWI Special World Interval

SWIFT Society for Worldwide Interbank Financial Telecommunications (banking network)

SWR Standing Wave Ratio

SX Simplex

SXS Step-by-Step

SYN Synchronous idle

sync Synchronous, Synchronize

SYNTRAN Synchronous Transmission (at 45 Mbit/s)

sys System

sysgen System Generation

SYU Synchronization Signal Unit

SYU or SU Synchronization Unit or Utility

SZ Seizure

T

T Tip, tera

T1 DS1 Digital system (1.544 Mbit/s)

T1 DM T1 Data Multiplexer

T1 WB4 T1 Data/voice multiplexer

TA Tape Armored, Test Access, Transmission Authenticator, Terminal Adapter

TA-182 AN/TA-182 signal converter

tab Table, Tabulate

TAC Telenet Access Controller, Technical Assistance Center, Test Access Control

TACI Test Access Control Interface

TACL Telecommunications Analysis Center Library (Rolm)

TACS Total Access Communication System

TAD Thrust Augmented improved Delta, Test Access Digroup

TADIL Tactical Digital Information Link

TADS Tactical Automatic Digital Switching

TAGIDE Submarine cable France-Portugal

TAI Temps Atomique International (international atomic time)

TALTC Test Access Line Termination Circuit

TAM Test Access Multiplexer

tan Tangent, Tandem office

TANE Telephone Assn. of New England

tanh Hyperbolic tangent

TAP Test Access Path

TAR Technical Action Request

TARS Turn Around Ranging Station

TAS Telephone Answering Service, Test Access Selector, Telecommunications Authority Singapore

TASC Telecommunications Alarm Surveillance and Control

TASCC Test Access Signaling Conversion Circuit

TASI Time Assignment Speech Interpolation

TASMAN Submarine cable Australia-Tasmania

TASO Television Allocations Study Organization

TASP Toll Alternatives Studies Program

TASS Trouble Analysis System or Subsystem

TAT Transatlantic Telephone cable

TAT 1-8 Submarine cables North America-Europe

TAU Test Access Unit

tbl Trouble, fault

TBS To Be Specified, later

TBU Terminal Buffer Unit

T&C Time and Charges

TC Toll Center, Terminal Controller, T-Carrier, Terminal Congestion, Test Code

TCA Time of Closest Approach, Tele-Communication Assn., toll completing (trunk)

TCAM Telecommunications Access Method

TCAS T-Carrier Administration System, Traffic Alert and Collision Avoidance System (radar)

TCC Telecommunications Coordinating Committee, Transmission Control Character, Through-Connected Circuit

TCE Telephone Company Engineered

TCF Technical Control Facility

TCG Test Call Generator

TCL Toll Circuit Layout

TCLR Toll Circuit Layout Record

TCM Test Call Module, Time Compression Multiplex, Terminal Charge Management, Trellis Code Modulation

TCMF Touch Calling Multifrequency

TCMS Toll Centering and Metropolitan Sectoring, Telephone Cost Management System

TCN Telecommunications Cooperative Network

TCOS Trunk Class Of Service

TCR Transient Call Register

TCS Total Communications System

TCSP Tandem Cross Section Program

TCT Toll Connecting Trunks

TCTS Trans-Canada Telephone System

TCU Teletypewriter Control Unit, Transmission Control Unit

TCXO Temperature compensated crystal oscillator

TD Temporarily Disconnected, Test Distributor, Transmitter-Distributor, Terminal Digit(s)

TDA Tunnel Diode Amplifier, Toll Dial Assistance, Telecoms Dealers Association (UK)

TDAS Traffic Data Administration System

TDD Telecoms Device for the Deaf, Time Division Duplexing

TDF Trunk Distribution Frame

TDM Time Division Multiplex, Telephone Directory Memory

TDMA Time-Division Multiple Access

TDMS Transmission Distortion Measuring Set, Telegraph Distortion Measuring Set

TDM-VDMA Time Division Multiplex-Variable Destination Multiple Access

TDNS Total Data Network System

TD-PSK Time-Differential Phase Shift Keyed system

TDR Temporarily Disconnected at subscriber's Request, Time Domain Reflectometer, Terminal Digit(s) Requested

TDRE Tracking and Data Relay Experiment

TDRS Tracking and Data Relay Satellite

TDS Tracking and Data acquisition Station

TE Trunk Expansion, Transverse Electric wave

T/E Test and Evaluation

TEC Total Electron Content, Telephone Engineering Center

tech Technician, Technical

TED Trunk Encryption Device

TEGAS Time Generation And Simulation

tel Telephone

telco Telephone Company or administration

telecon Teletypewriter Conference

TELENET Telenet Communication Corp.

TELEPAK Telephone Package

TELESAT Canadian domestic satellite system

TELETEX Super-telex service

TELEX automatic Teletypewriter Exchange service

telg Telegram, Telegraph

Telidon Canadian videotex system

TELITA Videotex services, Malaysia

TELPAK AT&T's private line bulk rate tariff discontinued in 1981

TELPAL Submarine cable Tel Aviv, Israel-Palo, Italy

TELSAM Telephone Service Attitude Measurement

TEM Transverse Electromagnetic wave, Transverse Electric Mode

temp Temperature

TENS Transcutaneous Electronic Nerve Stimulators

TEO Telephone Equipment Order

term Terminal

TES Time Encoded Speech

TF Trunk Frame, Test Frame

TFC Traffic, Transmission Fault Control

TFE Tetrafluorethylene

TFMS Trunk and Facilities Maintenance System

TFR Theoretical Final Route

TFS Traffic Forecasting System, Traffic Flow Security, Trunk Forecasting System

TFT Thin Film Transistor

TFTP Television Facility Test Position

tg Telegraph

TG Trunk Group, Tone Generator

TGB Termination barred

TGC Transmitter Gain Control

TGF Through Group Filter

TGID Trunk Group Identification

TGM Trunk Group Multiplexer

TGN Trunk Group Number

therm Thermistor

THF Tremendously High Frequency

THIR Temperature/Humidity Infra-Red radiometer

THL Trans Hybrid Loss

THP Terminal Handler Process

THZ Teraherz

TIAG Telecommunications Industry Advisory Group

TIC-TOC Telecoms Information Centre-Telecoms Office for Consumers (UK)

TID Traveling Ionospheric Disturbance

TIDF Trunk Intermediate Distribution Frame

TIE Time Interval Error

TIF Telephone Influence Factor

TIFO Text Interchange Format

TIG Telegram Identification Group

TIGER Telephone Information Gathering for Evaluation and Review

TIM Transmitter Intermodulation, Token/Net Interface Module (packet switching), Speaking Clock Service

TIMS Transmission Impairment Measuring Set, Telephone Information Management System

TIP Terminal Interface Processor, Terminal Interface Package

TIPS Terminal Interface Processors

TIQ Task Input Queue

TIRKS Trunk Integrated Records Keeping System

TIROS Television Infra-Red Observation Satellite

TIS Terminal Interface Subsystem

TJR Trunk & Junction Routing(-er)

TK Trunk equipment, Test desk

TKO Trunk Offer

TL Tie Line, Transmission Level

TLF Trunk Link Frame

TLK Talking, Talk

TLM Trouble Locating Manual

TLP Transmission Level Point

TLR Toll Line Release key

TLTP Trunk Line Test Panel

TLU Terminal Logic Unit

TLWS Trunk and Line Work Station

TM Transverse Magnetic wave, Time Modulation, Technical Manual, Timing Module

TMA Telecommunications Managers Association

TMAX Maximum time

TMC Transmission Maintenance Center

TMIN Minimum time

TMF Trunk Maintenance Files

TML Tandem Matching Loss

TMn Traffic Mix n($= 1 - 4$)

TMRS Traffic Measuring and Recording System

TMS Transmission Measuring Set, Telecommunications Message Switch, Time-Multiplexed switch

TMT Testing Methods & Techniques

TMU Transmission Message Unit

TN Tone, Twisted Nematic (liquid crystal display)

TNA Telex Network Adapter

TNDS Total Network Data System

TNL Terminal Net Loss

TNOP Total Network Operations Plan

TNS Transaction Network Service

TO Technical Order, Traffic Order

TOC Television Operating Center, Table Of Coincidences, Task-Oriented Costing

TOCC Technical and Operational Control Center, Technical and Operational Coordination Center

TOD Time of day

TOF Tone Off

TOLD Telecoms On Line Data (system)

TOLR Toll Restricted

TON Tone On

TONLAR Tone-Operated Net Loss Adjuser

TOPES Telephone Office Planning and Engineering System

TOP HAT Wal 2 diphase line code

TOPP Task-Oriented Plant Practice

TOPS Traffic Operator Position System

TOR Telegraph On Radio

TORC automatic Traffic Overflow Reroute Control

TOS Taken Out of Service, Time Ordered System, Tiros Operational Satellite, Temporarily Out of Service

TOSD Telephone Operations and Standards Division (Rural Electrification Administration)

TOT Transfer Of Technology

TOVS Tiros Operational Vertical Sounder

TOW Tube-launched, Optically-tracked, Wire-guided (missile)

tp Telephone

TP Toll Prefix, Terminal Pole, Toll Point, Test Position

TPA Telephone Pioneers of America, Toll Pulse Accepter

tpd, tpdb Third party data base

TPF Time Prism Filter

TPI Transmission Performance Index

TPL Terminal Per Line, Toll Pole Line

TPO Telecommunications Program Objective

TPP Telephony Preprocessor

TPR Teleprinter

TPS Terminal Per Station

TQC Technical Quality Control

tr, trsp Transpose

TR Transmission Report, Traffic Route, Technical Reference

T/R Transmit/Receive, Tip/Ring

TRAMPS Traffic Measure and Path Search

TRAN Computer Transmission Corp. of California

TRANSCAN Submarine cables, Canary Islands

TRANSEC Transmission Security

TRANSMIC French digital leased lines service

TRANSPAC French packet switched network

TRANSPAC 1, 2, 3 Submarine cables Hawaii-Japan-Philippines

TRANSPLAN Transaction Network Service Planning Model

TRAVIS Traffic Retrieval Analysis Validation and Information System

TRCC "T" carrier Restoration Control Center

TRD Timed Release Disconnect

TRF Tuned Radio Frequency

trfr Transfer

TRG Tip/Ring to Ground

TRIF Technical Requirements Industry Forum

TRL Trunk Register Link, Transistor-Resistor Logic

TROPIC Family of digital switching centers developed in Brazil

TRPF Transmit/Receive Parity Failure

TRS Toll Room Switch

TRT Traffic Route Testing, Tropical Radio Telegraph Company

TS Toll Switching (trunk), Time Switch, Time Slot

TSAC Time Slot Assignment Circuit

TSAU Time Slot Access Unit

TSB Twin Sideband

TSC Transit Switching Center, Transmitter Start Code, Test System Controller

TSF Telegraphie Sans Fil (radio [French]), Through Supergroup Filter

TSFS Trunk Servicing Forecasting System

TSG Timeslot Generator

TSI Timeslot Interchanger

TSL Total Service Life

TSO Telecommunications Service Order, Time Sharing Option, Telephone Service Observations

TSORT Transmission System Optimum Relief Tool

TSP Traffic Service Position

TSPS Traffic Service Positions System

TSPSCAP Traffic Service Position System real-time Capacity program

TSR Telecommunications Service Request

TSS Toll Switching System, Trunk Servicing System

TSS-C Transmission Surveillance System — Cable

TSSST (TS'T) Time-Space-Space-Space-Time

TST Time-Space-Time

TSTA Transmission Signaling and Test Access

TSTE Texas Society of Telephone Engineers

TSTPAC Transmission and Signaling Test Plan and Analysis Concept

TSU Tape Search Unit, Tandem Signal Unit, Test Signal Unit, Telephone Signal Unit

TSW or TS Time Switch

TS16 Time Slot 16

TT Temporarily Transferred, Teletypewriter Terminal, Traffic Tester

TTA Traffic Trunk Administration

TTB Toll Testboard

TTC Terminating Toll Center, Teletypewriter Center

TT&C Tracking, Telemetry and Control

TTF Transmission Test Facility

TTFN Slang ending to service messages (Ta Ta For Now, i.e., goodbye)

TTL (T²L) Transistor-Transistor Logic

TT/N Test Tone to Noise ratio

TTO Traffic Trunk Order

TTP Trunk Test Panel

TTS Teletypesetter

TTY Teletypewriter

TTYS Teletypesetters

TU Traffic Unit (= Erlang), Transmission Unit, Tape Unit, Timing Unit

TUCC Triangle University Computing Center

TUG Tape Unit Group

TUMS Temporary Usage Measured Service

TUP Telephony User Part

TUPS Technical User Performance Specifications (USTA)

TUR Traffic Usage Recorder

TV Television

TW Transit Working

TWDD Two-Way/Delay Dial

TWEB Transcribed Weather Broadcast

TWERLE Tropical Wind, Energy conversion, Reference Level experiment

TWID Two-Way/Immediate Dial

2PPAPM Two-Pulse Amplitude and Phase Modulation Modem

TWT Traveling Wave Tube, Traffic Work Table

TWWS Two-Way/Wink Start

TWX Teletypewriter exchange service, Now called Telex II

Tx Transmit

TX Terminating toll operator

TXD Telephone exchange (Digital)

TXE Telephone exchange (Electronic)

TXK Telephone exchange (crossbar)

TXS Telephone exchange (Strowger)

TYMNET Timeshare, Inc. Network

typ Typical

TZ Transmitter Zone

U

U Unclassified, Utility, Unit

UA User Agent

UAMPT African and Malagasy Postal and Telecommunications Union

UART Universal Asynchronous Receiver/Transmitter

UAX Unit Automatic exchange

UBA Unblocking Acknowledge

UBL Unblocking

UCC Universal Conference Circuit

UCD Uniform Call Distributor

UCP Update Control Process

UD Not known if the party will be there today

UDTS Universal Data Transfer Service

UG Underground

UHF Ultra-High Frequency (300-3000 MHz)

UIC User Interface Circuit

UIT French for ITU

UKB Universal Keyboard

UKPO United Kingdom Post Office

UL Underwriters' Laboratories, Inc., Utility Lead

ULA Uncommitted Logic Array

ULSI Ultra Large Scale Integrated circuit

UMS Universal special service order

UN Unknown, Unassigned, United Nations

UNA Universal Night Answering

unbal Unbalanced

UNCTAD United Nations Conference on Trade And Development

UNDRO United Nations Disaster Relief Office

UNESCO United Nations Educational, Scientific and Cultural Organization

UNICCAP Universal Cable Circuit Analysis Program

UNIPEDE International union of producers and distributors of electrical energy

UNITAR United Nations Institute for Training And Research

UPS Uninterruptible Power Supply, Upper Performance Score

URS Update Report System

URSI International scientific radio union

URTNA Union of national radio and television organizations of Africa

US Unit Separator

USART Universal Synchronous-Asynchronous Receiver-Transmitter

USASCII United States of America Standard Code for Information Interchange

USB Upper Sideband

USC Universal Service Circuit

μsec Microsecond

USITA (now USTA)

USNO United States Naval Observatory

USO Universal Service Order

USOA Uniform System Of Accounts

USOAR Uniform System of Accounts Revision

USOC Uniform Service Order Code

USP Usage Sensitive Pricing

USTA United States Telephone Association

USTSA United States Telephone Suppliers Assn.

UT Universal Time

UTC Coordinated universal time, Utilities Telecommunications Council

UTD   Universal Tone Decoder
UTG   Universal Tone Generator
UTI   International universal time
UTS   Update Transaction System
μμf   Micromicrofarad
UTV   Uncompensated Temperature Variation
UV   Ultraviolet
UX   Not expected today and do not know when party will be there

V

V   Volts
VA   Volt-Amperes
VAC   Vacant, Vacuum, volts-alternating, Value Added Carrier
VAD   Vapor-phase Axial Deposition
VADIS   Voice And Data Integrated System
VAI   Video Assisted Instruction
VAN   Value Added Network
VANSGL   Value Added Network Services General Licence (UK)
VAR   Variable, Variant, Varley, reactive volt-amperes
varactor   Variable reactor
VARC   Variable Axis Rotor Control system (wind power)
VAS   Vissr Atmospheric Sounder, Value Added Service
VB   Voice Bank
VBL   Vertical Block Line
VC   Virtual Channel, Virtual Call
VCA   Voice Connecting Arrangement
VCO   Voice Controlled Oscillator
VCXO   Voltage Controlled Crystal Oscillator
vdc   Volts, Direct Current
VDE   Verband Deutsche Elektrotechniker (German technical association)
VDS   Video Distribution System
VDT   Visual Display Terminal
VDU   Video Display Unit
VDX   Videotex
VE   Value Engineering
VEN   Virtual Equipment Number
VER   Verified, Verify, Verifying operator
vert   Vertical
VF   Voice Frequency
VFA   Volunteer Fire Alarm
VFCT   Voice Frequency Carrier Telegraph
VFD   Vacuum Fluorescent Display
VFFT   Voice Frequency Facility Terminal
VFL   Voice Frequency Line
VFO   Variable Frequency Oscillator
VFR   Visual Flight Rules
VFT   Voice Frequency Telegraphy
VFTG   Voice Frequency Telegraph

VG   Voice Grade
VGI   Voice Group Interface
VHF   Very High Frequency (30-300 MHz)
VHPIC   Very High Performance Integrated Circuits
VHRR   Very High Resolution Radiometer
VHSIC   Very High Speed Integrated Circuit
VIAS   Voice Intelligibility Analysis Set
VID   Video
VIDEOTEX   Interactive computer-controlled system using TV and phone
VIDF   Vertical side of Intermediate Distribution Frame
VIEWDATA   British Telecom's form of Videotex, now called Prestel
Vin   Voltage Input
VIP   Versatile Information Processor, Visual Information Projection
VIR   Vertical Interference Reference signal
VISSR   Visible and Infrared Spin Scan Radiometer
VITS   Vertical Insertion Test Signal
VIU   Voice Interface Unit
viz   Videlicet, namely, that is
VLBI   Very Long Base Interferometer
VLF   Very Low Frequency
VLSI   Very Large Scale Integration
vm   Voltmeter
v/m   Volts per Meter
VM   Virtual Memory
VMC   Variable Message Cycle
VMDF   Vertical side of Main Distribution Frame
VMOS   Vertical Metal Oxide Semiconductor
VMS   Voice Messaging Service, Voice Mailbox Service
VMX   Voice Message Exchange
VN   Verify Number if no answer
VNL   Via Net Loss
VNLF   Via Net Loss Factor
VNN   Vacant National Number
VOCODER   Voice Operated Coder
VODAS   Voice Operated Device, Anti-Singing
VOGAD   Voice Operated Gain Adjusting Device
VOLCAS   Voice Operated Loss Control And echo/Singing suppression circuit
VOM   Volt-Ohm Meter
VOR   VHF Omnidirectional Radio range
vox   Voice operated transmission
VPT   Voice Plus Telegraph
VRAM   Variable Rate Adaptive Multiplexing
VRC   Vertical Redundancy Check
VRI   Varistor
VS   Virtual Storage, Virtual System, Versus
VSAT   Very Small Aperture Terminals
VSB   Vestigal Sideband
VSF   Voice Store-and-Forward

VSLE Very Small Local Exchange
VSP Virtual Switching Point
VSPC Virtual Storage Personal Computing
VST Volume Sensitive Tariff
vswr Voltage Standing Wave Ratio
VT Vacuum Tube, Vertical Tabulation
VTAM Virtual Telecommunications Access Method
VTPR Vertical Temperature Profile Radiometer
VTR Video Tape Recording
V+TU Voice plus Teleprinter Unit
vtvm Vacuum Tube Voltmeter
VU Volume Unit

W

W Width, Watts, West
WA Wire Armored
WADS Wide Area Data Service
WAHT Weighted Average Holding Time
WAL 2 Second order Walsh function: diphase line code
WAN Wide Area Network
WARC World Administrative Radio Conference
WARC-BS WARC for Broadcast Satellites
WARC-MOB WARC for Mobile Services
WARC-MR WARC for Maritime Mobile Telecommunication
WARC-ORB WARC for usage of the geostationary satellite orbit
WARC-ST WARC for Space Telecommunications
WATS Wide Area Telephone Service
watt-hr Watt Hour(s)
WB Wet Bulb
WBFM Wideband Frequency Modulation
WBS Wideband System
WBSARC World Broadcasting-Satellite Administrative Radio Conference
WBVTR Wideband Video Tape Recorder
WC Wire Chief (test clerk), Western Cedar (utility pole)
WCC Wireline Common Carrier
WCS Writable Control Store
WCTP Wire Chief Test Panel
WDC World Data Center
wdg Windings
WDM Wavelength Division Multiplex
WDT A/B Watchdog Timer A/B
WECo Western Electric Co.
WEFAX Weather Facsimile
WES Wind Electric System
WFD Waveform Distortion
WFG Waveform Generator
WH We Have, ready with called party, Western Hemlock (utility pole)
whr Watt Hour(s)

W/I Within
WIS Wats Information System
WL Western Larch (utility pole)
WMC World Meteorological Center
WMO World Meteorological Organization
WMS World Magnetic Survey
WN Wrong Number
WNO Wrong number
WO Work Order
w/o Without
WP Western Pine (utility pole)
WPC World Power Conference
WPG Water Pipe Ground
wpm Words Per Minute
wrg Wrong
WRMI Slang confirmation in service message (We Really Mean It)
WRS Working transmission Reference System
WRTC Working Reference Telephone Circuit
WRU Who are you?
WS Wire Send, Work Station
WSD World Systems Division (of Comsat)
WSG Wired Shelf Group
w/t Radiotelegraphy
wt, wgt Weight
WT Will Talk
WTCI Western Tele-Communications Inc.
WU Western Union Telegraph Co.
WUI Western Union International
WW Wire-Wound, Wall-to-Wall
WWW World Weather Watch

X

X Cross, reactance, any number from 0 to 9
X.25 CCITT specification and protocols for public packet-switched networks
X 400 CCITT recommendation for public message handling systems
XA, Xarm Cross-Arm
XB, Xbar Crossbar
XBL Extension Bell
XBT CrossBar Tandem
XCS Ten call seconds
xcvr Transceiver
XD Ex-Directory, crossed
XFC transfer Charge
Xfer Transfer
xfmr Transformer
xmit Transmit
xmsn Transmission
xmtg Transmitting
xmtr Transmitter
XON Cross-office highway

x-on, x-off Transmitter on, transmitter off

XOR exclusive-or

XOS cross-Office Slot

XOW Express Order Wire

XPD Cross-Polar Discrimination

XPT cross-Point

XTC External Transmit Clock

XVR Exchange Voltage Regulator

Y

Y Admittance (in ohms)

YAG Yttrium Aluminum Garnet

YIG Yttrium Iron Garnet

YPS Yellow Pages Service

yr Year

Z

Z Impedance (in ohms)

Z Zulu time zone (Greenwich Mean Time)

2B+D ISDN term for two voice circuits plus one data circuit on the same two-wire physical loop.

3P Three Pole

3PDT Three Pole, Double Throw

3PST Three Pole, Single Throw

4P Four Pole

4PDT Four Pole, Double Throw

4WTS Four Wire Terminating Set

7D Seven Digit number

911 System Emergency call reporting system widely used in U.S.

999 System Nationwide emergency call system used in U.K.

0TLP Zero Transmission Level reference Point

APPENDIX B
SI UNITS AND
ELECTROMAGNETIC SPECTRUM

SI UNITS (Systeme International d'Unites)

SI (Systeme International d'Unites) is the internationally agreed upon system of units for expressing the values of physical quantities.

There are seven base SI units:

Quantity	Name of Unit	Symbol
length	meter	m
mass	kilogram	kg
time	second	s
electric current	ampere	A
thermodynamic temperature	kelvin	K
luminous intensity	candela	cd
amount of substance	mole	mol

There are two supplementary dimensionless units:

Quantity	Name of Unit	Symbol
angle	radian	rad
solid angle	steradian	sr

Other SI units derived from these base and/or supplementary units are:

Quantity	Name of Unit	Symbol
area	square meter	m^2
volume	cubic meter	m^3
frequency	hertz	Hz
density	kilogram per cubic meter	kg/m^3
velocity	meter per second	m/s
angular velocity	radian per second	rad/s
acceleration	meter per second squared	m/s^2
angular acceleration	radian per second squared	rad/s^2
force	newton	N (kg. m/s^2)
pressure	pascal (newton per square meter)	N/m^2
surface tension	newton per meter	N/m
dynamic viscosity	newton second per meter squared	N. s/m^2
kinematic viscosity, or	meter squared per	m^2/s
energy, work, quantity of heat	joule	J (N. m)
power	watt	W (J/s)
electric quantity, charge	coulomb	C (A. s)
electric potential, or potential difference	volt	V (W/A)
electric field strength	volt per meter	V/m
electric inductance	henry	H (V. s/A)
electric resistance	ohm	Ω(V/A) (H. Hz)
electric conductance	siemens	S (A/V)
magnetic field strength	ampere per meter	A/m
magnetic flux	weber	Wb (V. s)
magnetic permeability	henry per meter	H/m
ectric capacitance	farad	F (C/V) (A. s/V)
c permittivity	farad per meter	F/m
conductivity	watt per meter kelvin	W/m. K
t capacity	joule per kilogram kelvin	J/kg. K
	joule per kelvin	J/K
eat	joule per kilogram	J/kg
	candela (or nit) per square meter	cd/m^2

Quantity	Name of Unit	Symbol
luminous flux	lumen	lm(cd. sr)
illumination	lux	lx(lm/m²)

Multiples and submiltiples of these units are indicated by a standard range of prefixes as follows:

Factor	Prefix	Symbol	Factor	Prefix	Symbol
10^{12}	tera	T	10^{-3}	milli	m
10^{9}	giga	G	10^{-6}	micro	μ
10^{6}	mega	M	10^{-9}	nano	n
10^{3}	kilo	k	10^{-12}	pico	p
			10^{-15}	femto	f
			10^{-18}	atto	a

Quantities expressed in multiples of three are preferred, but in some special circumstances smaller multiples can be used:

Factor	Prefix	Symbol
10^{2}	hecto	h
10	deca	da
10^{-1}	deci	d
12^{-2}	centi	c

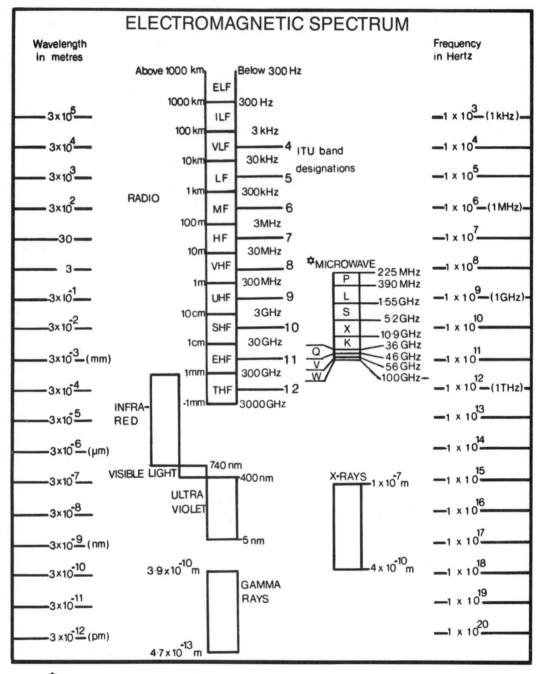

ELECTROMAGNETIC SPECTRUM

Wavelength in metres

Frequency in Hertz

Above 1000 km — Below 300 Hz

ELF

1000 km — 300 Hz
ILF
100 km — 3 kHz

3 x 10^5 — — 1 x 10^3 (1 kHz)

VLF — 4 ITU band designations
3 x 10^4 — 10 km — 30 kHz — 1 x 10^4

LF — 5
3 x 10^3 — 1 km — 300 kHz — 1 x 10^5

RADIO MF — 6
3 x 10^2 — 100 m — 3 MHz — 1 x 10^6 (1 MHz)

HF — 7
30 — 10 m — 30 MHz — 1 x 10^7

VHF — 8 ✿MICROWAVE
3 — 1 m — 300 MHz — 225 MHz — 1 x 10^8
P — 390 MHz
UHF — 9 L — 1·55 GHz — 1 x 10^9 (1 GHz)
3 x 10^-1 — 10 cm — 3 GHz S — 5·2 GHz
SHF — 10 X — 10·9 GHz — 1 x 10^10
3 x 10^-2 — 1 cm — 30 GHz K — 36 GHz
EHF — 11 Q — 46 GHz — 1 x 10^11
3 x 10^-3 (mm) — 1 mm — 300 GHz V — 56 GHz
W — 100 GHz —
THF — 12 — 1 x 10^12 (1 THz)
3 x 10^-4 — ·1 mm — 3000 GHz

INFRA-RED
3 x 10^-5 — — 1 x 10^13

3 x 10^-6 (μm) — — 1 x 10^14

VISIBLE LIGHT — 740 nm
3 x 10^-7 — 400 nm X·RAYS — 1 x 10^15
ULTRA VIOLET 1 x 10^-7 m — 1 x 10^16

3 x 10^-8 — — 1 x 10^16

3 x 10^-9 (nm) — 5 nm — 1 x 10^17

3 x 10^-10 — 3·9 x 10^-10 m 4 x 10^-10 m — 1 x 10^18

GAMMA RAYS
3 x 10^-11 — — 1 x 10^19

3 x 10^-12 (pm) — — 1 x 10^20

4·7 x 10^-13 m

✿The microwave band letters shown are widely used but have no officially agreed international status.